The Struggles and Dreams of
Robert Langer

Series in Structural Biology

Series Editor: Anders Liljas *(Lund University, Sweden)*

Published

Vol. 1 Structural Aspects of Protein Synthesis, 2nd ed.
by Anders Liljas and Måns Ehrenberg

Vol. 2 Found in Translation: Collection of Original Articles on Single-Particle Reconstruction and the Structural Basis of Protein Synthesis
by Joachim Frank

Vol. 3 Selected Papers of Michael G. Rossmann with Commentaries: The Development of Structural Biology
by Michael G. Rossmann

Vol. 4 From a Grain of Salt to the Ribosome: The History of Crystallography as seen Through the Lens of the Nobel Prize
edited by Ivar Olovsson, Anders Liljas and Sven Lidin

Vol. 5 The Struggles and Dreams of Robert Langer
by Robert Langer
edited by Shuguang Zhang

Forthcoming

Vol. 6 Structure and Action of Molecular Chaperones: Machines that Assist Protein Folding in the Cell
by Lila M Gierasch, Arthur L Horwich, Christine Slingsby, Sue Wickner and David Agard

Vol. 7 Curiosity and Passion for Science and Art
by Uwe B Sleytr

Vol. 8 Textbook of Structural Biology, 2nd ed.
by Anders Liljas, Lars Liljas, Miriam-Rose Ash, Göran Lindblom, Poul Nissen and Morten Kjeldgaard

Series in Structural Biology – Vol. 5

THE STRUGGLES AND DREAMS OF ROBERT LANGER

Editor

Shuguang Zhang

Massachusetts Institute of Technology, USA

World Scientific

NEW JERSEY • LONDON • SINGAPORE • BEIJING • SHANGHAI • HONG KONG • TAIPEI • CHENNAI • TOKYO

Published by

World Scientific Publishing Co. Pte. Ltd.

5 Toh Tuck Link, Singapore 596224

USA office: 27 Warren Street, Suite 401-402, Hackensack, NJ 07601

UK office: 57 Shelton Street, Covent Garden, London WC2H 9HE

Library of Congress Cataloging-in-Publication Data

Names: Langer, Robert S., author. | Zhang, Shuguang, Dr., editor.

Title: The struggles and dreams of Robert Langer / Robert Langer ; editor, Shuguang Zhang, Massachusetts Institute of Technology, USA.

Description: New Jersey : World Scientific, 2016. | Series: Series in structural biology ; vol. 5

Identifiers: LCCN 2015046967| ISBN 9789814749039 (hardcover : alk. paper) | ISBN 9814749036 (hardcover : alk. paper) | ISBN 9789814749046 (pbk. : alk. paper) | ISBN 9814749044 (pbk. : alk. paper)

Subjects: LCSH: Langer, Robert S. | Biomedical engineers--United States--Biography. | Biomedical engineering--United States.

Classification: LCC R856.25 .L36 2016 | DDC 610.28--dc23

LC record available at https://lccn.loc.gov/2015046967

British Library Cataloguing-in-Publication Data

A catalogue record for this book is available from the British Library.

Printed in Singapore

Foreword

Pathfinder pioneers of geographic explorations were often fearless, self-confident and highly motivated as they encountered endless dangers. They encountered numerous unexpected treacherous situations that seemed to be insurmountable, yet they overcame them. They faced repeated setbacks and, sometimes, near-death and real-death threats. They endured hardship and lack of support of all sorts; but they persevered. As a result, they occasionally encountered exhilarating discoveries of new vistas, new continents and new oceans beyond their dreams. In the long run, some of their explorations paid off and they are admired for their discoveries and their explorations benefited mankind.

Scientific pioneers often experience fates similar to geographic explorers. Their initial intuitive pursuits are often misunderstood; they may be ridiculed and humiliated; their grant applications and rigorous experimental research papers are rejected and sometimes, their jobs are on the line. But as they persevere, their discoveries and pioneering research gradually become appreciated and eventually widely adopted. Their discoveries have transformative effects, generate new knowledge, and stimulate new knowledge-based economy that benefits mankind. Some of them are eventually to be recognized and admired, and their early hardship and rejections all forgotten.

Robert Langer of the Massachusetts Institute of Technology is certainly one of the great path-breaking scientific pioneers. However, very few people know of his early hardships. In the 1970s, after the Middle East oil embargo, when almost all chemical engineers in the US worked in the petroleum industry with high salaries, he was struggling almost alone, overcoming many obstacles, and in the process, he created two new fields of biomaterials: polymer controlled macromolecular delivery and tissue engineering. A young scientist with no track record, he was subjected to ridicule and humiliation. He experienced numerous setbacks. His first nine NIH grant applications were rejected with harsh criticism; and his early papers were repeatedly turned down. His job as an assistant professor at MIT was on the line. (One senior faculty member at a department dinner blew smoke right into the young Langer's face, essentially telling him to look for another job. One professor said this seemed like a scene in a Fellini movie). However, as the late Alexander Rich best put it, "*Persistence is luck*." Langer persisted.

Robert Langer not only generated new knowledge in several fields; (a) designer biomaterials; (b) controlled and sustained molecular delivery; (c) tissue engineering; (d) new medical devices; and (e) angiogenesis inhibitors; but he also laid the foundation for several industries. His research revolutionized biomedicine and clinical treatments that benefit billions of people around the world. Steve Jobs was often credited for launching several new industries; Robert Langer contributed no less.

Nowadays, people all consider Robert Langer a genius and believe he achieved his fame effortlessly. Nothing could be further from the truth. Langer was not a prodigy, as most people assume. As he himself wrote in the Introduction, he was a happy ordinary child, not much different from other children of his era and neighborhood, but with a curious mind interested in math and chemistry. What set him apart from others is that he asked good questions, did not bother to follow the trends or to fit in with his peers academically. From early on, he had an open mind and a very big heart; he was interested in helping others to learn, and for them to stand on their own feet. He tutored high school math when he was a graduate student at MIT and was interested in teaching after graduation from MIT. Like several eminent people featured in Apple Computer's "*Think Different*" campaign (Albert Einstein, Thomas Edison, Buckminster Fuller, Martin Luther King Jr., Mahatma Gandhi, Pablo Picasso, Alfred Hitchcock, Amelia Earhart, Martha Graham, Frank Lloyd Wright and others), Langer thinks differently. After he earned his doctorate in Chemical Engineering from MIT, he did postdoctoral research in the Boston Children's Hospital. This was totally unheard of in 1970s. It was from there that he was inspired by a medical doctor and surgeon, his close mentor the late Dr. Judah Folkman.

In Folkman's lab, Langer not only single-handedly developed an elegant method using the rabbit eye assay for isolating and purifying angiogenesis factors, but he also isolated the first inhibitor of neovascularization. This simple assay system opened a floodgate for numerous angiogenesis stimulators and inhibitors to be isolated, including inhibitors of epidermal growth factor (EGF), fibroblast growth factor (FGF), vascular endothelial growth factor (VEGF), and many others. They have now been commercialized and have subsequently been used to treat millions of patients.

Because Langer was so unique and so early to lay the foundations for the new fields, there are over 1100 issued and pending patents and counting under his belt (more than anyone in history in the medical or chemical field), and over 1300 scientific publications and counting. His scientific citations on Google Scholar are ~200,000 and h-index 223 in early 2016 (for the obsessed administrative bean counters). Over 270 of his students are current faculty members in universities around the world. His pioneering research has been honored by numerous awards, including the Gairdner Prize of Canada, Prince of Asturias Award of Spain, Wilhelm Exner Medal of Austria, the Priestley Medal, the Kyoto Prize of Japan, the Breakthrough Prize in Life Sciences of US, the Rusnano Prize of Russia, the Queen Elizabeth Prize of UK, the Wolf Prize of Israel and the Kazemi Prize of Iran, and 28 honorary degrees. He has been inducted into four independent US National Academies of Sciences, Engineering, Medicine and Invention, and the Austrian Academy of Sciences as well as other academies.

However, despite his honors and busy schedule, he remains humble and down-to-earth, without a trace of arrogance or pretentiousness. He spends valuable time with his students including undergraduate students, postdocs and collaborators to discuss the latest advances in the fields; he exudes child-like excitement when new research results emerge. His lightning speed in replying to e-mail is legendary, even to people he has never met. He always honors his tightly arranged appointments, even with visitors and strangers.

Robert Langer is unique in another way. He loves science and his laboratory so deeply that he turns down all job offers with heavy administrative duties, including the director of centers, head of departments, dean of schools and presidents of research institutions, universities and companies. It is extremely rare to find such a humble human being in the current competitive environment.

In this collection of selected papers and the illuminating Introduction of Robert Langer, it is with great excitement and enjoyment that you are invited to read these selected papers and to explore the mind of a fearless knowledge pioneer, a humble and rare person who not only is a leading scientist and engineers but also a widely admired humanist.

Shuguang Zhang
Massachusetts Institute of Technology
Cambridge MA, USA
March 2016

Contents

Bachrach

Introduction

My Early Years

I grew up in Albany, New York. My father, Robert, ran a small liquor store. He was very smart. I remember he would do crossword puzzles very quickly and he had very broad knowledge. He worked very, very hard, but when he came home, he would play math games with me. My mother, Mary, was a homemaker. She took care of my sister, Kathy, and me. My mother worried a lot — in fact she still does — but she was and is one of the nicest people you could ever meet. I hope some of that rubbed off on me.

Among the gifts my parents got me when I was between the ages of 10 and 13 were Erector, microscope, and chemistry sets. I loved these sets. I would make robots and rocket launchers with the Erector set, watch shrimp hatch with the microscope, and I set up a little laboratory with the chemistry set in the basement of our small house, where I would mix chemicals together and watch them turn different colors. I would also make rubber, polymers, and other materials. I received other gifts as well, including baseball gloves, a basketball, and a football. I loved these too and I played a lot of sports with my friends in the neighborhood. I think I had a pretty regular and pretty happy childhood.

High School and College

I attended the Milne High School in Albany, New York from the ages of 12 to 17. I was very good in math, and pretty good in science; however, courses like English and French were very challenging for me. My father and my guidance counselor both suggested I become an engineer because I was good in math and science, and that was what I did. I applied to engineering schools and was fortunate to be accepted by Cornell University. In my freshman year, chemistry was by far my favorite subject and it was the only course where I did really well, so I decided to major in chemical engineering. I worked very hard at Cornell and I also got my first taste of teaching by being a teaching assistant in a course on heat and mass transfer. I loved it. When I graduated from Cornell, I received several job offers to run chemical plants, but I did not think I would be very good at it and it was not exciting to me, so I decided to apply to graduate school.

I applied to six graduate schools and chose to go to MIT. While at MIT, I did two things. The first was my doctoral thesis on enzymatic regeneration of adenosine triphosphate (ATP). The goal was to explore enzymes for synthetic purposes, and an energy source, ATP, was required. On the positive side, my advisor, Clark Colton, was very thorough; and one of the postdoctoral fellows in the lab, Colin Gardner, taught me how to do very careful, reproducible research. But, on the negative side, after the first year or two, I started to think that the research I was doing was not really important. I could not see how it would have great impact or change the world. That discouraged me about research. The second thing I did was get involved in a lot of educational outreach, particularly teaching underprivileged children. During my first year at MIT, I did a lot of tutoring in poor communities. During my second year, some people wanted to start a school for underprivileged high school students who had dropped out of public high schools. They asked me to help and we started the Group School. I helped create and chaired the math and science departments. My big goal was to make math and science interesting. I got MIT to fund a program to create a novel chemistry lab course where we would teach chemical principles by practical examples. For instance, making rubber illustrated a chemical reaction and making ice cream illustrated freezing point lowering because you need to lower the freezing point of water to make ice cream into a solid. I also created a number of games to help make math interesting. The school was very liberal and students were not required to take math or science. I remember that when I first joined, only 5 of 40 students signed up for math, but one year later 45 out of 50 did. This experience reinforced my love of teaching.

I finished graduate school with a doctorate in chemical engineering but I did not know what I wanted to do careerwise. I graduated in 1974 and at that time there was a big gas shortage. The price of gas went way, way up, and in many areas you had to wait in line at the gas station for hours to fill up your car. The consequence was that if you were a chemical engineer, you received a lot of job offers. In fact, nearly all of my classmates in the 1970s joined oil companies. There were many openings and that was really where the high-paying chemical engineering jobs were at that time. I got 20 job offers from oil companies — 4 from Exxon alone. I also received offers from Shell, Chevron, and others. One job interview made quite an impression on me. I went to this interview at Exxon in Baton Rouge and one of the engineers there said to me that if I could increase the yield of a particular chemical by about 0.1%, would not that be wonderful? He said that would be worth billions of dollars. I remember flying home to Boston that night, thinking to myself that I really did not want to do that.

What did I want to do? Well, from my college experiences I had this dream of using my background in chemistry and chemical engineering to improve people's lives. As I mentioned, I had spent a lot of my time as a graduate student starting a school for underprivileged high school kids and developing new chemistry and math curricula. One day, I saw an advertisement to be an assistant professor to develop a chemistry curriculum at City College in New York. So I wrote them a letter, but they did not write me back. But I liked that idea, so I found all the ads I could for an assistant professor position to develop chemistry curricula. I found about 40 such ads and wrote to all of them, but I do not think any of them replied.

Postdoctoral Research

Another way I thought I could help people was through health-related research. So I applied to a lot of hospitals and medical schools. None of them wrote me back either. Then one day Barry Bunow, one of the people in the lab where I worked, said I should write to a surgeon named Dr. Judah Folkman at Harvard. He said "sometimes he hires unusual people" and Dr. Folkman was kind enough to offer me a job. I took what, at that time, seemed to all chemical engineers like a huge risk and began doing postdoctoral work in a hospital. It might seem more common today, but at that time few, if any, chemical engineers had done postgraduate work in a surgery lab. Dr. Folkman told me about a problem he had been investigating — trying to figure out how blood vessels grow toward cancerous tumors in the body. Cancerous tumors grow because they induce blood vessels to grow into them. That way, they will receive nutrients and can grow much larger, but if you could prevent nutrients from getting to the tumor that might stop the tumor from growing. When I started working with Dr. Folkman, this concept was only theoretical, and many people did not agree with his concept. Moreover, this area of blood vessel growth was difficult to study. We realized that in order to solve this problem we would not only need to isolate a blood vessel inhibitor, which is often in the form of a large molecule, but, as has often been the case in the development of new medicines, discoveries were hampered because no bioassays existed. We would also need to develop a bioassay. We chose the cornea of a rabbit, because there are normally no blood vessels in the cornea. We put a tumor in the cornea in a way that would mimic what happens in the human body. Over 2–3 months, blood vessels grew from the edge of the cornea to the tumor. At that point, we wanted to stop those blood vessels from growing, but to do that we needed a controlled-release polymer that could deliver the different molecules I was isolating over a 2–3-month period. Since there were no such systems at that time, I had to try to develop one. My research in the drug delivery field started with this project.

My work was somewhat basic — to help in blood vessel research. However, from the standpoint of potential practical impact, something happened that I could not have anticipated. That was the advent of biotechnology and genetic engineering, where, for the first time, it became possible to create large molecules including peptides and proteins such as human growth hormones in a commercial way. But these molecules faced serious delivery challenges. Swallowing them did not work, because they were too large and would be destroyed by enzymes or acid in the stomach or intestines. They were also too large to use in a transdermal skin patch. If you tried injecting the molecules, they were quickly destroyed by enzymes. Delivering any of these molecules on a chronic basis would require a way to deliver them in an unaltered form, and yet protect them from harm. When we started this work, the conventional wisdom in the field was that it could not be done. Scientists simply felt it was not possible to slowly release these large molecules from biocompatible polymers, any more than a person could walk through a solid wall.

Against this background, I began working in the laboratory to see if I could come up with a way to make tiny systems that could deliver molecules for long times. Over two years of experimentation, I had found hundreds of different unsuccessful methods. Finally, I discovered one way to make it work. My students and I took water-repellent polymers and dissolved them in organic solvents, usually at extremely low temperatures, like –80°C. We added the molecules to them and slowly dried off the solvent. This is how we created small microspheres or even nanospheres. We published in the journal *Nature* that you could use this approach to release molecules of almost any size. These molecules could be released for more than 100 days in test tubes. Although the release rates were not constant in those initial studies, we later developed ways to ensure constant rates. For example, we were able to achieve a constant drug release for over 50 days. At first, our concepts were not well accepted, and many scientists ridiculed my work. For example, as a postdoc, I had not realized that being a member of the scientific community meant giving speeches at conferences. In 1976, I was asked for the first time to give a lecture. It was at the Midland Macromolecular Symposium in Midland, Michigan. I was 27 years old at the time, addressing distinguished elder polymer chemists and engineers, and I had never given a big talk before. In fact, the last talk I had given was in eighth grade, when I had to give a minute-and-a-half speech. The night before my eighth grade talk, I rehearsed four hours in my parents' bedroom in front of a mirror. The next day, I started to give the talk, but after one minute of speaking, I could not remember the next word and I froze. Eventually the teacher told me to sit down and gave me a very poor grade. I think it was an F.

When this Michigan talk came about many years later, I was very nervous and I stopped working two weeks in advance of it. I practiced my talk over and over into a tape recorder until, finally, the day of the lecture arrived. I got up, gave that talk, and I actually was pretty pleased by the end of it. I had not forgotten too much of what I had intended to say and I did not stammer too much. I thought that when I was done with that talk all these much older, distinguished chemists and engineers in the audience, being nice people, would want to encourage me, this young guy. But when I was done, a number of people gathered around me and they said, "We do not believe anything you have just said. We know that you cannot get these molecules that you are talking about to slowly diffuse through these polymers." It was not until several years later that other people began repeating what we did and then the question shifted to: How could this possibly happen? In fact, I spent a good part of my early career at MIT understanding how these polymer drug delivery systems functioned and trying to make them useful for different applications.

Becoming a Professor

Shortly after that talk, I tried to receive funding to support my research and wrote a number of grants. My first nine were turned down. I remember I wrote one grant to the National Institutes of Health (NIH) for some of my cancer research and the reviews were very, very negative. They not only turned me down, but they said, "Well, how could Dr. Langer do this? He's a chemical engineer. He does not know anything about biology and knows even less about cancer."

When I was done with my postdoctoral work, I applied for faculty positions in a number of chemical engineering departments. But I had trouble getting faculty jobs, because people felt that, at that time, what I was doing was not engineering. They thought it was more like biology. I ended up joining what was then the Nutrition and Food Science Department at MIT, but the year after I got the position, the chairman (who had hired me) left and a number of the senior faculty in the department decided to give me advice. They told me I should start looking for another job.

There I was, getting my grants turned down and people not believing in my research, and it appeared I would not even get promoted to associate professor. Fortunately, within the next few years, scientists in the pharmaceutical industry and at different universities started using some of the principles and techniques I had developed, and slowly things began to turn around.

Going back to my research in the 1970s, people could not understand how we could achieve long-lasting drug release from polymers. To answer this Rajan Bawa, one of my students, began to use the cryomicrotome, a low temperature cutting instrument, to cut thin sections through the polymers we used so we could visualize what was happening. We did a permeability study on a very thin slice of one of these polymers and found that if the molecules were 300 MW or greater, they were not able to diffuse from one side of the polymer section to the other. To understand the mechanism of release better, we then put a reddish molecule in the polymer matrix. We observed two separate phases and next we began the release process. After the release had gone on for a year, we cut another thin slice and found that what was left behind where the reddish molecule had been were holes. Since these holes had not existed before the addition of the molecule, they clearly had been made by the way we made the polymer systems. We then did a lot of imaging studies and found that these holes were interconnected. They also had a lot of very tight constrictions between them and were incredibly winding and tortuous, which accounts for the very long time it took for the molecule to get through. I sometimes like to explain this long release process by comparing it to driving a car through my hometown of Boston, which is often very tortuous. We then discovered that by adjusting these porous structures we could actually make the systems last anywhere from a day to more than three years, or any time period in between.

Moving Our Research Forward

One of my goals in doing laboratory work has been to move beyond conducting the work and publishing the results to applying that work to helping people. The first thing I did was to work with Dr. Folkman to use this polymer system in the bioassay I mentioned above to see if we could find a substance that could stop blood vessels from growing in the body. I had isolated many different substances and I tested all of them in 1000 different studies. All substances but one failed to stop the blood vessels from growing. We did over 20 different experiments with the successful substance and it was very exciting. For the first time, we saw that blood vessels growing toward a tumor were halted. They actually formed a zone of inhibition around the polymer that was releasing the inhibitor. That never had happened before in thousands of control experiments. We published this paper, which showed for the first time that inhibitors of blood vessels did exist, and also provided bioassays which could and were used to isolate future inhibitors in the journal *Science* in 1976. Today, many inhibitors of blood vessel growth have been isolated. They have been used to treat over 20 million patients who have different types of cancer, or eye diseases such as macular degeneration. Regulatory authorities like the US Food and Drug Administration (FDA) now say that there are four ways to treat cancer — radiation, chemotherapy, surgery, and stopping blood vessel growth (i.e. antiangiogenesis). Of course, many times several of these therapies are used together.

After these successful experiments, I wanted to move to the next logical step, which was getting the drug release system to work systemically (through the body) in animals. For this experiment, I worked with one of my graduate students, Larry Brown. The model system we first chose was diabetes. Our goal was to deliver insulin, which is a large molecule, to diabetic rats in order to lower their blood sugar. After making the rats diabetic, we made small polymer pellets, no bigger than aspirin tablets, containing insulin, and implanted them into the rats. We found that the pellets could lower the rats' blood sugar from 400 mg/dl, which is diabetic, to 100 mg/dl, which is normal, for over 100 days. This experiment showed that we could maintain biological activities of molecules like insulin for long time periods in animals. Our subsequent goal was to see if we could get this drug release process to work in humans. I will discuss that a little later.

The original controlled-release materials we developed were small particles. In many cases these were microspheres that one might inject under the skin and release molecules for a long time, like weeks or months. Another challenge is controlling the location of the drug. Here we created nanoparticles that you could inject into the bloodstream. Nanoparticles are critical for delivering significant payloads of any drug into cells, particularly newer potential drugs like siRNA that can shut off genes. But polymeric nanoparticles injected into the body were destroyed almost immediately by certain cells in the body called macrophages. This made their use essentially nonexistent. In a 1994 *Science* paper, we addressed these problems. We created nanoparticles that could circulate for hours in animals, be stable on the shelf for years, and not aggregate. These principles are being widely used by many scientists and companies to practice "nanomedicine."

Another area I started thinking about involved creating new polymer materials. Working in a hospital, it became clear to me that almost all polymers used in medicine were derived from household objects and their medical use was driven by clinicians who wanted to use them in medical areas based on their physical properties. For example, the polyether urethanes used in ladies' girdles are now used in artificial hearts because of their good flex life. The polyurethanes in mattress stuffings are used in breast implants. Yet, such an approach often leads to problems. Artificial hearts, for example, can cause clots to form when blood hits their surface, the ladies' girdle material, and these clots can cause strokes and death.

Against this background, I began thinking that we needed to find ways of solving medical problems other than to search for materials in everyday settings. As a chemical engineer, I believed that researchers could take an engineering design approach, asking the question "What do we really want in a biomaterial?" from the engineering, chemistry, and biology standpoints, and then synthesizing the materials from first principles. As a proof of principle, we decided to synthesize a new family of biodegradable polymers, namely polyanhydrides, for medical use. The first step was to select the monomers — the building blocks of the polymer. I asked my friend Mike Marletta, who was then at MIT, what monomers, from a list of substances we thought might make good polymers, would be safe in the human body. We then synthesized these polymers and discovered that by changing compositions we could make them last in the body anywhere from days to years. Then, with Henry Brem, who is now Chief of Neurosurgery at Johns Hopkins, we thought we could use these polymers to locally deliver drugs to treat brain cancer. I had to raise money for this project, so I wrote grants to NIH and other funding agencies. These grants were then reviewed by Study Sections composed of other professors. Their reviews were very negative. In our first grant in 1981 the reviewers said we would never be able to synthesize the polymers. However, Howie Rosen, one of my graduate students at the time, synthesized the polymers for his graduate thesis. He ultimately became president of the ALZA Corp., a very successful company, and was later elected as a member of the National Academy of Engineering. After Rosen accomplished the polymer synthesis, we sent the grant back for another review, and received the following reply: *The grant should still not be funded because the polymers will react with whatever drug you put in*. But several postdocs, including Kam Leong, now at Columbia University and a member of the National Academy of Engineering, and Robert Linhardt, who is the Constellation Professor at RPI, showed there was no reaction.

Then we returned the grant for yet another review, which came back with the comment that the polymers were fragile and would break. This time, two other postdocs in our laboratory — Avi Domb, who later became Chair of Medicinal Chemistry at the Hebrew University, and Edith Mathiowitz, who is now a Full Professor at Brown University and was elected to the National Academy of Inventors — addressed this problem. The revised grant was sent for evaluation, and this time the reviewers said it should not be funded because new polymers would not be safe to test on animals or people. But another graduate student, Cato Laurencin, who would become Dean of Medicine at the University of Connecticut as well as a member of the National Academy of Engineering and the National Academy of Medicine, showed the polymers are safe, supporting Mike Marletta's early predictions. Then we got the grant back again, but the reviewers said the drug would not diffuse far enough. But my graduate student Mark Saltzman, now Chair of Yale's Biomedical Engineering Department and a member of the National Academy of Medicine, showed it did. These kinds of negative reviews occurred for a long time, but in 1996 the FDA finally approved this treatment — the first time in more than 20 years that the FDA had approved a new treatment for brain cancer, and the first time they had approved the concept of polymer-based local chemotherapy. I should add that the funding that we obtained to support our research was based on broad patent coverage of polyanhydrides that we received from a company that licensed our patents. Nineteen years after their approval, these polymer systems are still used for treating brain cancer patients in over 30 countries, and they provided an entirely new paradigm in the drug delivery field, helping pave the way for drug-eluting stents and other local delivery systems.

I would now like to describe another area, the delivery of mammalian cells through materials, which our group has been working on for some time. My motivation for working on this project goes back more than 30 years of work we did with Boston surgeon Joseph P. "Jay" Vacanti, who is now chief of pediatric surgery at Massachusetts General Hospital and John Homans Professor of Surgery at Harvard Medical School. Among his patients were small children dying of liver failure whose lives depended on someone else dying and donating a liver for transplant. Not enough transplants, however, were available for treating these patients. The problem, of course, was not limited to liver disease, but extended to paralysis and to advanced diseases of nearly any organ or tissue.

Over 30 years ago, Jay and I came up with an idea for tissue engineering. This concept involves taking isolated dissociated cells from patients themselves or from a close relative. These cells include bone cells, cartilage cells, liver cells, intestinal cells, and urothelial cells. Today, they could also include stem cells. If these cells are injected at random, very little happens. It turns out that the cells are "smart," however, and if you put them close enough together, they can organize themselves and create structures. A group at UC Berkeley has shown that if you take mammary epithelial cells and put them close enough together *in vitro*, they can form acini and make milk. We wondered if we could make tissues by taking cells and putting them on a polymer template, in three dimensions, with the right media. We would grow the cells outside the body, on the right materials, and then ultimately return them to the body to make whatever tissue we wanted.

Let me now speculate about where this field might be in 30 years. A good example would be plastic surgery. Imagine a patient visiting a plastic surgeon and asking for a new nose. I believe that in 30 or 40 years patients might be able to go to a computer screen to select whatever nose they like, or create their own "designer" noses.

V. Prasad Shastri, one of my former postdocs and now a professor at the University of Freiburg, worked out a way to do just that using a novel fabrication process that could someday be adapted for computer-aided design techniques. This could also be done by three-dimensional printing. Shastri was able to fabricate a "regular" nose by using a polymer, lactic–glycolic acid, and forming it into the shape of a nose that was 97% porous. Pieces of the polymer could then be added or reshaped to create other kinds (or shapes) of noses that patients might desire. The donor cells for the nose could, for example, be taken from the patient's ear through a minimally invasive surgical procedure known as arthroscopy. The cells would be placed on a polymer scaffold and grown in a culture until the nose was fully formed. We are now using this procedure, though not on noses.

I would like to present three examples of tissue engineering using the principles I have discussed above. The first one is a form of replacement cartilage. Every year, about one million people need to replace cartilage in their bodies. We did some experiments with nude (hairless) animals, using their own cells on a scaffold to develop new cartilage that would resist rejection. For example, we reconstructed the skull of one animal and the cheek of another. If you opened up the animals to see the results, you would see pure-white glistening cartilage, with histologically the same characteristics as the original cartilage. However, this replacement cartilage is not as mechanically strong as the original. Thus, we still have work to do before we could provide, for example, new cartilage for a human arthritic knee. However, the development of artificial cartilage for such areas as cosmetic defects may be closer to becoming a reality. Jay sees patients who, for example, do not have ears. Linda Griffith (one of my former postdocs and now a professor of mechanical and biological engineering at MIT), Jay, and I received funding from the US Army to help create body parts like ears for soldiers who come back from Iraq or Afghanistan without such parts. We created an artificial human ear by making a scaffold in the form of an ear with polymer fibers, cartilage cells, and a matrix. It has not been put in humans, but Jay has tried it on animals and it is safe. Replacement cartilage has been tried on humans, however, in other forms. Jay first used it on a 12-year-old boy who had no ribs covering his heart. This boy, like many others his age, liked to play baseball, but he would have been at grave risk from something as ordinary as being hit in the chest by a stray ball. Jay operated on him and gave him a new chest created on a polymer scaffold using the boy's own cells. Now it is many years later and the boy is a grown man and is doing very well.

A second example is the development of new skin for burn victims. Let us consider an actual patient, a two-year-old boy who was badly burned. The clinicians took a polymer scaffold with neonatal skin fibroblasts and inserted it into the child at the time of injury. Three weeks later, new skin had started to form. Six months later, the burned areas were nearly invisible. This treatment was approved by the FDA for patients with skin ulcers or burns.

The final example is spinal cord repair. Erin Lavik, who is now a professor at Case Western Reserve University, led this work in our lab. He made a polymer scaffold that mimicked the human spinal cord and seeded them with neural stem cells. We then worked with neurosurgeon Ted Tang and stem cell specialist Evan Snyder to test this device in animals. Paraplegic rats that had difficulty moving their back legs were tested. We divided the rats into four groups: a treatment group that had the polymer stem cell scaffold inserted, and three control groups — one was given a sham operation, one implanted with stem cells only, and one with the polymer only. We then observed the rats for a year, doing numerous studies. At 100 days following the operation, we found that in the control groups the mean, about 40 rats, dragged their legs, and their paws were splayed in an awkward fashion. This is typical. In contrast, in the rats that were implanted with a polymer scaffold containing stem cells, we see results that are dramatically different from those of the control group. These rats can now bear their own weight and walk. Their paws are splayed in a much more normal fashion. This study was repeated on non-human primates by neurosurgeons Eric Woodard and John Slotkin and scientists at InVivo Therapeutics. Recently, the FDA gave the go ahead to start human clinical trials, and the first patient has just been treated.

Creating Companies and Products

I also wanted very much to have the inventions and materials we developed get to the point where they could help patients. This was very difficult because it takes a great deal of money and people to develop medical products. I began writing patents on our work. Today, we have licensed and sublicensed the patents to approximately 300 different companies, and I even helped start a number of companies with my students. I should add that when I wrote these patents and helped start these companies, many scientists looked down on it. They thought it was not a very good thing for a professor to do. But, today, these companies have made all kinds of products that treat patients with cancer, heart disease, and many other sicknesses. They have also created many thousands of jobs throughout the world.

Let me give a few examples. In the first example, I would like to expand on what I mentioned earlier — where we developed polymer systems that could slowly and continuously release large molecules. One of the challenges was getting a patent. In fact we filed a patent in 1976 and the Patent Office turned it down. I believe they turned it down five times between 1976 and 1981. My lawyer told me I should just give up, but I have never given up easily and I started to think about new ways in which we could get this patent allowed. The patent examiner said that what we had done was obvious, but I knew that was not true since, as I mentioned, scientists said it was impossible. I scoured the literature and I discovered there had been a paper published by five famous chemists and chemical engineers in 1979 that referred to our work by saying, "Generally the agent to be released is a relatively small molecule with a molecular weight no larger than a few hundred. One would not expect that macromolecules, e.g. proteins, could be released by such a technique because of their extremely small permeation rates through polymers. However, Folkman and Langer have reported some surprising results that clearly demonstrate the opposite." (Stannett, Koros, Paul, Baker, Lonsdale, *Adv. Poly. Sci.*, 1979.)

I showed this to our lawyer, who said, "Oh, this is very interesting," and he showed it to the patent examiner, who said, "I will allow this patent if Dr. Langer can get affidavits from the five chemists and engineers saying they really wrote this." All five scientists were kind enough to sign the affidavits and we got this very broad patent. We then licensed it to two very large companies — one in animal health and the other in human health. Both companies gave us grant money and they promised to do experiments to develop our invention, which I was very excited about. However, these companies were very large and what happened was that they would do one or two experiments and if these did not work right, they would give up. A few years later my friend and MIT colleague Alex Klibanov said, "Bob, why don't we start a company ourselves?" I was able to get these patents back and we started a little company called Enzytech, which later merged to become Alkermes. Alkermes has made long-acting injectable microspheres that have helped millions of patients suffering from diabetes, schizophrenia, alcoholism, narcotic addiction, and pituitary dwarfism.

In a second example, in 2001, I started Momenta Pharmaceuticals with Ram Sasisekharan and Ganesh Venkataraman, a former PhD student and a former postdoc, respectively. The technology in this case involved the use of enzymes known as heparinases for polysaccharide sequencing, which was described in papers that appeared in *Science* and other journals between 1982 and 1999, and protected by a number of patents. This technology is being employed to create new anticoagulants and other drugs for use in heart disease, cancer, and other indications. Momenta also used this polysaccharide sequencing approach to create the first FDA-approved generic, low-molecular-weight heparin, which is an anticoagulant. Today, this product has helped nearly 10 million patients, and it was the largest syringe launch in history. A second product is pending approval at the FDA, and others are in clinical trials. Furthermore, in 2008, there was a heparin crisis in which contaminated heparin obtained from China killed hundreds of people in 22 countries. Ram and I, as well as Bob Linhardt, another of my former postdocs, worked with Momenta, the FDA, and the Centers for Disease Control and Prevention in the United States to sequence the contaminated heparin, identify the contaminant, set new standards, and stop the deaths.

The above companies are examples of how we have been able to move discoveries from the lab to patients. We have been involved in launching some 30 companies. These companies are creating new aerosols to treat diseases like diabetes or Parkinson's; doing new types of genetic therapy for treating diseases like cancer or heart disease; developing new types of drug delivery microchips that can enable personalized medical therapies for treating osteoporosis and other diseases; developing new nanotechnology-based therapies for treating cancer and eye diseases, as well as vaccines; creating ultrafast nanotechnology-based diagnostics for detecting infectious diseases; and even developing new hair and skin care products. Polaris Ventures estimates that 2 billion people will be helped by products created by these companies.

My Family and Life Today

I should add a word or two about my personal life. I have been married for over 27 years to a wonderful, beautiful, and kind woman — my wife, Laura. I met Laura because she was the roommate of one of my postdocs and I had seen her running on the track where I also ran for exercise. I thought she was very beautiful, stimulating, smart, and nice. Laura has a bachelor's degree from Harvard and a PhD in neuroscience from MIT. Being married to someone with a scientific background has the added benefit that she knows the pressures I feel and the rewards I get from science. Being able to share that with her has been very wonderful. Laura is my best friend as well as my wife. I have also been very lucky because she is a very straightforward person. I always know exactly where I stand. I should add that a number of years ago my postdocs and students held a symposium for me, and one of them, Edith Mathiowitz, made a graph of my productivity. She asked everyone a question: Why is there a big inflection point in 1986? Her answer was, that was when Bob met Laura. We also have three wonderful children: Michael, Susan, and Sam. I feel very, very lucky, because being a husband, a father, and a scientist has enabled me to have a very complete and wonderful life.

Throughout all of this research, with its challenges and setbacks, I have been incredibly excited about the potential of science and engineering to make the world a better place and transform human health care. I also believe that we are only at the tip of the iceberg in creating materials for all types of applications that can profoundly relieve suffering and prolong life. I have also learned a great deal from the challenges and setbacks I have faced, and if there is any advice I might give to young people, it would be to dream big dreams; do not give up on those dreams; and recognize that conventional wisdom is not always correct.

I feel incredibly fortunate that I have had such a wonderful staff at MIT and such super students, postdocs, and collaborators. I have seen over 700 of my students get wonderful jobs, and over 270 are professors around the world. I view my students and postdocs as an extended family and I am so very proud of them. I would not be here today without having had so much support and help from so many people.

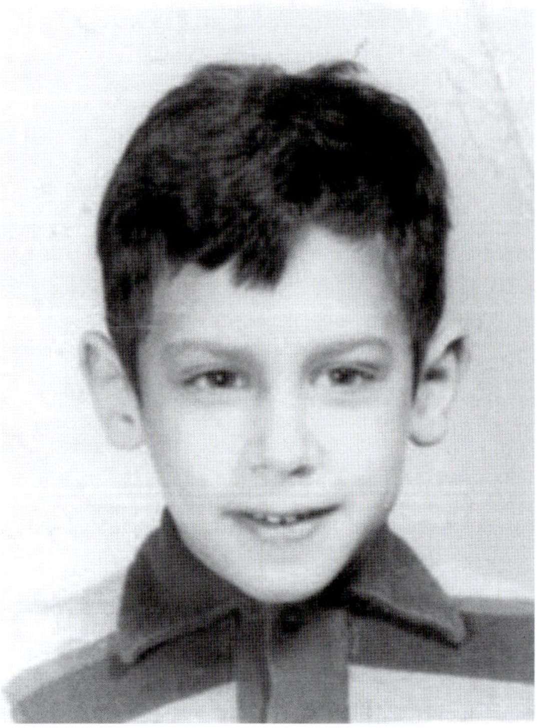

Robert Langer's childhood

Young Langer

Robert Langer's experiments

Bob Langer and his wife Laura

Langer and his young family

Langer and his young family

Bob Langer and his family

Bob Langer and his wife Laura

Cartilage Tissue Engineering
AFTER
2 weeks in culture

Langer in the laboratory

Langer in the laboratory

Langer with his mentor, the late Judah Folkman

Langer in his MIT office

Robert Langer's laboratory people

List of Publications

Drug Delivery (*Including Nanotechnology*)

Nature Vol. 263 October 28 1976

Polymers for the sustained release of proteins and other macromolecules

SINCE the first demonstration that silicone rubber[1] could be used as an implantable carrier for sustained delivery of low molecular weight compounds in animal tissues, various drug delivery systems have been developed. But except for the reports of Davis[2,3] and Gimbrone *et al.*[4], there has been little success in the development of slow release agents for large molecular weight compounds. Furthermore, the polymers used in those studies, polyvinylpyrrolidone and polyacrylamide, are often inflammatory in animal tissues and

Table 1 Host response to corneal polymer gel implants

Polymer	No. of tests	Inflammation* None	Mild	Significant
Polyacrylamide, 20%	10	20%	30%	50%
Polyvinylpyrrolidone, 20%	4	0	0	100%
Polyvinylalcohol (unwashed, 1%, 5%, 10%)	20	15%	70%	15%
Polyvinylalcohol (washed, 10%)	6	67%	33%	0
Hydron-S (with and without water in the polymer casting solution)	15	100%	0	0
Ethylene–vinyl acetate copolymer (unwashed)	20	40%	60%	0
Ethylene–vinyl acetate copolymer (washed)	20	100%	0	0

*The criteria were oedema, white cells and neovascularisation as observed through a stereomicroscope. A response was judged to be mild if any one of the three characteristics was detected, and significant if their presence was enough to cause corneal opacity.

usually permit only brief periods of sustained release. We now present a simple method for incorporating various proteins and other macromolecules into non-inflammatory polymers. Sustained release of biochemically active macromolecules has been demonstrated for periods exceeding 100 d.

Sterile polymer casting solutions were made by dissolving polymers in appropriate solvents: Hydron (a polymer of hydroxyethylmethacrylate) in absolute alcohol at 37 °C; ethylene–vinyl acetate copolymer (40% by weight vinyl acetate) in methylene chloride at 37 °C; and polyvinylalcohol by autoclaving in distilled water. Slow release pellets were made by placing a mixture of casting solution and the macromolecule to be released in a conical mould, 2–4 mm in diameter and 1.5 mm deep. Glass moulds were used for the ethylene–vinyl acetate copolymer because the polymer casting solution reacted with plastics causing adhesion. The moulds were dried under a mild vacuum (600 mmHg) overnight causing the solvent to evaporate leaving the macromolecule trapped within the polymer matrix. To retard further release of macromolecules, 'sandwiches' were sometimes made by coating the matrix with pure polymer in casting solution. Coating was achieved by dipping the polymer pellet in a puddle of pure polymer in casting solution and drying. In an alternative method of coating, pure polymer in casting solution was dried in the bottom of a mould, a middle layer of polymer and macromolecule was added and dried, and finally a top layer of pure polymer in casting solution was added and dried. Before use, the dried slow-release pellets were rehydrated by adding a drop of Ringer solution.

To study tissue response to the polymers, sterile polymer pellets ($1.5 \times 1.5 \times 0.5$ mm^3) were implanted in the eyes of rabbits, by aseptically creating a small pocket in the cornea[4]. Inflammation was judged by stereomicroscopy. The sensitivity of the cornea to inflammatory stimuli as well as its clarity and avascularity make it an excellent site to observe inflammatory characteristics: oedema, white-cell infiltration, and neovascularisation[5].

Ethylene–vinyl acetate copolymer, exhaustively washed in alcohol, and Hydron caused no inflammation in the cornea (Table 1). Histological sections made 3–4 weeks after implantation showed no foreign cells in the region surrounding the polymers or in other areas of the cornea. In contrast, polyacrylamide and polyvinylpyrrolidone, prepared according to Davis[3], were significantly inflammatory. Polyacrylamide has been used in our laboratory for several years[4], and inflammatory responses have been reduced by various washing techniques (R. Arensman and M. Gimbrone, unpublished). Unwashed polyvinylalcohol pellets also caused inflammation. Washing the polymer exhaustively with alcohol greatly reduced, but did not eliminate, the inflammatory response (Table 1).

Polyvinylalcohol, Hydron and ethylene–vinyl acetate copolymer were examined for the ability to release soybean trypsin inhibitor (molecular weight 21,000). The polymer–water ratio in the casting solution of polyvinylalcohol was important in the determination of release rates. The effects of various polyvinylalcohol concentrations in the casting solution on the rate of release of soybean trypsin inhibitor are summarised in Fig. 1. An increase in the polymer concentration in the casting solution significantly decreased the initial rate at which protein was released. A threshold level was apparently reached at about 10% polyvinylalcohol, because further increases in polymer concentration did little to retard release from the pellet. Water was also added to the Hydron–alcohol–protein casting solution. The greater the water–polymer ratio, the greater was the rate of protein release from the pellet.

Figure 1 illustrates two additional features of these slow release systems: (1) a large percentage of protein was released during the first hour of incubation, that is, 'burst effect'; (2) the rate of release was significantly retarded in polymer 'sandwiches'.

Figure 2*a* shows the release of soybean trypsin inhibitor from the three polymers. Protein was released from Hydron relatively rapidly, more slowly from polyvinylalcohol, and least rapidly from ethylene–vinyl acetate copolymer. The 'burst' was evident for all three systems.

Fig. 1 Release of soybean trypsin inhibitor from polyvinylalcohol. The percentage of polymer in the casting solution was varied. Protein concentration in all casting solutions was 12 mg ml^{-1}. Release characteristics were studied by incubating polymer pellets in small test tubes with lactated Ringer solution at 37 °C. Periodically, pellets were removed to new tubes containing fresh Ringer solution. Protein release from the pellets was assayed by the Lowry method[6]. Polyvinylalcohol; △, 1.2%; ○, 4.8%; ▲, 6%; ■, 10%; circle and cross, 20%; ●, 10% sandwich.

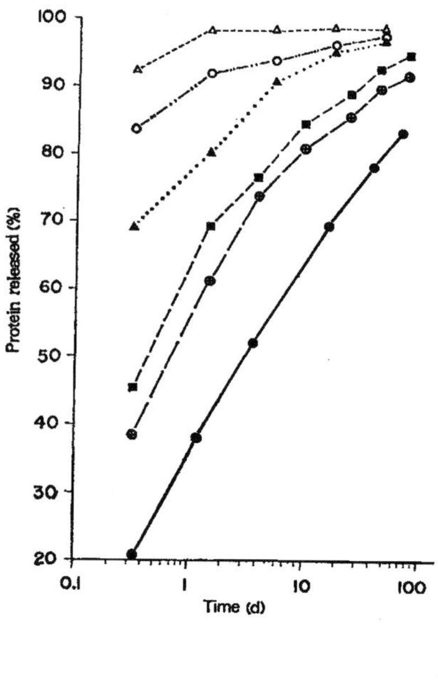

Nature Vol. 263 October 28 1976

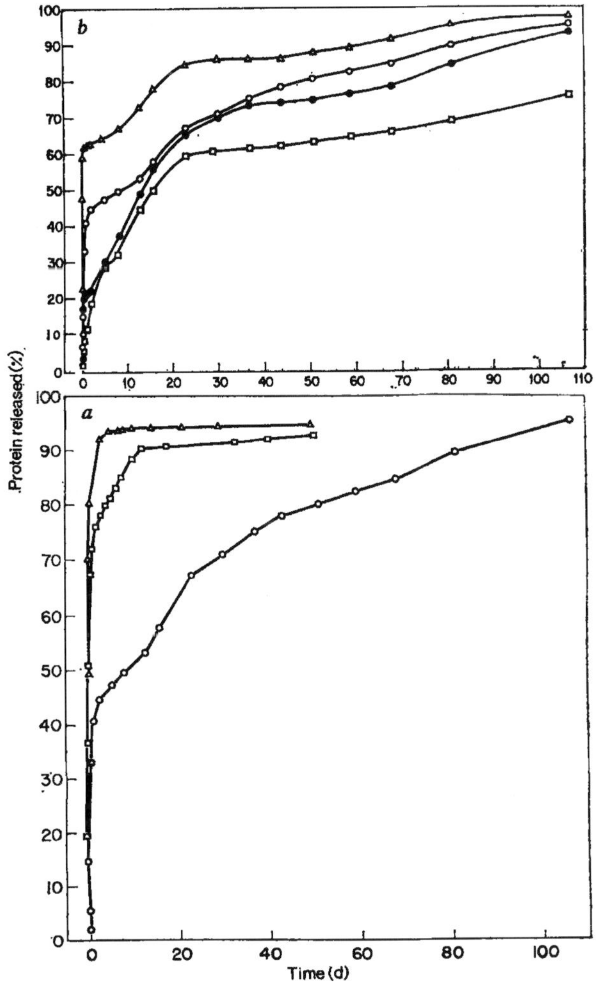

Fig. 2 *a*, Release of soybean trypsin inhibitor from Hydron (△), polyvinylalcohol (□) and ethylene–vinyl acetate copolymer 'sandwiches' (○). Protein concentration in all casting solutions was 50 mg ml^{-1}. Polymer concentrations in casting solutions were 12% Hydron, 10% polyvinylalcohol and 10% ethylene–vinyl acetate copolymer. These concentrations minimised initial rates of release from each of the individual polymer systems and still permitted viscosities low enough so that the solutions were easy to work with. *b*, Release of four proteins from ethylene–vinyl acetate copolymer 'sandwiches'. Protein and polymer concentrations in casting solutions were 50 mg ml^{-1} and 10%, respectively. Release was studied as described in Fig. 1. △, Lysozyme; ○, soybean trypsin inhibitor; ●, alkaline phosphatase; □, catalase.

Three other proteins, lysozyme (molecular weight 14,400), alkaline phosphatase (molecular weight 88,000), and catalase (molecular weight 250,000) were similarly studied. Their release profiles from ethylene–vinyl acetate copolymer are shown in Fig. 2*b*. Release rates approach zero-order kinetics over periods of 20–100 d. Using standard assays[7–9], we also found that this polymer released more than 100 ng per day of heparin, and ^{3}H-thymidine-labelled DNA (isolated from mouse SVT2 cells); and in excess of 100 μU per day of insulin for periods exceeding 40 d.

To determine whether the polymers were releasing biochemically active material *in vitro*, pellets containing 300 μg of lysozyme, soybean trypsin inhibitor, and alkaline phosphatase were placed in 20-ml volumes of Ringer at 37 °C. After 2 d of this incubation, each polymer was washed with 100 ml of Ringer for 5 min and placed on an agar diffusion slide described before[10–12]. Formation of zones of activity on these slides indicated the release of at least 100 ng per day of biochemically active material. When this cycle was repeated, new zones continued to form for periods exceeding 100 d with ethylene–vinyl acetate copolymer 'sandwiches', and for 14 and 4 d with polyvinylalcohol and Hydron 'sandwiches', respectively. Furthermore, when the specific activity of alkaline phosphatase was tested before and after incorporation into each of the three polymers, more than 80% of the enzyme escaping from the polymers was found to be active.

To determine whether biologically active macromolecules could be released *in vivo*, ethylene–vinyl acetate copolymer 'sandwiches' containing tumour angiogenesis factor (TAF), a tumour cell extract that induces neovascularisation[13], were tested. Pellets containing TAF were implanted into 20 rabbit corneas and in every case vessels sprouted from the corneal edge and grew towards the polymer. No oedema or white cells were detected. After 40 d, several of the polymers were removed and vessels disappeared within 21 d. When the same polymers were washed in Ringer solution and transplanted to new corneas, vascular responses were again observed, although rates of vessel growth were somewhat slower than with the initial implants. Also, pellets of polyvinylalcohol or ethylene–vinyl acetate copolymer containing TAF consistently produced neovascular responses when implanted on a 10- or 11-d-old chick chorioallantoic membrane. The polymers alone induced no such response.

These studies show that sustained release of proteins and other macromolecules from polymeric vehicles can be achieved over prolonged periods. The mechanism of release of large molecules from these polymers is not entirely clear. The mechanism, at least for ethylene–vinyl acetate copolymer, does not seem to depend on simple diffusion. The same macromolecules which were released so readily from this polymer after incorporation into its matrix, do not diffuse through a film of the pure polymer. Furthermore, when Dextran blue (molecular weight 2,000,000) was incorporated into ethylene–vinyl acetate copolymer, more than 90% was released in 3 d, in contrast to proteins such as alkaline phosphatase (molecular weight 88,000) where 90% release required nearly 100 d. It can be speculated that because this polymer is permeable to water, proteins deep within the polymer matrix may be driven to the surface by osmotic pressure or capillary action.

There are several possible applications. These polymers may be useful in the development of bioassays for various substances, including growth factors, chemotactic compounds and other informational molecules. The ability of ethylene–vinylacetate copolymer in casting solution to bind to plastics may facilitate new approaches for tissue culture experiments. It remains to be seen whether these polymers will eventually be clinically useful for long term delivery of macromolecules such as insulin, heparin or enzymes.

We thank Dr Patrick Wong of Alza Corporation for advice and polymers, and we thank Ken Tyler, Henry Brem, Dr David Tapper, Dr Michael Klein, Dr Kenneth Falterman, Dr Michael Klagsbrun, Germaine Grant, Von Gryska and Carl Cobb for assistance.

ROBERT LANGER
JUDAH FOLKMAN

Department of Surgery,
Children's Hospital Medical Center and
Harvard Medical School,
Boston, Massachusetts 02115

Received June 21; accepted August 30, 1976.

1 Folkman, J., and Long, D. M., *J. surg. Res.*, **4**, 139–142 (1964).
2 Davis, B. K., *Experientia*, **28**, 348 (1972).
3 Davis, B. K., *Proc. natn. Acad. Sci. U.S.A.*, **71**, 3120–3123 (1974).
4 Gimbrone, M. A., Jr, Cotran, R. S., Leapman, S. B., and Folkman, J., *J. natn. Cancer Inst.*, **42**, 314–427 (1974).
5 Fromer, C. H., and Klintworth, G. K., *Am. J. Path.*, **79**, 537–554 (1975).
6 Lowry, O. H., Rosebrough, N. J., Farr, A. L., and Randall, R. J., *J. biol. Chem.* **193**, 265–275 (1951).

7 Bittner, T., and Muir, H., *Analyt. Biochem.*, **4**, 330–334 (1962).
8 Furano, A. V., *Analyt. Biochem.*, **43**, 639–640 (1972).
9 Soeldner, J. S., and Slone, D., *Diabetes*, **14**, 771 (1965).
10 Carlsson, L., and Karlsson, B., *Experientia*, **28**, 990–991 (1972).
11 Schill, W. R., and Schumacher, G. F. B., *Analyt. Biochem.*, **46**, 502–533 (1972).
12 Schumacher, G. F. B., and Schill, W. R., *Analyt. Biochem.*, **48**, 9–26 (1972).
13 Folkman, J., *Cancer Res.*, **34**, 2109–2113 (1974).

Inhibition of Calcification of Bioprosthetic Heart Valves by Local Controlled-Release Diphosphonate

Abstract. *Bioprostheses fabricated from porcine aortic valves are widely used to replace diseased heart valves. Calcification is the principal cause of the clinical failure of these devices. In the present study, inhibition of the calcification of bioprosthetic heart valve cusps implanted subcutaneously in rats was achieved through the adjacent implantation of controlled-release matrices containing the anticalcification agent ethanehydroxydiphosphonate dispersed in a copolymer of ethylene-vinyl acetate. Prevention of calcification was virtually complete, without the adverse effects of retarded bone and somatic growth that accompany systemic administration of ethanehydroxydiphosphonate.*

Bioprostheses fabricated from porcine aortic valves treated with glutaraldehyde are widely used to replace diseased cardiac valves (*1–3*). More than 300,000 of these bioprostheses have been used in clinical implants since 1971 (*4–7*). Cuspal calcification is the most frequent cause of the clinical failure of these devices (*4–6*) and necessitates removal of the prostheses after 5 years in more than 50 percent of the children (*5*) and 5 to 10 percent of the adults receiving the valve implants (*2, 6, 7*). The pathogenesis of cuspal calcification is not completely understood (*7*), and there is no effective therapy for its prevention (*7*). Its pathophysiology can be reproduced with orthotopic valve replacements in sheep (*8*) or calves (*9*), or with subcutaneous cuspal implants in rabbits (*10*), mice (*11*), or rats (*12*). Subcutaneous implants are used because the time course of the mineral-

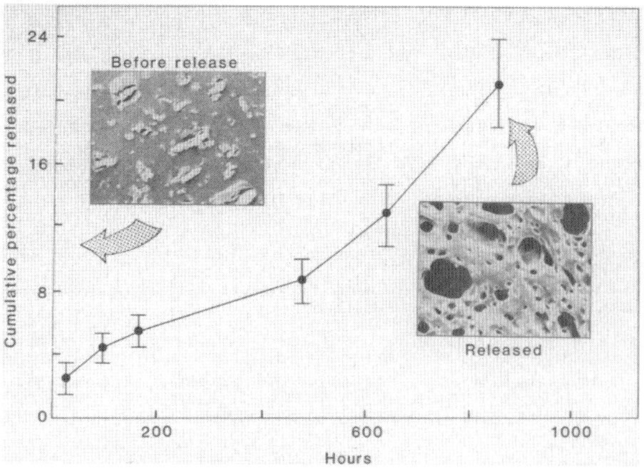

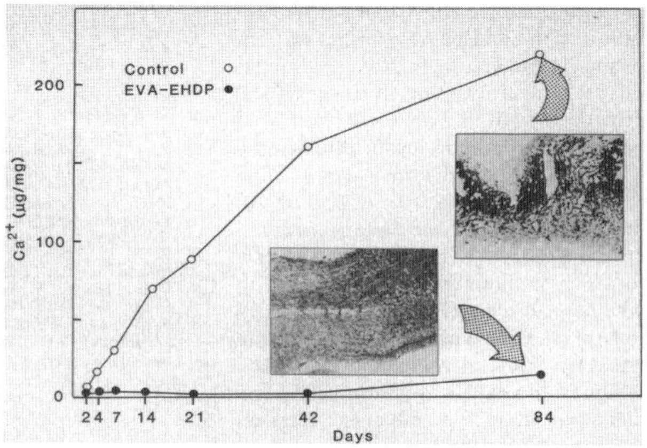

Fig. 1 (left). Cumulative percentage of EHDP released from an ethylene-vinyl acetate–EHDP matrix. Polymer matrices containing [^{14}C]EHDP were incubated at 37°C in 0.1*M* NaH_2PO_4, 7.40; released radioactivity was determined periodically. Inserts: scanning electron microscopy (ISI model DS-130) was performed on specimens before release and after 42 days of release in vitro. **Fig. 2 (right).** Ethylene-vinyl acetate (EVA)–EHDP inhibition of bioprosthetic valve cusp calcification. Calcification was assessed in cusps implanted subcutaneously next to 20 percent (by weight) ethylene-vinyl acetate–EHDP matrices or matrices containing only ethylene-vinyl acetate. Portions of retrieved cusps were fixed by the methods of Karnovsky (*22*) in cacodylate (0.2*M*)-buffered (*p*H 7.2) glutaraldehyde (2.5 percent) and formaldehyde (2.0 percent). Sections (3 µm) were stained with the von Kossa technique. Light microscopy (see inset) showed extensive intrinsic calcification at 84 days in a control specimen but no visible deposits in the implants treated with ethylene-vinyl acetate–EHDP.

ization is accelerated and the technique is simple (*12*).

In an effort to prevent this disease process, we administered the anticalcification agent ethanehydroxydiphosphonate (EHDP) directly into cusps implanted subcutaneously in rats. Local therapy was achieved with controlled-release matrices containing EHDP dispersed in a copolymer of ethylene-vinyl acetate.

Bioprosthetic heart valve cusps were prepared from porcine aortic valves that had been treated with glutaraldehyde (*12*) and stored in 0.2 percent glutaraldehyde (*p*H 7.4; 0.05*M* Hepes). Ethylene-vinyl acetate was obtained as a dried powder (ELVAX; 40 percent vinyl acetate; Du Pont), washed in alcohol and water, and dried (*13*). Ethanehydroxydiphosphonate was obtained as the disodium salt, both ^{14}C-labeled (9.2 mCi/mmol), and nonlabeled (Procter & Gamble). Rats used in these studies were 3-week-old male weanlings of the CD strain (Charles River).

Ethylene-vinyl acetate dissolved in methylene chloride (20 percent, weight to volume) was mixed with EHDP (20 percent by weight) and cast as hemispherical matrices (diameter, 1 cm) (*14*). This geometric structure provides nearly constant release rates for various test drugs (*14*). Before the release studies, the matrices were coated with ethylene-vinyl acetate in methylene chloride (15 percent, weight to volume) and air dried; a release aperture, 0.58 mm in diameter, was drilled into the flat face of the hemisphere.

Subcutaneous implants of ethylene-vinyl acetate–EHDP matrices were placed next to subcutaneous bioprosthetic cuspal (2 by 3 cm) implants in the rats (*12*). Control implants consisted of bioprosthetic cusps implanted next to ethylene-vinyl acetate hemispheres that did not contain EHDP. Animals were killed at intervals up to 84 days after implantation. Retrieved cusps were prepared for mineral analyses by hydrolysis with 6*N* HCl (*12*) and were analyzed for calcium by atomic absorption spectroscopy (*12*); the cusps were also analyzed for phosphorus (*15*). Blood specimens obtained by cardiac puncture at the end of the study were similarly analyzed for serum calcium. Potential adverse effects on bone development were assessed as follows. Representative femurs of the rats killed 84 days after receiving implants were fixed in neutral buffered formaldehyde solution, and the cortex was superficially decalcified to permit microtomy. The bones were embedded in glycomethacrylate, sectioned (3 µm) through the proximal femoral epiphyseal growth plate, and stained with hematoxylin and eosin to detect cellular abnormalities associated with diphosphonate toxicity (*16, 17*). The von Kossa technique was used to detect defective mineralization resulting from diphosphonate administration (*16, 17*).

Uniform, sustained release of EHDP [mean ± standard error of the mean (S.E.M.), 5.0 ± 0.4 µg/hour] was observed for more than 600 hours in vitro, with only 12 percent cumulative release by this time (Fig. 1). Scanning electron micrographs of unimplanted matrices of ethylene-vinyl acetate–EHDP showed a uniform distribution of EHDP particles ranging in size from 1 to 10 µm. After release of EHDP in vitro, scanning electron microscopy showed that canals had formed with dissolution of the EHDP (Fig. 1).

Minimal calcification was present in cusps removed from animals with hemisphere implants containing ethylene-vinyl acetate–EHDP. Control implants, in contrast, showed progressive extensive calcification (Fig. 2). Serum calcium levels were normal in both the group treated with EHDP for 84 days [Ca^{2+}, 9.5 ± 0.3 (S.E.M.) mg/dl; $n = 4$] and control groups (Ca^{2+}, 9.5 ± 0.3 mg/dl; $n = 3$). The body weight of the animals killed after 84 days of treatment with ethylene-vinyl acetate–EHDP [mean, 463 (± 48.3 standard deviation) g; $n = 3$] did not differ significantly from that of the controls [mean, 563 (± 23) g; $n = 4$], and histologic examination of representative von Kossa and hematoxylin-eosin sections of the femoral growth plates from the animals treated with ethylene-vinyl acetate–EHDP revealed no abnormalities in bone development.

Thus, local controlled release of EHDP from ethylene-vinyl acetate copolymer matrices directly into bioprosthetic cuspal implants inhibits calcification of the subcutaneous implants for up to 84 days without detectable untoward effects associated with diphosphonate therapy. The amount of calcification in the control cusps by 84 days was comparable to amounts occurring after several years of clinical implantation (*7*). Diphosphonates, including EHDP, inhibit calcification in bone and pathologic mineral deposits by binding to hydroxyapatite deposits and preventing further crystal growth (*16–18*). In an earlier study in which bioprosthetic cusp calcification was inhibited through the use of systemic EHDP, severe adverse effects on somatic growth and bone development occurred (*19*). However, we have now performed preliminary experiments in which short-term (2 weeks) administration of EHDP with an osmotic pump (ALZET 2001, Alza, Stanford, Calif.) inhibited the calcification during the implant period, with no adverse effects on bone development or somatic growth (data not shown). Technical constraints would prohibit the use of this type of osmotic device for long-term local therapy. The long-term (84 days) controlled-release system we have developed to inhibit cuspal calcification delivered EHDP at an approximate total body dosage of 6 μg/kg. Ethylene-vinyl acetate–EHDP matrices could potentially be incorporated into clinical valve prostheses during fabrication. In addition, ethylene-vinyl acetate–EHDP may be useful in controlling calcification in other types of prosthetic implants (*20, 21*).

ROBERT J. LEVY*
Department of Cardiology,
Children's Hospital,
Boston, Massachusetts 02115
JACQUELINE WOLFRUM
Department of Applied Biological Sciences and the Whitaker College of Health Sciences and Technology, Massachusetts Institute of Technology, Cambridge 02139
FREDERICK J. SCHOEN
Department of Pathology,
Brigham and Women's Hospital,
Boston, Massachusetts 02115
MARGUERITE A. HAWLEY
SALLY ANNE LUND
Department of Cardiology,
Children's Hospital
ROBERT LANGER
Department of Applied Biological Sciences and the Whitaker College of Health Sciences and Technology, Massachusetts Institute of Technology

References and Notes

1. A. Carpentier, G. Lemaigre, L. Robert, S. Carpentier, C. DuBost, *J. Thorac. Cardiovasc. Surg.* **58**, 467 (1969).
2. L. H. Cohn, G. H. Mudge, F. Pratter, J. J. Collins, Jr., *N. Engl. J. Med.* **304**, 258 (1981).
3. W. W. Angell, J. D. Angell, J. C. Kosek, *J. Thorac. Cardiovasc. Surg.* **83**, 493 (1982).
4. T. L. Spray and W. C. Roberts, *Am. J. Cardiol.* **40**, 319 (1977).
5. S. P. Sanders, R. J. Levy, M. D. Freed, W. I. Norwood, A. R. Castaneda, *ibid.* **46**, 429 (1980).
6. F. J. Schoen and C. E. Hobson, *Hum. Pathol.*, in press.
7. F. J. Schoen and R. J. Levy, *Cardiol. Clin.* **2**, 717 (1984).
8. Eidbo, R. J. Carroll, V. J. Ferrans, *Am. J. Cardiol.* **53**, 1388 (1984).
9. R. J. Levy, J. A. Zenker, W. F. Bernhard, *Ann. Thorac. Surg.* **36**, 187 (1983).
10. M. Fishbein *et al.*, *J. Thorac. Cardiovasc. Surg.* **83**, 602 (1982).
11. R. J. Levy, F. J. Schoen, and S. L. Howard, *Am. J. Cardiol.* **52**, 629 (1983).
12. R. J. Levy, F. J. Schoen, J. T. Levy, A. C. Nelson, S. L. Howard, L. J. Oshry, *Am. J. Pathol.* **113**, 143 (1983). Analytic-grade glutaraldehyde was obtained from Eastman-Kodak, Rochester, N.Y.
13. R. Langer, *Methods Enzymol.* **73**, 57 (1981).
14. D. S. T. Hsieh, W. D. Rhine, R. Langer, *J. Pharm. Sci.* **72**, 17 (1983).
15. P. S. Chen, T. Y. Toribara, H. Warner, *Anal. Chem.* **28**, 1756 (1956).
16. A. B. Gasser, D. B. Morgan, H. A. Fleisch, L. J. Richelle, *Clin. Sci.* **43**, 31 (1972).
17. R. Schenk, W. A. Merz, R. Muhlbauer, R. G. G. Russell, H. Fleisch, *Calcif. Tissue Res.* **11**, 196 (1973).
18. J. L. Meyer and G. H. Nancollas, *ibid.* **13**, 295 (1973).
19. R. J. Levy *et al.*, *Circulation* **68** (Suppl. 3), 395 (1983).
20. W. S. Pierce, J. H. Donachy, G. Rosenberg, *Science* **208**, 601 (1980).
21. J. B. Lian, R. J. Levy, W. F. Bernhard, M. Szycher, *Trans. Am. Soc. Artif. Intern. Organs* **46**, 429 (1981).
22. M. J. Karnovksy, *J. Cell Biol.* **27**, 137A (1965).
23. We thank E. Flynn for preparing the illustrations and L. D. Helstowski, L. Brown, and J. Kost for their assistance. This work was supported by NIH grants HL24463, HL20764, and GM26698.
* To whom requests for reprints should be sent.

11 September 1984; accepted 10 December 1984

An Enzymatic System for Removing Heparin in Extracorporeal Therapy

Abstract. *The need to fully heparinize patients undergoing extracorporeal therapy often leads to hemorrhagic complications. To enable heparinization of only the extracorporeal circuit, a blood filter containing immobilized heparinase was developed. This filter degraded 99 percent of heparin's anticoagulant activity within minutes in both canine and human blood.*

Extracorporeal devices perfused with blood, such as the artificial kidney and the pump-oxygenator, have been an important part of the clinician's armamentarium for many years. These devices all rely on the patient being heparinized to prevent clotting in the device. However, heparinization of the patient often leads to hemorrhagic complications (*1*, *2*). With the prospect of longer perfusion times with machines such as the membrane oxygenator (*3*), the drawbacks of systemic heparinization are multiplied. A number of approaches have been tested to solve this problem. They include (i) administering protamine to neutralize heparin (*4*), (ii) using antithrombotic drugs other than heparin (*5*), (iii) bonding heparin, albumin, or fibrinolytic enzymes to the extracorporeal device (*6*), and (iv) developing new blood-compatible materials for construction of the device (*7*). Although these approaches have led to some improvements, control of the heparin concentration in the blood remains a serious problem.

We suggest a novel approach that would permit full heparinization of blood entering any extracorporeal device but which would enable enzymatic elimination of the heparin before the blood returned to the patient. This approach consists of placing a blood filter containing immobilized heparinase—which degrades heparin into small polysaccharides—at the effluent of the extracorporeal device. Initial studies reported here to test the feasibility of this approach show that a conventional blood transfusion filter containing immobilized heparinase can remove the anticoagulant activity of clinically used amounts of heparin in several minutes in human blood and in dogs.

Heparinase was produced from *Flavobacterium heparinum* (*8*) and purified by cell sonication, nucleic acid precipitation with protamine sulfate, and hydroxylapatite chromatography (*9*). At this stage of purification, heparinase degrades 150 mg of heparin per milligram of protein per hour and is catalytically pure (free from contaminating sulfatases and glucuronidases) (*10*). Heparinase was covalently bound to Sepharose and washed to remove noncovalently bound enzyme by a variation (*9*) of the procedure of March *et al.* (*11*). There was no detectable leaching, as judged by protein concentration (*12*), of heparinase from the support over a 1-month period at 25°C.

Immobilization increases the stability of heparinase: at 4°, 25°, and 37°C the immobilized enzyme has half-lives of 5000, 1000, and 15 hours, respectively, compared with the native enzyme, which has half-lives of 125 and 30 hours and 1 hour at the same temperatures.

Active, immobilized heparinase and a control consisting of thermally inactivated, immobilized heparinase were loaded into two columns with a 1.5-ml bed volume. Heparin solutions were passed through both columns in stepwise fashion at concentrations of 15 to 75 μg/ml (clinically used heparin concentrations, having a specific activity of 0.15 U/μg, range from 5 to 25 μg/ml) (Fig. 1). As the heparin level in the input solutions increased, the difference in the heparin recovered at the outlets of the columns increased (Fig. 1). Even at 75 μg/ml, the heparin was largely removed by the column containing active heparinase, while the control column had no effect.

At flow rates above 1 ml/min the Sepharose packed tightly, preventing flow. Therefore we used a fluidized bed of Sepharose, obtained by pumping Sepha-

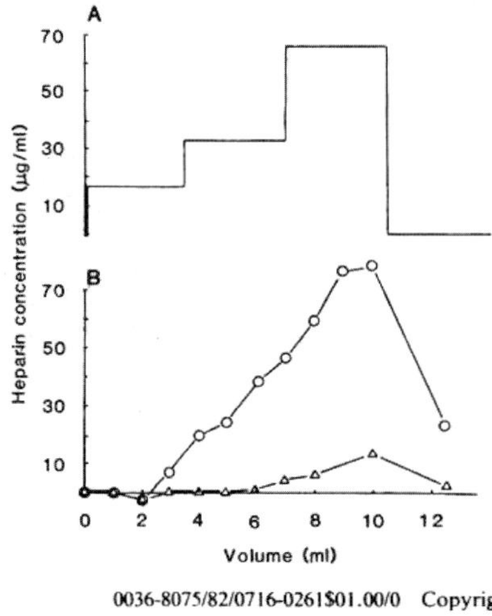

Fig. 1. (A) Input concentrations of heparin to both active and inactive columns, showing stepwise increases from 15 to 75 μg/ml. (B) Output concentrations of heparin from the column containing active heparinase (△) and the column containing heat-inactivated heparinase (○). Solutions of bovine serum albumin (60 mg/ml), salts, and heparin were passed through the columns (diameter, 1 cm) at a flow rate of 0.5 ml/min. The inactive heparinase was prepared by heating a suspension of active Sepharose-heparinase at 100°C for 30 minutes. The physiological concentrations of the nonheparin species in blood were used. Heparin concentration was measured by the azure assay (*8*, *17*).

0036-8075/82/0716-0261$01.00/0

rose through a U-tube interconnected to a 100-ml blood transfusion filter at a flow rate six times greater than the rate at which blood was being pumped through the same filter (Fig. 2).

Whole human blood, anticoagulated with 90 U of heparin per milliliter (an even higher concentration than is normally used clinically) (*1*), was passed through the filter at a flow rate of 50 ml/min. Blood samples were taken directly from the blood reservoir (sample port a in Fig. 2) before and after each of six passes of the entire blood volume through the filter. As shown in Fig. 3A, 60 percent of heparin's anticoagulant activity, as measured by activated partial thromboplastin time (APTT) (*13*), was destroyed in 2 minutes (one pass), and almost all the activity was gone after 6 minutes (three passes). To demonstrate that the change in APTT was due to a decrease in the heparin concentration and not to a nonspecific effect of immobilized heparinase on coagulation factors, we also used an azure A assay that detects heparin metachromatically (*14*). This assay also showed that heparin levels fell as blood passed through the filter (Fig. 3A). Control filters containing Sepharose without heparinase showed no binding or degradation of heparin in the blood, as judged by both assays.

In an in vivo experiment, two dogs were anesthetized and the carotid artery and internal jugular vein were cannulated with a Scribner shunt (*15*) connected to Silastic tubing joined directly to the heparinase reactor. Heparin (450 U/kg) was added and blood was sampled after it left the filter (sample port b). Three heparin assays were used: APTT, whole blood recalcification time (WBRT) (*16*), and azure A (*8, 17*). Within 2 minutes

Fig. 2. Diagram showing the immobilized heparinase reactor system. Two holes 9.5 mm in diameter were made in the top of a Bentley infusion filter (model PFT-100). Silicone rubber tubing (inner diameter, 0.8 cm; outer diameter, 1.1 cm; length, 27 cm) was inserted 10 cm into one hole and 3 cm into the other hole and the connections were sealed with a silicone sealant. A 40-cm silicone tube (inner diameter, 1.5 cm; outer diameter, 1.0 cm) with a sample port was clamped to the input of the filter. A medical grade Tygon tube (Cobe Hemaflow) from a kidney dialysis set with a bubble trap and a sampling septum was connected to the filter exit. The reactor was filled with saline and then purged of all air bubbles. A suspension of the Sepharose-heparinase was then slowly (10 to 15 ml/min) pumped into the reactor.

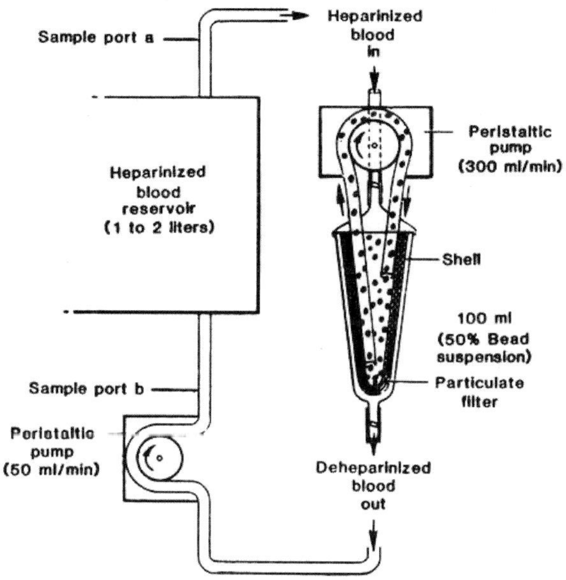

Fig. 3. (A) In vitro deheparinization of human blood. One liter of blood was obtained from healthy donors at the Children's Hospital Blood Bank, Boston. (The donors were advised not to eat during the morning of the donation or to take medication during the 3 days before donation.) While still fresh, the blood was heparinized (90 U/ml) and passed through the filter at a rate of 50 ml/min. The filter was placed between two 1-liter Fenwal blood transfer packs. After the entire blood volume from one pack had passed through the filter into the other pack (one pass), the packs were disconnected and their positions with respect to the filter were switched. The blood was sampled from the blood-containing pack before heparinization and after each of six passes through the filter. Samples were drawn into Vacutainers containing sodium citrate and centrifuged at 4500 rev/min for 20 minutes; the plasma was decanted from the formed components. Heparin in plasma was assayed for anticoagulant activity by APTT (△) (*13*) and for metachromatic activity by azure A (□) (*8, 17*). The reason that the azure A assay does not show zero heparin activity is that azure detects only heparin fragments of hexasaccharide size or larger. Fragments of this size are present [we determined this by high-pressure and gel permeation chromatography (*10*); digestion of heparin by heparinase in buffer also demonstrates this phenomenon], and these fragments are detected by azure A but do not possess anticoagulant activity as measured by APTT (*24*). (B) In vivo deheparinization of blood from an 11-kg female and a 22-kg male dog (blood volume, 80 cm^3/kg). A polyethylene catheter tip (No. 17 arterial, No. 19 venous) attached to a silicone rubber coupler tube was inserted into the carotid artery and internal jugular vein. The Tygon tubing from the filter (Fig. 2) was connected to these catheters. The dogs received heparin at a dose of 450 U/kg. Blood was drawn from the carotid artery through a peristaltic pump to the filter. Blood exiting the filter flowed through a sampling septum, then through a bubble trap containing a Dacron mesh thrombus filter, and was returned to the jugular vein. Samples were collected from sampling ports a and b in Vacutainers containing EDTA and analyzed immediately by Coulter counter at the Hematological Laboratory, Boston Brigham and Women's Hospital. The data were normalized to 0 percent activity before heparinization of the dog and 100 percent activity immediately after heparinization (●). Plasma samples to measure heparin in blood were prepared and assayed for anticoagulant activity by APTT [sample port a (▲) and sample port b (△)] and for metachromatic activity by azure A [sample port b (□)].

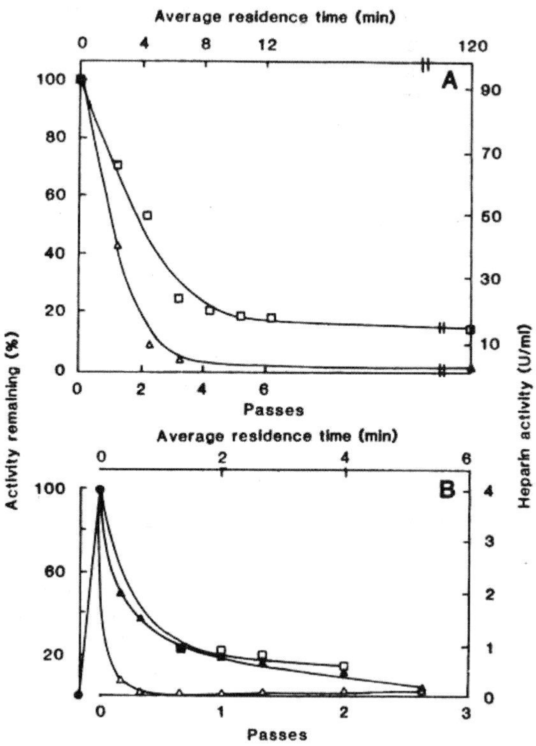

nearly all of heparin's anticoagulant activity measured by APTT was lost (Fig. 3B); the WBRT test gave identical results. The azure assay again demonstrated direct loss of heparin. Although it might be expected that removal of heparin would cause clotting, blood continued to flow unrestricted through the filter even after six passes. Blood samples were also taken at sample port a. After one pass there was an 80 percent loss of heparin by APTT (70 percent by azure) and after two passes there was a 90 percent loss by both assays (Fig. 3B). Heparin levels measured at sample port a were consistently higher than those measured at sample port b, presumably due to the mixing of blood in the dog. Control reactors without heparinase had no detectable effect on blood heparin levels in vivo (an in vivo half-life of 2 hours for heparin at this dosage was measured in the dog). The dogs still appeared healthy 3 months after the experiments. Blood was taken from the dogs 1 and 2 months after the experiments, and no antibodies to heparinase were detected by Ouchterlony precipitin tests (*18*). Products of the enzymatically degraded heparin were tested for cytotoxic and mutagenic effects on *Salmonella typhimurium* (*19*). No toxicity or mutagenicity was observed, even with concentrations 1000 times in excess of those we would anticipate clinically. In addition, heparin was neither toxic nor mutagenic by this assay.

At the end of the experiment, blood taken from the dogs showed no decrease in hematocrit, a 30 percent decrease in white blood cell count, and a 70 percent decrease in platelet count. These values are typical of those obtained for tests of extracorporeal circuits in dogs (*1*, *20*). To further improve the blood compatibility of immobilized heparinase, other heparinase supports and the use of a smaller filter should be examined. The latter is possible because heparinase has been purified to specific activities 70 times greater (with low recovery) than the enzyme used in this study by phosphocellulose ion-exchange chromatography and isoelectric focusing (*9*).

An immobilized heparinase filter could be used either at the end of a clinical procedure to eliminate heparin without the toxic effects of heparin-neutralizing substances such as protamine or continuously to prevent high levels of heparin from ever entering the patient. Blood filters as large as 2 liters are often used at the effluent of extracorporeal devices to remove microemboli. Heparinase might be bonded to the biomaterials of these filters. Unlike many proposed medical applications of immobilized enzymes such as asparaginase (*21*) and phenylalanine ammonia–lyase (*22*), immobilized heparinase would be used in procedures in which blood must enter extracorporeal circulation and in which existing biomaterials already interact with blood at the desired location (the end of the extracorporeal circuit). Thus this process may not require any additional apparatus or invasive procedure.

The heparinase filter may also be a prototype for other selective drug removal systems. Microbial selection techniques (*8*) could produce enzymes for use in filters to degrade potentially toxic drugs such as amphetamines or barbiturates. Such drug removal systems, like drug delivery systems (*23*), could prove useful in controlling drug concentrations in the blood.

ROBERT LANGER

Department of Nutrition and Food Science, Massachusetts Institute of Technology, and *Harvard-MIT Division of Health Sciences and Technology, Cambridge 02139*, and *Department of Surgery, Children's Hospital, Boston 02115*

ROBERT J. LINHARDT
STEVEN HOFFBERG
ANNETTE K. LARSEN
CHARLES L. COONEY

Department of Nutrition and Food Science, Massachusetts Institute of Technology

DAVID TAPPER

Department of Surgery, Children's Hospital Medical Center, and *Harvard Medical School, Boston*

MICHAEL KLEIN

Section of Pediatric Surgery, Mott Children's Hospital, and *University of Michigan, Ann Arbor 48109*

References and Notes

1. A. S. Gervin, *Surg. Gynecol. Obstet.* **140**, 789 (1975).
2. D. Basu, M. Gallus, J. Hirsh, J. Cade, *N. Engl. J. Med.* **287**, 324 (1972).
3. J. R. Fletcher, A. E. McKee, M. Mills, K. C. Snyder, C. M. Herman, *Surgery* **80**, 214 (1976).
4. J. Jastrzebski, P. Hilgard, M. K. Sykes, *Cardiovasc. Res.* **9**, 691 (1975).
5. E. Berglin, H. A. Hansson, A. C. Teger-Nilsson, G. William-Olsson, *Thromb. Res.* **9**, 81 (1976).
6. R. G. Mason, *Prog. Hemostasis Thromb.* **1**, 141 (1972).
7. W. J. Kolff and F. Stellwag, *Ann. N.Y. Acad. Sci.* **283**, 443 (1977).
8. P. M. Galliher, C. L. Cooney, R. Langer, R. J. Linhardt, *Appl. Environ. Microbiol.* **41**, 360 (1981).
9. R. Langer, R. J. Linhardt, M. Klein, M. M. Flanagan, P. M. Galliher, C. L. Cooney, in *Biomaterials: Interfacial Phenomena and Applications*, S. Cooper, A. Hoffman, N. Peppas, B. Rattner, Eds. (American Chemical Society, Washington, D.C., 1982), chap. 13, pp. 493–509.
10. R. J. Linhardt, G. L. Fitzgerald, C. L. Cooney, R. Langer, *Biochim. Biophys. Acta* **702**, 197 (1982).
11. S. C. March, I. Parikh, P. Cuatrecasas, *Anal. Biochem.* **60**, 149 (1974).
12. O. H. Lowry, N. J. Rosebrough, A. L. Farr, R. J. Randall, *J. Biol. Chem.* **193**, 265 (1951).
13. R. D. Langdel, in *Thrombosis and Bleeding Disorders*, N. U. Bang, F. K. Beller, E. Deutsch, E. F. Mammen, Eds. (Academic Press, New York, 1971).
14. C. P. Dietrich, *Biochem. J.* **108**, 647 (1968).
15. W. Quinton, D. Dillard, B. H. Scribner, *Trans. Am. Soc. Artif. Intern. Organs* **6**, 104 (1960).
16. C. H. Tsao, T. S. Galluzzo, M. Lo, K. G. Peterson, *Am. J. Clin. Pathol.* **71**, 17 (1979).
17. M. D. Klein, R. A. Drongowski, R. J. Linhardt, R. S. Langer, *Anal. Biochem.*, in press.
18. O. Ouchterlony and L. A. Nilsson, in *Handbook of Experimental Immunology*, D. M. Weir, Ed. (Oxford Univ. Press, Oxford, England, ed. 3, 1978), pp. 19.1–19.44.
19. T. R. Skopek, H. L. Liber, J. J. Krowelski, W. G. Thilly, *Proc. Natl. Acad. Sci. U.S.A.* **75**, 410 (1978).
20. J. C. F. deJong, C. Th. Smit Sibinga, Ch. R. H. Wildevuur, *Trans. Am. Soc. Artif. Intern. Organs* **21**, 40 (1975); J. C. F. deJong, J. M. Nelems, C. Th. Smit Sibinga, Ch. R. H. Wildevuur, *ibid.* **20**, 596 (1974); P. D. Richardson, *Bull. N.Y. Acad. Med.* **48**, 379 (1972).
21. L. Olanoff, K. Venkatasubramanian, F. R. Bernath, R. Joyeuse, *J. Biomed. Mater. Res.* **8**, 125 (1977).
22. C. M. Ambrus *et al.*, *Science* **201**, 837 (1978).
23. R. Langer, *Chem. Eng. Commun.* **6**, 1 (1980); *Chemtech* **12**, 98 (1982).
24. R. J. Linhardt, A. Grant, C. L. Cooney, R. Langer, *J. Biol. Chem.*, in press.
25. Supported by NIH grant GM 25810. We thank B. Andon and R. Drongowski for assistance.

25 January 1982; revised 13 April 1982

 0036-8075/82/0716-0263$01.00/0

Journal of Immunological Methods, 28 (1979) 193—197 193

Short communication

A SINGLE-STEP IMMUNIZATION BY SUSTAINED ANTIGEN RELEASE

IVAN PREIS and ROBERT S. LANGER [1]

Department of Surgery, Children's Hospital Medical Center, Boston, MA 02115 and Department of Nutrition and Food Science, Massachusetts Institute of Technology, Cambridge, MA 02139, U.S.A.

(Received 1 July 1978, accepted 26 March 1979)

An inert polymer pellet less than 1 mm in diameter, implanted subcutaneously in mice, releases free antigen continuously and enhances antibody formation for over 6 months. The immune response stimulated by sustained antigen delivery is comparable to the secondary response induced by the same total dose of antigen emulsified in complete Freund's adjuvant. The sustained release polymer implants, using antigens over a wide molecular weight range, proved effective in eliciting prolonged antibody formation. The antigens tested included bovine serum albumin, γ-globulin, and ribonuclease. Superior biocompatibility and release characteristics make the ethylene-vinyl acetate copolymer a promising method for the single-step induction of immunity.

INTRODUCTION

Since their introduction by Freund (1951) 28 years ago, the water-in-oil suspensions containing antigen have become standard immunological adjuvants. However, poor degradation of the mineral oil base restricts their use to experimental animals. Consequently, there has been an effort to find safer adjuvants with comparable effectiveness for use in human immunization (Hilleman, 1965). Although these antigen-dispersing vehicles enhance the antibody production if compared to injection of free material, they fall short of the level attained with the help of Freund's complete or incomplete (minus bacteria) adjuvants. We now present preliminary evidence that a sustained release polymer can provide a simple, safe and effective means of immunization. In this study, a single inert polymer pellet released antigen continuously and stimulated antibody formation for over 25 weeks when implanted subcutaneously in mice.

We have previously demonstrated that the ethylene-vinyl acetate

[1] Investigators desiring ethylene-vinyl acetate copolymer should write to Dr. Langer at the Department of Nutrition and Food Science, 16-309, Massachusetts Institute of Technology, Cambridge, MA 02139, U.S.A. To whom reprint requests should be addressed as well.

copolymer is non-inflammatory and is capable of releasing biochemically active macromolecules for periods exceeding 100 days in vivo and in vitro (Langer and Folkman, 1976). It can be implanted as a tiny pellet or may be injected with a spinal lumbar (No. 16) needle. These unique features offer the slow-release system for exploitation in many branches of immunological research. Here we report its adjuvant activity.

MATERIALS AND METHODS

The polymer pellets were made by mixing 5 mg of dry protein powder with 100 μl of methylene chloride containing (10%, w/v) ethylene-vinyl acetate copolymer (60% ethylene/40% vinyl acetate). The mixture was vacuum-dried, cut into 0.3 mm^3 pellets (100 μg protein/pellet), and coated with a thin film of pure polymer by dipping them in polymer-methylene chloride solution for 20 sec (Langer and Folkman, 1976). Aseptic technique was followed throughout. The proteins used were crystalline bovine albumin (Schwartz-Mann), ribonuclease-A (Sigma) from bovine pancreas, and bovine γ-globulin (Cohn Fraction II, Sigma).

C57 Black/6J mice (Jackson Laboratory, Bar Harbor, ME) were anesthetized with ether. Their backs were shaved and a small incision was made in the skin. Polymers were soaked in lactated Ringer's solution for 5 min and then implanted under the skin; one pellet was used per mouse. The pocket was closed with a Michel clip.

Two sets of experiments were conducted. In one set, the antibody response induced by bovine serum albumin (BSA) in the polymer was compared to the antibody response elicited by BSA in complete Freund's adjuvant (CFA). Eight mice received albumin-containing polymers. Four control animals were injected subcutaneously, at the same site where the polymers were implanted, with 50 μg BSA/ml phosphate-buffered saline (PBS) in CFA (1 : 1, v/v); a secondary boost was given 7 weeks later (100 μg antigen total).

In a second experiment, polymers with one of 3 different antigens were implanted into 3 separate groups of mice. Single polymer pellets containing 100 μg of either BSA, ribonuclease, or γ-globulin were used. Each test group consisted of 5 animals. Additional animals were implanted with polymer pellets without antigen.

Blood was collected weekly from the median tail artery. The sera from each group were combined, and the anti-protein serum antibody titer was determined by indirect hemagglutination. Fresh sheep red blood cells (SRBC) were washed 3 times for 10 min each in 0.9% saline at 2000 rev/min. Then 1 ml of packed cells was incubated at 25°C with equal volumes of protein (1 mg/ml) and $CrCl_3$ (1 mg/ml) in saline for 4 min. The cells were then washed 3 times for 10 min each in PBS at 2000 rev/min; the final concentration was adjusted to 0.75% (v/v). Twenty-five μl of PBS with 0.1% gelatin were layered in each microtiter well, followed by 25 μl of antiserum, and the serial dilutions were performed. Next, 25 μl of pre-sensitized SRBC were pipetted into each well; hemagglutination titers were read after 1 h.

Antibody formation was corroborated qualitatively by the Ouchterlony precipitin test (Campbell et al., 1970). Immunoglobulin class was determined at weeks 6, 10, 17 and 25 by immunodiffusion with rabbit anti-mouse globulin specific sera (Kishimoto and Ishizaka, 1971) and by immunoelectrophoretic analysis (Williams and Chase, 1971). Analysis of sera from control mice containing polymers without antigen consistently showed zero antibody titer.

RESULTS

Prolonged stimulation with antigen encapsulated in a polymer increased antibody production with time and did not induce tolerance (Mitchison, 1969). The antibodies formed were the IgG types. Mice implanted with polymer pellets showed no early decline in antibody formation, as was observed with the CFA-sponsored primary response in controls; antibody titer in the experimental group increased steadily, reaching a plateau after 9 weeks (Fig. 1). When, after 7 weeks, controls were given a booster dose in CFA, antibody formation of control and experimental groups was comparable until waning of the secondary response in controls. The titer in the experimental group diminished more slowly than did that of the controls, and was slightly higher during the final 3 months. Six months following implantation of the pellets, there was still antibody (+4) detectable in sera of all experimental animals (Fig. 1). When several mice containing polymer pellets were given a booster dose in CFA to test the generation of memory

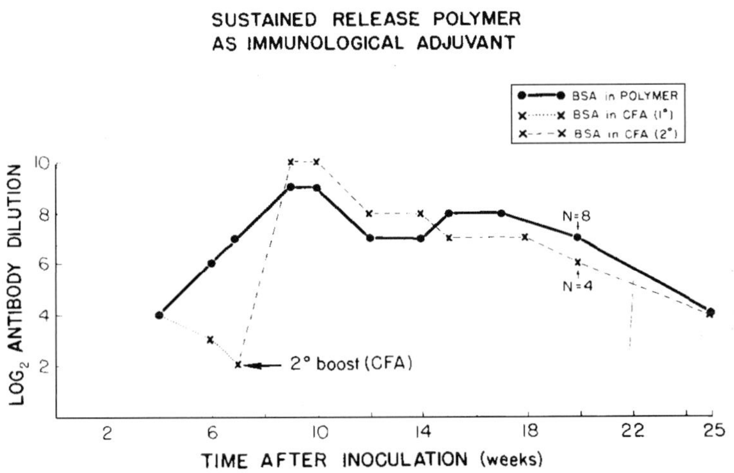

Fig. 1. Antibody titer as determined by indirect hemagglutination assay. Repeated quantitations of individual antibody titers differed from the titer of pooled antisera by $\pm 2 \log_2$ dilution units or less (S.D. = 1.16).

196

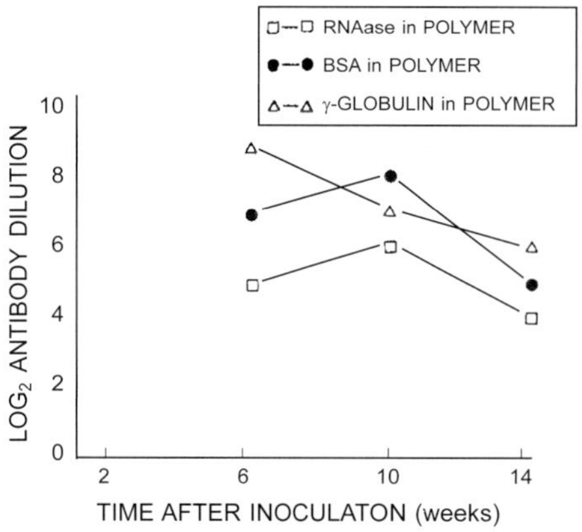

Fig. 2. Antibody titer as a function of time for 3 different antigens released by polymer. Each test group consisted of 5 mice.

cells, the hemagglutination assay 10 days later showed the antibody titer to be +7, indicating a secondary response.

The polymer capsule proved to be non-irritating throughout the 6-month study. No visible inflammatory reaction was seen at the implantation site at any time. When new BSA-containing polymers were implanted in mice at the end of the study, again no visible inflammation occurred.

In a second set of experiments, the ability of 3 different antigens of varying molecular weight to induce antibody formation was compared. All 3 antigens in the polymer elicited a prolonged immune response (Fig. 2). The strongest response was observed for γ-globulin (mol. wt. 158,000), followed by albumin (mol. wt. 68,000) and ribonuclease (mol. wt. 14,000).

DISCUSSION

The results demonstrate that an inert sustained-release polymer system can induce prolonged antibody formation. The polymer presumably acts as a repository for the antigen, protecting it from rapid degradation and allowing a small antigenic mass to be delivered continuously and effectively. In the present study, the antigen doses were always identical (100 μg), and the polymers were implanted subcutaneously for practical simplicity; other sites (e.g., peritoneum) or different doses may be more effective. Excellent biocompatibility and single-step introduction of the antigen could make this method of immunization preferable to immunological adjuvants such as CFA. Furthermore, the molecules entrapped in the polymer are released in unaltered form at controlled rates (Langer and Folkman, 1976), and immune cells are given ample time for recognition of true antigenic determinants. The rapid degradation of foreign protein is compensated by the constant diffusion of the immunogen from the polymer, thus insuring that antigen molecules are available in tissues for long periods. This may be particularly important when the quantity of immunogen is limited. In addition, valuable immunogens are not lost following implantation of the pellet, which can be removed, stored at 4° C, and reused in other animals.

The polymer proved to be an effective adjuvant for antigens over a wide molecular weight range. Although it is possible that the higher molecular weight substances are released more slowly, thereby causing prolonged, higher antibody titers, as shown in Fig. 2, previous studies (Langer and Folkman, 1976) have demonstrated that the rate of release of proteins from polymers is only weakly dependent on molecular weight. Studies in our laboratory (Langer and Folkman, 1976; Rhine et al., 1979) have shown that factors such as the type of polymer employed, the particle size of the antigen powders incorporated into the polymers, the ratio of polymer to antigen, and the thickness of the coating applied to the pellets are more critical factors in controlling release rates.

From the investigations described, it is evident that the gradual antigen-release system lends itself to experimental studies of immunological phenomena and may eventually represent a safe and efficient means for clinical immunization and toleragenesis.

ACKNOWLEDGEMENTS

We are grateful to Douglas Weber for valuable assistance, and we thank Dr. Judah Folkman for helpful comments and advice.

REFERENCES

Campbell, D.H., J.S. Garrey, N.E. Cremer and D.H. Sussdorf, 1970, Methods in Immunology, 2nd edn. (Benjamin, Inc., New York) p. 236.

Freund, J., 1951, Am. J. Clin. Pathol. 21, 645.

Hilleman, O.R., 1965, Am. J. Med. 38, 751.

Kishimoto, T. and K. Ishizaka, 1971, J. Immunol. 107, 1567.

Langer, R. and J. Folkman, 1976, Nature 263, 797.

Mitchison, N.A., 1969, in: Regulation of the Antibody Response, ed. B. Cinander (Thomas, Springfield, IL) p. 54.

Rhine, W.D., D.S.T. Hsieh and R.S. Langer, 1979, Polymer 20, 596.

Williams, C.A. and M.W. Chase (eds.), 1971, Methods in Immunology and Immunochemistry, Vol. III (Academic Press, New York) p. 174.

Journal of Membrane Science, 7 (1980) 333—350 333
Elsevier Scientific Publishing Company, Amsterdam — Printed in The Netherlands

CONTROL OF RELEASE KINETICS OF MACROMOLECULES FROM POLYMERS*

ROBERT LANGER**, DEAN S.T. HSIEH, WILLIAM RHINE and JUDAH FOLKMAN

Department of Nutrition and Food Science, M.I.T., Cambridge, Massachusetts 02139 (U.S.A.), the Department of Surgery, Children's Hospital Medical Center, Boston, Massachusetts 02115 (U.S.A.), and Harvard Medical School, Boston, Massachusetts 02115 (U.S.A.)

(Received March 29, 1980; accepted in revised form May 27, 1980)

Summary

Internal and external means for controlling the release rates of large molecules, such as proteins, from ethylene—vinyl acetate copolymer matrices are presented. Internal approaches include alteration of the polymer—drug design, such as changing drug loading and particle size, coating the matrix, or altering matrix geometry. Kinetic and microstructural analyses are discussed. Applications of these polymeric systems, for instance, in delivery of insulin for diabetes, improved immunization procedures, and in developing bioassays for informational macromolecules are considered. In addition, a new approach for externally controlling release rates of drugs using magnetism has been developed. Until now, drug delivery systems were capable of delivering drugs at either constant or decreasing rates. We sought a system that permitted delivery of increased doses on demand, and achieved this by incorporating magnetic particles and drugs into polymeric matrices. Drug release rates can then be increased by an appropriate application of an external magnetic field. Over a five-day period, the magnetic field was applied ten times and drug release rates increased by up to 100% each time. Initial results indicate that this system does not cause tissue damage.

Introduction

In recent years, the use of polymers as vehicles to deliver drugs for prolonged time periods has received widespread attention (see references 1—20 for reviews). However, most polymeric systems are only capable of continuously releasing low-molecular-weight compounds, and no system has been capable of releasing drugs at increased rates on demand. In this paper, we discuss the development of two unique systems: (1) systems that continuously release large molecules such as proteins, and (2) systems that contain drug and small magnetic beads, and which upon application of an external magnetic field release the drug at an increased level.

*Paper presented at the Symposium "Controlled-Release Membranes", Houston, Texas, March 27, 1980.
**To whom inquiries should be directed, at M.I.T.

0376-7388/80/0000—0000/ $02.50

Macromolecular delivery systems

Effective polymeric delivery systems for macromolecules could have considerable importance for high-molecular-weight drugs such as polypeptide hormones (e.g., insulin), antigens, heparin, and interferon. In many ways, macromolecular drugs are ideally suited to polymeric delivery systems because most have very short *in vivo* half lives [21]. Therefore, physiologic drug concentrations are almost totally dependent on the release of the drug from the polymer and not on the fate of the drug in the body. However, for many years the low permeability of conventional polymers to molecules over 600 daltons made it difficult to develop controlled-release delivery systems for macromolecules [22].

Davis first reported sustained release of macromolecules in 1972 [23, 24]. In his studies, polyacrylamide and polyvinylpyrrolidone were used as vehicles. However, these polymers were inflammatory and permitted only brief periods of sustained release [25]. In our laboratory, Gimbrone and co-workers also reported the use of polyacrylamide as a vehicle for sustained release of macromolecular tumor proteins [26]. Nevertheless, despite attempts to eliminate inflammation by exhaustive washing and other means, polyacrylamide still caused tissue damage [26].

In 1974, we focused our attention on several new polymers — poly-2-hydroxyethylmethacrylate (polyHEMA or Hydron®), polyvinylalcohol, and ethylene—vinyl acetate copolymer — and on new methods of drug incorporation to achieve sustained macromolecular release. In our original work [25], we found that if the above polymers were dissolved in appropriate solvents, mixed with powdered or dissolved macromolecules, and then dried, the resulting pellets were capable of continuous release. With ethylene—vinyl acetate copolymer, proteins were released in biochemically active form for over 100 days. Both ethylene—vinyl acetate copolymer (especially when exhaustively washed in absolute alcohol) and polyHEMA displayed excellent tissue biocompatibility. In the following sections we discuss our early work demonstrating the long-term release of macromolecules as well as more recent studies concerning (1) factors controlling release kinetics and (2) applications of these systems.

Demonstration of long-term release

In our original studies [25], slow-release pellets were made by first dissolving polymers in appropriate solvents: polyHEMA in absolute alcohol at 37°C; ethylene—vinyl acetate copolymer in methylene chloride at 37°C; and polyvinylalcohol by autoclaving in distilled deionized water. With ethylene—vinyl acetate copolymer, the macromolecules were in powdered form; for the other polymer systems, the macromolecules were either powdered or in aqueous solution. One hundred to 500 μl of polymer solution were mixed with the macromolecules to be released, and a small amount of the resultant mixture was placed in conical molds 2 mm in diameter and 1.5 mm deep. The molds were then dried under vacuum overnight, which caused the solvent to evaporate with the macromolecules trapped within the polymer matrix. The dried polymer was hydrated by adding a drop of lactated Ringer's solution, and gently removed from the mold. The entire procedure was simple, rapid, and performed at room temperature. Sterility was easily accomplished [27].

To examine kinetics, we incorporated different proteins into polymers and measured the rate of protein release using the Lowry protein assay [28]. We found that the rate of protein release depended on the polymer concentration in solution during casting and on the type of polymer used [25]. In initial trials with soybean trypsin inhibitor (M.W. 21,000), protein was released rapidly from polyHEMA, somewhat more slowly from polyvinylalcohol, and least rapidly from ethylene—vinyl acetate copolymer (Fig. 1). In a subsequent test, four different proteins, ranging in molecular weight from 14,000 to

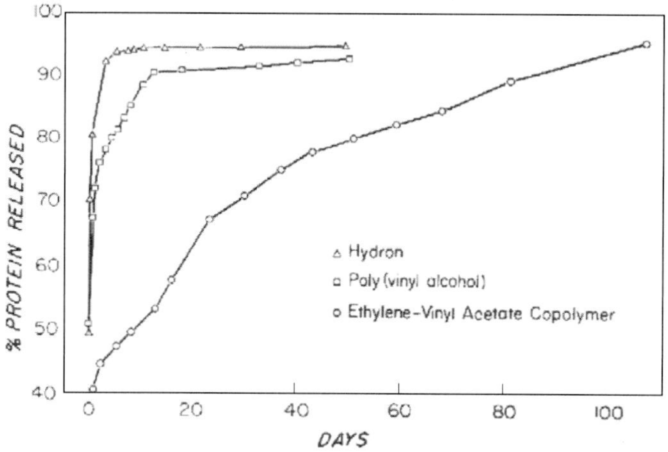

Fig. 1. Release of soybean trypsin inhibitor from coated polymer pellets. Protein concentrations in polymer solutions were all 50 mg/ml when cast [25].

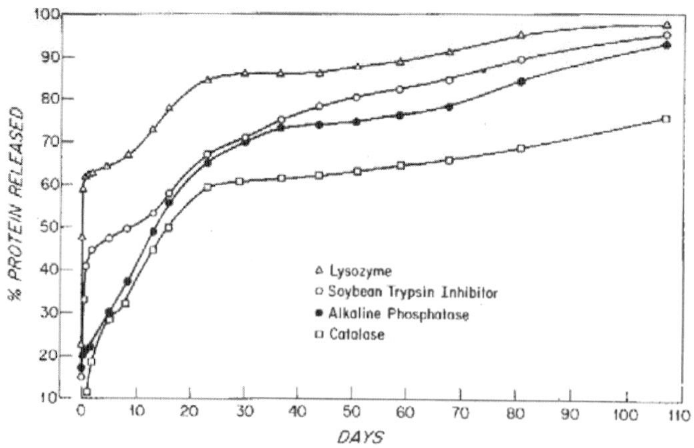

Fig. 2. Protein release from coated ethylene—vinyl acetate copolymer pellets. Protein concentrations in polymer solutions were all 50 mg/ml when cast [25].

250,000, were continuously released from ethylene—vinyl acetate copolymer for over 100 days (Fig. 2). Additional studies demonstrated that over 80% of the protein escaping from the polymers was in biochemically active form [27].

Reproducible procedures for fabricating polymeric delivery systems and factors controlling release kinetics

Although our original methods for producing polymeric delivery systems from ethylene—vinyl acetate copolymer permitted sustained release, the kinetics were often not reproducible; therefore, controlled release was not achieved. We concluded that the irreproducibility was a result of drug settling and redistribution during casting and drying, caused by the insolubility of the incorporated macromolecules in the polymer solvent. At room temperature, the drug migrated vertically and there also was visible lateral motion caused by currents (possibly thermal) in the mixture. We therefore developed a low-temperature casting and drying procedure that minimized this drug

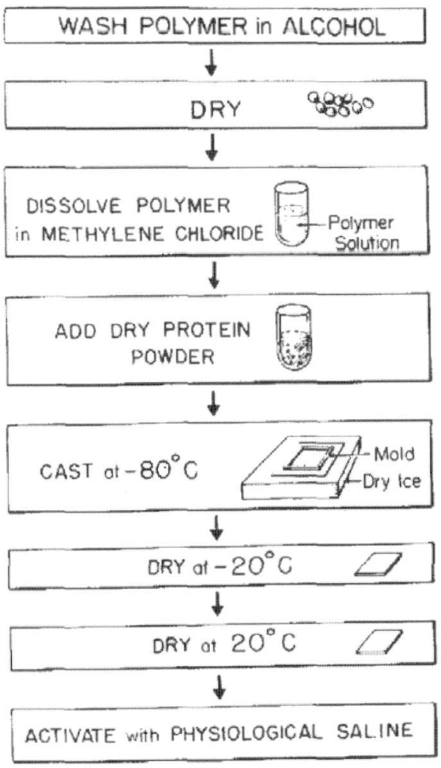

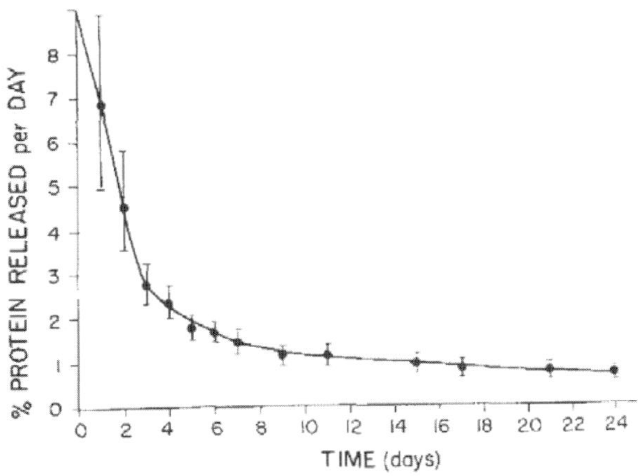

Fig. 3. Flow diagram describing the preparation of sustained-release polymers composed of ethylene—vinyl acetate copolymer. The alcohol wash is only necessary for removing impurities that cause minor inflammation in sensitive tissues.

Fig. 4. Reproducibility of release rates for 27 matrices prepared from three different slabs. The incorporated protein is bovine serum albumin at a particle size of 75—250 μm and a loading of 25% [29].

movement during matrix formation [29]. This method, illustrated in Fig. 3, produced polymer systems with markedly improved reproducibility of release kinetics. Using this technique, we then studied the effects of certain fabrication factors, including drug particle size, drug loading, and matrix coating, on macromolecular release.

Figure 4 depicts the reproducibility of release kinetics from matrices prepared at low temperature. Three slabs were prepared with bovine serum albumin, each was cut into nine 1 cm × 1 cm squares, and all 27 squares were measured for release for more than 24 days. The mean daily release rates and their standard deviations are graphed in Fig. 4. (Standard deviations of release rates from matrices prepared at room temperature were only within 75%.) In addition, more than 90% of the total release variance was due to variability within a single slab. Other studies using different proteins, particle sizes, and loadings yielded release reproducibilities consistent with the above observation, with one exception: at high loadings (>50%), the center square had a significantly slower release rate than the other eight squares from the same slab [29].

Two important factors contributing to macromolecular release were drug particle size and loading [29]. A series of experiments was conducted to examine the effects of particle size and loading on release rate. Examples of our findings are shown in Figs. 5 and 6. In general, we observed that increased

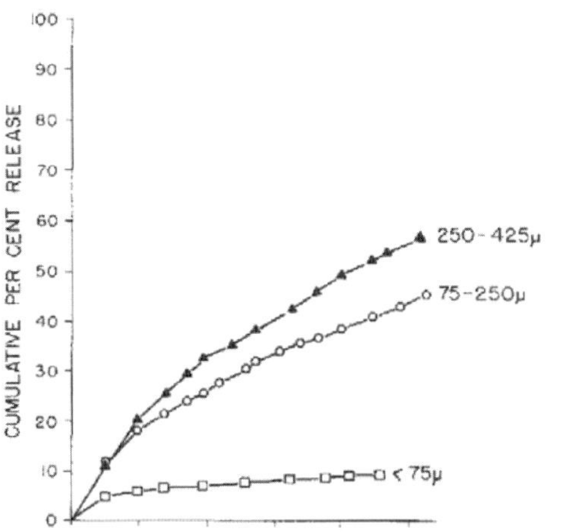

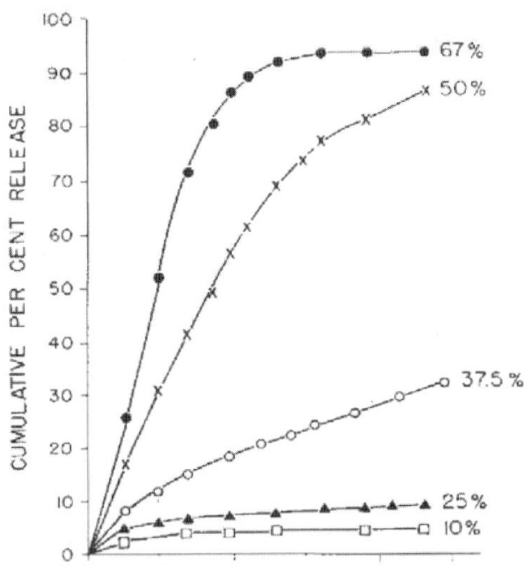

Fig. 5. Effect of particle size on the cumulative release of bovine serum albumin. The loading is 25% by weight. Each point represents the mean of at least eight samples. Standard deviations are all less than 10% of the respective mean values [29].

Fig. 6. Effect of loading on the cumulative release of bovine serum albumin. The particle size is less than 75 μm. Each point represents the mean of at least eight samples. Standard deviations are all less than 10% of the respective mean values [29].

particle size resulted in increased release rate. In the example shown, a change in particle size altered the release rate by a factor of six (Fig. 5).

Increases in drug loading also increased the release rate (Fig. 6). Not only did total drug release increase, but the cumulative percent of matrix drug released over time increased. As much as a fifty-fold increase in the slopes of percent release was caused by increasing drug loading from 10 to 50% by weight.

The effect of coating matrices with an additional layer of polymer was examined. Polymeric matrices, 50% insulin by weight, were prepared as described in Fig. 3. After the insulin—polymer matrices were dried and weighed, the squares were coated by dropping each into a vial containing 15 ml of polymer solution, either 5 or 20% (w/v) ethylene—vinyl acetate copolymer in methylene chloride. After 1 min in solution, the square was grasped on its edges with forceps, removed from solution, and held to dry at room temperature for 2 min. The matrix was dried for an additional 24 h at room temperature under mild vacuum.

As shown in Fig. 7, coating significantly affected drug release rates. Release kinetics decreased with increases in coating solution concentration. A sevenfold difference in release rate was observed between uncoated samples and samples coated with 20% polymer solution. Standard deviations of cumulative release values were within 15% of the means.

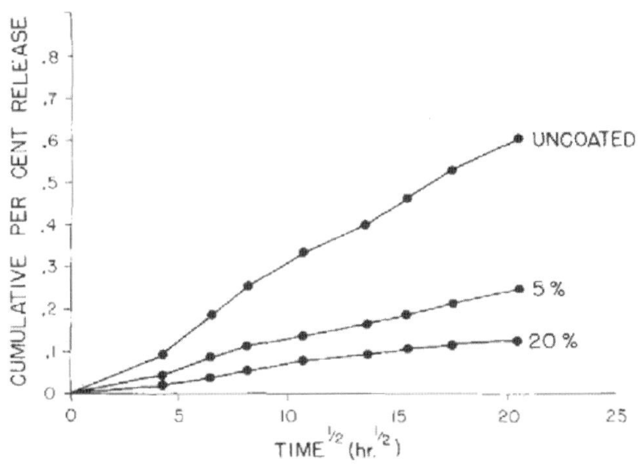

Fig. 7. Effect of coating on the cumulative release of insulin. Each point represents the mean of eight samples. Standard deviations are all less than 15% of the respective mean values. Release rates for insulin are slower than for other proteins because of insulin's lower solubility and particle size.

Applications of macromolecular polymeric delivery systems

We have explored a number of applications of these polymeric delivery systems. These applications include use as vehicles in bioassays or in creating chemical concentration gradients in chemotactic studies, continuous insulin delivery, or as an improved immunization method.

Biological assays: The most immediate application of macromolecular polymeric delivery systems has been in the development of biological assays for informational macromolecules. Such molecules (e.g., growth factors) are generally available only in small quantities and diffuse rapidly away from an implant site *in vivo* if injected or delivered by some other means. The polymers, however, maintain the macromolecule at an appropriate *in vivo* site and can provide macromolecules continuously for months, if desired. One example is the development of a bioassay for tumor neovascularization. It has been found that tumors elicit a diffusible material called tumor angiogenesis factor (TAF), which induces new blood vessels to grow from the host [30]. The new vessels are essential for continued tumor growth, and therefore purification of TAF is of fundamental interest. To develop an *in vivo* assay for TAF, we implanted ethylene—vinyl acetate copolymers in rabbit corneas. In over 200 experiments, slow-release pellets containing crude tumor extracts with TAF activity consistently elicited a neovascular response (Fig. 8). Furthermore, the rate at which vessels grew in the cornea depended on the quantity [27] and purity of TAF in the polymers.

In addition, these polymers have provided the basis for bioassays for growth factors derived from cartilage [31], the vitreous body [32], macrophages [33] and fibroblasts [34]. They also have been used in studies concerning the development of corneal ulcers [35] and in tests exploring the mechanism of vascular regression [36].

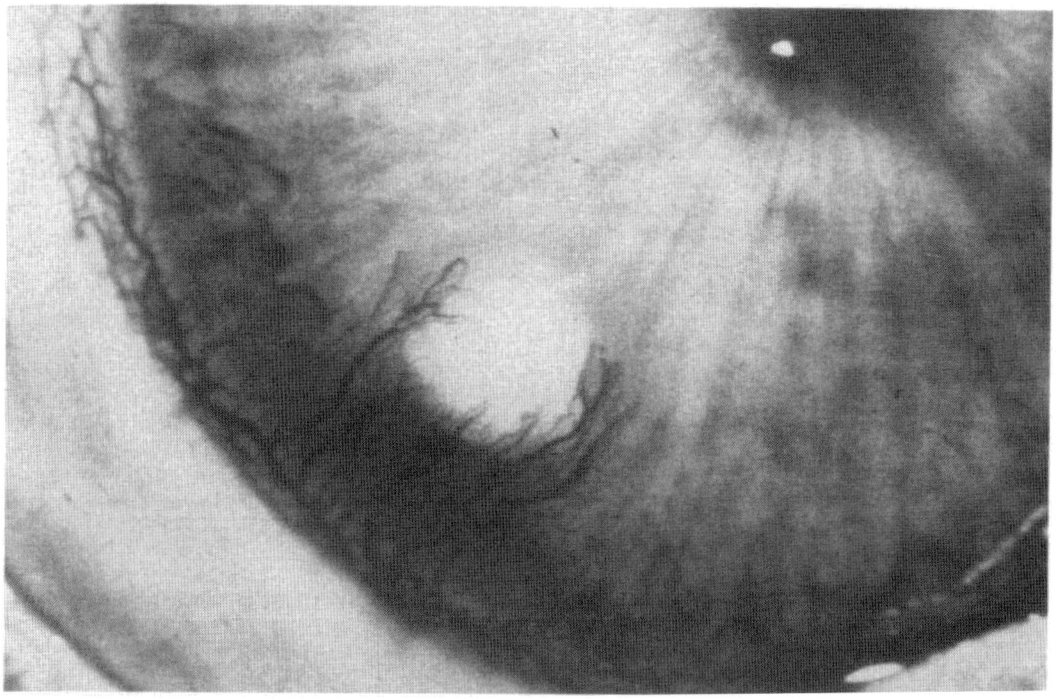

Fig. 8. Rabbit cornea 20 days after implantation of coated ethylene—vinyl acetate copolymer pellets containing tumor angiogenesis factor (TAF). Vessels from the corneal edge grew toward the polymer.

Chemotactic studies: Chemotaxis, or the movement toward or away from the source of a chemical compound, has been observed in a wide variety of organisms. Although numerous techniques have been employed to create concentration gradients to study this phenomenon, these methods are complicated and lack versatility [37–39]. We believed that controlled-release polymers might provide a simple and inexpensive means of creating concentration gradients to study chemotaxis. To examine this possibility, we studied the response of worms (planaria, *Dugesia dorotocephala*) to chemical gradients in an unstabilized aqueous medium. For this purpose, we adapted the T-maze system (Fig. 9) of Coward and Johannes, who have previously demonstrated a chemotactic response of the worms toward L-lysine, using synchronized infusion pumps to establish a chemical gradient [40]. In seventy tests, we found that the worms consistently migrated to the arm of the maze containing polymers impregnated with L-lysine ($P < 0.001$); they migrated with equal frequency to either arm of the maze if the polymers were empty or contained a control substance (L-glutamate) [41]. We have also found that these controlled-release polymers provide a simple method for investigating bacterial chemotaxis [41].

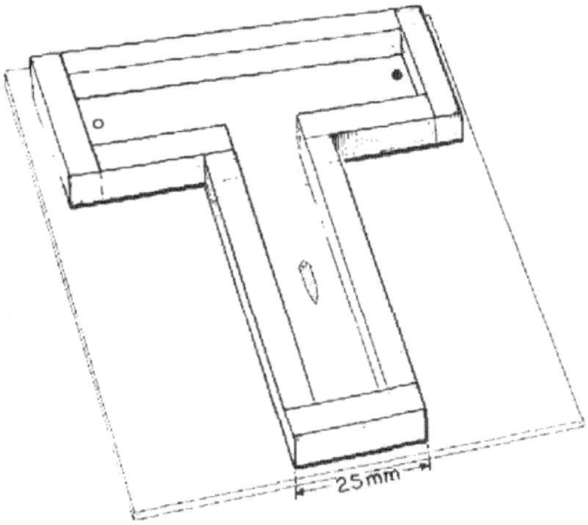

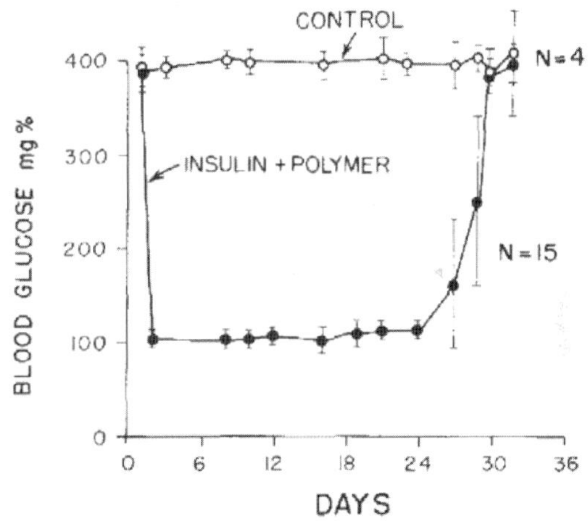

Fig. 9. T-maze used to assay chemotaxis of planaria. The black dot represents a polymer containing a chemotactic factor; the white dot indicates an empty polymer. The number of times a planaria swam to a particular arm of the T-maze was counted [41].

Fig. 10. Implantation of single insulin-containing polymers into diabetic rats. Controls were diabetic rats receiving empty polymers or no polymers [43].

Insulin delivery: Diabetics normally require insulin injections once or twice daily. Besides the inconvenience, this schedule of insulin administration results in abnormal peaks and valleys in blood glucose levels. Poor control of blood glucose may be responsible, in part, for diabetic complications such as blindness and heart and kidney disease [42].

The ability of controlled-release polymers containing insulin to maintain normoglycemia was tested in diabetic rats. Male rats (CD strain, Charles River Breeding Laboratories, 21 days old) weighing 150 to 200 g were made diabetic with streptozotocin. Ethylene—vinyl acetate copolymer discs (1.3 cm diameter, 1 mm high) containing 100 mg of bovine pancreatic insulin were placed subcutaneously in the lower abdomen; results are summarized in Fig. 10. Control animals with empty polymers or without polymers showed consistently high blood glucose levels. In contrast, the rats implanted with insulin polymers showed normal glucose levels for one month [43]. Furthermore, rats in the experimental groups gained weight at a normal rate, whereas growth of the controls was depressed as a result of the disease (Fig. 11).

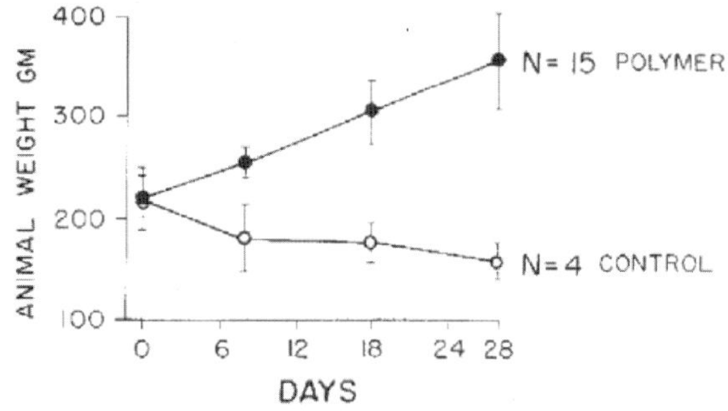

Fig. 11. Weights of diabetic rats containing insulin polymers and controls for the experiment shown in Fig. 10 [43].

Immunization procedures: Conventional methods for immunizing both animals and humans generally require multiple injections and often cause tissue irritation [44]. In a recent study, we considered the possibility that

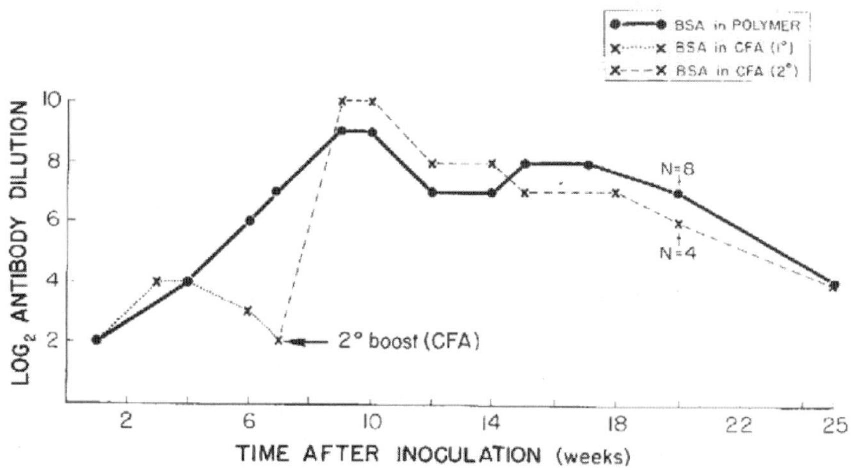

Fig. 12. Antibody titre in C57 black mice as a function of time, as determined by haemagglutination. The standard deviation on a given point is 1.16 [46]. The antibody titre induced by bovine serum albumin (BSA) in a single polymer is equal or superior to two injections of BSA in complete Freund's adjuvant (CFA).

controlled-release polymers for macromolecules might provide a simple, safe, and effective means of immunization by acting as a series of minishots [45]. Ethylene—vinyl acetate copolymer pellets containing 100 μg of bovine serum albumin were fabricated according to previously described methods [25] and implanted in C57 black mice. The immune response stimulated by sustained antigen delivery from one pellet was compared to a conventional method of immunization — two injections of bovine serum albumin emulsified in complete Freund's adjuvant [46]. The results, summarized in Fig. 12, indicated that the immune response induced by the antigen in polymer is comparable or superior to that induced by the same total dose of antigen using the conventional technique. Furthermore, a single polymer caused a sustained immune response for six months. No tissue irritation was observed with the controlled release system. This immunization procedure is effective for antigens other than albumin and over a wide range of molecular weights [47].

Magnetically modulated sustained release systems

One problem central to the field of controlled-release technology is that most vehicles so far developed display drug release rates that are either constant or decay with time. There has been no way to change or modulate the release rate on demand, once release has commenced.

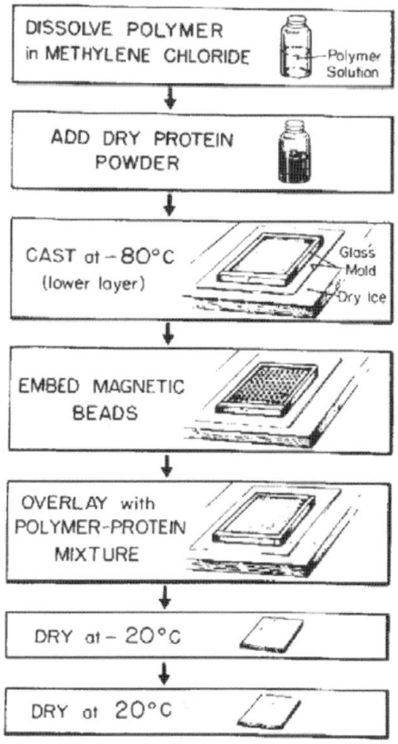

Fig. 13. Flow diagram describing the preparation of magnetic sustained release implants.

We recently developed a polymeric system capable of delivering macromolecules at increased rates on demand [48]. Small magnetic stainless steel beads (1.4 mm diameter) were embedded in the polymer along with the drug, using the procedure described in Fig. 13. The system was cast in three stages because direct incorporation of the magnetic beads resulted in their settling to the bottom of the polymer slab. A specially designed loading device was used to quickly and uniformly place the magnetic beads onto the polymer [48]. The appearance of the polymer slab under normal conditions and exposed to an X-ray to display the magnetic beads is shown in Fig. 14. Release rates were controlled by an oscillating external bar magnet (Fig. 15).

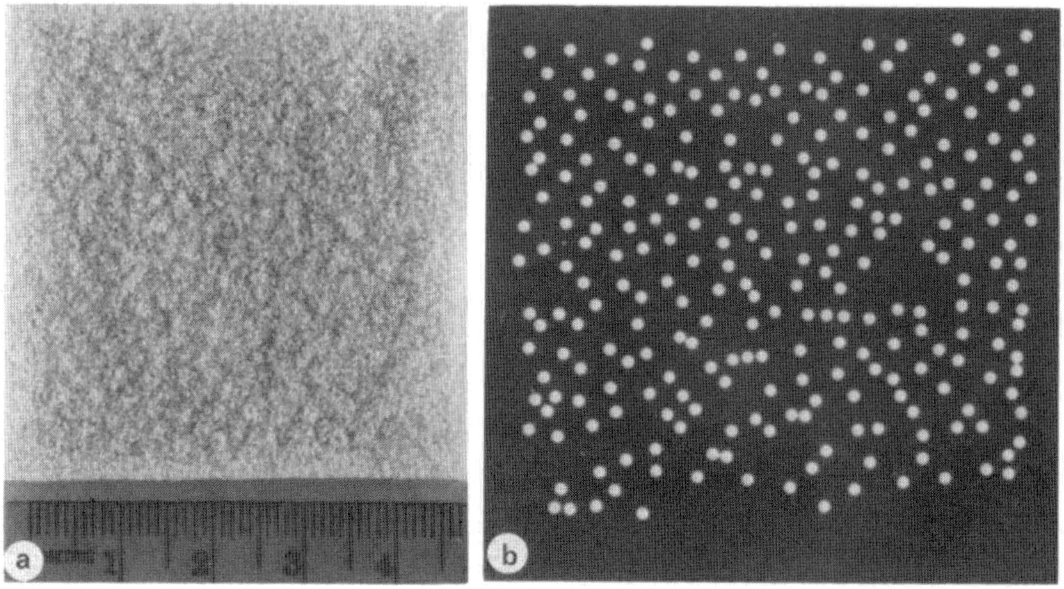

Fig. 14. (a) Photograph of ethylene—vinyl acetate copolymer slab containing drug (bovine serum albumin) and magnetic beads (magnification, 1.8×); (b) X-ray photograph of the same polymer slab (magnification, 2.2×) [48].

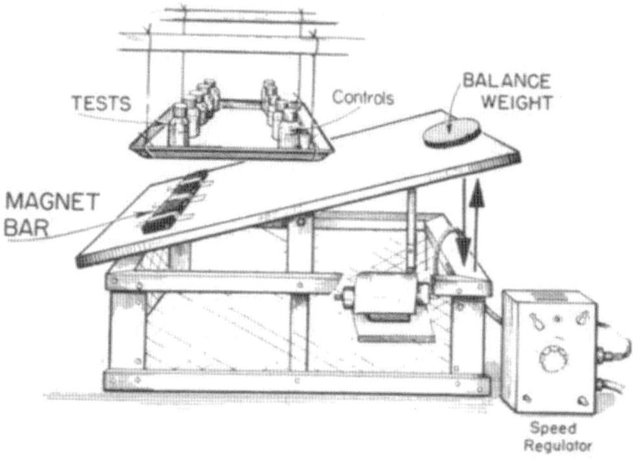

Fig. 15. Diagram of a motor driven see-saw rocker used as a magnetic triggering device.

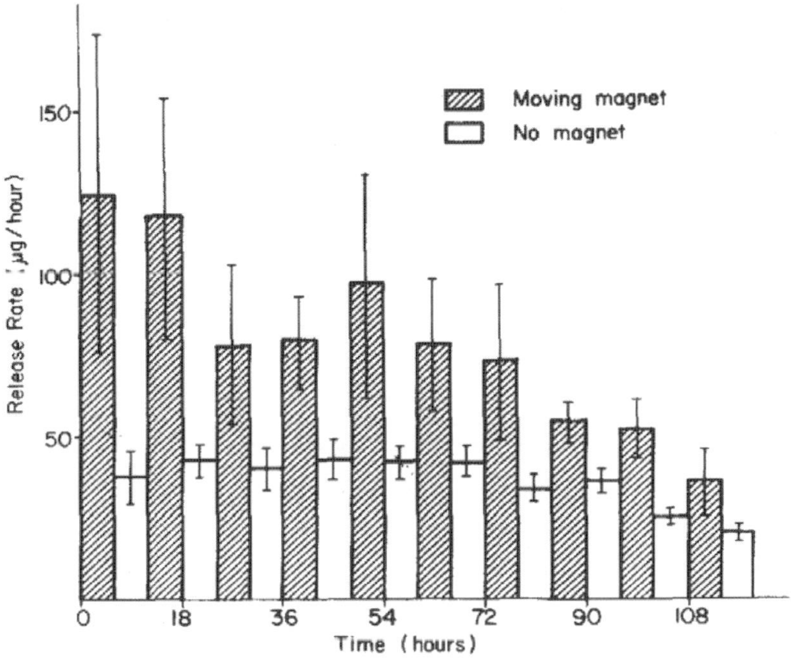

Fig. 16. Magnetically induced, modulated sustained release of bovine serum albumin from polymeric vehicles. Each histogram represents the average release rate of eight polymer squares (1 cm × 1 cm). This graph shows cyclic shifting between triggering (▨) and non-triggering (▭) for alternating six-hour periods over five days [48].

When exposed to the magnetic field, polymers released up to 100% more drug than when the magnetic field was discontinued (Fig. 16). Release rates returned to baseline levels when the magnetic field was discontinued. Increased rates of release on demand were repeatedly reproducible (see Discussion for possible mechanisms of release and applications of this system).

The validity of these findings was strengthened by four different control experiments: (1) To control for possible non-specific effects of the magnetic steel beads, polymer squares excised from the same slab as those used in the experiment in Fig. 16 were subjected to the same manipulations except that the bar magnet was absent. (2) To control for possible non-specific effects of the magnetic field (e.g., heating), polymer slabs were made without magnetic steel beads. (3) To control for any minor motion of the whole polymer, polymer squares containing drug and steel beads were exposed to intervals of shaking in the absence of the bar magnet. (4) To determine if the fluctuating magnetic field was an essential factor, the bar magnet was maintained in apposition to the polymer squares throughout alternating six-hour periods. In all four cases, these controls displayed baseline release rates that were the same as those of polymer squares not exposed to a magnet. No modulation was observed.

Biocompatibility was tested by implanting polymers ($1.5 \times 1.5 \times 1.5$ mm^3) containing steel beads into the rabbit cornea using previously described techniques [27]. No inflammation was detected [48].

This study demonstrated that release rates of drugs from polymers can be increased on demand by an external magnetic field. Although the first test experiment was conducted for a five-day period (Fig. 16), only 20% of the total drug was released. Further experiments have shown that these polymer systems are capable of releasing drugs for periods of three weeks or longer.

Discussion

The mechanism of release of macromolecules from polymers is currently under investigation in our laboratory. Because macromolecules are too large to diffuse through a pure film of ethylene—vinyl acetate copolymer [25] (Fig. 17a), we previously proposed that sustained release occurred via diffusion through channels in the matrix [29, 49]. In a recent study we were able to identify such channels [50] (Fig. 17b). The incorporation of the macromolecules during casting must introduce these channels through which dissolved drug can subsequently diffuse. In this case, observed differences in release rates for different proteins would be attributable to differences in protein properties (solubility, diffusivity) and matrix characteristics (porosity, tortuosity).

Our studies demonstrate the marked effect of drug particle size and loading on release kinetics. Release-rate increases caused by increases in particle size may result from the formation of larger channels or pores in the polymer matrix. Similarly, increased loadings may provide simpler pathways (lower tortuosity) and greater porosity for diffusion, both of which would facilitate the movement of water into, and proteins out of, the matrix.

Coating was also used to control the kinetics of macromolecular release; it may reduce the amount of macromolecule on the matrix surface and may decrease surface access of pores for channeled diffusion. It is assumed that the coating does not cause a complete barrier to form around the matrix because that would prevent any macromolecular diffusion into the surrounding medium [25].

One of the limitations of the present design of macromolecular delivery systems is that they do not release drug at a zero-order rate, but rather appear to display $(\text{time})^{1/2}$ kinetics (Figs. 4—6). However, we have recently suggested approaches based on geometric alteration of the polymeric matrices that should result in constant release rates [51]. Work is in progress to test these approaches.

In addition to the macromolecular delivery systems developed by Davis [23, 24] and our laboratory, several other approaches have recently been tested. One approach is the use of implantable pumps powered by osmotic pressure [52, 53] or other means [20, 54]. A second approach is the incorporation of macromolecules in bioerodible polymers. This has been successfully accomplished by Torchilin and coworkers using polyvinylpyrrolidone [55] and by Heller and Baker [56] using polymers of unreported origin. Other biodegradable polymers such as polylactic acid [57], lactic/glycolic acid co-

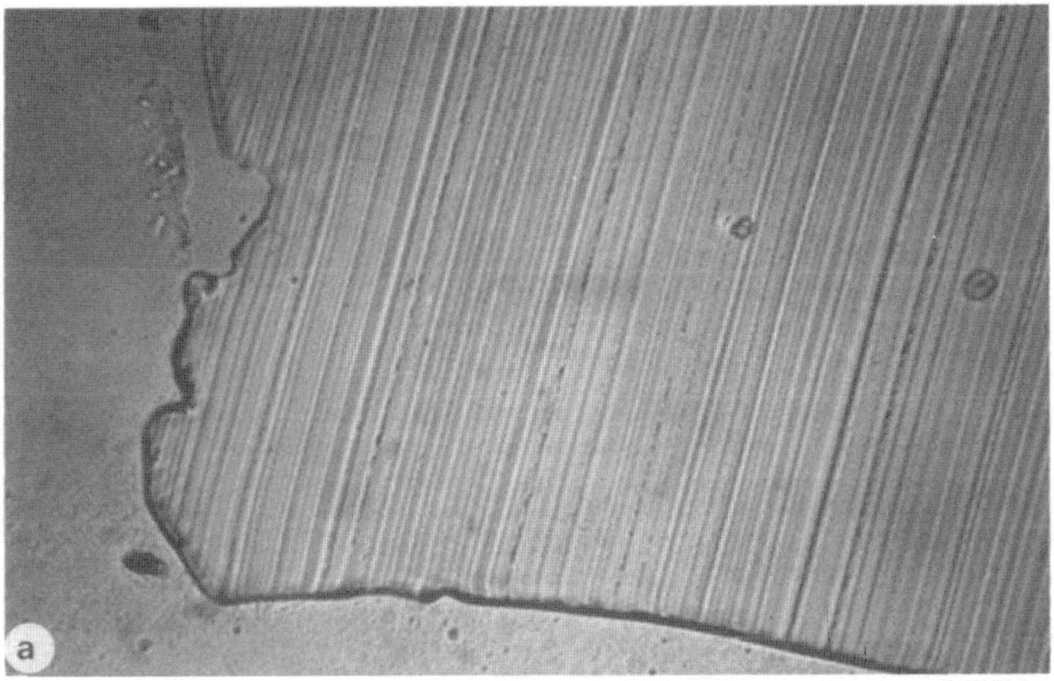

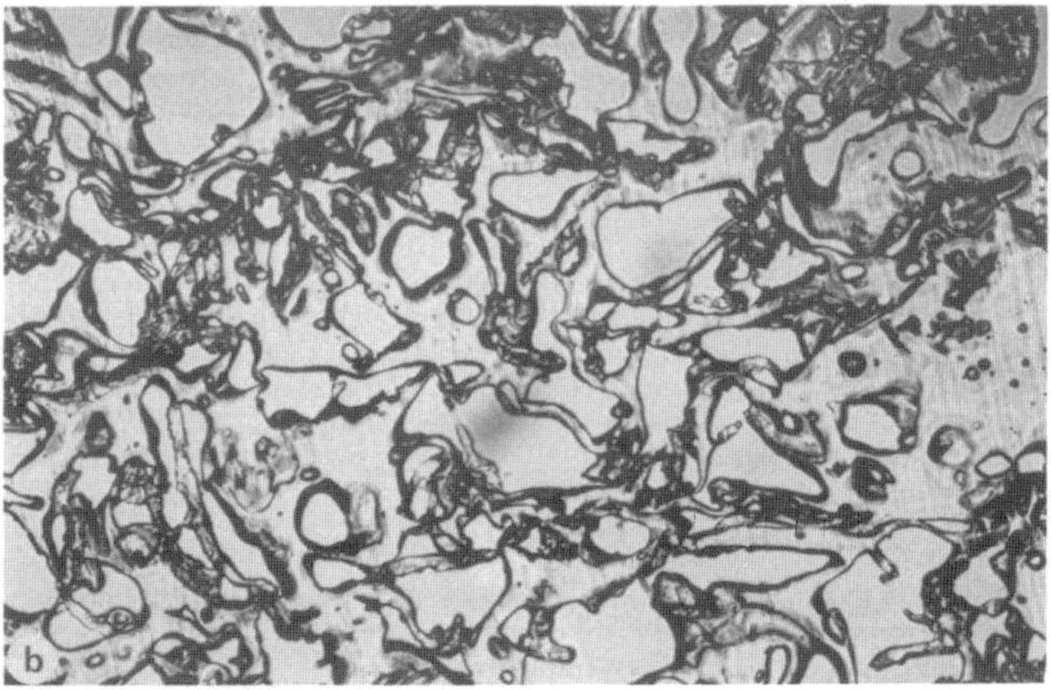

Fig. 17. Photomicrographs of 5 μm-thick sections obtained from mid-sections of polymer slabs at 125× (a) ethylene—vinyl acetate copolymer cast without drug; (b) ethylene—vinyl acetate copolymer originally cast with a macromolecular drug (bovine serum albumin) but with contents now totally released [50].

polymers [58], polyaminoacids [59], or polyorthoesters [60] may also prove useful for releasing macromolecules.

The possibility of delivering high molecular weight drugs may hold increasing importance in the future, particularly with the emergence of highly potent pharmaceuticals with new kinds of biological activity (e.g., interferon, enkephalins) [21]. Two recent studies have demonstrated new applications for macromolecular delivery systems. These applications are the sustained release of heparin for anticoagulation [61], and the sustained release of trasylol (M.W. 6500) from intrauterine devices to decrease menstrual bleeding [62].

The mechanism by which the magnetic field is able to induce increased release rates has not been established. We might speculate that the beads cause alternating compression and expansion of the matrix channels. Alternatively, the beads could strain the matrix and alter the integrity of the channels, which may facilitate diffusion.

Modulated delivery systems controlled by external means may ultimately improve the release pattern of certain drugs such as insulin. For example, studies have shown that excellent control of blood glucose can be achieved using large pumps that infuse insulin at a constant rate, but are capable of increasing the rate before meals [63–66]. Modulated release systems are also potentially useful in birth control as a means for delivering drug in correspondence with the menstrual cycle.

Alternative approaches for achieving modulated sustained release may also prove feasible. One example is the work of Heller and Trescony [67], who incorporated an enzyme (urease) which catalyzed a reaction which increased the local pH in a pH-sensitive bioerodible polymer together with the drug (hydrocortisone). In the absence of the enzyme's substrate (urea), drug release was relatively constant; when the substrate was present, the release rate increased.

Experiments on both macromolecular and modulated controlled-release systems are underway to improve our understanding of the critical parameters contributing to release kinetics and to develop mathematical models describing them. The studies conducted thus far demonstrate that these release systems may be made reproducibly, and can be potentially useful in a variety of applications.

Acknowledgements

This study was supported by NIH grant GM 26698. We would like to thank Barbara Haskell and Annette LaRocca for their help in editing and typing the manuscript.

References

1 R. Langer, Polymeric delivery systems for controlled drug release, Chem. Eng. Commun., 6 (1980) 1.

2 J.R. Robinson (Ed.), Sustained and Controlled Release Drug Delivery Systems, Marcel Dekker, New York, 1978.
3 D.R. Paul and F.W. Harris (Eds.), Controlled Release Polymeric Formulations, American Chemical Society Symposium Series 33, Washington, D.C., 1976.
4 A.C. Tanquary and R.E. Lacey (Eds.), Controlled Release of Biologically Active Agents, Vol. 47 of Advances in Experimental Medicine and Biology, Plenum Press, New York, 1974.
5 R.J. Kostelnik (Ed.), Polymeric delivery systems, Midland Macromolecular Symposium 5, Gordon and Breach, New York, 1978.
6 N.F. Cardarelli (Ed.), Controlled Release Pesticide Formulations, CRC Press, Boca Raton, Florida, 1976.
7 A. Kydoneius (Ed.), Controlled Release Technology, CRC Press, Boca Raton, Florida, 1980.
8 N.F. Cardarelli (Ed.), Controlled Release Pesticide Symposium, University of Akron, Ohio, 1974.
9 F.W. Harris (Ed.), Proceedings 1975 International Controlled Release Pesticide Symposium, Wright State University, Dayton, Ohio, 1975.
10 N.F. Cardarelli (Ed.), Proceedings of the 1976 Controlled Release Pesticide Symposium, Akron, Ohio, 1976.
11 H.B. Sher (Ed.), Controlled Release Pesticides, American Chemical Society Symposium Series Vol. 53, Washington, D.C., 1977.
12 R.F. Goulding (Ed.), Proceedings of the 1977 Controlled Release Pesticide Symposium, Oregon State University, Corvallis, Oregon, 1977.
13 G.A. Montemarano (Ed.), Proceedings of the 5th International Symposium on Controlled Release of Bioactive Materials, National Bureau of Standards, Gaithersburg, Maryland, 1978.
14 R.W. Baker (Ed.), Proceedings of the 6th International Symposium on Controlled Release of Bioactive Materials, Academic Press, New York, in press.
15 A. Zaffaroni, Controlled delivery therapy is here, Chemtech., 6 (1976) 756.
16 G.L. Flynn, S.H. Yalkowsky and T.J. Roseman, Mass transport phenomena and models: Theoretical concepts, J. Pharm. Sci., 63 (1974) 479.
17 A.S. Michaels, Applications of the theory of molecular transport in polymers to the design of controlled drug delivery systems, Polym. Prepr., 20: (1979) 1, 332.
18 V.T. Stannett, W.J. Koros, D.R. Paul, H.K. Lonsdale and R.W. Baker, Recent advances in membrane science and technology, Adv. Polym. Sci., 32 (1979) 69.
19 J. Heller, Controlled release of biologically active compounds from bioerodible polymers, Biomaterials, 1 (1980) 51.
20 P.J. Blackshear, Implantable drug-delivery systems, Sci. Am., 241 (1979) 66.
21 R. Langer, Bioavailability of macromolecular drugs and its control, in V.F. Smolen (Ed.), Bioavailability and the Pharmacokinetic Control of Drug Response, John Wiley and Sons, New York, in press.
22 R.W. Baker and H.K. Lonsdale, Controlled release: Mechanisms and rates, in A.C. Tanquary and R.E. Lacey (Eds.), Controlled Release of Biologically Active Agents, Vol. 47 of Advances in experimental biology and medicine, Plenum Press, New York, 1974, p. 15.
23 B.K. Davis, Control of diabetes with polyacrylamide implants containing insulin, Experientia, 28 (1972) 348.
24 B.K. Davis, Diffusion in polymer gel implants, Proc. Natl. Acad. Sci., USA, 71 (1974) 3120.
25 R. Langer and J. Folkman, Polymers for the sustained release of proteins and other macromolecules, Nature, 263 (1976) 797.
26 M.A. Gimbrone, Jr., R.S. Cotran, S.B. Leapman and J. Folkman, Tumor growth and neovascularization: An experimental model using the rabbit cornea, J. Nat. Cancer Inst., 52 (1974) 413.

27 R. Langer and J. Folkman, Sustained release of macromolecules from polymers, in R.J. Kostelnik (Ed.), Polymeric Delivery Systems, Midland Macromolecular Symposium 5, Gordon and Breach, N.Y., 1978, p. 175.
28 O.H. Lowry, N.J. Rosebrough, A.L. Farr and R.J. Randall, Protein measurement with the folin phenol reagent, J. Biol. Chem., 193 (1951) 265.
29 W.D. Rhine, D.S.T. Hsieh and R. Langer, Polymers for the sustained release of macromolecules: Methods to fabricate reproducible systems and control release kinetics, J. Pharm. Sci., 69 (1980) 265.
30 J. Folkman, E. Merler, C. Abernathy and G. Williams, Isolation of a tumor factor responsible for angiogenesis, J. Exp. Med., 133 (1971) 275.
31 R. Langer, H. Brem, K. Falterman, M. Klein and J. Folkman, Isolation of a cartilage factor that inhibits tumor neovascularization, Science, 193 (1976) 70.
32 S. Brem, I. Preis, R. Langer, H. Brem, J. Folkman and A. Patz, Inhibition of neovascularization by an extract derived from vitreous, Am. J. Ophthal., 84 (1977) 323.
33 P. Polverini, R. Cotran, M. Gimbrone and E. Unanue, Activated macrophages induce vascular proliferation, Nature, 269 (1977) 804.
34 D. Gospodarowicz, H. Bialecki and J.K. Thakaral, The angiogenic activity of the fibroblast and epidermal growth factor, Exp. Eye Res., 28 (1979) 501.
35 H. Conn, M. Berman, K. Kenyon, R. Langer and J. Gage, Stromal vascularization prevents corneal ulceration, Invest. Ophthal., 19 (1980) 362.
36 D. Ausprunk, K. Falterman and J. Folkman, The sequence of events in the regression of corneal capillaries, Lab. Invest., 38 (1978) 284.
37 S. Boyden, The chemotactic effect of mixtures of antibody and antigen on polymorphonuclear leukocytes, J. Exp. Med., 115 (1962) 453.
38 W.W. Tso and J. Adler, Negative chemotaxis in *escherichia coli*, J. Bacteriol., 118 (1974) 560.
39 S.J. Coward and R.E. Johannes, Amino acid chemoreception by the planaria *duqesia dorotocephala*, Comp. Biochem. Physiol., 29 (1969) 475.
40 S.J. Coward, R.C. Taylor and R.E. Johannes, A Simple flow system for studying chemoreception in aquatic organisms, J. Biol. Physiol., 9 (1967) 32.
41 R. Langer, M. Fefferman, P. Gryska and K. Bergman, A simple method for studying chemotaxis using sustained release of attractants from inert polymers, Can. J. Microbiol., 26 (1980) 274.
42 D. Job, E. Eschwege, C. Guyot-Argenton, J.P. Aubry and G. Tchobroutsky, Effect of multiple daily insulin injections on the course of diabetic retinopathy, Diabetes, 25 (1976) 463.
43 H.M. Creque, R. Langer and J. Folkman, One month sustained release of insulin from a polymer implant, Diabetes, 29 (1980) 37.
44 O.R. Hilleman, Immunologic, chemotherapeutic and interferon approaches to control of viral disease, Am. J. Med., 38 (1965) 751.
45 I. Preis and R. Langer, A single step immunization by sustained antigen release, J. Immunol. Methods, 28 (1979) 193.
46 J. Freund, The effect of paraffin oil and mycobacteria on antibody formation and sensitization, Am. J. Clin. Pathol., 21 (1951) 645.
47 R. Langer, Polymers for the sustained release of macromolecules: Their use in a single step method of immunization, in H. Van Vunakis and J.J. Langone (Eds.), Immunological Techniques, Methods in Enzymology, Academic Press, New York, in press.
48 D.S.T. Hsieh, R. Langer and J. Folkman, Mangetic modulation of release of macromolecules from polymers, Proc. Natl. Acad. Sci., in press.
49 R. Langer and J. Folkman, Controlled release of macromolecules from polymers, in E.P. Goldberg and A.K. Nakajima (Eds.), Biomedical Polymers, Academic Press, New York, in press.

50 R. Langer, W. Rhine, D. Hsieh and R. Bawa, Polymers for the sustained release of macromolecules: Applications and control of release kinetics, in R. Baker (Ed.), Controlled Release of Bioactive Materials, Academic Press, N.Y., 1980, p. 83.

51 W. Rhine, V. Sukhatme, D.S.T. Hsieh and R. Langer, A new approach to achieve zero-order release kinetics from diffusion controlled polymeric matrix systems, in R. Baker (Ed.), Controlled Release of Bioactive Materials, Academic Press, New York, 1980, p. 177.

52 F. Theeuwes, Elementary osmotic pump, J. Pharm. Sci., 64 (1975) 1987.

53 R. Cappoza, B. Eckenhoff and S.I. Yum, Design and performance of the implantable osmotic minipump, Biomed. Eng., 1 (1977) 281.

54 J.A. Parsons, D. Rothwell and J.F. Sharpe, A miniature syringe pump for continuous administration of drugs and hormones: The mill-hill infuser, Lancet, 8002 (1977) 77.

55 V.P. Torchilin, E.G. Tischenko, V.N. Svirnov and E.I. Chazov, Immobilization of enzymes on slowly soluble carriers, J. Biomed. Mat. Res., 11 (1977) 223.

56 J. Heller and R.W. Baker, Theory and practice of controlled drug delivery from bioerodible polymers, in R.W. Baker (Ed.), Proc. 6th International Symposium on Controlled Release of Bioactive Materials, Academic Press, New York, in press.

57 S. Yolles, J. Eldridge, T. Leafe, J. Woodland, D. Blake and F. Meyer, Long-acting systems for narcotic antagonists, in A.C. Tanquary and R.E. Lacey (Eds.), Controlled Release of Biologically Active Agents, Vol. 47 in Advances in Experimental Biology and Medicine, p. 177, Plenum Press, New York, 1974.

58 D.L. Wise, A.D. Schwope, S.E. Harrigan, D.A. McCarthy and J.F. Howes, Sustained delivery of a narcotic antagonist from lactic/glycolic acid copolymer implants, in R.J. Kostelnik (Ed.), Polymeric Delivery Systems, Midland Macromolecular Monograph 5, Gordon and Breach, New York, 1978, p. 75.

59 K.R. Sidman, A.D. Schwope, W.D. Steber, S.E. Rudolph and S.B. Poulin, Biodegradable, implantable sustained release systems based on glutamic acid copolymers, J. Membr. Sci., 7 (1980) 277.

60 R.C. Capozza, L. Sendelbeck and W.J. Balkenhol, Preparation and evaluation of a bioerodible naltrexone delivery system, in R.J. Kostelnik (Ed.), Polymeric Delivery Systems, Midland Macromolecular Monograph 5, Gordon and Breach, New York, 1978, p. 59.

61 C.D. Ebert, J.C. McRea and S.W. Kim, Controlled release of pharmacologically active agents for thrombus prevention: Prostaglandin and heparin releasing polymers, in R.W. Baker (Ed.), Controlled Release of Bioactive Materials, Academic Press, New York, 1980, p. 107.

62 M.E. Tuttle, R.W. Baker and L.E. Laufe, Slow release aprotinin delivery for control of intrauterine-device-induced hemorrhage, J. Membr. Sci., 7 (1980) 351.

63 J.L. Pickup, H. Keen, J.A. Parsons and K. Alberti, Continuous subcutaneous insulin infusion: An approach to normal glycemia, Brit. J. Med., 1 (1978) 204.

64 S. Genuth and P. Martin, Control of hyperglycemia in adult diabetics by pulsed insulin delivery, Diabetes, 26 (1977) 571.

65 G. Slama, M. Hautecouverture, R. Assan and G. Tchobroutsky, One to five days of continuous intravenous infusion on seven diabetic patients, Diabetes, 23 (1974) 732.

66 W.V. Tamborlane, R.S. Sherwin, M. Genel and P. Felig, Reduction to normal of plasma glucose in juvenile diabetes by subcutaneous administration of insulin with a portable infusion pump, New Engl. J. Med., 300 (1979) 573.

67 J. Heller and P.V. Trescony, Controlled drug release by polymer dissolution II. Enzyme-mediated device, J. Pharm. Sci., 68 (1979) 919.

Bioerodible polyanhydrides for controlled drug delivery

H. B. Rosen[a], J. Chang[b], G. E. Wnek[b], R. J. Linhardt[c,d] and R. Langer[c,e]

[a]Dept. of Chemical Engineering, [b]Dept. of Materials Science and Engineering, [c]Dept. of Nutrition and Food Science, Massachusetts Institute of Technology, Cambridge, MA 02139, USA; [d]Division of Medicinal Chemistry, College of Pharmacy, University of Iowa, Iowa City, Iowa 52242, USA; [e]Dept. of Surgical Research, Children's Hospital Medical Center, Boston, MA 02115, USA
(Received 9 July 1982; revised 4 October 1982)

Initial *in vitro* and *in vivo* (rat) studies using a poly(anhydride), poly[bis(*p*-carboxyphenoxy)methane], as a bioerodible matrix for controlled drug delivery are described. Drug delivery matrices fabricated from this material and containing cholic acid showed a period of nearly zero-order erosion kinetics during which this steroid was released at nearly the same constant rate.

Keywords: Drugs, drug delivery system, polymer, polyanhydride, bioerosion

Although controlled release of drugs can be accomplished by several mechanisms[1], biodegradation of an insoluble polymer carrier to soluble monomer units offers the advantage of eliminating the need for surgical removal of the device. Controlled release matrices composed of hydrophilic biodegradable polymers such as poly(lactic acid), poly(glycolic acid), and their copolymers generally erode in a homogenous manner[2,3]. This leads to a progressive loosening of the matrix which causes changes in both the permeability and mechanical strength of the devices during bioerosion[2,4]. It would be far more desirable if a matrix were to erode heterogeneously, from the surface first. Such erosion will lead to zero-order drug release provided that diffusional release is minimal and the overall shape of the device remains nearly constant; thus maintaining constant surface area[5]. To obtain a device that erodes heterogeneously, the polymer used should be hydrophobic yet contain water labile linkages. The only polymers designed for this purpose have been poly(orthoesters)[6]. However, because of the stability of the backbone bonds, these polymers erode slowly and require additives to promote biodegradation. The delivery systems containing additives, such as water soluble salts, swell considerably leading to diffusional release. It occurred to us that poly(anhydrides), which were originally synthesized as fibre forming polymers in the textile industry[7], but rejected because of their hydrolytic instability compared to polyesters of similar structure[7], might be sufficiently hydrolytically labile to produce heterogeneous erosion at rates suitable for controlled release applications. This paper reports the use of one of these polyanhydrides as a prototype for a new class of biodegradable drug delivery matrices.

Correspondence to Professor R. Langer at M.I.T.

METHODS

Poly[bis(*p*-carboxyphenoxy) methane] (PCPM), was selected as a prototype poly(anhydride) because its erosion rate in NaOH[8] suggested to us sufficient hydrolytic lability for drug delivery application. PCPM was synthesized by the method of Conix[8] from the mixed anhydride of bis(*p*-carboxyphenoxy)methane and acetic acid. The polycondensation was performed *in vacuo* at 195–210°C[8,9]. The *Tg* and *Tm* of the melt pressed polymer were determined on a Perkin–Elmer DSC-2 differential scanning calorimeter.

To formulate drug-free matrices, PCPM was ground using a Fisher Scientific Micro Mill and the resulting particles sized using sieves. PCPM particles (150–300 μm diameter) were melt-pressed between sheets of aluminum foil at 121.1°C and 8000 lbs. (representing from 22–81 kpsi) on a Carver Laboratory Press using shims to regulate device thickness. The matrix dimensions examined ranged from having face areas of 0.2 to 0.8 cm^2 and having thicknesses of 0.05 to 0.10 cm. Specific formulations are given in the legends of *Figures 1 and 2*.

To incorporate a drug, unlabelled cholic acid from Sigma Chemical Co. (15 mg) was dissolved in 10 ml of ethanol. Tritiated cholic acid (2.4-^{3}H) from New England Nuclear, in 125 μl of ethanol (2.42×10^8 DPM, S.A. = 3.72×10^{10} DPM/mg) was added to the unlabelled solution and the solution was dried *in vacuo*. The drug powder (10.5 wt %) was mixed with the PCPM particles and melt-pressed as above. The drug containing device was then sandwiched between two very thin (<1 wt %) drug free layers of ground (150–300 μm diameter) PCPM to eliminate the presence of surface exposed drug particles and then repressed.

In vitro erosion and drug release studies were performed by placing the PCPM devices weighing from 10–50 mg in glass scintillation vials containing 10 ml of

0.2 M Na phosphate buffer (pH 7.4) at 37° or 60°C. The buffer was periodically changed by removing the device from the vial and placing it in a vial of fresh buffer. The absorbance of the collected buffer solutions was measured on a Gilford spectrophotometer at 243 nm to detect the diacid monomer, bis(*p*-carboxyphenoxy) methane. Concentrations were determined from a standard curve constructed by measuring the absorbance at 243 nm of the pure diacid monomer at concentrations from 0 to 0.02 mg/ml. The polymer devices containing tritiated cholic acid were eroded and the buffer solutions were analyzed on a LS-230 Beckman scintillation counter.

In vivo erosion was studied using three portions cut from a single melt-pressed PCPM slab each with the dimensions 0.26 cm^2 face area ×0.0695 cm thick and weighing 24 mg. The surfaces of these devices were disinfected under u.v. light for 30 mins. Each square was implanted subcutaneously in the abdominal region of a Sprague–Dawley rat. An animal was sacrificed at the end of 21, 44, and 153 days and the polymer removed, dried, and weighed.

RESULTS

Poly[bis(*p*-carboxyphenoxy)methane] (PCPM), was chosen as a prototype to examine the usefulness of poly(anhydrides) in controlled release devices. PCPM, synthesized by an adaption of the method of Conix[8], was a yellow translucent amorphous solid with a *Tg* = 92°C (5°C higher than reported by Conix[8]) and no observable *Tm* (a sample with cholic acid incorporated did, however, show a *Tm* = 192°C due to cholic acid melting). The molecular weight was reported to be 40 000[8]. The polymer had broad carbonyl stretching vibrations in the infrared of 1780 and 1720 cm^{-1} characteristic of a poly(anhydride) and no OH stretching vibrations (i.e. no stretching between 3300 to 2500 (cm^{-1}) characteristic of the diacid monomer.

Devices were pressed at temperatures and pressures ranging from 93°C to 163°C and 22 kpsi to 81 kpsi. At temperatures below 120°C the polymer did not flow well giving devices with poor mechanical properties. At temperatures above 145°C the devices formed were brittle. There were no discernable differences between devices pressed over the range of 22 to 81 kpsi. The conditions chosen to melt press PCPM were 121°C and 22 kpsi giving devices with suitable mechanical properties and minimizing the possible induction of morphological changes within the polymer during pressing. PCPM hydrolysed completely leaving no insoluble residue. The u.v. spectra of the erosion product was identical to that of the diacid monomer bis(*p*-carboxyphenoxy)methane.

The erosion curves for drug-free PCPM slabs at both 37°C and 60°C are shown in *Figure 1*. Both curves are characterized by an induction period followed by a linear region of mass loss at a nearly constant rate. Throughout the erosion, the devices decreased in size while maintaining their physical integrity suggesting surface erosion is occurring.

The erosion profile was unaltered by scraping the polymer matrix to remove the surface layer; the same induction period was observed. When the polymer matrices that were pre-eroded for 50 hours at 60°C (until the induction period ended), were removed, vacuum dried, and returned to fresh buffer solutions at either 37°C or 60°C, near zero-order erosion of the samples began almost immediately (*Figure 2*).

Drug release from PCPM was investigated *in vitro* using cholic acid which because of its low u.v. absorb-

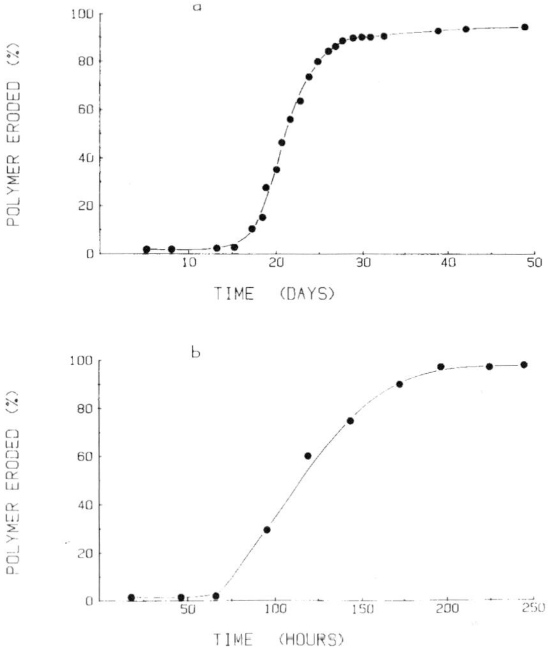

Figure 1 *Erosion curves for drug-free PCPM matrices in phosphate buffer at 37°C (a) and 60°C (b). Percent polymer eroded is 100x the cumulative mass eroded at each sample point divided by the total mass of the matrices eroded at 37° and 60°C which were 23 and 18 mg, respectively. The PCPM matrices eroded at 37°C and 60°C were pressed in the same rectangular mould and had the dimensions 0.24 cm^2 face area X 0.08 cm thick and 0.33 cm^2 face area X 0.05 cm thick, respectively*

ance at 243 nm did not interfere with the matrix erosion measurement. The erosion and release profiles were nearly zero-order and had similar slopes (*Figure 3*). Both the drug and the polymer had completely disappeared at nearly identical times.

The *in vivo* erosion of drug-free PCPM slabs showed a half-life of about 47 days or approximately five days shorter than *in vitro* erosion at 37°C in pH 7.4 buffer. After 153 days <1% of the polymer remained. The devices were only slightly encapsulated by tissue.

DISCUSSION

These initial studies using PCPM as a prototype poly(anhydride) suggest the suitability of this polymer class for use in bioerodible drug delivery systems. PCPM completely degrades to its monomer, under physiological conditions at rates useful for drug delivery applications. The erosion profile (*Figure 1*) is characterized by an induction period followed by a period of nearly linear zero-under erosion. An investigation of this undesirable induction period was conducted by scraping the surface layers from the polymer matrix. The fact that such treatment did not eliminate or shorten the induction period indicates that morphological surface changes, occurring during the melt pressing, are not the cause of this induction period. The induction period was, however, eliminated by pre-eroding followed by vacuum drying the device. The effectiveness of this pretreatment step suggests that two rate constants might control the rate of device erosion. The first is the rate of hydrolysis of the anhydride linkage and the second the rate of polymer dissolution. In order to begin

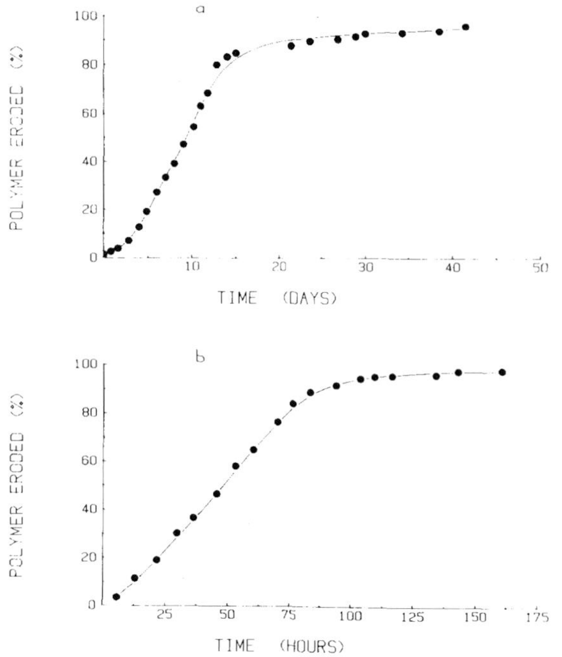

Figure 2 Erosion curves for the drug-free PCPM matrices which have been pre-eroded at 60°C for 50 h. Erosion took place in phosphate buffer at 37°C (a) and 60°C (b). Percent polymer erosion is 100x the cumulative mass eroded at each sample point divided by the total mass of the matrices eroded at 37° and 60°C which were 74 and 25 mg, respectively. The PCPM matrices eroded at 37°C and 60°C had the dimensions 0.57 cm^2 face area ×0.06 cm thick and 1.13 cm^2 face area X 0.05 cm thick, respectively

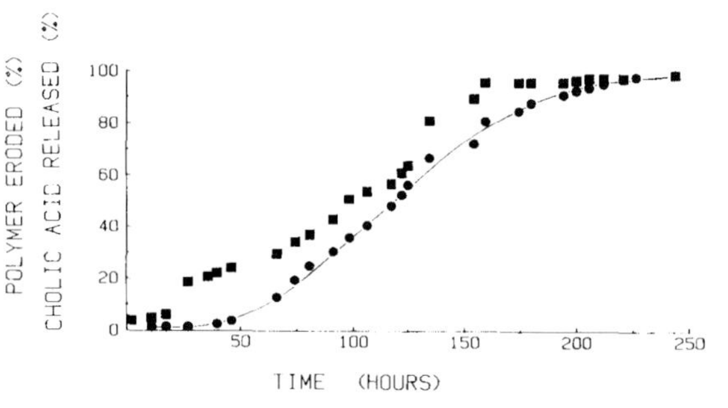

Figure 3 Erosion and release curves for a PCPM matrix containing cholic acid at 10.5 wt.% eroded in phosphate buffer at 60°C. PCPM erosion (●) is plotted as percent polymer erosion which equals 100X the cumulative mass eroded at each sample point divided by total mass of the matrix, 25 mg. Cholic acid release (■) is plotted as the percent cholic acid released which equals 100x the cumulative counts per minute at each sample point divided by the total number of counts per minute within the polymer matrix. The matrix has the dimensions 0.5 cm^2 face area × 0.06 cm thick

to measure monomer units in the buffer it seems likely that many anhydride linkages on the polymer's surface must be cleaved. This non-productive hydrolysis corresponds to the observed induction time. Pre-erosion of the device presumably decreases the surface polymer's chain length and in the subsequent erosion the anhydride hydrolysis rate and device dissolution rate become equivalent. Alternatively the observed induction period and its elimination by pre-erosion may be the result of an initially hydrophobic surface becoming increasingly hydrophilic as hydrolysis occurs. The rate of erosion would increase up to the point where it becomes limited by both the rate of hydrolysis and the rate of water penetration into the polymer.

The decrease in the device thickness throughout the erosion and the maintenance of the matrice's structural integrity as well as the nearly zero-order erosion kinetics suggests that heterogeneous surface erosion predominates. PCPM eroded *in vivo* slightly faster than its *in vitro* erosion in phosphate buffer at pH 7.4 at the same temperature. The thin, partially eroded, polymer slabs were difficult to completely remove from the rats and hence may be the cause for the slightly accelerated *in vivo* erosion rate measurement. Although no tissue irritation was apparent and only slight device encapsulation occurred within the test animals, the toxicity of the polymer matrix and monomer erosion products remains to be investigated. Currently, work is underway in this laboratory[9] to extend the types of drugs released from this matrix and to examine the synthesis of other novel poly(anhydrides) from naturally occurring diacid monomers which might reduce the potential toxicological problems associated with a bio-erodible drug delivery matrix.

ACKNOWLEDGEMENTS

The authors would like to thank Larry Brown, Gerald Fitzgerald and Ronald Siegel for assistance and Pamela Brown for preparing this manuscript. This study was supported by N.I.H. grant no. GM 26698. Gary E. Wnek is the recipient of a Dupont Young Faculty Award.

REFERENCES

1 Langer, R. and N.A. Peppas, Present and Future Applications of Biomaterials in Controlled Drug Release, *Biomaterials* 1981 **2**, 201–214

2 Heller, J. and R.W. Baker, Theory and Practice of Controlled Drug Delivery from Bioerodible Polymers, in *Controlled Release of Bioactive Materials.* (Ed. R.W. Baker), Academic Press, NYC 1980, pp 1–17 (1980)

3 Pitt, C.G., M.M. Gratzl, G.L. Kummel, J. Surles and A. Schindler, Aliphatic Polyesters II. The degradation of poly(DL-lactide), poly(ε-caprolactone), and their copolymers in vivo, *Biomaterials* 1981, **2**, 215–220 (1981)

4 Chu, C.C., Hydrolytic Degradation of Polyglycolic Acid: Tensile Strength and Crystallinity Study, *J. Appl. Polym. Sci.* 1981, **26** (5), 1727–34

5 Hopfenberg, H.B., Controlled Release from Erodible Slabs, Cylinders and Spheres, *Controlled Release Polymeric Formulations*, (Eds. D.R. Paul and F.W. Harris) *ACS Symposium Series* No. 33, 1976, pp 26–32

6 Heller, J., Penhale, D.W.H., Helwing, R.F., and Fritzinger, B.K. Release of Norethindrone from Polyacetals and Poly(Ortho Esters), *Polym. Eng. Sci.*, 1981, **21**, 727–731

7 Conix, A., Aromatic Polyanhydrides, A New Class of High Melting Fiber-Forming Polymers. *J. Polym. Sci.* 1958, **29**, 343–353

8 Conix, A., Poly[1,3-Bis(*p*-carboxyphenoxy)-propane anhydride], *Macromolecular Synthesis*, Vol. 2, (Ed. J.R. Elliot) Wiley, NYC 1966. pp 95–99

9 Rosen, H.B., Synthesis and Characterization of Bioerodible Polymers for Controlled Drug Release, *S.M. Thesis*, Massachusetts Institute of Technology, 1982

Effect of Ultrasound on Transdermal Drug Delivery to Rats and Guinea Pigs

D. Levy,* J. Kost,‡ Y. Meshulam,* and R. Langer§
**Israel Institute for Biological Research, Ness-Ziona, Israel; ‡Department of Chemical Engineering, Ben-Gurion University, Beer-Sheeva, Israel; §Department of Chemical Engineering, Massachusetts Institute of Technology, Cambridge, Massachusetts 02139*

Abstract

The effect of therapeutic range ultrasound (1 MHz) on skin permeation of D-mannitol, a highly polar sugar alcohol, inulin, a high molecular weight polysaccharide and physostigmine, a lipophilic anticholinesterase drug was studied in rats and guinea pigs. D-Mannitol and inulin are totally and rapidly excreted, once they have penetrated through the skin into the blood stream, permitting direct in vivo monitoring. For evaluating skin penetration of physostigmine the decrease of whole blood cholinesterase was measured. Ultrasound nearly completely eliminated the lag time usually associated with transdermal delivery of drugs. 3–5 min of ultrasound irradiation (1.5 W/cm^2 continuous wave or 3 W/cm^2 pulsed wave) increased the transdermal permeation of inulin and mannitol in rats by 5–20-fold within 1–2 h following ultrasound application. Ultrasound treatment also significantly increased ($P < 0.05$) the inhibition of cholinesterase during the first hour after application in both physostigmine treated rats and guinea pigs: while in control guinea pigs no significant inhibition of cholinesterase could be detected during the first 2 h after application of physostigmine, the ultrasound treated group showed a 15±5% (mean±SEM) decrease in blood cholinesterase 1 h after ultrasound application. For physostigmine-treated rats the level of cholinesterase inhibition 1 h after ultrasound application was 53±5% in the ultrasound-treated group and 35±5% in the controls.

Introduction

Topical administration of drugs has been used for many years, but only recently has controlled administration of systemic drugs through the skin (transdermal drug delivery) become a clinically accepted procedure. While the earliest indications that transdermal delivery might be useful date back to the 1960s (1), the first transdermal therapeutic system (for scopolamine), was introduced in 1981. Recently, several other transdermal systems (for nitroglycerine, estradiol, clonidine) have been developed (2). However, while transdermal delivery of drugs is an alluring concept, it has drawbacks that limit its scope. Relatively few drugs penetrate the skin at rates high enough to yield therapeutic levels in the plasma, and the lag time for establishing a stable concentration of the drug in the circulation may be several hours or even days (3). For example: transdermal systems for clonidine proposed for antihypertensive therapy have been reported to reach therapeutic steady state levels only after one to two days (3); scopolamine which is used for treating motion sickness is also effective only several hours after application (4). In order to enhance penetration of drugs through the skin it is possible to use chemical enhancers or to manipulate skin permeability by external means such as iontophoresis (5). External modulation of skin permeability may allow better control of drug release rates than the use of chemical enhancers, as it can be stopped and reapplied according to need. The actual rate of administration of a systemic drug from a transdermal therapeutic system is related to the rate of drug release from the therapeutic system and the rate of drug permeation through the skin (6). The lag time phase may also be attributed to both the skin and the transdermal system, as it takes some time for the drug to migrate from the therapeutic system and across the skin. We have previously shown that ultrasound can enhance the release rates of drugs through solid polymers (7), and in the current study, examined the effect of ultrasound on skin permeation of drugs with special attention to the lag time period. Ultrasound has previously been explored for treating localized skin conditions (8), and for delivering drugs to inflamed joints (9, 10). However, there is limited quantitative information on the effect of ultrasound on the delivery and the lag time phase of transdermally administered drugs.

In the current study we report that ultrasound nearly completely eliminates the lag time associated with transdermal delivery, and in some cases increases the overall drug penetration through the skin. Three different substances and two animal models, rats and guinea pigs, were used. The molecules evaluated were D-mannitol, a highly polar sugar alcohol, inulin, a large molecular weight polysaccharide, and physostigmine, a lipophilic molecule. D-Mannitol and inulin are totally and rapidly excreted (11, 12) permitting direct in vivo monitoring of the labeled drug once it penetrated through the skin into the blood circulation. Physostigmine is a reversible inhibitor of cholinesterase, used for its anticholinesterase action in glaucoma (13), in tricyclic antidepressant poisoning (13) and in the treatment of Alzheimer's disease (14). The biological effect of physostigmine was determined by measuring the decrease of whole blood cholinesterase of physostigmine-treated animals. The ultrasonic unit applied was a therapeutic ultrasound system approved for human use.

Methods

Materials. The test solutions were: D-[^{3}H]mannitol (New England Nuclear, Boston, MA), 1 mCi/ml in ethanol water solution 9:1, saturated with unlabeled D-mannitol (Sigma Chemical Co., St. Louis, MO); [^{3}H]inulin (New England Nuclear) mixed with unlabeled inulin and dissolved in warm water to make a 1-mCi/5 ml saturated solution; and physostigmine base (Sigma) in ethanol (15% or 30% wt/vol). We

Address reprint requests to Dr. Langer, Massachusetts Institute of Technology, E25-342, 77 Massachusetts Avenue, Cambridge, MA 02139.

Received for publication 17 August 1987 and in revised form 7 February 1989.

J. Clin. Invest.

0021-9738/89/06/2074/05 $2.00
Volume 83, June 1989, 2074–2078

did not use saturated solutions of physostigmine in order to avoid too high a cholinesterase inhibition in control animals.

In vitro studies. A transport cell composed of two compartments (48 ml) each separated by the membrane of interest was used to obtain the permeability data. The membranes evaluated were cellulose (Cuprophan) and microporous polypropylene films (Celgard; Celanese Corp., New York). The permeating molecules detected were: NaCl, KIO_3, $(NH_4)_2S_2O_8$, urea (Merck & Co., Inc., Rahway, NJ) and bovine serum albumin (BSA) (Sigma). In these experiments an ultrasonic probe (Vibra Cell 250; Sonics and Materials, Danbury, CT) was immersed into the upstream compartment and the solution was exposed to ultrasound. The downstream compartment solution was circulated (Ismatec pump 7620; Zurich, Switzerland) through a conductivity cell (El-Hamma EHT-24, TH-2400; EL-Hamma, Israel) to detect salt conductivity, or a flow-through spectrophotometer (Spectronic 1001; Bausch & Lomb, Inc., Rochester, NY) at 220 nm for BSA detection. Urea was detected by taking aliquots from the downstream compartment based on the assay developed by Marsh et al. (15). In several permeability experiments instead of an ultrasonic probe, a stirrer (RGL-100; Heidolph, Kelheim, West Germany) was immersed in the upstream compartment and permeability was evaluated at defined mixing rates. To evaluate the effect of temperature on permeability, experiments were performed in a thermostated environment set to defined temperatures (model 3550 incubator; Lab-Line Instruments, McGaw Park, IL). In several other permeability experiments the solution in the transport cell was degassed just before the permeability experiment (vacuum of 60 mm Hg for 15 min).

In vivo drug and ultrasound application. At 30 min before drug application the hair of dorsal test sites of the experimental animals (Sprague-Dawley rats, 200-300 g each, or albino guinea pigs of a local strain, Dunkin-Hardly origin, 300-400 g each) was closely clipped using an electric clipper, and the skin was washed with a mild detergent (Proseptyl Bactericide; Sicca, Holon, Israel) to effect wetting by the ultrasonic coupling gel (EMS Therasonic coupling medium). Test solutions of 20 μl (20 μCi) of D-[^{3}H]mannitol, 50 μl (11 μCi) of [^{3}H]inulin or 6–12 μl of physostigmine were applied to test sites on the upper back of rats or guinea pigs, (20 μl/kg of 15% wt/vol physostigmine in ethanol for rats, and 30 μl/kg of 30% wt/vol of physostigmine in ethanol for guinea pigs), followed by the ultrasonic gel. Ultrasound energy, 1.5 W/cm^2 continuous wave (CW),[1] or 3 W/cm^2 pulsed wave 4:1, was generated by an ultrasonic transducer (1 MHz Therasonic Double Five, model 14A; Electro Medical Supplies, Greenham, Wantage, Ltd., Oxfordshire, England) applied to the treated area for 3 min for a CW or 5 min for a pulsed mode (CW ultrasound was applied for a shorter time and lower intensity to avoid excess heating). Skin temperature was recorded immediately after ultrasound application by placing a temperature probe (Tele Thermometer model 46 TUC; Yellow Springs Instruments, Inc., Yellow Springs, OH) on the surface of the treated skin. The treated area was covered by Blenderm surgical tape (3M Corp., St. Paul, MN).

Permeability evaluation. The D-[^{3}H]mannitol and [^{3}H]inulin treated animals were housed in metabolic cages for the duration of the experiment or catheterized to obtain more reliable short term data. Urine was collected at several time points, 10 ml of scintillation fluor (Aquasol 2; New England Nuclear) was added, and the radioactivity was measured by liquid scintillation counting (Tri-Carb 4530; Packard Instruments, Downers Grove, IL).

For the catheterization, the rats were anesthetized with methoxyflurane (Pitman Moore Inc.), and their abdominal and cervical fur were shaved and cleansed. A 2-cm long midline abdominal incision was made through the abdominal musculature. The bladder was exposed, a nick was made in the bladder top, the catheter (PE-50, 7411; Intramedic Clay Adams, Parsippany, NJ) was inserted in the bladder and secured by 4-0 Dexon suture. After urine flow was verified, the abdominal musculature was sutured (4-0 Dexon) as well as the skin incision (4-0 nylon). Test solutions of D-[^{3}H]mannitol or [^{3}H]inulin were applied to the skin, followed by ultrasound application as described above. Urine was collected every 15–30 min for the first 2 h and at several additional times during the experiment, and the radioactivity was measured.

For evaluating the effect of ultrasound on the transdermal delivery of physostigmine, blood samples of physostigmine treated rats or guinea pigs were obtained at specific times by cardiac puncture and the decrease of whole blood cholinesterase (ChE) activity was assayed by the Johnson method (16).

Controls. Control animals in each experiment were treated by the procedure described above moving the ultrasonic applicator head over the treated area for 3–5 min, with the power level at the applicator set to zero.

To check for a possible decomposing effect of the ultrasound treatment on the applied D-mannitol, urine samples of the ultrasound treated rats and controls were compared by thin-layer chromatography on silica plates (Kieselgel 60; Merck) developed in *t*-butanol/pyridine/water (6:4:3) or in *t*-butanol/formic acid/water (33:50:17) and visualized by phosphomolybdic acid reagent. The developed silica gel plates were cut to several sections according to the locations of the spots, and the pieces were placed in scintillation vials. 1 ml of distilled water was added to each vial to dissolve the mannitol, then 10 ml of Aquasol scintillation liquid was added and the radioactivity measured.

We also checked the ultrasound effect on ChE inhibition of blood previously inhibited by physostigmine in vitro. 1-ml samples of heparinized blood collected from rats were incubated at 37°C in the presence of 5×10^{-9}–5×10^{-8} M physostigmine. After steady state was reached (~ 30 min), samples were placed in a water bath (room temperature) in which an ultrasonic probe was inserted for 5 min (3 W/cm^2 pulsed wave). The level of cholinesterase inhibition before and after the ultrasound treatment was recorded (Table I).

Histology. Rats exposed to ultrasound, as described above (5 min 3 W/cm^2, pulsed wave), were killed using an overdose of carbon dioxide. Sections (5 mm) were prepared using hematoxylin and eosin staining.

Statistical analysis. Values are expressed as mean±SEM. Student's *t* test was used for statistical analysis.

Results

In vivo the pronounced effect of ultrasound on transdermal permeability was observed for all three molecules studied. During the first 2 h the secretion rate of radiolabeled mannitol was 20-fold higher in the ultrasound treated group than in the controls (Fig. 1). In the inulin experiment (Fig. 2), the experimental group had five times more excreted radioactive drug than the control group during the first hour.

In order to eliminate the possibility that the excess radioactivity observed in the excretion of ultrasound-treated rats was

Table I. The Concentration Dependence of Inhibition of Rat Blood Cholinesterase by Physostigmine

Concentration of physostigmine	% of ChE inhibition	% of ChE inhibition after ultrasound pulsing
nM		
5	17±3	18±3
10	29±4	24±4
20	45±2	40±7
30	58±2	53±3

The blood samples were incubated with physostigmine for 30 min at 37°C, then placed in a water bath (room temperature) and pulsed with US probe for 5 min (3 W/cm^2 pulsed wave). Each value is the mean±SEM of three experiments.

1. *Abbreviations used in this paper:* ChE, cholinesterase; CW, continuous wave; US, ultrasound.

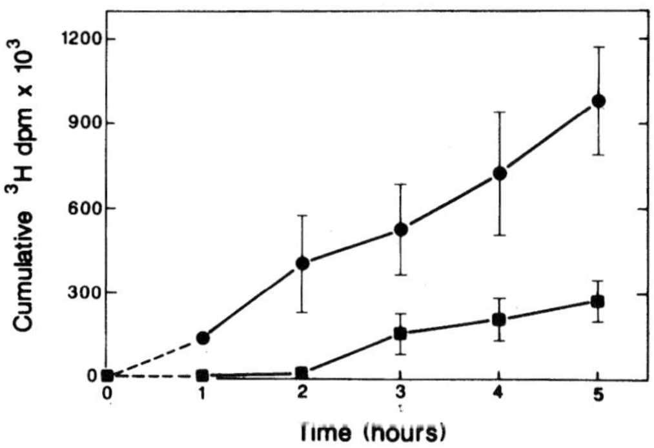

Figure 1. Rats' cumulative radioactive secretion after topical application of 20 μl of saturated solution of D-mannitol containing 20 μCi D-[^{3}H]mannitol. •, Ultrasound (US) treated rats. n = 4. (1.5 W/cm^2 CW, 3 min). ▪, Control rats. no US. n = 12. Values are mean±SEM.

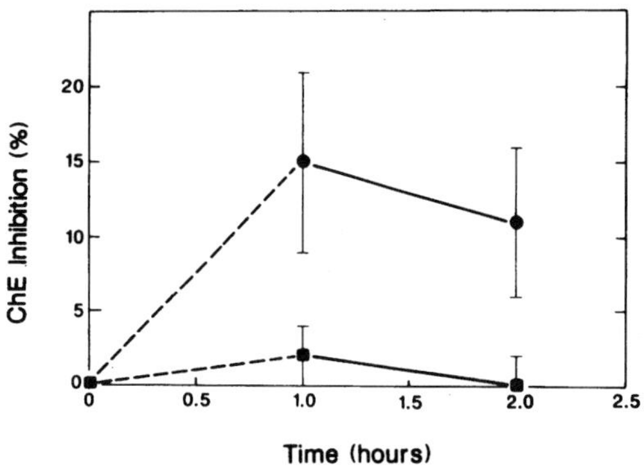

Figure 3. Whole blood ChE inhibition profile in guinea pigs after topical application of physostigmine (30% wt/vol in ethanol). •, Ultrasound treated animals. n = 4. (3 W/cm^2 pulsed wave 5 min). ▪, Control animals. no US. n = 4. Values are mean±SEM.

due to changes in urine flow rather than changes in skin permeation, the rats were catheterized, and the urine volume was recorded. The mean±SEM volume of urine secretion during the first 2 h was 1.50±0.25 ml for ultrasound-treated rats and 0.84±0.25 ml for the controls (in the mannitol experiment). The fact that the urination rate of the ultrasound-treated rats was less than twice that of the control rats is thus unable to account for the much higher changes in drug excretion.

The possibility of artifact in the amount of D-mannitol and inulin recovered due to tritium exchange with surrounding aqueous media was also evaluated. The media was assayed as described by Brown et al. (17). No tritium exchange was observed.

Thin-layer chromatography of urine excreted D-[^{3}H]-mannitol from ultrasound treated rats was identical to that of untreated rats. The main spot appeared to be D-[^{3}H]mannitol, indicating that the applied mannitol was not decomposed by ultrasound. We did not test the excreted [^{3}H]inulin, due to the very small amount secreted, (< 0.03% of the applied inulin was excreted in 5 h, as compared to ~ 2% of mannitol).

To estimate the permeation of physostigmine through the skin, we measured the decrease in blood ChE which depends on the concentration of physostigmine in the blood (Table I) (18). The level of ChE inhibition 1 h following ultrasound application was significantly higher ($P < 0.05$) for both rats and guinea pigs in the ultrasound treated animals than in controls. No significant inhibition of ChE could be detected during the first 2 h after application of physostigmine to control guinea pigs, while the ultrasound-treated group showed a 15±5% decrease in ChE activity 1 h after ultrasound application (Fig. 3). 5 h after ultrasound application, the level of ChE inhibition was 13±4% for both groups (not shown). For the physostigmine-treated rats, the level of ChE inhibition 1 h after application was 53%±5 in the ultrasound group and 35%±5 in the control (Fig. 4). To rule out the possibility of ultrasound itself affecting blood ChE activity, ultrasound energy (same conditions as for experimental animals) was applied to a dorsal site of drug-free animals and blood ChE activity was assayed before and after ultrasound treatment. Ultrasound treatment alone had no effect on the level of ChE inhibition in vivo, nor did it have any effect on the level of ChE inhibition of physostigmine-treated blood in vitro (Table I), thus indicating that the higher ChE inhibition is directly related to increased drug penetration. Furthermore, when ultrasound was applied to guinea pigs 1 h before physostigmine application, there was no difference in ChE inhibition between

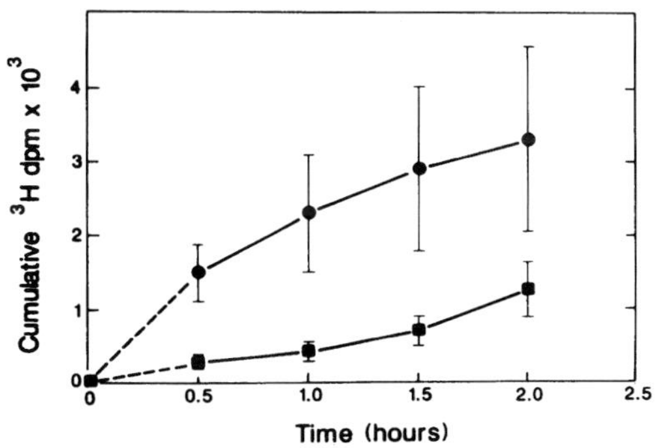

Figure 2. Rats' cumulative radioactive secretion after topical application of 50 μl of saturated solution of inulin containing 11 μCi [^{3}H]-inulin. •, Ultrasound treated rats. n = 4. (3 W/cm^2 pulsed wave, 5 min). ▪, Control rats. no US. n = 4. Values are mean±SEM.

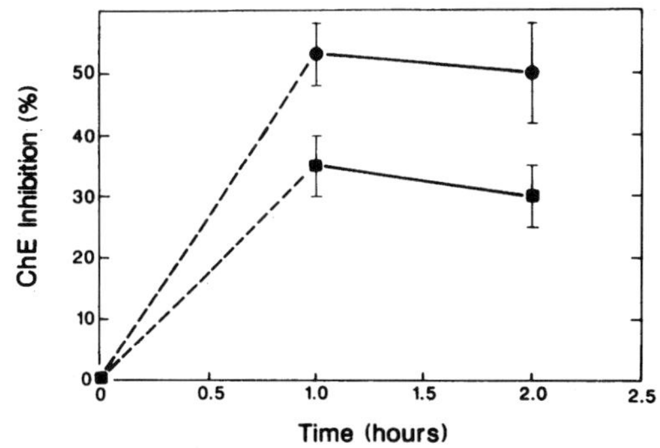

Figure 4. Whole blood ChE inhibition profile in rats after topical application of physostigmine (15% wt/vol in ethanol). •, Ultrasound treated rats. n = 7. (3 W/cm^2 pulsed wave, 5 min). ▪, Control rats. no US. n = 7. Values are mean±SEM.

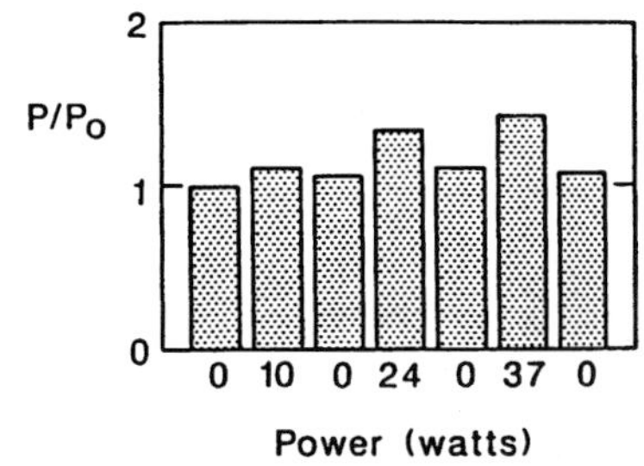

Figure 5. Effect of ultrasound power on KIO_3 permeability through polypropylene (Celgard 5511) membrane placed in a two compartment diffusion cell. P_0 is the initial permeability of the membrane, before ultrasound application. An ultrasonic probe was placed in the upstream component and samples were analyzed in the downstream compartment.

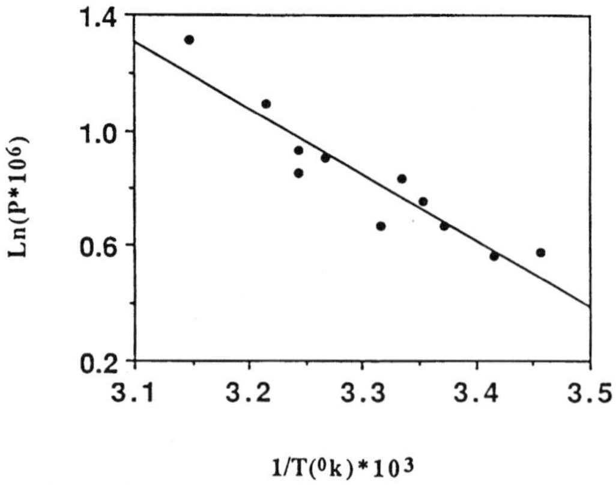

Figure 6. Permeability of urea through cuprophan membranes vs. temperature. Membranes were placed in a two compartment diffusion cell, which was located in a thermostatted environment. The constant temperatures in which the permeability was evaluated were in the range of 17–44°C.

the experimental group and the controls. The increase in surface skin temperature measured immediately after ultrasound application did not exceed 1–2°C.

Mechanistic studies. To gain insight into the mechanism of ultrasonically enhanced release, synthetic membranes were evaluated in an in vitro diffusion cell. The advantages of using such a defined system involving synthetic membranes rather than skin are that these membranes are very reproducible and their permeability characteristics are well characterized. Therefore individual effects contributing to ultrasound enhanced permeability can be accurately assessed. Studies were conducted to examine both the reversibility of the ultrasound effect and the mechanism of this effect. To evaluate the reversibility of the ultrasound phenomenon, experiments were performed in which the ultrasound was turned on and off alternatively at different power levels and the permeability was calculated. Fig. 5 shows the effect of ultrasound power on the ratio of permeabilities (while exposed to ultrasound to the permeability in unexposed solutions). As can be seen the effect is reversible. The permeabilities of the membranes return to their original values (before ultrasound application), when the output power of the ultrasound was set back to zero.

We considered three factors that might contribute to ultrasound enhanced permeability: mixing, cavitation, and temperature. To examine whether ultrasound might affect a boundary layer surrounding the membranes, and therefore cause higher permeability values, experiments were performed under controlled mixing rates. A boundary layer is a thin layer of fluid created when a fluid flows over a surface. The velocity in this layer decreases until it is zero at the surface. The boundary layer is an additional resistance for drug transport. The thickness of the layer is inversely proportional to the velocity of flowing fluid and therefore mixing could eliminate a boundary layer and its effect on permeability. Table II summarizes the effect of mixing on the urea permeability through cuprophan membranes. From 600 rpm and greater there is no effect of additional mixing on permeability. An additional factor that could be associated with ultrasound is cavitation (i.e. formation and collapse of gas cavities in the liquid creating shock waves). Pulsation of cavitation bubbles might affect the boundary layer as well as transport by convection. Therefore experiments were conducted in degassed buffer, where cavitation was minimized. The observed enhancement of urea permeability was much smaller: 23% (at 19.5°C and 40% duty cycle) compared to 55% in the original buffer at the same conditions and temperature. Heat liberation will eventually be produced within any medium exposed to ultrasonic energy. As mass diffusivity is a function of temperature, urea permeability experiments through cuprophan membranes were performed at defined constant temperatures. The permeability vs. temperature displayed Arrhenius behavior (Fig. 6).

Table II. The Effect of Mixing Rate on Urea Permeability through Cuprophan Membranes

Permeability $P \cdot 10^6$	Mixing
cm^2/s	*rpm*
1.6±0.067	0
1.73±0.095	300
1.88±0.143	600
1.788±0.156	900

Studies were conducted by placing the membrane in a two-compartment transport cell. A stirrer was placed in the upstream compartment. Urea was analyzed by taking samples from the downstream compartment. For details, see Methods.

Histology. Comparisons were made between rat skin exposed to ultrasound and controls not exposed to ultrasound. The sections of skin and associated underlying tissues appeared essentially the same and were within normal limits.

Discussion

One effect, due to ultrasound, observed in this study was the shortening of the lag time, normally associated with all forms of transdermal drug delivery. This was observed in all but the physostigmine-treated rats. The permeability of rat skin to physostigmine solution was relatively high, therefore the lag time phase was not observed perhaps because the sampling periods were not frequent enough to detect it. However, although the systemic effect (ChE inhibition) could be observed after 1 h in both ultrasound-treated and the control rats, the effect was larger in the former group.

The fact that 3–5 min of ultrasound irradiation have shortened the onset time of transdermally applied substances suggests that ultrasound can temporarily and reversibly alter the permeability of skin for both hydrophilic and lipophilic drugs. The reversibility of the ultrasound effect on skin was demonstrated by the observation that ultrasound irradiation of guinea pigs 1 h before drug application did not change the permeation rate of physostigmine through the skin. This is also supported by the observation that the penetration rate of physostigmine in guinea pigs, 5 h after ultrasound application, was the same for both ultrasound treated and controls, (as judged by the level of ChE inhibition) in contrast to the large difference in permeation rate observed right after ultrasound application (Fig. 3).

The small increase in surface skin temperature observed after ultrasound application (1–2°C), is not likely to cause dramatic changes in skin permeability (19, 20). However, ultrasound is known for its ability to cause deep penetrating hyperthermia which may increase diffusivity, solubility of drugs, vasodilation and blood flow, all of which may facilitate drug penetration.

The mechanistic in vitro studies showed that ultrasound can reversibly enhance the permeability of synthetic membranes (Fig. 5). The temperature increase of 1–2°C based on the Arrhenius correlation has a relatively very small effect on the permeability. In addition to temperature, other important phenomena involved are mixing and cavitation. Although it's difficult to be certain whether the data obtained in the in vitro experiments with synthetic polymeric membranes are extendable to the in vivo situation with skin, it is likely that since both involve diffusion through membranes, that those factors which ultrasound affects most significantly in vitro (e.g., mixing, cavitation) also play a significant role in vivo (21).

The observed phenomenon that short exposures of ultrasound can alter both the lag time and the permeability of transdermally administered substances suggests a number of future studies. Determining the appropriate ultrasonic frequencies, duty cycles, orientation, and energies to achieve desired effects for different drugs may greatly broaden the range of therapeutic agents that can be administered transdermally. Eliminating the lag phase associated with transdermal dosage forms may eventually broaden the use of these dosage forms to acute as well as chronic indications. Portable ultrasonic triggering devices must be developed, however, if this type of approach is to be generally usable. There has been considerable clinical interest in the use of iontophoresis to affect transdermal drug delivery (5). The current study indicates that ultrasound may also be of value in this regard.

Acknowledgments

The authors thank R. Chen, O. Lykin, and M. Goldriech for their technical assistance.

This study was supported by National Institutes of Health Grant CM26698 and a grant from the United States-Israel Binational Science Foundation (BSF), Jerusalem, Israel.

References

1. Maibach, H. I., and R. J. Feldmann. 1967. The effect of DMSO on percutaneous penetration of hydrocortisone and testosterone on man. *Ann. NY Acad. Sci. USA* 141:423–427.

2. Brown, L. R., and R. Langer. 1988. Transdermal delivery of drugs. *Annu. Rev. Med.* 39:221–229.

3. Arndts, D., and K. Arndts. 1984. Pharmacokinetics and pharmacodynamics of transdermally administered clonidine. *Eur. J. Clin. Pharmacol.* 26:79–85.

4. Price, N. M., L. G. Schmitt, G. McGuire, J. E. Shaw, and G. Trobough. 1981. Transdermal scopolamine in the prevention of motion sickness at sea. *Clin. Pharmacol. Ther.* 29:414–419.

5. Kari, B. 1986. Control of blood glucose levels in alloxan-diabetic rabbits by iontophoresis of insulin. *Diabetes.* 35:217–221.

6. Cleary, G. W. 1984. Transdermal controlled release systems. *In* Medical Applications of Controlled Release. Vol. 1. R. S. Langer and D. L. Wise, editors. CRC Press, Inc, Boca Raton, FL. 203–253.

7. Kost, J., K. Leong, and R. Langer. 1986. Ultrasonic modulated drug delivery systems. *In* Polymers in Medicine, Biology and Pharmaceutical Applications II. E. Chiellini, editor. Plenum Press, New York/London. 376–396.

8. McElnay, J. C., M. P. Mathews, R. Harland, and D. F. McCafferty. 1985. The effect of ultrasound on the percutanous absorption of lignocaine, *Br. J. Clin. Pharmacol.* 20:421–424.

9. Fellinger, K., and J. Schmid. 1954. Klinik und Therapie des chronischen, Gelenkreumatismus. Maudrich, Vienna, Austria. 549–554.

10. Skauen, D. M., and G. M. Zentner. 1984. Phonophoresis, *Int. J. Pharmacol.* 20:235–245.

11. Gutman, Y., C. W. Gattschalk, and W. T. Lassiter. 1965. Macropuncture study of inulin absorption in the rat kidney. *Science (Wash. DC).* 147:753–754.

12. Weiner, I. M., and G. H. Mudge. 1985. Diuretics and other agents employed in the mobilization of edema fluid. *In* The Pharmacological Basis of Therapeutics. 7th ed. L. Goodman, A. Gilman, T. W. Rall, and F. Murad, editors. McMillan Publishing Co., New York. 887–889.

13. Taylor, P. 1985. Anticholinesterase agents. *In* The Pharmacological Basis of Therapeutics. 7th ed. L. Goodman, A. Gilman, T. W. Rall, and F. Murad, editors. McMillan Publishing Co., New York. 110–129.

14. Thal, L. J., P. A. Fuld, D. M. Masur, N. S. Sharpless, and P. Davies. 1983. Oral physostigmine and lecithin improve memory in Alzheimer's disease, *Psychopharmacol. Bull.* 19:454–456.

15. March, W. H., B. Fingerhut, and H. Miller. 1965. Automated and manual direct methods for the determination of blood urea. *Clin. Chem.* 11:624–627.

16. Johnson, C. D., and R. L. Russell. 1975. A rapid simple radiometric assay for cholinesterase, suitable for multiple determinations. *Anal. Biochem.* 64:229–238.

17. Brown, L. R., C. L. Wei, and R. Langer. 1983. In vivo and in vitro release of macromolecules from polymeric drug delivery systems. *J. Pharmacol. Sci.* 72:1181–1185.

18. Aldridge, W. N., and E. Reiner. 1972. Enzyme Inhibitors as Substrates. North Holland Publishing Co., Amsterdam/London.

19. Knutson, K., R. O. Potts, D. B. Guzek, G. M. Golden, J. E. McKie, W. J. Lambert, and W. I. Higuchi. 1985. Macro and molecular physical-chemical considerations in understanding drug transport in the stratum corneum. *J. Controlled Release.* 2:67–87.

20. Flynn, G. L. 1985. Mechanism of percutanous absorption from physicochemical evidence. *In* Percutanous Absorption. R. L. Bronaugh, and H. I. Maibach, editors. Marcel Dekker, Inc., New York/Basel. 33.

21. Haar, G. R., and S. Daniels. 1981. Evidence for ultrasonically induced cavitation in vivo. *Phys. Med. Biol.* 26:1145–1149.

New Methods of Drug Delivery

ROBERT LANGER*

Conventional forms of drug administration generally rely on pills, eye drops, ointments, and intravenous solutions. Recently, a number of novel drug delivery approaches have been developed. These approaches include drug modification by chemical means, drug entrapment in small vesicles that are injected into the bloodstream, and drug entrapment within pumps or polymeric materials that are placed in desired bodily compartments (for example, the eye or beneath the skin). These techniques have already led to delivery systems that improve human health, and continued research may revolutionize the way many drugs are delivered.

IN THE LAST FEW YEARS, WE HAVE WITNESSED AN EXPLOSION in research aimed at creating new drug delivery systems. This research has been fueled by several developments. (i) Many drugs, both old pharmaceutical products and new molecular entities, can be administered in ways that not only improve safety and efficacy but, in some cases, permit new therapies. (ii) Newer and complex drugs such as proteins are becoming available through genetic engineering; the delivery of these drugs is often more complicated than that of more conventional drugs, necessitating novel delivery systems. (iii) There is an increasing awareness that drug release patterns (continuous versus pulsatile) significantly affect therapeutic responses. (iv) The overall expense to create a pharmaceutical that is a new molecular entity is at least $150 million; the lower cost to improve the delivery of an existing drug is sometimes seen as a better investment. This issue is exacerbated because drug patents expire after 17 years, and a new drug delivery system may permit continued benefits for the company producing it. (v) Advances in materials science and biotechnology are permitting the development of new physical and chemical methods of drug delivery. In this article, some of the methods being studied to deliver drugs are discussed.

The author is in the Department of Chemical Engineering, Harvard-MIT Division of Health Sciences and Technology, and Whitaker College of Health Sciences, Massachusetts Institute of Technology, Cambridge, MA 02139, and Department of Surgery, Children's Hospital, Boston, MA 02115. He is Kenneth J. Germeshausen Professor of Chemical and Biochemical Engineering at MIT.

*To whom correspondence should be addressed at the Massachusetts Institute of Technology, E25-342, Cambridge, MA 02139.

Chemical Modification

A drug may be chemically modified to selectively alter such properties as biodistribution, pharmacokinetics, solubility, or antigenicity. One example is drugs that are designed to cross a normally impermeable barrier. The blood brain barrier, which contains tight endothelial cell junctions and prevents most molecules from entering the central nervous system, has been the target of considerable research. Several experimental approaches have been developed, in which drugs are complexed to agents that enable them to cross this barrier (for example, by rendering the drug more lipophilic or coupling it to a molecule that has a specific transport mechanism) (*1*).

Drugs have also been attached to soluble macromolecules such as proteins, polysaccharides, or synthetic polymers via degradable linkages. This process alters the drug's size and other properties, resulting in different pharmacokinetics and biodistribution. One example involves coupling the antitumor agent neocarzinostatin to styrene–maleic acid copolymers (*2*). When this complex was injected intra-arterially into patients with hepatocellular carcinoma, decreases in α-fetoprotein levels and tumor size were observed. In animals, antitumor agents such as doxorubicin coupled to *N*-(2-hydroxypropyl) methacrylamide copolymers showed radically altered pharmacokinetics, resulting in reduced toxicity. The half-life of the drug in plasma and the drug levels in the tumor were increased while the concentrations in the periphery decreased (*3*).

An exciting approach for "targeting" drugs to specific cells involves linkage of a bioactive agent (drug, radioisotope, or toxin) to a monoclonal antibody. Antibody conjugates are now being studied in the treatment of cancer, septic shock, and acquired immunodeficiency syndrome (AIDS). There are several critical issues in the use of antibodies. With mouse antibodies, anaphylactic reactions frequently occur with repeated administration. Thus, ongoing research is directed toward producing human monoclonal antibodies or toward making mouse antibodies more human-like through the use of genetic engineering. This problem may be exacerbated for immunotoxins (antibody-toxins) because of the proteinaceous character of the toxin. Thus far, clinical usefulness of immunotoxins has been demonstrated in therapy regimens characterized by rapid pharmacokinetics, such as treatments for lymphoma and graft versus host disease, and extracorporeal treatments such as bone marrow purgings. The powerful killer potential of certain toxins, such as ricin or diphtheria toxin, makes immunotoxins an attractive approach if an appropriate antibody is available that can be internalized by desired cells (*4*). Antibody-radioisotopes act over a greater distance than immunotoxins. One requirement with such complexes is the availability of a suitable chelator that allows a kinetically stable binding of the radioisotope. The degradation of the linker structure between the chelator and antibody is also critical, since nondegradable structures may cause kidney and liver toxicity. Initial clinical results with certain beta-emitters have shown regression of lymphomas. Other critical issues in the use of an antibody are its affinity, specificity, size, and large-scale production; for cancer chemotherapy, tumor characteristics and blood flow are important considerations (*5*).

Polymers, such as polyethylene glycol (PEG), can be attached to drugs to either lengthen their lifetime or alter their immunogenicity. The polymers physically prevent cells and enzymes from attacking the drug. PEG-uricase reduced serum urate levels in patients with hyperuricemia and gout; PEG-asparaginase has been used for patients with leukemia, and PEG–adenosine deaminase has been used for patients with a severe combined immunodeficiency (*6*). Drug longevity and immunogenicity may also be affected by biological approaches, including protein engineering and altering glycosylation patterns.

Vesicles

Vesicles are microparticulates or colloidal carriers composed of substances such as proteins, lipids (for example, liposomes), carbohydrates, or synthetic polymers. Vesicles share some of the advantages of drug-macromolecular conjugates (altered pharmacokinetics and biodistribution) and make possible a potentially higher drug payload. Liposomes, the most widely studied of these vesicles, can be formulated with a variety of lipid compositions and structures and are potentially nontoxic, degradable, and nonimmunogenic. However, many liposomes exhibit poor stability during storage and use. Liposome stability may be improved by increasing the liposomal cholesterol content or synthesizing polymerizable liposomes, but biodegradability may then be diminished (*7*). Engineering issues such as large-scale lipid production and manufacturing of liposomes are also critical to the more widespread use of these vesicles.

In clinical studies, liposomal doxorubicin reduces side effects such as alopecia and nausea associated with the administration of the free drug yet permits a higher maximal tolerated dose and a reduction in cardiac toxicity of 86% (*8*). Liposomal amphotericin B is more effective than the free drug in treating immunocompromised cancer patients with fungal infections (*9*). Methods are also being studied to create liposomes that release more drug in response to specific stimuli such as heat, enzymes, polycations, light, or pH (*10*).

Vesicles may be "targeted" either passively or actively. Passive targeting involves the natural uptake by cells that scavenge foreign microparticulates such as reticuloendothelial cells, which are concentrated in tissues such as the liver or spleen, or circulating monocytes. Thus, liposomes have been used for delivering toxic agents, such as arsenic, to treat liver-specific parasitic diseases (for example, schistosomiasis) in animal models (*11*) with doses 0.1% of those of conventional regimens. Similarly, immunostimulating agents encapsulated in liposomes are taken up by monocytes, which then leads to enhanced killer cell activity. This approach is being tested in certain cancer treatments (*12*). Liposomes can also be used to deliver vaccines (*13*).

Active targeting generally involves placing a charge or recognition sequence (for example, from an antibody) onto the vesicle such that it is more rapidly taken up by certain cell types (such as cancer cells) than others. One difficulty with this approach is that reticuloendothelial cells also scavenge these vesicles. However, recent approaches for altering vesicle compositions, by coating them with surfactants or altering lipid compositions, may reduce the magnitude of this problem (*14*). Vesicles that contain magnetic microparticles have also been used to target drugs to specific locations in animal models via external magnetic fields (*15*).

Controlled Release Systems

Controlled release systems deliver a drug at a predetermined rate for a definite time period. In general, release rates are determined by the design of the system and are nearly independent of environmental conditions, such as pH. These systems can also deliver drugs for long time periods (days to years). Although vesicles or drug macromolecule conjugates may prolong release, optimal control is afforded if the drug is placed in a polymeric material or pump. Controlled release systems differ from older "sustained release" or "slow release" preparations that include complexes (to salts or ion-exchange resins), suspensions, emulsions, slowly dissolving coatings that do not dissolve in the stomach yet do dissolve in the intestine (enteric coatings), and compressed tablets. Generally, sustained-release systems emit drugs in less than a day, and environmental conditions influence release rates, which leads to patient to patient variations.

Controlled release systems provide advantages over conventional drug therapies. For example, after ingestion or injection of standard dosage forms, the blood level of the drug rises, peaks, and then declines. Since each drug has a therapeutic range above which it is toxic and below which it is ineffective, oscillating drug levels may cause alternating periods of ineffectiveness and toxicity. Although sustained release preparations attenuate the peaks and valleys, they do not eliminate them. In contrast, a controlled release preparation maintains the drug in the desired therapeutic range by a single administration. Other potential advantages of controlled release systems include (i) localized delivery of the drug to a particular body compartment, thereby lowering the systemic drug level; (ii) preservation of medications that are rapidly destroyed by the body (this is particularly important for biologically sensitive molecules such as proteins); (iii) reduced need for follow-up care; (iv) increased comfort; and (v) improved compliance.

Pumps are larger and more costly than polymeric systems and require surgery for implantation; however, they offer the advantage of very precise drug control and can release the drug directly into the bloodstream. In addition, some pumps are refillable. Both externally worn and implantable pumps have been developed. In both cases, the driving force is a pressure difference, which results in bulk flow of a drug solution through an orifice.

A common externally worn pressure-driven pump is the miniature syringe pump, in which the drug is delivered at a constant rate by a syringe barrel that moves at a constant velocity; the delivery rate is adjusted by altering either the drug concentration in the syringe or the barrel velocity. An implantable pressure-driven pump has been developed that uses a fluorocarbon propellant as a driving force. In this case, the pump controls a collapsible bellows, which divides the pump interior into two chambers, one containing the propellant and the other containing the drug solution. At body temperature, the vapor pressure exerted by the propellant forces the drug solution through a filter and flow regulator at a constant rate. Other pressure-driven pumps use piezoelectric disk benders or valves. Release rates can be externally regulated by the use of approaches such as telemetry. Pumps have been used in cancer therapy where a catheter extending from a pump is selectively inserted into a blood vessel feeding an organ such as the liver or brain to increase the delivery rate to the diseased organ while sparing the rest of the body. Pumps have also been used to release insulin, heparin, morphine, and other drugs (*16*).

Polymeric materials generally release drugs by the following mechanisms: (i) diffusion, (ii) chemical reaction, or (iii) solvent activation. There are two types of diffusion-controlled systems: reservoirs (Fig. 1A) and matrices (Fig. 1B). Chemical control is accomplished either by polymer degradation (Fig. 1C) or chemical cleavage of the drug from a polymer (Fig. 1D). Solvent activation involves either swelling of the polymer (Fig. 1E) or osmotic effects (Fig. 1, F and G).

One of the first clinically used controlled release polymer systems was the Ocusert, a reservoir system designed to improve therapy for glaucoma, one of the world's leading causes of blindness. The conventional treatment involved the use of pilocarpine eye drops (which reduce intraocular pressure) four times a day. The eye drops often caused side effects, and patient compliance was sometimes poor. The Ocusert delivers pilocarpine (20 or 40 μg/hour) continuously for 1 week and controls intraocular pressure with less drug and fewer side effects. It is placed in the lower eyelid's conjunctival cul-de-sac, where it floats in the tear film. Despite its advantages, the Ocusert never achieved widespread use, initially because of its expense and poor acceptance by older patients who were reluctant to adjust to this system and later because of the introduction of timolol, a drug that requires only two applications per day.

The use of polymers to deliver contraceptive steroids has been widely studied. Four types of systems have been examined: (i) subdermal reservoir implants composed of nondegradable polymers that release drug for over 5 years (for example, the Norplant); these systems, based on a seminal study of diffusion through silicone rubber (*17*), are approved for use in 15 countries; (ii) subdermal implants or injectable microspheres composed of degradable materials, such as lactic acid–glycolic acid copolymers, polycaprolactones, or cholesterol; (iii) steroid-releasing intrauterine devices, such as the Progestasert, an ethylene–vinyl acetate copolymer reservoir that contains a 3-day supply (38 mg) of the amount of progesterone normally taken orally, but which, since it delivers progesterone to its

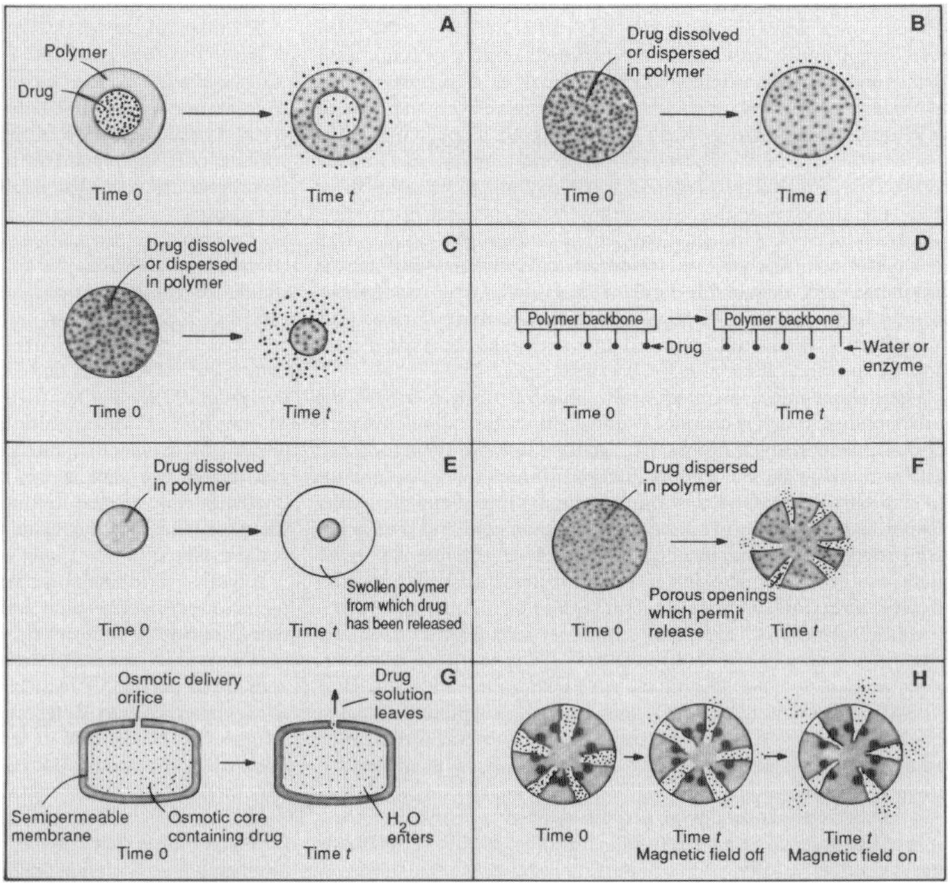

Fig. 1. Polymer release mechanisms. The most common release mechanism is diffusion, whereby the drug migrates from its initial position in the polymeric system to the polymer's outer surface and then to the body. Diffusion may occur through a reservoir (**A**), in which a drug core is surrounded by a polymer film, or in a matrix (**B**), where the drug is uniformly distributed through the polymeric system. Drugs can also be released by chemical mechanisms such as degradation of the polymer (**C**) or cleavage of the drug from a polymer backbone (**D**). Exposure to a solvent can also activate drug release. For example, the drug may be locked into place by polymer chains, and, upon exposure to environmental fluid, the outer polymer regions begin to swell, allowing the drug to move outward (**E**), or water may permeate a drug-polymer system as a result of osmotic pressure, causing pores to form and bringing about drug release (**F**). An attractive osmotic system that can provide constant release rates exists in the form of a pill that has a laser-drilled hole in the surface of a polymer coating (**G**). Some polymer systems can be externally activated to release more drug when needed, using forces such as magnetism (**H**). In this case, an external magnetic field causes polymer-embedded magnetic beads to "squeeze" drug-containing pores, forcing more drug out of a matrix. In all cases, dots represent drug, and in (H) the large dots represent magnetic beads. Combinations of the above mechanisms are possible. Release rates from polymer systems can be controlled by the nature of the polymeric material (for example, crystallinity or pore structure for diffusion-controlled systems; the lability of the bonds or the hydrophobicity of the monomers for chemically controlled systems) and the design of the system (for example, thickness and shape). The advantage of having systems with different release mechanisms is that each can accomplish different goals. For example, reservoir systems are able to produce near-constant release rates, whereas matrix systems are inexpensive to manufacture. Chemically controlled systems generally result in elimination of the polymer, whereas solvent-activated systems have release rates independent of pH (*50*).

target locally at a rate of approximately 65 μg/day, lasts for over 1 year; and (iv) vaginal rings, which are silicone reservoir systems used for 3 to 6 months; generally, for each monthly cycle they are inserted for 3 weeks and then are withdrawn for 1 week (*18*).

Tetracycline, incorporated into diffusion-controlled systems composed of ethylene–vinyl acetate copolymer or other substances, has been used to treat periodontal disease. When this controlled release system was placed in the periodontal pocket, significant reductions in bacterial counts and in the incidence of gingivitis were observed. Furthermore, because the systems are placed next to their target, treatment is accomplished with less than one-thousandth of the normal systemic dose (*19*).

A number of other controlled release systems are under study. These include localized release of diphosphonates (calcium chelators) to prevent heart valve calcification, dopamine or bromocriptine for potential treatment of Parkinson's disease, and bethanecol for potential treatment of Alzheimer's disease (*20*).

Controlled Release Systems for Peptides and Proteins

For many years, controlled-release systems were capable of slowly releasing drugs of only low molecular weight (<600). Large molecules such as proteins were not considered feasible candidates, because polypeptides were considered too large to slowly diffuse through most polymeric materials, even after swelling of the polymer. Large molecules could diffuse through highly porous membranes such as Millipore filters or certain gels such as polyacrylamide; however, in these cases, diffusion was generally too rapid to be of value and tissue damage was usually observed. The discovery that matrices of solid hydrophobic polymers containing powdered macromolecules enabled molecules of nearly any size to be released for over 100 days permitted controlled delivery of a variety of proteins, polysaccharides, and polynucleotides (*21*). Examples of polymers that perform in this way are nondegradable ethylene–vinyl acetate copolymer and degradable lactic acid–glycolic acid copolymers. Certain hydrogels such as poly(hydroxyethylmethacrylate) or poly(vinylalcohol) also work effectively but release proteins for shorter time periods than the above polymer systems.

The release mechanism generally involves movement of the polypeptide through a complex porous path in the polymer matrix. If the polymer erodes, this will affect the pore structure and accelerate the release. Factors influencing release rates include protein particle size and loading, protein solubility and molecular weight, polymer composition and molecular weight, and the dimensions and shape of the matrix (*22*). Polymer systems are now being used in animal studies to release many proteins, including insulin, growth factors, and angiogenesis inhibitors (*23*). The first Food and Drug Administration (FDA)–approved system for controlled release of a peptide, the Lupron Depot (injectable microspheres composed of lactic acid–glycolic acid copolymer and leuprolide acetate, and lasting 30 days) was recently introduced for the treatment of prostate cancer. Other polymeric systems for releasing similar drugs (*24*) are also under evaluation for treating endometriosis and other conditions.

A number of challenges in protein delivery remain. Foremost among these is that, when encapsulated proteins remain in the body for a long time, they may denature or aggregate as a result of exposure to moisture at 37°C. This can cause a loss of biological activity and possible changes in immunogenicity. Stabilization approaches being developed in protein chemistry (*25*) will be important for the success of some of these delivery systems. In one study that used solid proteins as a model, small amounts of added water induced aggregation of albumin, ovalbumin, glucose oxidase, and β-lactoglobulin. The aggregation as a function of added water went through a maximum with just 3 μl of water, causing 97% aggregation of 10 mg of albumin in 24 hours. At lower and higher water concentrations, aggregation was reduced. The aggregation mechanism was discovered to be intermolecular S–S bond formation through thiol-disulfide interchange. This, in turn, suggested rational strategies for protein stabilization, including modifying sulfhydryl residues, lyophilizing from acidic solutions, controlling moisture content, using appropriate additives, and developing specific polymer matrix compositions (*26*). In a study of ribonuclease, oxygen was responsible for protein aggregation (*27*).

Transdermal Controlled Release Systems

The skin is often considered a barrier that keeps all agents, including drugs, out of the body. However, a few drugs have just the right properties to penetrate the skin at appreciable rates and are potent enough so that only low doses are required. Furthermore, compared to the oral route, losses due to liver metabolism are reduced. The rate-limiting barrier to drug entry through the skin is the outermost skin layer, the stratum corneum, which is composed primarily of keratin and lipids. For a drug to penetrate the skin significantly, it should have a low molecular weight and appreciable solubility in both water and oil.

The first transdermal delivery system introduced clinically released scopolamine from patches (reservoir systems) to prevent nausea associated with motion sickness. After the patch has been applied, a 4- to 6-hour lag period is required for the drug to reach therapeutic concentrations. Because of the small amount of drug required (7 μg/hour over 3 days) and the high skin permeability of scopolamine, this system can be designed so that the device rather than the skin is rate-controlling. This minimizes patient to patient variations. The device is placed behind the ear because the permeability of the stratum corneum there is comparatively high, which further enables the device, rather than the skin, to provide the principal diffusion barrier.

The most widely used transdermal systems release nitroglycerin daily for the treatment of heart disease. These systems, first introduced in 1982, have annual sales of approximately $500 million. The amount of nitroglycerin absorbed is determined by the skin rather than the device; nitroglycerin patches of different sizes are available so that patients can select the desired dosages. However, the continuous delivery of nitroglycerin may create drug tolerance. The possibility of controlled intermittent delivery of nitroglycerin is being explored.

A weekly clonidine patch and a twice weekly estradiol patch are used to treat hypertension and estradiol deficiency (for postmenopausal females), respectively. There have been reports of local irritation with these systems, perhaps because of their longer application periods or because of the combined effects of bioadhesives, chemicals, and drugs used in the formulations. Transdermal systems for the delivery of testosterone, fentanyl, isosorbide dinitrate, nicotine, timolol, and antihistamines, although not yet clinically available, are under study.

The biggest challenge in transdermal delivery is to increase the variety of drugs that can be administered. Four approaches have been explored. Electrical means such as iontophoresis, which can drive charged molecules through the skin, have received considerable attention. It has been proposed that iontophoresis might allow the transdermal delivery of larger molecular weight drugs, such as insulin. Animal studies with insulin have not led to conclusive results; insulin permeation depends on the animal model, the type of current, and whether the stratum corneum has been removed (*28*). Nonetheless, clinical studies have shown that smaller peptides such as luteinizing hormone–releasing hormone (LHRH) can be delivered at increased rates (*29*). A second approach uses ultrasound to enhance transdermal drug permeation. Ultrasound also eliminates the lag times associated with transdermal drug delivery in animal models (*30*). Chemical modification provides a third approach: a lipophilic drug could be synthesized that penetrates the skin and is subsequently converted by epidermal enzymes into the original drug. Finally, penetration enhancers such as Azone, dimethyl sulfoxide, and dimethyl formamide have been used. However, extensive testing must be done to establish safety. It may be more useful to utilize agents used in FDA-approved topical formulations (for example, ethanol is used in the estradiol system to enhance penetration).

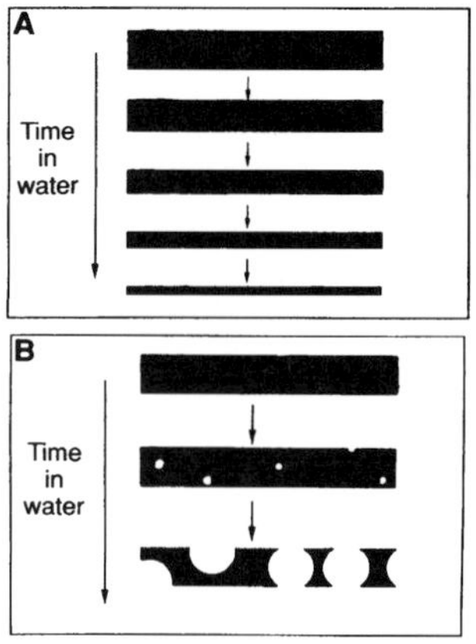

Fig. 2. Idealized diagram of polymer matrices displaying surface erosion (**A**) or bulk erosion (**B**).

Novel Degradable Polymers

Most materials used in medicine today were not designed for biomedical applications. For example, the polymers used in the artificial heart and dialysis tubing were originally used in ladies' girdles and sausage casings, respectively. These materials were chosen because they appeared, to some extent, to resemble the organs they were intended to replace. A significant challenge is to develop more rational approaches for creating improved materials for humans. This may be particularly important in the development of degradable polymers.

For such polymers, to maximize control over release, it is often desirable for a system to degrade only from its surface (Fig. 2A). [The only degradable polymers in common use, polyesters such as lactic acid–glycolic acid copolymers, display bulk (homogeneous) erosion (Fig. 2B), resulting in significant degradation in the matrix interior.] For surface-eroding systems, the drug release rate is proportional to the polymer erosion rate. This eliminates the possibility of dose dumping, improving device safety; release rates can be controlled by changes in system thickness and total drug content, facilitating device design. Achieving surface erosion requires that the degradation rate on the polymer matrix surface be much faster than the rate of water penetration into the matrix bulk. Efforts have begun to design such ideal polymers. Theoretically, the polymer should be hydrophobic but should have water-labile linkages connecting monomers.

It was proposed that, because of the lability of anhydride linkages, polyanhydrides would be a promising class of polymers. By varying the monomer ratios in polyanhydride copolymers, surface-eroding polymers lasting from 1 week to several years were designed and synthesized (*31*).

The possibility of implanting polyanhydride disks containing nitrosoureas for treating brain cancer after surgery is being explored. Normally, nitrosoureas are given intravenously (they have a half-life of 12 to 15 min and cause serious toxicity to several organs). By placing nitrosoureas in polyanhydrides, the drug is protected and its efficacy lasts approximately for the duration of the polymer lifetime (in this case, nearly 1 month). The polymer disks also deliver the drug locally to the brain, significantly reducing systemic toxicity. Surface erosion is desirable, for, if bulk erosion were to occur, uncontrolled amounts of this potentially toxic drug could be released during breakup of the matrix. These polymers have been shown to be safe in numerous animal models (*32*). Institutional Review Board approval was then obtained to conduct clinical trials with polyanhydrides at five U.S. hospitals. In 1987, the FDA approved these polyanhydrides for clinical trials. In an initial study of 21 patients, safety was demonstrated and patient lifetime was extended significantly beyond that afforded by conventional treatments (*33*). A phase-3 trial involving 32 hospitals is currently under way; over 100 patients have been treated.

Several different surface-eroding polyorthoester systems have been synthesized. In this case, additives are placed inside the polymer matrix, which causes the surface to degrade at a different rate than the rest of the matrix. Such a degradation pattern can occur because these polymers erode at very different rates, depending on pH, and the additives maintain the matrix bulk at a pH different from that of the surface. By varying the type and amount of additive, release rates can be controlled (*34*).

It may be desirable to have degradable polymers that consist of, and break down into, naturally occurring metabolites. Thus, new polyamino acids were synthesized in which L-amino acids or dipeptides were polymerized by nonamide bonds between functional groups (for example, esters) located on amino acid side chains. This approach permits the synthesis of biomaterials (for drug delivery systems, artificial organs, vascular grafts, or other prostheses) that are derived from nontoxic substances, which also have other desirable properties: (i) the incorporation of an anhydride linkage into the polymer backbone causes rapid degradability; (ii) an ester bond provides better film and fiber formation; and (iii) an imide or iminocarbonate bond improves mechanical strength (*35*).

One such polymer is being studied in vaccine delivery. Many adjuvants such as aluminum oxide or Freund's adjuvant rely on a simple "depot" effect, releasing antigen over a short period, from several hours to a few weeks. In earlier studies in mice and rabbits, prolonged release of small amounts of antigen from a nondegradable device resulted in sustained antibody production for over 6 months (*36*). Although these studies demonstrated the potential value of controlled release in immunization, it would be advantageous to use degradable systems to avoid implant retrieval. This concept is particularly attractive, because the polymer degradation products could be intentionally designed to have adjuvant properties, that is, an "engineered" polymer. This would permit the design of a system that could stimulate the immune response while simultaneously releasing antigen over long periods. Because of the adjuvanticity of L-tyrosine and its derivatives, a polymer consisting of tyrosine or a tyrosine derivative connected by hydrolyzable iminocarbonate bonds was synthesized. When this polymer was converted into small pellets, this system provided sustained adjuvanticity while simultaneously serving as an antigen repository. The release of antigen from a single tyrosine-based polyiminocarbonate pellet gave rise, in mice, to higher antibody titers than release of the same antigen dose from a control polyiminocarbonate pellet or from two injections of the antigen over 1 year (*37*).

Pulsatile Polymeric Controlled Release Systems

It would be desirable if polymeric systems could be designed to release increased levels of drug when needed; this would mimic the body's physiological processes. Both open-loop and closed-loop approaches are being studied. One open-loop system contains drug and small magnetic beads embedded in a polymer matrix (Fig. 1H). Release rates are enhanced when desired by an oscillating external magnetic field. Parameters that affect the release rate include the magnetic field frequency and strength, the polymer composition, and the strength and orientation of the polymer-embedded magnets. Application of the magnetic field causes up to 30-fold increases in release rates (*38*). Ultrasound can also be used to enhance drug release rates from polymers (*39*). Successful clinical implementation of the ultrasonic or magnetic systems will probably require the creation of small portable triggering devices (wristwatch size) that can be preprogrammed or activated manually when desired.

Several closed-loop polymeric systems are being developed. In one case intended for the increased release of insulin in the presence of excess glucose, glucose oxidase was immobilized within an insulin-containing polyamine membrane. Glucose oxidase converts glucose to gluconic acid; the acid protonates amine groups within the membrane. The electrostatic repulsion of the positively charged amine groups causes expansion of the membrane and increased delivery of insulin. As the physiologic glucose concentration decreases in response to the released insulin, the membrane contracts, decreasing the rate of insulin release (*40*). In another approach, glucose oxidase was immobilized to agarose beads contained within a polymer matrix. The acid formed when external glucose reacts with the immobilized enzyme lowers the pH, which changes the solubility of insulin and the diffusional driving force. Increased release rates to glucose challenges were observed in vitro and in diabetic rats (*41*). A third approach involves the synthesis of glycosylated insulin bound to concanavalin A (Con A). Con A is immobilized on Sepharose beads. The glycosylated insulin is displaced from Con A in response to glucose, which competes for the same binding sites. The rate of insulin release also depends on the binding affinity of the insulin derivative to Con A and can be influenced by the choice of saccharide group in glycosylated insulin. By encapsulating glycosylated insulin-bound Con A within a suitable polymer that is permeable to both glucose and insulin, it is possible to control glucose influx and insulin efflux (*42*). Critical issues with respect to each of these delivery systems are the stability of insulin and enzymes and the rapidity of movement (response time) of insulin from the polymer matrix to the circulation. Such systems may also benefit from ongoing research in biosensors (*43*).

Research is being conducted on self-triggered release of drugs such as narcotic antagonists in multicomponent systems involving erodible polymers, antibodies, and enzymes (*44*). Pulsatile systems involving pH-sensitive or temperature-sensitive polymers are also being studied, as are polymer systems that can be activated by light or electricity (*45*).

Conclusions and Future Directions

The studies discussed here show that carriers can affect drug level, location, longevity, and antigenicity. Although this technology is at an early stage, it has already made a significant clinical and commercial impact. This technology is not limited to medicine. Controlled release has been used for pet flea collars, pesticides, anti-fouling agents, fertilizers, and fragrances. Liposomes are used in cosmetics.

There are numerous challenges ahead. One area is the creation of bioadhesive polymers that could alter a drug's location when given orally (*46*). This could be particularly important for drugs that are absorbed only in certain segments of the gastrointestinal tract. Even more tantalizing, but more difficult, is delivering large and complex molecules such as proteins orally. Research on novel anatomical delivery pathways such as the nose or lung may also permit the delivery of a wider spectrum of drugs. Furthermore, an understanding of cell transport mechanisms may aid in cellular targeting (*47*).

Although this article has focused principally on specific carriers for pharmaceuticals, ongoing research in cell transplantation could be used to provide desired agents (*48*). The possibility of inserting genes into cells to produce desired entities is being explored (*49*).

Furthermore, continuous advances in biotechnology will have at least several major effects on drug delivery. First, novel complex drugs will be created that will be difficult to administer by conventional means. Second, approaches being developed in genetic engineering may enable the creation of new molecular constructs (for example, deletion mutants, hybrid proteins, and ligated gene fusion hybrids) with increased ability to achieve site-specific delivery. Finally, advances in materials science and chemical engineering should permit improved polymers, lipids, antibodies, and other substances to be synthesized, better understood, manufactured, and effectively used in drug delivery.

REFERENCES AND NOTES

1. N. Bodor and J. W. Simpkins, *Science* **221**, 65 (1983); A. K. Kumagai, J. B. Einsen, W. M. Pardridge, *J. Biol. Chem.* **262**, 15214 (1987); J. N. Jacob, G. W. Hesse, V. E. Shashoua, *J. Med. Chem.* **33**, 733 (1990).
2. Y. Matsumura and H. Maeda, *Cancer Res.* **46**, 6387 (1986).
3. J. Kopecek and R. Duncan, *J. Controlled Release* **6**, 315 (1987).
4. M. A. Till *et al.*, *Science* **242**, 1166 (1988); P. J. Martin *et al.*, *Bone Marrow Transplant.* **3**, 437 (1988).
5. T. R. Shockley *et al.*, in *Temporal Control of Drug Delivery*, W. J. M. Hrushesky, R. Langer, F. Theeuwes, Eds. (New York Academy of Sciences, New York, in press); R. Jain, *Cancer Res.* **48**, 2641 (1988); E. Tomlinson, *Adv. Drug Delivery Rev.* **1**, 87 (1987); E. Hurwitz, R. Kashi, R. Arnon, M. Wilchek, M. Sela, *J. Med. Chem.* **28**, 137 (1985).
6. M. S. Hershfield and R. H. Buckley, *N. Engl. J. Med.* **316**, 589 (1987); D. H. Ho *et al.*, *Drug Metab. Dispos.* **14**, 349 (1986); C. C. Chua *et al.*, *Ann. Intern. Med.* **109**, 114 (1988).
7. B. Hupfer, H. Ringsdorf, H. Schupp, *Makromol. Chem.* **82**, 247 (1981); S. L. Regen, B. Czech, A. Singh, *J. Am. Chem. Soc.* **102**, 6638 (1980).
8. J. Treat, A. R. Greenspan, A. Rahman, in *Liposomes in the Therapy of Infectious Diseases and Cancer*, G. Lopez-Berestein and I. J. Fidler, Eds. (Liss, New York, 1989), pp. 353–365.
9. G. Lopez-Berestein, *ibid.*, pp. 317–327.
10. M. Maeda, A. Kumano, D. A. Tirrell, *J. Am. Chem. Soc.* **110**, 7455 (1988); J. N. Weinstein, R. L. Magin, M. B. Yatvin, D. S. Zaharko, *Science* **204**, 188 (1979); P. Kibat, Y. Igari, M. Wheatley, H. Eisen, R. Langer, *FASEB J.* **4**, 2533 (1990); H. Ellens, J. Bentz, F. C. Szoka, *Biochemistry* **24**, 3099 (1985); S. Cohen, M. C. Bano, M. Chow, R. Langer, in preparation; C. Pidgeon and C. A. Hunt, *Photochem. Photobiol.* **37**, 491 (1983); M. B. Yatvin *et al.*, *Science* **210**, 1253 (1980).
11. C. Alving *et al.*, *Proc. Natl. Acad. Sci. U.S.A.* **75**, 2959 (1978).
12. E. S. Kleinerman, J. L. Murray, J. S. Snyder, J. E. Cunningham, I. J. Fidler, *Cancer Res.* **49**, 4665 (1989); G. Poste *et al.*, *ibid.* **42**, 1412 (1982).
13. C. R. Alving *et al.*, *Vaccine* **4**, 166 (1986); A. C. Allison and G. Gregoriadis, *Nature* **252**, 252 (1974).
14. A. Gabizon and D. Papahadjopoulos, *Proc. Natl. Acad. Sci. U.S.A.* **85**, 6949 (1988); L. Illum, S. S. Davis, R. H. Muller, E. Mak, P. West, *Life Sci.* **40**, 367 (1987); P. Arturson, T. Laakso, P. Edman, *J. Pharm. Sci.* **72**, 1415 (1983).
15. K. Widder, G. Flouret, A. Senyei, *J. Pharm. Sci.* **68**, 79 (1979); K. Mosbach and U. Schroder, *FEBS Lett.* **102**, 112 (1979).
16. M. V. Sefton, *CRC Crit. Rev. Biomed. Eng.* **14**, 201 (1987); H. Buchwald, T. D. Rohde, P. D. Schneider, R. L. Varco, P. J. Blackshear, *Surgery* **88**, 507 (1980); C. D. Saudek *et al.*, *N. Engl. J. Med.* **321**, 574 (1989).
17. J. Folkman and D. M. Long, *J. Surg. Res.* **4**, 139 (1964).
18. H. A. Nash, in *Medical Applications of Controlled Release*, R. Langer and D. Wise, Eds. (CRC Press, Boca Raton, FL, 1984), vol. 2, pp. 35–64.
19. J. M. Goodson, *ibid.*, pp. 115–138.
20. R. J. Levy *et al.*, *Science* **228**, 190 (1985); M. J. During *et al.*, *Ann. Neurol.* **25**, 351 (1989); M. Howard *et al.*, *J. Neurosurg.* **71**, 105 (1989).
21. R. Langer and J. Folkman, *Nature* **263**, 797 (1976).
22. R. Bawa, R. Siegel, B. Marasca, M. Karel, R. Langer, *J. Controlled Release* **1**, 259 (1985); T. Hsu and R. Langer, *J. Biomed. Mater. Res.* **19**, 445 (1985); Y. Ogawa, M. Yamamoto, S. Takada, H. Okada, T. Shimamoto, *Chem. Pharm. Bull.* **36**, 1502 (1988); W. M. Saltzman and R. Langer, *Biophys. J.* **55**, 163 (1989).
23. L. Brown, L. Siemer, C. Munoz, E. Edelman, R. Langer, *Diabetes* **35**, 692 (1986);

G. B. Silberstein and C. W. Daniel, *Science* **237**, 291 (1987); A. Lee and R. Langer, *ibid.* **221**, 1185 (1983).
24. L. M. Sanders *et al.*, *J. Pharm. Sci.* **73**, 1294 (1984); F. G. Hutchinson and B. J. A. Furr, in *Drug Carrier Systems*, F. H. D. Roerdink and A. M. Kroon, Eds. (Wiley, New York, 1989), pp. 111–130.
25. A. M. Klibanov, *Adv. Appl. Microbiol.* **29**, 1 (1983); D. B. Volkin and A. M. Klibanov, in *Protein Function: A Practical Approach*, T. E. Creighton, Ed. (IRL Press, Oxford, 1989), pp. 1–24.
26. R. Liu, R. Langer, A. M. Klibanov, *Biotech. Bioeng.*, in press.
27. M. J. Hageman, *Drug. Dev. Ind. Pharm.* **14**, 2047 (1988).
28. B. Kari, *Diabetes* **35**, 217 (1986); O. Siddiqui, Y. Sun, J. C. Liu, Y. W. Chien, *J. Pharm. Sci.* **76**, 341 (1987).
29. B. R. Meyer *et al.*, *Clinical Pharmacol. Ther.* **44**, 607 (1988).
30. D. Levy, J. Kost, Y. Meshulam, R. Langer, *J. Clin. Invest.* **83**, 2074 (1989).
31. H. Rosen, J. Chang, G. Wnek, R. Linhardt, R. Langer, *Biomaterials* **4**, 131 (1983); K. W. Leong, B. C. Brott, R. Langer, *Biomaterials* **4**, 131 (1983); *J. Biomed. Mater. Res.* **19**, 941 (1985); A. Domb, C. Gallardo, R. Langer, *Macromolecules* **22**, 3200 (1989).
32. K. W. Leong, P. D'Amore, M. Marletta, R. Langer, *J. Biomed. Mater. Res.* **20**, 51 (1986); C. Laurencin *et al.*, *ibid.*, in press; H. Brem *et al.*, *Selective Cancer Ther.* **5**, 55 (1989); R. J. Tamargo, J. J. Epstein, C. S. Reinhard, M. Chasin, H. Brem, *J. Biomed. Mater. Res.* **23**, 253 (1989).
33. H. Brem *et al.*, *J. Neurosurg.*, in press.
34. J. Heller, R. V. Sparer, G. M. Zenter, in *Biodegradable Polymers as Drug Delivery Systems*, M. Chasin and R. Langer, Eds. (Dekker, New York, 1990), pp. 121–161.
35. J. Kohn and R. Langer, *J. Am. Chem. Soc.* **109**, 817 (1987); A. Staubli, E. Ron, R. Langer, *ibid.* **112**, 4419 (1990).
36. I. Preis and R. Langer, *J. Immunol. Methods* **28**, 193 (1979); S. Niemi, J. Fox, L. Brown, R. Langer, *Lab. Anim. Sci.* **35**, 609 (1985).
37. J. Kohn *et al.*, *J. Immunol. Methods* **95**, 31 (1986).
38. E. Edelman, J. Kost, H. Bobeck, R. Langer, *J. Biomed. Mater. Res.* **19**, 67 (1985).
39. J. Kost, K. Leong, R. Langer, *Proc. Natl. Acad. Sci. U.S.A.* **86**, 7663 (1989).
40. J. Kost *et al.*, *J. Biomed. Mater. Res.* **19**, 1117 (1985).
41. F. F. Ghodsian, L. Brown, E. Mathiowitz, D. Brandenburg, R. Langer, *Proc. Natl. Acad. Sci. U.S.A.* **85**, 2403 (1988).
42. M. Brownlee and A. Cerami, *Science* **206**, 1190 (1979); S. W. Kim *et al.*, *J. Controlled Release* **11**, 193 (1990).
43. A. Heller, *Acc. Chem. Res.* **23**, 128 (1990).
44. C. G. Pitt, Z.-W. Gu, R. W. Hendren, J. Thompson, M. C. Wani, *J. Controlled Release* **2**, 363 (1985); J. Heller, *ibid.* **8**, 111 (1988).
45. R. A. Siegel, M. Falmarzian, B. A. Firestone, B. C. Moxley, *ibid.*, p. 179; L. Brannon-Peppas and N. A. Peppas, *ibid.*, p. 267; E. Mathiowitz and M. D. Cohen, *J. Membr. Sci.* **40**, 67 (1989); A. S. Hoffman, A. Afrassiabi, L. C. Dong, *J. Controlled Release* **4**, 213 (1986); Y. H. Bae, T. Okano, R. Hsu, S. W. Kim, *Makromol. Chem. Rapid Commun.* **8**, 481 (1987); P. E. Grimshaw, A. J. Grodzinsky, M. L. Yarmush, D. M. Yarmush, *Chem. Eng. Sci.* **44**, 827 (1989).
46. R. Gurny, J. M. Meyer, N. A. Peppas, *Biomaterials* **5**, 336 (1984); K. Park and J. R. Robinson, *Int. J. Pharm. Amst.* **19**, 107 (1984).
47. B. B. Kahn, M. J. Charron, H. F. Lodish, S. V. Cushman, J. S. Flier, *J. Clin. Invest.* **84**, 404 (1989); R. C. Kowal, J. Herz, J. L. Goldstein, V. Esser, M. S. Brown, *Proc. Natl. Acad. Sci. U.S.A.* **86**, 5810 (1989).
48. J. Vacanti *et al.*, *J. Pediatr. Surg.* **23**, 3 (1988); F. Lim and A. M. Sun, *Science* **210**, 908 (1980); P. Aebischer, S. R. Winn, P. M. Galletti, *Brain Res.* **448**, 364 (1988); R. M. Dawson, R. L. Broughton, W. T. K. Stevenson, M. V. Sefton, *Biomaterials* **8**, 360 (1987); S. Cohen *et al.*, *J. Am. Chem. Soc.*, in press.
49. J. M. Wilson *et al.*, *Science* **244**, 1344 (1989); J. R. Morgan, Y. Barrandon, H. Green, R. Mulligan, *ibid.* **237**, 1476 (1987).
50. N. A. Peppas, in *Controlled Drug Bioavailability, Drug Product Design and Performance*, V. F. Smolen and L. A. Ball, Eds. (Wiley, New York, 1984), vol. 1, pp. 203–237; R. Langer and N. A. Peppas, *J. Macromol. Sci. Rev. Macromol. Chem.* **23**, 61 (1983).
51. I thank H. Bernstein, S. Cohen, E. Edelman, A. Klibanov, and E. Tomlinson for their helpful reviews of this manuscript and J. Folkman for providing me with the inspiration to do research in this field. Supported by NIH grant GM 26698.

LETTERS TO NATURE

Antisense c-*myb* oligonucleotides inhibit intimal arterial smooth muscle cell accumulation *in vivo*

Michael Simons*†, Elazer R. Edelman‡§, Jean-Luc DeKeyser§, Robert Langer§ & Robert D. Rosenberg*†¶

* Department of Biology, ‡ Division of Health, Science and Technology, and § Department of Chemical Engineering, Massachusetts Institute of Technology, Cambridge, Massachusetts 02139, USA and † Department of Medicine, Harvard Medical School, Beth Israel Hospital, Boston, Massachusetts 02122, USA

SYNTHETIC antisense oligonucleotides have been used to dissect gene function *in vitro*. Technical difficulties prevented the use of this approach for investigating the effect of gene products *in vivo*. Here we report the use of local delivery of antisense c-*myb* oligonucleotide to suppress intimal accumulation of rat carotid arterial smooth muscle cells. Our results suggest that antisense oligonucleotides can be used to define the *in vivo* biological role of specific macromolecules in the blood vessel wall and could potentially serve as a new class of therapeutic agents for cardiovascular disorders.

The proto-oncogene c-*myb* is homologous to the transforming gene product of the avian myeloblastosis virus and was originally thought to be present only in haematopoietic cells[1,2]. But c-*myb* and its various family members are synthesized in diverse cell types and regulate cell growth and differentiation[3]. The proto-oncogene is involved in mitogen-induced proliferation of vascular smooth muscle cells (SMC) which constitutes a major pathway of atherogenesis[4]. Previous studies demonstrated that *in vitro* stimulation of SMC growth is followed by an elevation in c-*myb* messenger RNA levels and suppression of the message increase by antisense oligonucleotide halts cell proliferation[5,6].

We have used a rat carotid artery injury model to show that

¶ To whom correspondence should be addressed.

LETTERS TO NATURE

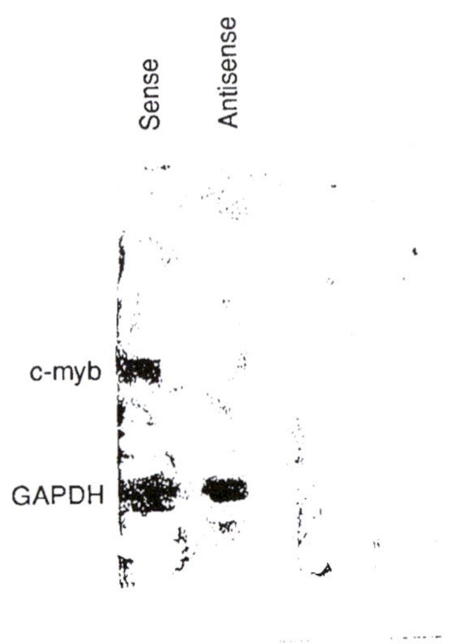

FIG. 1 Northern blot analysis of carotid artery c-*myb* mRNA levels. The levels of c-*myb* mRNA in the left carotid artery of sense (left lane) or antisense (right) oligonucleotide-treated rats. The left carotid artery in the neck was removed 2 weeks after balloon angioplasty from four rats treated with F127 pluronic gels containing 200 μg of sense or antisense oligonucleotide. The two antisense- and two sense-treated segments exposed to gel were individually pooled, frozen in liquid nitrogen, crushed, and total RNA was isolated in parallel from all animals as previously described[23]. Subsequently, poly(A)$^+$RNA was prepared in parallel from all tissues using Invitrogen Micro-Fast Track columns (Invitrogen, San Diego, CA). Ethidium stained gels showed no degradation of RNA samples as evidenced by intact ribosomal bands and mRNA smears centred at a molecular size of about 3–4 kilobases (kb). Samples of 10 μg of poly(A)$^+$RNA were subjected to electrophoresis on a 1% formaldehyde-agarose gel, transferred to a Gene Screen Plus membrane (New England Nuclear, Boston, MA) and hybridized with cDNA probes for c-*myb* and GAPDH as previously described[5]. The c.p.m. used in c-*myb* hybridization were 10 times those used for GAPDH hybridization. The blot was then exposed using a Betascope 603 Blot Analyzer (Betagen, Waltham, MA). The counts in the GAPDH band of the sense-treated sample were 6.5 times the counts in the c-*myb* band of the same sample, and the counts in the c-*myb* band of the antisense-treated sample were at background levels. The counts in the GAPDH band of the sense-treated sample were 35% greater than the counts in the GAPDH band of the antisense-treated sample. On the basis of the available data, the levels of c-*myb* and GAPDH mRNA in the sense-treated sample differ by about 65-fold assuming equivalent specific activity of cDNA probes, and the levels of c-*myb* mRNA in the antisense-treated animals must be less than 5% of the sense-treated controls.

local delivery of antisense c-*myb* phosphorothioate oligonucleotide suppresses SMC migration/growth *in vivo*. To this end, the left common carotid arteries of rats were subjected to balloon angioplasty which denudes endothelium and induces a highly reproducible intimal migration/proliferation of SMC over the entire length of the affected blood vessel[7,8]. The antisense oligonucleotide (GTGTCGGGGTCTCCGGGC) (position 4 to 22 of the mouse c-*myb* sequence) or the corresponding sense oligonucleotide (GCCCGGAGACCCCGACAC) were added to pluronic solutions at 1 mg ml^{-1} and kept at 4 °C. Immediately after balloon angioplasty, 200 μl of solution was applied to the carotid artery from which the adventia was stripped. The treated area constitutes about half the blood vessel and represents the portion that lies within the neck. On contact with tissues at 39 °C, the solution gelled instantaneously generating a translucent layer that enveloped the treated region. The wounds were closed immediately after application of gel, and the rats were returned to their cages. Inspection of additional animals revealed that pluronic gel disappears over 1–2 h.

We initially ascertained the effect of c-*myb* oligonucleotide in suppressing oncogene mRNA levels in the rat carotid artery 2 weeks after injury when the extent of SMC accumulation had reached a maximum. The treated portion of the blood vessel was surgically removed from five pairs of antisense and sense-treated rats. Northern blot analysis of one set of animals (Fig. 1) shows that injured carotid artery treated with antisense oligonucleotide exhibits no detectable levels of c-*myb* mRNA whereas injured carotid artery treated with sense oligonucleotide possesses significant amounts of oncogene mRNA. Examination of the injured carotid arteries of three additional sets of animals

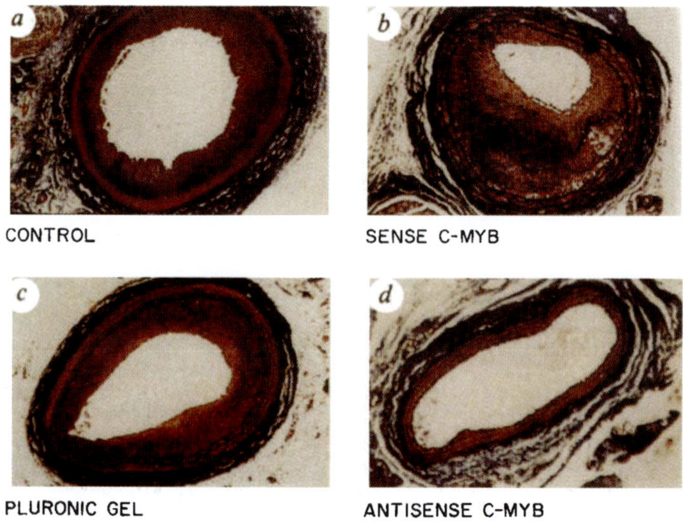

FIG. 2 Effect of antisense and sense c-*myb* oligonucleotides on neointimal formation in rat carotid arteries subjected to balloon angioplasty. Representative cross-sections from the carotid artery of an untreated rat (*a*), a rat treated with pluronic gel containing 200 μg sense oligonucleotide (*b*), a rat treated with pluronic gel (*c*), and a rat treated with pluronic gel containing 200 μg antisense oligonucleotide (*d*). (Mason trichrome, magnification ×80.) Male Sprague-Daley rats (average weight 500 g) were anaesthetized with Nembutal (4 mg per 100 g), and the left carotid artery of each animal was isolated by a midline cervical incision, suspended on ties and stripped of adventitia as previously described[24]. A 2F Fogarty catheter was introduced through the external carotid artery of each rat, advanced to the aortic arch, the balloon was inflated to produce moderate resistance to catheter movement and then gradually withdrawn to the entry point. The entire procedure was repeated three times for each animal. The oligonucleotides were added at a concentration of 1 mg ml^{-1} to 25% (w/v) solutions of F127 pluronic gels prepared as outlined by the manufacturer (BASF Wyandotte Corporation, Wyandotte, MI) and maintained at 4 °C. Prechilled pipettes and tips were used to apply the solution to the left carotid artery of each animal with generation of a gel surrounding the exposed blood vessel, and the wounds were then closed. At the time of killing (2 weeks later), the animals were anaesthetized with Nembutal and perfused with 150 cc normal saline under a pressure of 120 mm Hg. The carotid arteries were removed, fixed in 3% formalin, and processed for light microscopy in a standard manner.

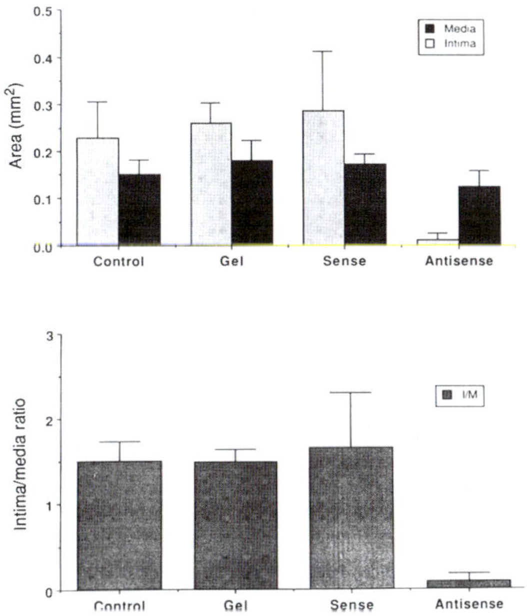

FIG. 3 Intimal and medial cross-sectional areas, as well as intimal/medial ratios of untreated and treated rat carotid arteries. The upper panel shows mean cross-sectional areas of the intimal (light bars) and medial (dark bars) regions of rat carotid arteries which were untreated, treated with pluronic gel, treated with pluronic gel containing 200 μg sense oligonucleotide, and pluronic gel containing 200 μg antisense oligonucleotide. The lower panel depicts the ratio of intimal to medial areas for the same four groups of animals. The data are provided as means ±1 standard deviation.

showed identical results. Analysis of one other set of animals revealed a minimal amount of oncogene mRNA in the antisense oligonucleotide-treated artery as compared with that exposed to sense oligonucleotide. The equivalent loadings of mRNA were documented by hybridization against a glyceraldehyde phosphate dehydrogenase (GAPDH) complementary DNA probe and examination of ethidium bromide-stained gels.

We then determined the effect of antisense oligonucleotide on intimal SMC accumulation 2 weeks after angioplasty in 22 rats who were treated with pluronic gels containing antisense oligonucleotide ($n = 8$); sense oligonucleotide ($n = 5$); no oligonucleotide ($n = 4$); or were untreated ($n = 5$). Morphological examination revealed that minimal intimal accumulation occurred with antisense oligonucleotide whereas extensive SMC accumulation was observed in the other treatment or control regimens (Fig. 2). The measurements of arterial segments document suppression of intimal SMC accumulation by antisense oligonucleotide without any apparent effect on medial SMC viability (Fig. 3). We also did experiments using an antisense two-base-pair mismatch oligonucleotide (GTGCCGGGGTCTTCGGGC) to substantiate further the specificity of the antisense oligonucleotide. Previous *in vitro* results show that antisense mismatch oligonucleotide does not reduce c-*myb* mRNA levels or SMC growth[5]. The *in vivo* data reveal that antisense mismatch oligonucleotide is unable to inhibit intimal SMC accumulation (the ratios of intimal to medial cross-sectional areas for mismatch oligonucleotide, sense oligonucleotide, and antisense oligonucleotide are 1.05 ± 0.14 ($n = 3$), 1.02 ± 0.07 ($n = 2$), and 0.16 ± 0.03 ($n = 2$), respectively). We have also used local single bolus intravascular delivery of oligonucleotides in rabbits during catheter-induced injury of iliac arteries and angiographically demonstrated significant reduction of SMC growth for up to 2 months after treatment.

Rat carotid arteries were examined to evaluate the lengthwise extent of suppression of SMC accumulation. Figure 4 shows a treated region of a rat carotid artery in the neck and an untreated segment of the same blood vessel in the chest. Similar patterns of SMC accumulation were noted with other antisense oligonucleotide-treated rats. The action of antisense oligonucleotide is apparently limited to the portion of carotid artery immediately in contact with gel. This supposition was confirmed in two additional rats where the extent of gel deposition was marked with silk ties which reveal a transition from minimal to extensive intimal accumulation over 3–4 mm of arterial length just proximal to application of gel (Fig. 4).

The injury of the arterial wall induces synthesis and/or release of numerous mitogens which bind to their own specific receptors eventually leading to migration/proliferation of SMC[9–12]. Recent attempts to inhibit this event used prolonged infusions of specific antibodies against individual mitogens, but were only partially successful, or had no effect[13,14]. An alternative strategy for suppressing the above process would be to inhibit production of a key intracellular intermediate of a common mitogenic pathway required by all growth factors. Data summarized in this communication indicate that c-*myb* expression is essential

FIG. 4 Spatial distribution of the antiproliferative effect of antisense oligonucleotides. Typical cross-sections (Mason trichrome, ×80) from treated and untreated segments of the left carotid artery of a rat are shown on the right and a longitudinal section of the transitional area between these zones (Mason trichrome, ×200) is depicted on the left. The shaded area represents the extent of application of pluronic gel.

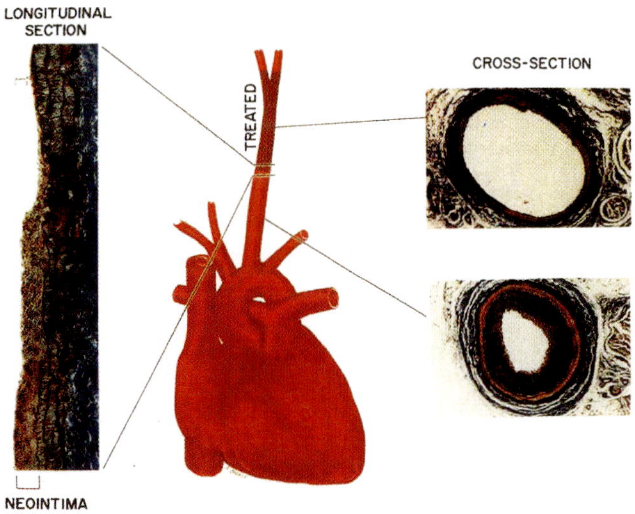

LETTERS TO NATURE

for the function of this common mitogenic pathway. We have previously demonstrated that the drug heparin and blood vessel wall heparan sulphates probably also inhibit proliferation of SMC by suppression of c-*myb*[15]. Earlier studies have revealed that suppression of SMC migration/growth for two to three days after arterial injury is sufficient to effect cell proliferation at 2 weeks[16]. Therefore, the action of antisense oligonucleotide could have been limited to a few days. But preliminary experiments with radioactively labelled phosphorothioate oligonucleotide indicate that the amounts of this component in the blood vessel wall remains roughly constant from 90 min to at least 72 h. The stability of phosphorothioate oligonucleotide to nuclease action[17] suggests that the effects of these agents may be manifest for a much longer time.

To the best of our knowledge, these studies constitute the first reported use of antisense oligonucleotide to inhibit synthesis of a normal gene product under *in vivo* conditions with a subsequent effect on a cellular process. Our positive results are probably due to local delivery of high concentrations of an antisense oligonucleotide modified to reduce *in vivo* degradation. The success of this strategy suggests that antisense oligonucleotides can be used *in vivo* to suppress other low abundance gene products segmentally in the blood vessel wall with subsequent determination of the resultant physiological effects. Furthermore, intimal SMC accumulation constitutes a major pathological event responsible for long-term failure of coronary and peripheral arterial bypass grafts as well as the development of stenosis following coronary artery angioplasty[18,19]. Several promising approaches have been described for suppressing this process, but none have proved clinically useful as yet[20–22]. The antisense oligonucleotides constitute a new class of therapeutic agents which can be locally targeted to inhibit specific gene products that result in suppression of intimal SMC accumulation, and possibly other acute pathologic processes. □

Received 15 May; accepted 30 July 1992.

1. Luscher, B. & Eisenman, R. N. *Genes Dev.* **4,** 2235–2241 (1990).
2. Anfossi, G., Gewirtz, A. M. & Calabretta, B. *Proc. natn. Acad. Sci. U.S.A.* **86,** 3379–3383 (1989).
3. Thompson, C. B., Challoner, P. B., Neiman, P. E. & Groudine, M. *Nature* **319,** 374–380 (1986).
4. Ross, R. *N. Engl. J. Med.* **314,** 488–500 (1986).
5. Simons, M. & Rosenberg, R. D. *Circ. Res.* **70,** 835–843 (1992).
6. Brown, K. E., Kindy, M. S. & Sonenshein, G. E. *J. biol. Chem.* **267,** 4625–4630 (1992).
7. Clowes, A. W., Reidy, M. A. & Clowes, M. M. *Lab. Invest.* **49,** 327–333 (1983).
8. Edelman, E. R., Adams, D. H. & Karnovsky, M. S. *Proc. natn. Acad. Sci. U.S.A.* **87,** 3773–3777 (1990).
9. Deuel, T. F. & Huang, J. S. *J. clin. Invest.* **74,** 669–676 (1984).
10. Linder, V., Lappi, D. A., Baird, A., Majack, R. A. & Reidy, M. A. *Circ. Res.* **68,** 106–113 (1991).
11. Jawien, A., Bowen-Pope, D. F., Lindner, V., Schwartz, S. M. & Clowes, A. M. *J. clin. Invest.* **89,** 507–511 (1992).
12. Hansson, G. K., Jonasson, L., Holm, J., Clowes, M. M. & Clowes, A. W. *Circ. Res.* **63,** 712–719 (1988).
13. Ferns, G. A. A. *et al. Science* **253,** 1129–1132 (1991).
14. Lindner, V. & Reidy, M. A. *Proc. natn. Acad. Sci. U.S.A.* **88,** 3739–3743 (1991).
15. Reilly, C. F., Kindy, M. S., Brown, K. E., Rosenberg, R. D. & Sonenshein, G. E. *J. biol. Chem.* **264,** 6990–6995 (1989).
16. Guyton, J. R., Rosenberg, R. D., Clowes, A. W. & Karnovsky, M. J. *Circ. Res.* **46,** 625–634 (1980).
17. Agrawal, S., Temsamani, J. & Tang, J. Y. *Proc. natn. Acad. Sci. U.S.A.* **88,** 7595–7599 (1991).
18. Dilley, R. J., McGeachie, J. K. & Prendergast, F. J. *Atherosclerosis* **63,** 99–107 (1987).
19. Nobuyoshi, M. *et al. J. Am. Coll. Cardiol.* **17,** 433–439 (1991).
20. Jonasson, L., Holm, J. & Hannson, G. K. *Proc. natn. Acad. Sci. U.S.A.* **85,** 2303–2306 (1988).
21. Powell, J. S. *Science* **245,** 186–189 (1989).
22. Pukac, L. A. *Am. J. Path.* **139,** 1501–1509 (1991).
23. Chomczynski, P. & Sacchi, N. *Analyt. Biochem.* **162,** 156–159 (1987).
24. Edelman, E. R., Nugent, M. A., Smith, L. T. & Karnovsky, M. J. *J. clin. Invest.* **89,** 465–473 (1992).

Biodegradable Long-Circulating Polymeric Nanospheres

Ruxandra Gref,* Yoshiharu Minamitake,†
Maria Teresa Peracchia,‡ Vladimir Trubetskoy,§
Vladimir Torchilin,§ Robert Langer‖¶

Injectable nanoparticulate carriers have important potential applications such as site-specific drug delivery or medical imaging. Conventional carriers, however, cannot generally be used because they are eliminated by the reticulo-endothelial system within seconds or minutes after intravenous injection. To address these limitations, monodisperse biodegradable nanospheres were developed from amphiphilic copolymers composed of two biocompatible blocks. The nanospheres exhibited dramatically increased blood circulation times and reduced liver accumulation in mice. Furthermore, they entrapped up to 45 percent by weight of the drug in the dense core in a one-step procedure and could be freeze-dried and easily redispersed without additives in aqueous solutions.

The development of intravenously administered carriers with blood circulation times long enough to continuously deliver drugs, imaging agents, or other entities to specific sites of action has been a major challenge. The desired features of such a carrier include (i) that the agent to be encapsulated comprises a reasonably high weight fraction (loading) of the total carrier system (for example, more than 30%), (ii) that the amount of agent used in the first step of the encapsulation process is incorporated into the final carrier (entrapment efficiency) at a reasonably high level (for example, more than 80%), (iii) the ability to be freeze-dried and reconstituted in solution without aggregation, (iv) biodegradability, (v) small size (less than 5 μm), and (vi) characteristics to prevent rapid clearance of the particles from the bloodstream. For example, albumin or galactose microspheres used in clinical studies for imaging are cleared from the blood within 20 s, preventing their use in many imaging applications (*1*). Some progress has been made in reducing the rapid clearance of microparticulates, for example, by the attachment of poloxamer or polysorbate to nondegradable polystyrene or polymethylmethacrylate particles or by the creation of liposomes or other carriers containing glycolipids, albumin, or derivatives of polyethylene glycol (PEG) (*2*). However, it has been difficult to develop particles displaying all of these qualities. We present here an approach to create degradable polymeric nanospheres that display, to a large extent, the above characteristics.

To form the core of the particles, we used biodegradable materials shown to be safe in the human body such as poly(lactic-co-glycolic acid) (PLGA), polycaprolactone (PCL), and their copolymers (*3*). By the adjustment of the chemical composition and molecular weight (MW) of a polymer, the degradation time of the core and release kinetics of the encapsulated agent can be controlled (*4*). To obtain a coating that might prevent opsonization and subsequent recognition by the macrophages of the reticulo-endothelial system, PEG was covalently attached to the nanosphere core by the synthesis of amphiphilic diblock copolymers, PEG-R, where R is one of the previously mentioned biodegradable materials. These diblock copolymers contain groups of ethylene glycol (–EG–EG–EG–) attached together, and this entire block is then bound to a second block of repetitive sequences of monomers such as lactic or glycolic acid or caprolactone (forming the more hydrophobic part of the copolymer, R). We then took advantage of the different solubilities of R (lipophilic) and PEG (hydrophilic) in aqueous and organic solutions to obtain a phase-separated structure.

Department of Chemical Engineering, Massachusetts Institute of Technology, Cambridge, MA 02139, USA.

*Present address: ENSIC-LCPM, 1 Rue Grandville, Nancy 54001, France.

†Present address: Suntory Limited, Ohra-Gun, Gunma-Ken 370-05, Japan.

‡Present address: Dipartimento Farmaceutico, Università degli Studi di Parma, Parma 43100, Italy.

§Permanent address: Department of Radiology, Massachusetts General Hospital-East, Charlestown, MA 02129, USA.

||Permanent address: Department of Chemical Engineering, Harvard–Massachusetts Institute of Technology (MIT) Division of Health Sciences, and Whitaker College of Health Sciences, MIT, Cambridge, MA 02139; and Department of Surgery, Children's Hospital, Boston, MA 02115, USA.

¶To whom correspondence should be addressed at MIT, E25-342, Cambridge, MA 02139, USA.

By using the diblock polymers, we were able to form nanospheres in a one-step procedure. First, PEG-R was dissolved in an organic solvent (methylene chloride, ethyl acetate, or acetone). Then an emulsion (oil in water) was formed in an aqueous phase by vortexing and sonicating. PEG is soluble in water but not in ethyl acetate or acetone. Conversely, R is not water-soluble but is highly soluble in the organic phase. Consequently, PEG migrates from the droplets' inner core to the water interface, and R remains inside the droplets during emulsification. Upon solvent evaporation, the nanosphere core solidifies. As a result of the amphiphilic nature of the diblock polymers, the use of other surfactants (not always biocompatible) can be avoided. The particles can then be recovered by centrifugation and lyophilized without protective additives. Lyophilized nanoparticles are easily redispersed in aqueous solutions. Size distribution measurements by quasi-elastic light scattering (QELS) before and after lyophilization established that aggregation does not occur.

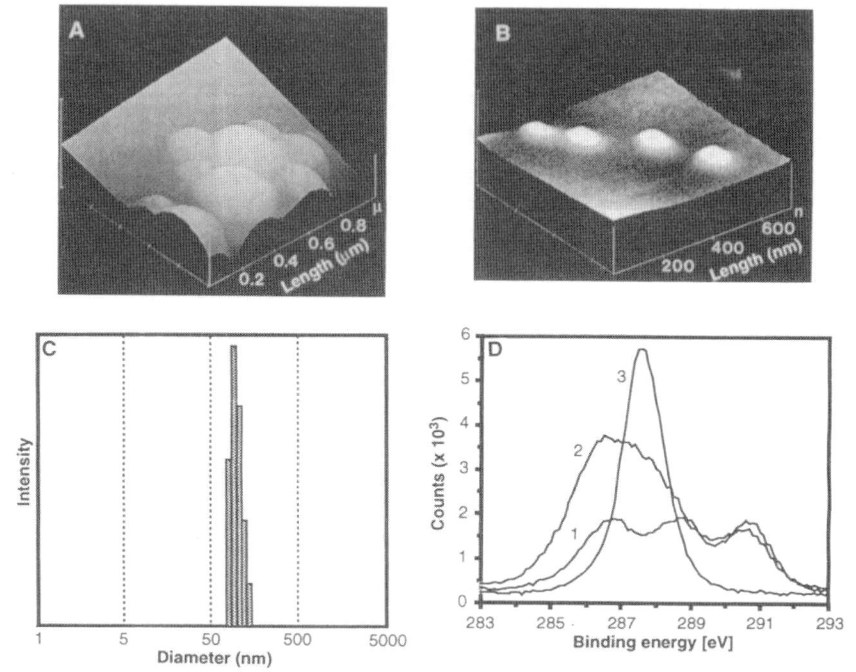

Fig. 1. (**A** and **B**) Images of nanospheres taken with an atomic force microscope (Nanoscope III, Digital Instruments, Santa Barbara, California) (scan rate, 2.0 Hz; number of samples, 512). The cantilever oscillates vertically (tapping mode) with a frequency of 350 kHz. (A) PLGA nanospheres (scan size, 1 μm; set point, 1.6 V; *z* axis, 250,000 nm per division); (B) PEG-PLGA nanospheres (MW of PEG, 5 kD; MW of PLGA, 45 kD; molar ratio lactic acid:glycolic acid in PLGA, 75:25; scan size, 712.9 nm; set point, 1.7 V; *z* axis, 100,000 nm per division). The nanospheres were prepared from diblock PLGA-PEG copolymers. These polymers were formed by the direct reaction of the terminal amine group from monoamine, monomethoxy PEG with PLGA (comporting a reactive carboxyl end group). They were also synthesized by the ring-opening polymerization at 114°C of lactide and glycolide in the presence of monomethoxy PEG (weight ratio, 36:9:5), by the use of stannous octoate as a catalyst (0.2 weight percent). Characterization of the polymers by ^{13}C-nuclear magnetic resonance showed that the molar ratio of lactic acid and glycolic acid in PLGA was in each case 3:1 and that the PEG content was approximately 10% by weight. To prepare the nanospheres, a solvent evaporation method (*9*) was modified for use as follows: The copolymer was first dissolved in an appropriate solvent (ethyl acetate or methylene chloride, 25 mg/2 ml). This solution was poured into 30 ml of distilled water, and an oil-in-water emulsion was formed by vortexing (30 s) and sonicating (1 min, 40-W output). The organic solvent was slowly removed by evaporation at room temperature, under gentle stirring for 2 hours. (**C**) Quasi-elastic light scattering of PEG-PLGA nanospheres showing unimodal size distribution (mean diameter of 140 nm). Instruments used were a Lexel Argon-ion laser (Fremont, California) (model BI-200SM), with a Brookhaven apparatus consisting of a goniometer and a 136-channel digital correlator and a signal processor. Measurements were made with a laser wavelength of 488 nm at a scattering angle of 90°. (**D**) Surface chemical analysis (XPS) of (trace 1) PLGA nanospheres, (trace 2) PEG-PLGA nanospheres (MW of PEG, 5 kD; MW of PLGA, 45 kD; and molar ratio of lactic acid:glycolic acid in PLGA, 75:25), and (trace 3) PEG 5 kD crystal powder. Data were collected by MgKα x-rays with a power of 300 W on a Perkin-Elmer 5100 apparatus.

The nanospheres' spherical shape was established by atomic force microscopy (Fig. 1, A and B), a nondestructive technique. The nanospheres are monodispersed, as determined by QELS (Fig. 1C) with a mean diameter between 90 and 150 nm, depending on the type of polymer, MW of PEG, and the organic solvent used.

Surface chemical analysis [x-ray photoelectron spectroscopy (XPS)] performed on lyophilized nanospheres confirmed that the major amount of PEG is concentrated within 5 nm of the outer layer of the nanospheres (Fig. 1D). By comparing XPS spectra before and after incubation in phosphate-buffered saline at 37°C, we found that only a minimal amount of PEG was detached (less than 5% in 24 hours), presumably because of the stability of the ester bond between PEG and R at pH 7.4.

To determine the efficacy of the PEG coating in altering organ distribution of the particles, PEG-coated and noncoated particles were injected into mice (four or five mice per group). Blood circulation time increases as the MW of PEG increases (Fig. 2A). This phenomenon can be explained by an increased thickness of the protective PEG layer, which prevents opsonization. The most striking results were seen in the reduction of liver uptake (Fig. 2B). Only 5 min after injection, 66% of noncoated particles were removed by the liver, while less than 30% of the 20-kD PEG-coated nanospheres were captured by the liver 2 hours after injection. After 5 hours, accumulation of 20-kD PEG-coated nanospheres by the liver still did not exceed 30%.

These dramatic body distribution differences were confirmed by gamma scintigraphy with ^{111}In-labeled particles (three mice per experiment) (5). After 15 min, only liver- and spleen-associated ^{111}In radioactivity was observed with PLGA nanospheres containing no PEG coating. However, high radioactivity was observed in the blood pool (heart and lung) when 20-kD PEG-coated nanospheres of the same size and prepared in the same way were injected, showing the persistence of these particles in the vascular compartment.

To study the encapsulation properties of PEG-coated nanospheres, lidocaine was used as a model drug. Drug-loaded nanospheres were prepared in the same one-step procedure (lidocaine was added to the organic phase, in which the copolymer was dissolved), freeze-dried, and easily dispersed in aqueous phases. High drug loadings (up to 45% by nanosphere weight) and entrapment efficiencies (more than 95% of the initial drug in the encapsulation solution was entrapped) were achieved. Similar loading capacities and entrapment efficiencies were observed for other drugs; for example, prednisolone was encapsulated with nearly identical efficacy. Depending on the properties of the drug (such as charge or hydrophobicity) to be encapsulated, it may be important to alter the solvent system or to use additives to achieve desirable characteristics.

Lidocaine is continually released in vitro over 14 hours (Fig. 3A). The release is fast because of the nanosphere's high surface area and small drug size. We observed that the higher the nanoparticle drug content, the slower the release. One explanation for this relation takes into account the possibility of drug crystallization inside the nanospheres. This hypothesis was supported by calorimetric (Fig. 3B) and x-ray diffraction studies (5). At low loading, lidocaine is present as a molecular dispersion inside the PLGA core. At high loading, a phase separation occurs, leading to the crystallization of part of the drug. The crystallized drug should dissolve and diffuse more slowly into the outer aqueous phase. Control of the drug-release kinetics can presumably be achieved by optimizing the chemical composition of the diblock polymers, the drug loading, and the particle size of the nanospheres (6).

The PEG-coated particles could be used in various ways. By the attachment of an appropriate protein such as transferrin to the surface of the particles (7), endocytosis of a particle containing DNA may be possible. Alternatively, antibodies could be attached to the PEG end group (8), potentially forming highly specific, targetable entities to desired tissues. With further study, these nanospheres may be useful in drug delivery, medical imaging, gene therapy, or a variety of other applications.

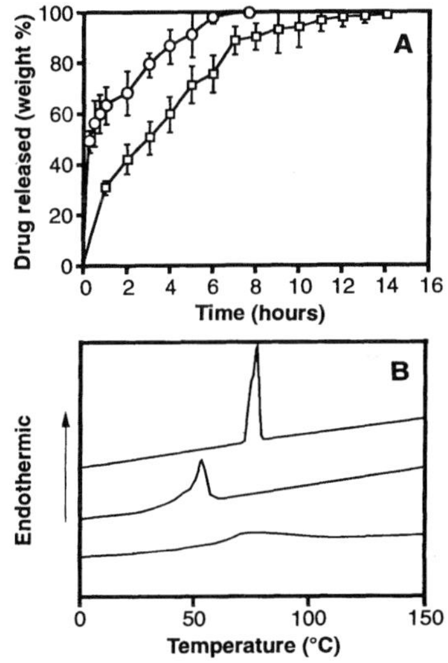

Fig. 3. (**A**) Comparison of lidocaine release kinetics from 20 kD PEG–80 kD PLGA nanospheres at (circles) 10 weight percent and (squares) 33 weight percent loading. Data are shown as mean ± SE obtained from three experiments per group. (**B**) Crystallinity of (top trace) lidocaine crystals (middle trace) 20 kD PEG–80 kD PLGA nanospheres at 33 weight percent and (bottom trace) 10 weight percent lidocaine loading were determined by the use of differential scanning calorimetry. Lidocaine crystals melt at 75°C. This melting point is depressed at 33 weight percent loading, indicating that part of the encapsulated drug exists in a small crystalline form. At lower loading (10 weight percent), practically no lidocaine crystals are observed, indicating that the drug is homogenously dispersed in the polymer matrix.

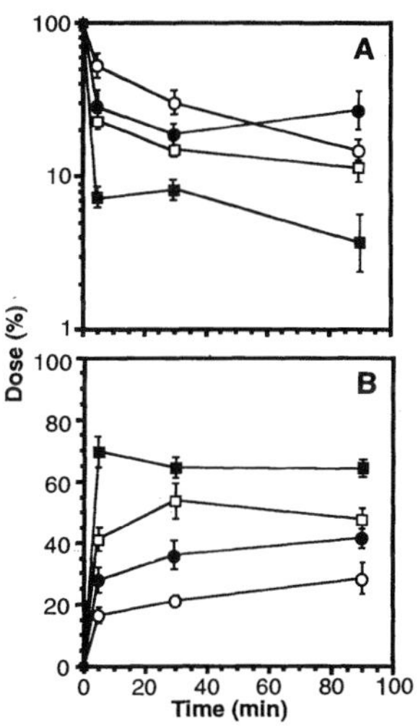

Fig. 2. Comparative biodistribution after injection of ^{111}In-labeled, noncoated PLGA and PEG-PLGA nanospheres (mean diameter of 140 ± 10 nm) in the tail vein of BALB/c mice (18 to 20 g). Each polymer used has a 36:9:5 weight ratio of lactide, glycolide, and PEG (5, 12, or 20 kD). The ^{111}In was chelated with diethylenetriamine pentaacetic acid stearyl amide (DTPA-SA) by incubation in hot ethanol (T = 70°C; concentration of DTPA-SA, 1 mg/ml) for 1 hour. This solution was added to the polymer dissolved in methylene chloride, and the nanospheres were prepared by emulsification in water (0.3 weight percent polyvinyl alcohol). The excess of non-entrapped label was washed away three times with Hanks buffer solution (HBS) by centrifugation (5 min, 50,000 rpm), and the supernatant was removed. Incubation of ^{111}In-labeled nanospheres in HBS or horse serum at 37°C showed less than 2% label loss during 4 hours (as assessed by the counting of radioactivity in the aqueous phase), suggesting that the hydrophobic stearyl moiety was firmly anchored to the particle core. Data were obtained from four mice per group and shown as mean ± SE. The nanospheres in the figures are indicated as follows: filled squares, PLGA; open squares, 5 kD PEG-PLGA; filled circles, 12 kD PEG-PLGA; and open circles, 20 kD PEG-PLGA. (**A**) Blood-associated radioactivity (percent of injected dose) determined by the gamma-counting of 0.1-ml blood samples. Injected dose, 0.35-ml suspension of nanospheres in HBS, 1 μCi. (**B**) Liver uptake of dose measured by the counting of organ samples. Injected dose, 0.35-ml suspension of nanospheres in HBS, 1 μCi.

REFERENCES AND NOTES

1. S. Gottlleb, A. Ernst, L. Litt, K. Q. Schwarz, R. S. Meltzer, *J. Am. Soc. Echocardiography* **3**, 238 (1990); J. R. Saphiro, S. A. Reisner, G. S. Lichtenberg, R. S. Meltzer, *J. Am. Coll. Cardiol.* **16**, 1603 (1990); K. Sugibayashi, Y. Morimoto, T. Nadai, Y. Kato, *Chem. Pharmacol. Bull.* **25**, 3433 (1977).
2. V. P. Torchilin *et al.*, *Fed. Am. Soc. Exp. Biol. J.* **6**, 2716 (1992); L. Illum, L. O. Jacobsen, R. H. Müller, R. Mak, S. S. Davis, *Biomaterials* **8**, 113 (1987); J. Lee, P. A. Martic, J. S. Tan, *J. Colloid Interface Sci.* **131**, 252 (1989); D. D. Lasic, F. J. Martin, A. Gabizon, S. K. Huang, D. Papahadjopoulos, *Biochim. Biophys. Acta* **1070**, 187 (1991); M. Donbrow, Ed., *Microcapsules and Nanoparticles in Medicine and Pharmacy* (CRC Press, Boca Raton, FL, 1992); T. M. Allen and C. Hansen, *Biochim. Biophys. Acta* **1068**, 133 (1991); R. H. Müller, K. H. Wallis, S. D. Tröster, J. Kreuter, *J. Controlled Release* **20**, 237 (1992); T. Verrecchia *et al.*, paper presented at VIII Journees Scientifiques du Groupe Thematique de Recherche sur les Vecteurs, Nancy, France, 6 to 10 December 1993; B. G. Müller and T. Kissel, *Pharm. Pharmacol. Lett.* **3**, 67 (1993); D. Bazile *et al.*, *Yanuzaigaku* **53**, 10 (1993).
3. A. Schindler *et al.*, *Contemp. Top. Polym. Sci.* **2**, 251 (1977); R. J. Kelly, *Rev. Surg.* **2**, 142 (1970).
4. S. W. Kim, R. V. Petersen, J. Feijen, in *Polymeric Drug Delivery Systems*, A. J. Ariens, Ed. (Academic Press, New York, 1992), p. 193.
5. R. Gref, Y. Minamitake, M. T. Peracchia, V. Torchilin, R. Langer, unpublished data.
6. S. J. Holland, B. J. Tighe, P. L. Gould, *J. Controlled Release* **4**, 155 (1986).
7. E. Wagner, M. Zenke, M. Cotten, H. Beug, M. L. Birnstiel, *Proc. Natl. Acad. Sci. U.S.A.* **87**, 3410 (1990).
8. G. Blume *et al.*, *Biochem. Biophys. Acta* **1149**, 180 (1993).
9. L. R. Beck, V. Z. Pope, D. R. Cowsar, D. H. Lewis, T. R. Tice, *Adv. Contracep. Delivery Syst.* **1**, 79 (1986).

10. We thank A. Milshteyn, E. L. Shaw, S. Cohen, A. Domb, G. Wolf, and R. Nanda for assistance. Supported by National Institutes of Health grants U01 CA52857 and GM 26698 and a Lavoisier grant from the French Foreign Affairs Ministry.

6 October 1993; accepted 21 January 1994

Large Porous Particles for Pulmonary Drug Delivery

David A. Edwards,* Justin Hanes, Giovanni Caponetti, Jeffrey Hrkach, Abdelaziz Ben-Jebria, Mary Lou Eskew, Jeffrey Mintzes, Daniel Deaver, Noah Lotan, Robert Langer*

A new type of inhalation aerosol, characterized by particles of small mass density and large size, permitted the highly efficient delivery of inhaled therapeutics into the systemic circulation. Particles with mass densities less than 0.4 gram per cubic centimeter and mean diameters exceeding 5 micrometers were inspired deep into the lungs and escaped the lungs' natural clearance mechanisms until the inhaled particles delivered their therapeutic payload. Inhalation of large porous insulin particles resulted in elevated systemic levels of insulin and suppressed systemic glucose levels for 96 hours, whereas small nonporous insulin particles had this effect for only 4 hours. High systemic bioavailability of testosterone was also achieved by inhalation delivery of porous particles with a mean diameter (20 micrometers) approximately 10 times that of conventional inhaled therapeutic particles.

Inhaled aerosols are effective therapeutic carriers for the treatment of respiratory inflammation (*1*), cystic fibrosis (*2*), and other lung disorders (*3*); they also offer potential for noninvasive systemic delivery of peptides and proteins (*4*). Local and systemic inhalation therapies can often benefit from a controlled release of the therapeutic agent (*5*), as is achievable with the use of biodegradable polymeric materials (*6*). Slow release from an inhaled therapeutic particle can prolong the residence of an administered drug in the airways or acini and can diminish the rate of a drug's appearance in the bloodstream (*7*). Also, patient compliance increases when dosage frequency is reduced (*7*).

The human lungs, however, have efficient means of removing deposited particles over periods ranging from minutes to hours. In the upper airways, ciliated epithelia contribute to the "mucociliary escalator" (*8*), by which particles are swept from the airways toward the mouth. In the deep lungs, an army of alveolar macrophages is capable of phagocytosing particles soon after their deposition (*9*). An effective slow-release inhalation therapy therefore requires a means of avoiding or suspend-

D. A. Edwards, A. Ben-Jebria, J. Mintzes, Department of Chemical Engineering, Pennsylvania State University, 204 Fenske Laboratory, University Park, PA 16802, USA.
J. Hanes, J. Hrkach, R. Langer, Department of Chemical Engineering, Massachusetts Institute of Technology, Cambridge, MA 02139, USA.
G. Caponetti, Department of Chemical Engineering, Massachusetts Institute of Technology, Cambridge, MA 02139, USA, and Department of Pharmacy, Universite di Parma, 43100 Parma, Italy.
M. L. Eskew, Environmental Resources Research Institute, Pennsylvania State University, Fenske Laboratory, University Park, PA 16802, USA.
D. Deaver, Department of Animal Science, Pennsylvania State University, 324 Henning Building, University Park, PA 16802, USA.
N. Lotan, Department of Biomedical Engineering, Technion–Israel Institute of Technology, Haifa 32000, Israel.

*To whom correspondence should be addressed. E-mail: dxe11@psuv.edu (D.A.E.), rlanger@mit.edu (R.L.).

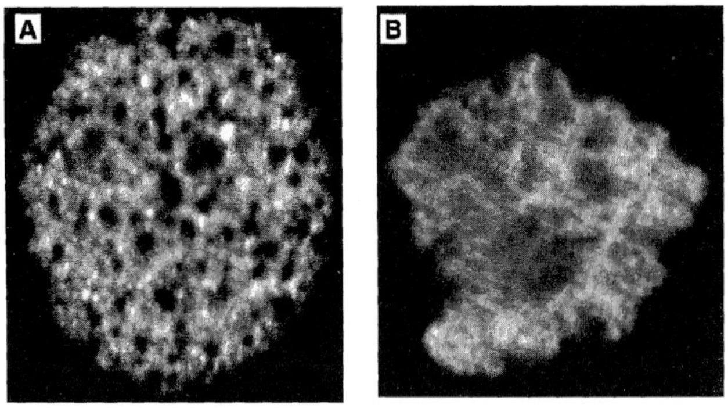

Fig. 1. Confocal microscopy images of (**A**) porous PLGA and (**B**) porous PLAL-Lys particles. Fluorescein isothiocyanate–dextran was encapsulated in the PLGA particle to render the pore spaces of the particle visible in the fluorescent confocal image. The PLAL-Lys particles were fluorescently labeled through the reaction of rhodamine isothiocyanate with lysine amine groups on the surface of the particles. The PLGA and PLAL-Lys particles are highly porous, as evidenced by the appearance of fluorescence throughout the particle structure.

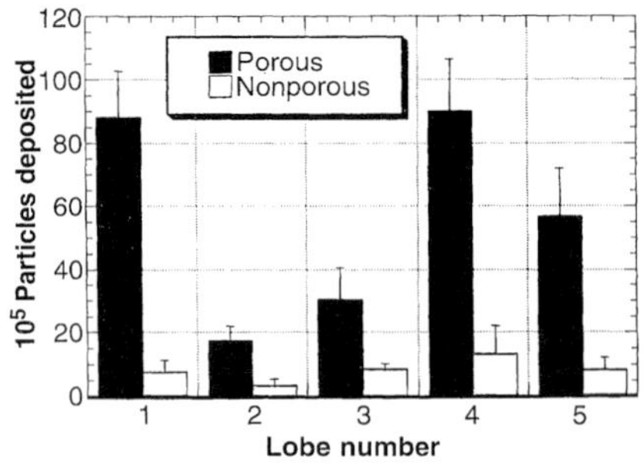

Fig. 2. Total particle recovery in rat lungs after bronchoalveolar lavage. Lobe numbers correspond to (1) left lung, (2) anterior, (3) median, (4) posterior, and (5) postcaval. For porous PLAL-Lys particles, d = 6.9 ± 4.2 μm and ρ = 0.1 g/cm^3. For nonporous PLA particles, d = 6.7 ± 3.2 μm and ρ = 0.94 g/cm^3. Means and SEs are based on n = 4.

ing the lungs' natural clearance mechanisms until encapsulated drugs have been effectively delivered.

Until now, therapeutic dry powder aerosols have been made with particle mass densities (ρ) of ~1 ± 0.5 g/cm^3 and mean geometric diameters (d) of <5 μm to avoid excessive deposition in the dry powder inhaler (DPI) and oropharyngeal cavity (*5*, *10*). Here, we show that very light particles (ρ < ~0.4 g/cm^3) with d > 5 μm can be deposited in the lungs. As a consequence of their large size and low mass density, porous particles can aerosolize from a DPI more efficiently than smaller nonporous particles, resulting in higher respirable fractions of inhaled therapeutics. Also by virtue of their size, large particles can avoid phagocytic clearance from the lungs until the particles have delivered their therapeutic dose; this attribute can be particularly useful for controlled-release inhalation therapies.

To assess the merits of large porous particles for pulmonary drug delivery, we encapsulated model therapeutics inside porous particles (Fig. 1A) composed of 50:50 poly(lactic acid-*co*-glycolic acid) (PLGA). Double- and single-emulsion solvent evaporation techniques (*11*) were used to prepare porous and nonporous PLGA particles, respectively. Porous and nonporous particles of similar aerodynamic diameter (*12*), loaded with ~15 weight % model therapeutic (testosterone), were aerosolized into a cascade impactor system (*13*) from a Spinhaler DPI for 30 s at an airflow rate of 28.3 liter/min. The cascade impactor provides an in vitro system for estimating the respirable fraction of a dry powder; it consists of a closed chamber within which flat plates are arranged perpendicular to the airflow, such that particles deposit stagewise in a manner reflective of their aerodynamic diameters. After deposition on the stages of the impactor, particles were collected (*13*) and total particle mass was assessed stagewise; the respirable fraction was determined as the percent of total particle mass exiting the DPI, recovered from the terminal, "respirable" stages of the impactor. Nonporous particles [d = 3.5 μm, ρ = 0.8 g/cm^3 (*14*)] exhibited a respirable fraction of 20.5 ± 3.5%, whereas 50 ± 10% of porous particles (d = 8.5 μm, ρ = 0.1 g/cm^3) were respirable, even though the aerodynamic diameters (*12*) of the two particle types are nearly identical. The large porous particles' high efficiency can be attributed to their smaller surface-to-volume ratio. Large particles aggregate less than small particles, all other considerations being equal (*15*, *16*); thus, while both have identical aerodynamic diameters, the large particles tend to exit the DPI more generally as single particles. The smaller particles aggregate more, leading to their deposition by gravity and inertia before reaching the "respirable" stages of the impactor.

To assess the influence of particle composition, we aerosolized a second type of porous particle (Fig. 1B), composed of poly(lactic acid-*co*-lysine-*graft*-lysine) (PLAL-Lys) (*11*). The PLAL-Lys particles exhibit some hygroscopicity, possibly a result of their lysine content, whereas the PLGA particles do not. Porous PLAL-Lys aerosols (d = 8.2 μm, ρ < 0.1 g/cm^3) exhibited an in vitro respirable fraction (57 ± 1.9%) similar to that of the porous PLGA particles (50 ± 10%), which suggests that absolute particle mass density, rather than particle chemistry or hygroscopicity, is the prime determinant of the relatively high respirable fractions observed for the large porous particles. The values of 50 ± 10% and 57 ± 1.9% for porous particles exceed comparable respirable fractions obtained in recent aerosolization studies (*15*) performed with mannitol (4 ± 0.3%) and recombinant human granulocyte colony-stimulating factor (blended with mannitol) (34 ± 2%) powders using a Spinhaler DPI at a similar airflow rate (30 liter/min).

To determine whether the relatively efficient in vitro aerosolization of large porous particles translates into improved respirable fractions in vivo, we aerosolized porous and nonporous particles into the airways of rats (*17*). During forced ventilation, rats were exposed to porous or nonporous particles; bronchoalveolar lavage was used to collect particles deposited in the trachea as well as in the airways and acini (*18*). The nonporous particles deposited primarily in the trachea (~79% of all particle mass that entered the trachea), whereas only 46% of the porous particle mass deposited in the trachea. Particles remaining in the rat lungs after bronchoalveolar lavage were obtained by careful dissection of the individual lobes of the lungs in subsequent experiments (*19*) (Fig. 2). The absolute number of porous particles remaining in the lungs was approximately an order of magnitude greater than the corresponding number of nonporous particles.

The role of low mass density in rendering large particles respirable can be understood in terms of the particles' mean aerodynamic diameter (*12*). Relatively large particles with high porosity have the same aerodynamic diameter as smaller, nonporous particles; these larger particles can enter the lungs because particle mass dictates the location of aerosol deposition in the lungs. The increased aerosolization efficiency of large, light particles lowers the probability of deposition losses before particle entry into the airways, thereby increasing the systemic bioavailability of an inhaled drug.

To test whether large particle size can increase systemic bioavailability, we encapsulated insulin into porous and nonporous polymeric particles. We designed the mass densities and mean diameters of the two particles such that they each had an aerodynamic diameter (~2 μm) suitable for deep lung deposition (*12*); the mean diameters of the porous and nonporous particles were >5 μm and <5 μm, respectively (Fig. 3, A to C). Identical masses of the porous or nonporous particles were administered to rats as an inhalation aerosol or injected subcutaneously (controls)

(*20*). Serum insulin concentrations were monitored as a function of time after inhalation or injection. For both porous (Fig. 3A) and nonporous (Fig. 3B) particles, blood levels of insulin reached high values within the first hour after inhalation. Only with large porous particles did blood levels of insulin remain elevated ($P < 0.05$) beyond 4 hours, with a relatively constant insulin release continuing to at least 96 hours ($0.04 < P < 0.2$). These results were confirmed by serum glucose values (Fig. 3C), which show falling glucose levels for the first 10 hours after inhalation of the porous insulin particles, followed by relatively constant low glucose levels for the remainder of the 96-hour period [for small nonporous insulin particles, initially suppressed glucose values rose after 24 hours (*21*)]. Similar biphasic release profiles of macromolecules from PLGA polymers have been reported (*22*). For the large porous particles, insulin bioavailability relative to subcutaneous injection (*23*) was 87.5%, whereas the small nonporous particles yielded a relative bioavailability of 12% after inhalation. By comparison, bioavailability (relative to subcutaneous injection) of insulin administered to rats as an inhalation liquid aerosol has been reported as 37.3% using a similar endotracheal method (*24*). The absolute bioavailability of insulin inhaled into rat lungs in the form of a lactose-insulin powder through a DPI connected to an endotracheal tube has been reported as 6.5% (*25*). The longest sustained insulin release previously reported (6 hours) was achieved using liposomes intratracheally instilled into rat lungs (*26*).

Given the short systemic half-life of insulin (11 min) (*27*) and the 12- to 24-hour time scale of particle clearance from the central and upper airways (*5*), the appearance of exogenous insulin in the bloodstream several days after inhalation appears to indicate that large porous particles achieve long, nonphagocytosed lifetimes in the deep lungs. To test this hypothesis, we lavaged (*18*) the lungs of rats immediately after inhalation of the porous and nonporous insulin particles as well as 48 hours after inhalation. For nonporous particles, 30 ± 3% of phagocytic cells contained particles immediately after inhalation, and 39 ± 5% contained particles 48 hours after inhalation. By contrast, only 8 ± 2% of phagocytic cells contained large porous particles immediately after inhalation, and 12.5 ± 3.5% contained particles 48 hours after inhalation. For small nonporous particles, 17.5 ± 1.5% of the phagocytic cell population contained three or more particles 48 hours after inhalation, compared with 4 ± 1% for large nonporous particles. Inflammatory response was also elevated with small nonporous particles; neutrophils represented 34 ± 12% of the phagocytic cell population 48 hours after inhalation of the small nonporous particles, compared with 8.5 ± 3.5% for large porous particles (alveolar macrophages represented 100% of cells measured immediately after inhalation). These results are consistent with those of in vitro experiments showing that phagocytosis of particles diminishes precipitously as particle diameter increases beyond 3 μm (*28*).

To further determine whether increased bioavailability correlates with increasing size of porous particles, we encapsulated a second model drug, testosterone, in porous particles of two different mean geometric diameters (10.1 and 20.4 μm). An identical mass of powder was administered to rats as an inhalation aerosol or as a subcutaneous injection (controls). Serum testosterone concentrations were monitored as a function of time after inhalation or injection (Fig. 3, D and E). Blood levels of testosterone remained well above background levels ($P < 0.05$) for 12 to 24 hours, even though the systemic half-life of testosterone is 10 to 20 min (*27*). Testosterone bioavailability relative to subcutaneous injection was 177% for the 20.4-μm-diameter particles (Fig. 3D) and 53% for the 10.1-μm-diameter particles (Fig. 3E). The increase in testosterone bioavailability with increasing size of porous particles is especially notable given that the mean diameter of the 20.4-μm particles is ~10 times that of nonporous conventional therapeutic particles (*5*, *10*). The relatively short time scale of testosterone release observed for both the inhalation and subcutaneous controls is near the in vitro time scale of release (several hours) reported elsewhere for 50:50 PLGA microparticles of similar size encapsulating a therapeutic substance (bupivacaine) of similar molecular weight and lipophilicity (*29*).

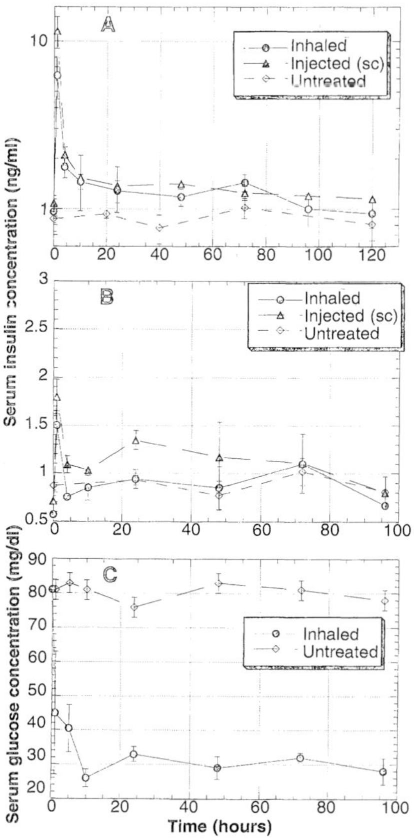

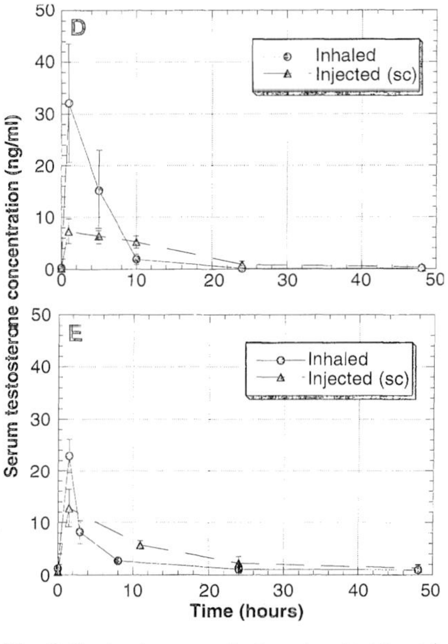

Fig. 3. Systemic concentrations in rats after administration of porous and nonporous therapeutic particles. (**A**) Serum insulin concentration after inhalation and subcutaneous injection of 9 mg of large porous insulin particles. No insulin particles were administered to nontreated controls. Porous particles contained insulin (20.0 weight %) and 50:50 PLGA (80.0 weight %) (*11*). Their mean *d* and d_{aer} values were 6.8 μm and 2.15 μm, respectively. Means and SEs are based on $n = 3$. (**B**) Serum insulin concentration after inhalation and subcutaneous injection of 9 mg of small nonporous insulin particles. No insulin particles were administered to nontreated controls. Nonporous particles contained insulin (10.0 weight %) and 50:50 PLGA (90.0 weight %) (*11*). Their mean *d* and d_{aer} values were 4.4 μm and 2.15 μm, respectively. Means and SEs are based on $n = 3$. (**C**) Serum glucose concentration after inhalation of 9 mg of large porous insulin particles or 9 mg of small nonporous insulin particles. No insulin particles were administered to nontreated controls. Means and SEs are based on $n = 3$. (**D**) Serum testosterone concentration after administration of 6 mg of porous testosterone particles ($d = 20.4$ μm) as an inhalation powder and as a subcutaneous control. Particles contained testosterone (15 weight %), 50:50 PLGA (76.5 weight %), and PLAL-Lys (8.5 weight %). For the dry powder, $\rho = 0.1$ g/cm^3. (**E**) Same as (D) but with smaller porous testosterone particles ($d = 10.1$ μm).

Porous particles comprising therapeutics and pharmaceutical excipients can easily be formed by spray-drying (*30*), rapid expansion of supercritical fluids (*31*), and other particle formation technologies. Hence, they can immediately address a variety of needs as therapeutic carriers for inhalation therapies. Their potential for high aerosolization efficiency, long-term drug release, and increased systemic bioavailability makes large porous particles especially attractive for systemic inhalation therapies.

REFERENCES AND NOTES

1. O. Einarsson *et al.*, *Ann. N.Y. Acad. Sci.* **762**, 89 (1995).
2. R. Niven, *Crit. Rev. Ther. Drug Carrier Syst.* **12**, 151 (1995).
3. S. J. Smith and J. A. Bernstein, in *Inhalation Aerosols*, A. J. Hickey, Ed. (Dekker, New York, 1996), pp. 233–269.
4. J. S. Patton and R. M. Platz, *Adv. Drug Delivery Rev.* **8**, 179 (1992).
5. I. Gonda, in *Topics in Pharmaceutical Sciences 1991*, D. J. A. Crommelin and K. K. Midha, Eds. (Medpharm Scientific, Stuttgart, 1992), pp. 95–115.
6. R. Langer, *Science* **249**, 1527 (1990).
7. I. Gonda, *Crit. Rev. Ther. Drug Carrier Syst.* **6**, 4273 (1990).
8. D. Pavia, in *Aerosols and the Lung: Clinical and Experimental Aspects*, S. W. Clarke and D. Pavia, Eds. (Butterworths, London, 1984), pp. 200–229.
9. M. B. Warheit and M. A. Hartsky, *Microscopy Res. Tech.* **26**, 412 (1993); J. D. Brain, in *The Reticuloendothelial System*, S. M. Reichard and J. Filkins, Eds. (Plenum, New York, 1985), pp. 315–331.
10. D. Ganderton, *J. Biopharm. Sci.* **3**, 101 (1992); A. Adjei and J. Garren, *Pharm. Res.* **7**, 565 (1990).
11. PLGA particles were made using the double- and single-emulsification solvent evaporation procedures. In the double-emulsification procedure, 300 μl of an aqueous solution was emulsified on ice into 4.0 ml of 50:50 PLGA polymer solution in methylene chloride (200 mg of polymer) by probe sonication. The first emulsion was poured into 100 ml of 1% aqueous polyvinyl alcohol (PVA) solution (average molecular weight 25 kD, 88% hydrolyzed) and homogenized (Silverson Machines, London, England) at 6000 rpm to form the double emulsion. The microspheres were continuously stirred for 3 hours to allow hardening, collected by centrifugation, washed several times with double-distilled water, and freeze-dried into a freely flowing powder. Particles (PLGA or blends of PLGA and PLAL-LYS) containing testosterone were made by dissolving the testosterone in the polymer solution. For the insulin particles, an aqueous solution of human insulin was emulsified with PLGA in the first emulsification. The amounts of water and insulin were decreased to diminish particle porosity, whereas insulin microsphere aerosolization efficiency was enhanced by the addition of L-α-phosphatidylcholine dipalmitoyl to the polymer solution [J. Hanes, thesis, Massachusetts Institute of Technology (1996)]. In the single-emulsification method, 200 mg of the copolymer 50:50 PLGA was dissolved in 2.8 ml of methylene chloride. The polymer solution was homogenized at 7500 rpm in 100 ml of 1% (w/v) aqueous PVA solution. The resulting dispersion was stirred using a magnetic stirrer for 3 hours until the methylene chloride was completely evaporated and the particles hardened. The particles were then isolated by centrifugation. The precipitate was washed three times with distilled water. Finally, the particles were freeze-dried (Labconco freeze dryer 8) for at least 48 hours. Particle sizing was performed using a Coulter Multisizer II (Coulter Electronics, Luton, England). PLGA particles prepared by the single-emulsification method are naturally less porous than those made by double emulsification. For example, PLGA microspheres with bulk mass densities as low as 0.05 g/cm^3 have been prepared by the double-emulsification technique [R. Jeyanthi, B. C. Thanoo, R. C. Metha, P. P. DeLuca, *J. Controlled Release* **38**, 235 (1996)], although these microspheres, at 35 to 140 μm, were inappropriately large for deep-lung inhalation. PLAL-Lys particles were also made using the single-emulsification technique [J. S. Hrkach, J. Ou, N. Lotan, R. Langer, *Macromolecules* **28**, 4736 (1995)]; these particles, apparently because of the cationic charge of the graft copolymer, are highly porous even when prepared by the single-emulsification method.
12. The aerodynamic diameter d_{aer} is related to the actual sphere diameter d by the formula
$$d_{aer} = d\sqrt{\rho} \quad (1)$$
(*5*). Maximal deposition of monodisperse aerosol particles in the alveolar region of the human lung (~60% of the inhaled particles) occurs for d_{aer} = ~3 μm (*32*). Inhaled particles with $d_{aer} > 3$ μm tend to deposit in the upper airways by gravity and inertia, whereas particles with $d_{aer} < 3$ μm tend to be exhaled. On the basis of Eq. 1, the actual diameter d (in micrometers) of porous particles that will exhibit maximum deep-lung deposition because of their small ρ is
$$d = 3/\sqrt{\rho} \quad (2)$$
(where $\rho < 1$). Hence, for such particles d is always greater than 3 μm. For example, porous particles with d = 8.5 μm and ρ = 0.1 g/cm^3 have approximately the same aerodynamic diameter (~2.7 μm) as nonporous particles with d = 3.5 μm and ρ = 0.8 g/cm^3.
13. We placed 20 mg of porous or nonporous microparticles, with encapsulated rhodamine as a fluorescent marker, in No. 1 hard gelatin capsules (Eli Lilly), loaded the capsules into a Spinhaler DPI (Fisons, Bedford, MA), and activated the DPI. Particles were aerosolized into a Mark I Andersen Impactor (Andersen Samplers, Atlanta, GA) from the Spinhaler device for 30 s at a flow rate of 28.3 liter/min. [Each plate of the impactor was previously coated with Tween 80 by immersing the plates in an acetone solution (5% w/v) and then evaporating the acetone in an oven at 60°C for 5 min.] After aerosolization and deposition, particles were collected from each stage of the impactor system and completely degraded in NaOH solution (0.2 N). After incubation at 37°C for 12 hours, the fluorescence of each solution was measured (wavelengths of 554 nm excitation, 574 nm emission).
14. Particle mass density was determined either by nonmercury porosimetry or tap density measurements (Micromeritics, Norcross, GA). The mass density of nonporous particles ranged from ~0.4 to 1.0 g/cm^3. The lower mass density limit (0.4 g/cm^3) is near that recently reported for spray-dried pharmaceutical powders (*15*), whereas the upper limit (1.0 g/cm^3) is similar to the mass density of liquid pharmaceutical aerosols [A. Adjei *et al.*, *Int. J. Pharm.* **107**, 57 (1994)] and other dry-powder aerosols [M. T. Vidgren, P. A. Vidgren, T. P. Paronen, *ibid.* **35**, 139 (1987)]. The mass density of porous particles was typically near or equal to 0.1 g/cm^3. Mass density variation from particle to particle within powders did not appear to introduce a substantial source of uncertainty. For example, using an API aerosizer, we measured the mean time-of-flight of the porous and nonporous particles described in the legend of Fig. 2. This gave a mean d_{aer} of 1.57 ± 2.41 μm for the porous particles and 5.82 ± 1.86 μm for the nonporous particles. Using Eq. 1 in (*12*), this indirectly yields a mass density ratio (porous to nonporous) of 0.1 ± 0.3. Given that the measured ratio of mass densities is 0.11 (Fig. 2), mass density variation within the two powders appears to be small (<10%).
15. D. L. French, D. A. Edwards, R. W. Niven, *J. Aerosol Sci.* **27**, 769 (1996).
16. W. I. Li *et al.*, *ibid.*, p. 1269.
17. Male Sprague-Dawley rats (150 to 200 g) were anesthetized using ketamine (90 mg/kg)-xylazine (10 mg/kg). Anesthetized rats were placed ventral side up on a surgical table provided with a temperature-controlled pad to maintain physiological temperature. Rats were cannulated above the carina with an endotracheal tube connected to a Harvard ventilator and were force-ventilated for 20 min at 300 ml/min. Fifty micrograms of porous (PLAL-Lys) or nonporous (PLA) microparticles were introduced into the endotracheal tube. After the period of forced ventilation, rats were killed and the lungs and trachea were separately washed using bronchoalveolar lavage (*18*). The lavage fluid was centrifuged (400*g*), and the pellets were collected and resuspended in 2 ml of phenol red–free Hanks' balanced salt solution (HBSS) (Gibco) without Ca^{2+} and Mg^{2+}. A 100-μl sample was removed for particle counting using a hemacytometer. The remaining solution was mixed with 10 ml of 0.4 N NaOH. After incubation at 37°C for 12 hours, the fluorescence of each solution was measured (wavelengths of 554 nm excitation, 574 nm emission).
18. Lungs were lavaged immediately (5 to 10 min) after inhalation. A tracheal cannula was inserted and tied into place, and the airways were washed with 10-ml aliquots of HBSS. The lavage procedure was repeated until a total volume of 30 ml was collected. In the phagocytic cell experiments, lavage fluid was centrifuged and the cell pellets were resuspended in HBSS for counting, differentiation of cell types, and measurement of phagocytosis. Engulfment of particles by phagocytic cells was determined by counting microscopically the number of particles incorporated per cell in wet mounts and in fixed and stained cytocentrifuge preparations. Reported numbers are based on the wet-mount experiments; similar numbers were obtained using the stained cytocentrifuge preparations.
19. The lobes were placed in separate petri dishes containing 5 ml of HBSS. Each lobe was teased through a 60-mesh screen to dissociate the tissue and was then filtered through cotton gauze to remove tissue debris and connective tissue. The petri dish and gauze were washed with an additional 15 ml of HBSS to maximize microparticle collection. Each tissue preparation was centrifuged and resuspended in 2 ml of HBSS, and the particles were counted in a hemacytometer.
20. Rats were anesthetized and cannulated as described (*17*) and were force-ventilated for 10 to 20 min at 300 ml/min. Rats received two types of aerosols through the endotracheal tube. After the period of forced ventilation, necks were sutured and rats were revived within 1 to 2 hours. Blood samples (300 μl) were periodically withdrawn from the tail vein over a period of 2 to 6 days. These samples were mixed with assay buffer, centrifuged, and the supernatant examined for the presence of (endogenous and exogenous) insulin or testosterone using radioimmunoassays (ICN Pharmaceuticals, Costa Mesa, CA). Glucose was measured using a colorimetric assay (Sigma). Control studies involved subcutaneous injection of the same amount of powder as was inhaled. The particles were injected into the scruff of the neck.
21. D. A. Edwards, A. Ben-Jebria, J. Mintzes, J. Hanes, J. Hrkach, R. Langer, G. Caponetti, M. L. Eskew, D. Deaver, N. Lotan, data not shown.
22. S. Cohen *et al.*, *Pharm. Res.* **8**, 713 (1991).
23. Bioavailability was determined as $(AUC_{inhaled}/AUC_{subcutaneous}) \times 100$, where AUC is area under the curve.
24. P. Colthorpe *et al.*, *Pharm. Res.* **9**, 764 (1992).
25. F. Komada *et al.*, *J. Pharm. Sci.* **83**, 863 (1994).
26. F. Y. Liu, Z. Shao, D. O. Kildsig, A. K. Mitra, *Pharm. Res.* **10**, 228 (1993).
27. A. R. Gennaro, Ed., *Remington: The Science and Practice of Pharmacy* (Mack, Easton, PA, ed. 19, 1995).
28. H. Kawaguchi *et al.*, *Biomaterials* **7**, 61 (1986); L. J. Krenis and B. Strauss, *Proc. Soc. Exp. Med.* **107**, 748 (1961); S. Rudt and R. H. Muller, *J. Controlled Release* **22**, 263 (1992).
29. P. Le Corre *et al.*, *Int. J. Pharm.* **107**, 41 (1994).
30. M. Sacchetti and M. M. Van Oort, in (*3*), pp. 337–384.
31. J. W. Tom and P. G. Debenedetti, *J. Aerosol Sci.* **22**, 555 (1991).
32. J. Heyder *et al.*, *ibid.* **17**, 811 (1986); D. A. Edwards, *ibid.* **26**, 293 (1995).

33. We thank J. Heidel and C. Perez de la Cruz for helpful technical assistance. Supported by an NSF CAREER Grant (D.E., J.M.) and NIH grants GM26698 (R.L.) and HD29125 (R.L., J.H.).

5 February 1997; accepted 8 May 1997

reviews

Drug delivery and targeting

Robert Langer

When a pharmaceutical agent is encapsulated within, or attached to, a polymer or lipid, drug safety and efficacy can be greatly improved and new therapies are possible. This has provided the impetus for active study of the design of degradable materials, intelligent delivery systems and approaches for delivery through different portals in the body.

Although delivery systems that could target drugs to specific body sites or precisely control drug release rates for prolonged times have long been dreamed of, only in recent years has the development of such systems become practical. Yet in a short time, new drug-delivery systems have had an impact on nearly every branch of medicine including cardiology, ophthalmology, endocrinology, oncology, pulmonology, immunology and pain management. Annual sales in the United States of advanced drug-delivery systems exceed $10 billion (Table 1) and are rising rapidly. Many principles used in drug delivery, such as the application of materials for long-term controlled release of encapsulated agents, have also been used for the delivery of pesticides, fertilizers, herbicides, fragrances, flavours and other substances, making controlled release an area of remarkably widespread application.

Potential advantages of improved drug delivery include: (1) continuous maintenance of drug levels in a therapeutically desirable range; (2) reduction of harmful side effects due to targeted delivery to a particular cell type or tissue; (3) potentially decreased amount of drug needed; (4) decreased number of dosages and possibly less invasive dosing, leading to improved patient compliance with the prescribed drug regimen; and (5) facilitation of drug administration for pharmaceuticals with short *in vivo* half-lives (for example peptides and proteins). Advantages 1 to 3 are additionally important because up to 15% of all hospital admissions, some 100,000 deaths and $136 billion in health care costs in the US each year can be attributed to adverse drug events[1,2]. Another 10% of hospital admissions are due to a lack of patient compliance (advantage 4)[2,3]. The above advantages must be weighed against the following concerns in the development of each particular drug-delivery system: (1) toxicity of the materials (or their degradation products) from which the drug is released, or other safety issues such as unwanted rapid release of the drug (dose dumping); (2) discomfort caused by the system itself or the means of insertion; and (3) expense of the system due to the drug encapsulation materials or the manufacturing process.

Polymer-based drug-delivery systems

There are three general mechanisms by which drugs are delivered from polymer or lipid systems: (1) diffusion of the drug species from or through the system; (2) a chemical or enzymatic reaction leading to degradation of the system, or cleavage of the drug from the system; and (3) solvent activation, either through osmosis or swelling of the system[4]. Advantages and disadvantages of these systems have been discussed previously[5]. A combination of mechanisms is possible.

Polymer-based systems have had an enormous impact on drug

Table 1 US sales of advanced drug-delivery systems (1996 and 1997)

Type of system	Sales (millions of $US) 1996	1997
Oral	6,066	7,178
Transdermal	1,525	1,701
Injectable/implantable polymeric systems	856	1,109
Ocular	12	17
Liposomes	107	208
Transmucosal (vaginal, buccal)	35	65
Lung delivery (nebulizers, metered-dose inhalers)	2,513	2,573
Nasal delivery	851	994
Total:	11,965	13,841

Sources: Cowen and Co., Boston, Massachusetts, USA; IMS America Plymouth Meeting, Pennsylvania, USA.

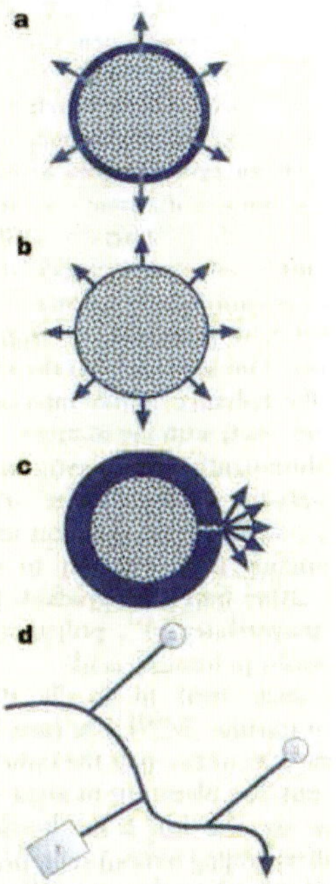

Figure 1 Examples of polymer-based delivery systems. In **a**–**c**, small dots represent drug, and arrows show the direction in which drug is released. **a**, Reservoir system in which drug diffuses through a polymer membrane (blue rim). This is the design for Norplant, Ocusert and a number of other systems. **b**, Matrix system in which the drug is evenly distributed through a polymer system. The drug can be released by diffusion through the polymer. Alternatively, drug release can occur by a combination of drug diffusion and polymer erosion. **c**, Osmotic system in which drug is pumped out through a laser-drilled hole. Water is attracted through the blue semipermeable membrane (the membrane is permeable to water but impermeable to drug) by osmosis due to the drug or a coencapsulated salt in the system. **d**, Polymeric drug conjugates. The curved line represents polymer. The bonds connecting drug (D) and polymer are cleavable inside the body. The targeting moiety, T, is optional.

therapies (Fig. 1). In one approach, the drug is physically entrapped inside a solid polymer that can then be injected or implanted in the body. Early forms of these systems involved non-degradable polymers (membrane-controlled diffusion) (Fig. 1a) such as silicone rubber, which could release low molecular mass lipophilic drugs for extremely long times[6]. This type of approach led to the development of Norplant, small silicone capsules containing contraceptives that are slowly released by diffusion through the polymer for 5 years. However, this approach does not permit the slow delivery of either ionic species or molecules with a relative molecular mass (M_r) over about 400 because they are not able to diffuse through such polymers. To address this problem, drugs were physically embedded in polymers at concentrations high enough to create a series of interconnecting pores through which the drug could subsequently slowly diffuse[7] (a type of matrix system, Fig. 1b). Early studies utilized ethylene–vinyl acetate copolymer or various hydrogels as model polymers. Subsequently, to develop biodegradable systems with such properties, lactic/glycolic acid copolymers were utilized. In this case, the combination of diffusion through pores as well as polymer matrix degradation allows control of release rates. This approach has provided the basis for injectable delivery systems lasting for 1 to 4 months for normally short-lived polypeptide hormones such as luteinizing hormone releasing hormone analogues[8] now used by about 300,000 patients annually for treating advanced prostate cancer, endometriosis or precocious puberty.

To permit the slow release of even larger molecules such as proteins, approaches for controlling release kinetics must also consider

reviews

the stability of proteins in the solid state (the state in which they generally exist in polymer matrices)[9]. Enhancement of stability is important because exposure to moisture in the body can lead to significant protein aggregation. Approaches are being developed to stabilize important molecules such as human growth hormone so that they can be released for a month from injectable microcapsules[10].

Although much effort has been focused on utilizing polymers that have a history of medical use and then adapting their microstructures to provide desired delivery rates, another approach is the intentional design of materials that solve specific drug-delivery problems. In fact, most biomedical research in the 1960s and 1970s focused on utilizing 'off-the-shelf' polymers designed for consumer applications and adapting these polymers for medical purposes. For example, materials used in the artificial heart were originally components used to make women's girdles[11]. Consider the fact that most degradable polymers used in injectable drug-delivery systems display bulk erosion that causes the polymer to dissolve throughout the entire matrix. This makes constant release rates complex to achieve and creates the possibility of dose dumping as the system eventually hydrolyses. To address these issues it would be desirable to design polymers that display surface erosion, analogous to a bar of soap dissolving. One strategy to achieve predominantly surface erosion involves synthesizing polymers that have hydrophobic monomer units connected by water-labile bonds. This approach has the advantage of keeping water away from the polymer matrix interior but allows controlled erosion when water reacts with the matrix surface. Polymer matrices that display predominantly surface erosion have been created by synthesizing hydrophobic polyanhydrides[12]. By the appropriate choice of monomers, polyanhydride matrices can be made to degrade over periods ranging from one day to many months or any time in between[13]. Other new biodegradable polymers being synthesized include polyorthoesters[14], polyphosphoesters[15], polyphosphazenes[16] and pseudo-polyamino acids[17].

Polyanhydride matrices have been used to locally deliver chemotherapeutic drugs such as carmustine (BCNU) to treat brain cancer[18]. In this case, the surgeon removes as much of the tumour as possible at the time of operation, but also places up to eight small polymer–drug wafers at the tumour site. The drug is slowly released from the polymer for 1 month to kill remaining tumour cells. Because the drug is delivered locally, harmful side effects that normally occur from systemic chemotherapy are minimized. One recent clinical trial showed that after 2 years, 31% of the treated patients were alive whereas only 6% in the control group survived[19]. In 1996, the United States Food and Drug Administration approved this treatment for patients with recurrent glioblastoma, the first new brain cancer therapy approved in over 20 years. A similar approach also in clinical trials has been used to locally release gentamycin to treat osteomyelitis[20].

In some situations, it would be useful to create materials that can be injected through an extremely small needle. Recent work has involved the synthesis of a thermosensitive biodegradable hydrogel consisting of blocks of poly(ethylene oxide) and poly(lactic acid)[21]. A bioactive agent is dissolved in the aqueous polymer phase at 45 °C (where the system is liquid) and the solution is injected at 37 °C (body temperature) whereupon it quickly solidifies. Another approach to achieve solidification of a liquid polymer solution involves first dissolving polymers in non-toxic organic solvents. The polymer solution is then injected, where exposure to water causes precipitation[22].

Polymers that are bioadhesive to mucosa are being explored to prolong drug residence time in the vagina, gastrointestinal tract and other body sites[23]. Materials containing functional groups that exhibit hydrogen bonding, such as certain hydrogels or polymers containing free carboxyl groups, are useful materials for this application[24]. Another goal is to design pharmacological activity into a polymer as well. One example is the incorporation of immune adjuvants, such as tyrosine derivatives, into polymers so that when the polymer degrades and a vaccine is slowly released an adjuvant is also released that stimulates the immune response further[25].

The concept of chemically binding (as opposed to physically entrapping) drugs to water-soluble polymers can confer new properties upon the drug such as decreased immunogenicity or tissue targeting. For example, the biological half-life of high molecular mass substances such as adenosine deaminase (ADA) and asparaginase has been lengthened and their immunogenicities reduced by binding these molecules to polyethylene glycol (PEG)[26]. This has led to new treatments for severe combined immunodeficiency disease associated with ADA deficiency and acute lymphoblastic leukaemia.

For tissue targeting, water-soluble non-immunogenic biocompatible polymers, which will either degrade or be eliminated by the kidney, are chemically linked to drugs, ideally through bonds that are cleaved once they reach their target (for example a tumour) (Fig. 1d). By changing the drug from a small to a large molecule, the biodistribution of the drug following intravenous administration will be different[27,28]. This approach has been used in several cancer chemotherapy strategies. The concept is that low molecular mass anticancer drugs will penetrate most tissues because they pass rapidly through cell membranes. Thus, the drug is quickly distributed throughout the body, with no selectivity toward tumour tissue. However, if the polymer–drug linkages are selected so that they are stable in the bloodstream, the polymer–drug conjugate will circulate for a longer time than just the drug itself because the high molecular mass polymer–drug can generally only gain entry to cells by endocytosis. Because most normal tissues have intact non-leaky microvasculatures, the polymer–drug accumulates to a greater extent in tumour tissue which has a leaky capillary bed[27–29]. This approach is known as passive targeting. One example involves *N*-(2-hydroxypropyl) methacrylamide (HPMA copolymer) conjugated to doxorubicin through a peptidyl linker that is cleaved by thiol-dependent proteases in lysosomes. Up to 70 times more doxorubicin accumulates in mouse melanoma tumours *in vivo* than in normal tissues. In addition, the maximum tolerated dose of the polymer–drug conjugate is 5–10 times higher than that of the free drug in animals and humans[27]. Another example involves injecting a styrene–maleic anhydride copolymer coupled to neocarzinostatin (SMANCS) into the hepatic artery to treat liver cancer[30]. Passive targeting may also be useful for modulating drug distribution after direct delivery to tissues[31].

Active targeting to specific tissues can be achieved by complexing the polymer–drug conjugate with a molecule (such as antibody, carbohydrate) that will be recognized by cell surface receptors in that tissue. One difficulty with this approach has been finding highly specific non-immunogenic targeting molecules. Nonetheless, progress is being made. For example, the HPMA copolymer drug conjugate has been targeted to the liver by adding galactose which is recognized by the hepatocyte cell surface asialoglycoprotein receptor[27]. These systems as well as HPMA–taxol conjugates are in clinical trials. Another approach involving polymeric drugs utilizes a polymeric (poly[N^5-(2-hydroxyethyl)-L-glutamine]) immunoconjugate of adriamycin and a targeting antibody such as a human IgM antibody directed against tumour-associated antigen. These immunoconjugates were designed to allow selective release of adriamycin in the acidic environment of the tumour through a new acid-labile maleamic acid linker[32].

Liposome-based delivery systems

Small lipid vesicles known as liposomes have been widely explored as drug carriers[33–35]. Encapsulating drugs within a vesicle, as opposed to attaching them to a single polymer chain, has the potential advantage of providing a higher drug-carrying capacity. However, to use liposomes effectively, a number of issues had to be overcome including shelf-life, difficulty in targeting specific tissues and rapid clearance by the phagocytic cells of the reticuloendothelial system (RES). A strategy to address this latter issue was to alter the liposome surface by adding hydrophilic substituents, such as PEG, that reduce RES uptake thereby prolonging liposome circulation time[33].

As the above liposome development problems began to be solved, the ability of liposomes to alter drug biodistribution, thereby leading

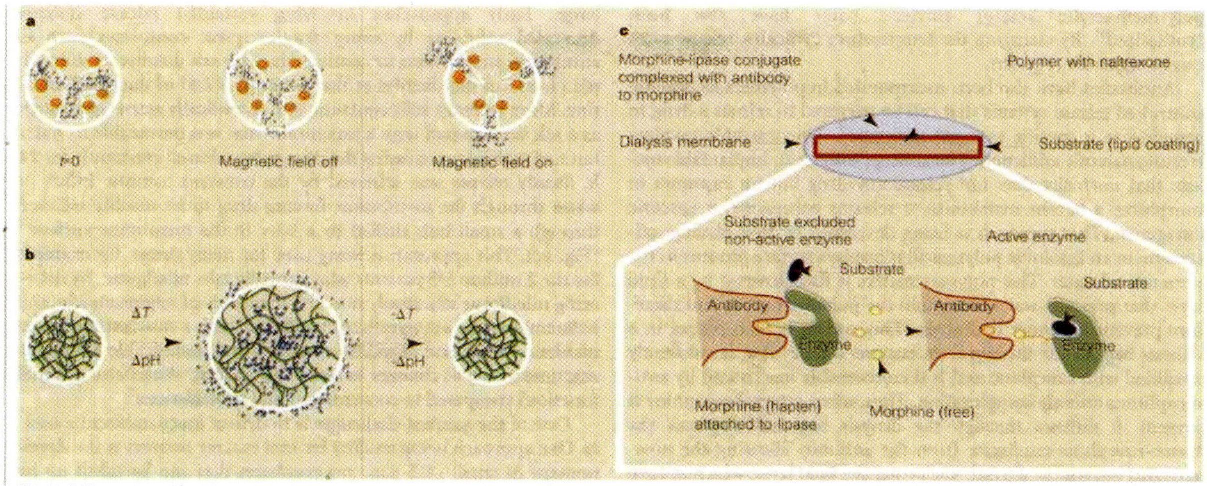

Figure 2 Examples of intelligent polymer delivery systems. See the main text for explanations of how these systems function. **a**, Magnetically controlled system. Large dots are magnetic beads. Small dots represent drug. **b**, pH- or temperature-controlled system (adapted from ref. 51). Small dots represent drugs. Lines represent polymer chains. **c**, Antibody-regulated system (adapted from ref. 56). Top panel, overall system; bottom panel, close-up of enzyme–antibody complex.

to decreases in drug toxicity, was utilized to create safer intravenous drug formulations. Liposomes loaded with duanorubicin and doxorubicin have now been approved by regulatory authorities for the treatment of HIV-associated Kaposi's sarcoma as has liposomal amphotericin B for fungal infections in cancer. The ability to achieve effective cellular targeting may depend on the ability to produce highly specific antibodies. For example, new antibodies coupled to liposomes such as those against the Her2 proto-oncogene, which plays an important role in the progression of breast and other cancers, are being explored for tumour targeting[36]. Liposomes that contain antibodies that recognize E-selectin, an endothelial-specific surface molecule, are being used to target vascular endothelial cells to achieve site-selective delivery in the cardiovascular system[37].

In some cases, polymeric drugs can be encapsulated within liposomes to beneficial effect. For example, *cis*-4-hydroxyproline (Chyp) prevents collagen accumulation when administered to animals and has been considered as a potential treatment for lung fibrosis. However, Chyp rapidly diffuses to many body sites causing toxicity by interfering with non-collagen protein synthesis. By coupling Chyp to a PEG–lysine copolymer, which prevents rapid diffusion and slowly releases Chyp, and encapsulating it into liposomes with amylopectin on their surface, which helps in lung targeting, collagen accumulation was prevented in the pulmonary arteries of hypoxic rats for over 1 week[38].

One area of intense investigation is the application of liposomes, particularly cationic liposomes, for gene therapy. Negatively charged DNA binds to positively charged liposomes, providing efficient transfection for many cell types[39]. For example, certain types of cationic liposomes can target endothelial cells in specific organs such as the lung[40]. Liposomes may also be useful for the delivery of DNA vaccines[41].

Polymer vesicles have also been investigated for their ability to achieve long circulation times and altered drug biodistribution. PEG has been conjugated to lactic/glycolic acid copolymers[42] or polyphosphazenes[43] and formed into nanospheres that circulate for long time periods. Albumin–heparin microspheres containing adriamycin have also shown improved biodistribution and reduced toxicity in animals[44]. In gene therapy, cationic nanospheres composed of chitosan[45], cationic polymer conjugates such as polyethyleneimine[46] or cationic starburst dendrimers made of poly(amido amine) polymers[47] may be useful in delivering DNA.

Intelligent delivery systems

In many cases, constant or decreasing drug release rates, as achieved with most drug-delivery systems, do not always mimic the body's natural pattern of providing chemicals. For example, for drugs such as insulin, pulsatile delivery is desirable.

An early approach to pulsatile delivery involved incorporating magnetic beads in an elastic polymer (Fig. 2a). When an oscillating magnetic field was applied (for example as may someday be achieved in a wristwatch-like device) more drug was released. This occurred reversibly and repeatedly over several months. The external magnetic field appears to cause alternative expansion and contraction of the drug-carrying pores. Key factors in terms of achieving pulsatile release are magnetic bead strength, magnetic field strength, polymer elasticity, magnetic bead size and polymer matrix structure[48].

Ultrasound also noticeably enhances transport of molecules entrapped within polymer release systems. The enhancing effect of ultrasound on drug release appears largely due to convection, generated by cavitation. Critical parameters are ultrasound frequency, molecular mass of the incorporated drug and polymer matrix structure (size of pores in the polymer network)[49].

Electric current has been used to control release rates from polymers. In one new approach, two water-soluble polymers, poly(ethyloxazoline) and either polymethacrylic acid or polyacrylic acid, form a solid because of intramolecular hydrogen bonding between the carboxylic and oxazoline groups. When an electric field is applied by attaching a cathode to the polymer matrix surface, hydroxyl ions are produced by electrolysis of water causing a local pH increase near the cathode surface, disrupting hydrogen bonding and causing disintegration of the polymer complex. If a drug such as insulin is placed in the insoluble complex it is released only when the electric field is applied[50].

Hydrogels are being widely studied for pulsatile delivery[51]. The electrostatic charge of a hydrogel (due for example to charged amino groups) can change in response to an environmental stimulus such as pH, leading to increased repulsive forces between polymer chains, which alters the diffusion rates of an entrapped molecule. One example involves a copolymer of polyhydroxylethyl methacrylate and *N,N*-dimethylaminoethyl methacrylate. Depending on the pH, the systems can either expand or squeeze tightly leading to an on–off mechanism to control the release rate of a drug encapsulated within the hydrogel (Fig. 2b). Such systems can also contain drugs such as insulin and biosensors such as glucose oxidase, which responds to changing external glucose concentrations by producing acidic byproducts. Thus, a change in glucose concentration is sensed by the hydrogel and insulin release is correspondingly altered[52]. Temperature-sensitive polymers such as poly-*N*-isopropylacrylamide[53,54] or

reviews

poly(methacrylic acid-g) ethylene oxide have also been synthesized[55]. By changing the temperature cyclically, release rates can be regulated (Fig. 2b).

Antibodies have also been incorporated in polymers to develop controlled release systems that can be triggered to release a drug in response to a specific external molecule[56]. One example involves treating narcotic addiction. The concept utilizes an implantable system that normally does not release any drug but on exposure to morphine, a heroin metabolite, it releases naltrexone, a narcotic antagonist. This approach is being developed by first placing naltrexone in an insoluble polymer that displays surface erosion in the presence of water. This polymer matrix is then covered by a lipid layer that prevents water entry into the polymer matrix and therefore prevents polymer hydrolysis. This system is then placed in a dialysis bag. Inside the bag is an enzyme (lipase) that is covalently modified with morphine and is then reversibly inactivated by anti-morphine antibody complexation. Thus, when external morphine is present, it diffuses through the dialysis bag and displaces the lipase–morphine conjugate from the antibody allowing the now-activated enzyme to degrade the protective lipid layer, which in turn permits polymer hydrolysis, thereby liberating the narcotic antagonist which subsequently diffuses out of the dialysis membrane into the body (Fig. 2c)

Delivery routes

Scientists have been studying how to get drugs into every part of the body with the goal of either local delivery to that site or non-invasive systemic delivery of drugs using that part of the body as a route to the systemic circulation.

Pulmonary. Local delivery to the lung has long been used in the treatment of respiratory diseases such as asthma and more recently for protein therapies such as DNase for cystic fibrosis. The deep part of the lung, which contains the alveoli, also has a number of potential advantages for systemic delivery of molecules including a large surface area, thin tissue lining and a limited number of proteolytic enzymes[57,58]. Most current lung delivery systems deliver drugs in liquid form and many incorporate chloroflurocarbon (CFC) propellants which may be environmentally dangerous. In addition, many of these systems do not deliver the drug reproducibly or efficiently; generally less than 10% of the drug is received by the lung from the device. Furthermore, repeated delivery every few hours is often necessary.

Recent advances in inhaler design involve non-CFC propellants and electronic breath actuation. New inhalers being designed control delivery by pushing liquid drug formulations through very tiny nozzles (2.5 μm diameter) at preprogrammed values of inspiratory flow rate and inhaled volume[59]. Approaches are also being developed to deliver fine dry-powder aerosols to the lung. An aerosol cloud is generated by compressing air into the drug powder inside the inhaler, thus breaking the powder into tiny (1–5 μm) particles that are capable of reaching the deep part of the lung. These newer inhalers reproducibly deliver 20–50% of the drug to the lung[60].

In spite of these advances, pulmonary drug delivery efficiency often remains relatively low and repeated administration continues to be necessary. To address these issues, the design of the aerosol particles themselves may be important. One recent approach has been to design large (5–20 μm) highly porous particles with extremely low densities. By lowering their density, the aerodynamics of the particles are altered making it possible for unusually large particles to enter the lungs through an airstream. Increasing particle size leads to decreased particle aggregation, creating greater inhalation efficiency, as well as decreased phagocytosis by alveolar macrophages. The decreased phagocytosis can result in sustained drug release. Insulin can be delivered in animals using these particles for over 4 days from a single inhaled dose[61].

Oral. Oral administration of small molecules is by far the most common approach for delivery of pharmaceuticals. However, maintaining a steady drug release rate for 12–24 h has long been a challenge. Early approaches involving sustained release systems decreased solubility by using drug/excipient complexes such as emulsions, suspensions or coatings that did not dissolve at stomach pH (1–5) but did dissolve at the higher pH (7.4) of the small intestine. More recently, pills containing an osmotically active agent such as a salt were coated with a membrane that was permeable to water but not to the drug, causing the drug to be released constantly for 24 h. Steady release was achieved by the constant osmotic influx of water through the membrane, forcing drug to be steadily released through a small hole drilled by a laser in the membrane surface[62] (Fig. 1c). This approach is being used for many drugs, for example for the 2 million US patients who annually take nifedipine. By delivering nifedipine at a steady rate, the frequency of angina attacks and ischaemic events are significantly reduced, with a substantially lower incidence of adverse vasodilatory effects and undesirable metabolic reactions (such as changes in serum potassium, cholesterol or renal function) compared to conventional oral formulations[2].

One of the greatest challenges is to deliver macromolecules orally. One approach being studied for oral vaccine delivery is the development of small (≤5 μm) microspheres that can be taken up by intestinal Peyer's patches and, to a lesser extent, enterocytes[63–66]. Although many materials such as poly(lactic/glycolic acid) have been studied as delivery vehicles there is usually a low uptake (less than 1%) by Peyer's patches. Lipophilic polymers led to higher uptake but most materials used to date are not degradable[63]. To develop lipophilic particles, liposomes were used orally, but were polymerized to prevent their destruction by bile acids. Over 3% uptake by Peyer's patches has been achieved using polymerized liposomes and this has been increased to 10% by covalently attaching ligands with an affinity for Peyer's patch M cells, such as ulex europaeus agglutinin, to the liposome surfaces[67].

One approach to achieving oral delivery of complex molecules, including genes, has been the development of polyanhydride microspheres, which display strong adhesive interactions with the intestinal mucosa and cellular lining. Low molecular mass drugs such as dicumarol, as well as larger molecules such as insulin and even genes, can be taken up in animals using this type of approach, presumably because the microspheres maintain contact with the intestinal epithelium for long times. Microscopic evidence has shown that the microspheres actually penetrate the epithelium, through and between cells[68].

Another approach involves designing low molecular mass carrier molecules that weakly bind to proteins and apparently carry them across the gastrointestinal epithelium. The carrier molecules, such as hydroxybenzylaminophenylbutyric acid[69] reversibly destabilize the native protein state favouring a partially unfolded conformation that can potentially diffuse more readily across the gastrointestinal epithelium into the systemic circulation. The unfolded form of the protein satisfies the definition of the protein's molten-globule state. The transport efficiency of these carriers for specific molecules correlates with their destabilizing potency which can be measured by techniques such as differential scanning calorimetry. Different carrier molecules appear to be specific for different proteins. This approach has been used for the oral delivery of interferon, calcitonin and human growth hormone in animals[70] and is currently in clinical trials with heparin.

The targeting of protein drugs to the colon, which has much lower levels of protease activity than other portions of the gastrointestinal tract, is also being studied. To achieve this, polymers that will not degrade in the intestine before entry into the colon are being synthesized. For example, by constructing polymers with enzymatically degradable azoaromatic crosslinks, azoreductases commonly found in high concentrations in the colon but much lower concentrations elsewhere can degrade the polymer, allowing the entrapped drug to escape. It may be necessary not only to protect the drug from proteolytic enzymes but also to coincorporate penetration enhancers into the oral formulation[71]. One of the challenges for this and all the above approaches to oral protein delivery will be achieving bioavailabilities high enough and reproducibilities consistent enough for practical use in humans.

Transdermal. The skin acts effectively as an impenetrable barrier for most drugs. However, small lipophilic drugs can cross the skin at low flux rates and this provides a means for delivering drugs that are destroyed by the liver when taken orally. Transdermal patches have been developed for seven different drugs lasting from 1 to 7 days (Table 2) and have been useful in producing new therapies and reducing adverse drug effects. For example, 2 years after being on transdermal nicotine patches for 12 weeks, 4 times as many patch users still do not smoke compared to patients who received placebos[72]. Over 700,000 US smokers have given up smoking because of these patches. Transdermal oestradiol patches are now being used annually for 1.3 million US patients; in contrast to oral formulations, these are not associated with liver damage. Transdermal clonidine, nitroglycerin and fentanyl patches also exhibited fewer adverse effects than conventional oral dosage forms[2].

To make transdermal delivery of other molecules practical, electrical approaches such as iontophoresis, which involves the usage of low-voltage pulses for long time periods, are being explored[73]. Iontophoresis can provide enhanced transport for some low molecular mass molecules such as pain medications and even decapeptides. To improve fluxes, high-voltage pulses for short time periods (milliseconds) have been used to induce pores temporarily in the skin (electroporation)[74], enabling the delivery of larger highly charged molecules including heparin[75] and oligonucleotides through human cadaver skin (the most commonly used model for human skin transport).

Ultrasound, particularly at low frequencies, can greatly enhance the flux of high molecular mass substances through the skin, presumably by temporarily disordering the lipid bilayers in the skin's outermost layer, the stratum corneum, which provides the principal barrier to diffusion. Using ultrasound, fluxes up to 5,000 times normal have been achieved and molecules the size of insulin or larger have been non-invasively transported through human skin models[76].

Other routes. The nose represents an additional portal for the delivery of large molecules. Using bioadhesive chitosan microspheres, the residence time of different macromolecules in the nose has been prolonged[77]. In addition, positively charged chitosan may interact with nasal epithelial cells, temporarily loosening up tight junctions in the nasal mucosa[78], causing enhanced permeability. This approach is in clinical trials for insulin.

Delivery to the brain has been a particularly difficult challenge because of the blood–brain barrier (BBB), which is composed of endothelial cells forming tight junctions that permit transport only of low molecular mass lipid-soluble molecules, with the exception of a few select peptides and nutrients that cross through specific transporting mechanisms. Although brain implants such as those discussed earlier permit localized delivery of encapsulated agents including proteins[79], it would also be useful to have less invasive means of delivering drugs, and of delivering them throughout the entire brain. Approaches to BBB transport include: BBB disruption using intracarotid administration of vasoactive molecules such as bradykinin or hypertonic solutions; the creation of prodrugs in which water-soluble drugs are attached by cleavable bonds to lipid-soluble carriers such as dihydropyridine; or attaching the drug to a vector that will transport across the BBB, such as the 83-14 murine monoclonal antibody to the human insulin receptor[80]. One approach that has moved to advanced clinical trials involves the use of a bradykinin agonist known as RMP-7 to aid in the delivery of anticancer agents such as carboplatin that normally do not cross the BBB[81].

The vaginal route has also been widely used for drug delivery, most notably in the form of vaginal rings for contraception. Generally, these systems are used over 6 months and are periodically removed monthly for 1-week periods. The vagina is a primary infectious site for many sexually transmitted pathogens as well as the entry site for sperm. Recently, the possibility was explored that antibodies delivered directly to the vagina can provide significant protection against sexually transmitted diseases, as well as pregnancy. By designing a polymeric vaginal delivery device that delivers antibodies for long time periods, a continuous supply of antibodies to the vagina was provided[82].

Table 2 Currently marketed transdermal drug-delivery systems

Drug	Indication	Duration (days)
Scopolamine	Motion sickness	3
Nitroglycerin	Angina	1
Nicotine	Smoking cessation	1
Clonidine	Hypertension	7
Fentanyl	Pain	1
Oestradiol	Oestrogen replacement	3.5–7
Testosterone	Male hormone replacement	1

The eye is also an important site for continuous delivery[83]. One early example was the development of the Ocusert, an ethylene-vinyl acetate copolymer-controlled release system that slowly delivers pilocarpine for 1 week to treat glaucoma. The Ocusert is inserted into the conjuctiva. However, because many patients do not like to insert objects into their eye, eye drops that could gel in the eye because of a pH or temperature change might be a useful area of research. Many important ocular diseases, such as macular degeneration and diabetic retinopathy are located in the back of the eye; it would be of value to develop systems that could target this area. Although not directly placed in the back of the eye, the Vitrasert polymer system, which is placed in the vitreous and slowly delivers gangcyclovir, is useful in treating AIDS patients with cytomegalovirus retinitis[84].

Many diseases are confined to specific anatomic sites in the body. By physically applying polymer systems directly at these sites, these diseases can potentially be locally treated, increasing both safety and efficacy. For example, in vascular disease, the ability to locally deliver heparin[85] or antisense oligonucleotides[86] to blood vessels has been shown to prevent smooth muscle cell proliferation, leading to potentially new approaches to treat restenosis. Another application has been the delivery of tissue plasminogen activator or urokinase in biodegradable polymers to prevent post-surgical adhesions in rat models. The polymers used in this case are applied as a liquid and then photopolymerized in the presence of external light so that they not only solidify but adhere to the tissue surface to which they are applied[87]. Local polymeric delivery of calcium chelating agents such as bisphosphonates has been shown to be useful in preventing calcification of bioprosthetic heart valves[88]. The local delivery of antibiotics to the gums of patients with periodontal disease greatly reduces the systemic doses of these antibiotics and provides a new way of non-surgically treating this disease[89].

Discussion

Drug delivery is a remarkably interdisciplinary field. Important contributions have come from material scientists, engineers, biologists, pharmaceutical scientists and others who have developed important concepts and brought them to clinical application. Certainly one activity that will take place in the next 10 to 20 years is the clinical introduction and evaluation of many of the new delivery systems discussed above. At the same time, advances in many other scientific areas will aid in the development of new drug-delivery systems. Progress in immunology and human genomics should lead to a greater insight into the type of targeting molecules that can be used to achieve site-specific drug delivery. Advances in combinatorial chemistry are already being used to create new biomaterials and may enable large numbers of new biomaterials to be screened more rapidly[90]. Progress in microelectronics and nanotechnology may someday lead to tiny robots that may be able to travel through the bloodstream and perform both physical and chemical functions. The understanding of transport phenomena through various portals in the body such as the intestine, lung and skin may lead to new drug-delivery strategies. Delivery to sites that are not easily accessi-

reviews

ble such as the sinuses or nerves may also be important. In addition, the development of mathematical models that can predict delivery performance[91] will facilitate the design of various delivery systems.

Finally, the design of ideal carriers for applications such as gene therapy will be extremely important. The possibility of designing materials that can simultaneously avoid the RES, target specific cell types, be taken up by those cell types in such a way as not to destroy their DNA, and then made to release the DNA unharmed so that the DNA can travel to the nucleus will be an important challenge. Nonetheless, with the progress being made in biology, chemistry, biomaterials, engineering and pharmaceutical sciences, this field should have a bright and rapidly evolving future. □

Robert Langer is at the Department of Chemical Engineering, Massachusetts Institute of Technology, 45 Carlton Street, Building E25, Cambridge, Massachusetts 02139, USA.

1. Classen, D. C., Pestotnik, S. L., Evans, S. R., Lloyd, J. F. & Burke, J. P. Adverse drug events in hospitalized patients. *J. Am. Med. Assoc.* **277**, 301–306 (1997).
2. Cramer, M. P.& Saks, S. R. Value of controlled release dosage forms. *PharmacoEconomics* **5**, 482–504 (1994).
3. McKenney, J. M. & Harrison, W. L. Drug related hospital admissions. *Am. J. Hosp. Pharm.* **33**, 792–795 (1976).
4. Langer, R. New methods of drug delivery. *Science* **249**, 1527–1533 (1990).
5. Langer, R. Polymeric delivery systems for controlled drug release. *Chem. Eng. Commun.* **6**, 15–48 (1980).
6. Folkman, J. & Long, D. The use of silicone rubber as a carrier for prolonged drug therapy. *Surg. Res.* **4**, 139–142 (1964).
7. Langer, R. & Folkman, J. Polymers for the sustained release of proteins and other macromolecules. *Nature* **263**, 797–800 (1976).
8. Ogawa, Y., Yamamoto, M., Okada, H., Yashiki, T. & Shimamoto, T. A new technique to efficiently entrap leuprolide acetate into microcapsules of polylactic acid and copoly(lactic/glycolic acid). *Bull. Chem. Pharm.* **36**, 1095–1103 (1988).
9. Costantino, H., Langer, R. & Klibanov, A. Solid-phase aggregation of proteins under pharmaceutically relevant conditions. *J. Pharm. Sci*, **83**, 1662–1666 (1994).
10. Johnson, O. L. *et al.* A month-long effect from a single injection of microencapsulated human growth hormone. *Nature Med.* **3**, 795–799 (1996).
11. Peppas, N. A. & Langer, R. New challenges in biomaterials. *Science* **263**, 1715–1720 (1994).
12. Tamada, J. A. & Langer, R. Erosion kinetics of hydrolytically degradable polymers. *Proc. Natl Acad. Sci. USA* **90**, 552–556 (1993).
13. Leong, K. W., Brott, B. C. & Langer, R. Bioerodible polyanhydrides as drug-carrier matrices: I. characterization, degradation and release characteristics. *J. Biomed. Mater. Res.* **19**, 941–955 (1985).
14. Ng, S. Y., Vandamme, T., Taylor, M. S. & Heller, J. Synthesis and erosion studies of self-catalyzed poly(ortho ester)s. *Macromolecules* **30**, 770–772 (1997).
15. Kadiyala, S., Lo, H., Ponticiallo, M. S. & Leong, K. in *Biomedical Applications of Synthetic Biodegradable Polymers* (ed. Hollinger, J.) 33–57 (CRC Press, Boca Raton, 1995).
16. Allcock, H. R. Design and synthesis of new biomaterials via macromolecules substitution. *Ann. N.Y. Acad. Sci.* **831**, 13–31 (1997).
17. James, K. & Kohn, J. in *Controlled Drug Delivery: Challenges and Strategies* (ed. Park, K) 389-403 (American Chemical Society, Washington DC, 1997).
18. Brem, H. *et al.* Placebo-controlled trial of safety and efficacy of intraoperative controlled delivery by biodegradable polymers of chemotherapy for recurrent gliomas. *Lancet* **345**, 1008–1012 (1995).
19. Valtonen, S. *et al.* Interstitial chemotherapy with carmustine-loaded polymers for high-grade gliomas: a randomized double-blind study. *Neurosurgery* **41**, 44–49 (1997).
20. Laurencin, C. T. *et al.* Bioerodible poly(anhydrides) for antibiotic drug delivery: *in vivo* osteomyelitis treatment in a rat model system. *J. Orthopaed. Res.* **11**, 256–262 (1993).
21. Jeong, B., Bae, Y. H., Lee, D. S. & Kim, S. W. Biodegradable block copolymers as injectable drug-delivery systems. *Nature* **388**, 860–862 (1997).
22. Dunn, R. *et al.* Sustained release of cisplatin in dogs from an injectable implant delivery system. *J. Bioact. Compat. Polym.* **11**, 287–300 (1996).
23. Peppas, N. A. & Sahlin, J. J. Hydrogels as mucoadhesive and bioadhesive materials: a review. *Biomaterials* **17**, 1553–1561 (1996).
24. Lenaerts, V. & Gurny, R. (eds) *Bioadhesive Drug Delivery Systems* (CRC Press, Boca Raton, 1990).
25. Kohn, J. *et al.* Single-step immunization using a controlled release, biodegradable polymer with sustained adjuvant activity. *J. Immunol. Methods* **95**, 31– 38 (1986).
26. Burnham, N. L. Polymers for delivering peptides and proteins. *Am. J. Hosp. Pharm.* **51**, 210–218 (1994).
27. Duncan, R., Dimitrijevic, S. & Evagorou, E. G. The role of polymer conjugates in the diagnosis and treatment of cancer. *STP Pharma Sci.* **6**, 237–263 (1996).
28. Putnam, D. & Kopecek, J. Polymer conjugates with anticancer activity. *Adv. In Polym. Sci.* **122**, 59-123 (1995).
29. Murakami, Y., Tabata, Y. & Ikada, Y. Tumor accumulation of poly(ethylene glycol) with different molecular weights after intravenous injections. *Drug Deliv.* **4**, 23–31 (1997).
30. Maeda, H. SMANCS and polymer-conjugated macromolecular drugs: advantages in cancer chemotherapy. *Adv. Drug Deliv. Rev.* **6**, 181–202 (1991).
31. Dang, W., Colvin, O. M., Brem, H. & Saltzman, W. M. Covalent coupling of methotrexate to dextran enhances the penetration of cytotoxicity into a tissue-like matrix. *Cancer Res.* **54**, 1729–1735 (1994).
32. Hoes, C. J. T. *et al.* Synthesis and biological evaluation of immunoconjugates of adriamycin and a human IgM linked by poly[N^5-(2-hydroxyethyl)-L-glutamine]. *J. Contr. Rel.* **38**, 245–266 (1996).
33. Park, J. W., Hong, K., Kirpotin, D., Papahadjopoulos, D. & Benz, C. C. Immunoliposomes for cancer treatment. *Adv. Pharmacol.* **40**, 399–435 (1997).
34. Gregoiraidis, G. Engineered liposomes for drug delivery: progress and problems. *Trends Biotechnol.* **13**, 527–537 (1995).
35. Torchilin, V. P. Immunoliposomes and PEGylated immunoliposomes: possible use for targeted delivery of imaging agents. *Immunomethods* **4**, 244–258 (1994).
36. Kirpotin, D. *et al.* Sterically stabilized anti-HER2 immunoliposomes; design and targeting to human breast cancer cells *in vitro*. *Biochemistry* **36**, 66–75 (1997).
37. Spragg, D. D. *et al.* Immunotargeting of liposomes to activated vascular endothelial cells: a strategy for site-selective delivery in the cardiovascular system. *Proc. Natl Acad. Sci. USA* **94**, 8795–8800 (1997).
38. Poiani, G. J. *et al.* Polymeric carrier of proline analogue with antifibrotic effect in pulmonary vascular remodeling. *Am. J. Respir. Crit. Care Med.* **155**, 1384–1390 (1997).
39. Li, S. *et al.* DC-Chol lipid system in gene transfer. *J. Contr. Rel.* **39**, 373–381 (1996).
40. McLean, J. W. *et al.* Organ-specific endothelial cell uptake of cationic liposome-DNA complexes in mice. *Am. Physiol. Soc.* **273**, H387–H404 (1997).
41. Gregoriadis, G., Saffie, R. & de Souza, J. B. Liposome-mediated DNA vaccination. *FEBS Lett.* **402**, 107–110 (1997).
42. Gref, R. *et al.* Biodegradable long-circulating polymeric nanospheres. *Science* **263**, 1600–1603 (1994).
43. Vandorpe, J. *et al.* Long circulating biodegradable poly(phosphazene) nanoparticles with poly(phosphazene)-poly(ethyleneoxide) copolymer. *Biomaterials* **18**, 1147–1152 (1997).
44. Cremers, H. F. M. *et al.* Improved distribution and reduced toxicity of adriamycin bound to albumin-heparin microspheres. *Int. J. Pharm.* **120**, 51–61 (1995).
45. Leong, K. W. *et al.* DNA-polycation nanospheres as non-viral gene delivery vehicles. *J. Cont. Rel.* **53**, 183–193 (1998).
46. Boussif, O. *et al.* A versatile vector for gene and oligonucleotide transfer into cells in culture and *in vivo* polyethylenimine. *Proc. Natl Acad. Sci. USA* **92**, 7297–7301 (1995).
47. Tang, M. X. & Szoka, F. C. The influence of polymer structure on the interactions of cationic polymers with DNA and morphology of the resulting complexes. *Gene Ther.* **4**, 823–832 (1997).
48. Edelman, E., Brown, L. & Langer, R. In vitro and in vivo kinetics of regulated drug release from polymer matrices by oscillating magnetic fields. *J. Biomed. Mater. Sci.* **21**, 339–353 (1987).
49. Lavon, I. & Kost, J. Mass transport enhancement by ultrasound in non-degradable polymeric controlled release systems. *J. Contr. Rel.* (in the press).
50. Kwon, I. C., Bae, Y. H. & Kim, S. W. Electrically erodible polymer gel for controlled release of drugs. *Nature* **354**, 291–293 (1991).
51. Peppas, N. A. Hydrogels and drug delivery. *Curr. Opin. Colloid . Interface Sci.* **2**, 531–537 (1997).
52. Goldraich, M. & Kost, J. Glucose-sensitive polymeric matrices for controlled drug delivery. *Clin. Mater.* **13**, 135–142 (1993).
53. Chen, G. & Hoffman, A. S. Graft copolymers that exhibit temperature-induced phase transitions over a wide range of pH. *Nature* **373**, 49–53 (1995).
54. Yoshida, R. *et al.* Comb-type grafted hydrogels with rapid deswelling response to temperature changes. *Nature* **374**, 240–242 (1995).
55. Lowman, A. M. & Peppas, N. A. Analysis of the complexation/decomplexation phenomena in graft copolymer networks. *Macromolecules* **30**, 4959–4965 (1997).
56. Heller, J. in *Controlled Drug Delivery: Challenges and Strategies* (ed. Park, K.) 127–146 (American Chemical Society, Washington DC, 1997).
57. Niven, R. W. Delivery of biotherapeutics by inhalation aerosol. *Crit. Rev. Therap. Drug Carrier Syst.* **12**, 151–231 (1995).
58. Patton, J. S. Mechanisms of macromolecule absorption by the lungs. *Adv. Drug Deliv. Rev.* **19**, 3–36 (1996).
59. Schuster, J., Rubsamen, R., Lloyd, P. & Lloyd, J. The AERx aerosol delivery system. *Pharm. Res.* **14**, 354–357 (1997).
60. Service, R. F. Drug delivery takes a deep breath. *Science* **277**, 1199–1120 (1997).
61. Edwards, D. A. *et al.* Large porous aerosols for pulmonary drug delivery. *Science* **276**, 1868–1871 (1997).
62. Florence, A. T. & Jani, P. U. Novel oral drug formulations: their potential in modulating adverse effects. *Drug Safety* **10**, 233–266 (1994).
63. Eldridge, J. H. *et al.* Controlled vaccine release in the gut-associated lymphoid tissues: I. orally administered biodegradable microspheres target the Peyer's patches. *J. Contr. Rel.* **11**, 205–214 (1990).
64. Desai, M. P., Labhasetwar, V., Amidon, G. L. & Levy, R. J. Gastrointestinal uptake of biodegradable microparticles: effect of particle size. *Pharm. Res.* **13**, 1838–1845 (1996).
65. Florence, A. T. The oral absorption of micro- and nanoparticulates: neither exceptional nor unusual. *Pharm. Res.* **14**, 259–266 (1997).
66. Tabata, Y., Inoue, Y. & Ikada, Y. Size effect on systemic and mucosal immune responses induced by oral administration of biodegradable microspheres. *Vaccine* **14**, 1677–1685 (1996).
67. Chen, H. & Langer, R. Lectin-bearing polymerized liposomes as potential oral vaccine carriers. *Pharm. Res.* **13**, 1378–1383 (1996).
68. Mathiowitz, E. *et al.* Biologically erodable microspheres as potential oral drug delivery systems. *Nature* **386**, 410–414 (1997).
69. Leone-Bay, A. *et al.* 4-[4-(2-Hydroxybenzoyl)amino]phenyl-butyric acid as a novel oral delivery agent for recombinant human growth hormone. *J. Med. Chem.* **39**, 2571–2578 (1996).
70. Milstein, S. J. *et al.* Partially unfolded proteins efficiently penetrate cell membranes: Implications for oral drug delivery. *J. Contr. Rel.* **53**, 259–267 (1998).
71. Yeh, P. Y., Berenson, M. M., Samawitz, W. S., Kopeckova, P. & Kopechek, J. Site-specific drug delivery and penetration enhancement in the gastrointestinal tract. *J. Contr. Rel.* **36**, 109–124 (1995).
72. Henningfield, J. E. Nicotine medications for smoking cessation. *N. Engl. J. Med.* **333**, 1196–1203 (1995).
73. Merino, V., Kalia, Y. N. & Guy, R. H. Transdermal therapy and diagnosis by iontophoresis. *Trends Biotechnol.* **15**, 288–290 (1997).
74. Prausnitz, M. R., Bose, V. G., Langer, R. & Weaver, J. C. Electroporation of mammalian skin: a new mechanism to enhance transdermal drug delivery. *Proc. Natl Acad. Sci. USA* **90**, 10504–10508 (1993).
75. Prausnitz, M., Edelman, E. R., Gimm, J. A., Langer, R. & Weaver, J.C. Transdermal delivery of heparin by skin electroporation. *Bio/Technology* **13**, 1205–1209 (1995).
76. Mitragotri, S., Blankschtein, D. & Langer, R. Ultrasound-mediated transdermal protein delivery. *Science* **269**, 850–853 (1995).
77. Aspden, T. J. *et al.* Chitosan as a nasal delivery system: the effect of chitosan solutions on *in vitro* and *in vivo* mucociliary transport rates in human tubinates and volunteers. *J. Pharm. Sci.* **86**, 509–513 (1997).
78. Artursson, P., Lindmark, T., Davis, S. S. & Illum, L. Effect of chitosan on the permeability of monolayers of intestinal epithelial cells (Caco-2). *Pharm. Res.* **11**, 1358–1361 (1994).
79. Sherwood, J. K., Zeitlin, L., Whaley, K. J., Cone, R. A. & Saltzman, M. Controlled release of antibodies for longterm topical passive immunoprotection of female mice against genital herpes. *Nature Biotechnol.* **14**, 468–472 (1996).
80. Pardridge, W. M. Drug delivery to the brain. *J. Cerebral Blood Flow Metab.* **17**, 713–731 (1997).
81. Bartus, R. T. *et al.* Controlled modulation of BBB permeability using the bradykinin agonist, RMP-7. *Exp. Neurol.* **142**, 14–28 (1996).
82. Mitra, A. (ed.) *Ophthalmic Drug Delivery Systems* 223–262 (Marcel Dekker, New York, 1993).
83. Krewson, C. E. & Saltzman, W. M. Transport and elimination of recombinant human NGF during long-term delivery to the brain. *Brain Res.* **727**, 169–181 (1996).
84. Musch, D. C. *et al.* Treatment of cytomegalovirus retinitis with a sustained-release gancyclovir implant. *N. Engl. J. Med.* **337**, 83–90 (1997).
85. Rodgers, C. & Edelman, E. R. Controlled release of heparin reduces neointimal hyperplasia in stented rabbit arteries: ramifications for local therapy. *J. Intervent. Cardiol.* **5**, 195–202 (1992).
86. Simons, M., Edelman, E. R., DeKeyser, J.-L., Langer, R. & Rosenberg, R. D. Antisense c-myb oligonucleotides inhibit arterial smooth muscle proliferation *in vivo*. *Nature* **359**, 67–70 (1992).
87. Hill-West, J. L., Dunn, R. D. & Hubbell, J. A. Local release of fibrinolytic agents for adhesion prevention. *J. Surg. Res.* **59**, 759–763 (1995).
88. Levy, R. J. *et al.* Inhibition of calcification of bioprosthetic heart valves by local controlled release diphosphonate. *Science* **228**, 190–192 (1985).
89. Goodson, J. M. Life on a string: development of the tetracycline fiber delivery system. *Techn. Health Care* **4**, 269–282 (1996).
90. Brocchini, S., James, K., Tangpasuthadol, V. & Kohn, J. A combinatorial approach for polymer design. *J. Am. Chem. Soc.* **119**, 4553–4554 (1997).
91. Narasimhan, B. & Peppas, N. A. in *Controlled Drug Delivery: Challenges and Strategies* (ed. Park, K.) 529–557 (American Chemical Society, Washington DC, 1997).

Acknowledgements. I would like to thank J. Kost, G. Cowley, H. Rosen, S. Mitragotri, D. Putnam, L. Niklason, I. Sanderson, J. Hrkach and S. Cannizzaro for their assistance. R.L.'s research in drug delivery is supported by the NIH.

A controlled-release microchip

John T. Santini Jr*, Michael J. Cima† & Robert Langer*

* *Department of Chemical Engineering,* † *Department of Materials Science and Engineering, Massachusetts Institute of Technology, Cambridge, Massachusetts 02139, USA*

Much previous work in methods of achieving complex drug-release patterns has focused on pulsatile release from polymeric materials in response to specific stimuli[1], such as electric[2–5] or magnetic[6,7] fields, exposure to ultrasound[7,8], light[9] or enzymes[10], and changes in pH[11] or temperature[12–14]. An alternative method for achieving pulsatile release involves using microfabrication technology to develop active devices that incorporate micrometre-scale pumps, valves and flow channels to deliver liquid solutions[15,16]. Here we report a solid-state silicon microchip that can provide controlled release of single or multiple chemical substances on demand. The release mechanism is based on the electrochemical dissolution of thin anode membranes covering microreservoirs filled with chemicals in solid, liquid or gel form. We have conducted proof-of-principle release studies with a prototype microchip using gold and saline solution as a model electrode material and release medium, and we have demonstrated controlled, pulsatile release of chemical substances with this device.

Controlled release from our microchip involves no moving parts. Release from a particular reservoir is initiated by applying an electric potential between the anode membrane covering that reservoir and a cathode. Fig. 1a shows a cut-away portion of a prototype microchip containing reservoirs filled with the chemical to be released. The devices used in this study were 17 mm by 17 mm by 310 μm and contained 34 reservoirs. Device size could be reduced to <2 mm, depending on the particular application. As a point of reference, a device of the size used in these studies (17 mm) has enough surface area to accommodate over 1,000 reservoirs.

Devices for this study were fabricated by a sequential process using silicon wafers and microelectronic processing techniques[17,18] including ultraviolet photolithography, chemical vapour deposition (CVD), electron beam evaporation and reactive ion etching. Each device contained reservoirs that extended completely through the wafer. The reservoirs were square pyramidal in shape (that is, they contained one large and one small square opening, see Fig. 1b), had a volume of ~25 nl, and were sealed on the small square end (50 × 50 μm) by a 0.3-μm-thick, gold membrane anode. Gold was chosen as a model membrane material because it is easily deposited and patterned, has a low reactivity with other substances and resists spontaneous corrosion in many solutions over the entire pH range. However, the presence of a small amount of chloride ion creates an electric potential region which favours the formation of soluble gold chloride complexes[19]. Holding the anode potential in this corrosion region enables reproducible gold dissolution. Potentials below this region are too low to cause appreciable corrosion, whereas potentials above this region result in gas evolution and formation of a passivating gold oxide layer that causes corrosion to slow or stop. Other metals such as copper or titanium tend to dissolve spontaneously under these conditions or do not form soluble materials on application of an electric potential. Although it is used as a model compound in these studies, gold has also been shown to be a biocompatible material[20].

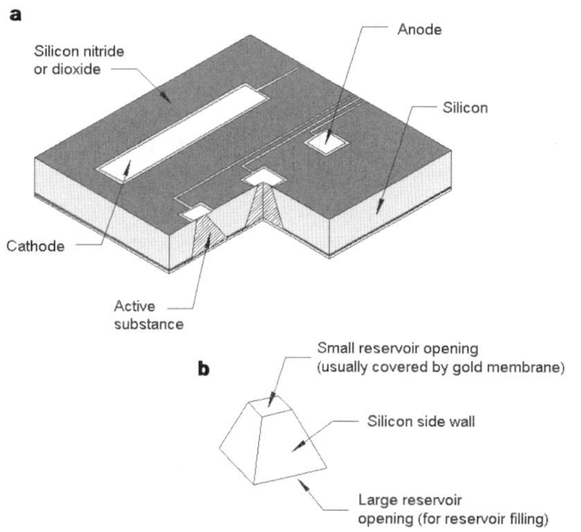

Figure 1 A prototype microchip for controlled release showing the shape of a single reservoir. **a**, Fabrication of these microchips began by depositing ~0.12 μm of low stress, silicon-rich nitride on both sides of prime grade, (100) silicon wafers using a vertical tube reactor. The silicon nitride layer on one side of the wafer was patterned by photolithography and electron cyclotron resonance (ECR) enhanced reactive ion etching (RIE) to give a square device (17 × 17 mm) containing 34,480-μm-square reservoirs. The silicon nitride served as an etch mask for potassium hydroxide solution at 85 °C, which anisotropically etched square pyramidal reservoirs (**b**) into the silicon along the (111) crystal planes until the silicon nitride film on the opposite side of the wafer was reached. The newly fabricated silicon nitride membranes completely covered the 40–60 μm square openings of the reservoir. Gold electrodes (0.3 μm thick, with a 0.01-μm chromium adhesion layer) were deposited and patterned over the silicon nitride membranes by electron beam evaporation and lift-off. A 0.6-μm layer of plasma enhanced chemical vapour deposition silicon dioxide was deposited at 350 °C over the entire electrode containing surface. The silicon dioxide located over portions of the anode, cathode and bonding pads were etched with ECR-enhanced RIE to expose the underlying gold film; this technique was then used to remove the thin silicon nitride and chromium membranes located in the reservoir underneath the gold anode.

Some portions of the electrodes must be protected from unwanted corrosion by an adherent, non-porous coating that isolates the electrode materials from the surrounding electrolyte. Silicon dioxide (SiO_2) was used as a model protective coating because its physical properties can be tailored to a particular application by selecting appropriate processing conditions. For example, SiO_2 deposited by CVD at low temperatures (<100 °C) tends to be porous and non-adherent in solution, whereas high-temperature (>700 °C) deposition or annealing results in denser SiO_2 but may also lead to thermal grooving and void formation in gold films[21]. We used plasma-enhanced CVD at moderate temperatures (350 °C) to address these problems and produce SiO_2 films possessing adequate density and adhesion with negligible gold void formation. Light microscopy revealed no evidence of corrosion following the applied potential in those areas of the gold electrodes covered by SiO_2, except in those regions immediately adjacent to the exposed gold membrane.

The gold membrane anodes of each microchip were examined for defects (such as pinholes) using reflected and transmission light microscopy after completion of the entire fabrication process. Reservoirs covered by defect-free, gold membrane anodes were considered viable for use in chemical-release experiments and were loaded with model chemical substances, sodium fluorescein and $^{45}Ca^{2+}$ (as $CaCl_2$), selected for ease of detection. Inkjet printing, in combination with a computer-controlled alignment apparatus[22], is capable of depositing as little as 0.2 nl of a liquid or gel solution of known concentration into each reservoir. Microsyringe pumps also have the ability to deposit nanolitre quantities of solutions into device reservoirs. We found that these two filling methods work well and prevent any leaking of solutions or gels, such as our aqueous solutions composed of 15–25% (by volume) liquid polymer (polyethylene glycol, relative molecular mass $M_r = 200$) and the model release compounds. The water quickly evaporates after injection leaving only liquid polymer and model compound in the reservoir. The reservoirs are then covered with squares of a thin adhesive plastic and sealed with a waterproof epoxy. The gold membranes do not appear to be fragile even though the liquid material in a filled reservoir can exert a stress on the gold membrane anode due to capillary forces. Calculations indicate that these capillary forces are nearly two orders of magnitude smaller than that required to rupture a 0.3-μm-thick gold membrane, which can withstand pressures up to 3 lb in^{-2}. In addition, prototype devices have been stored for over a year without any known effects on operation or reliability.

The objective of initial release experiments was to determine if pulsatile release of a single compound could be obtained from a microchip device. Release was achieved from a reservoir of a prototype device immersed in phosphate-buffered saline (PBS) by applying a potential of +1.04 volts with respect to a saturated

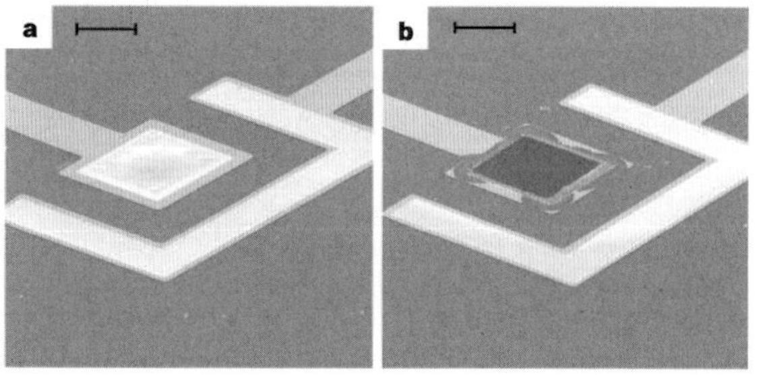

Figure 2 Removal of an anode membrane to initiate release from a reservoir. **a**, **b**, Scanning electron micrographs of a gold membrane anode covering a reservoir are shown before (**a**) and after (**b**) the application of +1.04 V w.r.t. SCE for several seconds in PBS. (Scale bar, 50 μm.)

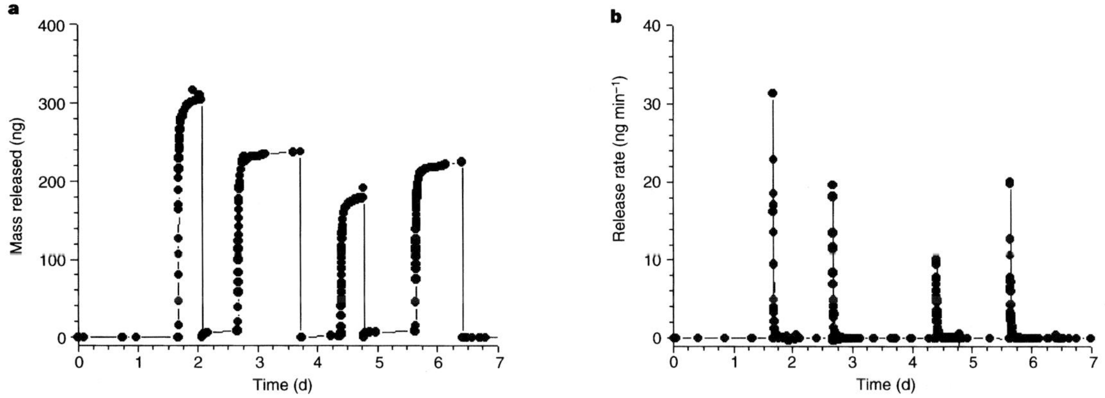

Figure 3 Pulsatile release of a single substance from a microchip device. **a**, The total mass of sodium fluorescein released into PBS over a period of several days is shown for each of four reservoirs. This release study was conducted in PBS stirred with a magnetic stirring bar at room temperature. The device was submerged in the PBS for >36 h before the first release to ensure that there was no leakage from any of the loaded reservoirs. The tip of the SCE was placed within 1–3 mm of the anode surface immediately before application of the electric potential. Release of fluorescein was obtained on demand by applying a potential of +1.04 V w.r.t. SCE for up to 30 s to each anode. The SCE was removed immediately after the potential was applied. Samples (3 ml) were taken every few minutes from the release medium, analysed for dye content in a fluorimeter, and replaced with an equal volume of fresh PBS solution. The entire release medium (100 ml) was replaced with fresh solution several hours after each release. The release rate for the same experiment is shown in **b**. The release rate was calculated by taking the amount of fluorescein released between two time points and dividing it by the time elapsed, starting after the device was immersed in the PBS.

calomel reference electrode (+1.04 V w.r.t. SCE) to the gold anode covering that reservoir. The use of a reference electrode ensured that the potential of the gold membrane anode remained in the gold corrosion region during activation. Figures 2a and b show a gold membrane anode before and after a potential was applied in PBS. Release of the fluorescent model compound, sodium fluorescein, from a reservoir was detected by fluorescence spectroscopy within 1–2 minutes after the potential was applied (Fig. 3).

Subsequent release experiments were designed to determine if the independent release of multiple compounds could be obtained from a single device. Reservoirs containing one of two model compounds were opened by applying +1.04 V w.r.t. SCE to the corresponding anode immersed in saline solution without phosphate buffer. The pulsatile release of $^{45}Ca^{2+}$ ions and sodium fluorescein over a period of several hours was achieved, showing that multiple compounds can be released independently from a single microchip device (Fig. 4).

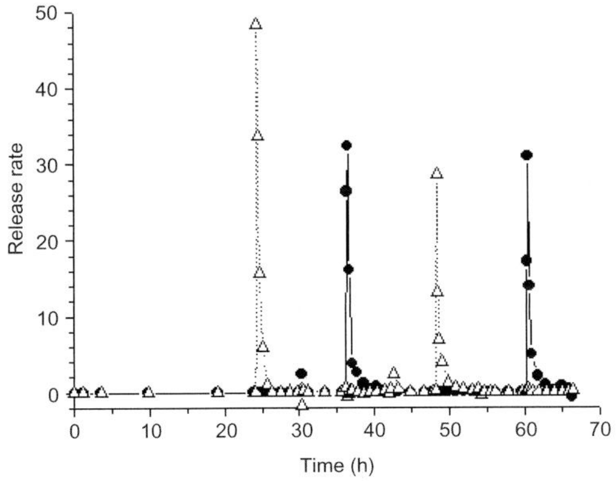

Figure 4 Pulsatile release of multiple substances from a single microchip device. The release rate of $^{45}Ca^{2+}$ ions (open triangles; vertical scale in units of $5 \times$ nCi min^{-1}) and sodium fluorescein (filled circles; in units of ng min^{-1}) into 0.145 M NaCl solution over several hours is shown for each reservoir. The experimental procedures and release-rate calculations were the same as in Fig. 3, with the addition that samples (1 ml) were taken every several minutes from the release medium, analysed for radionuclide content in a scintillation counter, and replaced with an equal volume of fresh saline solution.

The release studies demonstrate that the activation of each reservoir can be controlled individually, creating a possibility for achieving many complex release patterns. Varying amounts of chemical substances in solid, liquid or gel form can be released into solution in either a pulsatile manner, a continuous manner, or a combination of both, either sequentially or simultaneously from a single device. Such a device has additional potential advantages including small size, quick response times and low power consumption. In addition, all chemical substances to be released are stored in the reservoirs of the microchip itself, creating a possibility for the future development of autonomous devices. A microbattery, multiplexing circuitry and memory could be integrated directly onto the device, allowing the entire device to be mounted onto the tip of a small probe, implanted, swallowed, integrated with microfluidic components to develop a 'laboratory-on-a-chip', or incorporated into a standard electronic package, depending on the particular application. Some fields where this device may become useful include medical diagnostics, analytical chemistry, chemical detection, industrial process monitoring and control, combinatorial chemistry and microbiology. Future work will focus on improving reservoir filling methods and determining the factors that control the release of material from opened reservoirs. Gold and silicon were used as model materials for the prototype device in these proof-of-principle release studies. In the future, proper selection of biocompatible device materials may result in the development of an autonomous, controlled-release implant ('pharmacy-on-a-chip') or a highly controllable tablet ('smart tablet') for drug delivery applications. □

Received 25 August; accepted 16 November 1998.

1. Kost, J. & Langer, R. Responsive polymeric delivery systems. *Adv. Drug Delivery Rev.* **6,** 19–50 (1991).
2. Kwon, I. C., Bae, Y. H. & Kim, S. W. Electrically erodible polymer gel for controlled release of drugs. *Nature* **354,** 291–293 (1991).

3. Bae, Y. H., Kwon, I. C. & Kim, S. W. in *Polymeric Drugs and Drug Adminsitration* (ed. Ottenbrite, R. M.) 98–110 (Am. Chem. Soc., Washington DC, 1994).
4. Miller, L. L. Electrochemically controlled release of drug ions from conducting polymers. *Mol. Cryst. Liq. Cryst.* **160,** 297–301 (1988).
5. Hepel, M. & Fijalek, Z. in *Polymeric Drugs and Drug Adminsitraiton* (ed. Ottenbrite, R. M.) 79–97 (Am. Chem. Soc., Washington DC, 1994).
6. Edelman, E. R., Kost, J., Bobeck, H. & Langer, R. Regulation of drug release from polymer matrices by oscillating magnetic fields. *J. Biomed. Mater. Res.* **19,** 67–83 (1985).
7. Kost, J. & Langer, R. in *Pulsed and Self-Regulated Drug Delivery* (ed. Kost, J.) 3–16 (CRC, Boca Raton, 1990).
8. Kost, J., Leong, K. & Langer, R. Ultrasound-enhanced polymer degradation and release of incorporated substances. *Proc. Natl Acad. Sci. USA* **86,** 7663–7666 (1989).
9. Mathiowitz, E. & Cohen, M. D. Polyamide microcapsules for controlled release. V. Photochemical release. *J. Membr. Sci.* **40,** 67–86 (1989).
10. Fischel-Ghodsian, F., Brown, L., Mathiowitz, E., Brandenburg, D. & Langer, R. Enzymatically controlled drug delivery. *Proc. Natl Acad. Sci. USA* **85,** 2403–2406 (1988).
11. Siegel, R. A., Falamarzian, M., Firestone, B. A. & Moxley, B. C. pH-controlled release from hydrophobic/polyelectrolyte copolymer hydrogels. *J. Control. Release* **8,** 179–182 (1988).
12. Bae, Y. H., Okano, T., Husu, R. & Kim, S. W. Thermo-sensitive polymers as on-off switches for drug release. *Makromol. Chem. Rapid Commun.* **8,** 481–485 (1987).
13. Hoffman, A. S., Afrassiabi, A. & Dong, L. C. Thermally reversible hydrogels: II. Delivery and selective removal of substances from aqueous solutions. *J. Control. Release* **4,** 213–222 (1986).
14. Okano, T., Bae, Y. H. & Kim, S. W. in *Pulsed and Self-Regulated Drug Delivery* (ed. Kost, J.) 17–45 (CRC, Boca Raton, 1990).
15. Gravesen, P., Branebjerg, J. & Jensen, O. S. Microfluidics - a review. *J. Micromech. Microeng.* **3,** 168–182 (1993).
16. Shoji, S. & Esashi, M. Microflow devices and systems. *J. Micromech. Microeng.* **4,** 157–171 (1994).
17. Wolf, S. & Tauber, R. N. *Silicon Processing for the VLSI Era* Vol. 1 *Process Technology* (Lattice Sunset Beach, California, 1986).
18. Jaeger, R. C. *Introduction to Microelectronic Fabrication* (Addison-Wesley, Reading, Massachusetts, 1988).
19. Frankenthal, R. P. & Siconolfi, D. J. The anodic corrosion of gold in concentrated chloride solutions. *J. Electrochem. Soc.* **129,** 1192–1196 (1982).
20. Merchant, B. Gold, the noble metal and the paradoxes of its toxicology. *Biologicals* **26,** 49–59 (1998).
21. Jiran, E. & Thompson, C. V. Capillary instabilities in thin, continuous films. *Thin Solid Films* **208,** 23–28 (1992).
22. Wu, B. M. in *Microstructural Control During Three Dimensional Printing of Polymeric Medical Devices* Thesis, 25–40 MIT (1997).

Acknowledgements. We thank A. Göpferich and M. Llabres for help in the early stages of the project, and K. Jensen, C. Thompson and R. Latanision for discussions about materials and electrochemical aspects of this project. All fabrication work was carried out at the Microsystems Technology Laboratory at MIT. This work was partially supported by the US NSF.

Correspondence and requests for materials should be addressed to R.L. (e-mail: rlanger@mit.edu).

Multi-pulse drug delivery from a resorbable polymeric microchip device

AMY C. RICHARDS GRAYSON[1,†], INSUNG S. CHOI[2], BETTY M. TYLER[3], PAUL P. WANG[3], HENRY BREM[3], MICHAEL J. CIMA[1] AND ROBERT LANGER*[4]

[1]Department of Materials Science and Engineering, Massachusetts Institute of Technology, 77 Massachusetts Avenue, Cambridge, Massachusetts 02139, USA
[2]Department of Chemistry, School of Molecular Science – BK21, and Center for Molecular Design and Synthesis, Korea Advanced Institute of Science and Technology, Daejeon 305-701, Korea
[3]School of Medicine and Department of Neurosurgery, Johns Hopkins University, 817 Hunterian, 725 North Wolfe Street, Baltimore, Maryland 21205, USA
[4]Department of Chemical Engineering, Massachusetts Institute of Technology, 77 Massachusetts Avenue, Bldg E25-342, Cambridge, Massachusetts 02139, USA
†Present address: School of Chemical and Biomolecular Engineering and the Biomedical Engineering Program, 120 Olin Hall, Cornell University, Ithaca, New York 14853, USA
*e-mail: rlanger@mit.edu

Published online: 19 October 2003; doi:10.1038/nmat998

Controlled-release drug delivery systems have many applications, including treatments for hormone deficiencies and chronic pain. A biodegradable device that could provide multi-dose drug delivery would be advantageous for long-term treatment of conditions requiring pulsatile drug release. In this work, biodegradable polymeric microchips were fabricated that released four pulses of radiolabelled dextran, human growth hormone or heparin *in vitro*. Heparin that was released over 142 days retained on average 96 ± 12% of its bioactivity. The microchips were 1.2 cm in diameter, 480–560 μm thick and had 36 reservoirs that could each be filled with a different chemical. The devices were fabricated from poly(L-lactic acid) and had poly(D,L-lactic-*co*-glycolic acid) membranes of different molecular masses covering the reservoirs. A drug delivery system can be designed with the potential to release pulses of different drugs at intervals after implantation in a patient by using different molecular masses or materials for the membrane.

The method by which a drug is delivered can have a significant effect on the drug's therapeutic efficacy[1]. Conventional drug delivery systems such as implants and some oral delivery systems typically produce a sharp initial increase in drug concentration to a peak above the therapeutic range, followed by a fast decrease in concentration to a level below the therapeutic range. The time spent in the optimum concentration range for therapeutic effect is therefore short[2]. Initial concentration peaks can pose a serious risk of toxicity and related complications for potent drugs. Additionally, accurate and repeatable dosing of medication to patients may be difficult to achieve because of patient non-compliance. The field of controlled-release drug delivery therefore initially focused on achieving a constant release of drug over long periods with minimal influence by outside factors. However, for certain molecules or drugs, sustained or continuous release is not optimal. For example, insulin[3] and hormones of the anterior pituitary gland such as growth hormone and gonadotropin-releasing hormone (GnRH) are secreted by the human body in a pulsatile fashion[4]. It is desirable for therapies that treat deficiencies or disorders of these molecules to mimic the pulsatile fashion in which the molecules are naturally released from the body. Similarly, for multi-dose vaccines, repeated delivery of the antigen is desired for optimal antibody response.

Previous work in the area of controlled release has focused on approaches such as tablets[5–7], pumps[8–10], implants (Gliadel[11,12], for example), injectable microspheres (including Lupron Depot[12]) and patches[13]. Typically, these devices are capable of achieving continuous (tablets, patches or implants) or pulsatile (pumps) release of drugs over periods ranging from minutes to months or years. Matrix-type systems, such as Lupron Depot and Gliadel, are composed of a mixture of a polymer and the drug to be released. Changing the component polymer, polymer molecular mass, excipients or drug loading can significantly change both the drug's stability and the kinetics of drug release, owing to the physical interdependence of the release formulation and the drug formulation.

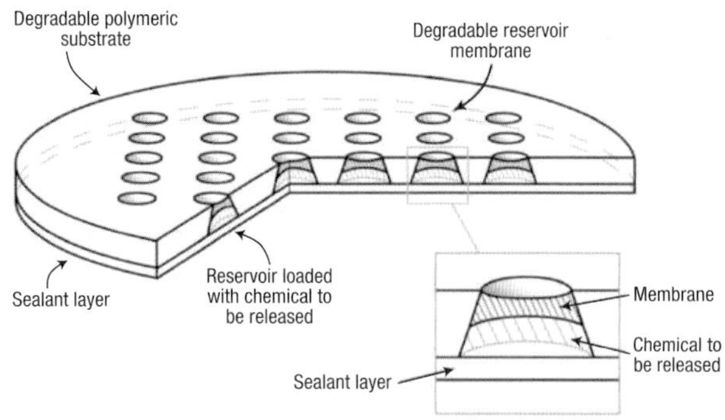

Figure 1 Diagram of polymeric microchip device. The main body of the device is composed of a reservoir-containing substrate that is fabricated from a degradable polymer. Truncated conical reservoirs in the substrate are loaded with the chemical to be released, and sealed with polymeric degradable reservoir membranes on one end and a sealant layer (polyester tape) on the opposite end. Inset, close-up of a reservoir, reservoir membrane, sealant layer and chemical to be released.

Several different strategies have been investigated for triggering the pulsatile release of drugs from polymeric implants[14], such as changes in pH (refs 15–17) or temperature[18–20], exposure to ultrasound[21,22], enzymes[23] or light[24], or changes in electric[25–29] or magnetic[21,30] fields or molecules present in the human body, such as antigens[31] or water. A newer approach uses microfabricated silicon devices for the controlled release of drugs[32]. These devices are activated by the application of a small voltage to a drug-containing reservoir, and potentially allow the release of hundreds of individual doses of drugs or chemicals (liquids and solids) from a single device. However, these devices are not self-contained or biodegradable, and require components such as a power supply or timer.

Pulsatile release from polymeric drug-delivery systems in the absence of an external trigger has been investigated more recently, and approaches involving multilayer compression-moulded devices[33–35], beads of various diameters[36] or novel hydrogels[37] have shown promising results.

We designed, developed and tested a biodegradable polymeric microchip device to investigate the feasibility of achieving multi-pulse drug release from a polymeric system over periods of several months without requiring the application of a stimulus to trigger the drug release. A prototype device is shown in a cut-away diagram in Fig. 1. This device was designed to release chemicals at different times on the basis of the characteristics of the reservoir membranes that affect their degradation rate, such as the material used, the molecular mass, the composition or the thickness. Separation of the release formulation (the reservoir membranes) from the drug formulation (which is loaded

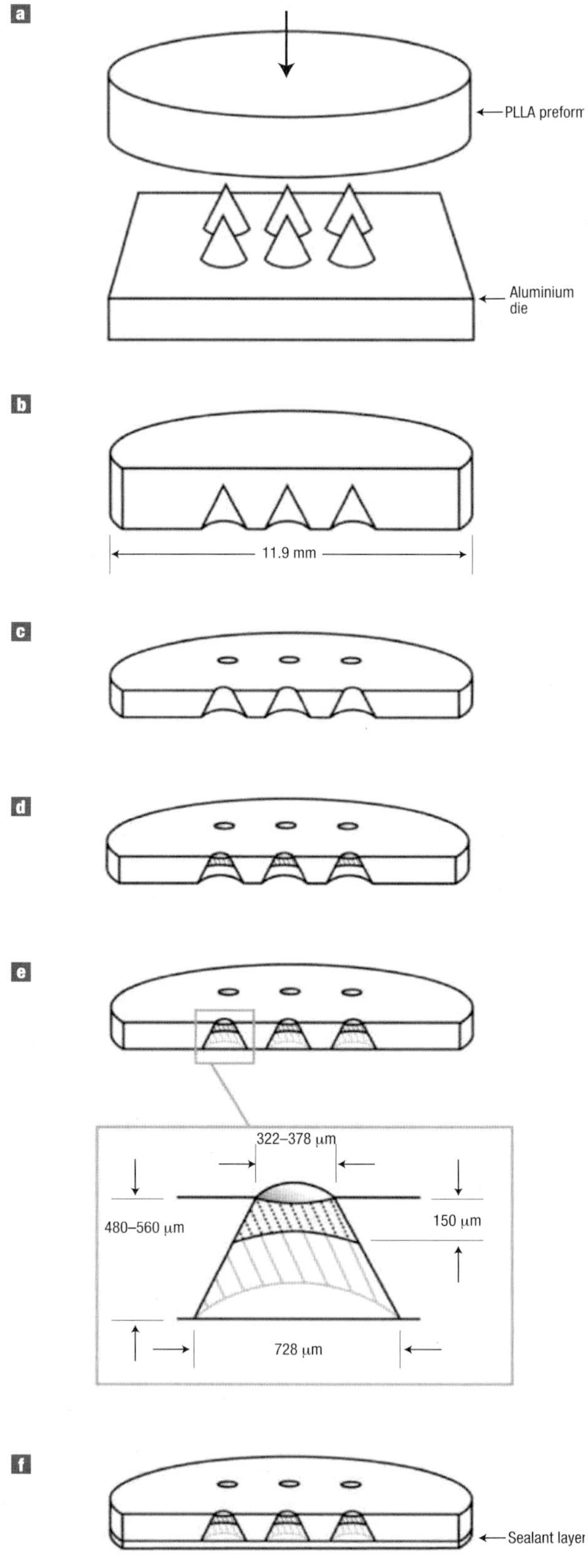

Figure 2 Diagram of fabrication process for polymeric microchip device. **a**, The poly(L-lactic acid) (PLLA) preform is compression-moulded on an aluminum die plate at elevated temperature to form reservoirs in the device substrate. The initial compression-moulding step at room temperature to form a partly dense, disk-shaped preform is not shown. **b**, Diagram of device substrate after high-temperature compression-moulding, showing conical reservoirs. **c**, Diagram of device substrate after polishing to truncate the conical reservoirs. **d**, Device substrate after the formation of reservoir membranes by microinjection. **e**, Device after loading the reservoirs with the chemicals to be released by microinjection, and enlargement of reservoir showing typical dimensions of reservoir and membrane. **f**, Device after sealing and completion of the fabrication process.

into the reservoirs behind the membranes) might enable greater flexibility in adapting this system for a desired application than is currently achievable with existing methods.

Proof-of-principle studies were conducted with prototype devices fabricated from poly(L-lactic acid) (PLLA) that used poly(lactic-*co*-glycolic acid) (PLGA) reservoir membranes of various molecular masses to control the time at which various chemicals (dextran, heparin and human growth hormone (HGH)) were released from the devices. PLGA copolymers with a 50:50 ratio of lactic acid to glycolic acid were selected as the materials from which to fabricate the reservoir membranes. This ratio of monomers provides degradation times suitable for an *in vitro* study of a few weeks to months[38,39]. A range of relative molecular masses (M_r values) was chosen (4,400 (PLGA4.4), 11,000 (PLGA11), 28,000 (PLGA28) and 64,000 (PLGA64)), on the basis of results from previous studies in our laboratory, to provide membrane degradation rates and chemical releases from the devices over a period of a few months.

We developed a new approach to fabricating devices and loading them with chemicals by means of two compression-moulding steps and two microinjection steps, as shown in Fig. 2. The devices were about 11.9 mm in diameter and between 480 and 560 µm thick. PLLA was chosen for the main body of the devices because of its long biodegradation time and its well-characterized biocompatibility and biodegradation[39–47]. The long degradation time enables the chemicals to be completely released from the device before the main body of the microchip degrades, minimizing the possibility of dose dumping. Prototype devices had 36 reservoirs of a truncated conical shape, and each reservoir had a volume of 120–130 nl.

The objective of our initial release experiments was to determine whether multiple pulses of molecules could be reproducibly obtained from these microchip devices *in vitro*. Figure 3 shows the cumulative amounts of ^{14}C-methylated dextran, ^{3}H-heparin or ^{125}I-HGH (preliminary results) released over time from different devices. Each device had a total of four reservoirs that were loaded equally with one of the radiolabelled molecules. Each reservoir was sealed with a membrane that was fabricated from one of the PLGA4.4, PLGA11, PLGA28 and PLGA64 copolymers. The results presented in Fig. 3 show four clear pulses of drug from the devices as each of the reservoir membranes sequentially degraded and opened. Preliminary results from other studies conducted in our laboratory suggest that water uptake and swelling of the polymer membranes provides a driving force for the opening or rupturing of the membranes and subsequent chemical release, which is counterbalanced by the mechanical strength of the polymer. Polymers with a higher molecular mass seem to retain their mechanical strength for longer periods, leading to chemical release at later times. The total cumulative amounts of initial loading recovered, as well as the average release times and standard deviations from reservoirs with different membranes, are summarized in Table 1. The pulses were almost exactly equal in magnitude. The larger cumulative amount and greater noise in the data for the device loaded

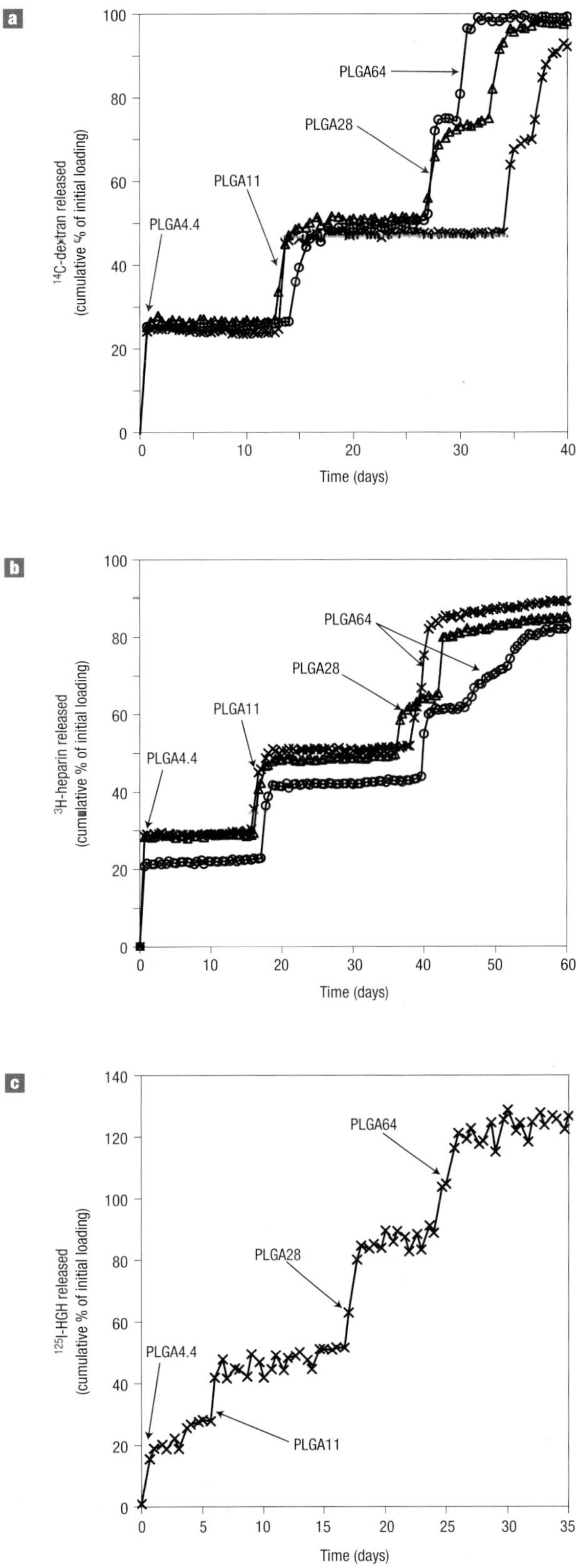

Figure 3 Cumulative percentage of initial loading released from microchip devices *in vitro*. Each symbol (triangle, cross or circle) in a panel represents data collected for a different device. Each device had a total of four reservoirs that were loaded with radiolabelled molecules and sealed with a membrane, and each reservoir had a different membrane that was fabricated from one of the PLGA4.4, PLGA11, PLGA28 or PLGA64 copolymers. The release times of the chemicals from the reservoirs increased as the molecular mass of the reservoir membrane polymers was increased, as shown by the arrows indicating the opening of each type of membrane on the devices. Experiments were conducted in saline solution at 28–33 °C *in vitro*. Devices represented by triangles are included for comparison[49]. **a**, Cumulative release results for devices loaded with ^{14}C-dextran. **b**, Cumulative release results for devices loaded with ^{3}H-heparin. **c**, Cumulative release results for a device loaded with ^{125}I-HGH.

Table 1 Summary of release times and amounts of chemicals recovered from microchips

Device loading	Temperature (°C)	Membrane material				Percentage of initial loading recovered
		PLGA4.4	PLGA11	PLGA28	PLGA64	
^{14}C-dextran	28–33	1 ± 0	14 ± 1	30 ± 4	33 ± 3	92–99 at 40 days
^{3}H-heparin	28–33	1 ± 0	17 ± 1	38 ± 2	43 ± 3	83–89 at 60 days
^{125}I-HGH	28–33	1	6	17	25	127 at 35 days
^{14}C-dextran and ^{3}H-heparin	28–33		15		37	92 at 54 days
		1		32		79 at 54 days
^{3}H-heparin	25	0.7 ± 0.6	44 ± 4	135 ± 4		68–77 at 143 days

Results are average release times ± standard deviations (in days) from reservoirs with different membranes, and cumulative percentages of initial loading released from microchips loaded with various molecules. The release conditions were *in vitro* in all cases.

with ^{125}I-HGH are due to the greater background noise present for ^{125}I, which is a gamma emitter.

Further studies were performed to demonstrate the delivery of more than one chemical from single devices. Four devices were each loaded with both ^{14}C-dextran and ^{3}H-heparin. A total of four reservoirs were loaded with the chemicals and sealed with reservoir membranes.

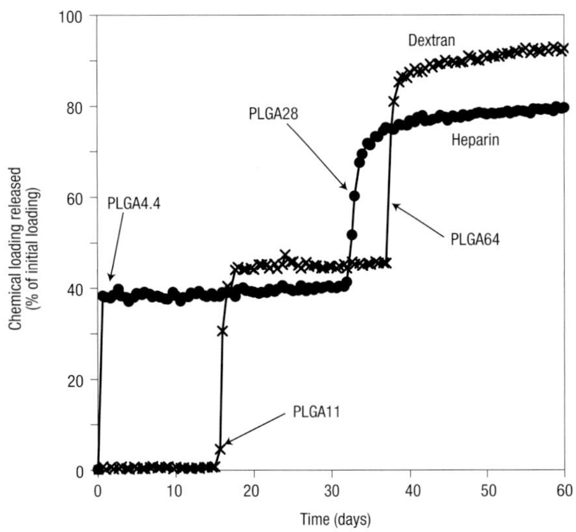

Figure 4 Cumulative percentage of initial loading released from microchip device *in vitro*. Release results are shown for a representative device that was loaded with both ^{14}C-dextran (crosses) and ^{3}H-heparin (circles). Similarly to the results obtained for devices that were loaded with only one chemical, the release times of the chemicals from the reservoirs increased as the molecular mass of the reservoir membrane polymers was increased, as shown by the arrows indicating the opening of each type of membrane on the devices. Experiments were conducted in saline solution at 28–33 °C *in vitro*.

Reservoirs having PLGA4.4 and PLGA28 membranes were equally loaded with ^{3}H-heparin, whereas reservoirs having PLGA11 and PLGA64 membranes were loaded equally with ^{14}C-dextran. Figure 4 shows the percentage of cumulative loading of ^{14}C-dextran and ^{3}H-heparin released from a representative device. The release times and cumulative loading recovered for this device are listed in Table 1. Each chemical was released in two pulses that were almost exactly equal. The mathematical method used by the scintillation counter to separate the counts due to each isotope, as well as the calibration based on a quench curve, might account for the fact that the measured amount of each chemical is less than 100%. The release times observed for the membranes on this device agree quite well with those obtained for the devices that were loaded with single chemicals discussed above, with the exception of the reservoir loaded with ^{3}H-heparin and having a PLGA28 membrane, which opened slightly earlier than expected.

It is well known that molecular stability is a critical issue for the clinical utility of drug delivery devices. The optimization of a drug formulation, with the addition of various stabilizers, is often required. The bioactivity of heparin delivered from devices during an *in vitro* release study was therefore characterized. The cumulative amount of ^{3}H-heparin detected from three devices over the course of 145 days is shown in Fig. 5. Each device had one reservoir sealed with a membrane made from PLGA4.4, one reservoir with a PLGA11 membrane, and one with a PLGA64 membrane. Average release times and amounts of heparin recovered for each device are summarized in Table 1. The release times observed for this set of devices were slightly delayed compared with those seen at higher temperatures. The lower temperature at which these later experiments were performed (25 °C in comparison with 28–33 °C for the earlier studies) most probably slowed the degradation rate of the membrane polymers, although no significant broadening of the range of release times was observed.

The results of heparin anti-Factor XA activity assays are summarized in Table 2. The assays indicated that on average, 102 ± 14% (standard error of the mean) of the heparin that was present after the PLGA11 membranes had opened remained active. Even after 136–142 days, 96 ± 12% of the heparin present in the saline solution after the PLGA64 membranes had opened was still biologically active. Other studies on the long-term stability of heparin showed that at pH 7.0 and 100 °C, heparin remained stable over about 500 hours (21 days), and retained 80–90% of its activity as measured by anti-Factor XA and

Table 2 Measured activity of heparin released from polymeric microchip devices in vitro

Device	Reservoirs with PLGA11 membranes			Reservoirs with PLGA64 membranes		
	Release day	Assay day	Heparin activity (%)	Release day(s)	Assay day	Heparin activity (%)
1	43	43	129	132–136	136	110
2	40–41	41	85	139–143	142	73
3	48–52	49	91	133–136	136	106

The heparin activity (measured as the percentage of the total heparin present in the release medium at a given time that remained biologically active) was determined by anti-Factor XA inhibition.

anti-Factor IIA amidolytic methods. However, the measured activity showed an extremely large decrease at 1,000 hours to only about 6% of the initial activity[48]. The results in Table 2 show that the heparin released from the polymeric microchips did not undergo significant degradation and was still able to inhibit Factor XA after 142 days *in vitro* at 25 °C.

In summary, we report here the demonstration of multiple pulses of chemicals *in vitro* from a novel polymeric microchip device that is resorbable. Preliminary results obtained *in vivo* have shown pulsatile release analogous to that demonstrated *in vitro* (see Supplementary Information). This device brings together many of the appealing characteristics of current drug delivery systems, such as biodegradability and the ability to achieve pulsatile release, while increasing the number of chemicals and/or doses that could be achieved from a single polymeric device without the application of a stimulus to trigger drug release. Separation of the reservoir membranes from the chemicals within the reservoirs potentially allows greater independence of the formulations that control drug release and drug stability. This independence of the two formulations might in turn provide greater flexibility in device design than is currently available with other controlled release drug delivery systems.

These devices offer a method for molecular or pharmaceutical delivery, with the opportunity to tailor the release characteristics for specific applications by simply varying the attributes of the device (size, polymer), reservoirs (number, volume) and membranes (thickness, molecular mass, material, copolymer ratio). Superposition of the release profiles from individual reservoirs contained on the device could enable the construction of complex release profiles that have the potential to provide both pulsatile and continuous release of different chemicals. This device could be implanted for clinical applications and could enable the patterned delivery of multiple potent drugs. A similar strategy could be employed for an oral delivery device fabricated from different materials.

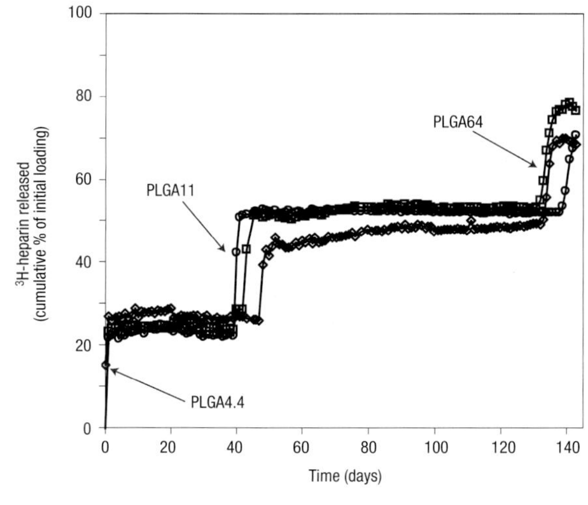

Figure 5 Cumulative percentage of initial ^{3}H-heparin loading released from microchip devices *in vitro*. Each symbol (circle, square or diamond) represents data collected for a different device. We again observed that the release times of the chemicals from the reservoirs increased as the molecular mass of the reservoir membrane polymers was increased, as shown by the arrows indicating the opening of each type of membrane on the devices, and that there was relatively small variation in the range of release times for reservoirs having a given type of membrane. Experiments were conducted in saline solution at 25 °C *in vitro*.

METHODS

DEVICE FABRICATION

Poly(L-lactic acid) (PLLA, relative molecular mass 194,000 (M_r 194K), T_m = 176 °C, Medisorb 100 L; Alkermes) microchips were fabricated by a two-step process on a Carver Lab Press, model C. PLLA powder was compression-moulded at room temperature in a cylindrical die to form a preform, which was subsequently moulded at high temperature on an aluminum die plate having conical protrusions to form reservoirs in the substrate. Substrate material was removed from one side of the device with a rotary polisher to truncate the small end of the reservoirs, and device thicknesses after polishing ranged from 480 to 560 μm.

Membranes were prepared in the device reservoirs by the microinjection of 12% v/v solutions of PLGA in 1,1,1,3,3,3-hexafluoro-2-propanol (Sigma-Aldrich). This solvent was selected for its ability to dissolve the entire family of poly(glycolic acid) (PGA), poly(lactic acid) (PLA) and PLGA homopolymers and copolymers. Although the amount of residual solvent is expected to be low in the reservoir membranes (see Supplementary Information), other cosolvents such as chloroform or dichloromethane could also be used for PLGA or PLA membranes. Poly(D,L-lactic-*co*-glycolic acid) polymer powders with M_r values of 4.4K, 11K, 28K and 64K (Medisorb® 5050 DL 1A, 2A, 3A and 4A, respectively; Alkermes) were used to fabricate membranes. Microinjected volumes ranged from 116.4 to 195.7 nl, for predicted final membrane thicknesses of 148–156 μm on devices loaded with ^{14}C-dextran, ^{3}H-heparin and ^{125}I-HGH.

Devices were suspended on glass slides and dried in a vacuum oven at 74 °C and 6.8 kPa vacuum for 48 hours after microinjection of the membrane solutions. The membranes did not touch the bottom of the oven during drying. Membranes were inspected after drying for defects at ×80 magnification with a PZMT Trinocular stereomicroscope. Selected reservoirs having defect-free membranes were loaded with a solution of the chemical or chemicals to be released. The chemical solutions were dried for 24 hours at room temperature and pressure. Devices were sealed opposite the membranes with Ideal 9144 Masking Tape (American Biltrite, Lowell, Massachusetts).

IN VITRO RELEASE STUDIES

Devices were microinjected with solutions of ^{14}C-methylated dextran (M_r 70K), unlabelled bovine heparin sodium salt (both from Sigma-Aldrich), ^{3}H-heparin sodium salt or ^{125}I-HGH (both from PerkinElmer Life Sciences) in deionized water for release studies. Devices were affixed to glass microscope slides for stability and put in screw-cap jars containing 20 ml stirred 1× PBS at 28–33 °C. Samples of the release medium were taken twice daily and analysed on a scintillation counter for radiolabel detection.

ARTICLES

Devices that were prepared for *in vitro* heparin release and bioactivity studies each had three reservoirs loaded with a combination of ^{3}H-heparin sodium salt (0.2918 mCi mg^{-1}; PerkinElmer Life Sciences) and unlabelled bovine heparin sodium salt. Devices were affixed to the bottom of six-well polystyrene tissue culture dishes and agitated in 4 ml PBS on an orbital shaker at 60 r.p.m. and 25 °C. Samples of the release medium were taken once daily and analysed on a scintillation counter for radiolabel detection. The amount of therapeutic (biologically active) heparin was quantified with a Sigma Diagnostics Accucolor heparin kit and a UV/visible spectrophotometer (Dynatech MR5000 or Molecular Devices Spectramax 384 Plus) at 405 nm after a reservoir release was detected.
See Supplementary Information for more complete experimental procedures.

Received 27 May 2003; accepted 15 September 2003; published 19 October 2003.

References

1. Bakken, E. E. & Heruth, K. Temporal control of drugs: an engineering perspective. *Ann. N.Y. Acad. Sci.* **618**, 422–427 (1991).
2. Langer, R. New methods of drug delivery. *Science* **249**, 1527–1533 (1990).
3. Matthews, D. R., Lang, D. A., Burnett, M. A. & Turner, R. C. Control of pulsatile insulin secretion in man. *Diabetologia* **24**, 231–237 (1983).
4. Santini, J. T. Jr, Richards, A. C., Scheidt, R., Cima, M. J. & Langer, R. Microchips as controlled drug-delivery devices. *Angew. Chem. Int. Edn* **39**, 2396–2407 (2000).
5. Wu, B. M. *et al.* Solid free-from fabrication of drug delivery devices. *J. Control. Release* **40**, 77–87 (1996).
6. Katstra, W. E. *et al.* Oral dosage forms fabricated by Three Dimensional Printing™. *J. Control. Release* **66**, 1–9 (2000).
7. Rowe, C. W. *et al.* Multimechanism oral dosage forms fabricated by Three Dimensional Printing™. *J. Control. Release* **66**, 11–17 (2000).
8. Santoro, N. Efficacy and safety of intravenous pulsatile gonadotropin-releasing hormone: Lutrepulse for injection. *Am. J. Obstet. Gynecol.* **163**, 1759–1764 (1990).
9. Creasy, G. W. & Jaffe, M. E. Pulsatile delivery systems. *Ann. N.Y. Acad. Sci.* **618**, 548–557 (1991).
10. Creasy, G. W. & Jaffe, M. E. Endocrine/reproductive pulsatile delivery systems. *Adv. Drug Deliv. Rev.* **6**, 51–56 (1991).
11. Sampath, P. & Brem, H. Implantable slow-release chemotherapeutic polymers for the treatment of malignant brain tumors. *J. Moffitt Cancer Center* **5**, 130–137 (1998).
12. Murray, L. (ed.) *Physicians' Desk Reference* 57th edn (Thomson PDR, Montvale, New Jersey, 2003).
13. Brown, L. & Langer, R. Transdermal delivery of drugs. *Annu. Rev. Med.* **39**, 221–229 (1988).
14. Sershen, S. & West, J. Implantable, polymeric systems for modulated drug delivery. *Adv. Drug Deliv. Rev.* **54**, 1225–1235 (2002).
15. Siegel, R. A., Falamarzian, M., Firestone, B. A. & Moxley, B. C. pH-controlled release from hydrophobic/polyelectrolyte copolymer hydrogels. *J. Control. Release* **8**, 179–182 (1988).
16. Kim, J. H., Kim, J. Y., Lee, Y. M. & Kim, K. Y. J. Controlled release of riboflavin and insulin through crosslinked poly(vinyl alcohol)/chitosan blend membrane. *Appl. Polym. Sci.* **44**, 1923–1828 (1992).
17. Gutowska, A. *et al.* Squeezing hydrogels for controlled oral drug delivery. *J. Control. Release* **48**, 141–148 (1997).
18. Bae, Y. H., Okano, T., Hsu, R. & Kim, S. W. Thermo-sensitive polymers as on-off switches for drug release. *Makromol. Chem., Rapid Commun.* **8**, 481–485 (1987).
19. Hoffman, A. S., Afrassiabi, A. & Dong, L. C. Thermally reversible hydrogels. II. delivery and selective removal of substances from aqueous solutions. *J. Control. Release* **4**, 213–222 (1986).
20. Okano, T., Bae, Y. H. & Kim, S. W. in *Pulsed and Self-Regulated Drug Delivery* (ed. Kost, J.) 17–45 (CRC Press, Boca Raton, Florida, 1990).
21. Kost, J. & Langer, R. in *Pulsed and Self-Regulated Drug Delivery* (ed. Kost, J.) 3–16 (CRC Press, Boca Raton, Florida, 1990).
22. Kost, J., Leong, K. & Langer, R. Ultrasound-enhanced polymer degradation and release of incorporated substances. *Proc. Natl Acad. Sci. USA* **86**, 7663–7666 (1989).
23. Fischel-Ghodsian, F., Brown, L., Mathiowitz, E., Brandenburg, D. & Langer, R. Enzymatically controlled drug delivery. *Proc. Natl Acad. Sci. USA* **85**, 2403–2406 (1988).
24. Mathiowitz, E. & Cohen, M. D. Polyamide microcapsules for controlled release. V. photochemical release. *J. Membr. Sci.* **40**, 67–86 (1989).
25. Kwon, I. C., Bae, Y. H. & Kim, S. W. Electrically erodible polymer gel for controlled release of drugs. *Nature* **354**, 291–293 (1991).
26. Bae, Y. H., Kwon, I. C. & Kim, S. W. in *Polymeric Drugs and Drug Administration* (ed. Ottenbrite, R. M.) 98–110 (American Chemical Society, Washington, DC, 1994).
27. Miller, L. L. Electrochemically controlled release of drug ions from conducting polymers. *Mol. Cryst. Liq. Cryst.* **160**, 297–301 (1998).
28. Hepel, M. & Fijalek, Z. in *Polymeric Drugs and Drug Administration* (ed. Ottenbrite, R. M.) 79–97 (American Chemical Society, Washington, DC, 1994).
29. Schwendeman, S. P., Amidon, G. L. & Levy, R. J. Determinants of the modulated release of antiarrhythmic drugs by iontophoresis through polymer membranes. *Macromolecules* **26**, 2264–2272 (1993).
30. Edelman, E. R., Kost, J., Bobeck, H. & Langer, R. Regulation of drug release from polymer matrices by oscillating magnetic fields. *J. Biomed. Mater. Res.* **19**, 67–83 (1985).
31. Suzuki, Y. *et al.* A new drug delivery system with controlled release of antibiotic only in the presence of infection. *J. Biomed. Mater. Res.* **42**, 112–116 (1998).
32. Santini, J. T. Jr, Cima, M. J. & Langer, R. A controlled-release microchip. *Nature* **397**, 335–338 (1999).
33. Jiang, H. L. & Zhu, K. J. Pulsatile protein release from a laminated device comprising of polyanhydrides and pH-sensitive complexes. *Int. J. Pharm.* **194**, 51–60 (2000).
34. Qiu, L. Y. & Zhu, K. J. Design of a core-shelled polymer cylinder for potential programmable drug delivery. *Int. J. Pharm.* **219**, 151–160 (2001).
35. Göpferich, A. Bioerodible implants with programmable drug release. *J. Control. Release* **44**, 271–281 (1997).
36. Kikuchi, A., Kawabuchi, M., Sugihara, M., Sakurai, Y. & Okano, T. Pulsed dextran release from calcium-alginate beads. *J. Control. Release* **47**, 21–29 (1997).
37. Misra, G. P. & Siegel, R. A. New mode of drug delivery: long term autonomous rhythmic hormone release across a hydrogel membrane. *J. Control. Release* **81**, 1–6 (2002).
38. Lu, L., Garcia, C. A. & Mikos, A. In vitro degradation of thin poly(DL-lactic-co-glycolic acid) films. *J. Biomed. Mater. Res.* **46**, 236–244 (1999).
39. Anderson, J. M. & Shive, M. S. Biodegradation and biocompatibility of PLA and PLGA microspheres. *Adv. Drug Deliv. Rev.* **28**, 5–24 (1997).
40. Grizzi, I., Garreau, H., Li, S. & Vert, M. Hydrolytic degradation of devices based on poly(DL-lactic acid) size-dependence. *Biomaterials* **16**, 305–311 (1995).
41. Tamada, J. A. & Langer, R. Erosion kinetics of hydrolytically degradable polymers. *Proc. Natl Acad. Sci. USA* **90**, 552–556 (1993).
42. Gilding, D. K. & Reed, A. M. Biodegradable polymers for use in surgery—polyglycolic/poly(lactic acid) homo- and copolymers: 1. *Polymer* **20**, 1459–1464 (1979).
43. Kulkarni, R. K., Moore, E. G., Hegyeli, A. F. & Leonard, F. Biodegradable poly(lactic acid) polymers. *J. Biomed. Mater. Res.* **5**, 169–181 (1971).
44. Göpferich, A. Mechanisms of polymer degradation and erosion. *Biomaterials* **17**, 103–114 (1996).
45. Miller, R. A., Brady, J. M. & Cutright, D. E. Degradation rates of oral resorbable implants (polylactates and polyglycolates): rate modification with changes in PLA/PGA copolymer ratios. *J. Biomed. Mater. Res.* **11**, 711–719 (1977).
46. Vert, M., Mauduit, J. & Li, S. Biodegradation of PLA/GA polymers: increasing complexity. *Biomaterials* **15**, 1209–1213 (1994).
47. Vert, M., Li, S. M. & Garreau, H. Attempts to map the structure and degradation characteristics of aliphatic polyesters derived from lactic and glycolic acids. *J. Biomater. Sci. Polym. Edn* **6**, 639–649 (1994).
48. Jandik, K. A., Kruep, D., Cartier, M. & Linhardt, R. J. Accelerated stability studies of heparin. *J. Pharm. Sci.* **85**, 45–51 (1996).
49. Shawgo, R. S., Grayson, A. C. R., Li, Y. & Cima, M. J. BioMEMS for drug delivery. *Curr. Opin. Solid State Mater. Sci.* **6**, 329–334 (2002).

Acknowledgements

This work was supported by the National Institutes of Health Bioengineering Research Partnership grant R24-AI47739-02. A.C.R.G. thanks the American Association of University Women and the National Science Foundation for fellowship funding. I.S.C. thanks the Korea Science and Engineering Foundation and the Brain Korea 21 project.
Correspondence and requests for materials should be addressed to R.L.
Supplementary Information accompanies the paper on www.nature.com/naturematerials

Competing financial interests

The authors declare competing financial interests: details accompany the paper on www.nature.com/naturematerials

Targeted nanoparticle-aptamer bioconjugates for cancer chemotherapy *in vivo*

Omid C. Farokhzad*†‡§, Jianjun Cheng‡¶‖, Benjamin A. Teply*‡¶, Ines Sherifi*¶, Sangyong Jon, Philip W. Kantoff††, Jerome P. Richie‡‡, and Robert Langer†§¶**

Departments of *Anesthesiology and ‡‡Urology, Brigham and Women's Hospital, Harvard Medical School, 75 Francis Street, Boston, MA 02115; †Massachusetts Institute of Technology–Harvard Center for Cancer Nanotechnology Excellence, 77 Massachusetts Avenue, Cambridge, MA 02139; ¶Department of Chemical Engineering, Massachusetts Institute of Technology, 77 Massachusetts Avenue, Cambridge, MA 02139; **Department of Life Science, Gwangju Institute of Science and Technology, Oryoung-dong, Buk-gu, Gwangju 500-712, Korea; and ††Lank Center for Genitourinary Oncology, Dana–Farber Cancer Institute, Harvard Medical School, 44 Binney Street, Dana 1230, Boston, MA 02115

Contributed by Robert Langer, March 3, 2006

Targeted uptake of therapeutic nanoparticles in a cell-, tissue-, or disease-specific manner represents a potentially powerful technology. Using prostate cancer as a model, we report docetaxel (Dtxl)-encapsulated nanoparticles formulated with biocompatible and biodegradable poly(D,L-lactic-*co*-glycolic acid)-*block*-poly(ethylene glycol) (PLGA-*b*-PEG) copolymer and surface functionalized with the A10 2′-fluoropyrimidine RNA aptamers that recognize the extracellular domain of the prostate-specific membrane antigen (PSMA), a well characterized antigen expressed on the surface of prostate cancer cells. These Dtxl-encapsulated nanoparticle-aptamer bioconjugates (Dtxl-NP-Apt) bind to the PSMA protein expressed on the surface of LNCaP prostate epithelial cells and get taken up by these cells resulting in significantly enhanced *in vitro* cellular toxicity as compared with nontargeted nanoparticles that lack the PSMA aptamer (Dtxl-NP) ($P < 0.0004$). The Dtxl-NP-Apt bioconjugates also exhibit remarkable efficacy and reduced toxicity as measured by mean body weight loss (BWL) *in vivo* [body weight loss of 7.7 ± 4% vs. 18 ± 5% for Dtxl-NP-Apt vs. Dtxl-NP at nadir, respectively (mean ± SD); $n = 7$]. After a single intratumoral injection of Dtxl-NP-Apt bioconjugates, complete tumor reduction was observed in five of seven LNCaP xenograft nude mice (initial tumor volume of ≈300 mm^3), and 100% of these animals survived our 109-day study. In contrast, two of seven mice in the Dtxl-NP group had complete tumor reduction with 109-day survivability of only 57%. Dtxl alone had a survivability of only 14%. Saline and nanoparticles without drug were similarly nonefficacious. This report demonstrates the potential utility of nanoparticle-aptamer bioconjugates for a therapeutic application.

docetaxel | prostate cancer | targeted delivery | prostate-specific membrane antigen | poly(D,L-lactic-*co*-glycolic acid) (PLGA)

There has been a substantial interest in developing localized therapeutic options for treatment of early-stage cancer that have reduced toxicity. For example, transperineal ultrasound guided prostate brachytherapy with radioactive ^{125}I or ^{103}Pd seeds represents an attractive therapeutic option for low- and intermediate-risk prostate cancer (PCa) patients (1), and its use has increased from 4% in 1993–1995 to 22% in 1999–2001 (2). Despite the rapid adoption of this therapeutic modality, complications still occur, including erectile dysfunction (33–53%) (3), urinary retention (15–32%) (4, 5), and severe radiation-induced bowel injury (1%) (6). Moreover, brachytherapy may fail to eradicate localized prostate cancer, resulting in local recurrence (7).

Nanotechnology approaches where a constant dose of chemotherapy is delivered directly to cancer cells over an extended period may result in alternative or complementary therapeutic options for patients with early-stage cancer. The challenge lies in the design of nanoparticles (NPs) that are specifically and differentially taken up by the targeted cells and release their payload over an extended period to achieve a clinical response (8–10). Using PCa as a model cancer and the following design criteria, we aimed to develop drug-encapsulated NPs for PCa targeting. First, we aimed to develop NPs using biodegradable and biocompatible components that were previously approved by the Food and Drug Administration (FDA) for a clinical use. We believe that the use of FDA-approved components may facilitate the translation of these vehicles into clinical practice. We chose poly(D,L-lactic-*co*-glycolic acid) (PLGA) as the controlled release polymer system because its safety in clinical use has been well established, initially as a biomaterial in Vicryl sutures (11) and later as excipients for sustained release of parenteral drugs (12, 13). Second, we aimed to develop NPs that could be surface functionalized with nucleic acid ligands [aptamers (Apts)] for targeted delivery and uptake in a cell-specific manner. Aptamers are DNA or RNA oligonucleotides that, through intramolecular interactions, fold into unique tertiary conformations capable of binding to target antigens with high affinity and specificity, analogous to antibodies (14, 15). We chose Apts as targeting molecules because this class of ligands is nonimmunogenic and exhibits remarkable stability in a wide range of pH (≈4–9), temperature, and organic solvents without loss of activity (16, 17). Furthermore, Apt synthesis does not rely on biological systems and is an entirely chemical process that can decrease batch-to-batch variability when production is scaled up. These characteristics are in contrast to antibodies that may be immunogenic or more labile, and their large-scale biological production is prone to significant batch-to-batch variability (18, 19). Third, we aimed to develop NPs that resist uptake by tissue macrophages and by nontargeted cells, thus increasing their residence time at the site of administration. We chose to develop poly(ethylene glycol) (PEG)-functionalized NPs because we had previously shown that pegylated polymeric NPs are considerably more effective against systemic clearance than similar particles without PEG (20, 21). PEG has also been used to improve the pharmacokinetic properties of liposomes (22), macromolecules (23), and small molecule drugs (24). Fourth, we aimed to develop NPs that demonstrate differential cytotoxicity against PCa *in vitro* and *in vivo* using a chemotherapeutic agent currently in clinical use for the management of PCa. Docetaxel (Dtxl), when used systemically, can prolong the survival of patients with hormone-resistant PCa (25, 26). We postulated that controlled release of Dtxl targeted to PCa cells may result in enhanced cytotoxicity and antitumor efficacy, making it a potential therapeutic modality for the management of localized prostate cancer. The combination of the above

Conflict of interest statement: No conflicts declared.

Abbreviations: PCa, prostate cancer; Dtxl, docetaxel; NP, nanoparticle; Apt, aptamer; PLGA, poly(D,L-lactic-*co*-glycolic acid); PEG, poly(ethylene glycol); PSMA, prostate-specific membrane antigen; H&E, hematoxylin/eosin; BWL, body weight loss.

‡O.C.F., J.C., and B.A.T. contributed equally to this work.

§To whom correspondence may be addressed. E-mail: ofarokhzad@partners.org or rlanger@mit.edu.

‖Present address: Department of Materials Science and Engineering, University of Illinois at Urbana–Champaign, 1304 West Green Street, Urbana, IL 61801.

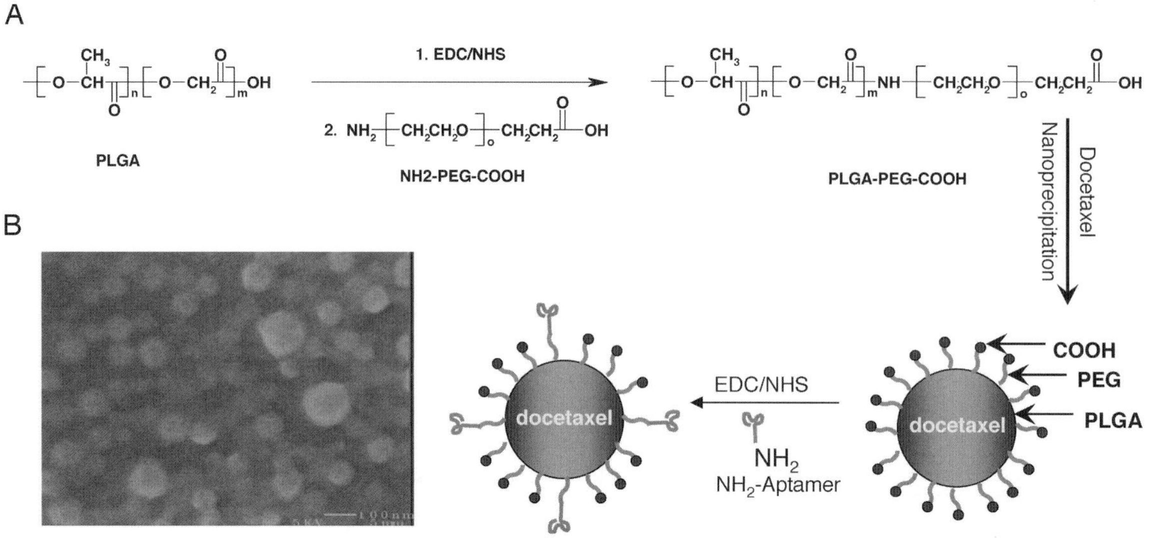

Fig. 1. Development of Dtxl-encapsulated pegylated PLGA NP-Apt bioconjugates. (*A*) Schematic representation of the synthesis of PLGA-PEG-COOH copolymer and strategy of encapsulation of Dtxl. We developed Dtxl-encapsulated, pegylated NPs by the nanoprecipitation method. These particles have a negative surface charge attributable to the carboxylic acid on the terminal end of the PEG. The NPs were conjugated to amine-functionalized A10 PSMA Apt by carbodiimide coupling chemistry. (*B*) Representative scanning electron microscopy image of resulting Dtxl-encapsulated NPs is shown. EDC, 1-ethyl-3-(3-dimethylaminopropyl)-carbodiimide; NHS, *N*-hydroxysuccinimide.

design criteria may facilitate the translation of therapeutically effective NP-Apt bioconjugates into clinical practice.

We had previously encapsulated rhodamine-labeled dextran (as a model drug) within NPs formulated with poly(D,L-lactic acid) (PLA)-*b*-PEG block copolymer and surface functionalized these NPs with nuclease-stabilized A10 2′-fluoropyrimidine RNA Apts (27) that recognize the extracellular domain of the prostate-specific membrane antigen (PSMA) (28–31). PSMA is a well characterized antigen expressed on the surface of PCa cells that participates in membrane recycling and becomes internalized through ligand-induced endocytosis (32, 33). Our data demonstrated that these fluorescently labeled, targeted NP-Apt bioconjugates differentially bound and got taken up by LNCaP prostate epithelial cells, which express the PSMA protein efficiently and with high specificity. No binding or uptake was detected in PC3 prostate epithelial cells, which do not express the PSMA protein (34, 35).

Herein, we developed Dtxl-encapsulated, pegylated PLGA NP-Apt bioconjugates that bind to the PSMA protein on the surface of PCa cells. We assessed the efficacy of these bioconjugates *in vitro* and *in vivo* using a LNCaP xenograft nude mouse model of PCa. We now report an example of NP-Apt bioconjugates that exhibit significant anticancer efficacy without the systemic toxicity that is common to chemotherapeutics.

Results and Discussion

Development of NPs. We used the nanoprecipitation method (36) to encapsulate Dtxl within PLGA-*b*-PEG block copolymer with a terminal carboxylic acid group (PLGA-PEG-COOH) and developed Dtxl-encapsulated, pegylated PLGA NPs [153.3 ± 13.9 nm (mean ± SD); $n = 10$] (Fig. 1*A*). The hydrophilic PEG group facilitates the presentation of the carboxylic acid on the NP surface. Additionally, the PEG group decreases nonspecific biofouling of particles *in vivo* (37) and minimizes the particle uptake by nontargeted cells, including their premature clearance by the mononuclear phagocytic system (21, 38). The presence of the carboxy-modified PEG on the NP surface also results in a negative surface charge [ζ potential, -42 ± 1 mV (mean ± SD); $n = 3$] that may decrease nonspecific interaction of the negatively charged Apts with the NP surface, thus preserving Apt conformation and binding characteristics (34). The surface of these NPs was functionalized with the A10 PSMA Apt (27) to allow differential uptake by targeted PCa cells. The surface morphology and size distribution of NPs were evaluated by scanning electron microscopy (SEM) (Fig. 1*B*). The functionalization of the NPs with Apt resulted in an ≈15-nm increase in particle size, presumably attributed to the presence of Apts on the NP surface.

In Vitro **Cellular Cytotoxicity Assays (MTT Assays).** We examined the *in vitro* differential cytotoxicity of Dtxl-encapsulated, pegylated PLGA NP-Apt bioconjugates (Dtxl-NP-Apt) vs. similar NPs lacking the A10 PSMA Apt (Dtxl-NP) using LNCaP cells, which express the PSMA protein. Because Dtxl is a hydrophobic and membrane-permeable drug and because PSMA has been shown

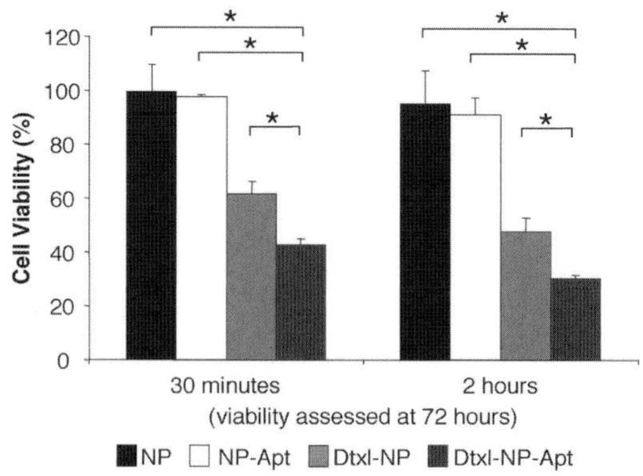

Fig. 2. MTT assay to determine the differential cytotoxicity of Dtxl-encapsulated NP-Apt bioconjugates (Dtxl-NP-Apt), Dtxl-encapsulated NPs lacking the A10 PSMA Apt (Dtxl-NP), control NP-Apt bioconjugates without Dtxl (NP-Apt), and control NPs without Dtxl (NP) after incubation with LNCaP prostate epithelial cells. NPs were incubated with cells for 30 min (*Left*) or 2 h (*Right*), and the cells were subsequently washed and incubated in media for a total of 72 h before assessing cell viability in each group. *, Significance by ANOVA at 95% confidence interval.

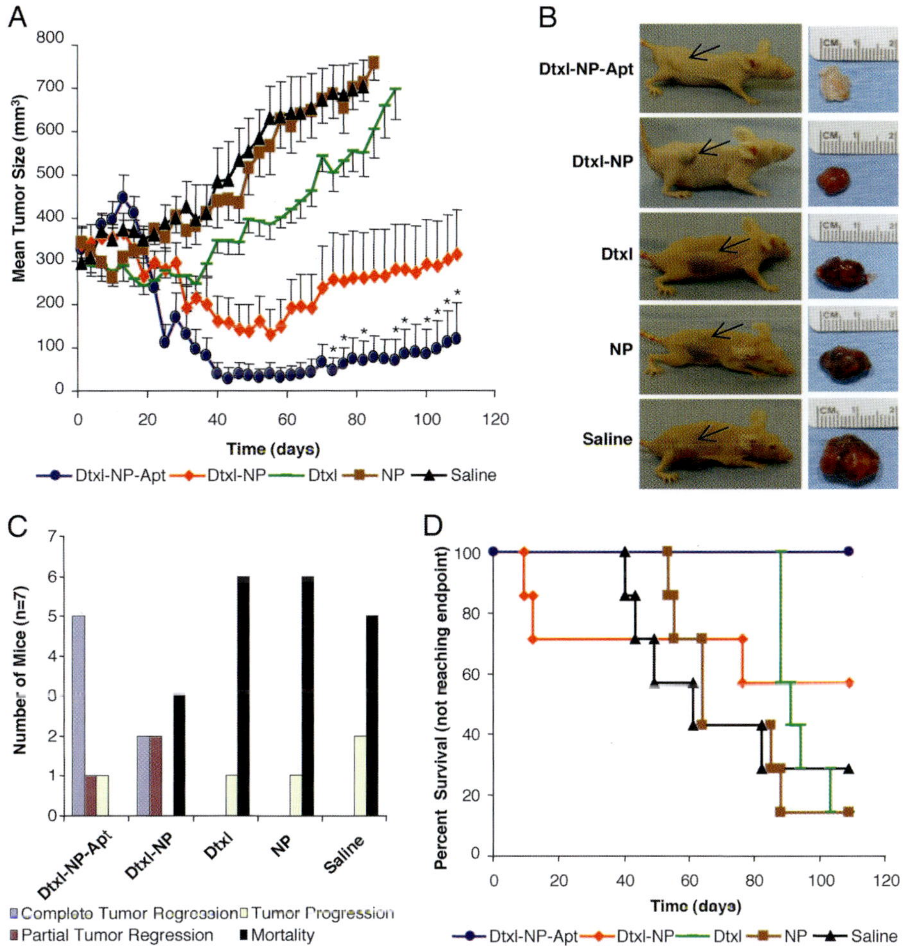

Fig. 3. Comparative efficacy study in LNCaP s.c. xenograft nude mouse model of PCa. (*A*) PCa was induced in mice by implanting LNCaP prostate epithelial cells s.c. in the flanks of nude mice and allowing the tumors to develop to appreciable size over 21 days (≈300 mm³). The comparative efficacy study of single intratumoral injection (day 0) of (*i*) saline (black); (*ii*) pegylated PLGA NP without drug (NP, brown); (*iii*) emulsified Dtxl (Dtxl, green), 40 mg/kg; (*iv*) Dtxl-encapsulated NPs (Dtxl-NP, red), 40 mg/kg; or (*v*) Dtxl-encapsulated NP-Apt bioconjugates (Dtxl-NP-Apt, blue), 40 mg/kg was evaluated over 109 days and demonstrated that targeted NPs are significantly more efficacious in tumor reduction as compared with other groups. Data represent mean ± SEM of seven mice per group. *, Data points for the Dtxl-NP-Apt group that were statistically significant compared with all other groups by ANOVA at 95% confidence interval. (*B*) Representative mouse at end point for each group is shown (*Left*) alongside images of excised tumors (*Right*). For the Dtxl-NP-Apt group, which achieved complete tumor regression, the scar tissue and underlying skin at the site of injection are shown. Black arrows point to the position of the implanted tumor on each mouse. (*C*) Plot of outcomes for each of the treatment groups divided into four categories: complete tumor regression (blue), incomplete tumor regression (red), tumor growth (yellow), and mortality (black). Two Dtxl-NP animals experienced >20% weight loss on days 9 and 12 after dosing and were euthanized. One mouse in each of the Dtxl and saline groups was euthanized late in the study for excessive weight loss likely attributable to large tumor load. (*D*) The Kaplan–Meier survival curve demonstrates that 100% of the Dtxl-NP-Apt group was alive on day 109, whereas the other groups had animals reaching our study end points and were euthanized on various days throughout the study period (end points defined as tumor load of 800 mm³ or BWL >20%).

to be constitutively endocytosed in LNCaP cells (33), we designed a study to best assay Dtxl toxicity due to uptake of the NPs and subsequent intracellular Dtxl release. We incubated Dtxl-NP and Dtxl-NP-Apt groups with LNCaP cells for 30 min or 2 h to allow time for specific particle uptake and further incubated in medium for a total of 72 h before the measurement of cell viability by MTT assay. This duration of incubation was consistent with our previous report that fluorescently labeled pegylated PLA NP-Apt bioconjugates that bound to the PSMA protein were efficiently taken up by LNCaP cells within 2 h (34). Our data suggest that Dtxl-NP-Apt bioconjugates are significantly more cytotoxic as compared with control Dtxl-NP lacking the A10 PSMA Apt [cell viability of 61 ± 5% vs. 42 ± 2% at 30 min and 48 ± 5% vs. 30 ± 1% at 2 h for Dtxl-NP vs. Dtxl-NP-Apt, respectively (mean ± SD); $P < 0.004$] (Fig. 2). The observed toxicity in the Dtxl-NP group may be due to nonspecific uptake of particles or alternatively may be attributed to release of Dtxl in the medium and subsequent cellular uptake of Dtxl during the 30-min or 2-h incubation before the wash step. To eliminate the possibility that the PSMA Apt or NPs formulated with the PLGA-PEG-COOH copolymer system were responsible for the cytotoxicity, we performed similar MTT assays with NPs and NP-Apt bioconjugates, which did not encapsulate Dtxl. These studies confirmed the lack of cellular cytotoxicity in the absence of Dtxl (Fig. 2).

***In Vivo* Efficacy and Toxicity Studies Using Mouse LNCaP Xenograft Model of PCa.** We next evaluated the efficacy of our bioconjugates using xenograft models of PCa developed by s.c. injection of LNCaP cells in the flank of BALB/c nude mice (39). After tumors had developed to ≈300 mm³, we performed comparative efficacy studies by dividing animals into five groups ($n = 7$) in a way to minimize weight and tumor size differences among the groups. Using the previously reported maximal tolerated dose (MTD) of 40 mg/kg for i.v. Dtxl (40) as a reference point, the following regimens were administered by a single intratumoral injection: (*i*) saline; (*ii*) pegylated PLGA NP without drug (NP); (*iii*) emulsified Dtxl, 40 mg/kg; (*iv*) Dtxl-NP, 40 mg/kg; or (*v*) Dtxl-NP-Apt, 40 mg/kg. The tumor size and body weight were then monitored for 109 days. The results showed that a single intratumoral administration of Dtxl-NP-Apt bioconjugates is significantly more efficacious in tumor reduction as compared with nontargeted NPs and controls (Fig. 3 *A* and *B*). One reason for this enhanced efficacy may be that the targeted particles are designed to bind to the PSMA proteins on PCa cells, thus possibly delaying clearance from the site of the tumor. If the targeted NPs are internalized after binding to PSMA proteins, as demonstrated (34), then subsequent intracellular delivery of Dtxl may contribute to the cytotoxicity and enhanced efficacy of this group in tumor reduction.

For each control group of saline, NP, and Dtxl, the treatment did

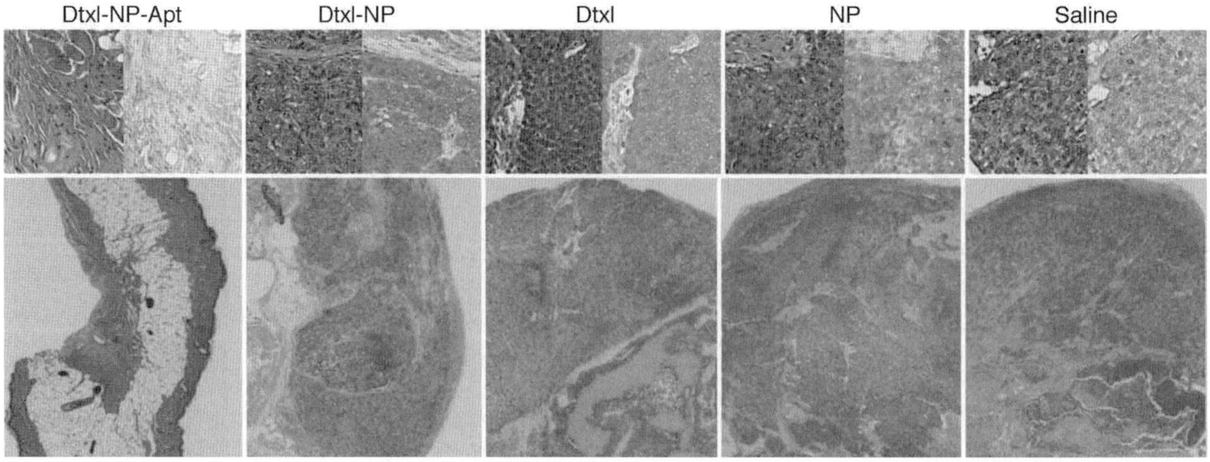

Fig. 4. Histological staining of the excised tumors in the (*i*) saline, (*ii*) pegylated PLGA NP without drug (NP), (*iii*) emulsified Dtxl (Dtxl), (*iv*) Dtxl-encapsulated NPs (Dtxl-NP), or (*v*) Dtxl-encapsulated NP-Apt bioconjugates (Dtxl-NP-Apt) was evaluated by an independent pathologist. The larger images (*Lower*) are H&E staining of representative specimens at ×20 magnification. The smaller images (*Upper*) are H&E (*Left*) and PSMA (*Right*) staining of consecutive sections for each group at ×50 magnification. All specimens except those obtained from the Dtxl-NP-Apt-treated mice were positive for PSMA staining [dark brown horseradish peroxidase stain]. The Dtxl-NP-Apt staining confirmed the absence of residual tumor and presence of scar and adipose tissue.

not show any long-term efficacy, and the mean tumor sizes at the end of the study for the groups were 786 ± 7 mm^3, 775 ± 25 mm^3, and 741 ± 40 mm^3, respectively (mean ± SEM; $n = 7$). None of the animals of the saline and NP groups exhibited tumor regression. Overall five of seven animals in the saline group and six of seven animals in the NP group reached end point (defined as excessive tumor load of >800 mm^3 or body weight loss (BWL) of >20%) during the 109-day study duration. The remaining three animals in these two groups failed to reach the end point during the study, an observation that is consistent with the well documented slow rate of LNCaP tumor growth, which is also characteristic of PCa growth in humans. Six of the seven animals in the Dtxl cohort reached the end point. The difference in the final mean tumor size or survival time for the Dtxl group compared with the saline and NP groups was not statistically significant (ANOVA at 95% confidence interval).

The Dtxl-NP-Apt-treated group demonstrated the most dramatic efficacy: the final mean tumor load was 119 ± 84 mm^3 (mean ± SEM; $n = 7$, significantly smaller than all other groups by ANOVA at 95% confidence interval). In total, five of seven treated animals experienced complete tumor reductions on days 25, 25, 31, 37, and 40. The remaining two animals also exhibited regression of their tumor size after the initial dosing, and one of these two animals experienced progression after the initial regression and had a tumor size of 567 mm^3 by the end of the study (Fig. 3*C*). All mice in the Dtxl-NP-Apt group survived the 109-day study duration. The Dtxl-NP group was also more efficacious than the Dtxl, NP, and saline control groups, but significantly less efficacious when compared with the Dtxl-NP-Apt group. The mean tumor size at end point was 315 ± 103 mm^3 (mean ± SEM; $n = 7$). Two complete tumor reductions were observed on days 31 and 43, and the tumor size in one animal reached end point on day 73. The study demonstrated that, after a single administration, the Dtxl-NP-Apt group was most efficacious against LNCaP PCa tumors, resulting in a better survival than other treatment groups (Fig. 3*D*).

We next performed histological staining of the excised tumors and the tissue at the injection sites, and the slides were evaluated by an independent pathologist (Fig. 4). The complete reduction of tumor and presence of fibrotic tissue in the median tumors of the Dtxl-NP-Apt group was confirmed by hematoxylin/eosin (H&E) staining and negative PSMA staining, consistent with the elimination of LNCaP tumor mass. In the remaining groups, the H&E and PSMA staining of the median tumors at end point demonstrated positive PSMA staining, with areas of hemorrhagic necrosis consistent with the presence of tumors (Fig. 4).

We assessed toxicity of each group *in vivo* by analyzing their effect on the WBC count and BWL. The assessment of the WBC count in all five groups at the end point confirmed no evidence of leukopenia or associated toxicities (Fig. 5*A*). The NP and saline groups demonstrated a BWL trend that roughly paralleled their tumor load. These control groups did not experience any acute depression in body weight after dosing, which is consistent with the absence of drug in the formulations. One animal in each of the saline and Dtxl groups experienced >20% BWL, respectively, on day 44 (755 mm^3) and day 95 (539 mm^3), possibly due to tumor load of these animals. The Dtxl-NP group had two animals with excessive BWL (>20%), which were euthanized on days 9 and 12. The average weights of the other animals in the Dtxl-NP cohort returned to predosing levels by day 26. At nadir, the Dtxl-NP group exhibited a maximal mean BWL of 18 ± 5% (mean ± SD; $n = 7$) on day 12 (Fig. 5*B*). The Dtxl mean BWL at nadir was 17 ± 3% (mean ± SD; $n = 7$) on day 8 and by day 19, the mean body weight had recovered to predosing levels. A statistically significant smaller maximal mean BWL of 7.7 ± 4.0% (mean ± SD; $n = 7$) was observed for the Dtxl-NP-Apt group on day 6, with recovery to its original mean body weight on day 19 (Fig. 5*B*). The observed acute BWL and subsequent recovery after dosing of the Dtxl-NP-Apt and Dtxl-NP groups may represent bulk degradation of PLGA, resulting in a burst followed by slower continued release of Dtxl over time. This release pattern is characteristic of the PLGA controlled release polymer system and allows for the presence of the Dtxl over an extended period at the site of administration. Our *in vitro* drug release assays demonstrate that ≈20% of the drug is yet to be released after 1 month of incubation in an aqueous solution (data not shown).

One possible explanation for significantly enhanced efficacy and a relatively smaller BWL in the Dtxl-NP-Apt group as compared with the Dtxl-NP group may be that the former group is expected to get internalized into tumor cells with subsequent intracellular release of the drug. The latter may release the drug in the extracellular space, causing systemic absorption and distribution, increased toxicity, and decreased efficacy. The nontargeted NPs may also diffuse away from the tumor site and release the drug after concentrating in other organs, possibly underlying the early mor-

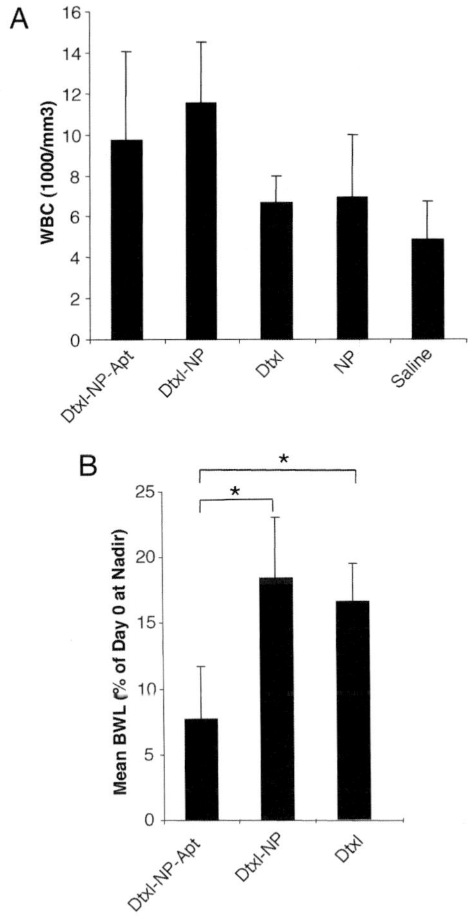

Fig. 5. Evaluation of treatment toxicity. (*A*) WBC counts at the experimental end point were within the normal range for all groups and confirmed the absence of persistent hematologic toxicity. (*B*) Mean BWL after dosing of mouse in each group is shown. *, Significant difference by ANOVA at 95% confidence interval.

bidity in two animals of the cohort. This explanation would be consistent with our own and other investigators' *in vitro* findings that PSMA is constitutively endocytosed and that the targeting of this molecule results in the intracellular delivery of the PSMA ligands (33, 34, 41). For example, the J591 antibody, which recognizes the extracellular domain of the PSMA, has been shown to specifically bind to LNCaP cells and get rapidly internalized and accumulate in the endosomes through a clathrin-mediated pathway (33). Additionally, we have shown that the A10 PSMA Apt-functionalized NPs are taken up by LNCaP cells whereas similar NPs lacking the PSMA Apt remain extracellular (34). These data, however, were generated by using *in vitro* models, and any *in vivo* correlation would need further evaluation.

It is also possible that the differential toxicity of the Dtxl-NP-Apt bioconjugates is due to their interaction with the tumor microenvironment in such a way as to lead to enhanced efficacy and/or decreased toxicity through a mechanism that is independent of their binding to the PSMA protein on the PCa cells. For example, it has been shown that both particle size and charge play a role in passive tumor targeting of systemically administered NPs and that cationic (42) and smaller (43) particles are better at concentrating in the tumors vs. similarly designed anionic and larger particles. Although we are not aware of previous studies suggesting that these parameters are also important in clearance of particles from tumor after intratumoral delivery, it is nonetheless possible that the presence of Apts on the surface of NPs may have altered the surface charge or size of the Dtxl-NP-Apt bioconjugates in a way to lead to a lower rate of lymphatic or systemic clearance of these particles from the tumor interstitium, leading to enhanced efficacy as compared with the Dtxl-NP group.

A thorough evaluation of these alternative possibilities to explain the dramatic efficacy of Dtxl-NP-Apt bioconjugate vs. other groups will require similar *in vivo* efficacy studies using LNCaP xenograft models that are deficient in PSMA expression or development and testing of similar Dtxl-NP-Apt bioconjugates using random Apts that lack specificity for the PSMA protein.

Summary. We had previously developed proof of concept drug delivery vehicles that were composed of polymeric NPs and Apts for targeted delivery and uptake by PCa cells (34). Herein, we have demonstrated *in vitro* and *in vivo* efficacy of NP-Apt bioconjugates against cancer cells. These bioconjugates have the advantage that the materials used in the development of the NPs were approved by the Food and Drug Administration for a prior clinical use and that the targeting molecules used in their development are small, relatively stable, nonimmunogenic, and easy to synthesize, which together may facilitate the translation of these bioconjugates into clinical practice. We postulate that a similar approach may be used to develop therapeutic and diagnostic NP-Apt bioconjugates for other important human diseases.

Materials and Methods

Materials. All chemicals were obtained from Sigma-Aldrich unless otherwise noted. PLGA [inherent viscosity 0.20 dl/g in hexafluoroisopropanol (HFIP)] with acid end groups was purchased from Absorbable Polymers International (Pelham, AL). The heterofunctional PEG polymer with a terminal amine and carboxylic acid functional groups (NH_2-PEG-COOH) was custom synthesized (molecular weight = 3,400; Nektar Therapeutics, San Carlos, CA).

Synthesis of PLGA-PEG-COOH Block Copolymer. The PLGA-COOH and NH_2-PEG_{3400}-COOH polymers were used to synthesize PLGA-*b*-PEG copolymer with terminal carboxylic acid groups (PLGA-PEG-COOH). PLGA-COOH was preactivated to its succinimide by using 1-ethyl-3-(3-dimethylaminopropyl)-carbodiimide (EDC) and *N*-hydroxysulfosuccinimide (sulfo-NHS) and then reacted with NH_2-PEG-COOH. The resulting PLGA-PEG-COOH was characterized by ^{1}H-NMR (400 MHz, $CDCl_3$), δ 5.19 (m br, $-C(O)CH(CH_3)O-$), 4.79 (m br, $-C(O)CH_2O-$), 3.62 (m, $-OCH_2CH_2-$), 3.43 (m, $-OCH_2CH_2-$), 1.54 (m, $-C(O)CH(CH_3)O-$).

Development of Dtxl-Encapsulated NPs. Dtxl-encapsulated NPs were prepared by using the nanoprecipitation method. Briefly, PLGA-PEG-COOH (10 mg/ml) and Dtxl (0.5 mg/ml) were dissolved in acetonitrile and together mixed dropwise into water, giving a final polymer concentration of 3.3 mg/ml. The NPs were stirred for 1 h, and the remaining organic solvent was removed in a rotary evaporator at reduced pressure. The NPs were centrifuged at 10,000 × *g* for 15 min and washed with deionized water, and the size (in nanometers) and surface charge (ζ-potential in millivolts) of NPs were evaluated by Quasi-elastic laser light scattering (QELS) by using a ZetaPALS dynamic light-scattering detector (15 mW laser, incident beam = 676 nm; Brookhaven Instruments, Holtsville, NY). Surface morphology and size were also determined by high-resolution scanning electron microscopy (JEOL 6320FV). Dtxl content in the NPs for calculation of dosing quantity was determined on an Agilent (Palo Alto, CA) 1100 HPLC equipped with a pentafluorophenyl column (Curosil-PFP, 250 × 4.6 mm, 5μ; Phenomenex, Torrance, CA) by using a UV detector at 227 nm.

NP-Apt Conjugation. The PLGA-PEG-COOH NP suspension (≈10 μg/μl in DNase RNase-free water) was incubated with 400 mM 1-ethyl-3-(3-dimethylaminopropyl)-carbodiimide and 100 mM *N*-hydroxysuccinimide for 15 min at room temperature with gentle stirring. The resulting *N*-hydroxysuccinimide-activated particles were covalently linked to 5′-NH_2 modified A10 PSMA Apts (2% weight compared with polymer concentration). The resulting NP-Apt were washed, resuspended in PBS, and used immediately.

***In Vitro* Cytotoxicity Assays.** The prostate LNCaP cells were grown in 12-well plates in RPMI medium 1640 (American Type Culture Collection), supplemented with 100 units/ml aqueous penicillin G, 100 μg/ml streptomycin, and 10% FBS at concentrations to allow 70% confluence in 24 h (i.e., 40,000 LNCaP cells per cm^2). On the day of experiments, cells were washed with prewarmed PBS and incubated with prewarmed phenol-red reduced OptiMEM media for 30 min before the addition of 0.6 μg/ml Dtxl-NP or Dtxl-NP-Apt. The control cells were incubated with NP and NP-Apt without Dtxl. Cells were incubated for 30 min or 2 h at 37°C and washed three times with PBS (100 μl), and fresh growth media were replaced in the plates. The cells were incubated for a total of 72 h. Cell viability was assessed colorimetrically with MTT reagent (Invitrogen). One-way ANOVA with Fisher's least significant difference (LSD) post hoc comparisons at 95% confidence interval was used for statistical comparisons.

***In Vivo* Efficacy Studies.** Animals were cared for under the supervision of the Massachusetts Institute of Technology Division of Comparative Medicine and in compliance with the Principles of Laboratory Animal Care of the National Institutes of Health. PSMA-expressing xenograph flank tumors were induced in 8-week-old BALB/c nude mice by s.c. injection of 3 million LNCaP cells suspended in 1:1 media and matrigel. After 3 weeks when tumors had reached ≈300 mm^3, mice were divided into five groups of seven mice, minimizing weight and tumor size differences. Tumor-bearing nude mice were treated by intratumoral injection of emulsified Dtxl (40 mg/kg), Dtxl-NP-Apt (40 mg/kg), Dtxl-NP (40 mg/kg), NP without drug (NP), or saline. After dosing, the mice were monitored for weight and implanted tumor size daily for 2 weeks and every 3 days thereafter. If BWL persisted beyond 20% of predosing weight, the animals were euthanized. The length and width of the tumors were measured by digital calipers, calculating tumor volume by the following formula: $(\text{width}^2 \times \text{length})/2$. Mice were monitored for a maximum of 109 days, until the tumor was completely regressed or until the tumor volume exceeded 800 mm^3, for which the mice were euthanized for excessive tumor load. For animals that were euthanized because of tumor load or BWL, the tumor size at the time of euthanasia was used for the purpose of mean tumor size calculation. Initial volume of the tumors averaged 328 mm^3. Average body weight for mice in the study was 19.2 g. One-way ANOVA with Fisher's LSD post hoc comparisons at 95% confidence interval was used for statistical comparisons.

Histology. After antigen retrieval, formalin-fixed and paraffin-embedded (FFPE) tissue slides from median tumors in each group were incubated with biotin-labeled A10 PSMA Apt in 1 ml of PBS in the presence of 5× molar excess of tRNA and 0.2% BSA for 30 min at 37°C. Slides were washed three times with PBS and incubated with horseradish peroxidase (HRP)-conjugated streptavidin for 5 min, washed three times with PBS, incubated with the peroxidase substrate, washed twice with PBS, mounted, and analyzed by light microscopy. H&E staining was performed by the Massachusetts Institute of Technology Division of Comparative Medicine.

We thank Drs. Chris Cannizzaro, Etgar Levy-Nissenbaum, Y. Z. Wang, Kris Wood, Ali Khademhosseini, Robert Ross, and Meredith Regan for helpful discussions and Dr. Arlin Rogers for evaluation of histological slides. This work was supported by National Institutes of Health (NIH)/ National Cancer Institute Grant CA 119349, by NIH/National Institute of Biomedical Imaging and Bioengineering Grant EB003647, and by the David Koch Cancer Research Fund.

1. Grimm, P. D., Blasko, J. C., Sylvester, J. E., Meier, R. M. & Cavanagh, W. (2001) *Int. J. Radiat. Oncol. Biol. Phys.* **51,** 31–40.
2. Cooperberg, M. R., Moul, J. W. & Carroll, P. R. (2005) *J. Clin. Oncol.* **23,** 8146–8151.
3. Teloken, C. (2001) *Cancer Control* **8,** 540–545.
4. Han, B. H., Demel, K. C., Wallner, K., Ellis, W., Young, L. & Russell, K. (2001) *J. Urol.* **166,** 953–957.
5. Bucci, J., Morris, W. J., Keyes, M., Spadinger, I., Sidhu, S. & Moravan, V. (2002) *Int. J. Radiat. Oncol. Biol. Phys.* **53,** 91–98.
6. Shah, S. A., Cima, R. R., Benoit, E., Breen, E. L. & Bleday, R. (2004) *Dis. Colon Rectum* **47,** 1487–1492.
7. Peschel, R. E. & Colberg, J. W. (2003) *Lancet Oncol.* **4,** 233–241.
8. Ferrari, M. (2005) *Nat. Rev. Cancer* **5,** 161–171.
9. Langer, R. (1998) *Nature* **392,** 5–10.
10. Langer, R. (2001) *Science* **293,** 58–59.
11. Saunders, R. A. & Helveston, E. M. (1979) *Ophthalmic Surg.* **10,** 13–18.
12. Caron, P. (2002) *Ann. Endocrinol. (Paris)* **63,** 2S19–2S24.
13. Guerin, C., Olivi, A., Weingart, J. D., Lawson, H. C. & Brem, H. (2004) *Invest. New Drugs* **22,** 27–37.
14. Ellington, A. D. & Szostak, J. W. (1990) *Nature* **346,** 818–822.
15. Tuerk, C. & Gold, L. (1990) *Science* **249,** 505–510.
16. Nimjee, S. M., Rusconi, C. P. & Sullenger, B. A. (2005) *Annu. Rev. Med.* **56,** 555–583.
17. Wilson, C. & Szostak, J. W. (1998) *Chem. Biol.* **5,** 609–617.
18. Cheifetz, A. & Mayer, L. (2005) *Mt. Sinai J. Med.* **72,** 250–256.
19. Chester, K., Pedley, B., Tolner, B., Violet, J., Mayer, A., Sharma, S., Boxer, G., Green, A., Nagl, S. & Begent, R. (2004) *Tumour Biol.* **25,** 91–98.
20. Gref, R., Minamitake, Y., Peracchia, M. T., Domb, A., Trubetskoy, V., Torchilin, V. & Langer, R. (1997) *Pharm. Biotechnol.* **10,** 167–198.
21. Gref, R., Minamitake, Y., Peracchia, M. T., Trubetskoy, V., Torchilin, V. & Langer, R. (1994) *Science* **263,** 1600–1603.
22. Moghimi, S. M. & Szebeni, J. (2003) *Prog. Lipid Res.* **42,** 463–478.
23. Christie, R. J. & Grainger, D. W. (2003) *Adv. Drug Deliv. Rev.* **55,** 421–437.
24. Pechar, M., Ulbrich, K., Subr, V., Seymour, L. W. & Schacht, E. H. (2000) *Bioconjug. Chem.* **11,** 131–139.
25. Petrylak, D. P., Tangen, C. M., Hussain, M. H., Lara, P. N., Jr., Jones, J. A., Taplin, M. E., Burch, P. A., Berry, D., Moinpour, C., Kohli, M., *et al.* (2004) *N. Engl. J. Med.* **351,** 1513–1520.
26. Tannock, I. F., de Wit, R., Berry, W. R., Horti, J., Pluzanska, A., Chi, K. N., Oudard, S., Theodore, C., James, N. D., Turesson, I., *et al.* (2004) *N. Engl. J. Med.* **351,** 1502–1512.
27. Lupold, S. E., Hicke, B. J., Lin, Y. & Coffey, D. S. (2002) *Cancer Res.* **62,** 4029–4033.
28. Leach, F. (2004) *Cancer Biol. Ther.* **3,** 559–560.
29. Murphy, G. P., Elgamal, A. A., Su, S. L., Bostwick, D. G. & Holmes, E. H. (1998) *Cancer* **83,** 2259–2269.
30. Thomas, T. P., Patri, A. K., Myc, A., Myaing, M. T., Ye, J. Y., Norris, T. B. & Baker, J. R., Jr. (2004) *Biomacromolecules* **5,** 2269–2274.
31. Slovin, S. F. (2005) *Expert Opin. Ther. Targets* **9,** 561–570.
32. Ghosh, A. & Heston, W. D. (2004) *J. Cell. Biochem.* **91,** 528–539.
33. Liu, H., Rajasekaran, A. K., Moy, P., Xia, Y., Kim, S., Navarro, V., Rahmati, R. & Bander, N. H. (1998) *Cancer Res.* **58,** 4055–4060.
34. Farokhzad, O. C., Jon, S., Khademhosseini, A., Tran, T. N., Lavan, D. A. & Langer, R. (2004) *Cancer Res.* **64,** 7668–7672.
35. Farokhzad, O. C., Khademhosseini, A., Jon, S., Hermmann, A., Cheng, J., Chin, C., Kiselyuk, A., Teply, B., Eng, G. & Langer, R. (2005) *Anal. Chem.* **77,** 5453–5459.
36. Chorny, M., Fishbein, I., Danenberg, H. D. & Golomb, G. (2002) *J. Controlled Release* **83,** 389–400.
37. Burchenal, J. E., Deible, C. R., Deglau, T. E., Russell, A. J., Beckman, E. J. & Wagner, W. R. (2002) *J. Thromb. Thrombolysis* **13,** 27–33.
38. Owens, D. E., III, & Peppas, N. A. (2006) *Int. J. Pharm.* **307,** 93–102.
39. Horoszewicz, J. S., Leong, S. S., Kawinski, E., Karr, J. P., Rosenthal, H., Chu, T. M., Mirand, E. A. & Murphy, G. P. (1983) *Cancer Res.* **43,** 1809–1818.
40. Vanhoefer, U., Cao, S., Harstrick, A., Seeber, S. & Rustum, Y. M. (1997) *Ann. Oncol.* **8,** 1221–1228.
41. Rajasekaran, S. A., Anilkumar, G., Oshima, E., Bowie, J. U., Liu, H., Heston, W., Bander, N. H. & Rajasekaran, A. K. (2003) *Mol. Biol. Cell* **14,** 4835–4845.
42. Campbell, R. B., Fukumura, D., Brown, E. B., Mazzola, L. M., Izumi, Y., Jain, R. K., Torchilin, V. P. & Munn, L. L. (2002) *Cancer Res.* **62,** 6831–6836.
43. Kong, G., Braun, R. D. & Dewhirst, M. W. (2000) *Cancer Res.* **60,** 4440–4445.

REVIEW ARTICLE

Nanocarriers as an emerging platform for cancer therapy

Nanotechnology has the potential to revolutionize cancer diagnosis and therapy. Advances in protein engineering and materials science have contributed to novel nanoscale targeting approaches that may bring new hope to cancer patients. Several therapeutic nanocarriers have been approved for clinical use. However, to date, there are only a few clinically approved nanocarriers that incorporate molecules to selectively bind and target cancer cells. This review examines some of the approved formulations and discusses the challenges in translating basic research to the clinic. We detail the arsenal of nanocarriers and molecules available for selective tumour targeting, and emphasize the challenges in cancer treatment.

DAN PEER[1†], JEFFREY M. KARP[2,3†], SEUNGPYO HONG[4†], OMID C. FAROKHZAD[5], RIMONA MARGALIT[6] AND ROBERT LANGER[3,4*]

[1]Immune Disease Institute and Department of Anesthesia, Harvard Medical School, Boston, Massachusetts 02115, USA; [2]HST Center for Biomedical Engineering, Department of Medicine, Brigham and Women's Hospital, Harvard Medical School, Cambridge, Massachusetts 02139, USA; [3]Harvard-MIT Division of Health Sciences and Technology, Massachusetts Institute of Technology, Cambridge, Massachusetts 02139, USA; [4]Department of Chemical Engineering, Massachusetts Institute of Technology, Cambridge, Massachusetts 02139, USA; [5]Laboratory of Nanomedicine and Biomaterials and Department of Anesthesiology, Brigham and Women's Hospital, Harvard Medical School, Boston, Massachusetts 02115, USA; [6]Department of Biochemistry, George S. Wise Faculty of Life Sciences, and the Center for Nanoscience and Nanotechnology, Tel Aviv University, Tel Aviv, 69978, Israel
†These authors contributed equally to this review
***e-mail: rlanger@mit.edu**

Cancer remains one of the world's most devastating diseases, with more than 10 million new cases every year[1]. However, mortality has decreased in the past two years[2] owing to better understanding of tumour biology and improved diagnostic devices and treatments. Current cancer treatments include surgical intervention, radiation and chemotherapeutic drugs, which often also kill healthy cells and cause toxicity to the patient. It would therefore be desirable to develop chemotherapeutics that can either passively or actively target cancerous cells. Passive targeting exploits the characteristic features of tumour biology that allow nanocarriers to accumulate in the tumour by the enhanced permeability and retention (EPR) effect[2]. Passively targeting nanocarriers first reached clinical trials in the mid-1980s, and the first products, based on liposomes and polymer–protein conjugates, were marketed in the mid-1990s. Later, therapeutic nanocarriers based on this strategy were approved for wider use (Table 1) and methods of further enhancing targeting of drugs to cancer cells were investigated. Active approaches achieve this by conjugating nanocarriers containing chemotherapeutics with molecules that bind to overexpressed antigens or receptors on the target cells. Recent reviews provide perspective on the use of nanotechnology as a fundamental tool in cancer research and nanomedicine[3,4]. Here we focus on the potential of nanocarriers and molecules that can selectively target tumours, and highlight the challenges in translating some of the basic research to the clinic.

PASSIVE AND ACTIVE TARGETING

Nanocarriers encounter numerous barriers en route to their target, such as mucosal barriers and non-specific uptake[5,6]. To address the challenges of targeting tumours with nanotechnology, it is necessary to combine the rational design of nanocarriers with the fundamental understanding of tumour biology (Box 1).

General features of tumours include leaky blood vessels and poor lymphatic drainage. Whereas free drugs may diffuse non-specifically, a nanocarrier can extravasate (escape) into the tumour tissues via the leaky vessels by the EPR effect[7] (Fig. 1). The increased permeability of the blood vessels in tumours is characteristic of rapid and defective angiogenesis (formation of new blood vessels from existing ones). Furthermore, the dysfunctional lymphatic drainage in tumours retains the accumulated nanocarriers and allows them to release drugs into the vicinity of the tumour cells. Experiments using liposomes of different mean size suggest that the threshold vesicle size for extravasation into tumours is ~400 nm (ref. 8), but other studies have shown that particles with diameters <200 nm are more effective[5,8–10].

Although passive targeting approaches form the basis of clinical therapy, they suffer from several limitations. Ubiquitously targeting cells within a tumour is not always feasible because some drugs cannot diffuse efficiently and the random nature of the approach makes it difficult to control the process. This lack of control may induce multiple-drug resistance (MDR) — a situation where chemotherapy treatments fail patients owing to resistance of cancer cells towards one or more drugs. MDR occurs because transporter proteins that expel drugs from cells are overexpressed on the surface of cancer cells[4,11,12]. Expelling drugs inevitably lowers the therapeutic effect and cancer cells soon develop resistance to a variety of drugs. The passive strategy is further limited because certain tumours do not exhibit

REVIEW ARTICLE

Box 1 Rational design of nanocarriers for cancer therapy

Nanocarriers can offer many advantages over free drugs. They:

- protect the drug from premature degradation;
- prevent drugs from prematurely interacting with the biological environment;
- enhance absorption of the drugs into a selected tissue (for example, solid tumour);
- control the pharmacokinetic and drug tissue distribution profile;
- improve intracellular penetration.

For rapid and effective clinical translation, the nanocarrier should:

- be made from a material that is biocompatible, well characterized, and easily functionalized;
- exhibit high differential uptake efficiency in the target cells over normal cells (or tissue);
- be either soluble or colloidal under aqueous conditions for increased effectiveness;
- have an extended circulating half-life, a low rate of aggregation, and a long shelf life.

Table 1 Representative examples of nanocarrier-based drugs on the market

Compound	Commercial name	Nanocarrier	Indications
Styrene maleic anhydride-neocarzinostatin (SMANCS)	Zinostatin/Stimalmer	Polymer–protein conjugate	Hepatocellular carcinoma
PEG-L-asparaginase	Oncaspar	Polymer–protein conjugate	Acute lymphoblastic leukemia
PEG-granulocyte colony-stimulating factor (G-CSF)	Neulasta/PEGfilgrastim	Polymer–protein conjugate	Prevention of chemotherapy-associated neutropenia
IL2 fused to diphtheria toxin	Ontak (Denilelukin diftitox)	Immunotoxin (fusion protein)	Cutaneous T-cell lymphoma
Anti-CD33 antibody conjugated to calicheamicin	Mylotarg	Chemo-immunoconjugate	Acute myelogenous leukemia
Anti-CD20 conjugated to yttrium-90 or indium-111	Zevalin	Radio-immunoconjugate	Relapsed or refractory, low-grade, follicular, or transformed non-Hodgkin's lymphoma
Anti-CD20 conjugated to iodine-131	Bexxar	Radio-immunoconjugate	Relapsed or refractory, low-grade, follicular, or transformed non-Hodgkin's lymphoma
Daunorubicin	DaunoXome	Liposomes	Kaposi's sarcoma
Doxorubicin	Myocet	Liposomes	Combinational therapy of recurrent breast cancer, ovarian cancer, Kaposi's sarcoma
Doxorubicin	Doxil/Caelyx	PEG-liposomes	Refractory Kaposi's sarcoma, recurrent breast cancer, ovarian cancer
Vincristine	Onco TCS	Liposomes	Relapsed aggressive non-Hodgkin's lymphoma (NHL)
Paclitaxel	Abraxane	Albumin-bound paclitaxel nanoparticles	Metastatic breast cancer

the EPR effect, and the permeability of vessels may not be the same throughout a single tumour[13].

One way to overcome these limitations is to programme the nanocarriers so they actively bind to specific cells after extravasation. This binding may be achieved by attaching targeting agents such as ligands — molecules that bind to specific receptors on the cell surface — to the surface of the nanocarrier by a variety of conjugation chemistries[9]. Nanocarriers will recognize and bind to target cells through ligand–receptor interactions, and bound carriers are internalized before the drug is released inside the cell (Fig 1). In general, when using a targeting agent to deliver nanocarriers to cancer cells, it is imperative that the agent binds with high selectivity to molecules that are uniquely expressed on the cell surface. Other important considerations are outlined below.

To maximize specificity, a surface marker (antigen or receptor) should be overexpressed on target cells relative to normal cells. For example, to efficiently deliver liposomes to B-cell receptors using the anti-CD19 monoclonal antibody (mAb), the density of receptors should be in the range of 10^4–10^5 copies per cell. Those with lower density are less effectively targeted[14]. In a breast cancer model, a receptor density of 10^5 copies of ErbB2 receptors per cell was necessary to improve the therapeutic efficacy of an anti-ErbB2-targeted liposomal doxorubicin relative to its non-targeted counterpart[15].

The binding of certain ligands to their receptors may cause receptor-mediated internalization, which is often necessary if nanocarriers are to release drugs inside the cell[16–18]. For example, a more significant therapeutic outcome was achieved when immunoliposomes targeted to human blood cancer (B-cell lymphoma) were labelled with an internalizing anti-CD19 ligand rather than a non-internalizing anti-CD20 ligand[19]. In contrast, targeting nanocarriers to non-internalizing receptors may sometimes be advantageous in solid tumours owing to the bystander effect, where cells lacking the target receptor can be killed through drug release at the surface of the neighbouring cells, where carriers can bind[20].

It is generally known that higher binding affinity increases targeting efficacy. However, for solid tumours, there is evidence that high binding affinity can decrease penetration of nanocarriers due to a 'binding-site barrier', where the nanocarrier binds to its target so strongly that penetration into the tissue is prevented[16,21]. In addition to enhanced affinity, multivalent binding effects (or avidity) may also be used to improve targeting. The collective binding in a multivalent interaction is much stronger than monovalent binding. For example, dendrimer nanocarriers conjugated to 3–15 folate molecules showed a 2,500–170,000-fold enhancement in dissociation constants (K_D) over free folate when attaching to folate-binding proteins immobilized on a surface. This was attributed to the avidity of the multiple folic acid groups on the periphery of the dendrimers[22].

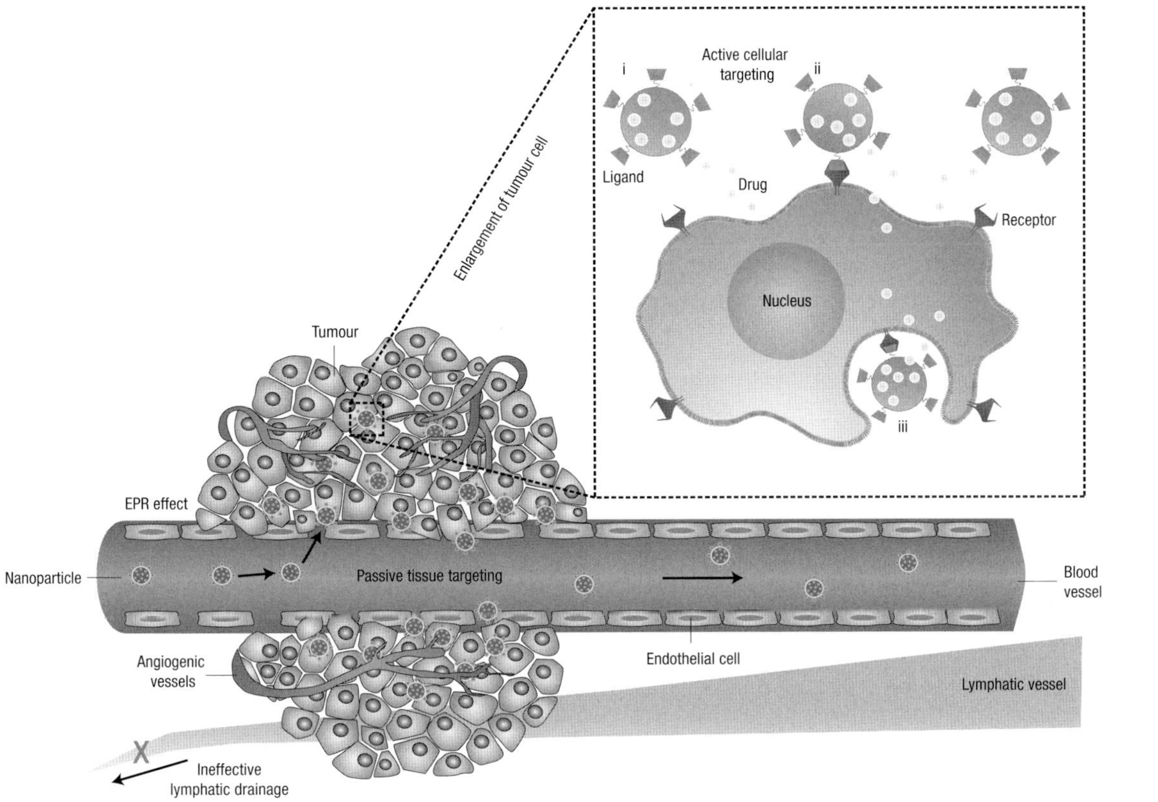

Figure 1 Schematic representation of different mechanisms by which nanocarriers can deliver drugs to tumours. Polymeric nanoparticles are shown as representative nanocarriers (circles). Passive tissue targeting is achieved by extravasation of nanoparticles through increased permeability of the tumour vasculature and ineffective lymphatic drainage (EPR effect). Active cellular targeting (inset) can be achieved by functionalizing the surface of nanoparticles with ligands that promote cell-specific recognition and binding. The nanoparticles can (i) release their contents in close proximity to the target cells; (ii) attach to the membrane of the cell and act as an extracellular sustained-release drug depot; or (iii) internalize into the cell.

TYPES OF TARGETING AGENTS

Targeting agents can be broadly classified as proteins (mainly antibodies and their fragments), nucleic acids (aptamers), or other receptor ligands (peptides, vitamins, and carbohydrates).

Targeting cancer with a mAb was described by Milstein in 1981[23]. Over the past two decades, the feasibility of antibody-based tissue targeting has been clinically demonstrated (reviewed in refs 24,25) with 17 different mAbs approved by the US Food and Drug Administration (FDA)[26]. The mAb rituximab (Rituxan) was approved in 1997 for treatment of patients with non-Hodgkin's lymphoma — a type of cancer that originates in lymphocytes[27]. A year later, Trastuzumab (Herceptin), an anti-HER2 mAb that binds to ErbB2 receptors, was approved for the treatment of breast cancer[28]. The first angiogenesis inhibitor for treating colorectal cancer, Bevacizumab (Avastin), an anti-VEGF mAb that inhibits the factor responsible for the growth of new blood vessels, was approved in 2004[29]. Today, over 200 delivery systems based on antibodies or their fragments are in preclinical and clinical trials[16,30]. Recent developments in the field of antibody engineering have resulted in the production of antibodies that contain animal and human origins such chimeric mAbs, humanized mAbs (those with a greater human contribution), and antibody fragments.

Antibodies may be used in their native state or as fragments for targeting (Fig. 2a). However, use of whole mAbs is advantageous because the presence of two binding sites (within a single antibody) gives rise to a higher binding avidity. Furthermore, when immune cells bind to the Fc portion of the antibody, a signalling cascade is initiated to kill the cancer cells. However, the Fc domain of an intact mAb can also bind to the Fc receptors on normal cells, as occurs with macrophages. This may lead to increased immunogenicity — the ability to evoke an immune response — and liver and spleen uptake of the nanocarrier. An additional advantage of whole/intact antibodies is their ability to maintain stability during long-term storage. Although antibody fragments including antigen-binding fragments (Fab), dimers of antigen-binding fragments ($F(ab')_2$), single-chain fragment variables (scFv) and other engineered fragments are less stable than whole antibodies, they are considered safer when injected systemically owing to reduced non-specific binding[16,30]. To rapidly select antibodies or their fragments that bind to and internalize within cancer cells, phage display libraries that involve a high throughput approach may be used[31,32]. This method generates a multitude of potentially useful antibodies that bind to the same target cells but to different epitopes (a part of a macromolecule that is recognized by antibodies; one receptor may have several epitopes that will be recognized by multiple antibodies). For example, through a selective

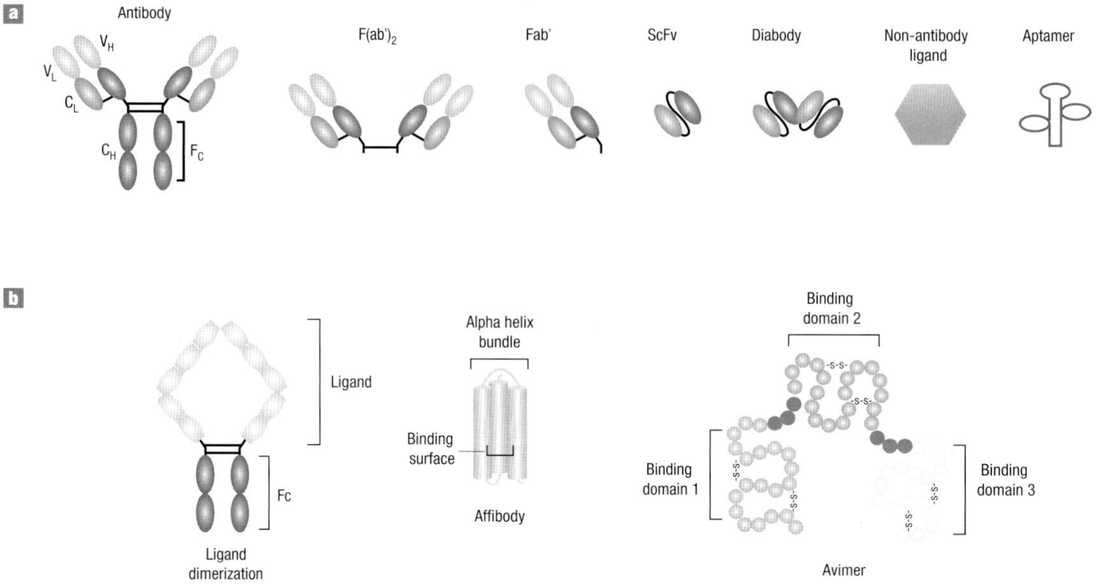

Figure 2 Common targeting agents and ways to improve their affinity and selectivity. **a**, The panel shows a variety of targeting molecules such as a monoclonal antibody or antibodies' fragments, non-antibody ligands, and aptamers. The antibody fragments F(ab)$_2$ and Fab are generated by enzymatic cleavage whereas the Fab , scFv, and bivalent scFv (diabody) fragments are created by molecular biology techniques. V_H: variable heavy chain; V_L: variable light chain; C_H: constant heavy chain; C_L: constant light chain. Non-antibody ligands include vitamins, carbohydrates, peptides, and other proteins. Aptamers can be composed of either DNA or RNA. **b**, Affinity and selectivity can be increased through ligand dimerization or by screening for conformational-sensitive targeting agents such as affibodies, avimers and nanobodies, as well as intact antibodies and their fragments.

process, scFv antibodies have been identified for superior binding and internalization properties for prostate cancer cells[33].

It is possible to increase the efficacy of antibodies by conjugating a therapeutic agent directly to it for targeted delivery. For example, in 2000, the chemotherapeutic drug, calicheamicin, which is conjugated with the anti-CD33 antibody (marketed under the trade name Mylotarg), was the first clinically approved formulation that targets cancerous cells. Others include Zevalin and Bexxar, which use anti-CD20 antibodies to target radioisotopes to cancer cells (Table 1). Although the efficacy of these therapies has been proven, lethal side effects have been observed, likely due to non-specific binding[34] between the targeting agent and non-target moieties on the cell surface. Another reason could be the interaction of the targeting agent with its target expressed on non-cancerous cells. For example, BR96-doxorubicin — an immunoconjugate linked with doxorubicin and comprising an antibody that targets and binds to the Lewis-Y antigen (expressed on 75% of all breast cancers) — demonstrated significant anti-tumour activity in mouse tumour models. BR96-doxorubicin showed lower toxicity than that resulting from doxorubicin alone and it was efficacious in these animal models[35]. However, in dogs, an acute enteropathy (pathology of the intestine) was observed presumably due to binding of the conjugate to Lewis-Y-related antigens expressed by non-targeted gastrointestinal epithelial cells. In Phase II human clinical studies, BR96-doxorubicin immunoconjugates had limited anti-tumour activity and caused severe gastrointestinal toxicity, leading to termination of the study[36].

Although using genomics and proteomics technology to choose appropriate targets is an active area of research, to date no clinically effective targets have been identified. Creating new technologies to enhance selectivity and targeting efficacy with existing targets seem more promising. For example, fusion proteins can be created by combining two or more genes to produce a new protein with desired properties. Antibodies can be engineered so they bind to their target with high affinity, and using molecular biology techniques, it is possible to design protein-based ligand mimetics based on the structure of a receptor. Dimerization of proteins or peptides can increase ligand affinity through divalency — two simultaneous binding events, usually involving concurrent binding of a protein or a peptide to the two Fc domains of an antibody (Fig 2b). For example, dimerization of a low-affinity scFv (also known as diabody) against the ErbB2, led to enhanced tumour localization in a mouse tumour model[37].

It is also possible to increase binding affinity and selectivity to cell surface targets by engineering proteins that detect a specific conformation of a target receptor. In a recent *in vivo* study using a fusion protein consisting of an scFv antibody fragment to target and deliver small interfering RNA (siRNA) to lymphocytes — a type of white blood cell — a 10,000-fold increased affinity for the target receptor, integrin LFA-1, was observed[18]. Integrin LFA-1 is usually present in a low-affinity non-adhesive form on naïve leukocytes (white blood cells that are not activated by cancer cells or pathogens that enter the body), but converts to the high-affinity adhesive form through conformational changes on activation of the immune system. Therefore, targeting the high-affinity form of LFA-1 enables drugs to be selectively delivered to the activated and adhesive leukocytes. New classes of targeting molecules can be engineered to target specific conformations. These include small protein domains, known as affibodies, that can be engineered to bind specifically to different target proteins in a conformational-sensitive manner. Other small proteins that act like antibodies — called avimers — are used to bind selectively to target receptors through multivalent effects. Nanobodies, which are heavy-chain antibodies engineered to one tenth of the size of an intact antibody with a missing light chain, have been used to

Table 2 Examples of nano-based platforms and their current stage of development for use in cancer therapy

Type of carrier and mean diameter (nm)	Drug entrapped or linked	Current stage of development	Type of cancer (for clinical trials)	References
Polymer–drug conjugates (6–15)	Doxorubicin, Paclitaxel, Camptothecin, Platinate, TNP-470	12 products under clinical trials (Phases I–III) and *in vivo*	Various tumours	Reviewed in 3, 61
Liposomes (both PEG and non-PEG coated) (85–100)	Lurtotecan, platinum compounds, Annamycin	Several products in clinical trials (Phases I–III) and *in vivo*	Solid tumours, renal cell carcinoma, mesothelioma, ovarian and acute lymphoblastic leukaemia	Reviewed in 9
Polymeric nanoparticles (50–200)	Doxorubicin, Paclitaxel, platinum-based drugs, Docetaxel	Several products are in clinical trials (Phases I–III) and *in vivo*	Adenocarcinoma of the oesophagus, metastatic breast cancer and acute lymphoblastic leukemia	5, 91, 100, 101
Polymersomes (~100)	Doxorubicin, Paclitaxel	*In vivo*		73, 74
Micelles (lipid based and polymeric) (5–100)	Doxorubicin	Clinical trials (Phase I)	Metastatic or recurrent solid tumours refractory to conventional chemotherapy	77, 92, 102
	Paclitaxel	Clinical trials (Phase I)	Pancreatic, bile duct, gastric and colonic cancers	
	Platinum-based drugs (carboplatin/ cisplatin), Camptothecin, Tamoxifen, Epirubicin	*In vivo* and *in vitro*		Reviewed in 75
Nanoshells (Gold-silica) (~130)	No drug (for photothermal therapy)	*In vivo*		37, 103
Gold nanoparticles (10–40)	No drug (for photothermal ablation)	*In vivo*		104
Nanocages (30–40)	No drug	Chemistry, structural analysis and *in vitro*		90, 105
Dendrimers (~ 5)	Methotrexate	*In vitro / in vivo*		46, 86
Immuno-PEG-liposomes (100)	Doxorubicin	Clinical trials (Phase I)	Metastatic stomach cancer	76
Immunoliposomes (100–150)	Doxorubicin, platinum-based drugs, Vinblastin, Vincristin, Topotecan, Paclitaxel	*In vivo*		Reviewed in 9, 106
Immunotoxins, Immunopolymers, and fusion proteins (3–15)	Various drugs, toxins	Clinical trials (Phases I–III)	Various types of cancer	Reviewed in 16, 17, 61

bind to carcinoembryonic antigen (CEA), a protein used as a tumour marker[38–40] (Fig. 2b).

In addition to the rational design of antibodies, high-throughput approaches have been used to generate targeting agents such as aptamers, which are short single-stranded DNA or RNA oligonucleotides selected *in vitro* from a large number of random sequences (~10^{14}–10^{15}). Aptamers are selected to bind to a wide variety of targets, including intracellular proteins, transmembrane proteins, soluble proteins, carbohydrates, and small molecule drugs. Several aptamers have also been developed to bind specifically to receptors on cancer cells, and thus may be suitable for nanoparticle-aptamer conjugate therapy[41]. For example, docetaxel (Dtxl)-encapsulated nanoparticles whose surface is modified with an aptamer that targets the antigen on the surface of prostate cancer cells, were delivered with high selectivity and efficacy *in vivo*[42].

Growth factor or vitamin interactions with cancer cells represent a commonly used targeting strategy, as cancer cells often overexpress the receptors for nutrition to maintain their fast-growing metabolism. Epidermal growth factor (EGF) has been shown to block and reduce tumour expression of the EGF receptor, which is overexpressed in a variety of tumour cells such as breast and tongue cancer[43]. Additionally, based on the same idea, the vitamin folic acid (folate) has also been used for cancer targeting because folate receptors (FRs) are frequently overexpressed in a range of tumour cells including ovarian, endometrial and kidney cancer[44]. Transferrin (Tf) interacts with Tf receptors (TfRs), which are overexpressed on a variety of tumour cells (including pancreatic, colon, lung, and bladder cancer) owing to increased metabolic rates[45]. Direct coupling of these targeting agents to nanocarriers containing chemotherapies such as drugs has improved intracellular delivery and therapeutic outcome in animal tumour models[46–48]. One challenge with targeting receptors whose expression correlates with metabolic rate, such as folate and Tf, is that these receptors are also expressed in fast-growing healthy cells such as fibroblasts, epithelial and endothelial cells. This could lead to non-specific targeting and subsequently decrease the effectiveness of the drug and increase toxicity[49].

The use of peptides as targeting agents — including arginine-glycine-aspartic acid (RGD), which is the ligand of the cell adhesion integrin $\alpha_v\beta_3$ on endothelial cells — results in increased intracellular drug delivery in different murine tumour models[50,51]. However, RGD also binds to other integrins such as $\alpha_5\beta_1$ and $\alpha_4\beta_1$ and therefore is not specific to cancer cells, which may limit its use. In addition to cell surface antigens, extracellular matrices (ECMs) overexpressed in tumours, such as heparin sulphate, chondroitin sulphate, and hyaluronan (HA), may also serve as effective targets for specific ECM receptors[52,53]. Coating liposomes with HA improves circulation time and enhances targeting to HA receptor-expressing tumours *in vivo*[54,55].

THE ARSENAL OF NANOCARRIERS

Nanocarriers are nanosized materials (diameter 1–100 nm) that can carry multiple drugs and/or imaging agents. Owing to their high surface-area-to-volume ratio, it is possible to achieve high ligand density on the surface for targeting purposes. Nanocarriers can also be used to increase local drug concentration by carrying the drug within and control-releasing it when bound to the targets. Currently, natural and synthetic polymers and lipids are typically used as drug delivery vectors; clinically approved formulations are listed in Table 1. The family of nanocarriers includes polymer

REVIEW ARTICLE

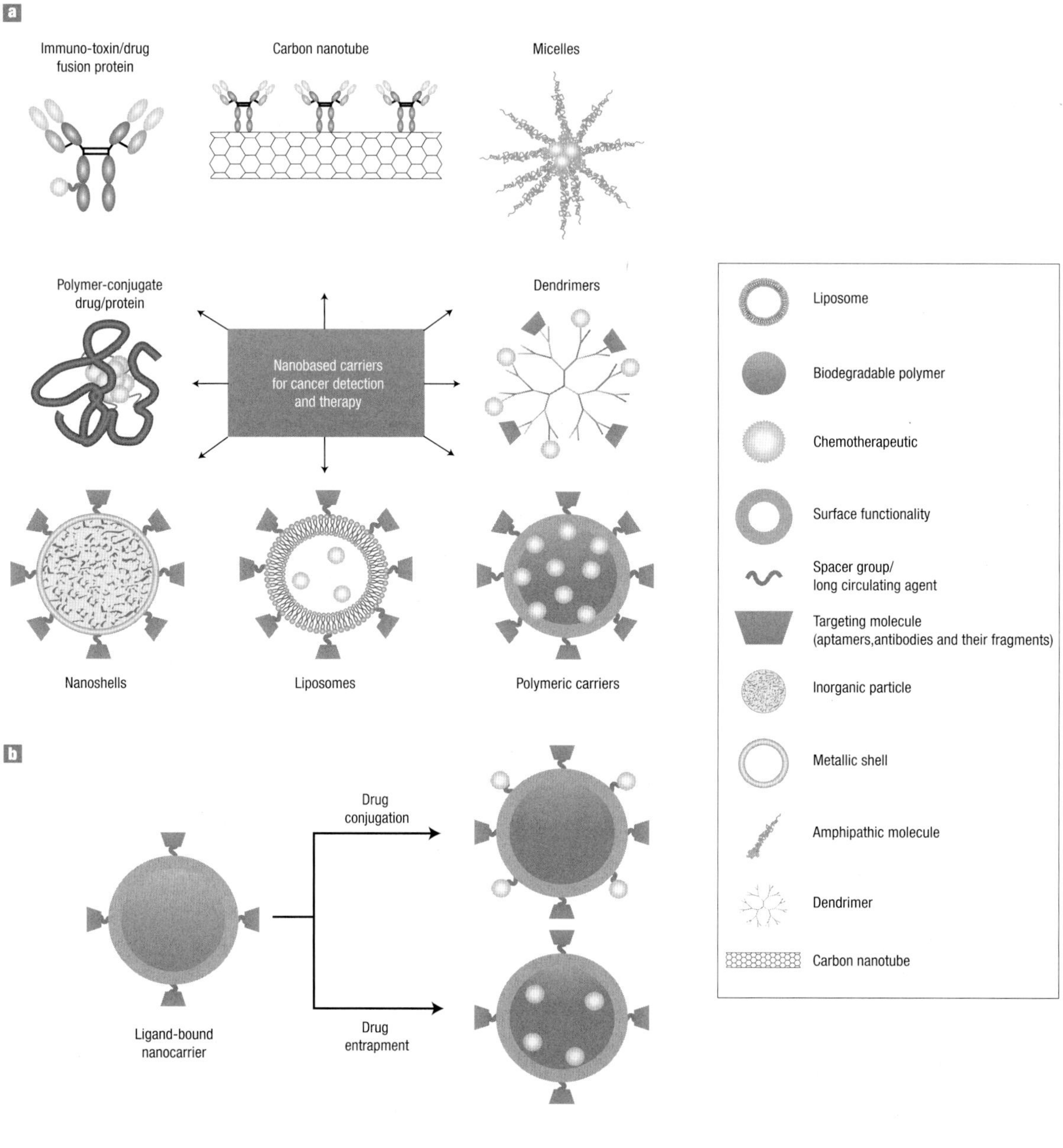

Figure 3 Examples of nanocarriers for targeting cancer. **a**, A whole range of delivery agents are possible but the main components typically include a nanocarrier, a targeting moiety conjugated to the nanocarrier, and a cargo (such as the desired chemotherapeutic drugs). **b**, Schematic diagram of the drug conjugation and entrapment processes. The chemotherapeutics could be bound to the nanocarrier, as in the use of polymer–drug conjugates, dendrimers and some particulate carriers, or they could be entrapped inside the nanocarrier.

conjugates, polymeric nanoparticles, lipid-based carriers such as liposomes and micelles, dendrimers, carbon nanotubes, and gold nanoparticles, including nanoshells and nanocages (Fig. 3a). These nanocarriers have been explored for a variety of applications such as drug delivery, imaging, photothermal ablation of tumours, radiation sensitizers, detection of apoptosis, and sentinel lymph-node mapping[3,4,56] (Table 2).

To date, at least 12 polymer–drug conjugates have entered Phase I and II clinical trials (Table 2 and Fig. 3a) and are especially useful for targeting blood vessels in tumours. Examples include anti-endothelial immunoconjugates, fusion proteins[57–59], and caplostatin, the first polymer-angiogenesis inhibitor conjugates[60]. Polymers that are chemically conjugated with drugs are often considered new chemical entities (NCEs) owing to a distinct pharmacokinetic profile from that of the parent drug. Despite the variety of novel drug targets and sophisticated chemistries available, only four drugs (doxorubicin, camptothecin, paclitaxel, and platinate) and four polymers (N-(2-hydroxylpropyl)methacrylamide (HPMA) copolymer, poly-*L*-glutamic acid, poly(ethylene glycol) (PEG), and Dextran) have been repeatedly used to develop polymer–drug conjugates[3,61].

Polymers are the most commonly explored materials for constructing nanoparticle-based drug carriers. One of the earliest reports of their use for cancer therapy dates back to 1979[62] when adsorption of anticancer drugs to polyalkylcyanoacrylate nanoparticles was described. Couvreur *et al.* revealed the release mechanism of the drugs from the polymer in calf serum, followed by tissue distribution and efficacy studies in a tumour model[63]. This work laid the foundation for the development of doxorubicin-loaded nanoparticles that were tested in clinical trials in the mid-1980s[64]. Polymeric nanoparticles can be made from synthetic polymers, including poly(lactic acid) (PLA) and poly(lactic co-glycolic acid)[65], or from natural polymers such as chitosan[66] and collagen[67] and may be used to encapsulate drugs without chemical modification. The drugs can be released in a controlled manner through surface or bulk erosion, diffusion through the polymer matrix, swelling followed by diffusion, or in response to the local environment. Several multifunctional polymeric nanoparticles are now in various stages of pre-clinical and clinical development[4,56,68,69]. Concerns arising from the use of polymer-based nanocarriers include the inherent structural heterogeneity of polymers, reflected, for example, in a high polydispersity index (the ratio of the weight-and-number-average molecular weight (M_w/M_n)). There are, however, a few examples of polymeric nanoparticles that show near-homogenous size distribution[70].

Lipid-based carriers have attractive biological properties, including general biocompatibility, biodegradability, isolation of drugs from the surrounding environment, and the ability to entrap both hydrophilic and hydrophobic drugs. Through the addition of agents to the lipid membrane or by the alteration of the surface chemistry, properties of lipid-based carriers, such as their size, charge, and surface functionality, can easily be modified. Liposomes, polymersomes, and micelles represent a class of amphiphile-based particles. Liposomes are spherical, self-closed structures formed by one or several concentric lipid bilayers with inner aqueous phases. Today, liposomes are approved by regulatory agencies to carry a range of chemotherapeutics[26,71,72] (Table 1).

Polymersomes have an architecture similar to that of liposomes, but they are composed of synthetic polymer amphiphiles, including PLA-based copolymers[73,74] (Table 2). However, as with polymer therapeutics, there are still no clinically approved strategies that use active cellular targeting for lipid-based carriers.

Micelles, which are self-assembling closed lipid monolayers with a hydrophobic core and hydrophilic shell, have been successfully used as pharmaceutical carriers for water-insoluble drugs (Table 2)[75]. They belong to a group of amphiphilic colloids that can be formed spontaneously under certain concentrations and temperatures from amphiphilic or surface-active agents (surfactants) (Fig. 3a). An example of a polymeric micelle under clinical evaluation is NK911, which is a block copolymer of PEG and poly(aspartic acid). NK911, which consists of a bound doxorubicin fraction (~45%) (Fig. 3b) and a free drug[76], was evaluated for metastatic pancreatic cancer treatment. Another carrier is NK105, a micelle containing paclitaxel, was evaluated for pancreatic, colonic and gastric tumour treatment[77].

Lipid-based carriers pose several challenges, which represent general issues in the use of other targeted nanocarriers such as polymeric nanoparticles. For example, upon intravenous injection, particles are rapidly cleared from the bloodstream by the reticuloendothelial defence mechanism, regardless of particle composition[78,79]. Moreover, instability of the carrier and burst drug release, as well as non-specific uptake by the mononuclear phagocytic system (MPS), provides additional challenges for translating these carriers to the clinic.

Given their long history, liposome-based carriers serve as a classic example of the challenges encountered in the development of nanocarriers and the solutions that have been attempted. For example, PEG has been used to improve circulation time by stabilizing and protecting micelles and liposomes from opsonization — a plasma protein deposition process that signals Kupffer cells in the liver to remove the carriers from circulation[75,80]. However, Daunosome and Myocet are examples of clinically used liposomes (80–90 nm in diameter) without PEG coating that have been reported to exhibit enhanced circulation times, although to a lesser degree than PEGylated liposomes such as Doxil/Caelyx (Table 1).

In addition to rapid clearance, another challenge is the fast burst release of the chemotherapeutic drugs from the liposomes. To overcome this phenomenon, doxorubicin, for example, may be encapsulated in the liposomal aqueous phase by an ammonium sulphate gradient[81]. This method achieves a stable drug entrapment with negligible drug leakage during circulation, even after prolonged residence in the blood stream[82]. In clinical practice, liposomal systems have shown preferential accumulation in tumours, via the EPR effect, and reduced toxicity of their cargo (Tables 1 and 2). However, long-circulating liposomes may lead to extravasation of the drug in unexpected sites. The most commonly experienced clinical toxic effect from the PEGylated liposomal doxorubicin is palmar-plantar erythrodysesthesia (PPE), also called the hand-foot syndrome. PPE — a dermatologic toxicity reaction seen with high doses of many types of chemotherapy — can be addressed by changing the dosing and scheduling of the treatment[83]. Other challenges facing the use of liposomes in the clinic include the high production cost, fast oxidation of some phospholipids, and lack of controlled-release properties of encapsulated drugs.

To achieve temporal release of two drugs, polymers and phospholipids can be combined as a single delivery agent (polymer core/lipid shell). After locating at a tumour site through the EPR effect, the outer phospholipid shell releases an anti-angiogenesis agent, and the inner polymeric nanoparticle subsequently releases a chemotherapy agent in response to local hypoxia — shortage of oxygen. This strategy led to reduced toxicity and enhanced anti-metastatic effects in two different mouse tumour models, emphasizing the advantages of a mechanism-based design for targeted nanocarriers[84].

Organic nanoparticles include dendrimers, viral capsids and nanostructures made from biological building blocks such as proteins. Abraxane is an albumin-bound paclitaxel nanoparticle formulation approved by the FDA in 2005 as a second-line treatment for metastatic breast cancer. Abraxane was designed to address insolubility problems encountered with paclitaxel. Its use eliminates the need for toxic solvents like Cremophor EL (polyoxyethylated castor oil), which has been shown to limit the dose of Taxol that can be administered[85].

Dendrimers are synthetic, branched macromolecules that form a tree-like structure whose synthesis represents a relatively new field in polymer chemistry. Polyamidoamine dendrimers have shown promise for biomedical applications because they (1) can be easily conjugated with targeting molecules, imaging agents, and drugs, (2) have high water solubility and well-defined chemical structures, (3) are biocompatible, and (4) are rapidly cleared from the blood through the kidneys, made possible by their small size (<5 nm), which eliminates the need for biodegradability. *In vivo* delivery of dendrimer–methotrexate conjugates using multivalent targeting results in a tenfold reduction in tumour size compared with that achieved with the same molar concentration of free systemic methotrexate[22,46]. This work provided motivation for further pre-clinical development, and a variety of dendrimers are now under investigation for cancer treatment and are extensively reviewed elsewhere[86,87]. Although

REVIEW ARTICLE

promising, dendrimers are more expensive than other nanoparticles and require many repetitive steps for synthesis, posing a challenge for large-scale production.

Inorganic nanoparticles are primarily metal based and have the potential to be produced with near monodispersity. Inorganic materials have been extensively studied for magnetic resonance imaging and high-resolution superconducting quantum interference devices[88]. Inorganic particles may also be functionalized to introduce targeting molecules and drugs. Specific types of recently developed inorganic nanoparticles include nanoshells and gold nanoparticles.

Nanoshells (100–200 nm) may use the same carrier for both imaging and therapy (Table 2). They are composed of a silica core and a metallic outer layer. Nanoshells have optical resonances that can be adjusted to absorb or scatter essentially anywhere in the electromagnetic spectrum, including the near infrared region (NIR, 820 nm, 4 W cm^{-2}), where transmission of light through tissue is optimal. Absorbing nanoshells are suitable for hyperthermia-based therapeutics, where the nanoshells absorb radiation and heat up the surrounding cancer tissue. Scattering nanoshells, on the other hand, are desirable as contrast agents for imaging applications. Recently, a cancer therapy was developed based on absorption of NIR light by nanoshells, resulting in rapid localized heating to selectively kill tumours implanted in mice. Tissues heated above the thermal damage threshold displayed coagulation, cell shrinkage and loss of nuclear staining, which are indicators of irreversible thermal damage, whereas control tissues appeared undamaged[37,89].

A similar approach involves gold nanocages which are smaller (<50 nm) than the nanoshells. These gold nanocages (Table 2) can be constructed to generate heat in response to NIR light and thus may also be useful in hyperthermia-based therapeutics[90]. Unlike nanoshells and nanocages, pure gold nanoparticles (Table 2) are relatively easy to synthesize and manipulate. Non-specific interactions that cause toxicity in healthy tissues may impede the use of many types of nanoparticles, but using inorganic particles for photo-ablation significantly limits non-specific toxicity because light is locally directed. However, inorganic particles may not provide advantages over other types of nanoparticles for systemic targeting of individual cancer cells because they are not biodegradable or small enough to be cleared easily, resulting in potential accumulation in the body, which may cause long-term toxicity.

THE CHALLENGES OF MULTIDRUG RESISTANCE

The delivery of drugs through targeted nanocarriers that are internalized by cells provides an alternative route to diffusion of drugs into cells. This approach may allow targeted carriers to bypass the activity of integral membrane proteins, known as MDR transporters, which transport a variety of anticancer drugs out of the cancer cell and produce resistance against chemotherapy[11]. The molecular basis of cancer drug resistance is complex and has been correlated to elevated levels of enzymes that can neutralize chemotherapeutic drugs. More often, however, it is due to the overexpression of MDR transporters that actively pump chemotherapeutic drugs out of the cell and reduce the intracellular drug doses below lethal threshold levels. Because not all cancer cells express the MDR transporters, chemotherapy will kill only drug-sensitive cells that do not or only mildly express MDR transporters, while leaving behind a small population of drug-resistant cells that highly express MDR transporters. With tumour recurrence, chemotherapy may fail because residual drug-resistant cells dominate the tumour population.

Among the MDR transporters, the most widely investigated proteins are: P-glycoprotein (also referred to as MDR1 or ABCB1); the multidrug resistance associated proteins (MRPs), of which the most studied is the MRP1 (or ABCC1); and the breast cancer resistance protein (ABCG2). These proteins have different structures, but they share a similar function of expelling chemotherapy drugs from the cells[12]. Several studies have demonstrated the possibility of using nanocarriers to bypass the MDR transporters. SP1049C is a non-ionic (pluronic or also known as poloxamer) block copolymer composed of a hydrophobic core and hydrophilic tail that contains doxorubicin. SP1049C has been shown to circumvent p-glycoprotein-mediated drug resistance in a mouse model of leukaemia and is now under clinical evaluation[91,92]. Folate receptor-mediated cell uptake of doxorubicin-loaded liposomes into an MDR cell line was shown to be unaffected by P-glycoprotein (Pgp)-mediated drug efflux, in contrast to the uptake of free doxorubicin[93]. In an attempt to reverse MDR, vincristine-loaded lipid nanoparticles conjugated to an anti-Pgp mAb (MRK-16), showed greater cytotoxicity in resistant human myelogenous leukaemia cell lines than control non-targeted particles — a response attributed to the inhibition of the Pgp-mediated efflux of vincristine by MRK-16[94]. Additional reports have addressed the challenge of MDR using polymer therapeutics[95], polymeric nanoparticles[96], lipid nanocapsules[97] and micelles[98] within cell lines or in mouse tumour models. Combination treatments with targeted nanocarriers for selective delivery of drugs and MDR pump inhibitors will likely address some of the problems posed by resistant tumours.

INTO THE FUTURE

The choice of an appropriate nanocarrier is not obvious, and the few existing comparative studies are difficult to interpret because several factors may simultaneously affect biodistribution and targeting. In addition, developing suitable screening methodologies for determining optimal characteristics of nanocarriers remains elusive. Therefore, successful targeting strategies must be determined experimentally on a case-by-case basis, which is laborious. In addition, systemic therapies using nanocarriers require methods that can overcome non-specific uptake by mononuclear phagocytic cells and by non-targeted cells. It is also not clear to what extent this is possible without substantially increasing the complexity of the nanocarrier and without influencing commercial scale-up. Improved therapeutic efficacy of targeted nanocarriers has been established in multiple animal models of cancer, and currently more than 120 clinical trials are underway with various antibody-containing nanocarrier formulations[99]. For the clinician, in addition to enhancing confidence through the ability to image the type and location of the tumour, it is imperative to construct appropriate therapeutic regimens. When targeting cell surface markers presents a significant challenge, as in the case for solid tumours, targeting tumour vasculature or the extracellular matrix surrounding the tumour microenvironment may be necessary. In the case of circulating cancer cells, as in leukaemia and lymphoma, a therapy that targets surface antigens with high affinity and includes a carrier with a long circulating half-life may be the most efficacious. Similar to combination drug strategies that may be personalized to optimize treatment regimens, oncologists in the near future may be presented with the ability to choose specific nanocarrier/targeting molecule combinations which could lead to improved therapeutic outcomes and reduced costs.

Although we are still far from Nobel Prize winner Paul Ehrlich's 'magic bullet', many believe that we will soon enter an era in which nanocarrier-based approaches will represent an important modality within therapeutic and diagnostic oncology.

doi:10.1038/nnano.2007.387

References

1. Stewart, B. W. & Kleihues, P. *World Cancer Report* (World Health Organization Press, Geneva, 2003).
2. *Cancer Facts & Figures 2007* (American Cancer Society, Atlanta, 2007).
3. Duncan, R. Polymer conjugates as anticancer nanomedicines. *Nat. Rev. Cancer* **6**, 688–701 (2006).
4. Ferrari, M. Cancer nanotechnology: opportunities and challenges. *Nat. Rev. Cancer* **5**, 161–171 (2005).

5. Couvreur, P. & Vauthier, C. Nanotechnology: Intelligent design to treat complex disease. *Pharm. Res.* **23,** 1417–1450 (2006).
6. Alonso, M. J. Nanomedicines for overcoming biological barriers. *Biomed. Pharmacother.* **58,** 168–172 (2004).
7. Matsumura, Y. & Maeda, H. A new concept for macromolecular therapeutics in cancer-chemotherapy — Mechanism of tumoritropic accumulation of proteins and the antitumor agent smancs. *Cancer Res.* **46,** 6387–6392 (1986).
8. Yuan, F. *et al.* Vascular-permeability in a human tumor xenograft — Molecular-size dependence and cutoff size. *Cancer Res.* **55,** 3752–3756 (1995).
9. Torchilin, V. P. Recent advances with liposomes as pharmaceutical carriers. *Nat. Rev. Drug Discov.* **4,** 145–160 (2005).
10. Hobbs, S. K. *et al.* Regulation of transport pathways in tumor vessels: role of tumor type and microenvironment. *Proc. Natl Acad. Sci. USA* **95,** 4607–4612 (1998).
11. Gottesman, M. M., Fojo, T. & Bates, S. E. Multidrug resistance in cancer: Role of ATP-dependent transporters. *Nat. Rev. Cancer* **2,** 48–58 (2002).
12. Peer, D. & Margalit, R. Fluoxetine and reversal of multidrug resistance. *Cancer Lett.* **237,** 180–187 (2006).
13. Jain, R. K. Barriers to drug-delivery in solid tumors. *Sci. Am.* **271,** 58–65 (1994).
14. de Menezes, D. E. L., Pilarski, L. M. & Allen, T. M. In vitro and in vivo targeting of immunoliposomal doxorubicin to human B-cell lymphoma. *Cancer Res.* **58,** 3320–3330 (1998).
15. Park, J. W. *et al.* Anti-HER2 immunoliposomes: enhanced efficacy attributable to targeted delivery. *Clin. Cancer Res.* **8,** 1172–1181 (2002).
16. Allen, T. M. Ligand-targeted therapeutics in anticancer therapy. *Nat. Rev. Cancer* **2,** 750–763 (2002).
17. Pastan, I., Hassan, R., FitzGerald, D. J. & Kreitman, R. J. Immunotoxin therapy of cancer. *Nat. Rev. Cancer* **6,** 559–565 (2006).
18. Peer, D., Zhu, P., Carman, C. V., Lieberman, J. & Shimaoka, M. Selective gene silencing in activated leukocytes by targeting siRNAs to the integrin lymphocyte function-associated antigen-1. *Proc. Natl Acad. Sci. USA* **104,** 4095–4100 (2007).
19. Sapra, P. & Allen, T. M. Internalizing antibodies are necessary for improved therapeutic efficacy of antibody-targeted liposomal drugs. *Cancer Res.* **62,** 7190–7194 (2002).
20. Allen, T. M. Long-circulating (sterically stabilized) liposomes for targeted drug-delivery. *Trends Pharmacol. Sci.* **15,** 215–220 (1994).
21. Adams, G. P. *et al.* High affinity restricts the localization and tumor penetration of single-chain Fv antibody molecules. *Cancer Res.* **61,** 4750–4755 (2001).
22. Hong, S. *et al.* The binding avidity of a nanoparticle-based multivalent targeted drug delivery platform. *Chem. Biol.* **14,** 107–115 (2007).
23. Warenius, H. M., Galfre, G., Bleehen, N. M. & Milstein, C. Attempted targeting of A monoclonal-antibody in a human-tumor xenograft system. *Eur. J. Cancer Clin. Oncology* **17,** 1009–1015 (1981).
24. von Mehren, A. G., Weiner L. M. Monoclonal antibody therapy for cancer. *Annu. Rev. Med.* **54,** 343–369 (2003).
25. Weiner, L. M. & Adams, G. P. New approaches to antibody therapy. *Oncogene* **19,** 6144–6151 (2000).
26. Gabizon, A. A. Pegylated liposomal doxorubicin: metamorphosis of an old drug into a new form of chemotherapy. *Cancer Invest.* **19,** 424–436 (2001).
27. James, J. S. & Dubs, G. FDA approves new kind of lymphoma treatment. *AIDS Treat. News* **284,** 2–3 (1997).
28. Albanell, J. & Baselga, J. Trastuzumab, a humanized anti-HER2 monoclonal antibody, for the treatment of breast cancer. *Drugs Today* **35,** 931–946 (1999).
29. Ferrara, N. VEGF as a therapeutic target in cancer. *Oncology* **69** (Suppl. 3), 11–16 (2005).
30. Carter, P. Improving the efficacy of antibody-based cancer therapies. *Nat. Rev. Cancer* **1,** 118–129 (2001).
31. Marks, J. D. Selection of internalizing antibodies for drug delivery. *Methods Mol. Biol.* **248,** 201–208 (2004).
32. Marks, J. D. *et al.* Human-antibody fragments specific for human blood-groups antigens from a phage display library. *Bio-Technol.* **11,** 1145–1149 (1993).
33. Liu, B., Conrad, F., Cooperberg, M. R., Kirpotin, D. B. & Marks, J. D. Mapping tumor epitope space by direct selection of single-chain Fv antibody libraries on prostate cancer cells. *Cancer Res.* **64,** 704–710 (2004).
34. Arnold, D. M. *et al.* Systematic review: efficacy and safety of rituximab for adults with idiopathic thrombocytopenic purpura. *Ann. Intern. Med.* **146,** 25–33 (2007).
35. Trail, P. A. *et al.* Cure of xenografted human carcinomas by Br96-doxorubicin immunoconjugates. *Science* **261,** 212–215 (1993).
36. Tolcher, A. W. *et al.* Randomized phase II study of BR96-doxorubicin conjugate in patients with metastatic breast cancer. *J. Clin. Oncology* **17,** 478–484 (1999).
37. Hirsch, L. R. *et al.* Nanoshell-mediated near-infrared thermal therapy of tumors under magnetic resonance guidance. *Proc. Natl Acad. Sci. USA* **100,** 13549–13554 (2003).
38. Silverman, J. *et al.* Multivalent avimer proteins evolved by exon shuffling of a family of human receptor domains. *Nat. Biotechnol.* **23,** 1556–1561 (2005).
39. Cortez-Retamozo, V. *et al.* Efficient cancer therapy with a nanobody-based conjugate. *Cancer Res.* **64,** 2853–2857 (2004).
40. Nord, K. *et al.* Binding proteins selected from combinatorial libraries of an alpha-helical bacterial receptor domain. *Nat. Biotechnol.* **15,** 772–777 (1997).
41. White, R. R., Sullenger, B. A. & Rusconi, C. P. Developing aptamers into therapeutics. *J. Clin. Invest.* **106,** 929–934 (2000).
42. Farokhzad, O. C. *et al.* Targeted nanoparticle-aptamer bioconjugates for cancer chemotherapy in vivo. *Proc. Natl Acad. Sci. USA* **103,** 6315–6320 (2006).
43. Sanfilippo, J. S. *et al.* Quantitative analyses of epidermal growth factor receptors, HER-2/neu oncoprotein and cathepsin D in nonmalignant and malignant uteri. *Cancer* **77,** 710–716 (1996).
44. Antony, A. C. The biological chemistry of folate receptors. *Blood* **79,** 2807–2820 (1992).
45. Prost, A. C. *et al.* Differential transferrin receptor density in human colorectal cancer: A potential probe for diagnosis and therapy. *Int. J. Oncol.* **13,** 871–875 (1998).
46. Kukowska-Latallo, J. F. *et al.* Nanoparticle targeting of anticancer drug improves therapeutic response in animal model of human epithelial cancer. *Cancer Res.* **65,** 5317–5324 (2005).
47. Iinuma, H. *et al.* Intracellular targeting therapy of cisplatin-encapsulated transferrin-polyethylene glycol liposome on peritoneal dissemination of gastric cancer. *Int. J. Cancer* **99,** 130–137 (2002).
48. Ishida, O. *et al.* Liposomes bearing polyethyleneglycol-coupled transferrin with intracellular targeting property to the solid tumors in vivo. *Pharm. Res.* **18,** 1042–1048 (2001).
49. Ekblom, P., Thesleff, I., Lehto, V. P. & Virtanen, I. Distribution of the transferrin receptor in normal human-fibroblasts and fibro-sarcoma cells. *Int. J. Cancer* **31,** 111–117 (1983).
50. Li, J. *et al.* Fusion protein from RGD peptide and Fc fragment of mouse immunoglobulin G inhibits angiogenesis in tumor. *Cancer Gene Ther.* **11,** 363–370 (2004).
51. Ruoslahti, E. Cell adhesion and tumor metastasis. *Princess Takamatsu Symp.* **24,** 99–105 (1994).
52. Peer, D. & Margalit, R. Tumor-targeted hyaluronan nanoliposomes increase the antitumor activity of liposomal Doxorubicin in syngeneic and human xenograft mouse tumor models. *Neoplasia* **6,** 343–353 (2004).
53. Hu, Z., Sun, Y. & Garen, A. Targeting tumor vasculature endothelial cells and tumor cells for immunotherapy of human melanoma in a mouse xenograft model. *Proc. Natl Acad. Sci. USA* **96,** 8161–8166 (1999).
54. Peer, D. & Margalit, R. Loading mitomycin C inside long circulating hyaluronan targeted nano-liposomes increases its antitumor activity in three mice tumor models. *Int. J. Cancer* **108,** 780–789 (2004).
55. Eliaz, R. E. & Szoka, F. C. Jr. Liposome-encapsulated doxorubicin targeted to CD44: a strategy to kill CD44-overexpressing tumor cells. *Cancer Res.* **61,** 2592–2601 (2001).
56. LaVan, D. A., McGuire, T. & Langer, R. Small-scale systems for in vivo drug delivery. *Nat. Biotechnol.* **21,** 1184–1191 (2003).
57. Arap, W., Pasqualini, R. & Ruoslahti, E. Cancer treatment by targeted drug delivery to tumor vasculature in a mouse model. *Science* **279,** 377–380 (1998).
58. Schraa, A. J. *et al.* Targeting of RGD-modified proteins to tumor vasculature: A pharmacokinetic and cellular distribution study. *Int. J. Cancer* **102,** 469–475 (2002).
59. Halin, C. *et al.* Enhancement of the antitumor activity of interleukin-12 by targeted delivery to neovasculature. *Nat. Biotechnol.* **20,** 264–269 (2002).
60. Satchi-Fainaro, R. *et al.* Targeting angiogenesis with a conjugate of HPMA copolymer and TNP-470. *Nat. Med.* **10,** 255–261 (2004).
61. Satchi-Fainaro, R., Duncan, R. & Barnes, C. M. in *Polymer Therapeutics II: Polymers as Drugs, Conjugates and Gene Delivery Systems* Vol. 193 (eds Satchi-Fainaro, R. & Duncan, R.) 1–65 (Springer-Verlag, Berlin, 2006).
62. Couvreur, P., Kante, B., Roland, M. & Speiser, P. Adsorption of anti-neoplastic drugs to polyalkylcyanoacrylate nanoparticles and their release in calf serum. *J. Pharm. Sci.* **68,** 1521–1524 (1979).
63. Couvreur, P. *et al.* Tissue distribution of anti-tumor drugs associated with polyalkylcyanoacrylate nanoparticles. *J. Pharm. Sci.* **69,** 199–202 (1980).
64. Couvreur, P., Kante, B., Grislain, L., Roland, M. & Speiser, P. Toxicity of polyalkylcyanoacrylate nanoparticles II: Doxorubicin-loaded nanoparticles. *J. Pharm. Sci.* **71,** 790–792 (1982).
65. Hrkach, J. S., Peracchia, M. T., Domb, A., Lotan, N. & Langer, R. Nanotechnology for biomaterials engineering: Structural characterization of amphiphilic polymeric nanoparticles by H-1 NMR spectroscopy. *Biomaterials* **18,** 27–30 (1997).
66. Calvo, P., RemunanLopez, C., VilaJato, J. L. & Alonso, M. J. Chitosan and chitosan ethylene oxide propylene oxide block copolymer nanoparticles as novel carriers for proteins and vaccines. *Pharm. Res.* **14,** 1431–1436 (1997).
67. Elsamaligy, M. S. & Rohdewald, P. Reconstituted collagen nanoparticles, a novel drug carrier delivery system. *J. Pharm. Pharmacol.* **35,** 537–539 (1983).
68. Moses, M. A., Brem, H. & Langer, R. Advancing the field of drug delivery: taking aim at cancer. *Cancer Cell* **4,** 337–341 (2003).
69. Farokhzad, O. C. & Langer, R. Nanomedicine: Developing smarter therapeutic and diagnostic modalities. *Adv. Drug Deliv. Rev.* **58,** 1456–1459 (2006).
70. Guo, R. *et al.* Synthesis of alginic acid-poly[2-(diethylamino)ethyl methacrylate] monodispersed nanoparticles by a polymer-monomer pair reaction system. *Biomacromolecules* **8,** 843–850 (2007).
71. Gabizon, A. A. Stealth liposomes and tumor targeting: one step further in the quest for the magic bullet. *Clin. Cancer Res.* **7,** 223–225 (2001).
72. Safra, T. *et al.* Pegylated liposomal doxorubicin (doxil): reduced clinical cardiotoxicity in patients reaching or exceeding cumulative doses of 500 mg/m^2. *Ann. Oncol.* **11,** 1029–1033 (2000).
73. Ahmed, F. *et al.* Shrinkage of a rapidly growing tumor by drug-loaded polymersomes: pH-triggered release through copolymer degradation. *Mol. Pharm.* **3,** 340–350 (2006).
74. Discher, D. E. & Ahmed, F. Polymersomes. *Annu. Rev. Biomed. Eng.* **8,** 323–341 (2006).
75. Matsumura, Y. *et al.* Phase I clinical trial and pharmacokinetic evaluation of NK911, a micelle-encapsulated doxorubicin. *Brit. J. Cancer* **91,** 1775–1781 (2004).
76. Kato, K. *et al.* Phase I study of NK105, a paclitaxel-incorporating micellar nanoparticle, in patients with advanced cancer. *J. Clin. Oncol* **24** (suppl.), 2018 (2006).
77. Torchilin, V. P. Micellar nanocarriers: Pharmaceutical perspectives. *Pharm. Res.* **24,** 1–16 (2007).
78. Brigger, I., Dubernet, C. & Couvreur, P. Nanoparticles in cancer therapy and diagnosis. *Adv. Drug Deliv. Rev.* **54,** 631–651 (2002).
79. Kreuter, J. & Higuchi, T. Improved delivery of methoxsalen. *J. Pharm. Sci.* **68,** 451–454 (1979).
80. Papahadjopoulos, D. *et al.* Sterically stabilized liposomes - improvements in pharmacokinetics and antitumor therapeutic efficacy. *Proc. Natl Acad. Sci. USA* **88,** 11460–11464 (1991).
81. Haran, G., Cohen, R., Bar, L. K. & Barenholz, Y. Transmembrane ammonium-sulfate gradients in liposomes produce efficient and stable entrapment of amphipathic weak bases. *Biochim. Biophys. Acta* **1151,** 201–215 (1993).
82. Gabizon, A. A., Shmeeda, H. & Zalipsky, S. Pros and cons of the liposome platform in cancer drug targeting. *J. Liposome Res.* **16,** 175–183 (2006).
83. Lorusso, D. *et al.* Pegylated liposomal doxorubicin-related palmar-plantar erythrodysesthesia ('hand-foot' syndrome). *Ann. Oncol.* (2007).
84. Sengupta, S. *et al.* Temporal targeting of tumour cells and neovasculature with a nanoscale delivery system. *Nature* **436,** 568–572 (2005).
85. Damascelli, B. *et al.* Intraarterial chemotherapy with polyoxyethylated castor oil free paclitaxel, incorporated in albumin nanoparticles (ABI-007). *Cancer* **92,** 2592–2602 (2001).
86. Gillies, E. R. & Frechet, J. M. J. Dendrimers and dendritic polymers in drug delivery. *Drug Discov. Today* **10,** 35–43 (2005).

87. Malik, N. *et al.* Dendrimers: Relationship between structure and biocompatibility in vitro, and preliminary studies on the biodistribution of I-125-labelled polyamidoamine dendrimers in vivo. *J. Control. Release* **65,** 133–148 (2000).
88. Morawski, A. M., Lanza, G. A. & Wickline, S. A. Targeted contrast agents for magnetic resonance imaging and ultrasound. *Curr. Opin. Biotechnol.* **16,** 89–92 (2005).
89. Loo, C., Lowery, A., Halas, N., West, J., Drezek, R. Immunotargeted nanoshells for integrated cancer imaging and therapy. *Nano Lett.* **5,** 709–711 (2005).
90. Chen, J. *et al.* Gold nanocages: Bioconjugation and their potential use as optical imaging contrast agents. *Nano Lett.* **5,** 473–477 (2005).
91. Danson, S. *et al.* Phase I dose escalation and pharmacokinetic study of pluronic polymer-bound doxorubicin (SP 1049C) in patients with advanced cancer. *Brit. J. Cancer* **90,** 2085–2091 (2004).
92. Batrakova, E. V. *et al.* Anthracycline antibiotics non-covalently incorporated into the block copolymer micelles: In vivo evaluation of anti-cancer activity. *Brit. J. Cancer* **74,** 1545–1552 (1996).
93. Goren, D. *et al.* Nuclear delivery of doxorubicin via folate-targeted liposomes with bypass of multidrug-resistance efflux pump. *Clin. Cancer Res.* **6,** 1949–1957 (2000).
94. Matsuo, H. *et al.* Possibility of the reversal of multidrug resistance and the avoidance of side effects by liposomes modified with MRK-16, a monoclonal antibody to P-glycoprotein. *J. Control. Release* **77,** 77–86 (2001).
95. Duncan, R., Vicent, M. J., Greco, F. & Nicholson, R. I. Polymer-drug conjugates: towards a novel approach for the treatment of endrocine-related cancer. *Endocrine-Relat. Cancer* 12, S189–S199 (2005).
96. Wong, H. L. *et al.* A new polymer-lipid hybrid nanoparticle system increases cytotoxicity of doxorubicin against multidrug-resistant human breast cancer cells. *Pharm. Res.* **23,** 1574–1585 (2006).
97. Garcion, E. *et al.* A new generation of anticancer, drug-loaded, colloidal vectors reverses multidrug resistance in glioma and reduces tumor progression in rats. *Mol. Cancer Ther.* **5,** 1710–1722 (2006).
98. Lee, E. S., Na, K. & Bae, Y. H. Doxorubicin loaded pH-sensitive polymeric micelles for reversal of resistant MCF-7 tumor. *J. Control. Release* **103,** 405–418 (2005).
99. Sapra, P. & Allen, T. M. Ligand-targeted liposomal anticancer drugs. *Prog. Lipid Res.* **42,** 439–462 (2003).
100. Moghimi, S. M. Recent developments in polymeric nanoparticle engineering and their applications in experimental and clinical oncology. *Anticancer Agents Med. Chem.* **6,** 553–561 (2006).
101. Lee, K. S. *et al.* Multicenter phase II trial of Genexol-PM, a Cremophor-free, polymeric micelle formulation of paclitaxel, in patients with metastatic breast cancer. *Breast Cancer Res. Treat.* (2007).
102. Nakanishi, T. *et al.* Development of the polymer micelle carrier system for doxorubicin. *J. Control. Release* **74,** 295–302 (2001).
103. Hirsch, L. R. *et al.* Metal nanoshells. *Ann. Biomed. Engin.* **34,** 15–22 (2006).
104. Sokolov, K. *et al.* Real-time vital optical imaging of precancer using anti-epidermal growth factor receptor antibodies conjugated to gold nanoparticles. *Cancer Res.* **63,** 1999–2004 (2003).
105. Chen, J. Y. *et al.* Facile synthesis of gold-silver nanocages with controllable pores on the surface. *J. Am. Chem. Soc.* **128,** 14776–14777 (2006).
106. Kontermann, R. E. Immunolliposomes for cancer therapy. *Curr. Opin. Mol. Ther.* **8,** 39–45 (2006).

Acknowledgments

We would like to acknowledge Shiladitya Sengupta for critically reviewing the manuscript and Maeve Cullinane for helpful discussions. This work was supported by federal funds NIH/NCI CA119349, NIH/NIBIB EB 003647, and NIH R01-EB000244. The content is solely the responsibility of the authors and does not necessarily represent the official view of the NIH.

Competing financial interests

The authors declare competing financial interests: details accompany the full-text HTML version of the paper at www.nature.com/naturenanotechnology.

RESEARCH ARTICLE

DRUG DELIVERY

First-in-Human Testing of a Wirelessly Controlled Drug Delivery Microchip

Robert Farra,[1]* Norman F. Sheppard Jr.,[1] Laura McCabe,[1] Robert M. Neer,[2] James M. Anderson,[3] John T. Santini Jr.,[4] Michael J. Cima,[5] Robert Langer[6]

The first clinical trial of an implantable microchip-based drug delivery device is discussed. Human parathyroid hormone fragment (1–34) [hPTH(1–34)] was delivered from the device in vivo. hPTH(1–34) is the only approved anabolic osteoporosis treatment, but requires daily injections, making patient compliance an obstacle to effective treatment. Furthermore, a net increase in bone mineral density requires intermittent or pulsatile hPTH(1–34) delivery, a challenge for implantable drug delivery products. The microchip-based devices, containing discrete doses of lyophilized hPTH(1–34), were implanted in eight osteoporotic postmenopausal women for 4 months and wirelessly programmed to release doses from the device once daily for up to 20 days. A computer-based programmer, operating in the Medical Implant Communications Service band, established a bidirectional wireless communication link with the implant to program the dosing schedule and receive implant status confirming proper operation. Each woman subsequently received hPTH(1–34) injections in escalating doses. The pharmacokinetics, safety, tolerability, and bioequivalence of hPTH(1–34) were assessed. Device dosing produced similar pharmacokinetics to multiple injections and had lower coefficients of variation. Bone marker evaluation indicated that daily release from the device increased bone formation. There were no toxic or adverse events due to the device or drug, and patients stated that the implant did not affect quality of life.

INTRODUCTION

Implantable medical devices are routinely used in many medical specialties, including cardiology, orthopedics, and neurology. Devices such as pacemakers, joint replacements, and pain pumps perform an electronic, mechanical, or fluidic function to help patients return to a healthier anatomical or physiological state. Over the past decade, device manufacturers have incorporated chemicals or drugs into medical implants with the objective to improve efficacy and reduce morbidity. Drug-eluting stents, for example, reduce in-stent restenosis when compared with bare-metal stents (*1*). The U.S. Food and Drug Administration (FDA) has defined products that combine devices, drugs, or biological products as "combination products." Other approved combination products include drug-releasing transdermal patches, absorbable sponges or meshes impregnated with antibiotics, and bone grafts consisting of protein solution with an absorbable structure or scaffold.

One class of combination products featuring on-demand drug release capabilities was first described by Santini *et al.*, who developed a microchip with many reservoirs containing discrete doses of drug (*2–4*). However, adapting the microchip technology for human use posed significant challenges. First, hermetic sealing of each reservoir at or near room temperature was critical to prevent degradation of the drug's composition. A compression welding process was developed to provide a long-term hermetic seal (*5*). Second, a reliable means to protect and expose the contents of each reservoir on command was required. An impermeable, thin metallic membrane was developed as an integral component of the reservoir. This membrane can be removed by electrothermal ablation (*6*). The drug is then released in a controlled, pulsatile manner. Third, aseptic filling and lyophilization of clinical doses of a drug in the microchip needed to be developed (*7, 8*). Implanted drug delivery systems based on the multireservoir microchip—with all of these optimized features—are particularly well suited for delivery of polypeptides based on a predefined or even improvised dosing schedule. Furthermore, despite the microchip's capability to deliver drugs in vitro, once implanted into the body, a fibrous, collagen-based membrane can develop around the device (*9–11*). The presence of this fibrous capsule may affect the resulting pharmacokinetics (PK) by slowing systemic absorption because the drug needs to diffuse across the membrane. One of the aims of this study was to determine the clinical relevance of this capsule.

Human parathyroid hormone fragment (1–34) [hPTH(1–34)] is used to treat osteoporosis. Osteoporosis is an imbalance in bone resorption and bone formation processes, where the resulting loss of bone mineral density and disrupted bone microarchitecture lead to an increase in fractures. The World Health Organization estimates that 9 million osteoporotic fractures occur annually worldwide, with a significant contribution to disability rates (*12*). The total cost for treatment of these fractures in the United States in 2015 is projected to be more than $20 billion (*13*). There are two classes of drugs used to treat osteoporosis: bone resorption inhibitors, such as estrogens, bisphosphonates, and calcitonin, and anabolic agents, such as human parathyroid hormone [hPTH(1–84)] and teriparatide [hPTH(1–34)], the hormone's 34–amino acid N-terminal fragment. In 2002, the FDA approved Eli Lilly and Company's teriparatide (U.S. and European Union trade names FORTEO and FORSTEO, respectively), which contains hPTH(1–34) as the active pharmaceutical ingredient. This drug is indicated to treat both men and postmenopausal women with osteoporosis who are at high risk for fracture. There were about 50,000 teriparatide users in the United States in 2010 (*14*).

[1]MicroCHIPS, Inc., Waltham, MA 02451, USA. [2]Endocrine Unit, Massachusetts General Hospital, Harvard Medical School, Boston, MA 02114, USA. [3]Department of Pathology, Case Western Reserve University, Cleveland, OH 44106, USA. [4]On Demand Therapeutics Inc., Tyngsboro, MA 01879, USA. [5]Department of Materials Science and Engineering, Koch Institute for Integrative Cancer Research, Massachusetts Institute of Technology, Cambridge, MA 02139, USA. [6]Department of Chemical Engineering, Koch Institute for Integrative Cancer Research, Massachusetts Institute of Technology, Cambridge, MA 02139, USA.
*To whom correspondence should be addressed. E-mail: rfarra@mchips.com

Continuous hPTH(1–34) administration promotes osteoclast activity, with resultant bone loss (*15*, *16*). Conversely, intermittent or pulsatile delivery of hPTH(1–34) provides anabolic therapy because it stimulates osteoblast cellular activity (bone formation) more than osteoclast cellular activity, thus increasing bone mass and bone mineral density (*17*). Subcutaneous injections of 20- to 40-µg doses of hPTH (1–34) administered daily for up to 2 years have resulted in a decrease in the incidence of fractures and have an acceptable safety profile (*18*, *19*). Teriparatide has therefore become an accepted drug for increasing bone mass and reconstituting bone structure and strength, but has poor patient compliance because of the need for daily subcutaneous injections (*20*).

This paper describes the first human trial of an implantable drug delivery device based on microchip reservoir technology that is wirelessly programmable over the Medical Implant Communications Service (MICS) band to deliver an anti-osteoporosis drug at precise times (Fig. 1, A to C). The primary objective of this clinical trial was to assess the PK of hPTH(1–34) released from implantable devices in vivo in patients with osteopenia or osteoporosis, with the PK after development of a fibrous capsule of specific interest. Safety measures included evaluation of the biological response to the implant and monitoring indicators of drug toxicity. Secondary objectives were to assess the bioactivity of the drug on the basis of changes in serum markers of bone formation and resorption, and to evaluate the reliability and reproducibility of releasing peptide from the device. The result of this effort was the demonstration of a programmable implant that was able to deliver hPTH(1–34) at scheduled intervals, with PK similar to multiple subcutaneous injections and without the pain and burden of daily injections.

RESULTS

Clinical trial design

The clinical trial for the microchip-based drug delivery device focused on assessing the PK profiles of drug released from the implant encapsulated with fibrous tissue in comparison to subcutaneous injections of FORSTEO. hPTH(1–34) releases were therefore initiated 8 weeks after implantation to ensure a fully developed tissue capsule. In addition to assessing the PK parameters, bone formation markers were evaluated to determine improvement to bone formation owing to hPTH(1–34) dosing from the implanted device. Comparator injections of FORSTEO were scheduled after device-mediated dosing was complete. These subcutaneous injections were administered at two doses: 20 and 40 µg. The 20-µg doses were administered before explanting the device, whereas the 40-µg doses were administered after explanting these devices as part of an amendment to the original protocol. The detailed protocol and study rationale are further described in Materials and Methods. Figure 1D summarizes the sequence and timing of the protocol.

hPTH(1–34) PK

PK profiles were obtained by measuring hPTH(1–34) in venous blood samples drawn periodically during the 6 hours after hPTH(1–34) release. For each subject (n = 7), hPTH(1–34) delivery from the implant was characterized by four PK profiles on days 60, 65, 70, and 84 (Fig. 2A), whereas two PK profiles each were determined for 20- and 40-µg injections of FORSTEO on days 91 and 96 and days 131 and 138, respectively (Fig. 2B). (The 40-µg dose was administered as two sequential 20-µg injections from the delivery pen and is hereinafter denoted "2 × 20 µg.") The interpatient differences in maximum concentration were attributed to the differences in patient weight (table S1). The average concentration profiles from all implant

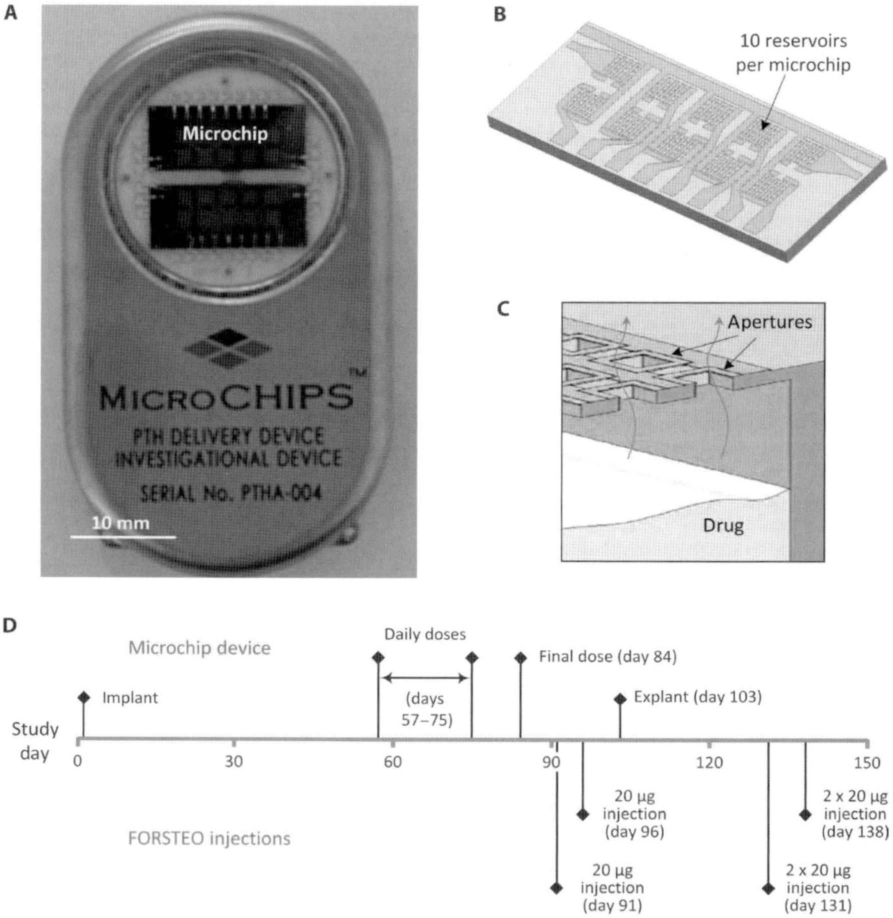

Fig. 1. The microchip-based drug delivery device and overview of study design. (**A** and **B**) Microchip-based hPTH(1–34) drug delivery device (54 mm × 31 mm × 11 mm, $l \times w \times h$) (A) containing two microchips with 10 reservoirs each (13.0 mm × 5.4 mm × 0.5 mm, $l \times w \times h$) (B). (**C**) Schematic cross section of microchip assembly showing drug releasing from one reservoir. (**D**) Timeline of study events.

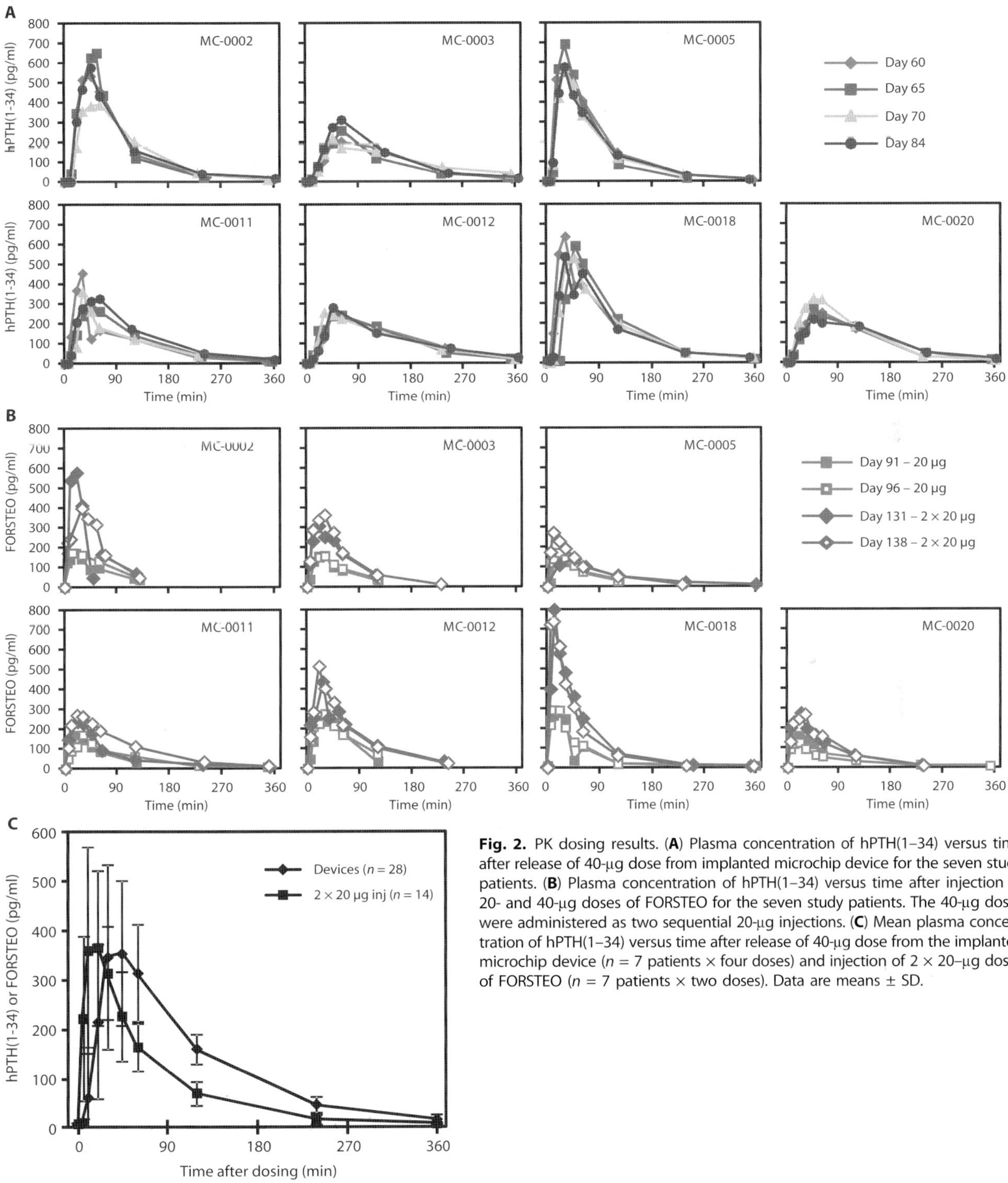

Fig. 2. PK dosing results. (**A**) Plasma concentration of hPTH(1–34) versus time after release of 40-µg dose from implanted microchip device for the seven study patients. (**B**) Plasma concentration of hPTH(1–34) versus time after injection of 20- and 40-µg doses of FORSTEO for the seven study patients. The 40-µg doses were administered as two sequential 20-µg injections. (**C**) Mean plasma concentration of hPTH(1–34) versus time after release of 40-µg dose from the implanted microchip device (n = 7 patients × four doses) and injection of 2 × 20–µg doses of FORSTEO (n = 7 patients × two doses). Data are means ± SD.

deliveries and all 2 × 20–μg subcutaneous injections are presented in Fig. 2C.

Noncompartmental analysis was conducted to determine the PK parameters characterizing hPTH(1–34) administration. These parameters included C_{max} (observed peak concentration), T_{max} (time to peak concentration), AUC (area under the curve), and $T_{1/2}$ (terminal half-life). The averages of these parameters by study day and by patient are presented in Tables 1 and 2A. The resulting intrapatient PK parameters of the microchip-implanted devices were reproducible for PK days 60, 65, 70, and 84 (Table 2A and Fig. 2A): coefficient of variance for C_{max} ranged from 7 to 22% and the coefficient of variance for AUC ranged from 2 to 20%. The corresponding intrapatient PK parameters for the FORSTEO injections are summarized in Table 2B and Fig. 2B. The coefficients of variance for the 2 × 20–μg injections for C_{max} ranged from 2 to 45%, and the coefficient of variance for AUC ranged from 2 to 34%. These data show that the coefficients of variance for C_{max} and AUC were lower for the device releases than for the subcutaneous injections. Similarly, the coefficients of variance for T_{max} and $T_{1/2}$ were lower for the implant releases than for the subcutaneous injections.

The average PK parameters for different modes of drug administration, based on this clinical study and published literature (*21*), are summarized in Table 3. PK parameter comparisons between microchip device–mediated release and subcutaneous injection can also be evaluated as a ratio of each delivery mode (average PK parameter for implant releases divided by average PK parameter for injections). Ratios of averages for the implant releases to the 40-μg FORTEO injections were as follows: C_{max} = 88%, AUC = 98%, T_{max} = 77% (Table 3). These ratios indicate that the resulting average implant C_{max} was 12% lower than that for the injection. The average implant AUC was about equal (within 2%) to that of the injections. The average time to maximum concentration for the implant was 1.3-fold faster than the time for the injections.

Ratios comparing the average values for the implant release to the 2 × 20–μg injections were also calculated: C_{max} = 101%, AUC = 157%, T_{max} = 196%, $T_{1/2}$ = 132% (Table 3). The average implant C_{max} was about equal (within 1%) to that of the 2 × 20–μg injections. The average implant AUC was about 1.6-fold higher than that of the 2 × 20–μg injections. The average time to maximum concentration for the implant was about twice that of the 2 × 20–μg injections. The average terminal half-life for the implant dosing was about one-third longer than that of the 2 × 20–μg injections.

Bone markers

Changes in bone formation and bone resorption were evaluated by measuring serum type I collagen propeptide (P1NP) and type I collagenolysis fragment (CTX), respectively. Blood samples to assess these two bone markers were obtained from each patient at screening, 2 weeks after device implant, on the first day of hPTH(1–34) dosing (day 57), and at each of the eight PK profile procedures (days 60, 65, 70, 84, 91, 96, 131, and 138). Serum P1NP progressively increased during the period of daily dosing (days 57 to 75) from the implant (Fig. 3A). The mean increase between the first day of dosing and the 14th day of dosing was 143% (P = 0.01, pairwise t test). An increase in P1NP is consistent with anabolic increase in bone formation, which is essential to increasing bone mineral density. The significance of increasing P1NP during device dosing confirms that the hPTH(1–34) delivery from the device was clinically effective. Pairwise t tests between

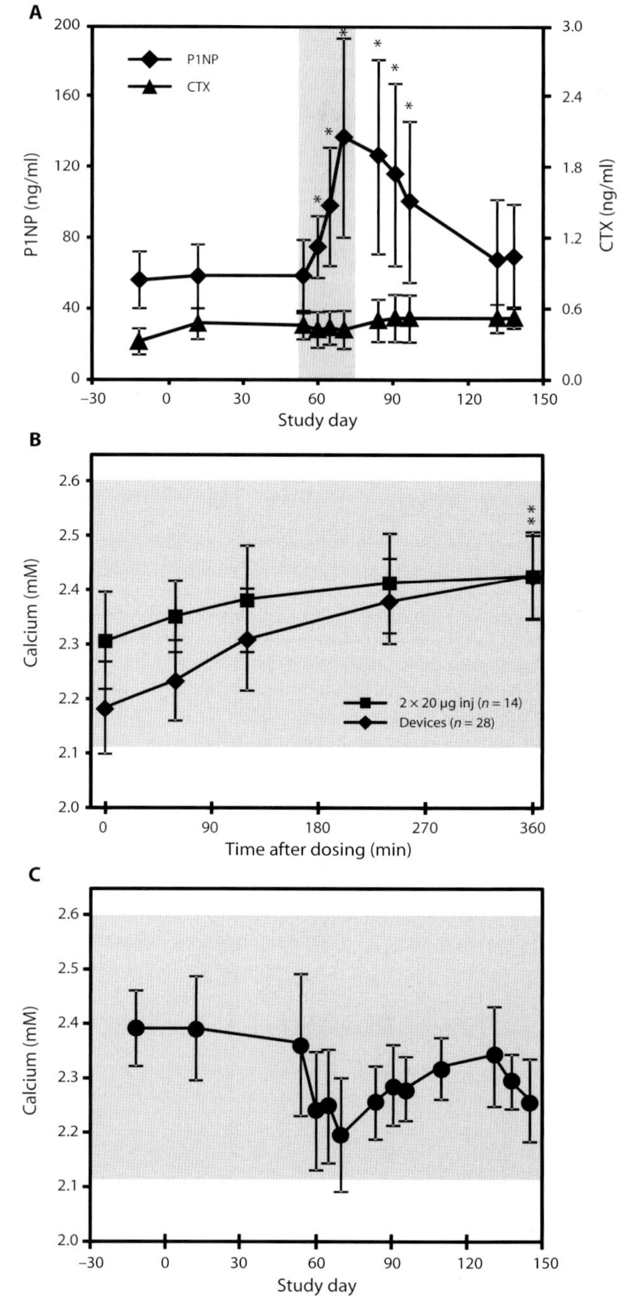

Fig. 3. Bone marker and calcium measurements. (**A**) P1NP and CTX bone marker concentrations in serum before, during, and after implant-mediated drug dosing. The shaded area encloses the 2 weeks during which individual 40-μg doses of hPTH(1–34) were released from the implant once daily. Data are means ± SD (n =7). *P < 0.05 compared to day −12 (screening visit), pairwise t test. (**B**) Serum calcium levels during dosing from the microchip implant (n = 28) and the 2 × 20–μg injections (n = 14). *P < 0.05 compared to predose calcium levels, pairwise t test. (**C**) Baseline serum calcium levels over the course of the study (n = 7). The shaded area in (B) and (C) represents range of normal serum calcium values. Data are means ± SD.

the average P1NP levels at screening (about 2 weeks before implantation) and those at study days 60, 65, 70, and 84 all had a *P* value of less than 0.05. P1NP levels began to fall after completion of the daily dosing from the implant (day >75). The P1NP levels at days 91 and 96, which were the 20-μg dosing days, had a *P* value of less than 0.05 (pairwise *t* test). The marker for bone resorption, serum CTX, was normal and did not vary during the interval of hPTH(1–34) dosing (Fig. 3A).

Table 1. PK parameters throughout the device dosing period. Data are means ± SD (n = 7).

Day	C_{max} (ng/ml)	T_{max} (min)	AUC (ng·min/ml)	$T_{1/2}$ (min)
60	410 ± 175	44 ± 13	44 ± 10	66 ± 16
65	426 ± 209	49 ± 10	44 ± 9	64 ± 20
70	378 ± 133	41 ± 11	43 ± 9	75 ± 27
84	405 ± 154	45 ± 12	46 ± 6	76 ± 16

Table 2. Results of PK parameters by patient. (A) Doses delivered by the implant (n = 4). (B) Doses delivered as 2 × 20–μg injections (n = 2). Data are means ± SD.

Patient	C_{max} (ng/ml)	T_{max} (min)	AUC (ng·min/ml)	$T_{1/2}$ (min)
		A. Implant [hPTH(1–34)]		
MC-0002	538 ± 111	51 ± 8	51 ± 2	51 ± 11
MC-0003	249 ± 49	56 ± 8	36 ± 4	89 ± 28
MC-0005	598 ± 67	32 ± 3	48 ± 4	48 ± 7
MC-0011	353 ± 79	41 ± 14	36 ± 7	69 ± 4
MC-0012	255 ± 19	45 ± 12	41 ± 1	91 ± 8
MC-0018	575 ± 51	39 ± 10	57 ± 2	68 ± 5
MC-0020	266 ± 43	49 ± 8	40 ± 3	72 ± 12
		B. Injection (FORSTEO)		
MC-0002	489 ± 127	26 ± 6	26 ± 1	39 ± 13
MC-0003	335 ± 39	25 ± 7	24 ± 5	44 ± 5
MC-0005	206 ± 93	28 ± 26	18 ± 1	64 ± 19
MC-0011	255 ± 23	21 ± 1	27 ± 9	77 ± 7
MC-0012	476 ± 56	23 ± 4	37 ± 1	61 ± 4
MC-0018	770 ± 43	11 ± 17	41 ± 3	45 ± 2
MC-0020	273 ± 5	27 ± 5	22 ± 1	49 ± 6

Safety assessment

Safety assessment was based on evaluating the biological response to the implant and the absence of any indications for drug toxicity. The laboratory assessment of drug toxicity indicated that the drug and device combination were safe, and there were no abnormalities reported as part of the adverse event/serious adverse event procedures.

Uric acid, creatinine, and blood urea nitrogen (BUN) values were measured during the dosing period to assess kidney function (table S2). No change in mean uric acid concentration was observed (P > 0.05, pairwise t test). The average concentrations of creatinine and BUN were also unchanged over the course of the study (both P > 0.05, pairwise t test), providing evidence of the safety of the treatment on kidney function. Routine liver panel assessments (aspartate transaminase, alanine transaminase, γ-glutamyl transpeptidase, alkaline phosphatase, total protein, and albumin) were also performed at dosing visits to ensure normal liver function (table S3). All results were within the normal limits.

Serum total calcium was measured in venous blood samples drawn immediately preceding and 1, 2, 4, and 6 hours after each dose delivered from the implant and after each subcutaneous injection of FORSTEO. The baseline value (t = −5 min) during the dosing period represented about the 24-hour time point with respect to the dose delivered on the previous day. A transient increase of serum calcium that is expected after dosing should return to the patient's typical range within 24 hours. Calcium levels increased slightly after dosing but remained below the upper limit of normal, which is 2.6 mM (*22*) (Fig. 3B). The increase in serum calcium levels at the 6-hour time point was statistically significant (P < 0.05, pairwise t test) compared to the predose calcium levels from both the implant and the FORSTEO injections (2 × 20 μg). Calcium levels 18 or more hours after teriparatide administration were all lower than baseline serum calcium measured before any teriparatide was administered (Fig. 3C).

Tissue capsule histology

Potential local effects of drug delivery were assessed for abnormal tissue reactions at the time of explantation by macroscopic evaluation of the capsule tissue surrounding the implant (Fig. 4, A and B) and by histological and cellular morphological analysis of this tissue (Fig. 4, C to H). The histology of the excised tissue capsules indicated that six of the seven capsules showed a normal wound-healing response, with few inflammatory cells (scores of 0 and 1 on a scale of 0 to 5, with 0

Table 3. Average PK parameters for hPTH(1–34) from the microchip device compared to 2 × 20–μg and single 20-μg FORSTEO injections. Data are means ± SD. ND, not determined.

Drug, method of delivery	Dose (μg)	Number of samples	C_{max} (pg/ml)	T_{max} (min)	$AUC_{0\text{-}last}$ (ng·min/ml)	$T_{1/2}$ (min)	Reference
hPTH(1–34), implant	40	28	405 ± 161	45 ± 11	44 ± 8	70 ± 20	This study
FORSTEO, injection	2 × 20	14	400 ± 194	23 ± 10	28 ± 9	53 ± 15	This study
FORTEO, injection*	40	34	460 (146–875)	58 (40–91)	46 (17–69)	ND	(*21*)
FORSTEO, injection	20	14	192 ± 55	22 ± 6	14 ± 4	55 ± 16	This study
FORTEO, injection	20	22	151 ± 57	32 ± 15	10 ± 4	90 ± 107	(*21*)

*Range shown in parentheses.

being no visible cells in field of view at 400×). These cells included neutrophils, lymphocytes, and macrophages as identified by histological and cellular morphological analysis (fig. S1). The seventh tissue capsule showed a higher amount of macrophages (scores of 0 to 3) but was still considered a normal wound-healing response (*23*). Microscopic examination showed minimal, if any, evidence of giant cells, edema, congestion, necrosis, hemorrhage, or granulation tissue cells. Histology showed no signs of degeneration, bacterial infection, or malignancy in any of the capsules (fig. S1).

The tissue capsule thickness over the microchip where the drug was released was consistent with the capsule thickness over the entire device. The capsule thickness over the microchip varied across the patients from 0.2 to 0.7 mm (Fig. 4, E and F). The average distance to the neovascularization bed across all patients was 0.1 mm, with a minimum and maximum distance of 0.05 and 0.35 mm, respectively. The relative size of the neovascularization bed was scored by assessing the width of the area from the implant/tissue interface to the unaffected areas that had the characteristics of normal tissue and normal vascularity.

Device functionality

The devices were inspected after explant to determine whether the reservoirs had opened properly by noting the number of apertures opened per reservoir. A total of 132 doses of drug were delivered from the seven devices. The electrothermal activation process was designed to simultaneously remove the 20 membranes that were sealing each of the reservoir's 20 apertures. Of the 132 reservoirs opened, 116 had all 20 membranes cleared. The number of membranes successfully removed from the remaining 16 reservoirs ranged from 9 to 19. Only three of these reservoirs were opened on days when the PK was assessed. The number of apertures opened for these three releases were 19, 16, and 15; however, only the latter case appeared to affect the resulting PK (patient MC-0002, day 70), as indicated by a slightly decreased C_{max} (Fig. 2A). The instances of partially opened reservoirs were limited to two of the seven devices. These have been correlated to high-resistance connections on a particular lot of printed circuit boards and were not design-related.

DISCUSSION

PK profile of implant versus injection

The primary objective of this study was to determine the PK profiles of hPTH(1–34) when delivered from multireservoir microchip implants after a fibrous capsule was formed around the implant. The PK profiles were assessed for reproducibility (day to day, interpatient, and intrapatient) and also compared with subcutaneous injections of FORSTEO. A key factor in assessing the bioequivalence of a new drug formulation is the correspondence of the PK. The FDA and the European Medicines Agency require the PK profiles once released from the device to be within 80 to 125% of the approved drug's PK values (*24*, *25*). The results obtained from this first-in-human microchip clinical trial indicated that the release profile of 40-μg doses of hPTH(1–34) from a wirelessly controlled microchip implant device was comparable to the profile of subcutaneous injections of 2 × 20–μg FORSTEO and was bioequivalent to the profile of single injections of 40 μg of FORTEO (Table 3). The release of hPTH(1–34) in plasma further showed the pulsatile profile required for anabolic response.

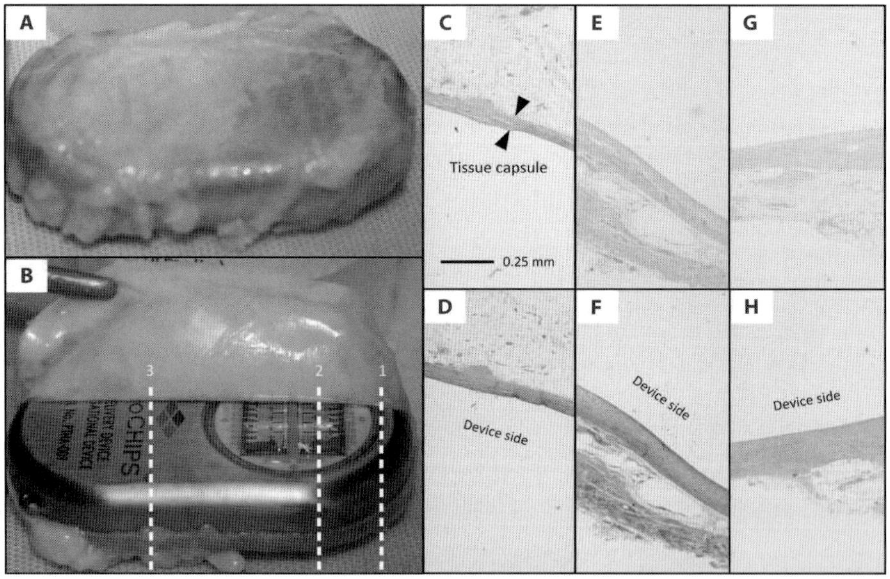

Fig. 4. Tissue histology results from a representative patient, MC-0012. (**A** and **B**) Two representative macroscopic images of the tissue capsule surrounding the device after explantation. (**C** to **H**) Micrographs of the tissue capsule from each patient consisted of three total images from both the dorsal (antenna, toward skin) and the ventral (microchip, toward muscle) sides. Top row, H&E stain; bottom row, Masson's trichrome stain. (C and D) Cross-section 1 at the microchip and titanium interface (dorsal). (E and F) Cross-section 2 over the microchip (ventral). (G and H) Cross-section 3 over the titanium case (ventral).

Drug absorption from the implant appeared to be slower than the 2 × 20–μg injections, but faster than the single 40-μg injections, based on observed T_{max} values. There was no evidence that such differences in absorption rate altered skeletal responses to hPTH(1–34), using P1NP as an indicator of bone formation. Here, the changes in serum calcium, CTX, and P1NP during implant therapy qualitatively and quantitatively mimic those reported previously during daily subcutaneous injections of FORSTEO (*26*). Because bioequivalence has been demonstrated here, future implant designs should be sized for the same dose size as that of the approved dose for injection. Furthermore, the 2 × 20–μg injections indicated that the T_{max} and $T_{1/2}$ values were about equal to those of a single 20-μg injection, rather than those of a 40-μg injection. These differences may be attributed to the method of administration of the 2 × 20–μg injections. Two injections versus one drug bolus may increase the surface area, resulting in absorption into the bloodstream, similar to that of a single 20-μg injection. Conversely, drug releases from the implants were reproducible within each patient and more reproducible than the subcutaneous injections. PK data from the microchip implant had lower coefficients of variance than the injections. A possible explanation for this was that the implant environment was more consistent from dose to dose than subcutaneous injections, because the needle location would have varied with each administration, whereas the implant resides in a stable environment.

Biological markers of bone turnover are used in the treatment of osteoporosis to monitor efficacy and improve fracture risk assessment. Two markers were monitored over the course of the study: P1NP, a widely accepted bone formation marker and a predictor of long-term increase in bone mass, and the bone resorption marker CTX. Daily hPTH(1–34) released from the microchip implant device progressively and statistically increased P1NP. When daily dosing was terminated, P1NP levels decreased back to levels observed at the start of the trial. Daily hPTH(1–34) released from the microchip implant device did not increase CTX. The increase in P1NP and the constancy of CTX are indications that the hPTH(1–34) dosing increased bone formation instead of bone resorption, as expected. Furthermore, serum calcium levels and markers for liver and kidney function remained within the clinically accepted ranges.

Histology of the explanted fibrous tissue capsule formed around the microchip implant device was consistent in terms of tissue thickness and composition with capsules of other biocompatible devices, such as pacemakers. The capsule thickness over the microchip varied across the seven patients, but there were no observable differences in capsule histology for sections directly over the microchip (where the Pt/Ti membranes were electrothermally removed to release the drug) versus sections contacting other inactive components of the implant. This observation implies that the membrane opening and drug release did not alter the capsule locally and is further evidence of the tolerability and biocompatibility of the drug and device.

Clinical and biological effects of the implant

This study has demonstrated the clinical viability of the microchip-based implantable drug delivery device. The device and hPTH(1–34) drug combination were biocompatible, had no adverse immune reaction, and were well tolerated and accepted by the patients in this study. The microchip successfully protected and released each dose precisely as programmed. Furthermore, the resulting PK profiles from the implant—even through the fibrous tissue capsule surrounding the implant—were comparable to the PK profiles of multiple subcutaneous injections. These dose releases were anabolic, because evidence showed an increase in the bone formation marker P1NP.

A benefit of this microchip device is that it can be both implanted and explanted in a physician's office using a local anesthetic. For this study, all of the patients' surgical incisions healed normally after both the implant and the explant surgical procedures. Discomfort associated with each procedure required only acetaminophen or ibuprofen taken for no more than 2 days after surgery. The microchip implant device was well tolerated by the patients on the basis of surveys conducted throughout the study (table S4). The patients responded favorably, indicating that they would repeat the procedure to implant such a device again, they were satisfied with the implant location, the implant site was comfortable, and they tolerated the size of the implant.

Missed drug doses and the few partial reservoir openings were due to printed circuit board fabrication (high-resistance or open connections) and did not affect the overall outcome of the trial. The one device that did not release any drug contained a faulty component in the membrane activation circuitry required to release the drug. Specific tests to assess the membrane activation circuitry will become part of the manufacturing inspection process and will ensure proper operation of implanted devices. Also, although our device contained only 20 reservoirs (for up to 20 doses), a microchip-based device containing a larger number of reservoirs will be needed to deliver daily doses of proteins, peptides, and other drugs over the course of 1 or more years. These implants will provide effective, scheduled treatment for patients without the disadvantages associated with injection-based drug administration. The cost of an annual drug delivery implant is expected to be equal to that of other electronic implants, such as pacemakers and implantable cardioverter defibrillators. Indications that will benefit from this implant will be those that require frequent, scheduled dosing, such as anabolic osteoporosis treatment requiring daily teriparatide injections, and multiple sclerosis treatment requiring injections of interferon-β1a every 48 hours to decrease the frequency of exacerbations and to improve physical abilities. Alternatively, one can envision use of such a device to deliver potent drugs on demand in an acute situation.

Despite the current limitations, a microchip-based drug delivery device has several advantages, including custom PK to achieve desired efficacy, as well as the ability to achieve injection-like PK profiles without repeated needle injections. In addition, the device can be implanted in various body compartments for more localized delivery to maximize delivery to target tissue while minimizing the systemic side effects. Future applications involving closed-loop control may be possible because the implant can be triggered to release a drug based on feedback from sensors in the body to achieve needed therapy or drug delivery. Patients will further benefit from the programmable and automatic drug delivery functionality without fear of overdosing and underdosing, and will be 100% compliant without having to intervene or remember to take their medication.

MATERIALS AND METHODS

Study design

The clinical trial was conducted in Denmark in accordance with the principles of the Declaration of Helsinki from the International Conference on Harmonization. A clinical research organization (CRO), Center for Clinical and Basic Research (CCBR)–Synarc, was contracted to facilitate regulatory approval, patient recruiting and management, and overall study execution. The trial was approved by the CRO's Research Ethics Committee and the Danish Medicines Agency. The trial was registered in the European Clinical Trials Database (EudraCT, number 2010-020040-35), and a MedDRA account was established to record adverse events. All patients provided informed consent.

Patients

Patients (n = 7) participating in the study were osteoporotic postmenopausal women between the ages of 65 and 70, in good health, with a body mass index ranging from 18.5 to 30.2, and with normal thyroid function. Patient weights varied from 45.3 to 80.7 kg (table S1). Dual-energy x-ray absorptiometry was used to screen for osteoporosis. Subjects who had taken bone-active drugs in the preceding 6 months were excluded from the study, as were those with any illness relating to bone or calcium metabolism (for example, kidney and urinary tract stones, hepatitis, or compromised immune function).

Potential subjects with active implantable medical devices were also excluded. Patients were asked to participate in periodic surveys to assess noticeable pain and their satisfaction level with an implantable drug delivery device. A total of four surveys were conducted with a 100% response rate. An eighth patient was enrolled in the study, but on-board diagnostics for that device reported that the drug was not released. All results from the eighth patient were excluded from the analysis.

Implantable device

The implantable drug delivery system consisted of the implant and a PC–based programmer. The implant (Fig. 1A) integrated two drug-containing microchip assemblies on the surface of a titanium housing that contained control and communication electronics. Each microchip assembly (Fig. 1, B and C) contained 10 individual 40-μg doses of lyophilized hPTH(1–34) formulation, for a total of 20 doses per device. A drug dose could be released immediately upon receipt of a command from the programmer or at a prespecified time in the future. The programmer, operating in the Medical Implant Communication Service (MICS) band, wirelessly transmitted instructions, such as dose scheduling, to the implant. The bidirectional communications link permitted the upload of implant status information, including dose delivery confirmation and battery voltage.

Devices and drug were manufactured with systems consistent with Good Manufacturing Practices. The assembled and packaged devices were sterilized by ethylene oxide gas at a maximum temperature of 36°C. Device verification tests for active implants (including ISO 14708, IEC 60601-1-2, FCC 47CFR 95, ETSI EN 301 839-1, -2, -3) were conducted before trial submission and approval to ensure functionality, safety, and electrical and radio-frequency compliance. The device was classified as a tissue-contacting implant for long implant duration (≥30 days). All biocompatibility tests were performed in compliance with ISO 10993 and the FDA Blue Book Memorandum G95. The drug was tested for toxicity, stability, and impurities. Nonclinical tests were conducted per Good Laboratory Practice standards by Toxikon Corp. Electrical, emissions, safety, and transportation testing were conducted by a certified laboratory (Intertek).

Microchip

The key component of the microchip assembly was a 13.0 mm × 5.4 mm × 0.5 mm ($l \times w \times h$) silicon chip having 10 individually addressable, 600-nl reservoirs. The microchip's reservoirs were filled with drug solution and lyophilized (see "Drug formulation"). A mating chip made of silicon was then joined to the silicon chip to hermetically seal the individual reservoirs by a room temperature compression welding process. The tissue-contacting face of the reservoirs was perforated with twenty 0.1-mm-diameter apertures, each covered by a composite membrane of titanium and platinum. Circuit traces, connecting the 20 membranes and wired to the internal electronics, provided the path for a current pulse to ablate individual membranes and to expose their reservoir's contents to tissue fluid surrounding the device. The silicon microchips were fabricated by Micralyne Inc.

Drug formulation

We formulated an hPTH(1–34) solution to meet the requirements of the multireservoir microchip delivery device. The requirements included (i) concentrated (>50 mg/ml) solutions to fill the microreservoirs and achieve a clinically relevant dose, (ii) rapid dissolution (seconds) and release of the dosage form to mimic pulsatile PK of a subcutaneous injection, and (iii) stability at 37°C for the duration of the study.

The active pharmaceutical ingredient, hPTH(1–34) acetate lyophilizate (PolyPeptide Laboratories), was prepared by adding 150 mg of hPTH (1–34) to 1670 μl of an aqueous solution of 4.4 M glacial acetic acid, 0.20 M citric acid, and 0.39 M histidine. After filtering through a 0.2-μm filter, a custom robotic system was used to aseptically dispense 500 nl into each microchip reservoir. The microchips were placed into a lyophilizer (Genesis 25 EL, VirTis Inc.) to remove water and acetic acid, leaving a solid dosage form. The hPTH(1–34) content of each dose, determined by high-performance liquid chromatography analysis, was 40 ± 2 μg.

The microchip assemblies were sealed with a cold compression weld and then tested for hermeticity. Reservoirs that did not pass the hermeticity specification (3×10^{-10} atm cm^3/s) were opened in sterile saline to release the drug. The microchip assemblies were then attached to the devices, resulting in the following number of reservoirs per device per patient: patient MC-0018 device had 20 doses; patients MC-0002, MC-0003, MC-0011, MC-0012, and MC-0020 devices had 19 doses each; patient MC-0005 device had 17 doses. This resulted in a total of 132 doses for the study.

Surgical procedures

The drug delivery device was implanted in a surgeon's office during an outpatient visit. The implant location was the subcutaneous space of the abdomen, just below the waistline. Patients were given injections of lidocaine as a local anesthetic. A 2.5-cm-long incision was made through the dermis followed by blunt dissection to create a pocket of equal size to the device. Each device was placed in the pocket with the microchip facing the muscle fascia and was anchored with two suture loops to minimize micromotion in the subcutaneous space. The sutures used to anchor the device were nonabsorbable polypropylene, and the incision was approximated with a nylon suture. Patients were instructed to take acetaminophen or ibuprofen to manage postsurgical pain. The condition of the implant site was documented during follow-up visits with a physician at 1 and 2 weeks after surgery.

The device explant procedure was also performed during an outpatient visit at the surgeon's office. The devices and their encapsulating fibrous tissue were removed under local anesthetic. The explanted tissue and device were placed in 10% buffered formalin in preparation for histological examination. Follow-up visits took place at about 1 and 6 weeks after explant.

Study procedure

Figure 1D summarizes the sequence and timing of the study. Eight weeks were allowed to pass after device implantation before drug release was initiated to ensure formation of a stable fibrous capsule around the implant. The implant delivered up to 19 daily doses of 40 μg of hPTH(1–34) per patient during days 57 to 75. The final implant dose was released about 1 week later, on day 84. The first of these doses was administered while the patient was under observation at the clinic. Four of the remaining doses were delivered while the patient was at the clinic for PK analysis (days 60, 65, 70, and 84). All other doses were released automatically at a predetermined time under control of the implant. Two doses of teriparatide (FORSTEO, Eli Lilly) were administered on days 91 and 96 to determine the comparative PK of a subcutaneous injection. The microchip was explanted on day 103. Two additional FORSTEO PK analyses were carried out on days

131 and 138 (after explant), during which patients were given two sequential injections (2 × 20 μg) of FORSTEO for a total dose of 40 μg.

PK determination

hPTH(1–34) PK analyses were carried out in the clinic. The patients' vital signs were measured and a peripheral intravenous line was inserted. A baseline blood sample was drawn within 5 min of dosing. A command was sent wirelessly to the microchip implant to release the drug. For subcutaneous injections, FORSTEO was administered with the manufacturer's injector pen. The 40-μg FORSTEO doses were accomplished by injecting a second dose at the same site without removing the pen. These are denoted 2 × 20 μg in the Results section. Blood samples for hPTH(1–34) determination were then collected at 5, 10, 20, 30, 45, 60, 120, 240, and 360 min after the dose. Serum calcium kinetics were determined with samples drawn at –5, 60, 120, 240, and 360 min.

Blood sampling and assays

Laboratory procedures conducted in Denmark were managed by CCBR-Synarc Research Laboratory. The concentration of hPTH(1–34) in the plasma samples was measured by Intertek ALTA Analytical Laboratory (San Diego, CA) with High Sensitivity ELISA (enzyme-linked immunosorbent assay) kit (Immutopics). Method validation for the kit was conducted according to FDA Good Laboratory Practices. The assay was qualified to a lower limit of quantitation (LLOQ) of 7.5 pg/ml, with a cross-reactivity to endogenous hPTH(1–84) of less than 6.5 weight percent (wt %) [for assay specifications for cross-reactivity, see (*27*)]. Endogenous PTH levels were measured at screening and found to be within the normal range (15 to 65 pg/ml) for all seven patients, so any contribution of hPTH(1–84) to the response was expected to be below the LLOQ. Synarc Research Laboratory (Rødovre, Denmark) conducted the sample analyses for serum calcium, P1NP, and CTX. The serum calcium test was qualified to an LLOQ of 0.25 mM. The LLOQs for P1NP and CTX were 5 ng/ml and 0.010 nM, respectively.

Tissue capsule histology

Tissue samples were excised at the time of device explant. The device within the capsule was stored in 10% buffered formalin for about 7 weeks before analysis. The devices were shipped to Toxikon Corp., and histological analysis was performed on each capsule. Once the device was removed from the capsule, three cross sections were taken in the following locations: the edge of the microchip, the center of the microchip, and over the titanium case. The sections were embedded in paraffin, and histologic slides were prepared and stained with both hematoxylin and eosin (H&E) and Masson's trichrome.

Biological reaction was assessed by light microscopic histological analysis for the presence of an inflammatory response (polymorphonuclear cells, lymphocytes, plasma cells, macrophages, giant cells, and necrosis). The healing response was scored by the amount of neovascularization and fibrosis. Neovascularization was identified by the distinct morphological appearance of blood vessels, and fibrosis was identified by the distinct morphological appearance and pattern of collagen deposition. The ventral (toward the skin) and dorsal (toward the muscle) sides of the capsule were scored separately. Scores were based on a five-point scale (fig. S1).

Statistical analysis

Summary statistics (mean, SD, coefficient of variance), the identification of maximum hPTH(1–34) concentration (C_{max}) and the time to reach C_{max} (T_{max}) from tabulated data, and the determination of the area under the PK curve (AUC) by application of the linear trapezoidal rule were performed with Microsoft Excel (version 12.0). Values of hPTH(1–34) that were below the LLOQ were not included in the PK analysis.

SUPPLEMENTARY MATERIALS

www.sciencetranslationalmedicine.org/cgi/content/full/4/122/122ra21/DC1
Fig. S1. Overview of the histological analysis performed on the tissue capsule for each patient.
Table S1. Patient history collected at screening visit (day –12).
Table S2. Results of routine kidney panel assessment (*n* = 7 patients).
Table S3. Results of routine liver panel assessment (*n* = 7 patients).
Table S4. Survey questions and responses conducted throughout the study.

REFERENCES AND NOTES

1. G. W. Stone, S. G. Ellis, D. A. Cox, J. Hermiller, C. O'Shaughnessy, J. T. Mann, M. Turco, R. Caputo, P. Bergin, J. Greenberg, J. J. Popma, M. E. Russell; TAXUS-IV Investigators, A polymer-based, paclitaxel-eluting stent in patients with coronary artery disease. *N. Engl. J. Med.* **350**, 221–231 (2004).
2. J. T. Santini Jr., M. J. Cima, R. Langer, A controlled-release microchip. *Nature* **397**, 335–338 (1999).
3. A. C. Richards Grayson, I. S. Choi, B. M. Tyler, P. P. Wang, H. Brem, M. J. Cima, R. Langer, Multi-pulse drug delivery from a resorbable polymeric microchip device. *Nat. Mater.* **2**, 767–772 (2003).
4. J. H. Prescott, S. Lipka, S. Baldwin, N. F. Sheppard Jr., J. M. Maloney, J. Coppeta, B. Yomtov, M. A. Staples, J. T. Santini Jr., Chronic, programmed polypeptide delivery from an implanted, multireservoir microchip device. *Nat. Biotechnol.* **24**, 437–438 (2006).
5. J. R. Coppeta, K. Shelton, N. F. Sheppard Jr., D. Snell, C. M. B. Santini, Compression and cold weld sealing methods and devices, U.S. Patent 2006/0115323 A1 (2006).
6. J. M. Maloney, S. A. Uhland, B. F. Polito, N. F. Sheppard Jr., C. M. Pelta, J. T. Santini Jr., Electrothermally activated microchips for implantable drug delivery and biosensing. *J. Control. Release* **109**, 244–255 (2005).
7. J. H. Prescott, T. J. Krieger, S. Lipka, M. A. Staples, Dosage form development, in vitro release kinetics, and in vitro–in vivo correlation for leuprolide released from an implantable multi-reservoir array. *Pharm. Res.* **24**, 1252–1261 (2007).
8. E. R. Proos, J. H. Prescott, M. A. Staples, Long-term stability and in vitro release of hPTH(1-34) from a multi-reservoir array. *Pharm. Res.* **25**, 1387–1395 (2008).
9. J. M. Anderson, Cardiovascular device retrieval and evaluation. *Cardiovasc. Pathol.* **2**, 199–208 (1993).
10. W. G. Brodbeck, M. MacEwan, E. Colton, H. Meyerson, J. M. Anderson, Lymphocytes and the foreign body response: Lymphocyte enhancement of macrophage adhesion and fusion. *J. Biomed. Mater. Res. A* **74**, 222–229 (2005).
11. L. Perry, F. Karp, K. Hauch, B. D. Ratner, Explanted pacemakers: Observations of the long-term foreign body response. *J. Undergrad. Res. Bioeng.* **7**, 13–21 (2007).
12. WHO Scientific Group on the Assessment of Osteoporosis at Primary Health Care Level, *Summary Meeting Report, Brussels, Belgium, 5-7 May 2004* (WHO Press, Geneva, Switzerland, 2007).
13. R. Burge, B. Dawson-Hughes, D. H. Solomon, J. B. Wong, A. King, A. Tosteson, Incidence and economic burden of osteoporosis-related fractures in the United States, 2005–2025. *J. Bone Miner. Res.* **22**, 465–475 (2007).
14. Eli Lilly and Co., *Annual Report and Proxy Statement*; http://investor.lilly.com/annuals.cfm
15. J. M. Hock, I. Gera, Effects of continuous and intermittent administration and inhibition of resorption on the anabolic response of bone to parathyroid hormone. *J. Bone Miner. Res.* **7**, 65–72 (1992).
16. H. Dobnig, R. T. Turner, The effects of programmed administration of human parathyroid hormone fragment (1–34) on bone histomorphometry and serum chemistry in rats. *Endocrinology* **138**, 4607–4612 (1997).
17. C. A. Frolik, E. C. Black, R. L. Cain, J. H. Satterwhite, P. L. Brown-Augsburger, M. Sato, J. M. Hock, Anabolic and catabolic bone effects of human parathyroid hormone (1-34) are predicted by duration of hormone exposure. *Bone* **33**, 372–379 (2003).
18. R. M. Neer, C. D. Arnaud, J. R. Zanchetta, R. Prince, G. A. Gaich, J. Y. Reginster, A. B. Hodsman, E. F. Eriksen, S. Ish-Shalom, H. K. Genant, O. Wang, B. H. Mitlak, Effect of parathyroid hormone (1-34) on fractures and bone mineral density in postmenopausal women with osteoporosis. *N. Engl. J. Med.* **344**, 1434–1441 (2001).
19. C. P. Jerome, D. B. Burr, T. Van Bibber, J. M. Hock, R. Brommage, Treatment with human parathyroid hormone (1-34) for 18 months increases cancellous bone volume and improves

trabecular architecture in ovariectomized cynomolgus monkeys (*Macaca fascicularis*). *Bone* **28**, 150–159 (2001).

20. K. Taylor, D. T. Gold, P. Miller, P. Chen, M. Wong, K. Krohn, Teriparatide therapy in a community setting: Persistence and use of other osteoporosis medications in DANCE, paper presented at the 30th Annual Meeting of the American Society for Bone and Mineral Research, 2008.
21. H. Y. Ahn, *Clinical Pharmacology and Biopharmaceutics Review* (Food and Drug Administration, Center for Drug Evaluation and Research, Silver Spring, MD, 2002); http://www.accessdata.fda.gov/drugsatfda_docs/nda/2002/21-318_FORTEO_BioPharmr.pdf
22. A. Kratz, M. Ferraro, P. M. Sluss, K. B. Lewandrowski, Case records of the Massachusetts General Hospital. Weekly clinicopathological exercises. Laboratory reference values. *N. Engl. J. Med.* **351**, 1548–1563 (2004).
23. J. M. Anderson, Biological responses to materials. *Annu. Rev. Mater. Res.* **31**, 81–110 (2001).
24. Food and Drug Administration, Center for Drug Evaluation and Research, *Guidance for Industry: Statistical Approaches to Establishing Bioequivalence* (Food and Drug Administration, Silver Spring, MD, 2001); http://www.fda.gov/downloads/Drugs/GuidanceComplianceRegulatoryInformation/Guidances/UCM070244.pdf.
25. European Medicines Agency, Committee for Medicinal Products for Human Use, *Guideline on the Investigation of Bioequivalence* (European Medicines Agency, London, UK, 2009).
26. S. J. Glover, R. Eastell, E. V. McCloskey, A. Rogers, P. Garnero, J. Lowery, R. Belleli, T. M. Wright, M. R. John, Rapid and robust response of biochemical markers of bone formation to teriparatide therapy. *Bone* **45**, 1053–1058 (2009).
27. Immutopics Inc., High Sensitivity Human PTH (1-34) ELISA Kit; http://www.immutopics.com/pdf/directional-inserts/60-3900.pdf.

Acknowledgments: We thank C. S. Teglbjærg and the staff at CCBR-Synarc in Denmark for their assistance in conducting the clinical trial. **Funding:** This study was sponsored by MicroCHIPS, Inc. **Author contributions:** All authors have contributed in one or more of the following ways: development of the microchip-based device, protocol development, conducting portions of the study, and data collection (R.F., N.F.S., and L.M.); data analysis (R.F., N.F.S., L.M., R.M.N., and J.M.A.); and preparing the manuscript (R.F., N.F.S., L.M., R.M.N., J.M.A., J.T.S., M.J.C., and R.L.). **Competing interests:** R.F., N.F.S., and L.M. are employees of MicroCHIPS Inc. R.L. is a board member and R.M.N. and M.J.C. are paid consultants of MicroCHIPS, Inc. N.F.S., J.T.S., M.J.C., and R.L. hold patents in various aspects of the microchip. The authors declare that their spouses, partners, or children have no financial relationships that may be relevant to the submitted work, and none of the authors have any nonfinancial interests that may be relevant to the submitted work. **Data and materials availability:** European Clinical Trials Database (EudraCT, number 2010-020040-35). All reasonable requests for collaboration or clinical testing of the device will be fulfilled provided that a written agreement is executed in advance between MicroCHIPS, Inc. and the requester (and his or her affiliated institution). Inquiries should be directed to the corresponding author.

Submitted 29 September 2011
Accepted 15 December 2011
Rapid Publication (accepted manuscript): 16 February 2012
10.1126/scitranslmed.3003276

Citation: R. Farra, N. F. Sheppard Jr., L. McCabe, R. M. Neer, J. M. Anderson, J. T. Santini Jr., M. J. Cima, R. Langer, First-in-human testing of a wirelessly controlled drug delivery microchip. *Sci. Transl. Med.* **4**, 122ra21 (2012).

ARTICLES

nature biotechnology

Combinatorial discovery of polymers resistant to bacterial attachment

Andrew L Hook[1,9], Chien-Yi Chang[2,9], Jing Yang[1,9], Jeni Luckett[2], Alan Cockayne[2], Steve Atkinson[2], Ying Mei[3,8], Roger Bayston[4], Derek J Irvine[5], Robert Langer[3,6,7], Daniel G Anderson[3,6,7], Paul Williams[2,10], Martyn C Davies[1,10] & Morgan R Alexander[1,10]

Bacterial attachment and subsequent biofilm formation pose key challenges to the optimal performance of medical devices. In this study, we determined the attachment of selected bacterial species to hundreds of polymeric materials in a high-throughput microarray format. Using this method, we identified a group of structurally related materials comprising ester and cyclic hydrocarbon moieties that substantially reduced the attachment of pathogenic bacteria (*Pseudomonas aeruginosa, Staphylococcus aureus* and *Escherichia coli*). Coating silicone with these 'hit' materials achieved up to a 30-fold (96.7%) reduction in the surface area covered by bacteria compared with a commercial silver hydrogel coating *in vitro*, and the same material coatings were effective at reducing bacterial attachment *in vivo* in a mouse implant infection model. These polymers represent a class of materials that reduce the attachment of bacteria that could not have been predicted to have this property from the current understanding of bacteria-surface interactions.

Healthcare-associated infection is widely acknowledged as the most frequent adverse event in hospitals. It has been estimated that 80% of the infections acquired in hospitals involve biofilms[1]. Biofilms are surface-associated bacterial communities within which bacteria have up to 1,000 times higher resistance to antimicrobials and host defenses compared with their planktonic counterparts[2,3]. Biofilms present on the surface of medical devices provide the bacterial inocula for disease and can also serve as reservoirs of plasmids carrying antibiotic-resistance genes[4,5]. Most strategies for reducing biofilm-associated infections focus on the modification of existing materials that are used to manufacture in-dwelling medical devices by the incorporation of antibiotics[6,7] or other antimicrobials, such as silver salts, nitrofurazone, chlorhexidine, polymerized quaternary ammonium surfactants, antibacterial peptides and anionic nanoporous hydrogels[8–15]. These approaches aim to kill bacterial cells that attach to a material, but we believe that greater efficacy in preventing biofilm formation might be achieved by the development of new materials that are inherently resistant to biofilm formation[16]. Previous attempts to use an anti-attachment strategy include poly(ethylene glycol) brushes[17] and zwitterionic polymers[18,19]. An important problem that has restricted the development of materials resistant to bacterial attachment and growth is the poor understanding of the interactions that bacteria make with surfaces, which is required for *ab initio* materials design.

To overcome this constraint, we have developed a high-throughput approach to study bacterial attachment to hundreds of materials in a parallel assay format that utilizes polymer microarrays[20,21]. A microarray format has previously proven useful for identifying polymeric materials that support stem cell attachment and proliferation[22–25]. In this study polymer microarrays were adapted for the combinatorial development of materials resistant to bacterial attachment and growth (**Fig. 1**). *In vitro* testing has allowed us to identify a novel class of structurally related polymers that are resistant to bacterial attachment, outperforming commercially available silver-containing coatings. *In vivo* testing of the most promising of our materials has demonstrated the potential of these materials as coatings to reduce device-centered infection.

RESULTS

A library of 22 acrylate monomers was selected from those available commercially to provide wide chemical diversity, including ethylene glycol chains of various lengths, fluoro-substituted alkanes, linear and cyclic aliphatic, aromatic and amine moieties. To generate a large combinatorial space, we mixed each of 16 monomers (**Fig. 1a**, 1–16) as the major component with each of another 6 monomers (**Fig. 1a**, A–F) at ratios of 100:0 (homopolymer, 6 repeats), 90:10, 85:15, 80:20, 75:25 and 70:30 to create 576 monomer solutions. The combined monomer solutions were printed in triplicate as 300-μm diameter spots with a height of ~20 μm onto a poly (hydroxyl ethylmethacrylate) (pHEMA)-coated microscope slide where they were photopolymerized by a free radical mechanism

[1]Laboratory of Biophysics and Surface Analysis, University of Nottingham, Nottingham, UK. [2]School of Molecular Medical Sciences, University of Nottingham, Nottingham, UK. [3]Department of Chemical Engineering, Massachusetts Institute of Technology, Cambridge, Massachusetts, USA. [4]School of Clinical Sciences, Queen's Medical Centre, University of Nottingham, Nottingham, UK. [5]School of Chemistry, University of Nottingham, Nottingham, UK. [6]David H. Koch Institute for Integrative Cancer Research, Massachusetts Institute of Technology, Cambridge, Massachusetts, USA. [7]Harvard-MIT Division of Health Science Technology, Massachusetts Institute of Technology, Cambridge, Massachusetts, USA. [8]Present address: Clemson-MUSC Bioengineering program, Clemson University, Charleston, South Carolina, USA. [9]These authors contributed equally to this work. [10]These authors jointly directed this work. Correspondence should be directed to M.R.A. (morgan.alexander@nottingham.ac.uk).

Received 11 May 2012; accepted 26 June; published online 12 August 2012; doi:10.1038/nbt.2316

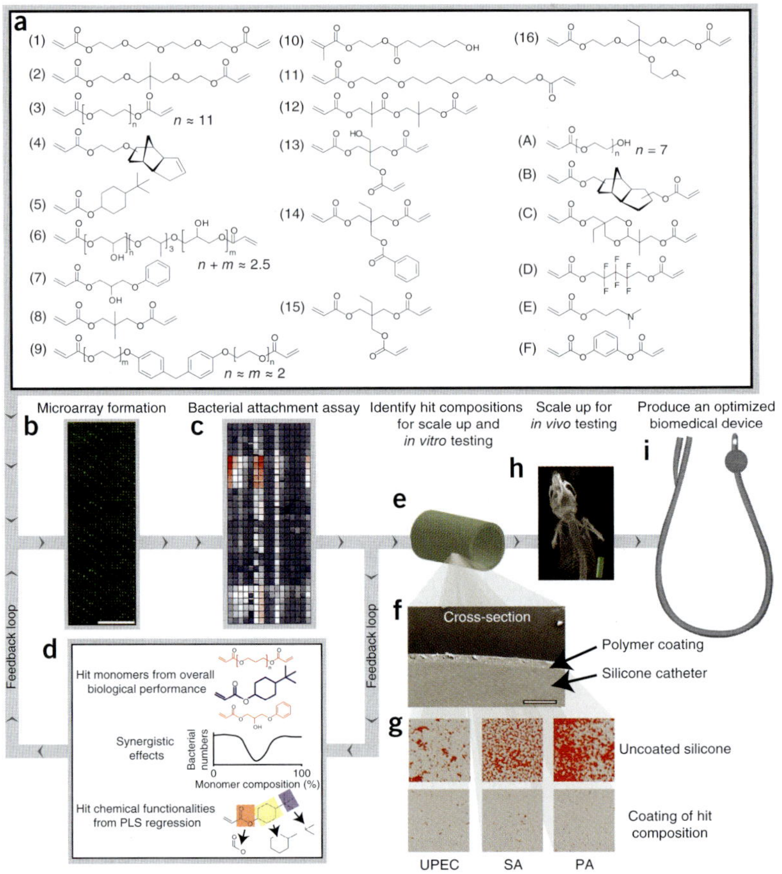

Figure 1 Schematic of the approach used to identify hit materials resistant to bacterial attachment and scale-up of hit materials. (**a**) The chemical structures of the monomers. (**b–f**) Outline of the strategy utilized for identifying hit material composition. (**b**) Fluorescence scanning image of the first-generation microarray after incubation with *P. aeruginosa* for 72 h. Three replicate arrays were present on each glass slide. Scale bar, 10 mm. (**c**) Intensity map of the bacterial data from the first-generation array. (**d**) The feedback loop used to select monomer compositions for a second-generation array. The biological performance of each monomer, the identification of hit material compositions that have synergistic effects and the results from the PLS regression of ToF-SIMS spectra and bacterial data were used to inform the composition of the second-generation array. (**e**) Schematic representation of the hit material composition scaled up. (**f**) An SEM image of the cross-section of a silicone catheter coated with a hit polymer (thickness, 20–25 μm). Scale bar, 100 μm. (**g**) Confocal images of SYTO17-stained biofilm for the three pathogens studied (*P. aeruginosa* (PA), *S. aureus* (SA), UPEC) from coated and uncoated silicone catheters. Each image is 160 × 160 μm. (**h**) Microcomputed tomography scan of a coated catheter implanted subcutaneously into a murine model. The catheter has been colored green in this image. (**i**) A schematic representation of a catheter coated with the hit composition.

to form the first-generation combinatorial polymer microarray (**Fig. 1b**). Hits from this first-generation polymer microarray were used to formulate the second-generation array and so on.

Bacterial microarray screening

The pathogens *P. aeruginosa*, *S. aureus* and uropathogenic *E. coli* (UPEC)[26] were transformed with plasmids expressing GFP to facilitate a high-throughput screen with a fluorescence intensity readout (**Fig. 2a**). These species were chosen as representatives of both Gram-positive (*S. aureus*) and Gram-negative (*P. aeruginosa* and UPEC) pathogens that are frequently found in medical device–associated infections[27,28]. The polymer microarrays were incubated with a suspension of planktonic bacteria for each of the pathogens separately for 3 d (72 h) and revealed a wide range of resistance to bacterial attachment (**Fig. 2b–d**). To simulate *in vivo* conditions, such as those encountered in the urinary tract, we conditioned one microarray with artificial urine for 72 h before incubation with UPEC and, in another experiment, we replaced the RPMI-1640 medium with artificial urine for a 72-h incubation with UPEC (**Fig. 2e,f**). Biomineralization was observed on some polymers as opaque deposits after conditioning in artificial urine and after culture in artificial urine in the absence of buffering (**Supplementary Fig. 1**), which stimulated greater bacterial attachment in all cases. The time period chosen provided a sufficiently stringent assay for the identification of a set of 'hit' materials that were resistant to bacterial attachment (**Supplementary Fig. 2**). The fluorescence signal (F) (Online Methods, equation (1)) was used to quantify the amount of bacterial attachment for each bacterial strain. A linear correlation ($R^2 = 0.93$) between F and surface coverage (by area), as determined by confocal measurements of biofilm after incubation with UPEC, confirmed the utility of F as a reliable estimate of bacterial attachment (**Supplementary Fig. 3**). In cases where polymers were formed from monomers that had opposing influences on bacterial attachment, a linear correlation between systematic variations in material composition and F was observed (**Supplementary Fig. 4**). For some polymers the biological behavior of the copolymer was superior to that of the homopolymers of the two constituent monomers. In one such example, the F of *P. aeruginosa* (F_{PA}) on the copolymer 13:E (85:15) was $1.0 \times 10^6 \pm 0.2 \times 10^6$ arbitrary units (AU), which was lower than the F_{PA} from the homopolymer of monomer 13 ($2.1 \times 10^6 \pm 0.2 \times 10^6$ AU) and the F_{PA} on the homopolymer of monomer E ($29.8 \times 10^6 \pm 1.1 \times 10^6$ AU) (**Supplementary Fig. 5**). This synergistic effect was observed for 11 of the 96 monomer pairs explored in the first-generation array and in all cases led to greater resistance to bacterial attachment, demonstrating that screening for hit materials using a copolymer library could produce copolymers with unexpected properties.

Varying levels of attachment were observed for the three bacterial species to the same polymer surfaces (**Supplementary Fig. 6**). This finding is unsurprising given the different surface properties and macromolecular surface composition of Gram-positive and Gram-negative bacteria[29] and the diverse attachment mechanisms they use, which can involve surface proteins, flagella, fimbriae[30] and exopolysaccharides[31].

To assess the resistance of any given material to attachment by diverse bacterial species and exposure conditions, and as a guide to select hit monomers for further study, we developed a composite

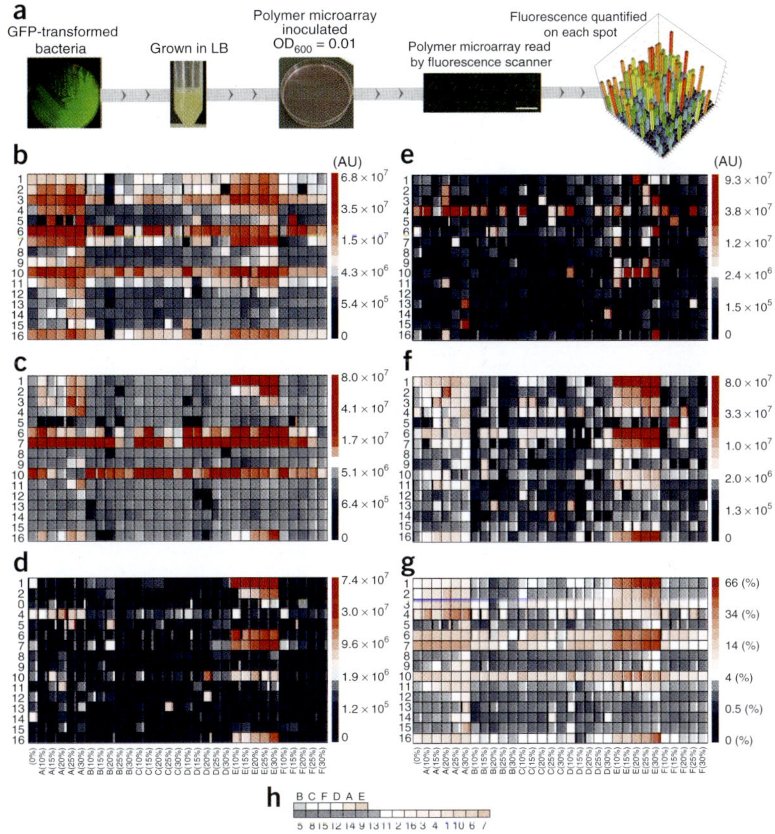

Figure 2 Slide inoculation procedure and 'heatmaps' of bacterial fluorescent intensity determined from first-generation slides (**a**) Bacterial attachment assay procedure. (**b–f**) Intensity maps of *F* measured for each bacterial strain on the first-generation array after 72-h incubation; *P. aeruginosa* (**b**), *S. aureus* (**c**), UPEC (**d**), UPEC grown in artificial urine (**e**) and UPEC grown on an artificial urine–conditioned slide (**f**). (**g**) Intensity map of the ι obtained for each material in the array, given as a percentage of the maximum bacterial numbers for all strains at all culture conditions. The major monomers are listed on the *y*-axis whereas the composition of the minor monomers is shown on the *x*-axis. The large shaded area within each outlined area indicates the mean value, and the mean ± 1 s.d. unit is presented in the narrow columns to the right (plus) and left (minus) of the mean, $n = 3$. Key to right. A square-root scale is applied to the intensity indicators. (**h**) The average ι for all materials containing a specific monomer, ranked from lowest to highest. The major and minor monomers were considered separately. The color next to each monomer is indicative of that monomer's mean ι and is colored by the same intensity scale as in **g**.

bacterial attachment parameter. The *F* value from each strain on each polymer composition, including UPEC incubated on artificial urine–conditioned slides and UPEC incubated in artificial urine, was normalized to the maximum *F* on the slide and averaged over all three replicates to produce an estimate of bacterial attachment value for all conditions and pathogens, called iota (ι) and defined in equation (2). A high value of ι indicated high bacterial attachment and therefore a poor material performance and vice versa. We plotted the value of ι in an intensity map for all first-generation polymer arrays, revealing which materials were least resistant to attachment by all three bacterial strains (red) and which were most resistant to all three (blue) (**Fig. 2g**). To compare the performance of each monomer used to fabricate the polymers, we determined the average ι from each of the materials containing a given monomer. The monomers were ranked by their average ι value, hence according to their resistance to bacterial attachment (**Fig. 2h**). Monomers 5 and B produced materials with the lowest average ι, that is, the most resistant to attachment, whereas 7 and E produced materials with the highest average ι.

High-throughput surface characterization of microarrays

To investigate the effect of polymer surface properties on bacterial attachment, we carried out high-throughput surface characterization (HT-SC) of the polymer microarray[32]. Techniques that probe the outermost surface of materials were used, including X-ray photoelectron spectroscopy (XPS) for quantitative elemental and functional analysis, atomic force microscopy (AFM) for topographical characterization, time-of-flight secondary ion mass spectrometry (ToF-SIMS) for molecular characterization and water contact angle (WCA) measurement to probe the surface wettability[20,32–34]. The influence of each property on bacterial attachment was assessed for all three bacterial strains separately. No correlation was identified between bacterial attachment and surface elemental composition (**Supplementary Fig. 7**), WCA or roughness (**Supplementary Fig. 8**), for the 496 materials studied in the first-generation array. ToF-SIMS analysis, coupled with the chemometrics technique of partial least squares (PLS) regression, has been demonstrated as a powerful technique for correlating surface chemistry represented in mass spectra with a univariate data set such as WCA[35,36] or stem cell attachment[24,25]. It was applied here to search for correlations between the surface chemistry of the array and the *F* values from each strain and to identify important surface moieties for bacterial attachment.

The PLS regression model produced by this analysis successfully predicted the *F* values for *P. aeruginosa* and *S. aureus* from the ToF-SIMS spectra (**Fig. 3a**), as evidenced by the linear relationship between the predicted and experimental *F* values (**Fig. 3b–d**), with an R^2 value of 0.68 and 0.76 for the two bacterial species, respectively. The PLS model for *S. aureus* gave a good prediction of the bacterial attachment for all polymers except those containing monomer 1, 6 or 10. No correlation was identified for UPEC ($R^2 = 0.28$). Only 17% of the materials on the array had a F_{UPEC} value >1% of $F_{\text{UPEC}max}$ compared with 97% and 96% for *P. aeruginosa* and *S. aureus*, respectively. The low attachment of this strain on the materials screened made it difficult to compute a suitable PLS regression model; thus, this lack of a correlation does not exclude the possibility of a dependence of UPEC bacterial attachment on surface chemistry.

The successful prediction of attachment by *P. aeruginosa* and *S. aureus* from the ToF-SIMS spectra demonstrates that the attachment of these species is dependent on the polymer surface chemistry. More specifically, the influence of each of the hundreds of

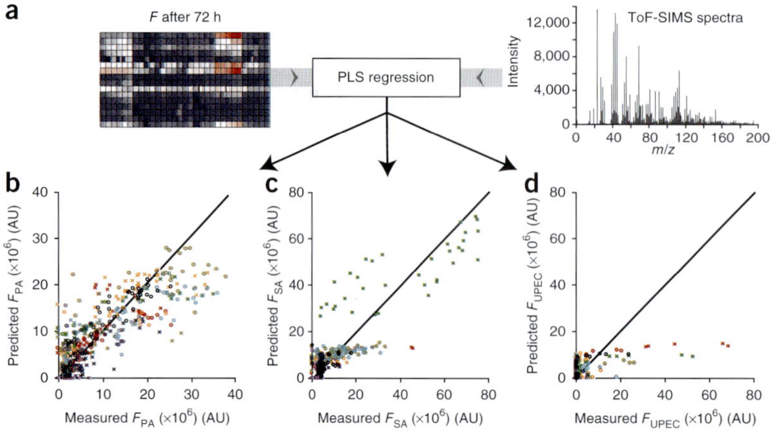

e

Ion	Positive RC_{PA} (×10^7)	Positive RC_{SA} (×10^7)	Ion	Negative RC_{PA} (×10^7)	Negative RC_{SA} (×10^7)
$C_2H_5O^+$	8.1	−0.2	C_2H^-	−7.8	−5
$C_3H_7O^+$	7.4	1.4	C_4H^-	−3.3	−1.6
$C_6H_5O^-$	4.4	26.9	$C_3H_3^+$	−2.5	−0.8
$C_6H_{11}O_3^-$	4.4	1.3	$C_2H_3^+$	−2.3	−1.6
$C_5H_9^+$	3.6	1	$C_8H_{15}^+$	−2.2	−1.1
$C_2H_3O^+$	3.1	0.1	$C_4H_7^+$	−2.2	−2.4
$C_3H_3O^+$	0.6	2.7	$C_3H_7^+$	−2.1	−2
$C_6H_7O^+$	0.4	1.8	$C_4H_5^+$	−1.7	−1.1
$C_6H_6O^+$	0.2	2.2	C_6H^-	−1.7	−0.8
$C_2H_3O_2^-$	0.1	2.2	CHO_2^-	−1.5	−2.1
$C_4H_5O_2^-$	0.1	2	$C_4H_7O^-$	−1.5	−0.4
$C_3H_3O_2^-$	0	4.5	$C_2H_5^+$	−1.4	−1
$C_7H_7O^+$	0	2.6	$C_4H_3^+$	−1.1	−0.4
$C_6H_5^+$	−0.5	5.4	$C_3H_5^+$	−0.5	−1.5
			$C_7H_5O_2^-$	0	−1.4
			$C_7H_5O^+$	1.4	−1.9

Figure 3 Correlation of the surface chemistries represented in the ToF-SIMS. (**a**) Schematic depiction of the PLS regression model used to predict the biological performance of materials by correlating *F* with the ToF-SIMS spectra. (**b–d**) The predicted bacterial attachment determined from the PLS regression model for *P. aeruginosa* (PA) (R^2 = 0.68) (**b**), *S. aureus* (SA) (R^2 = 0.76) (**c**), and UPEC (R^2 = 0.28) (**d**). The $y = x$ line is drawn as a guide. Polymers are grouped according to the major monomer 1 (×), 2 (○), 3 (×), 4 (○), 5 (×), 6 (○), 7 (×), 8 (○), 9 (×), 10 (○), 11 (×), 12 (○), 13 (×), 14 (○), 15 (×), 16 (○). (**e**) The key ions identified to be important by ToF-SIMS PLS regression analysis for the surface attachment of both *P. aeruginosa* (PA) and *S. aureus* (SA). The regression coefficient (RC) for each ion is also shown from the regression analysis with each bacterium separately. The RCs have been shaded according to their value (red, positive; blue, negative; white, neutral).

ions in the SIMS spectra on bacterial attachment is quantified by the regression coefficient, where a positive coefficient indicates that the ion in question promotes attachment, whereas a negative one indicates that the surface chemistry giving rise to the secondary ion is resistant to bacterial attachment (**Fig. 3e**). The surface chemical moieties assigned to secondary ion fragments with the highest PLS regression coefficients are shown in **Figure 3e** for both *P. aeruginosa* and *S. aureus*. In general, the presence of hydrocarbon secondary ions in spectra from materials was correlated with high resistance to bacterial attachment and oxygen-containing ions from certain pendant groups were observed to be correlated with low resistance to bacterial attachment for both *P. aeruginosa* and *S. aureus*. In particular, secondary ions consistent with cyclic carbon groups (C_4H^-, C_6H^-), ester groups (CHO_2^-), the tertiary butyl moiety ($C_4H_7^+$) and ions from aliphatic groups ($C_2H_3^+$, $C_2H_5^+$, $C_3H_7^+$) were correlated with lower bacterial attachment for both pathogens. Ions from ethylene glycol groups ($C_2H_3O^+$, $C_2H_3O_2^-$), and hydroxyl-containing fragments ($C_4H_5O_2^-$, $C_6H_{11}O_3^-$) correlated with higher bacterial attachment. These results point to the influence of moieties from particular monomers that are associated with the biological performance of the resultant polymer. Bacterial attachment is lower in monomers with the cyclic carbon environments as in monomers 4 and B, the relatively higher density of ester groups in materials formed from the triacrylate monomers 13 and 15, the tertiary butyl group of monomer 5 and the dimethyl hydrocarbon segment on monomers 2, 8 and 12. Bacterial attachment is greater where the ethylene glycol group is surface expressed as in polymers from monomers 1, 9, 16 and A, and where the hydroxyl group is present as in monomers 6, 7 and 10.

To determine the monomer composition most resistant to attachment by all three bacterial pathogens, we formulated a second-generation array that focused on the constituents from the first-generation array most resistant to bacterial attachment (**Fig. 1d**) but with greater variation in composition for each monomer pair. Monomer 4 was included as it contained the cyclic hydrocarbon moieties identified by the ToF-SIMS ions, C_4H^- and C_6H^-, and revealed to be associated with reduced bacterial attachment by the PLS regression analysis. The top four hit monomers were selected (15, 5, 8 and B) and mixed with each other and monomer 4 at ratios of x:(1 − x) where x varied from 10 to 90. This array contained 145 different materials (four replicates of each) plus four materials selected as positive controls that in the first-generation array were least resistant to bacterial attachment. Attachment after 3 d of incubation with all three bacterial pathogens, including UPEC incubated with artificial urine-conditioned slides, was assessed for the array and the ι for each material determined (**Fig. 4**).

Scale up from microarray spots

The six copolymer compositions (and four corresponding homopolymers) with the least bacterial attachment were chosen from the bacterial screen of the second-generation array as hit materials for further study. Scale-up from microarray spots to 8- to 10-mm diameter samples (termed coupons) was conducted to investigate the scalability of the polymers' physicochemical properties and biological performance (**Supplementary Figs. 9** and **10**). Initial scale-up experiments included monomer 15 (and other triacrylates), but inclusion of this monomer resulted in the formation of brittle materials that were observed to crack during the solvent extraction step. This was consistent with the formation of a highly cross-linked polymer and, thus, monomer 15 was removed from subsequent scale-up experiments. The area covered by the bacteria was at least twofold lower on all scaled-up materials compared with the positive control (monomer 7 homopolymer) (**Supplementary Fig. 9a**). Confocal microscopy coupled with live and/or dead staining (**Supplementary Fig. 10**) on the scaled-up polymer coupons revealed that, in common with the control polymer, a mixture of live and dead bacteria was attached to the materials. To simulate the more challenging *in vivo* conditioning by urine that is likely to occur on urinary catheters, *P. aeruginosa*, *S. aureus* and UPEC were incubated with polymer coupons preconditioned with artificial urine. For all three strains a greater than threefold reduction in the area of the sample covered by bacteria was observed for this assay compared with the positive control (monomer 7 homopolymer)

ARTICLES

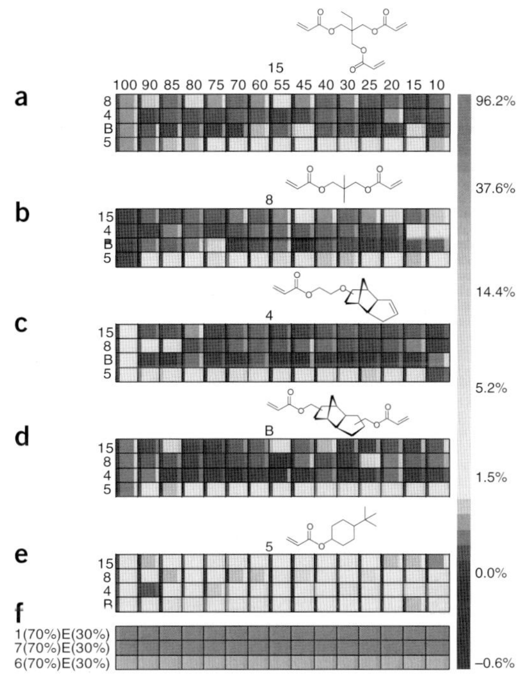

Figure 4 ι determined for all materials represented in the second-generation array. Five monomers from the first-generation array were mixed pairwise at 14 different ratios (15, 5, 8 and B were mixed with each other and monomer 4 at ratios of x:(1 − x) where x varied from 10 to 90). The content of the major monomer is indicated along the x-axis: (**a**) 15, (**b**) 8, (**c**) 4, (**d**) B and (**e**) 5 (**Fig. 1a**). The co-monomer is indicated in the y-axis. (**f**) The ι determined for high-bacterial-attachment polymers included in the second-generation array from the first-generation array. A square-root scale is applied to the color bar to highlight the low ι measured for this array. The mean of each value is shown as the large shaded area within each outlined rectangle and the mean ± 1 s.d. unit is presented in the narrow columns to the right (plus) and left (minus) of the mean, n = 4.

(**Supplementary Fig. 9b**), suggesting that these materials are resistant to bacterial attachment in the presence of urine components. ToF-SIMS of the polymer coupons indicates that the aliphatic pendant group on monomers 4 and 8 was surface enriched compared with the equivalent microarray spots, which suggests that in some instances scaling up from a microarray spot to a coupon modifies the material surface chemistry and biological performance (**Supplementary Figs. 9** and **11**).

Development of medical device coatings

To establish the potential of these hit materials as effective coatings for medical devices, we dip-coated oxygen plasma activated silicone catheters with monomer solution followed by UV curing (**Fig. 1e**) and compared bacterial attachment with silicone catheters and a commercially available, state-of-the-art, silver-containing coating. The four homopolymers of the hit monomers, six hit copolymer formulations and polymers from monomers 6, 7 and 1:E (70:30) (positive controls) were coated along the luminal and abluminal surfaces of the catheters (**Fig. 1f,g**). Scanning electron microscopy (SEM) was used to verify the presence of the coatings (**Supplementary Fig. 12**). The coated catheters were incubated with *P. aeruginosa*, *S. aureus* or UPEC for 72 h. Representative confocal images from the coated catheters after bacterial incubation are shown in **Figure 5a** and the area of biofilm coverage was quantified for each species. The corresponding ι values, normalized to the biofilm area coverage as measured on silicone alone, are shown in **Figure 5b**.

For all three pathogens, the coated catheters had a substantially lower bacterial coverage than the silicone elastomer. Several of the hit acrylate materials were superior to the silver-hydrogel (BactiGuard)-coated latex catheter (Bardex), which has been shown to be clinically effective at reducing catheter-associated asymptomatic bacteriuria when compared with standard latex catheters[28]. It should be noted that silver-hydrogel catheters are thought to be resistant to bacterial attachment as a consequence of the toxicity of silver ions to the bacteria, whereas our hit polymer coatings prevent attachment. The lowest area coverage for *P. aeruginosa* was achieved on the homopolymer formed from monomer 4, which had 1.2% ± 0.5% of the surface populated with bacteria. This represents a 28-fold reduction compared with the silicone catheter, and a 17-fold reduction compared with the BactiGuard-coated Bardex catheter. For *S. aureus* the best performing copolymer was formed from monomers B:5 (70:30) and resulted in a bacterial surface area coverage of 0.5% ± 0.3% , which represents a 67-fold reduction compared to silicone and a 30-fold reduction compared to BactiGuard. For UPEC the best performing copolymer was B:4 (90:10), which resulted in 1.2% ± 0.6% of the area populated with bacteria, representing a ninefold reduction compared to silicone and a sixfold reduction compared to BactiGuard. The homopolymer of B produced the catheter coating with the lowest ι of 2.0% ± 1.0%, and was the best material at preventing the attachment of all three different pathogens. The proportion of the surface covered for this coating was 2.3% ± 1.3%, 1.0% ± 0.4% and 1.5% ± 0.7% for *P. aeruginosa*, *S. aureus* and UPEC, respectively. This amounted to a 12-fold reduction in the composite parameter ι compared with an uncoated silicone catheter, and sevenfold lower than the BactiGuard-coated Bardex catheter.

There are several other different strategies that can be found in the literature to reduce bacterial attachment to surfaces that are not yet commercially available; for example, zwitterionic coatings function by preventing attachment[18,19]. Comparison with the performance of our hits to these literature reports is difficult as experimental methodologies vary widely. Attachment of *P. aeruginosa* (PA01) after 3 h incubation on a poly(sulfobetaine methacrylate)-grafted surface has been reported to be lower by 25-fold compared to that on glass[19], and in an incubation time closer to ours, a poly(carboxybetaine methacrylate)-grafted surface had 11-fold lower attachment of *P. aeruginosa* after 96-h bacterial exposure compared to glass[18].

Mechanism for reduction of bacterial coverage

The absence of a correlation between bacterial attachment and the contact angle or roughness of the materials studied (**Supplementary Fig. 8**) suggests that the interaction of bacteria with the methacrylate or acrylate library cannot be explained simply by hydrophobic interactions or roughness only, as was previously invoked to explain bacterial performance on self-assembled monolayers, stainless steel and certain polymers[37]. It should be noted that the bacterial cell incubation experiments were not conducted under flow conditions, thus, this assay did not assess any advantage gained from the shielding from shear that increased roughness could provide.

The ability to predict bacterial attachment from the chemistry of the materials as represented by the ToF-SIMS spectra (**Fig. 3b–d**) confirms that the bacteria-material interaction is dependent on surface chemistry. The PLS regression highlighted the ethylene glycol moieties and hydroxyl groups employed in these libraries for promoting the attachment of *P. aeruginosa* and *S. aureus*. The greater

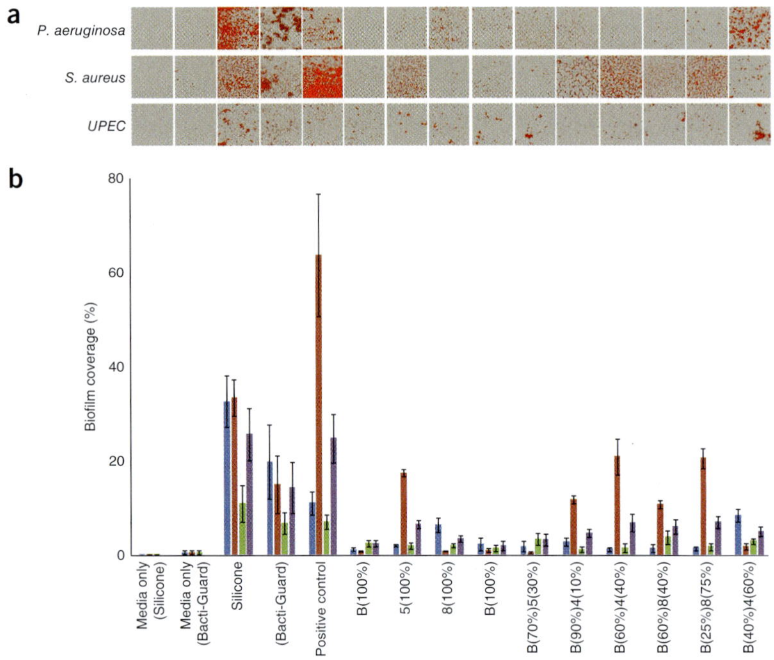

Figure 5 Proportion of surface covered by bacteria on catheters coated with hit polymers after 72 h incubation with planktonic bacteria. (**a**) Confocal microscopy images of *P. aeruginosa*, *S. aureus* and UPEC stained with SYTO17 growing on coated catheters, uncoated silicone and silicone treated with media without bacteria as a controls. Each image is 160 × 160 μm. (**b**) Quantification of bacterial area coverage for *P. aeruginosa* (blue), *S. aureus* (red), UPEC (green) and ι (purple) from confocal images of each sample. Area coverage was normalized to the coverage on silicone. The error bars represent mean ± 1 s.d. unit, $n = 5$. The composition of the positive control was specific to the bacterial strain used: *P. aeruginosa* = 6(100%), *S. aureus* = 7(100%), UPEC = 1(70%)E(30%). The fluorescence signal observed on the media-only controls is due to auto-fluorescence of the substrate and translates to a small erroneous (<0.5%) estimate of area of surface covered with bacteria.

attachment associated with surface hydroxyl groups suggests a role for hydrogen bonding, which may be through an interaction with the lipopolysaccharides, lipoteichoic acids or exopolysaccharides present on the bacterial cell surface[31]. For materials with resistance to bacterial attachment, the PLS regression analysis identified hydrophobic moieties such as aromatic and aliphatic carbon groups when accompanied by the weakly polar ester groups. By comparison with polystyrene, a purely hydrophobic material that is well known to support bacterial attachment (**Supplementary Fig. 13**)[38], the role of the ester group and the weakly amphiphilic structure of these hit polymers appear to be important for reduced bacterial attachment. Antibacterial behavior has previously been reported for zwitterionic materials[19] where disparate chemical properties are presented in close proximity on the molecular scale analogous to the weak amphiphiles identified in this study. To further investigate the role of weakly amphiphilic acrylate materials in preventing bacterial attachment, we produced a third polymer microarray that contained homopolymers of 15 methacrylate or acrylate monomers that had aliphatic, cyclic or aromatic pendant groups. F for *P. aeruginosa*, *S. aureus* and UPEC and ι for each material is shown in **Supplementary Figure 14**. The relatively high attachment of *P. aeruginosa* provided the greatest discrimination between the materials. The six materials with the lowest F_{PA} all contained cyclic or aromatic hydrocarbon groups. In contrast, the six materials with the highest F_{PA} all contained linear

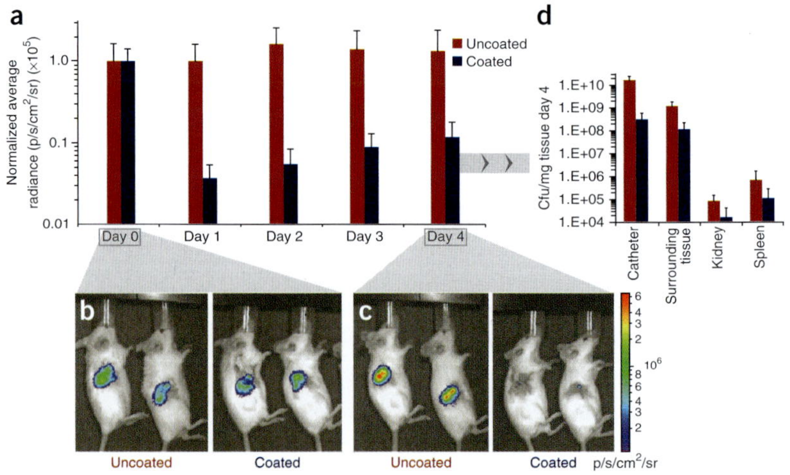

Figure 6 *In vivo* performance of hit polymer. Catheters dip-coated with hit polymer 4 copolymerized with DEGMA were implanted subcutaneously in mice. Mice were inoculated after 1 d with *S. aureus Xen29*, which were injected into the center of the tube. (**a**) The bioluminescence at the infection site was measured on the day of inoculation (day 0) and for the next 4 d. Error bars, ± 1 s.d. unit; $n = 9$ (three separate experiments, each comprising three control catheter mice and three polymer-coated catheter mice). The difference in bioluminescence between coated and uncoated samples from day 1 to 4 was confirmed to 99.5% confidence (*t*-test). Bioluminescence was normalized to the output at day 0 and all measurements have had the background luminescence subtracted, measured from an implanted catheter prior to inoculation. (**b**,**c**) Luminescence images with overlaid brightfield images of mice implanted with both uncoated (left) and coated (right) catheter segments on day 0 (**b**) and day 4 (**c**). (**d**) At day 4 the mice were euthanized, the insertion site plus the kidneys and spleen harvested, and the number of bacteria on the tissue determined. Error bars, ± 1 s.d. unit; $n = 3$ (taken from one group of three). The difference in cfu counts between coated and uncoated samples was confirmed to 99% confidence for the catheter and surrounding tissue, to 90% confidence for the kidneys and to 75% confidence for the spleen (*t*-test).

aliphatic carbon pendant groups, suggesting that the presence of ring structures is a determining factor for preventing bacterial attachment to methacrylate/acrylate polymers (**Supplementary Fig. 14**).

Confocal microscopy did not detect a significant number of either dead or live bacteria on the hit materials after 72 h incubation with planktonic bacteria (**Fig. 5a**), indicating that the mechanism behind the low attachment is prevention of bacterial attachment rather than a mechanism involving killing. Consistent with this conclusion, growth curves, which quantify bacterial numbers in the presence of the materials, showed no inhibition by the hit polymers for the bacterial strains used (**Supplementary Fig. 15**). In addition live and/or dead staining of the very low coverage of UPEC biofilm that were found revealed both live and dead cells present within the biofilm, which is typical of normal biofilms (**Supplementary Fig. 10**). Furthermore, there is no evidence of cytotoxicity to human cells of the polymers as the materials have been shown to support the culture of delicate embryonic stem cell lines[24].

The resistance of the lead acrylates discovered in this study to attachment of bacteria likely depends on the ability of bacterial cells to sense and respond to their immediate environment. This may be a consequence of the individual cells sensing the nature of the polymer surface by their cell envelope–associated sensory proteins or by specific surface structures such as flagella and pili involved in near-surface movement[39]. It is also possible that the bacterial population collectively responds to the polymer through quorum sensing (bacterial cell-to-cell communication) mechanisms[40], such that the lack of bacterial attachment occurs through this population-dependent decision-making process. In both scenarios the mechanism may be more complex than simple physicochemical interactions between the bacteria and surface. Specific examples include the Rcs sensor kinase, which controls the expression of a number of *E. coli* genes in response to growth on a solid surface[41]. Furthermore, production of the *P. aeruginosa* exopolysaccharides Pel and Psl are both under quorum-sensing control[42]. Thus, it is highly likely that bacterial responses to surfaces are more sophisticated than is currently appreciated.

In vivo assessment of hit material

The *in vivo* environment is more realistic than *in vitro* assessment of candidate materials with regards to eventual clinical application. We therefore carried out a subcutaneous foreign body infection model to test the efficacy of one of the hit materials (**Fig. 1h**). A copolymer of monomer 4 and di(ethylene glycol) methyl ether methacrylate (DEGMA) amenable to the dip-coating methodology was prepared by catalytic chain transfer polymerization (measured properties of the resultant polymer are shown in **Supplementary Table 1**). The DEGMA was included to tune the polymer's mechanical properties for *in vivo* use without compromising the polymer's resistance to bacterial attachment (**Supplementary Figs. 16 and 17**) and to exploit the ability of the oligo(ethylene glycol) moiety to reduce protein fouling[43] in the high protein–containing environment.

Both control silicone and dip-coated silicone catheters were implanted subcutaneously into mice and inoculated with bioluminescent *S. aureus* Xen29 by injection into the catheter lumen after 1 d. Immediately following inoculation similar bioluminescence was observed for both coated and uncoated catheters (**Fig. 6**). After 1 d, we observed more than a tenfold reduction in bioluminescence on the coated catheters compared with the uncoated silicone, a difference that persisted for 4 d (**Fig. 6a–c**). Bioluminescence requires the bacteria to be respiring aerobically, thus, it will not detect bacteria that are viable but are either dormant or growing anerobically. To investigate whether bacterial numbers on coated catheters were reduced, we euthanized the mice on day 4 and quantified the numbers of bacteria at the infection sites on the catheter and in the surrounding tissues, including kidneys and spleen. Bacterial numbers were reduced by nearly two orders of magnitude on the coated catheter compared to the uncoated catheter, and an order of magnitude reduction in bacterial numbers was observed in the tissue surrounding the implant, and in the kidneys and the spleen, suggesting a reduced amount of systemic bacteria (**Fig. 6e**). The reduction in bioluminescence on the coated catheter in the live animal demonstrates that the coating successfully reduced bacterial *in vivo* survival in the presence of the host defenses compared to the silicone control. We interpret the persistence of bacteria on the silicone catheter as indicative of bacteria that were able to attach to the silicone and form biofilms, thereby avoiding clearance by the host defenses. These observations suggest that the performance of the hit polymer identified *in vitro* translates to the challenging *in vivo* environment. Sufficiently small diameter (2.7 mm), silver-containing, hydrogel-coated catheters were not commercially available for the mouse model used in this study, preventing a direct comparison of the *in vivo* performance of the polymer in this animal study with the same controls as the *in vitro* experiments.

DISCUSSION

We have developed a method for the discovery of polymeric materials resistant to bacterial attachment using high-throughput surface characterization and chemometrics, which has identified simple chemical moieties that reduce bacterial adhesion to surfaces. The new class of materials discovered, which could not have been predicted from the current understanding of bacteria-material interactions, are resistant to bacterial attachment. An *in vitro* comparison after a 3-d incubation revealed up to 67-fold less bacterial attachment and biofilm formation compared with a commercial uncoated medical grade silicone, and a 30-fold lower bacterial attachment and biofilm formation than a commercial silver hydrogel coating. The resistance of one of these coatings to bacterial attachment was demonstrated *in vivo* using a mouse foreign-body infection model, which confirms the potential of the hit materials as coatings for indwelling biomedical devices. In contrast to the use of coatings that incorporate molecules that are toxic to and kill pathogens, the anti-attachment mechanism described here does not place selective evolutionary pressure on organisms to develop antimicrobial resistance.

By further developing this high-throughput screening approach to include other bacterial species (including freshly isolated clinical strains and pathogens specific to particular niches and different growth conditions), new polymeric materials may be found that are resistant to colonization by combinations of pathogens for medical devices, water purification systems, food preparation surfaces and utensils, or any other surface where bacterial adhesion is problematic.

METHODS

Methods and any associated references are available in the online version of the paper.

Note: Supplementary information is available in the online version of the paper.

ACKNOWLEDGMENTS

Funding from the Wellcome Trust (grant no. 085245 and support from N. Shepherd) and the Medical Research Council UK (for the *in vivo* work; grant no. G0802525) is gratefully acknowledged. M. Alexander gratefully acknowledges the Royal Society for the provision of his Wolfson Research Merit Award. Assistance with ToF-SIMS measurements from D. Scurr is kindly acknowledged. Assistance with the preparation of polymer for *in vivo* studies by E. Eaves, N. Nguyen and J. Li is kindly acknowledged.

AUTHOR CONTRIBUTIONS
M.R.A., M.C.D., P.W., R.L. and D.G.A. conceived the high-throughput strategy. Y.M., A.L.H. and J.Y. prepared the microarrays. J.Y. coated catheters with hit polymers. J.Y. and A.L.H. performed the HT-SC. C.-Y.C. performed the biological assays with support from S.A. D.J.I. prepared polymer for *in vivo* assays. J.L. and A.C. performed the *in vivo* assays. A.L.H., C.-Y.C., J.Y., M.R.A., M.C.D., P.W., R.B., D.G.A. and R.L. contributed to the analysis of data and the writing of the manuscript.

COMPETING FINANCIAL INTERESTS
The authors declare no competing financial interests.

Published online at http://www.nature.com/doifinder/10.1038/nbt.2316.
Reprints and permissions information is available online at http://www.nature.com/reprints/index.html.

1. Davies, D. Understanding biofilm resistance to antibacterial agents. *Nat. Rev. Drug Discov.* **2**, 114–122 (2003).
2. Costerton, J.W., Stewart, P.S. & Greenberg, E.P. Bacterial biofilms: a common cause of persistent infections. *Science* **284**, 1318–1322 (1999).
3. Smith, A.W. Biofilms and antibiotic therapy: is there a role for combating bacterial resistance by the use of novel drug delivery systems? *Adv. Drug Deliv. Rev.* **57**, 1539–1550 (2005).
4. Danese, P.N. Antibiofilm approaches: prevention of catheter colonization. *Chem. Biol.* **9**, 873–880 (2002).
5. Krol, J.E. *et al.* Increased transfer of a multidrug resistance plasmid in *Escherichia coli* biofilms at the air-liquid interface. *Appl. Environ. Microbiol.* **77**, 5079–5088 (2011).
6. Darouiche, R.O. *et al.* A comparison of two antimicrobial-impregnated central venous catheters. *N. Engl. J. Med.* **340**, 1–8 (1999).
7. Raad, I. *et al.* Central venous catheters coated with minocycline and rifampin for the prevention of catheter-related colonization and bloodstream infections. A randomized, double-blind trial. *Ann. Intern. Med.* **127**, 267–274 (1997).
8. Yorganci, K., Krepel, C., Weigelt, J.A. & Edmiston, C.E. Activity of antibacterial impregnated central venous catheters against Klebsiella pneumoniae. *Intensive Care Med.* **28**, 438–442 (2002).
9. Jaeger, K. *et al.* Efficacy of a benzalkonium chloride-impregnated central venous catheter to prevent catheter-associated infection in cancer patients. *Chemotherapy* **47**, 50–55 (2001).
10. Greenfeld, J.I. *et al.* Decreased bacterial adherence and biofilm formation on chlorhexidine and silver sulfadiazine-impregnated central venous catheters implanted in swine. *Crit. Care Med.* **23**, 894–900 (1995).
11. Guay, D.R. An update on the role of nitrofurans in the management of urinary tract infections. *Drugs* **61**, 353–364 (2001).
12. Caillier, L. *et al.* Synthesis and antimicrobial properties of polymerizable quaternary ammoniums. *Eur. J. Med. Chem.* **44**, 3201–3208 (2009).
13. Darouiche, R.O., Mansouri, M.D., Gawande, P.V. & Madhyastha, S. Efficacy of combination of chlorhexidine and protamine sulphate against device-associated pathogens. *J. Antimicrob. Chemother.* **61**, 651–657 (2008).
14. Li, P. *et al.* A polycationic antimicrobial and biocompatible hydrogel with microbe membrane suctioning ability. *Nat. Mater.* **10**, 149–156 (2011).
15. Costa, F., Carvalho, I.F., Montelaro, R.C., Gomes, P. & Martins, M.C.L. Covalent immobilization of antimicrobial peptides (AMPs) onto biomaterial surfaces. *Acta Biomater.* **7**, 1431–1440 (2011).
16. Monds, R.D. & O'Toole, G.A. The developmental model of microbial biofilms: ten years of a paradigm up for review. *Trends Microbiol.* **17**, 73–87 (2009).
17. Holmes, P.F. *et al.* Surface-modified nanoparticles as a new, versatile, and mechanically robust nonadhesive coating: suppression of protein adsorption and bacterial adhesion. *J. Biomed. Mater. Res. A* **91A**, 824–833 (2009).
18. Cheng, G. *et al.* Zwitterionic carboxybetaine polymer surfaces and their resistance to long-term biofilm formation. *Biomaterials* **30**, 5234–5240 (2009).
19. Cheng, G., Zhang, Z., Chen, S.F., Bryers, J.D. & Jiang, S.Y. Inhibition of bacterial adhesion and biofilm formation on zwitterionic surfaces. *Biomaterials* **28**, 4192–4199 (2007).
20. Hook, A.L. *et al.* High-throughput methods applied in biomaterial development and discovery. *Biomaterials* **31**, 187–198 (2010).
21. Pernagallo, S., Wu, M., Gallagher, M.P. & Bradley, M. Colonising new frontiers-microarrays reveal biofilm modulating polymers. *J. Mater. Chem.* **21**, 96–101 (2011).
22. Anderson, D.G., Levenberg, S. & Langer, R. Nanoliter-scale synthesis of arrayed biomaterials and application to human embryonic stem cells. *Nat. Biotechnol.* **22**, 863–866 (2004).
23. Mei, Y. *et al.* Mapping the interactions among biomaterials, adsorbed proteins, and human embryonic stem cells. *Adv. Mater.* **21**, 2781–2786 (2009).
24. Mei, Y. *et al.* Combinatorial development of biomaterials for clonal growth of human pluripotent stem cells. *Nat. Mater.* **9**, 768–778 (2010).
25. Yang, J. *et al.* Polymer surface functionalities that control human embryoid body cell adhesion revealed by high-throughput surface characterization of combinatorial material microarrays. *Biomaterials* **31**, 8827–8838 (2010).
26. Berger, H., Hacker, J., Juarez, A., Hughes, C. & Goebel, W. Cloning of the chromosomal determinants encoding hemolysin production and mannose-resistant hemagglutination in *Escherichia coli*. *J. Bacteriol.* **152**, 1241–1247 (1982).
27. Hidron, A.I. *et al.* Antimicrobial-resistant pathogens associated with healthcare-associated infections: annual summary of data reported to the national healthcare safety network at the Centers for Disease Control and Prevention, 2006–2007. *Infect. Control Hosp. Epidemiol.* **29**, 996–1011 (2008).
28. Schumm, K. & Lam, T.B. Types of urethral catheters for management of short-term voiding problems in hospitalised adults. *Cochrane DB. Syst. Rev.* **16**, CD004013 (2008).
29. Katsikogianni, M. & Missirlis, Y.F. Concise review of mechanisms of bacterial adhesion to biomaterials and of techniques used in estimating bacteria-material interactions. *Eur. Cell. Mater.* **8**, 37–57 (2004).
30. O'Toole, G., Kaplan, H.B. & Kolter, R. Biofilm formation as microbial development. *Annu. Rev. Microbiol.* **54**, 49–79 (2000).
31. Flemming, H.C. & Wingender, J. The biofilm matrix. *Nat. Rev. Microbiol.* **8**, 623–633 (2010).
32. Urquhart, A.J. *et al.* High-throughput surface characterisation of a combinatorial material library. *Adv. Mater. (Deerfield Beach Fla.)* **19**, 2486–2491 (2007).
33. Davies, M.C. *et al.* High-throughput surface characterization: a review of a new tool for screening prospective biomedical material arrays. *J. Drug Target.* **18**, 741–751 (2010).
34. Hook, A.L. *et al.* Polymers with hydro-responsive topography identified using high-throughput AFM of an acrylate microarray. *Soft Matter* **7**, 7194–7197 (2011).
35. Taylor, M. *et al.* Partial least squares regression as a powerful tool for investigating large combinatorial polymer libraries. *Surf. Interface Anal.* **41**, 127–135 (2009).
36. Urquhart, A.J. *et al.* TOF-SIMS analysis of a 576 micropatterned copolymer array to reveal surface moieties that control wettability. *Anal. Chem.* **80**, 135–142 (2008).
37. Dalton, H.M. & March, P.E. Molecular genetics of bacterial adhesion and biofouling. in *Handbook of Bacterial Adhesion: Principles, Methods, and Applications* (eds. An, Y.H. & Friedman, R.J.) 43–51 (Humana Press, 2000).
38. Ong, Y.L., Razatos, A., Georgiou, G. & Sharma, M.M. Adhesion forces between *E. coli* bacteria and biomaterial surfaces. *Langmuir* **15**, 2719–2725 (1999).
39. Conrad, J.C.C.J.C. *et al.* Flagella and pili-mediated near-surface single-cell motility mechanisms in P. aeruginosa. *Biophys. J.* **100**, 1608–1616 (2011).
40. Williams, P., Winzer, K., Chan, W.C. & Camara, M. Look who's talking: communication and quorum sensing in the bacterial world. *Philos. Trans. R. Soc. Lond. B. Biol. Sci.* **362**, 1119–1134 (2007).
41. Ferrieres, L. & Clarke, D.J. The RcsC sensor kinase is required for normal biofilm formation in *Escherichia coli* K-12 and controls the expression of a regulon in response to growth on a solid surface. *Mol. Microbiol.* **50**, 1665–1682 (2003).
42. Gilbert, K.B., Kim, T.H., Gupta, R., Greenberg, E.P. & Schuster, M. Global position analysis of the Pseudomonas aeruginosa quorum-sensing transcription factor LasR. *Mol. Microbiol.* **73**, 1072–1085 (2009).
43. Harder, P., Grunze, M., Dahint, R., Whitesides, G.M. & Laibinis, P.E. Molecular conformation in oligo(ethylene glycol)-terminated self-assembled monolayers on gold and silver surfaces determines their ability to resist protein adsorption. *J. Phys. Chem. B.* **102**, 426–436 (1998).

Lipid-derived nanoparticles for immunostimulatory RNA adjuvant delivery

David N. Nguyen[a,1], Kerry P. Mahon[b], Ghania Chikh[c], Phillip Kim[d], Hattie Chung[b], Alain P. Vicari[c], Kevin T. Love[b], Michael Goldberg[b], Steve Chen[b], Arthur M. Krieg[c], Jianzhu Chen[b], Robert Langer[a,b,d], and Daniel G. Anderson[a,b,d,2]

[a]Division of Health Sciences and Technology, Massachusetts Institute of Technology, Cambridge, MA 02142; [b]David H. Koch Institute for Integrative Cancer Research, Massachusetts Institute of Technology, Cambridge, MA 02142; [c]Pfizer Vaccines Research Ottawa, Pfizer Canada Inc., Ottawa, ON, Canada; and [d]Department of Chemical Engineering, Massachusetts Institute of Technology, Cambridge, MA 02142

Contributed by Robert Langer, January 4, 2012 (sent for review October 11, 2011)

The specific activation of Toll-like receptors (TLRs) has potential utility for a variety of therapeutic indications including antiviral immunotherapy and as vaccine adjuvants. TLR7 and TLR 8 may be activated by their native ligands, single-stranded RNA, or by small molecules of the imidazoquinoline family. However the use of TLR7/8 agonists for in vivo therapy is limited by instability, in the case of RNA, or systemic biodistribution and toxicity in the case of small molecule agonists. We hypothesized that unique lipid-like materials, termed "lipidoids," could be designed to efficiently deliver immunostimulatory RNA (isRNA) to TLR-expressing cells to drive innate and adaptive immune responses. A library of lipidoids was synthesized and screened for the ability to induce type I IFN activation in human peripheral blood mononuclear cells when combined with isRNA oligonucleotides. Effective lipidoid-isRNA nanoparticles, when tested in mice, stimulated strong IFN-α responses following subcutaneous injection, had robust antiviral activity that suppressed influenza virus replication, and enhanced antiovalbumin humoral and cell-mediated responses when used as a vaccine adjuvant. Further, we demonstrate that whereas all immunological activity was MyD88-dependent, certain materials were found to engage both TLR7-dependent and TLR7-independent activity in the mouse suggestive of cell-specific delivery. These lipidoid formulations, which are materials designed specifically for delivery of isRNA to Toll like receptors, were superior to the commonly used *N*-[1-(2,3-dioleoyloxy)propyl]-*N,N,N*-trimethylammonium methylsulfate–RNA delivery system and may provide new tools for the manipulation of TLR responses in vitro and in vivo.

innate immunity | dendritic cell | drug delivery | high-throughput screening

The development of vaccine adjuvants has focused on innate immune activation, which is an important early aspect of the protective immune response (1–5). A first step in triggering innate immunity is the activation of pattern recognition receptors (PRRs) that allow for identification of pathogens without the need for prior education of an adaptive response. The Toll-like receptors (TLR) recognize conserved structures among a diverse group of pathogens (2). Nucleic acids can be recognized by TLRs 7, 8, and 9, which comprise a closely related subfamily whose expression differs by species and cell type, and whose function is compartmentalized to the endosome (2, 6, 7). RNA molecules that directly stimulate innate immune responses through mechanisms such as the TLR pathways have been functionally termed as immunostimulatory RNA (isRNA) (8). In humans, TLR7 is highly expressed and functional in plasmacytoid dendritic cells (PDCs) and B cells, whereas TLR8 expression is localized mostly to monocytes, myeloid dendritic cells (MDCs), and monocyte-derived dendritic cellss (moDCs) (9). The engagement of TLRs 7 and 8 results in a characteristic type I interferon response (e.g., IFN-α), promotion of an antiviral state with induction of IFN-stimulated genes, and suppression of viral replication (5, 10–13). TLR activation also leads to a coordinated Th1-biasing cytokine profile (1) that increases immune surveillance of cancer (7) and stimulates adaptive immunity by providing the necessary "danger signals" for efficient production of T-cell-mediated responses and class switching to high-affinity antibodies (1, 14).

Recent advances in characterizing and producing defined antigens has led to a plethora of vaccine targets, but protein vaccination alone lacks the immunostimulatory danger signals present in live attenuated or inactivated pathogen preparations (5, 15). By increasing coupling of innate and adaptive responses (4), introducing TLR stimulation as a vaccine adjuvant may be useful for improving immune responses to vaccines. However, clinical use of synthetic small molecule TLR7 and TLR8 agonists, such as imiquimod, in cancer and infectious disease (3, 5) has been restricted to topical application to avoid systemic distribution, exuberant activation of innate immune responses (3, 16), and off-target activity (17). Targeted delivery of nucleic acid agonists may allow for localized activation of specific cells without the side effects of systemic small molecule agonists. Utilization of unprotected small single-stranded RNAs (ssRNAs), the natural ligands for TLR7/8 (6, 18), is impractical due to low stability, nuclease degradation (19–21), and the requirement of endosomal uptake for induction of TLR7/8 responses (11, 18, 22). Drug delivery systems have been shown to enhance both the efficiency and immunostimulatory side effects of siRNAs (23–26) yet to our knowledge no systems designed specifically for delivery of isRNAs to TLRs have been reported.

We hypothesized that controlled delivery of isRNA to TLR7 or TLR8 with synthetic nanoparticles could mimic the robust immune responses triggered by viral infection through efficient and localized activation of innate immune responses. Here we develop lipid-like materials, termed "lipidoids" (27), specifically capable of inducing robust isRNA-mediated TLR stimulation. We detail how lipidoid-isRNA nanoparticles can be synthesized, screened, and formulated to enhance humoral and cell-mediated immunity. Further, we discuss evidence supporting a mechanism of cell-specific targeting and MyD88-dependent activation of TLRs by lipioid-isRNA nanopartcles in vitro and in vivo.

Results

In Vitro Screening of Lipidoids Capable of isRNA Delivery. A preliminary library of lipidoid molecules was generated by combinatorial addition of alkyl-acrylate or alkyl-acrylamide tails to primary or

Author contributions: D.N.N., K.P.M., G.C., P.K., H.C., A.P.V., S.C., A.M.K., J.C., R.L., and D.G.A. designed research; D.N.N., K.P.M., G.C., P.K., H.C., A.P.V., K.T.L., and S.C. performed research; K.P.M., K.T.L., and M.G. contributed new reagents/analytic tools; D.N.N., G.C., P.K., H.C., A.P.V., and D.G.A. analyzed data; and D.N.N., K.P.M., G.C., R.L., and D.G.A. wrote the paper.

The authors declare no conflict of interest.

[1]To whom correspondence may be addressed at: Massachusetts Institute of Technology, Building 76, Room 653, 500 Main Street, Cambridge, Massachusetts 02142. E-mail: dgander@mit.edu.

[2]Present address: Stanford University School of Medicine, Stanford, CA 94305

See Author Summary on page 5150 (volume 109, number 14).

This article contains supporting information online at www.pnas.org/lookup/suppl/doi:10.1073/pnas.1121423109/-/DCSupplemental.

secondary amine-containing cores (Fig. S1*A*) following a simple solvent-free synthesis (Fig. S1 *B* and *C*). Lipidoid compounds are named according to the tail composition (designated by two letters) followed by number of tails if purified (in parentheses) and then by the amine core (designated by number) as described in Fig. S1. The 96 different lipidoid synthesis reactions (Table S1) included a variety of carbon alkyl tail lengths and amine-core structures. Lipidoid–RNA complexes were formed at four different mass ratios of lipidoid to RNA (15, 10, 5, 2.5 to 1) and screened in vitro for delivery in human peripheral blood mononuclear cells (PBMCs) using a high-throughput cell-based assay that detects type I IFN (28). The short single-stranded guanine-uracil (GU)-rich R-006 isRNA oligonucleotide has been shown to be an active TLR7 and TLR8 agonist inducing production of IFN-α and Th1-type cytokines across multiple species including human and mouse (29). To control for direct lipidoid-mediated immunostimulatory activity or toxicity, we also performed negative screening against the control R-1263 sequence ssRNA oligonucleotide, which has similar structure to R-006 but exhibits low TLR7 and TLR8 activity (29).

Many of the lipidoid–RNA complexes tested exhibited some level of immunostimulatory activity (Fig. 1 and Fig. S2). Almost 900 different conditions were tested that included different lipidoids with isRNA or control RNA at various lipidoid/RNA ratios. Of these, 106 conditions generated activity greater than half that of lipofectamine (L2K), which was used to normalize activity across donor PBMC batches. Although too toxic for in vivo administration, L2K has proven to be an effective transfection agent in vitro and a useful delivery screening benchmark (27). Excluding lipidoid formulations that were immunostimulatory with both the R-006 and R-1263 sequences, 14 distinct lipidoid materials were found to have activity equal to or greater than that of L2K combined with R-006. The 100-core diamine (Fig. S1*A*) was highly represented in this active subset. Of the seven 100-core lipidoid products tested, five enabled efficient delivery of R-006 including the top three lipidoid compounds at any weight ratio (Fig. 1, Fig. S2, and Fig. S3*A*).

Second-Generation Lipidoids Based on the 100-core. The 100-core has two primary amines that each can be substituted with up to two alkyl tails each for a total of four possible substitutions. A second-generation set of 100-core lipidoids incorporating short chain lengths and mixed alkyl chains was synthesized and evaluated for potential to deliver isRNA. We focused on three- and four-tail versions of the 100-core based on prior work indicating that fully (n) and partially (n-1)-substituted lipidoids exhibit the most efficient siRNA delivery (27). These 100-core lipidoids were synthesized by stepwise substitution of each primary amine (Fig. S1*D*) with a mixture of ND and NA tails, further identified as lipidoid "I" (I-3, three tails, and I-4, four-tails), or a mixture of ND and LD tails, further identified as lipidoid "II" (Fig. 2). Purified lipidoids I-3 and II-3 were soluble in sodium acetate whereas lipidoids I-4 and II-4 were insoluble in aqueous solutions. Upon repeat screening in vitro, lipidoids I and II exhibited high levels of isRNA delivery that were RNA specific and greater than that of ND(5)-98-1 (Fig. S3*B*), a lipidoid material previously shown to have high capacity for siRNA delivery in vitro and in vivo (27). Thus, the 100-core and second-generation derivatives were further investigated for in vivo isRNA delivery.

In Vivo Optimization of Lipidoid–RNA Nanoparticles. Previous work has demonstrated that formulation parameters and lipidoid: RNA (L/R) ratio can affect in vivo performance of lipidoid nanoparticles (23, 26, 30). To increase solubility and stability, lipidoids at 10 : 1, 11.5 : 1, or 15 : 1 L/R ratio were formulated with poly(ethylene-glycol) (PEG) and cholesterol (Ch) and extruded through 80 nm pores to generate nanoparticles. The particles were then either lyophilized or dialyzed to remove ethanol. Lyophilized nanoparticles under the conditions tested were found to have a heterogeneous size distribution into the micron range and aggregated after resuspension in HBSS (Table S2). Dialyzed nanoparticles remained stable in size below 200 nm for up to 1 mo (Table S3).

Nanoparticles of multiple different lipidoids encapsulating R-006 RNA were screened by s.c. injection in BALB/c mice, and the time course of IFN-α, interferon gamma-induced protein 10 (IP-10), and IL-6 induction was monitored over 24 h (Fig. S4 *A* and *B*). These cytokines were selected as markers of PDC (IFN-α and IP-10) and MDC (IL-6) activation (31, 32) and have been previously observed in response to R-006 RNA (29). *N*-[1-(2,3-dioleoyloxy)propyl]-*N,N,N*-trimethylammonium methylsulfate (DOTAP) has previously been used to facilitate cellular uptake and to investigate isRNA effects in vivo (18, 21). Lipidoids I-4 and II-4 consistently outperformed DOTAP and ND(5)-98-1. In direct comparisons, the dialyzed formulations of second-generation 100-core lipidoids I-4 and II-4 at a 15 : 1 L/R ratio had the highest propensity for cytokine induction (Fig. 3). Furthermore, T, B, and NK cells were all activated in response to both I-4 and II-4 lipidoid nanoparticles (Fig. S4*C*). Both I-4 and II-4 nanoparticles exhibited dose-dependent immunostimulation following s.c. administration (Fig. S4*D*). Thus, dialyzed lipidoid–RNA nanoparticles (LRNP) of lipidoids I-4 and II-4 at an L/R ratio of 15 : 1 were used for all subsequent in vivo investigations.

Increased Innate and Adaptive Immune Responses. We investigated clinically relevant endpoints in two mouse models of innate immune activation addressing antiviral and vaccine adjuvant properties. Because LRNP-II induced the greater amount of IFN-α secretion, these nanoparticles were tested in an established in vivo model of influenza infection previously shown to be sensitive to isRNA-triggered interferon and cytokine responses (23). LRNP-II were formulated with R-006, R-1263, or without RNA,

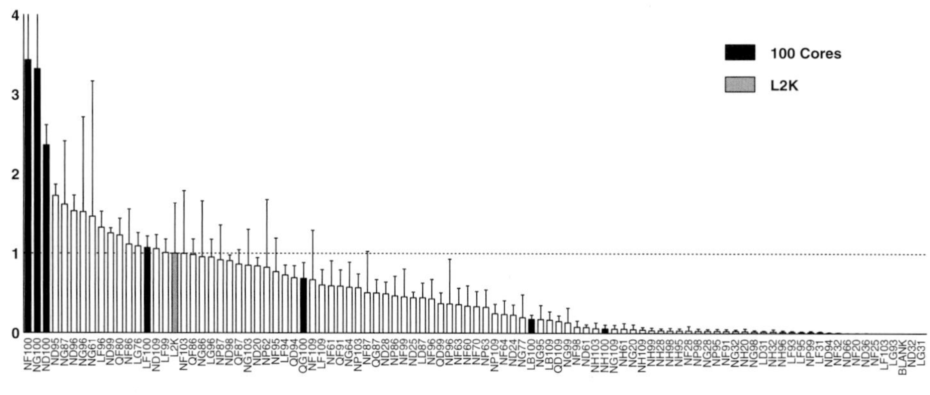

Fig. 1. Initial screening of lipidoid library for isRNA delivery. The highest relative type I IFN activity per unique compound, with either active or control RNA at any weight ratio, is shown for 96 lipidoids. All lipidoids were screened for isRNA delivery to human PBMCs in vitro independently with 200 μg of either immunostimulatory R-006 or control R-1263 comprising over 900 unique transfection experiments. Type I interferon activity was normalized for each batch of PBMCs to activity of L2K complexed with R-006 (gray bar, dotted line). 100-core lipidoids highlighted by solid black bars. Error bars represent standard deviation, $n = 4$.

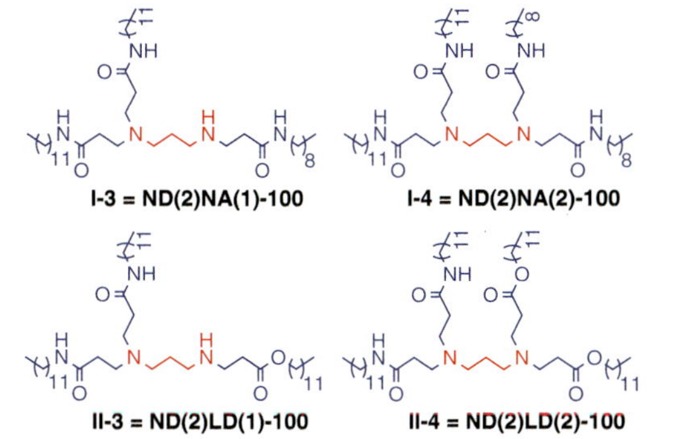

Fig. 2. Structures of second-generation lipidoids based on 100 core. Second-generation lipidoids were designed based on the 100 core (red). An ND(2)-100 precursor was substituted with the 10-carbon alkyl-acrylamide NA (lipidoid I) or the 12-carbon alkyl-acrylate LD (lipidoid II) and purified into single isomer components with either three or four total tails.

and injected s.c. into 129sv mice. After 9 h, mice were challenged intranasally with a supralethal dose (12,000 pfu) of influenza A/PR8 virus. Lung titers were measured 24 h postchallenge and found to be reduced 10-fold by isRNA delivery (Fig. 4*A*) relative to untreated control. Further, this antiviral prophylaxis was only observed for the immunostimulatory R-006 RNA, whereas neither control R-1263 nanoparticles nor empty nanoparticles without RNA (II-4 + Blank) resulted in significant reduction of viral titer.

We further investigated the potential of both LRNP-I and LRNP-II as adjuvants of vaccination with the model protein antigen chicken ovalbumin (Ova). C57Bl/6 mice were vaccinated by i.m. injections with Ova protein mixed with comparison adjuvants represented by LRNPs, DOTAP–RNA complexes, or CpG oligodeoxynucleotide (ODN) 1826, which is a mouse-specific B-class TLR9 agonist previously shown to adjuvant Ova antigen responses (33); two different pan-species B-class CpG ODN have shown enhancement of antibody titers against various antigens in clinical trials (5, 34, 35) and one is now in phase 3 testing in a hepatitis B vaccine (Heplisav, Dynavax). Immunization with LRNPs increased the magnitude of Ova-specific IgG by 3 to 4 log orders compared to immunization with protein alone (Fig. 4*B*). A significant increase in total IgG was observed for LRNP-I with both R-006 and R-1263, but for LRNP-II only immunization with R-006 RNA resulted in statistically greater levels of IgG antibody. Vaccination with either LRNP resulted in an increase in both IgG1 and IgG2c subclasses compared to protein alone (Fig. S5*A*). The increase in Th1-biased IgG2c subclass was notable compared to immunization with Ova protein alone although not as striking as it was for Ova mixed with CpG-ODNs, which have previously been shown to induce a strong Th1 bias (36, 37). LRNP adjuvants also increased absolute IgG2c titers, but preserved the Th2-associated IgG1 bias of protein vaccination alone.

More strikingly, LRNP also greatly enhanced cell-mediated immune responses to Ova. Both LRNP-I+R-006 and LRNP-II +R-006 induced greater numbers of splenic Ova-antigen-specific $CD8^+$ T cells than either CpG ODN 1826 or DOTAP+R-006 complexes, and all LRNP formulations increased antigen-specific $CD8^+$ T cells to levels greater than that with Ova vaccination alone (Fig. 4*C*). Using CpG ODN on its own as used here may be suboptimal for adjuvant effect as greater immunogenicity has been demonstrated in mice when CpG ODN is combined with an additional adjuvant possessing delivery properties (34, 37, 38). At low dose (10 μg) of LRNP-II nanoparticles, the increase in percentage of reactive CD8+ positive T cells was significantly greater with the R-006 RNA than R-1263. However, for LRNP-I the percentage of reactive CD8+ positive T cells was large but not significantly different between active R-006 and control R-1263 RNA. In vitro restimulation of splenocytes from vaccinated animals resulted in large increases in the Ova-specific secretion of the Th1-biasing cytokines IFN-γ and IL-2 (Fig. S5*B*). Increases of the Th2-associated cytokines IL-10 and IL-4 were also observed.

Characterization of Innate Immune Activation. LRNP-II exhibited 10-fold greater IFN-α and IP-10 responses than LRNP-I nanoparticles, which potently activated IL-6 following s.c. injection (Fig. S4*D*). We hypothesized that the lipidoid-specific cytokine profiles observed were due to preferential uptake and activation of different cell types following s.c. administration. We compared delivery of isRNA to primary monocytes or PDCs isolated from human PBMCs using lipidoids I-3 and II-3, the three-tailed versions of I and II that had greater isRNA activity in vitro (Fig. S3*B*). Lipidoid I-3 complexed with R-006 induced type I interferons from isolated CD14+ cells (i.e., monocytes), but not from isolated PDCs (Fig. S6*B*). Conversely, lipidoid II-3 complexed with R-006 induced large amounts of type I interferon production in PDCs but considerably less from CD14+ cells (monocytes).

After i.v. injection in the 129sv strain of mouse, LRNPs activated more robust responses than the s.c. route, and isRNA specificity was observed in stimulation of serum IFN-α and IP-10 with almost no activity of the control R-1263 RNA (Fig. S6*A*). However, LRNP formulated with R-1263 did induce IL-6 when injected i.v. raising concerns for increased nonspecific activation with i.v. injection, and this observation was most pronounced at the earliest time points with LRNP-I. We thus further investigated directly the role of TLRs in differential recognition of LRNPs.

The TLRs utilizing the common MyD88 pathway are TLR1, TLR2, TLR5, TLR6, TLR7, TLR8, and TLR9 (11). Because knock-out mice for all the TLRs were not readily available, we investigated TLR-specificity of LRNP-I in vitro in a human cell line stably expressing human TLRs. HEK293 cell lines stably expressing human TLRs 2, 3, 4, 5 or 6 were incubated with LRNP + R-006 without observation of any TLR-mediated activity above background (Fig. S6*C*). HEK293 cells stably transfected with TLR8, and TLR7 to a lesser extent, exhibited dose-dependent activation by LRNP-I incorporating R-006 RNA (Fig. S6*D*).

LRNP were also investigated in knock-out mouse models of the MyD88, TLR4, TLR7, and TLR9 genes and compared to wild-type (WT) controls on matched backgrounds (Fig. 5). Production of IFN-α, IP-10, and IL-6 were all dependent on MyD88 signaling for both LRNP-I and LRNP-II. However, only IFN-α and IP-10 production were also dependent upon TLR7. In the C57BL/6 strain, we observed that IL-6 production in response to LRNP-II was RNA specific and completely dependent upon TLR7. However, IL-6 activation in the same strain by LRNP-I appears to be only partially TLR7-dependent at a 50 μg RNA

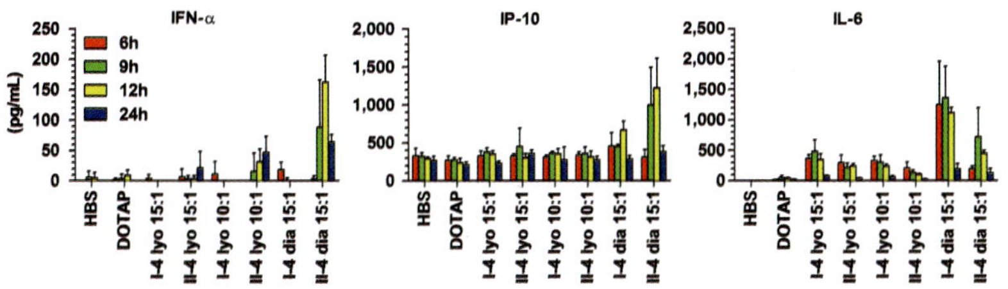

Fig. 3. In vivo screening for activation of innate immune responses following injection of formulated lipidoid–RNA nanoparticles. Lipidoid–RNA nanoparticles formulated with 100 ug R-006 RNA were injected s.c. in BALB/c mice ($n = 3$ or 4). R-006 formulated with DOTAP and a mock injection with HBSS were included as controls. Blood was collected at 6, 9, 12, and 24 h following injection and indicated cytokines were measured by ELISA.

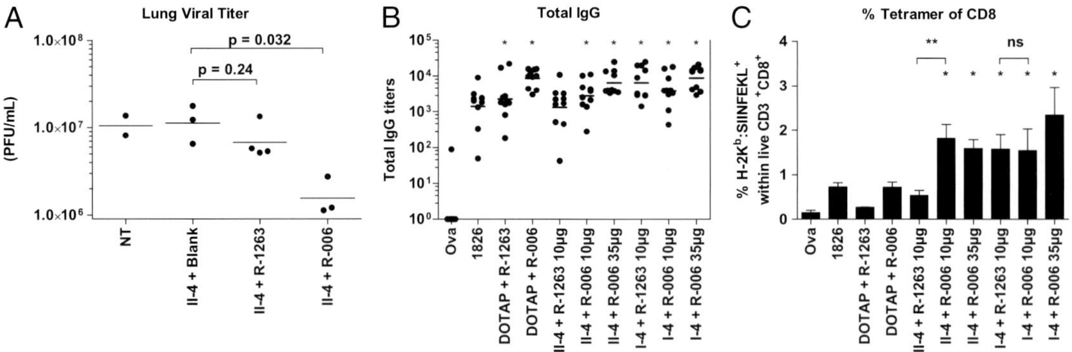

Fig. 4. In vivo adjuvant activity of formulated lipidoid–RNA nanoparticles (*A*) Prophylaxis against influenza infection. Lipidoid II-4 was formulated with R-006 RNA, R-1263 RNA, or without RNA (II-4 + blank) as a control and 50 µg RNA basis was injected s.c. in 129Sv mice. Lung viral titer was measured 24 h after challenge with 12,000 pfu of influenza (PR8) virus. Individual titers shown with geometric mean for each group represented by a bar. *p* values indicated for two-way *t*-test. (*B–C*) Adjuvant activity in chicken ovalbumin protein vaccination of C57Bl/6 mice together with lipidoid I-4 or lipidoid II-4 nanoparticles. Vaccination was also done with DOTAP-formulated RNA, CpG ODN 1826 (TLR9 agonist), or without adjuvant. Immunization was performed i.m. on days 0, 14, and 21, and samples were collected at day 28. (*B*) Total serum Ova-specific IgG. Individual titers shown with geometric mean for each group represented by a bar. (*C*) Spleen $CD8^+$ T cells specific for the SIINFEKL peptide fragment of the Ova protein. * $p = 0.05$ by one-way ANOVA with Dunn's Multiple Comparison posttest for means with unequal variance compared to Ova group ** $p = 0.05$ by one-way ANOVA with Dunn's Multiple Comparison posttest.

dose i.v. Further, IL-6 activity was completely abolished regardless of RNA without MyD88 signaling (Fig. 5). Incomplete reduction in IL-6 for LRNP-I + R-006 nanoparticles was observed in TLR − /− mice, and no inhibition was observed for LRNP-I + R-1263 nanoparticles. TLR4 and TLR9 were not required for cytokine production for either LRNP with either RNA, indicating that neither LPS nor immunostimulatory DNA contamination was responsible for IL-6 induction.

Given the differences in vaccine responses observed for LRNP-I compared to LRNP-II, we further investigated LRNP-I adjuvant properties in mice deficient in MyD88. Antigen-specific responses following immunization revealed a marked dependence upon MyD88 for increasing both cell-mediated and humoral immunity. Compared to wild-type controls, the humoral response of $MyD88^{-/-}$ mice exhibited reduced generation of total IgG titer for CpG ODN, lipidoid nanoparticles, and DOTAP–RNA complexes but not Ova antigen alone or polyI:C (Fig. S7*A*). $MyD88^{-/-}$ mice also exhibited complete loss of class switching to IgG2c (Fig. S7*B*) except for polyI:C. Additionally, a diminished cell-mediated response in $MyD88^{-/-}$ mice was observed with reduced antigen-specific cytotoxic T lymphocyte activity (Fig. S7*C*) and overall decreased numbers of antigen-specific T cells (Fig. S7*D*).

Discussion

Many groups have focused on strategies to chemically or physically alter nucleic acids such as siRNAs (25, 26, 39–41) to increase serum stability or render them immuno-silent; however, to our knowledge the development of delivery systems to intentionally enhance isRNA activity has not been reported. Here we develop a library of lipidoid materials for delivery of isRNA. These materials were able to condense isRNA molecules into nanoparticles and deliver isRNA molecules into endosomes of immune cells where they could trigger TLR-mediated responses. The RNA delivery systems reported herein have been specifically designed to incorporate these functionalities for the purpose of isRNA delivery.

Analysis of the chemical structures of lipidoids most efficient for isRNA delivery revealed that the isRNA delivery functionality was correlated with certain amine-core structures. Materials based upon the same cores (e.g., 100, 86, 87) exhibited similar properties (Fig. 1 and Fig. S2). It has been previously theorized that charge interaction is responsible for the nucleic acid packaging, release, and the endosomal escape properties of polyamines and positively charged lipids (25, 42). Following particle uptake, we hypothesize that the chemical properties of the available amines influence the intracellular location of release of the isRNA in such a manner that the isRNA payload remains within the endosome. Materials highly efficient for DNA delivery and RNA interference likely facilitate endosomal escape and thus may be poor agents for delivery of isRNA to TLRs in the endosome. Consistent with this hypothesis, lipioids previously found useful for delivery of double-stranded small interfering RNAs (27) (e.g., ND(5)-98-1) were not as useful for delivery of isRNA. Structures based upon the diamine core 100 were the most efficient in isRNA delivery across a variety of tail lengths. Additionally, lipidoids based upon the structurally and chemically similar 86 and 87 cores were highly immunostimulatory even with control RNA (Fig. S2*A*) possibly due to toxicity or direct TLR interaction.

Whereas the core material appears to be the major determinant of isRNA activity, it is clear that the alkane tails also influence delivery properties of the formulated nanoparticle (Fig. S3*A*) including cell-specific targeting. Tail chemistry can also influence the ability of a lipid to destabilize a cellular or endosomal membrane thus modifying rate of endosomal retention and release (25, 30). Further, our experience with lipidoids has shown that different tail-lengths or mixtures can result in unexpectedly enhanced functionality (43, 44). Thus, by screening and iteration based upon these design principles, we were able to explore a large chemical/functional space resulting in two novel 100-core lipidoids with mixed tails (I and II) highly efficient for isRNA delivery (Fig. 3, Fig. S4).

Although they both contain the same 100-amine core, lipidoids I and II exhibited differential patterns of innate immune activation suggestive of delivery to different cell types or different signaling pathways. The type I IFN and IP-10 production pattern associated with LRNP-II may be indicative of delivery of R-006 to TLR7 in PDCs. In humans and mice IFN-α is mostly produced by PDCs in responses to ssRNA activation of TLR7 although MDCs and monocyte-lineage cells can also secrete IFN-α with less potency (9, 45). IP-10 (CXCL10) is also produced primarily by PDCs through TLR7 in a ssRNA-dependent manner (6, 18, 32). However, IL-6 is primarily produced by MDCs, monocytes, and moDCs upon stimulation (9), though it can also be produced by murine PDCs in response to viral ssRNA (6, 18). Thus the activity of lipidoid I in terms of high IL-6 production is consistent with isRNA delivery preferentially to moDCs or MDCs. This hypothesis is further supported by the observation of cell-type-specific isRNA-mediated activation in vitro in human peripheral blood PDCs and monocytes (Fig. S6*B*).

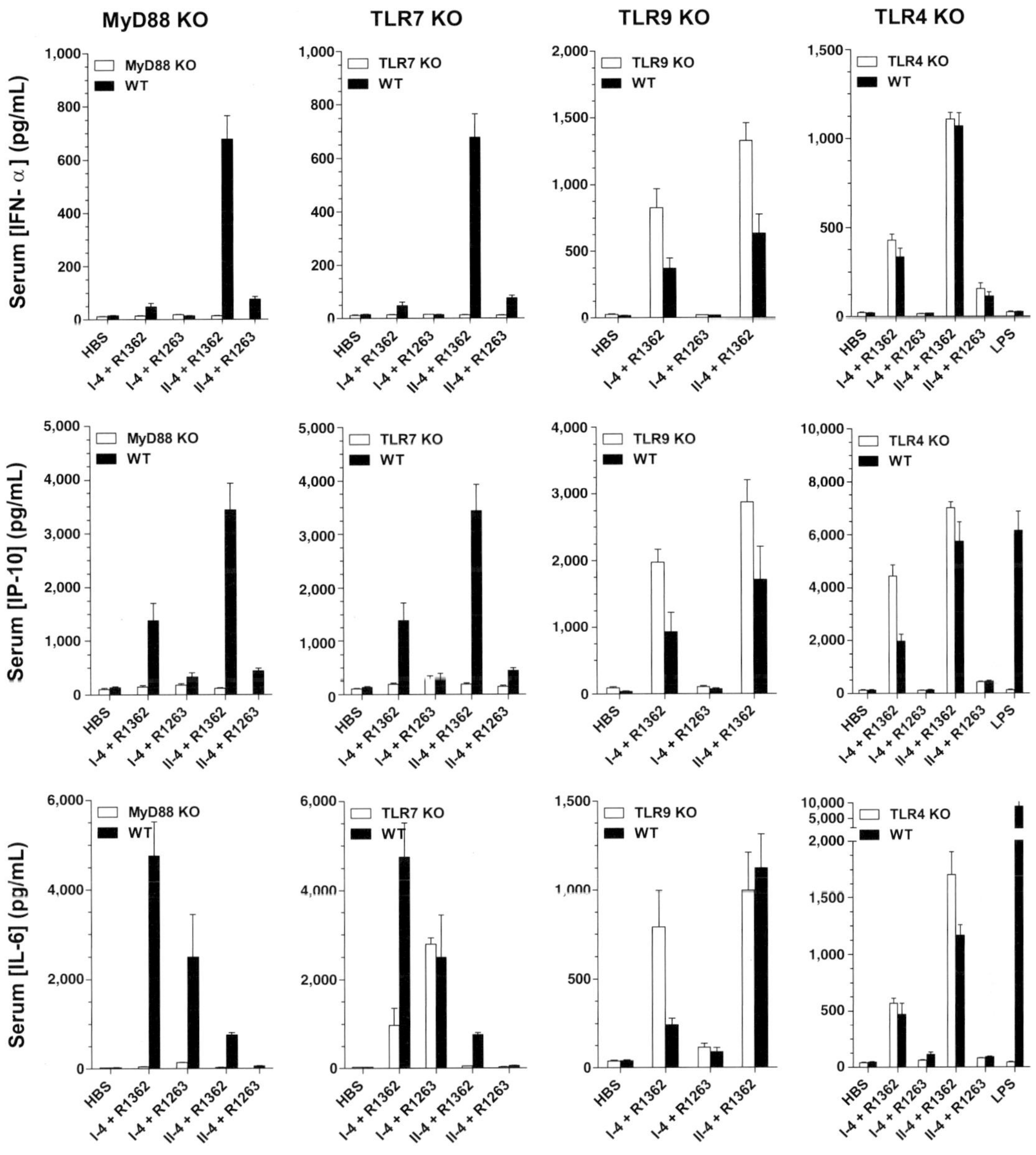

Fig. 5. Characterization of cytokine response from lipidoid–RNA nanoparticles in vivo. Lipidoids I-4 and II-4 were formulated with the strong TLR7/8 agonist R-006 RNA and the weak TLR7/8 agonist R-1263 RNA as control. Serum cytokine response of IFN-α (*Top*), IP-10 (*Middle*), and IL-6 (*Bottom*) at 6 h compared to strain background-matched controls following i.v. injection of 50 μg RNA into MyD88$^{-/-}$ or TLR7$^{-/-}$ mice, or 30 μg RNA into TLR9$^{-/-}$ or TLR4$^{-/-}$ mice; LPS = lipopolysaccharide; (n = 3 or 4).

The LRNP may activate more than one TLR. In the mouse, cell-specific segregation of TLR7 is less distinct than it is in humans (8, 11, 18); TLR7 is well-expressed in mouse MDCs and moDCs. Both IFN-α and IP-10 production were clearly TLR7- and MyD88-dependent in mice (Fig. 5) confirming an isRNA-specific TLR7-mediated mechanism. Recently it has been shown that mouse TLR8, previously thought to be nonfunctional (6, 18, 19, 46), can be activated under certain conditions (29, 47, 48). We hypothesize that TLR8 may play a partial role in LRNP-I isRNA-mediated activation of IL-6. In support of this hypothesis both TLR7 and TLR8 showed clear activity in response to LRNP-I stimulation in a human cell culture TLR-overexpression model (Fig. S6*D*) whereas the remaining MyD88-dependent TLRs were not stimulated (Fig. S6*C*). The IL-6 production induced by lipidoid-mediated isRNA delivery in TLR7-deficienct mice but not MyD88-deficient mice (Fig. 5) also indicates a possible role for isRNA stimulation of another MyD88-dependent receptor although no TLR8$^{-/-}$ mouse model was available to directly test the role of TLR8 responses to LRNPs in vivo. Together these data indicate roles for TLR7 and another MyD88-dependent function, possibly TLR8, for the induction of IL-6 by LRNP-I.

Prophylactic administration of LRNP-I resulted in suppression of influenza virus replication in the mouse lung (Fig. 4*A*) specific to the immunostimulatory activity of R-006 RNA. These data indicate that s.c. injections of lipidoid-encapsulated isRNA can achieve sufficient systemic innate immune stimulation to combat viral infection. Because they activate the same receptors as viruses resulting in a variety of cytokines and interferons, lipidoid isRNA delivery may resemble the innate response to infection by RNA viruses. Imiquimod, a TLR7/8 agonist, has seen extensive clinical use (3, 5) as a topical chemotherapeutic and antiviral. Experimental therapies employing TLR7 agonists have also demonstrated some clinical benefit in viral hepatitis (49), however, adverse immunologic events related to widespread biodistribution were observed at therapeutic doses (50, 51) including unacceptable toxicity in a human trial of oral imiquimod (52). No adverse effects were observed in mice injected with LRNPs in this study, though long-term and dose-response effects of LRNP administration need to be investigated. A restricted delivery approach such as LRNPs may limit the systemic distribution by localizing initiation of immune responses to the application site and draining lymph nodes as well as targeting uptake to specific cell types. Thus the endogenous antiviral-like response activated by LRNPs may be an attractive alternative therapy for conditions currently treated with recombinant IFN-α regimens.

The development of vaccine adjuvants that increase both humoral and cell-mediated immune responses by targeting TLRs represents a promising approach for the field of vaccine development (3, 5, 10). A hallmark of this cellular immunity is expansion of antigen-specific $CD8^+$ T cells, which was increased by immunization with either LRNP-I or LRNP-II (Fig. 4*C*). LRNP adjuvants increased secretion of Th1-biasing cytokines upon restimulation (Fig. S5*B*), which are associated with control of intracellular pathogens (2) and cancers (7). Efficient T-cell activation with multifunctional quality can lead to memory responses that confer protective immunity for long periods of time (1, 14, 53). Further studies will be necessary to evaluate the memory function and quality of T-cell responses to protein vaccines adjuvanted by LRNPs.

Class switching and production of IgG2c in B-cells and efficient antigen-specific T-cell induction requires activation of MyD88 (54). If the adjuvant mechanism of LRNP-I was completely due to activation of TLR-dependent innate immune responses, then these responses would be expected to be dependent upon MyD88. However, comparison of vaccine responses in wild-type and MyD88-deficient mice showed that although loss of MyD88 markedly reduced both cell-mediated and humoral immune responses, some adjuvant activity was maintained particularly with respect to generation of (non-IgG2c) IgG antibodies (Fig. S7*B*) indicating an additional role of a non-MyD88-dependent adjuvant effect. Commonly used adjuvants including alum, Freund's adjuvant, and squalene increase immune responses by both enhancing innate immune responses (4) and sustaining antigen availability and presentation in draining lymph nodes (55). Thus, in addition to providing isRNA-mediated innate signals, lipidoid nanoparticles may also enhance the delivery of protein antigens or TLR agonists to tissue-resident and lymphatic dendritic cells (55) thereby increasing vaccine responses.

In conclusion, combinatorial chemistry and high-throughput screening methods have enabled development of libraries of materials for isRNA delivery that provide for specific activation of immune responses. We report the development of unique delivery systems specifically designed to deliver isRNA in the context of a vaccine adjuvant or for use in immune prophylaxis against viral infection. Here we have directly shown LRNPs to be more efficient for isRNA delivery than currently available agents such as lipofectamine, DOTAP, or even the previously reported lipidoid ND(5)-98-1 (23, 27). Immunostimulatory RNA delivery to PDCs using formulations such as LRNP-II may be useful for antiviral applications, whereas delivery using LRNP-I may be most useful for activating monocytes and MDCs to enhance vaccine responses. Our screening revealed two promising candidate materials for future development, and further refinement of lipidoid design and nanoparticle formulation techniques may lead to even more specific and robust effects. Further development of such vehicles for controlled TLR7 and 8 stimulation, including additional optimization and safety assessment in larger animals, may have future clinical utility.

Materials and Methods

A summary of experimental techniques is presented here. For full details and methods, please see the *SI Materials and Methods*.

Lipidoid Synthesis and High-Throughput Screening for isRNA Delivery. Lipidoids were synthesized in a combinatorial fashion as depicted in Fig. S1 as previously described (27) by reacting primary and secondary amine-containing cores with alkyl-acrylate or alkyl-acrylamide tails. A complete list of crude lipidoids screened is found in Table S1. Second-generation lipidoids based on the 100-core diamine were synthesized in a four-step process (Fig. S1*D*) and renamed lipidoids I and II for clarification. Lipidoid products were complexed with RNA in sodium acetate at 15, 10, 5, and 2.5 : 1 mass ratios of lipid to RNA. RNAs R-006 or R-1263 were fully phosphorothioate-modified, 20-base, ssRNA synthesized by Coley Pharmaceuticals with sequences as previously described (29). PBMCs were isolated from donor-blind buffy-coat packs obtained from the Massachusetts General Hospital blood bank. Complexes were added to PBMCs, plated at 5×10^5 cells/well, for a final RNA concentration of 200 ng RNA per well (1 μg/mL, approximately 140 nM). Additionally, PBMCs were incubated with R-006 and L2K, according to manufacturers siRNA transfection protocols, to control for donor PBMC variability in maximum type I IFN secretion capacity. After 16–20 h of incubation, PBMC supernatants were stored at −80 °C for later quantification of type I interferon activity using a high-throughput cell-based detection assay as previously described (28).

Formulation and Characterization of Lipidoid–RNA Nanoparticles. Purified lipidoid was codissolved in ethanol with cholesterol and C16 mPEG 2000 ceramide at a 15:0.8:7 mass ratio (L:C:P) in a mixture of ethanol and sodium acetate. Lipidoid/Ch/PEG were added to RNA at a 15, 11.5, or 10:1 mass ratio (L:R). Lipidoid–RNA nanoparticles were then extruded through a double 200 nm membrane and then twice through a double 80 nm membrane (Whatman) on a Northern Lipids extrusion system at 40 °C. Prior to injection, nanoparticles were either dialyzed at 3,500 molecular weight cutoff in HBSS or lyophilized with 10 mg/mL sucrose. Final RNA concentration was determined by modified Ribogreen (Invitrogen) assay. No bacterial endotoxin was detected by limulus amebocyte lysate assay (Lonza) in any batches of nanoparticles.

In Vivo Characterization of Innate Immune Responses. All animal studies, except for influenza prophylaxis studies, were conducted at Coley Pharmaceuticals (now Pfizer Canada) under the approval of the institutional care committees and in accordance with the guidelines set forth by the Canadian Council on Animal Care. Mouse studies of influenza prophylaxis were conducted at the Massachusetts Institute of Technology (Cambridge, MA), where animals were cared for according to the guidance of the Division of Comparative Medicine. Animals were monitored for the duration of all experiments for adverse behavior or decrease in normal activity that might indicate toxicity or adverse inflammatory responses. Lipidoid–RNA nanoparticles were formulated with either R-006 or R-1263 RNA at 15:1 ratio (lipidoid to RNA), dialyzed to remove ethanol, and diluted in HBSS prior to injection under isofluorane anesthesia. Lipidoid nanoparticles were injected s.c. or by i.v. tail vein injection in 129Sv mice at indicated doses. Serum samples were analyzed for levels of IFN-α and cytokines at indicated time points. TLR-mediated responses were investigated following i.v. injection in TLR-deficient or genetic-background-matched WT controls. Prophylaxis of influenza infection models was conducted as previously described (23). Briefly, lipidoid–RNA nanoparticles were injected s.c. at 50 μg RNA per mouse. After 9 h, mice were infected by intranasal instillation with 12,000 pfu influenza A virus A/PR/8/34 (PR8). Lungs viral titer was quantified at 24 h after infection. Adjuvant studies of lipidoid I and II isRNA nanoparticles were performed by mixing indicated doses of lipidoid-formulated RNA with 20 μg chicken Ova immediately prior to i.m. injection into C57Bl/6 mice or $MyD88^{-/-}$ on the C57Bl/6 background. Mice were dosed at days 0, 14, and 21, then blood and splenocytes were collected at day 28. Plasma anti-Ova total IgG, IgG1, and IgG2c were determined by sandwich ELISA. Splenocytes were assessed using a chromium release assay for cytotoxic T cells and flow-cytometry-based tetramer analysis for quantifying antigen-specific T cells as previously described (56).

ACKNOWLEDGMENTS. This work was supported in part by grants from the National Institutes of Health EB00244 (to R.L. and D.G.A.) and AI56267 (to J.C.) and from the National Institute of Allergy and Infectious Diseases Contract HHSSN266200400044C (to G.C. and A.M.K.).

1. Iwasaki A, Medzhitov R (2004) Toll-like receptor control of the adaptive immune responses. *Nat Immunol* 5:987–995.
2. Zhang SY, et al. (2007) Human Toll-like receptor-dependent induction of interferons in protective immunity to viruses. *Immunol Rev* 220:225–236.
3. Meyer T, Stockfleth E (2008) Clinical investigations of Toll-like receptor agonists. *Expert Opin Investig Drugs* 17:1051–1065.
4. Pashine A, Valiante NM, Ulmer JB (2005) Targeting the innate immune response with improved vaccine adjuvants. *Nat Med* 11(Suppl 4):S63–68.
5. Kanzler H, Barrat FJ, Hessel EM, Coffman RL (2007) Therapeutic targeting of innate immunity with Toll-like receptor agonists and antagonists. *Nat Med* 13:552–559.
6. Heil F, et al. (2004) Species-specific recognition of single-stranded RNA via toll-like receptor 7 and 8. *Science* 303:1526–1529.
7. Schon MP, Schon M (2008) TLR7 and TLR8 as targets in cancer therapy. *Oncogene* 27:190–199.
8. Hornung V, et al. (2005) Sequence-specific potent induction of IFN-alpha by short interfering RNA in plasmacytoid dendritic cells through TLR7. *Nat Med* 11:263–270.
9. Jarrossay D, Napolitani G, Colonna M, Sallusto F, Lanzavecchia A (2001) Specialization and complementarity in microbial molecule recognition by human myeloid and plasmacytoid dendritic cells. *Eur J Immunol* 31:3388–3393.
10. Borden EC, et al. (2007) Interferons at age 50: Past, current and future impact on biomedicine. *Nat Rev Drug Discov* 6:975–990.
11. Akira S, Takeda K (2004) Functions of toll-like receptors: Lessons from KO mice. *C R Biol* 327:581–589.
12. Cristofaro P, Opal SM (2006) Role of Toll-like receptors in infection and immunity: Clinical implications. *Drugs* 66:15–29.
13. Honda K, Taniguchi T (2006) IRFs: Master regulators of signalling by Toll-like receptors and cytosolic pattern-recognition receptors. *Nat Rev Immunol* 6:644–658.
14. Steinman RM, Banchereau J (2007) Taking dendritic cells into medicine. *Nature* 449:419–426.
15. Delgado MF, et al. (2009) Lack of antibody affinity maturation due to poor Toll-like receptor stimulation leads to enhanced respiratory syncytial virus disease. *Nat Med* 15:34–41.
16. Marshak-Rothstein A (2006) Toll-like receptors in systemic autoimmune disease. *Nat Rev Immunol* 6:823–835.
17. Romagne F (2007) Current and future drugs targeting one class of innate immunity receptors: The Toll-like receptors. *Drug Discov Today* 12:80–87.
18. Diebold SS, Kaisho T, Hemmi H, Akira S, Reis e Sousa C (2004) Innate antiviral responses by means of TLR7-mediated recognition of single-stranded RNA. *Science* 303:1529–1531.
19. Lan T, et al. (2007) Stabilized immune modulatory RNA compounds as agonists of Toll-like receptors 7 and 8. *Proc Natl Acad Sci USA* 104:13750–13755.
20. Scheel B, et al. (2006) Therapeutic anti-tumor immunity triggered by injections of immunostimulating single-stranded RNA. *Eur J Immunol* 36:2807–2816.
21. Eberle F, et al. (2008) Modificationsin small interfering RNA that separate immunostimulation from RNA interference. *J Immunol* 180:3229–3237.
22. Sioud M (2005) Induction of inflammatory cytokines and interferon responses by double-stranded and single-stranded siRNAs is sequence-dependent and requires endosomal localization. *J Mol Biol* 348:1079–1090.
23. Nguyen DN, et al. (2009) Drug delivery-mediated control of RNA immunostimulation. *Mol Ther* 17:1555–1562.
24. Robbins M, et al. (2008) Misinterpreting the therapeutic effects of small interfering RNA caused by immune stimulation. *Hum Gene Ther* 19:991–999.
25. Schroeder A, Levins CG, Cortez C, Langer R, Anderson DG (2010) Lipid-based nanotherapeutics for siRNA delivery. *J Intern Med* 267:9–21.
26. Whitehead KA, Langer R, Anderson DG (2009) Knocking down barriers: advances in siRNA delivery. *Nat Rev Drug Discov* 8:129–138.
27. Akinc A, et al. (2008) A combinatorial library of lipid-like materials for delivery of RNAi therapeutics. *Nat Biotechnol* 26:561–569.
28. Nguyen DN, et al. (2009) A novel high-throughput cell-based method for integrated quantification of type I interferons and in vitro screening of immunostimulatory RNA drug delivery. *Biotechnol Bioeng* 103:664–675.
29. Forsbach A, et al. (2008) Identification of RNA sequence motifs stimulating sequence-specific TLR8-dependent immune responses. *J Immunol* 180:3729–3738.
30. Semple SC, et al. (2010) Rational design of cationic lipids for siRNA delivery. *Nat Biotechnol* 28:172–176.
31. Barchet W, et al. (2005) Dendritic cells respond to influenza virus through TLR7- and PKR-independent pathways. *Eur J Immunol* 35:236–242.
32. Megjugorac NJ, Young HA, Amrute SB, Olshalsky SL, Fitzgerald-Bocarsly P (2004) Virally stimulated plasmacytoid dendritic cells produce chemokines and induce migration of T and NK cells. *J Leukoc Biol* 75:504–514.
33. Davila E, Celis E (2000) Repeated administration of cytosine-phosphorothiolated guanine-containing oligonucleotides together with peptide/protein immunization results in enhanced CTL responses with anti-tumor activity. *J Immunol* 165:539–547.
34. Halperin SA, et al. (2006) Comparison of the safety and immunogenicity of hepatitis B virus surface antigen co-administered with an immunostimulatory phosphorothioate oligonucleotide and a licensed hepatitis B vaccine in healthy young adults. *Vaccine* 24:20–26.
35. Vollmer J, Krieg AM (2009) Immunotherapeutic applications of CpG oligodeoxynucleotide TLR9 agonists. *Adv Drug Deliv Rev* 61:195–204.
36. Malyala P, O'Hagan DT, Singh M (2009) Enhancing the therapeutic efficacy of CpG oligonucleotides using biodegradable microparticles. *Adv Drug Deliv Rev* 61:218–225.
37. Wilson KD, de Jong SD, Tam YK (2009) Lipid-based delivery of CpG oligonucleotides enhances immunotherapeutic efficacy. *Adv Drug Deliv Rev* 61:233–242.
38. Cooper CL, Angel JB, Seguin I, Davis HL, Cameron DW (2008) CPG 7909 adjuvant plus hepatitis B virus vaccination in HIV-infected adults achieves long-term seroprotection for up to 5 years. *Clin Infect Dis* 46:1310–1314.
39. Schlee M, Hornung V, Hartmann G (2006) siRNA and isRNA: Two edges of one sword. *Mol Ther* 14:463–470.
40. Sioud M (2007) RNA interference and innate immunity. *Adv Drug Deliv Rev* 59:153–163.
41. Weinstein S, Peer D (2010) RNAi nanomedicines: Challenges and opportunities within the immune system. *Nanotechnology* 21:232001.
42. Putnam D (2006) Polymers for gene delivery across length scales. *Nat Mater* 5:439–451.
43. Love KT, et al. (2010) Lipid-like materials for low-dose, in vivo gene silencing. *Proc Natl Acad Sci USA* 107:1864–1869.
44. Mahon KP, et al. (2010) Combinatorial approach to determine functional group effects on lipidoid-mediated siRNA delivery. *Bioconjug Chem* 21:1448–1454.
45. Asselin-Paturel C, Brizard G, Pin J J, Briere F, Trinchieri G (2003) Mouse strain differences in plasmacytoid dendritic cell frequency and function revealed by a novel monoclonal antibody. *J Immunol* 171:6466–6477.
46. Jurk M, et al. (2002) Human TLR7 or TLR8 independently confer responsiveness to the antiviral compound R-848. *Nat Immunol* 3:499–499.
47. Gorden KK, et al. (2006) Oligodeoxynucleotides differentially modulate activation of TLR7 and TLR8 by imidazoquinolines. *J Immunol* 177:8164–8170.
48. Martinez J, Huang X, Yang Y (2010) Toll-like receptor 8-mediated activation of murine plasmacytoid dendritic cells by vaccinia viral DNA. *Proc Natl Acad Sci USA* 107:6442–6447.
49. Horsmans Y, et al. (2005) Isatoribine, an agonist of TLR7, reduces plasma virus concentration in chronic hepatitis C infection. *Hepatology* 42:724–731.
50. Fidock MD, et al. (2011) The innate immune response, clinical outcomes,and ex vivo HCV antiviral efficacy of a TLR7 agonist (PF-4878691). *Clin Pharmacol Ther* 89:821–829.
51. Fletcher S, Steffy K, Averett D (2006) Masked oral prodrugs of toll-like receptor 7 agonists: A new approach for the treatment of infectious disease. *Curr Opin Investig Drugs* 7:702–708.
52. Savage P, et al. (1996) A phase I clinical trial of imiquimod, an oral interferon inducer, administered daily. *Br J Cancer* 74:1482–1486.
53. Seder RA, Darrah PA, Roederer M (2008) T-cell quality in memory and protection: Implications for vaccine design. *Nat Rev Immunol* 8:247–258.
54. Barr TA, Brown S, Mastroeni P, Gray D (2009) B cell intrinsic MyD88 signals drive IFN-gamma production from T cells and control switching to IgG2c. *J Immunol* 183:1005–1012.
55. Tritto E, Mosca F, De Gregorio E (2009) Mechanism of action of licensed vaccine adjuvants. *Vaccine* 27:3331–3334.
56. Chikh G, et al. (2009) Synthetic methylated CpG ODNs are potent in vivo adjuvants when delivered in liposomal nanoparticles. *Int Immunol* 21:757–767.

RESEARCH ARTICLE

NANOMEDICINE

Transepithelial Transport of Fc-Targeted Nanoparticles by the Neonatal Fc Receptor for Oral Delivery

Eric M. Pridgen,[1,2]* Frank Alexis,[3,4]* Timothy T. Kuo,[5] Etgar Levy-Nissenbaum,[3,4] Rohit Karnik,[6] Richard S. Blumberg,[5] Robert Langer,[1,2,4] Omid C. Farokhzad[3,4]†

Nanoparticles are poised to have a tremendous impact on the treatment of many diseases, but their broad application is limited because currently they can only be administered by parenteral methods. Oral administration of nanoparticles is preferred but remains a challenge because transport across the intestinal epithelium is limited. We show that nanoparticles targeted to the neonatal Fc receptor (FcRn), which mediates the transport of immunoglobulin G antibodies across epithelial barriers, are efficiently transported across the intestinal epithelium using both in vitro and in vivo models. In mice, orally administered FcRn-targeted nanoparticles crossed the intestinal epithelium and reached systemic circulation with a mean absorption efficiency of 13.7%*hour compared with only 1.2%*hour for nontargeted nanoparticles. In addition, targeted nanoparticles containing insulin as a model nanoparticle-based therapy for diabetes were orally administered at a clinically relevant insulin dose of 1.1 U/kg and elicited a prolonged hypoglycemic response in wild-type mice. This effect was abolished in FcRn knockout mice, indicating that the enhanced nanoparticle transport was specifically due to FcRn. FcRn-targeted nanoparticles may have a major impact on the treatment of many diseases by enabling drugs currently limited by low bioavailability to be efficiently delivered though oral administration.

INTRODUCTION

Nanoparticles (NPs) have the potential to make an impact on the treatment of many diseases, including cancer, cardiovascular disease, and diabetes. Many NP-based therapeutics are now entering clinical trials or have been approved for use (*1*, *2*), including targeted polymeric NPs (*3*) (clinicaltrials.gov identifier: NCT01478893) based on technologies that we have described previously (*4*). However, the impact of NPs in the clinic may be limited to a narrow set of indications because NP administration is currently restricted to parenteral methods. Many diseases that could benefit from NP-based therapeutics require frequent administration. Alternate routes of administration, particularly oral, are preferred because of the convenience and compliance by patients (*5*). Intestinal absorption of NPs is highly inefficient because the physicochemical parameters of NPs prevent their transport across cellular barriers such as the intestinal epithelium (*6*). To improve the absorption efficiency of NPs and to make the oral administration of NPs practical in the clinic, new strategies are necessary to overcome the intestinal epithelial barrier.

The neonatal Fc receptor (FcRn) mediates immunoglobulin G (IgG) transport across polarized epithelial barriers (*7*, *8*). It was discovered as the receptor in the neonatal intestine that transports IgG in breast milk from mother to offspring (*9*). However, FcRn is expressed into adulthood at levels similar to fetal expression in the apical region of epithelial cells in the small intestine and diffusely throughout the colon (*10*). FcRn is also expressed in the vascular endothelium, blood-brain barrier, kidneys, liver, lungs, and throughout the hematopoietic system (*11*, *12*). FcRn interacts with the Fc portion of IgG in a pH-dependent manner, binding with high affinity in acidic (pH <6.5) but not physiological environments (pH ~7.4) (*13*). The intracellular trafficking of the IgG-FcRn complex has been conclusively demonstrated in the rat intestine using IgG Fc labeled with 1.4-nm gold as a contrast agent for electron tomography (*14*). The studies revealed that Fc is transported through a complex pathway involving a network of entangled tubular and irregular vesicles to reach the basolateral surface of the cell.

We hypothesized that targeting NPs to FcRn using IgG Fc fragments would allow orally administered NPs to be transported across the intestinal epithelium in rodents (Fig. 1). In acidic sections of the intestine, such as the duodenum and portions of the jejunum (*15*), Fc fragments conjugated to NPs [Fc-targeted NPs (NP-Fc)] will bind to FcRn at the apical surface of absorptive epithelial cells, leading to receptor-mediated endocytosis (*16*). NP-Fc could also be taken up by fluid-phase pinocytosis. During intracellular trafficking, NP-Fc and FcRn in the same acidic endosome compartments will bind with high affinity. FcRn can then guide bound NP-Fc through a transcytosis pathway, avoiding lysosomal degradation (*17*). On the basolateral side, exocytosis results in exposure to a neutral pH environment in the lamina propria, causing the release of NP-Fc (*18*). NP-Fc can then diffuse through the lamina propria and enter systemic circulation.

Fc fusion proteins have been used to overcome biological barriers: Fc fused with erythropoietin was measured in nonhuman primates (*19*) and humans (*20*) after pulmonary administration, indicating that using the FcRn is a valid transport pathway in humans, and follicle-stimulating hormone fused with Fc reached circulation after oral delivery in newborn rats (*21*). However, NPs offer several potential advantages over fusion proteins including (i) transport of many drug molecules with each transcytosis event; (ii) protection of drug molecules, particularly protein therapeutics, through encapsulation; and (iii) mitigation of the need for drug modification. Recently, 20- to 50-nm fluorescent NPs conjugated to IgG Fc were transported across an in vitro airway epithelial

[1]Department of Chemical Engineering, Massachusetts Institute of Technology, Cambridge, MA 02139, USA. [2]The David H. Koch Institute for Integrative Cancer Research, Cambridge, MA 02139, USA. [3]Laboratory of Nanomedicine and Biomaterials, Department of Anesthesiology, Brigham and Women's Hospital, Harvard Medical School, Boston, MA 02115, USA. [4]MIT-Harvard Center for Cancer Nanotechnology Excellence, Cambridge, MA 02139, USA. [5]Division of Gastroenterology, Department of Medicine, Brigham and Women's Hospital, Harvard Medical School, Boston, MA 02115, USA. [6]Department of Mechanical Engineering, Massachusetts Institute of Technology, Cambridge, MA 02139, USA.
*These authors contributed equally to this work.
†Corresponding author. E-mail: ofarokhzad@zeus.bwh.harvard.edu

cell monolayer (*22*). Yet, no studies, to our knowledge, have used drug-encapsulated NPs to target the FcRn for oral drug delivery applications.

Here, we developed polymeric NPs surface-modified with Fc to target FcRn, resulting in transepithelial transport both in vitro and in vivo. NP-Fc demonstrated enhanced transport specifically mediated by FcRn across an intestinal epithelial monolayer in vitro. In mice, NP-Fc were imaged crossing the intestinal epithelium and entering the lamina propria after oral administration. NP-Fc were also measured in several organs after oral administration, indicating that NP-Fc were able to reach systemic circulation. Finally, NP-Fc encapsulating insulin as a model NP-based therapeutic for diabetes were administered orally and elicited a hypoglycemic response. These results demonstrate that NPs targeted to the FcRn are capable of crossing an epithelial barrier in vivo using a transcytosis pathway, making NP-based oral drug delivery formulations readily suitable to deliver therapeutics to the systemic circulation.

RESULTS

Preparation of Fc-targeted NPs

NPs were formed from biodegradable and biocompatible poly(lactic acid)–*b*-poly(ethylene glycol) (PLA-PEG) block copolymers. PLA is a biodegradable polymer used in many U.S. Food and Drug Administration–approved products and forms the NP core owing to its hydrophobicity. PEG is a biocompatible polymer that remains on the NP surface owing to its hydrophilic nature and forms the NP corona. PLA-PEG was synthesized using ring-opening polymerization with a free terminal maleimide group (PLA-PEG-MAL) to conjugate the Fc portion of IgG.

Studies of NPs have shown enhanced intestinal uptake as particle size decreases (*23*, *24*). However, these previous studies on particle size have mostly focused on nonspecific uptake by M cells in the Peyer's patches as opposed to specifically targeting the NPs to a transcytosis pathway such as the FcRn-mediated pathway. Here, the nanoprecipitation self-assembly method (Fig. 2A) (*25*) was used to generate particles with a mean hydrodynamic diameter of 55 nm and a polydispersity of 0.05 (Fig. 2B). Polyclonal IgG Fc fragments were covalently conjugated to PEG using maleimide-thiol chemistry. 2-Iminothiolane was used to modify the Fc with thiol groups (Fc-SH). Fc-SH was incubated with NPs for conjugation (Fig. 2A).

The amount of Fc conjugated to the NPs was measured for both Fc-SH and unmodified Fc (Fig. 2C). Unmodified Fc resulted in a lower ligand density than Fc-SH, indicating minimal nonspecific interactions between Fc and the NP surface and that the unbound Fc was successfully separated from NPs using centrifugal filtration. The ligand density for Fc-SH was 32-fold higher than that for unmodified Fc, indicating that Fc was bound on the NP surface. In addition, the hydrodynamic diameter of the NPs increased from 55 to 63 nm after Fc conjugation (Fig. 2B)—an increase consistent with the hydrodynamic diameter of IgG Fc (~3 nm) (*26*). The surface charge showed only a minor change from −4.3 ± 0.4 mV for NPs to −5.6 ± 1.1 mV for NP-Fc (mean ± SD, $n = 3$; $P > 0.05$, Student's t test).

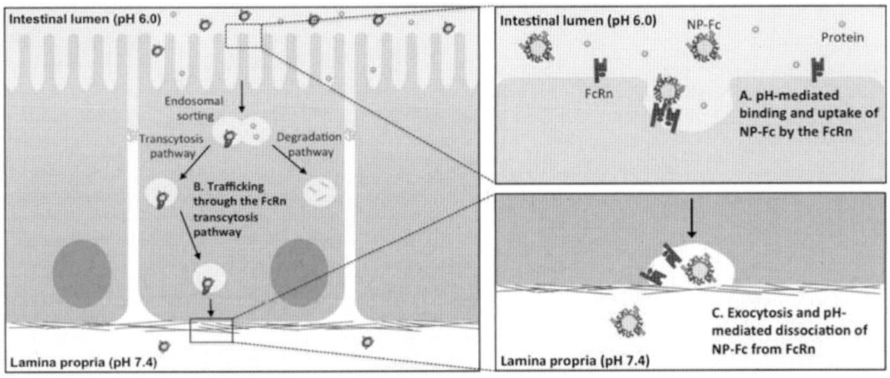

Fig. 1. Schematic of Fc-targeted NP transport across the intestinal epithelium by the FcRn through a transcytosis pathway. (**A**) IgG Fc on the NP surface binds to the FcRn on the apical side of absorptive epithelial cells under acidic conditions in the intestine. (**B**) NP-Fc are then trafficked across the epithelial cell through the FcRn transcytosis pathway in acidic endosomes. (**C**) Upon exocytosis on the basolateral side of the cell, the physiological pH causes IgG Fc to dissociate from the FcRn, and NP-Fc are free to diffuse through the intestinal lamina propria to the capillaries or lacteal and enter systemic circulation.

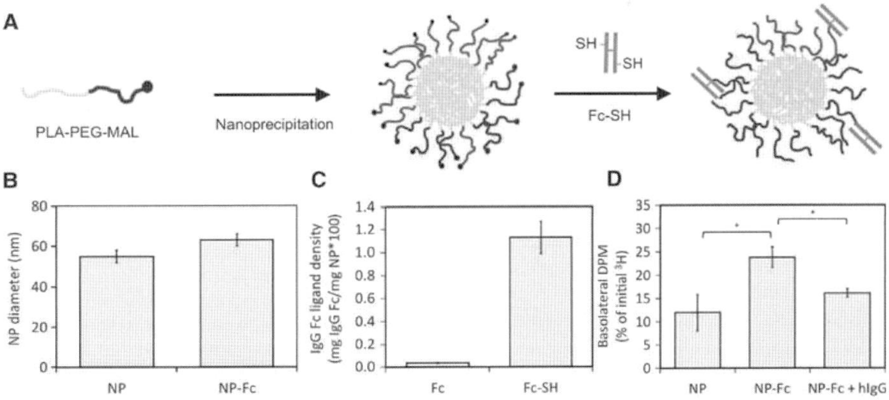

Fig. 2. NP assembly, characterization, and in vitro transepithelial transport. (**A**) Schematic of NP-Fc assembly. NPs consist of a biodegradable PLA core for drug encapsulation and a PEG surface coating for particle stability and to reduce phagocytic uptake. NPs were formed using the nanoprecipitation self-assembly method (*25*) and surface-modified with IgG Fc for FcRn targeting. (**B**) Dynamic light scattering measurements for nontargeted NPs and NP-Fc. Data are means ± SD ($n = 3$). (**C**) IgG Fc ligand density on the NP surface with (Fc-SH) and without (Fc) thiol modification of the IgG Fc. Data are means ± SD ($n = 3$). (**D**) In vitro transepithelial transport of nontargeted NPs, NP-Fc, and NP-Fc with an excess of human IgG Fc as a blocking agent for FcRn. Data are expressed as mean basolateral ^{3}H disintegrations per minute (DPM) as a percentage of the initial amount of ^{3}H (±SEM; $n = 4$ wells per group). *$P < 0.05$, two-tailed Student's t test.

RESEARCH ARTICLE

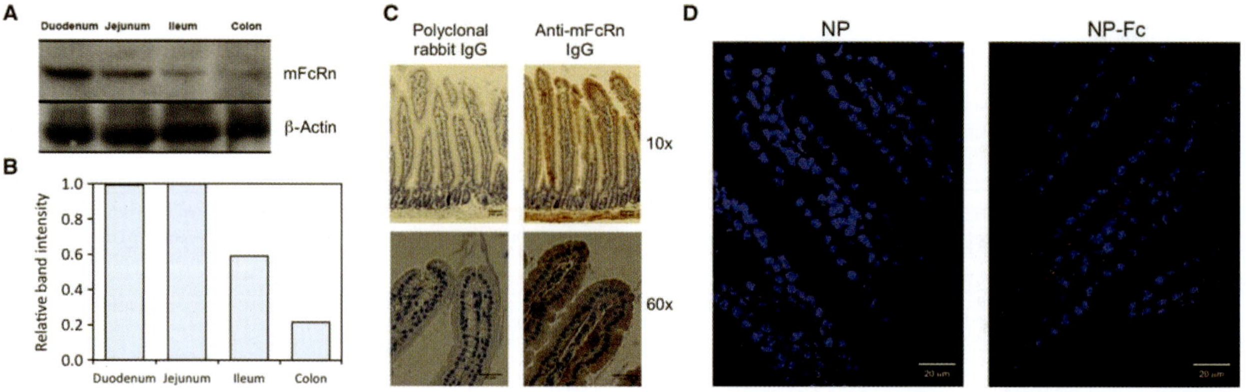

Fig. 3. FcRn expression and NP intestinal uptake in mice. (**A**) Western blot of mouse FcRn (mFcRn) in mouse intestinal tissue. (**B**) Quantification of Western blot band intensity in (A). The relative band intensity was calculated as the ratio of mouse FcRn to β-actin. (**C**) Immunohistochemistry on sections of mouse duodenum. Mouse FcRn appears brown. The negative control tissue was stained with polyclonal IgG. (**D**) Fluorescently labeled NPs (red) were administered to fasted wild-type (WT) mice by oral gavage, and the intestines were collected for sectioning and imaging 1.5 hours later. The panels are confocal fluorescence images of 12-μm sections of mouse duodenum. Cell nuclei were stained with 4′,6-diamidino-2-phenylindole (DAPI) (blue). Images are representative of $n = 3$ mice.

In vitro transepithelial transport

In vitro NP transepithelial transport studies were conducted using an epithelial cell monolayer model with Caco-2 cells, a human epithelial colorectal adenocarcinoma cell line typically used as a model of the intestine for drug permeability testing. Caco-2 cells endogenously express human FcRn and human β_2-microglobulin, and have previously been used for IgG transcytosis studies (*8, 27*). ^{3}H-labeled NPs were used to measure transport across the Caco-2 monolayer. Using this system with a pH gradient established from the apical to basolateral side of the Caco-2 polarized monolayer (to mimic the physiological pH of the duodenum and enhance apical binding), we measured the transcytosis of nontargeted NPs (control) and NP-Fc. ^{3}H measurements for NP-Fc on the basolateral side were twofold greater than nontargeted NPs after 24 hours (Fig. 2D), indicating that Fc on the NP surface significantly enhanced transepithelial transport in vitro. When NP-Fc was combined with a 50-fold excess of free IgG Fc as a blocking agent for the FcRn transcytosis pathway, NP-Fc transport was significantly reduced, indicating that the enhanced transport of NP-Fc is largely mediated by the FcRn (Fig. 2D).

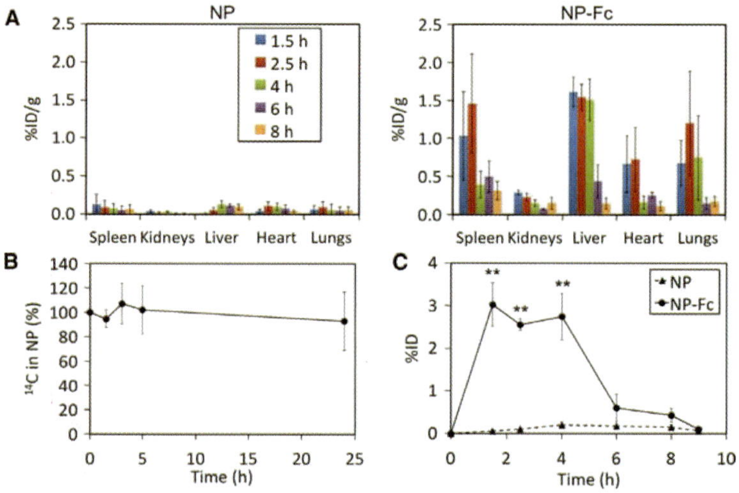

Fig. 4. NP absorption and biodistribution in mice. (**A**) Biodistribution of ^{14}C-labeled nontargeted NPs and NP-Fc after oral administration to fasted WT mice. Data are mean percent initial dose (%ID) per gram of tissue ± SEM ($n = 5$ mice per time point). (**B**) Release of ^{14}C from ^{14}C-labeled NPs in PBS at 37°C. Data are means ± SD for $n = 4$ release experiments. (**C**) Total absorbed ^{14}C over time for nontargeted NPs and NP-Fc after administration by oral gavage. Data are mean %ID measured in all of the organs added together ± SEM ($n = 5$ mice per time point). $^{**}P < 0.01$ for comparison of nontargeted NPs and NP-Fc at respective time point, two-tailed Student's *t* test.

FcRn expression in mice

FcRn expression throughout the entire small and large intestine of wild-type mice was confirmed by Western blot (Fig. 3A). Quantification of band intensity showed that FcRn expression decreased in the distal sections of the small intestine and the colon (Fig. 3B). Immunohistochemistry showed that FcRn was localized to the epithelium of the intestinal villi of the duodenum of wild-type mice (Fig. 3C).

In vivo absorption and biodistribution

In vivo transport of NP-Fc across the intestinal epithelium was visualized using fluorescently labeled NPs. Fasted wild-type mice were orally administered the fluorescently labeled NPs, and sections of the duodenum were collected and analyzed by confocal fluorescence microscopy. Figure 3D shows representative images for both nontargeted NPs and NP-Fc. For the nontargeted NPs, the fluorescence from the NPs was not observed in the villi, suggesting that the NPs were unable to enter the villi. However, for the NP-Fc, fluorescence was observed inside the villi on the basolateral side of the epithelial cells, indicating that NP-Fc crossed the intestinal epithelium and entered the lamina propria (Fig. 3D).

The biodistribution and absorption efficiency of both targeted and nontargeted NPs were quantitatively measured by radiolabeling the NPs with ^{14}C. Figure 4A shows the biodistribution of nontargeted NPs and NP-Fc over the course of 8 hours after oral administration to fasted wild-type mice. For the nontargeted NPs, a small amount of ^{14}C was measured in the organs. By contrast, a large amount of ^{14}C was measured in the spleen, kidneys, liver, and lungs for NP-Fc, indicating that NP-Fc entered systemic circulation after oral administration and reached several organs known to express FcRn (*11*). The ^{14}C in the organs was transient, peaking at 2.5 hours after delivery and clearing from the organs at later time points (Fig. 4A). To ensure that the ^{14}C remained with the NPs over the course of the experiment, we measured the release of ^{14}C from the NPs and observed no release over 24 hours (Fig. 4B).

The total ^{14}C absorbed over time was calculated by summing the ^{14}C measured in each of the organs (spleen, kidneys, liver, heart, and lungs) at a specific time point (Fig. 4C). Significantly higher amounts of ^{14}C were absorbed for NP-Fc at 1.5, 2.5, and 4 hours compared with nontargeted NPs, indicating that Fc targeting enhanced absorption. The area under the curve (AUC) was used to calculate the oral absorption efficiency, which was 1.2 ± 0.2%*hour for nontargeted NPs and 13.7 ± 1.3%*hour for NP-Fc (mean ± SEM, n = 5 mice per time point; P < 0.01, Student's t test). This difference in AUC suggests that IgG Fc targeting was responsible for an 11.5-fold increase in NP absorption.

Oral delivery of insulin to wild-type mice

NP-Fc encapsulating insulin were developed as a model NP-based therapeutic for diabetes that could be orally administered and evaluated for eliciting a pharmacologic response. Insulin NPs (insNPs) were prepared using nanoprecipitation, resulting in an insulin load of 0.5% (w/w). The nanoprecipitation method allowed the particle size to remain small (mean hydrodynamic diameter = 57 nm) while still allowing insulin encapsulation. The release of insulin from insNPs in phosphate-buffered saline (PBS) at 37°C demonstrated a strong burst release in the first hour followed by a controlled release (Fig. 5A). The insulin release profile was advantageous because it allowed all of the insulin to be delivered before complete clearance of the particles after 10 hours. To determine the bioactivity, we collected and injected the insulin released from insNPs into fasted wild-type mice through the tail vein. The bioactivity was measured by monitoring the blood glucose and comparing the response to an equivalent dose of free insulin solution (3.3 U/kg). The released insulin generated a hypoglycemic response in mice (Fig. 5B), indicating that the encapsulated insulin was bioactive after release from the insNPs.

The hypoglycemic response generated after oral administration of the targeted insNPs (insNP-Fc) was tested using fasted wild-type mice and compared with the efficacies of nontargeted insNPs, free insulin, and NP-Fc without insulin (Fig. 5C). The free insulin administered orally did not generate a glucose response in the mice [unlike the free insulin injected into the tail vein that was able to generate a glucose response (Fig. 4B)]. The NP-Fc without insulin was also unable to generate a glucose response. The glucose response generated by nontargeted insNPs was not different from that generated by the control groups at any time point. However, insNP-Fc caused a significant hypoglycemic response in the mice, reducing the glucose during the first 10 hours after administration (Fig. 5C). This is consistent with the biodistribution (Fig. 4A) and insulin release data (Fig. 5A), which demonstrated that the particles were cleared and the insulin was released within 10 hours, respectively. The blood glucose level then increased and was similar to that of the control groups by 15 hours. The insNP-Fc insulin dose required to generate the hypoglycemic response was 1.1 U/kg, which is clinically relevant (*28*) and lower than other oral insulin delivery systems that require 10 to 100 U/kg to generate a glucose response (*29*). When compared with the glucose response from free insulin administered by tail vein injection (Fig. 5B), the orally

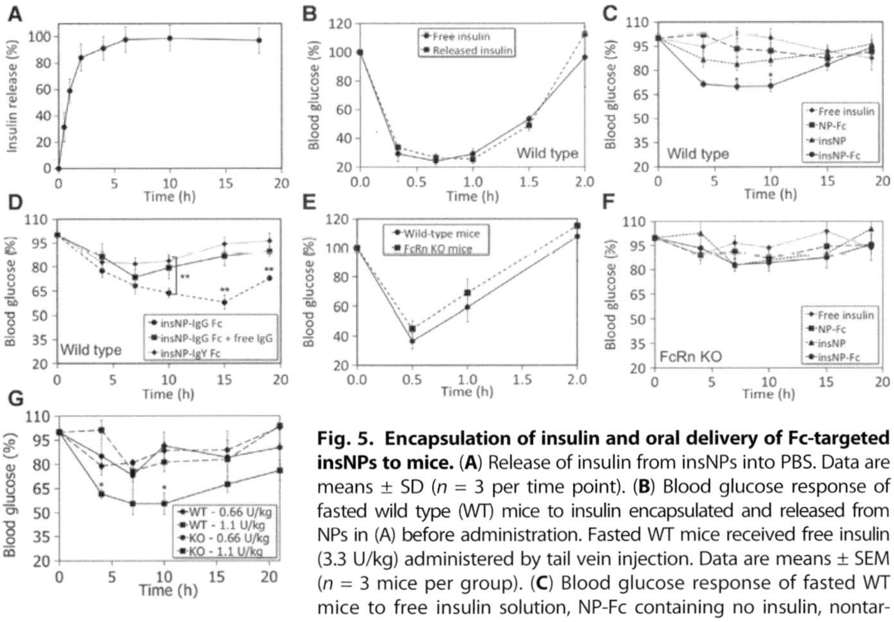

Fig. 5. Encapsulation of insulin and oral delivery of Fc-targeted insNPs to mice. (**A**) Release of insulin from insNPs into PBS. Data are means ± SD (n = 3 per time point). (**B**) Blood glucose response of fasted wild type (WT) mice to insulin encapsulated and released from NPs in (A) before administration. Fasted WT mice received free insulin (3.3 U/kg) administered by tail vein injection. Data are means ± SEM (n = 3 mice per group). (**C**) Blood glucose response of fasted WT mice to free insulin solution, NP-Fc containing no insulin, nontargeted insNPs, and insNP-Fc, each administered by oral gavage. Data are means ± SEM (n = 6 mice per group). *P < 0.05 for comparison of nontargeted insNPs and insNP-Fc at corresponding time points, two-tailed Student's t test. (**D**) Blood glucose response of fasted WT mice to insNP-Fc, insNP-Fc administered concurrently with excess of IgG Fc, and insNPs with chicken IgY Fc fragments, each administered by oral gavage. Data are means ± SEM (n = 5 mice per group). **P < 0.01 for comparison between insNP-Fc with insNP-Fc + free IgG Fc at the 15- and 19-hour time points and between insNP-IgG Fc and insNP-IgY Fc at the 10-, 15-, and 19-hour time points using a two-tailed Student's t test. (**E**) Blood glucose response to equivalent insulin doses (3.3 U/kg) administered by tail vein injection into fasted WT and FcRn knockout (KO) mice. Data are means ± SEM (n = 3 mice per group). (**F**) Fasted FcRn KO mice blood glucose response to free insulin solution, NP-Fc containing no insulin, nontargeted insNPs, or insNP-Fc, each administered by oral gavage. Data are means ± SEM (n = 5 mice per group). (**G**) Fasted WT and FcRn KO mice were dosed by oral gavage with nontargeted insNPs and insNP-Fc at two different doses. *P < 0.05 for comparison between insNP-Fc at 1.1 U/kg and each of the other groups at corresponding time points, two-tailed Student's t test.

administered insNP-Fc resulted in a prolonged (15 versus 1.5 hours) hypoglycemic response (Fig. 5C).

To demonstrate that the enhanced hypoglycemic response generated by insNP-Fc was due specifically to the IgG Fc ligand on the NP surface, we tested several additional control groups. In the first control, insNP-Fc was administered concurrently with a 50-fold excess of free IgG Fc. In the second control, chicken IgY Fc fragments were used instead of human IgG Fc fragments conjugated to insNPs. [Chicken IgY is the functional equivalent of IgG in nonmammalian species such as birds, but does not bind to mouse FcRn (*30*).] Both control groups had hypoglycemic responses that were significantly less than that with insNP-Fc (Fig. 5D), indicating that the use of the IgG Fc as a targeting ligand was responsible for the enhanced hypoglycemic response.

Oral delivery of insulin to FcRn knockout mice

The role of FcRn in NP transepithelial transport was tested by repeating the efficacy experiment using FcRn knockout mice. FcRn knockout mice had the same insulin sensitivity as the wild-type mice, so the same insulin dose was used for both strains (Fig. 5E). In contrast to the results in the wild-type mice (Fig. 5C), insNP-Fc did not generate a hypoglycemic response significantly different from the other three groups in the FcRn knockout mice (Fig. 5F). In these FcRn knockout mice, the response generated by insNP-Fc resembled the response generated by nontargeted insNPs in the wild-type mice (Fig. 5C), suggesting that the benefit gained from using Fc was specifically due to FcRn.

In vivo glucose response dose dependency

The glucose response in both wild-type and FcRn knockout mice was evaluated for dose dependency using two different doses of insNP-Fc: 0.66 and 1.1 U/kg (Fig. 5G). For the FcRn knockout mice, there was no difference in the glucose response between the two doses of insNP-Fc. However, for the wild-type mice, the glucose response was significantly greater at the higher dose of insNP-Fc, suggesting a possible dose dependence in the wild-type mice.

DISCUSSION

For many diseases, oral administration of therapeutics is the standard of care because daily therapy is required. In some cases, chronic diseases, such as cancer, that have been treated in the clinic with intravenous infusions are now increasingly being treated with oral therapeutics because patients prefer the convenience of oral administration relative to parenteral administration. For NP-based therapeutics to be a practical treatment for many diseases, NP formulations appropriate for oral administration are necessary. The most difficult barrier to the effective oral administration of NPs is the intestinal epithelium, which limits the absorption of NPs. To date, there is no practical solution to this problem.

There have been many attempts to develop oral drug delivery systems that overcome this barrier (*29*). For example, permeation enhancers have been used to open tight junctions to allow both paracellular and transcellular transport of drugs across the epithelium (*31*). Mucoadhesive biomaterials have been used to increase the retention time and local concentration of drugs near the apical surface of epithelial cells (*32*). Many oral NPs have been engineered for uptake by M cells in the Peyer's patches, although this limits the surface area available for absorption and exposes NPs to underlying immune cells (*33*). Finally, a few NP formulations have targeted cell receptors, but they still suffer from low bioavailability and require high oral drug dosages (*29*, *34*).

We have taken a novel approach to address this problem by developing drug-encapsulated NPs capable of targeting FcRn for transepithelial transport and enhanced NP intestinal absorption after oral administration. FcRn has been shown to mediate the transcytosis of IgG across several epithelial and endothelial barriers (*11*), and more recently, the transcytosis of fluorescent NPs was demonstrated across a monolayer of Calu-3 airway epithelial cells in vitro after adsorption of Fc on the NP surface (*22*). Harnessing the transcytosis pathway to cross the intestinal epithelium offers the advantage of leaving intact the integrity of the epithelial barrier, avoiding potential safety issues and adverse effects associated with manipulating the permeability of the intestine for paracellular or transcellular transport. An additional advantage of targeting the FcRn is that this receptor is expressed throughout the intestine, providing a significant increase in the available absorption surface area for NP-Fc, which is in contrast with other drug delivery systems that target only a specific portion of the intestine such as the Peyer's patches (*6*).

We have demonstrated that targeting FcRn for transepithelial transport by modifying the NP surface with IgG Fc provides a successful approach for the oral delivery of NPs. Using a mouse model, we demonstrated that FcRn enabled the NP-Fc to cross the intestinal epithelium and reach systemic circulation. The use of IgG Fc to target the NPs to FcRn resulted in absorption efficiency at least 11.5-fold higher than that for nontargeted NPs, but the efficiency could potentially be higher because some ^{14}C could have been absorbed in other tissues not measured in these studies. Furthermore, insNPs targeting FcRn, a model NP-based therapeutic for the treatment of diabetes, were able to generate a hypoglycemic response after oral administration at a clinically relevant insulin dose that is significantly lower than the doses required by other drug delivery systems designed for oral insulin delivery (*29*). Moreover, rodents significantly down-regulate FcRn expression in the intestine after weaning (*35*), but humans continue to express FcRn into adulthood. Therefore, although NP-Fc can increase transepithelial transport in mice, the transport in humans could potentially be even more efficient.

There are some potential limitations to this technology that need to be investigated. For example, FcRn is expressed in many tissues in addition to the intestine. Using IgG Fc to target the FcRn could potentially result in uptake in tissues other than the targeted tissue. However, FcRn expression in tissues such as the vascular endothelium, blood-brain barrier, and lungs could also allow the technology to be expanded for drug delivery across other cellular barriers. Immunological responses to Fc on the NP surface need to be investigated as well. The presence of IgG Fc may result in faster blood clearance of the NPs because of interactions with Fc receptors. NP-Fc may also generate immunological responses and thus could potentially be used for oral vaccine applications (*36*).

This technology may have a major impact on the treatment of several diseases by enabling NP-based therapies to be orally administered. In addition, the encapsulation of drugs or biologics that are currently limited by low bioavailability into NPs that target FcRn may enable markedly more efficient oral delivery of the therapies. The path forward for translation of this technology to the clinic is to test the NP-Fc in larger animal models such as pigs and nonhuman primates. Some key parameters that need to be understood for success in the clinic are the ideal release profile for an oral insulin formulation and how absorption varies based on diet and patient-to-patient variability.

MATERIALS AND METHODS

Study design

This study aimed to develop targeted Fc-functionalized NPs capable of binding to the FcRn protein, and evaluate whether this approach enabled the transcytosis of the FcRn-NP complex across the intestinal epithelial barrier using both in vitro and in vivo models. Fc-targeted NPs were first tested using an in vitro intestinal epithelium model and the transport across the epithelial barrier was quantified by ^{3}H-labeling of the NPs. Fc-targeted NPs were then tested in vivo using fluorescently-labeled NPs to observe transport across the intestinal epithelium and ^{14}C-labeled NPs to quantify the biodistribution and absorption efficiency after oral administration. Finally, Fc-targeted NPs containing insulin were used for plasma glycemic modulation studies in both wild-type and FcRn knockout mice as a proof-of-concept that targeted NPs were capable of oral delivery of bioactive protein therapeutics. For the imaging and ^{14}C experiments, the mice were selected based on weight and then randomly assigned to different treatment groups. For the plasma glycemic modulation studies, the mice were selected based on initial fasted glucose values and then randomly assigned to different treatment groups. The studies were not blinded. Sample sizes (n = 5 to 6 per group) for all of the experiments were sufficient to detect statistically significant differences between treatment groups, with all measurements included in analyses. Endpoints for glucose experiments were determined by total fasting time, which was limited to 24 h.

Polymer synthesis

D,L-Lactide (Sigma-Aldrich) and MAL-PEG-OH (JenKem Technology) were used to synthesize PLA-PEG-MAL by ring-opening polymerization. D,L-Lactide (3 g, 20.8 mmol) and MAL-PEG-OH (544 mg, 0.16 mmol) were dissolved in 15 ml of anhydrous toluene in a round-bottom flask. Tin(II) ethylhexanoate (38 mg, 0.09 mmol) was then added. The flask with condenser was placed in an oil bath, purged with nitrogen for 10 min, heated to 120°C, and reacted overnight while 4°C water circulated through the condenser. Toluene was then evaporated, and the polymer was precipitated in a 50:50 (v/v) mixture of ice-cold methanol and diethyl ether and vacuum-dried. The PLA-PEG-MAL was characterized by ^{1}H nuclear magnetic resonance (400 MHz), δ = 5.28 to 5.11 [br, -OC-CH(CH$_3$)O- in PLA], 3.62 (s, -CH$_2$CH$_2$O- in PEG), 1.57 to 1.45 (br, -OC-CHCH$_3$O- in PLA). With gel permeation chromatography, polymer M_n = 12.5 kD with M_w/M_n = 1.47 relative to polystyrene standards. PLA (inherent viscosity 0.50 dl/g) with terminal carboxylate groups (Lactel) was conjugated to ^{14}C doxorubicin (PerkinElmer) (PLA-^{14}C) and Alexa Fluor 647 (AF647) hydrazide tris(triethylammonium) salt (Invitrogen) (PLA-AF647) were prepared the same way. PLA (30 mg, 0.83 μmol) in 1 ml of dimethylformamide (DMF) was reacted with *N*-hydroxysuccinimide (NHS) (0.5 mg, 4.2 μmol) in the presence of 1-ethyl-3-[3-dimethylaminopropyl] (0.8 mg, 4.2 μmol) overnight. PLA-NHS was precipitated in a 50:50 (v/v) mixture of ice-cold methanol and diethyl ether and vacuum-dried. For PLA-^{14}C, PLA-NHS (18 mg, 0.5 μmol) was mixed overnight with ^{14}C doxorubicin (0.27 mg, 0.5 μmol) in 1 ml of DMF. For PLA-AF647, PLA-NHS (50 mg, 1.4 μmol) was mixed overnight with AF647 (1.5 mg, 1.4 μmol) in 1 ml of DMF. Both polymers were precipitated in a mixture of 50:50 (v/v) ice-cold methanol and diethyl ether and vacuum-dried.

Synthesis and characterization of NP-Fc

To prepare the NP-Fc, we dissolved 3 mg of PLA-PEG-MAL in 300 μl of acetonitrile and added it dropwise to 1.5 ml of water. The solution was mixed for 2 hours, and the NPs were purified by filtration with Millipore Amicon Ultra 100,000 NMWL (nominal molecular weight limit). The NPs were washed twice with water and twice with PBS containing 5 mM EDTA. Concurrently, 86 μg of purified human polyclonal IgG Fc prepared by papain digestion (Bethyl Laboratories) or 95 μg of chicken IgY Fc (Jackson ImmunoResearch Laboratories) in PBS containing 5 mM EDTA was reacted with 0.48 μl of 2-iminothiolane (5 mg/ml) (Traut's Reagent) for 1 hour. The modified Fc was added to the NPs and mixed for 1 hour to allow conjugation at 4°C. The NP-Fc were washed with PBS using Millipore Amicon Ultra 100,000 NMWL. The conjugation of IgG Fc to the NP surface was measured with a protein bicinchoninic acid (BCA) assay from Lamda Biotech. Particle diameter and surface charge (ζ potential) were measured with dynamic light scattering with a Brookhaven Instruments ZetaPALS.

In vitro transcytosis studies

Transepithelial transport studies used Transwell plates (Costar) with a Caco-2 [American Type Culture Collection (ATCC)] cell density of 5.5 × 10^4 in medium [ATCC-formulated Eagle's minimum essential medium with aqueous penicillin G (100 U/ml), streptomycin (100 U/ml), and fetal bovine serum (20%)]. On the day of the transport experiment, the medium was changed to Hanks' balanced salt solution (HBSS) (pH 6.5) in the apical chamber and HBSS (pH 7.4) in the basolateral chamber and allowed to equilibrate for 1 hour at 37°C and 5% CO_2. Before and at the end of the experiment, the monolayer integrity was checked by measuring the transepithelial resistance (TEER) with a Millicell-ERS (Millipore). TEER values were 900 to 1000 Ω/cm^2 for wells used in transport experiments, and the TEER remained constant throughout the experiments. ^{3}H-labeled NPs were prepared by blending 50 μg of ^{3}H–poly(lactic-*co*-glycolic acid) (PLGA) (PerkinElmer) with 1 mg of PLA-PEG-MAL in 100 μl of acetonitrile before nanoprecipitation in 500 μl of water and then washed in HBSS (pH 6.5) before the transport experiment. The apical solution was then replaced with a solution of 100 μg of ^{3}H-labeled NPs or NP-Fc in 250 μl of HBSS (pH 6.5). The NP formulations were incubated for 24 hours before measuring the ^{3}H content in the basolateral chamber. The basolateral solution was collected and added to a Hionic-Fluor scintillation cocktail (PerkinElmer) before analysis with a Packard Tri-Carb Scintillation Analyser. At the end of the experiment, the TEER was measured again to verify monolayer integrity. For the IgG Fc blocking experiment, a 50× molar excess of IgG Fc relative to Fc on the NP-Fc surface was added concurrently with NP-Fc to the apical chamber, and the ^{3}H content in the basolateral chamber was measured after 24 hours.

Western blot

Sections of small intestine and colon were removed from wild-type mice after euthanasia. Intestinal epithelial cells were isolated (*37*, *38*), and the protein was extracted. Protein concentrations in the extracts were determined with the BCA assay. Extracts were resolved on a 12% SDS–polyacrylamide gel electrophoresis gel under reducing conditions. Proteins were transferred onto a nitrocellulose membrane. The membrane was blocked with 5% non-fat milk, probed with rabbit anti-mouse FcRn (Santa Cruz Biotechnology) for 1 hour, and then incubated with goat anti-rabbit IgG–horseradish peroxidase (HRP) (Santa Cruz Biotechnology). All blocking, incubations, and washes used PBS-T (PBS with 0.05% Tween 20). Detection was by chemiluminescence. Band intensity was quantified with ImageJ.

Immunohistochemistry

Small intestine tissues were harvested and fixed in 10% formalin overnight. After ethanol dehydration, tissues were paraffin-embedded and cut into 8-μm-thick sections, mounted on slides, and dried overnight. The tissues were then rehydrated with xylene and ethanol. Endogenous peroxidase activity, endogenous biotin, and nonspecific proteins were blocked with 3% H_2O_2, avidin blocking agent, and 10% goat serum, respectively. The samples were incubated with polyclonal rabbit IgG or anti-mouse FcRn IgG (Santa Cruz Biotechnology) primary antibody overnight and then incubated with biotinylated anti-rabbit secondary antibodies (Santa Cruz Biotechnology) and with streptavidin-HRP, developed with DAPI, and mounted with hematoxylin counterstain.

In vivo fluorescence imaging

All animal studies were conducted under the supervision of the Massachusetts Institute of Technology (MIT) Division of Comparative Medicine in compliance with the National Institutes of Health (NIH) *Guide for the Care and Use of Laboratory Animals*. Wild-type BALB/c mice (Charles River Laboratories) (n = 3) were fasted overnight before gavage. Fluorescently labeled NPs were prepared by blending 100 μg of PLA-AF647 with 1 mg of PLA-PEG-MAL in 100 μl of acetonitrile before nanoprecipitation in 500 μl of water. Fluorescently labeled NPs and NP-Fc were washed three times in water until the flow-through was clear and suspended in 200 μl of water (7 ml/kg). The suspension was then administered to the mice by oral gavage. After 1.5 hours, the mice were euthanized. Duodenum tissue sections were frozen in Tissue-Tek OCT with liquid nitrogen. Cross sections of the tissue were obtained with a Leica CM1900 cryostat with a thickness of 12 μm. The tissue was air-dried overnight and then stained with ProLong Gold (Life Technologies) antifade reagent with DAPI. Fluorescent images were obtained with a Zeiss LSM 710 NLO scanning confocal microscope under oil immersion at 40× magnification.

In vivo biodistribution

To prepare ^{14}C-labeled NPs, we blended 450 μg of PLA-^{14}C with 3 mg of PLA-PEG-MAL in 300 μl of acetonitrile before nanoprecipitation in 1.5 ml of water. NPs were washed twice with water and twice with PBS. ^{14}C release was measured by preparing a batch of ^{14}C NPs in PBS (pH 7.4) and dividing the batch equally into 500-μg samples for incubation at 37°C. At each time point, samples were collected, washed three times with PBS with Millipore Amicon Ultra 100,000 NMWL, and then added to 15 ml of Hionic Fluor scintillation cocktail. The activity was counted with a Packard Tri-Carb Scintillation Analyser. For the in vivo biodistribution experiments, 6- to 12-week-old wild-type mice were fasted 8 hours before oral gavage of 1.5 mg (0.1 μCi per mouse) of ^{14}C-labeled NPs and NP-Fc in PBS (7 ml/kg). Groups of mice (n = 5 mice) were euthanized at each time point, and the spleen, kidneys, liver, lungs, and heart were harvested. Each organ was placed directly in a scintillation vial except for the liver, which was homogenized, and ~100 mg was analyzed. Each organ was solubilized in 2 ml of Solvable (PerkinElmer) for 12 hours at 60°C and then decolored with 200 μl of 0.5 M EDTA (Invitrogen) and 200 μl of 30% hydrogen peroxide (Fisher Scientific) for 1 hour at 60°C. The activity was counted in 15 ml of Hionic-Fluor scintillation cocktail with a Packard Tri-Carb Scintillation Analyser. To determine 100% dose, we counted vials of 500 μg of NPs and NP-Fc in 15 ml of Hionic-Fluor scintillation cocktail. For the oral absorption efficiency, total ^{14}C counted in all tissues was added at each time point. The AUC of total absorbed ^{14}C versus time was calculated with the trapezoid method and divided by the initial dose to determine the oral absorption efficiency.

Insulin encapsulation and release

insNPs were prepared by blending 900 μg of human recombinant insulin (Sigma-Aldrich), 450 μg of PLGA (50:50 glycolide/lactide, inherent viscosity of 0.20 dl/g) with terminal carboxylate groups (Lactel), and 3 mg of PLA-PEG-MAL together in 525 μl of dimethyl sulfoxide before nanoprecipitation in 2.1 ml of water. Free insulin was removed with Millipore Amicon Ultra 100,000 NMWL by washing with water and with PBS. Insulin encapsulation was measured by heating the NPs to 60°C for 30 min, and insulin was quantified with a BCA assay or insulin ELISA (enzyme-linked immunosorbent assay) kit (Millipore). Insulin release from the NP was measured by dividing a batch of insNPs equally into 24-kD dialysis units (Pierce) and incubating at 37°C in PBS (pH 7.4). At each time point, three samples of insNPs were collected, washed with PBS with Millipore Amicon Ultra 100,000 NMWL, heated to 60°C for 30 min, and measured for insulin with a BCA assay.

Insulin activity

Insulin activity was measured by preparing insNPs and allowing insulin release for 2 hours at 37°C in PBS (pH 7.4). Wild-type mice were fasted for 8 hours, and the mice (n = 3 per group) used were chosen so that the mean initial blood glucose levels were the same for each group. Released insulin (3.3 U/kg) was administered to the fasted mice by tail vein injection. An equivalent dose of free insulin by mass was administered by tail vein injection to another group of fasted mice. The blood glucose level was measured with the Contour blood glucose monitor (Bayer).

In vivo efficacy

Six- to 12-week old wild-type or FcRn knockout (*39*) mice (Jackson Laboratories) were fasted for 8 hours. Mice (n = 5 to 6) were chosen per group such that the mean initial blood glucose levels were the same per group. insNPs or insNP-Fc (150 or 250 μg) (insulin dose: 0.66 or 1.1 U/kg) in PBS (7 ml/kg) were administered to the mice by oral gavage. For controls, free insulin (1.1 U/kg) and 250 μg of NP-Fc without insulin in PBS (7 ml/kg) were administered by oral gavage. For the excess IgG Fc control, 250 μg of insNP-Fc was formulated with 50× molar excess of IgG Fc in PBS (7 ml/kg) before administration to the mice by oral gavage. The blood glucose level was measured as described above.

Statistical analysis

All data are presented as means with either SD or SEM as indicated. Statistical significance was determined by a two-tailed Student's t test (α = 0.05) assuming equal variance.

REFERENCES AND NOTES

1. M. E. Davis, J. E. Zuckerman, C. H. J. Choi, D. Seligson, A. Tolcher, C. A. Alabi, Y. Yen, J. D. Heidel, A. Ribas, Evidence of RNAi in humans from systemically administered siRNA via targeted nanoparticles. *Nature* **464**, 1067–1070 (2010).
2. A. Z. Wang, R. Langer, O. C. Farokhzad, Nanoparticle delivery of cancer drugs. *Annu. Rev. Med.* **63**, 185–198 (2012).
3. J. Hrkach, D. Von Hoff, M. M. Ali, E. Andrianova, J. Auer, T. Campbell, D. De Witt, M. Figa, M. Figueiredo, A. Horhota, S. Low, K. McDonnell, E. Peeke, B. Retnarajan, A. Sabnis, E. Schnipper, J. J. Song, Y. H. Song, J. Summa, D. Tompsett, G. Troiano, T. Van Geen Hoven, J. Wright, P. LoRusso, P. W. Kantoff, N. H. Bander, C. Sweeney, O. C. Farokhzad, R. Langer, S. Zale, Preclinical development and clinical translation of a PSMA-targeted docetaxel nanoparticle with a differentiated pharmacological profile. *Sci. Transl. Med.* **4**, 128ra39 (2012).
4. O. C. Farokhzad, J. Cheng, B. A. Teply, I. Sherifi, S. Jon, P. W. Kantoff, J. P. Richie, R. Langer, Targeted nanoparticle-aptamer bioconjugates for cancer chemotherapy in vivo. *Proc. Natl. Acad. Sci. U.S.A.* **103**, 6315–6320 (2006).

5. M. M. Borner, P. Schoffski, R. de Wit, F. Caponigro, G. Comella, A. Sulkes, G. Greim, G. J. Peters, K. van der Born, J. Wanders, R. F. de Boer, C. Martin, P. Fumoleau, Patient preference and pharmacokinetics of oral modulated UFT versus intravenous fluorouracil and leucovorin: A randomised crossover trial in advanced colorectal cancer. *Eur. J. Cancer* **38**, 349–358 (2002).
6. M. Goldberg, I. Gomez-Orellana, Challenges for the oral delivery of macromolecules. *Nat. Rev. Drug Discov.* **2**, 289–295 (2003).
7. M. Yoshida, S. M. Claypool, J. S. Wagner, E. Mizoguchi, A. Mizoguchi, D. C. Roopenian, W. I. Lencer, R. S. Blumberg, Human neonatal Fc receptor mediates transport of IgG into luminal secretions for delivery of antigens to mucosal dendritic cells. *Immunity* **20**, 769–783 (2004).
8. B. L. Dickinson, K. Badizadegan, Z. Wu, J. C. Ahouse, X. Zhu, N. E. Simister, R. S. Blumberg, W. I. Lencer, Bidirectional FcRn-dependent IgG transport in a polarized human intestinal epithelial cell line. *J. Clin. Invest.* **104**, 903–911 (1999).
9. F. W. Brambell, The transmission of immune globulins from the mother to the foetal and newborn young. *Proc. Nutr. Soc.* **28**, 35–41 (1969).
10. E. J. Israel, S. Taylor, Z. Wu, E. Mizoguchi, R. S. Blumberg, A. Bhan, N. E. Simister, Expression of the neonatal Fc receptor, FcRn, on human intestinal epithelial cells. *Immunology* **92**, 69–74 (1997).
11. T. T. Kuo, V. G. Aveson, Neonatal Fc receptor and IgG-based therapeutics. *MAbs* **3**, 422–430 (2011).
12. K. Baker, S. W. Qiao, T. T. Kuo, V. G. Aveson, B. Platzer, J. T. Andersen, I. Sandlie, Z. Chen, C. de Haar, W. I. Lencer, E. Fiebiger, R. S. Blumberg, Neonatal Fc receptor for IgG (FcRn) regulates cross-presentation of IgG immune complexes by $CD8^{-}CD11b^{+}$ dendritic cells. *Proc. Natl. Acad. Sci. U.S.A.* **108**, 9927–9932 (2011).
13. M. Raghavan, L. N. Gastinel, P. J. Bjorkman, The class I major histocompatibility complex related Fc receptor shows pH-dependent stability differences correlating with immunoglobulin binding and release. *Biochemistry* **32**, 8654–8660 (1993).
14. W. He, M. S. Ladinsky, K. E. Huey-Tubman, G. J. Jensen, J. R. McIntosh, P. J. Björkman, FcRn-mediated antibody transport across epithelial cells revealed by electron tomography. *Nature* **455**, 542–546 (2008).
15. D. Lalezari, Gastrointestinal pH profile in subjects with irritable bowel syndrome. *Ann. Gastroenterol.* **25**, 333–337 (2012).
16. D. C. Roopenian, S. Akilesh, FcRn: The neonatal Fc receptor comes of age. *Nat. Rev. Immunol.* **7**, 715–725 (2007).
17. B. L. Dickinson, S. M. Claypool, J. A. D'Angelo, M. L. Aiken, N. Venu, E. H. Yen, J. S. Wagner, J. A. Borawski, A. T. Pierce, R. Hershberg, Ca^{2+}-dependent calmodulin binding to FcRn affects immunoglobulin G transport in the transcytotic pathway. *Mol. Biol. Cell* **19**, 414–423 (2008).
18. R. J. Ober, C. Martinez, X. Lai, J. Zhou, E. S. Ward, Exocytosis of IgG as mediated by the receptor, FcRn: An analysis at the single-molecule level. *Proc. Natl. Acad. Sci. U.S.A.* **101**, 11076–11081 (2004).
19. A. J. Bitonti, J. A. Dumont, S. C. Low, R. T. Peters, K. E. Kropp, V. J. Palombella, J. M. Stattel, Y. Lu, C. A. Tan, J. J. Song, A. M. Garcia, N. E. Simister, G. M. Spiekermann, W. I. Lencer, R. S. Blumberg, Pulmonary delivery of an erythropoietin Fc fusion protein in non-human primates through an immunoglobulin transport pathway. *Proc. Natl. Acad. Sci. U.S.A.* **101**, 9763–9768 (2004).
20. J. A. Dumont, A. J. Bitonti, D. Clark, S. Evans, M. Pickford, S. P. Newman, Delivery of an erythropoietin-Fc fusion protein by inhalation in humans through an immunoglobulin transport pathway. *J. Aerosol Med.* **18**, 294–303 (2005).
21. S. C. Low, S. L. Nunes, A. J. Bitonti, J. A. Dumont, Oral and pulmonary delivery of FSH–Fc fusion proteins via neonatal Fc receptor-mediated transcytosis. *Hum. Reprod.* **20**, 1805–1813 (2005).
22. D. Vllasaliu, C. Alexander, M. Garnett, M. Eaton, S. Stolnik, Fc-mediated transport of nanoparticles across airway epithelial cell layers. *J. Control. Release* **158**, 479–486 (2012).
23. A. Awaad, M. Nakamura, K. Ishimura, Imaging of size-dependent uptake and identification of novel pathways in mouse Peyer's patches using fluorescent organosilica particles. *Nanomedicine* **8**, 627–636 (2012).
24. M. P. Desai, V. Labhasetwar, G. L. Amidon, R. J. Levy, Gastrointestinal uptake of biodegradable microparticles: Effect of particle size. *Pharm. Res.* **13**, 1838–1845 (1996).
25. U. Bilati, E. Allémann, E. Doelker, Nanoprecipitation versus emulsion-based techniques for the encapsulation of proteins into biodegradable nanoparticles and process-related stability issues. *AAPS PharmSciTech* **6**, E594–E604 (2005).
26. J. K. Armstrong, R. B. Wenby, H. J. Meiselman, T. C. Fisher, The hydrodynamic radii of macromolecules and their effect on red blood cell aggregation. *Biophys. J.* **87**, 4259–4270 (2004).
27. X. Liu, L. Ye, G. J. Christianson, J. Q. Yang, D. C. Roopenian, X. Zhu, NF-κB signaling regulates functional expression of the MHC class I-related neonatal Fc receptor for IgG via intronic binding sequences. *J. Immunol.* **179**, 2999–3011 (2007).
28. E. Cochran, C. Musso, P. Gorden, The use of U-500 in patients with extreme insulin resistance. *Diabetes Care* **28**, 1240–1244 (2005).
29. M. C. Chen, K. Sonaje, K. J. Chen, H. W. Sung, A review of the prospects for polymeric nanoparticle platforms in oral insulin delivery. *Biomaterials* **32**, 9826–9838 (2011).
30. E. J. Israel, D. F. Wilsker, K. C. Hayes, D. Schoenfeld, N. E. Simister, Increased clearance of IgG in mice that lack β_2-microglobulin: Possible protective role of FcRn. *Immunology* **89**, 573–578 (1996).
31. N. Salama, N. Eddington, A. Fasano, Tight junction modulation and its relationship to drug delivery. *Adv. Drug Deliv. Rev.* **58**, 15–28 (2006).
32. J. Smart, The basics and underlying mechanisms of mucoadhesion. *Adv. Drug Deliv. Rev.* **57**, 1556–1568 (2005).
33. M. Shakweh, G. Ponchel, E. Fattal, Particle uptake by Peyer's patches: A pathway for drug and vaccine delivery. *Expert Opin. Drug Deliv.* **1**, 141–163 (2004).
34. S. Jain, V. V. Rathi, A. K. Jain, M. Das, C. Godugu, Folate-decorated PLGA nanoparticles as a rationally designed vehicle for the oral delivery of insulin. *Nanomedicine* **7**, 1311–1337 (2012).
35. M. G. Martín, S. V. Wu, J. H. Walsh, Ontogenetic development and distribution of antibody transport and Fc receptor mRNA expression in rat intestine. *Dig. Dis. Sci.* **42**, 1062–1069 (1997).
36. L. Ye, R. Zeng, Y. Bai, D. C. Roopenian, X. Zhu, Efficient mucosal vaccination mediated by the neonatal Fc receptor. *Nat. Biotechnol.* **29**, 158–163 (2011).
37. C. Booth, J. A. O'Shea, in *Culture of Epithelial Cells* (Wiley-Liss Inc., New York, 2002), pp. 303–335.
38. D. Pan, A. Das, D. Liu, R. S. Veazey, B. Pahar, Isolation and characterization of intestinal epithelial cells from normal and SIV-infected rhesus macaques. *PLoS One* **7**, e30247 (2012).
39. D. C. Roopenian, G. J. Christianson, T. J. Sproule, A. C. Brown, S. Akilesh, N. Jung, S. Petkova, L. Avanessian, E. Y. Choi, D. J. Shaffer, P. A. Eden, C. L. Anderson, The MHC class I-like IgG receptor controls perinatal IgG transport, IgG homeostasis, and fate of IgG Fc coupled drugs. *J. Immunol.* **170**, 3528–3533 (2003).

Acknowledgments: We thank the W. M. Keck Microscope Facility at the Whitehead Institute, in particular W. Salmon for her assistance with the fluorescence microscopy imaging. We thank B. Tang for his help with the confocal microscopy image analysis. **Funding:** Supported in part by a Koch–Prostate Cancer Foundation Award in Nanotherapeutics (R.L. and O.C.F.); the National Cancer Institute Center of Cancer Nanotechnology Excellence at MIT-Harvard (U54-CA151884); a National Heart, Lung, and Blood Institute Program of Excellence in Nanotechnology Award (contract HHSN268201000045C); NIH grant EB000244; and NIH R01 grant EB015419-01. R.S.B. is supported by NIH DK53056 and Harvard Digestive Disease Center grant DK0034854. E.M.P. is supported by a National Defense Science and Engineering Graduate research fellowship and a Center of Cancer Nanotechnology Excellence graduate research fellowship (5 U54 CA151884-02). **Author contributions:** E.M.P. and F.A. designed and performed the NP synthesis and characterization, in vitro experiments, and in vivo experiments; analyzed the data; did the statistical analysis; and wrote the manuscript. O.C.F. designed the experiments, analyzed the data, and wrote the manuscript. R.L., R.K., and R.S.B. designed the experiments, analyzed the data, and critically reviewed the manuscript. T.T.K. and E.L.-N. analyzed the data and critically reviewed the manuscript. **Competing interests:** O.C.F., R.K., and R.L. have financial interests in BIND Therapeutics, Selecta Biosciences, and Blend Therapeutics, which are developing NP therapeutics. O.C.F., R.L., F.A., E.M.P., and T.T.K. are inventors on the patent application related to this technology. BIND and Selecta have an exclusive license from Brigham and Women's Hospital (BWH) and MIT for this technology. **Data and materials availability:** Materials are readily available and will be provided under the material transfer policies of the MIT and BWH.

Submitted 16 July 2013
Accepted 19 September 2013
Published 27 November 2013
10.1126/scitranslmed.3007049

Citation: E. M. Pridgen, F. Alexis, T. T. Kuo, E. Levy-Nissenbaum, R. Karnik, R. S. Blumberg, R. Langer, O. C. Farokhzad, Transepithelial transport of Fc-targeted nanoparticles by the neonatal Fc receptor for oral delivery. *Sci. Transl. Med.* **5**, 213ra167 (2013).

nature nanotechnology

ARTICLES

PUBLISHED ONLINE: 11 MAY 2014 | DOI: 10.1038/NNANO.2014.84

In vivo endothelial siRNA delivery using polymeric nanoparticles with low molecular weight

James E. Dahlman and Carmen Barnes *et al.**

Dysfunctional endothelium contributes to more diseases than any other tissue in the body. Small interfering RNAs (siRNAs) can help in the study and treatment of endothelial cells *in vivo* by durably silencing multiple genes simultaneously, but efficient siRNA delivery has so far remained challenging. Here, we show that polymeric nanoparticles made of low-molecular-weight polyamines and lipids can deliver siRNA to endothelial cells with high efficiency, thereby facilitating the simultaneous silencing of multiple endothelial genes *in vivo*. Unlike lipid or lipid-like nanoparticles, this formulation does not significantly reduce gene expression in hepatocytes or immune cells even at the dosage necessary for endothelial gene silencing. These nanoparticles mediate the most durable non-liver silencing reported so far and facilitate the delivery of siRNAs that modify endothelial function in mouse models of vascular permeability, emphysema, primary tumour growth and metastasis.

The vascular system releases factors into the bloodstream, changes the expression of specific receptors, modifies intercellular junctions, and regulates the immune response[1]. Endothelial cells also mediate biological functions such as endocytosis and metabolism[2]. Because these processes relate fundamentally to physiology, dysfunctional endothelium promotes more disease than any other tissue in the body[3]. Yet, modulating the behaviour of endothelial cells *in vivo* remains challenging, particularly in cases that require the inhibition of multiple endothelial genes.

RNAi-mediated modification of gene expression has the potential to improve disease treatment and *in vivo* studies of complex biological processes. However, its utility is limited by inefficient systemic delivery, with the exception of ionizable lipids and lipid-like compounds termed lipidoids, which reduce hepatic gene expression by 50% after injection of 0.01 mg kg^{-1} siRNA (refs 4–7). In contrast, efficient endothelial gene silencing without the transfection of hepatocytes has remained challenging. Although cationic lipids have been reported to deliver siRNA to endothelial cells, these endothelial delivery systems require cumulative doses of up to 7.5 mg kg^{-1} to achieve robust gene silencing[8–13].

Nanoformulations based on polymeric materials have delivered siRNA to hepatocytes and melanoma[14,15]. Unlike lipid-based nanoparticles, polymer–nucleic acid nanoparticles condense via multivalent interactions, leading to significantly different physical stabilities. One polymer class that has been investigated as a gene delivery material is polyethyleneimine (PEI)[16]. Although nanoparticles made from high-molecular-weight PEI ($M_w \approx 25{,}000$ Da) have delivered nucleic acids, they are associated with off-target effects[17]. In contrast, nanoformulations from low-molecular-weight PEI ($M_w \approx 600$ Da) are relatively well tolerated but cannot facilitate siRNA delivery[17,18].

Here, we report an siRNA–nanoparticle formulation that reduces endothelial gene expression by over 90% at a dose of 0.10 mg kg^{-1} and by 50% at doses as low as 0.02 mg kg^{-1}. This formulation, termed 7C1, differs from traditional lipid-based nanoparticle formulations because it can deliver siRNA to lung endothelial cells at low doses without substantially reducing gene expression in pulmonary immune cells, hepatocytes or peritoneal immune cells. We demonstrate that 7C1-mediated endothelial gene silencing affects function *in vivo* by using our nanoformulation to modify mouse models of vascular permeability, emphysema, lung tumour growth and lung metastasis.

Efficient siRNA delivery to endothelium *in vitro* and *in vivo*

A diverse library of epoxide-modified lipid–polymer hybrids were synthesized (Fig. 1a, Supplementary Fig. 1a). The compounds were tested for their ability to reduce gene expression in HeLa cells at four different lipid:siRNA mass ratios (2.5:1, 5:1, 10:1 and 15:1) (Supplementary Fig. 1b). HeLa cells expressing Firefly and Renilla luciferase were chosen as a first-pass screen for these 2,000 nanoparticles formulated with siLuciferase because of the cost-effectiveness of the assay[5]. We defined a successful nanoparticle as one that silenced Firefly luminescence by more than 70%, but decreased Renilla luminescence by less than 25%. Although only 0.9% of compounds in the library were successful at a mass ratio of 2.5, 6.5% were successful when the mass ratio was 15 (Table 1). We then measured Firefly luminescence as a function of lipids bound to successful PEI_{600} compounds. Luminescence decreased with the number of lipids bound (Supplementary Fig. 1c). A subset of formulations tested in HeLa cells were tested for their ability to deliver siRNA to human (HMVEC) and murine (bEnd.3) endothelial cells *in vitro*. The most effective compound, termed 7C1 based on its composition, reduced target mRNA expression by more than 85% in HeLa, HMVECs and bEnd.3 cells at a dose of 30 nM (Fig. 1b). Interestingly, 7C1 efficacy did not change with mass ratio in HeLa cells (Supplementary Fig. 1d). Only 2 nM was required to reduce target gene expression in bEnd.3 cells by 50%, and 7C1 did not affect bEnd.3 cell morphology or induce apoptosis *in vitro* at doses as high as 133 nM (Supplementary Fig. 1e,f).

7C1 was synthesized by reacting C_{15} epoxide-terminated lipids with PEI_{600} at a 14:1 molar ratio, and was formulated with $C_{14}PEG_{2000}$ to produce nanoparticles (diameter between 35 and 60 nm) that were stable in PBS solution for at least 40 days (Fig. 1c–e, Supplementary Fig. 1g–i). Particles formed multilamellar vesicles rather than the periodic aqueous compartments containing siRNA that make up stable nucleic-acid lipid particle formulations[19] (Fig. 1d, Supplementary Fig. 1j). Because particle charge at different values of pH can affect delivery by modifying interactions with

*A full list of authors and their affiliations appears at the end of the paper.

Table 1 | Per cent of compounds reducing Firefly luminescence more than 70% while not reducing Renilla luminescence more than 25% as a function of lipid:siFire mass ratio.

Compound:siFire mass ratio	Successful nanoparticles (%)
2.5	0.9
5	0.7
10	3.5
15	6.5

Table 2 | Intravenous dose required to reduce target gene mRNA expression by 50% *in vivo* (mg kg^{-1}) (ED_{50}).

Gene	Lung	Heart	Kidney
VEcad	0.02	0.04	0.08
ICAM2	0.02	0.08	0.15
Tie2	0.04	0.12	0.12
VEGFR-2*	0.05	0.10	0.25
Tie1*	0.05	0.10	0.15

*ED_{50} value calculated from multigene silencing experiment (Fig. 2d).

serum proteins, the zeta potential was measured for 7C1 formulated with siRNA at blood physiological pH (7.4) and pK_a (ref. 6; Fig. 1f). While 7C1 formed electrically neutral particles at pH 7.4, its pK_a was 5.0. This value is different than the pK_a of particles optimized for hepatocyte delivery[20].

We investigated the serum kinetics and biodistribution of 7C1 siRNA nanoparticles *in vivo*. 7C1 was complexed with Alexa647-tagged siRNA and injected intravenously. After 1 h, skin tissue whole-mounted for confocal microscopy showed colocalization between 7C1 and endothelial cells, suggesting endothelial cells endocytosed 7C1 *in vivo* (Fig. 1g). Endothelial cell uptake was confirmed by an increase in Alexa647 mean fluorescence intensity in endothelial cells sorted from pulmonary tissue 1 h after injection with 7C1 formulated with Alexa647-tagged siRNA (Fig. 2g). 7C1 serum kinetics was then measured. 7C1 serum concentration decreased by 50% within 20 min of intravenous injection, indicating the formulation was rapidly cleared or endocytosed (Fig. 1h). To investigate 7C1 biodistribution, Cy5.5 fluorescence was quantified 4 and 24 h after injection (when 7C1 serum concentrations were negligible) (Fig. 1i). Renal fluorescence was high, indicating that the kidneys aid in the clearance of siRNA delivered by 7C1. *In vitro*, HMVECs take up 7C1 via caveolae- and clathrin-mediated endocytosis (Fig. 2a).

To confirm that endothelial localization resulted in functional siRNA delivery, mRNA was measured after 7C1 was formulated with siRNAs targeting genes expressed primarily by endothelial cells. 7C1 was formulated with siRNA targeting the gene ICAM-2 and injected with a dose of 0.6 mg kg^{-1} on day one and four. Five days later, skin was analysed with confocal microscopy and flow cytometry (Supplementary Fig. 2a,b). ICAM-2 expression on lymph node and omentum endothelial cells was also measured with flow cytometry (Supplementary Fig. 2b). ICAM-2 expression decreased compared with PBS- and siLuciferase- (termed siCntrol) treated mice. 7C1 was then formulated with 0.60, 0.05, 0.02 or 0.007 mg kg^{-1} si-ICAM2 and injected once (Fig. 2b). Because ICAM-2 is principally expressed by endothelial cells in all major tissue beds, this assay detects endothelial gene silencing in any organ[21]. 7C1 reduced ICAM-2 mRNA expression in the pulmonary, cardiovascular and renal endothelium by 50% at doses of 0.02, 0.08 and 0.15 mg kg^{-1}, respectively (Fig. 2b, Table 2). To ensure efficient delivery was not limited to siICAM-2, we measured gene silencing in mice treated with siRNA targeting VE-cadherin (VEcad), a junctional protein whose expression is limited to endothelium. Cardiovascular, renal and pulmonary VEcad mRNA decreased by 50% at doses of 0.04, 0.08 and less than 0.02 mg kg^{-1}, respectively (Fig. 2c, Table 2). 7C1-siVEcad also reduced VEcad protein levels in whole-lung homogenates at 0.03 mg kg^{-1} (Fig. 2d). We then investigated whether reduced VEcad protein levels increased vascular permeability. Compared with mice treated with PBS or siCntrol, the extravasation of Evans Blue Dye out of the pulmonary vasculature increased 2.5-fold seven and fourteen days after a single 0.6 mg kg^{-1} injection of siVEcad (Fig. 2e). These data demonstrate that 7C1 facilitates the most efficient non-liver siRNA delivery reported to date.

Because *in vivo* multigene silencing requires highly efficient delivery, it has been limited to hepatocytes[5]. 7C1 silenced five endothelial genes (Tie1, Tie2, VEcad, VEGFR-2 and ICAM-2) concurrently. Three days following intravenous injection with a total dose of 0.25 mg kg^{-1}, target mRNA of all five genes decreased between 60% and 80% in pulmonary vasculature (Fig. 2f, Supplementary Fig. 2c–e). Target gene expression remained constant after siCntrol was injected (total dose of 2.0 mg kg^{-1}), suggesting that the reduced mRNA levels were due to RNAi. To our knowledge, this is the first demonstration of multi-gene silencing in endothelial cells *in vivo*.

Efficient delivery also facilitates durable gene silencing, because the duration of gene silencing is generally dose-dependent[5]. mRNA silencing was measured as a function of time after a 0.6 mg kg^{-1} injection of siICAM-2 (Fig. 2g). Pulmonary ICAM-2 mRNA expression initially decreased by 92% and remained suppressed by between 73% and 85% for 21 days. By contrast, cardiovascular and renal endothelial ICAM-2 expression increased continually over the first 28 days, reaching 50% of initial expression after day ten, again suggesting less efficient endothelial cell delivery in these vascular beds compared with lung endothelium (Fig. 2g). We then measured gene silencing in different organs after modifying the particle size and 7C1:siRNA mass ratio (see Supplementary Section '7C1 biophysical optimization'; Supplementary Fig. 2f–i)). 7C1 was complexed with siRNA targeting the endothelial specific gene Tie2 (ref. 21). In all cases, the most potent delivery was measured in pulmonary endothelial cells (Supplementary Fig. 2f–i).

Although others have reported siRNA delivery to the lung, functional gene silencing required doses much higher than 0.02 mg kg^{-1}. Because the relative silencing in different cell types dictates the type of *in vivo* models nanoparticles can be used to study, we studied 7C1 biodistribution and silencing in pulmonary epithelial cells, haematopoietic cells, T cells and B cells (Fig. 3a–c). We also measured gene silencing in hepatocytes and peritoneal immune cells, two cell types that have been preferentially transfected by lipid nanoparticles (Fig. 3d–f).

We quantified the uptake of Alexa647-labelled siRNA delivered by 7C1. One hour after a 1.0 mg kg^{-1} injection, lungs were digested into a single cell suspension and labelled with antibodies. Flow cytometry revealed that Alexa647 median fluorescent intensity was significantly higher in endothelial cells than in pulmonary epithelial cells, haematopoietic cells, T cells and B cells (Fig. 3a). Twenty-four hours after injection, endothelial cell uptake decreased. Although the significance of the decreased signal is unknown, it could result from fluorophore cleavage. We then used flow cytometry to simultaneously quantify ICAM-2 protein expression in pulmonary endothelial cells, haematopoietic cells, T cells and B cells. Three days following injection of 0.30, 0.20, 0.10 or 0.05 mg kg^{-1}, pulmonary endothelial cell ICAM-2 median fluorescent intensity decreased between 60% and 68% compared with cells from siCntrol treated mice (Fig. 3b). ICAM-2 median fluorescent intensity did not decrease in pulmonary haematopoietic cells, T cells or B cells. The relative delivery to lung endothelium and epithelium was then quantified with siRNA targeting Integrinß1 (Fig. 3c). Two days after injection, lung were digested before epithelial and endothelial cells were sorted into separate tubes with fluorescence activated cell sorting.

NATURE NANOTECHNOLOGY DOI: 10.1038/NNANO.2014.84 ARTICLES

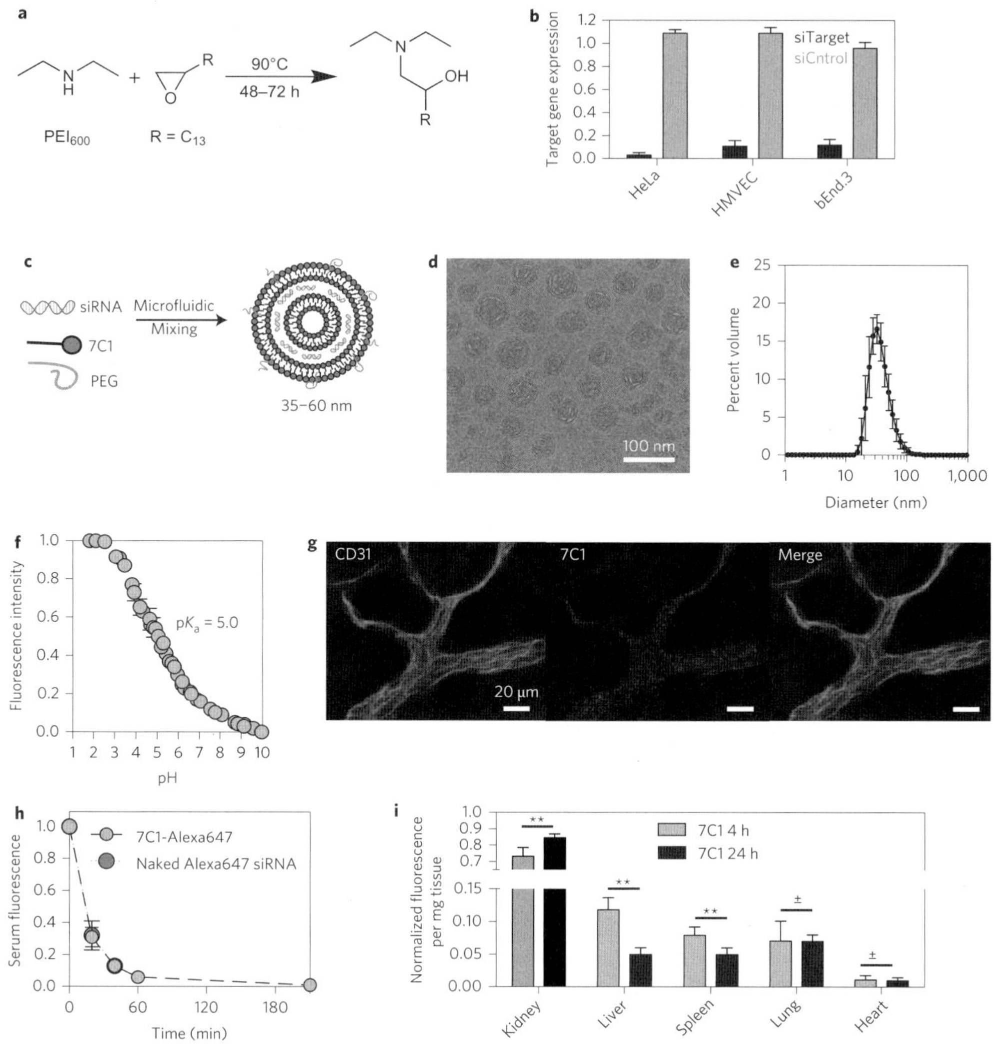

Figure 1 | 7C1 synthesis, characterization and *in vivo* biodistribution. a, 7C1 synthesis scheme. **b**, Target gene expression 24 h following 30 nM treatment with siRNA in human cervical carcinoma (HeLa), human primary endothelial (HMVEC) and murine endothelial (bEnd.3) cells. HeLa target gene expression was measured as Firefly luminescence in HeLa cells expressing Luciferase that were treated with siRNA targeting luciferase. bEnd.3 and HMVEC target gene expression was measured as Tie2 mRNA levels following treatment with siRNA targeting Tie2. **c**, 7C1 formulation scheme. 7C1 nanoparticles were mixed with $C_{14}PEG_{2000}$ and siRNA in a high-throughput microfluidic chamber as previously described[30]. **d**, 7C1 internal structure characterized by cryo-TEM. Dark bands indicate lipid layers and light bands indicate regions with siRNA. **e**, Average 7C1 hydrodynamic diameter, measured by dynamic light scattering, and weighted by volume ($N = 20$ formulations). **f**, 6,P-toluidinylnapthalene-2-sulfonate (TNS) fluorescence of formulated 7C1 nanoparticles as a function of pH (used to measure 7C1 pK_a). **g**, Representative confocal image of Alexa647-tagged siRNA complexed to 7C1 1 h after intravenous injection. CD31 is a ubiquitous marker for endothelium. **h**, Serum Cy5.5 concentration following injection with 7C1-Cy5.5 siRNA or naked Cy5.5 siRNA. **i**, Cy5.5 fluorescence per mg tissue after injection with 7C1-Cy5.5 siRNA. Tissues were removed after injection and weighed individually. Cy5.5 intensity was normalized to each individual tissue. Time points were selected to measure systemic siRNA accumulation after Cy5.5 was cleared from serum. $N = 4$–5 mice per group. In all cases, data are shown as mean ± s.d. $^{**}P < 0.005$, $^{\pm}P > 0.75$.

The purity of the sorted cells was confirmed by measuring cell-specific markers using RT-PCR (PCR with reverse transcription, Supplementary Fig. 2j). Compared with siCntrol-treated cells, endothelial cell Integrinß1 mRNA decreased by between 70% and 82%, while epithelial cell mRNA did not change substantially (Fig. 3c). These data indicate that, at these doses, 7C1 preferentially delivers siRNA to pulmonary endothelial cells.

We then analysed whether 7C1 delivered siRNA to hepatocytes, which are preferentially targeted by many lipid nanoparticle formulations (Fig. 3d). We measured the expression of a hepatocyte-specific gene Factor 7 (F7) after injection with the highly potent siF7[5]. F7 serum concentration remained constant after siF7 was injected at a dose of 1.5 mg kg^{-1} and was only reduced by 35% after an injection of 2.0 mg kg^{-1}, but the positive control lipid nanoparticle Hepat01 decreased F7 expression by 95% after a 0.30 mg kg^{-1} dose (Fig. 3d). To confirm that 7C1 reduced endothelial gene silencing without silencing hepatocyte gene expression, 7C1 was simultaneously complexed with siF7 and siTie2 (Fig. 3e).

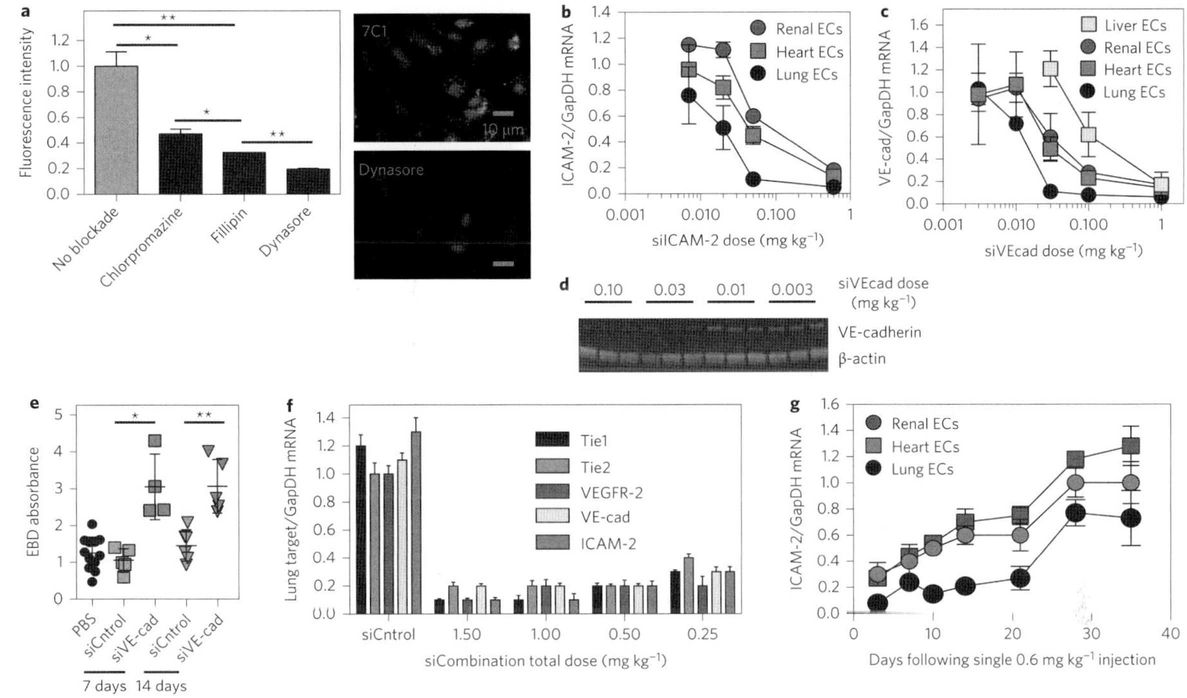

Figure 2 | 7C1 delivers siRNA to endothelial cells. a, Alexa647 fluorescence uptake in HMVEC cells following 7C1-Alexa647 treatment in the presence of small molecules inhibiting clathrin (Chlorpromazine), caveolin (Fillipin) and both endocytotic pathways (Dynasore). Representative images of cells treated with 7C1 or Dynasore are shown. **b**, ICAM-2/GapDH mRNA ratios (normalized to PBS-treated mice) following intravenous injection of 7C1-siICAM-2. **c**, VE-cadherin/GapDH mRNA ratios (normalized to PBS-treated mice) following intravenous injection of 7C1-siVEcad. **d**, VE-cadherin and β-actin protein expression following treatment with 7C1 siVEcad. **e**, Evans Blue Dye (EBD) pulmonary absorbance 7 and 14 days following a 0.6 mg kg^{-1} injection of 7C1-siVEcad. **f**, Target/GapDH mRNA ratios (normalized to PBS-treated mice) following injection of 7C1 formulated with siCntrol or five siRNAs targeting ICAM-2, Tie2, VE-cadherin, VEGFR2 or Tie1, respectively (siCombination). **g**, ICAM-2/GapDH mRNA levels as a function of time following a 0.6 mg kg^{-1} injection of siICAM-2. Data shown as mean $\pm$ s.d. $N = 4$–5 mice per group, $^{*}P < 0.05$, $^{**}P < 0.005$.

Because efficacy can vary with the molar ratio of PEG and cholesterol, we performed this two-gene experiment with particles formulated with different 7C1:cholesterol:$C_{14}PEG_{2000}$ molar ratios. Two formulations reduced lung Tie2 mRNA by nearly 90% after a single 0.15 mg kg^{-1} dose, but did not reduce *F7* expression (Fig. 3e).

Intravenously injected particles may also transfect peritoneal immune cells, especially $CD11b^{+}$ monocytes and macrophages[22]. CD45 median fluorescent intensity was quantified in immune cells isolated from the peritoneal cavity following intravenous injection of 2.0 mg kg^{-1} 7C1 formulated with an siRNA targeting CD45 (siCD45)[22] (Fig. 3f). CD45 protein expression remained constant in macrophages, B cells, T cells and dendritic cells following treatment with 7C1 (Fig. 3f). By contrast, CD45 expression decreased in macrophages cells following treatment with the positive control lipid nanoparticle C12-200. Taken together, these data indicate that 7C1 does not deliver siRNA to hepatocytes or peritoneal immune cells in healthy BL/6 mice at any dose tested.

Endothelial RNAi affects multiple animal models

Whether RNA delivery could change endothelial function was then investigated in mouse models of emphysema, primary tumour growth and lung metastasis. Emphysema is characterized by decreased pulmonary surface area, which reduces gas transport, causing dyspnea and cough[23,24]. Along with macrophage-mediated protease imbalance, decreased VEGF and VEGFR-2 expression has been documented in the lungs of patients with chronic obstructive pulmonary disease. Moreover, genetic Cre-lox mediated deletion of VEGF causes emphysema in mice[24]. VEGF receptor blockade with SU5416 promotes emphysema in rodents, leading to decreased alveolar surface to volume ratios, increased lung volume and increased distance between alveolar walls (termed the mean linear intercept)[23]. Because SU5416 simultaneously inhibits VEGFR-1 and VEGFR-2, we characterized the pulmonary phenotype following VEGFR-2 specific silencing. VEGFR-2 silencing induced emphysema-like changes, indicated by decreased alveolar surface to volume ratios and increases in lung volume and mean linear intercept compared with siCntrol-treated mice (Fig. 3g–i,k). These changes were not due to infiltration of myeloid cells (Fig. 3j). These results suggest VEGFR-2 specific silencing is sufficient to induce emphysema-like phenotypes in mice and that systemic endothelial cell RNAi can be used to investigate the role of endothelial gene function *in vivo*.

The therapeutic effect of endothelial RNAi on primary tumour growth was then investigated in a Lewis lung carcinoma model. Previous work has demonstrated that antibodies targeting VEGFR-1 and Dll4 can reduce primary tumour growth through disrupted or non-productive angiogenesis, respectively[25,26]. In particular, targeting VEGFR-1, which is expressed on endothelial cells and pro-angiogenic myeloid cells, reduced tumour progression, metastasis and formation of a pre-metastatic niche[27,28]. However, monoclonal antibodies may function differently from RNAi-based methods, which inhibit both extracellular and intracellular signalling[29]. To investigate whether therapeutic deletion of VEGFR-1 and Dll4, both of which have intracellular signalling components,

NATURE NANOTECHNOLOGY DOI: 10.1038/NNANO.2014.84 ARTICLES

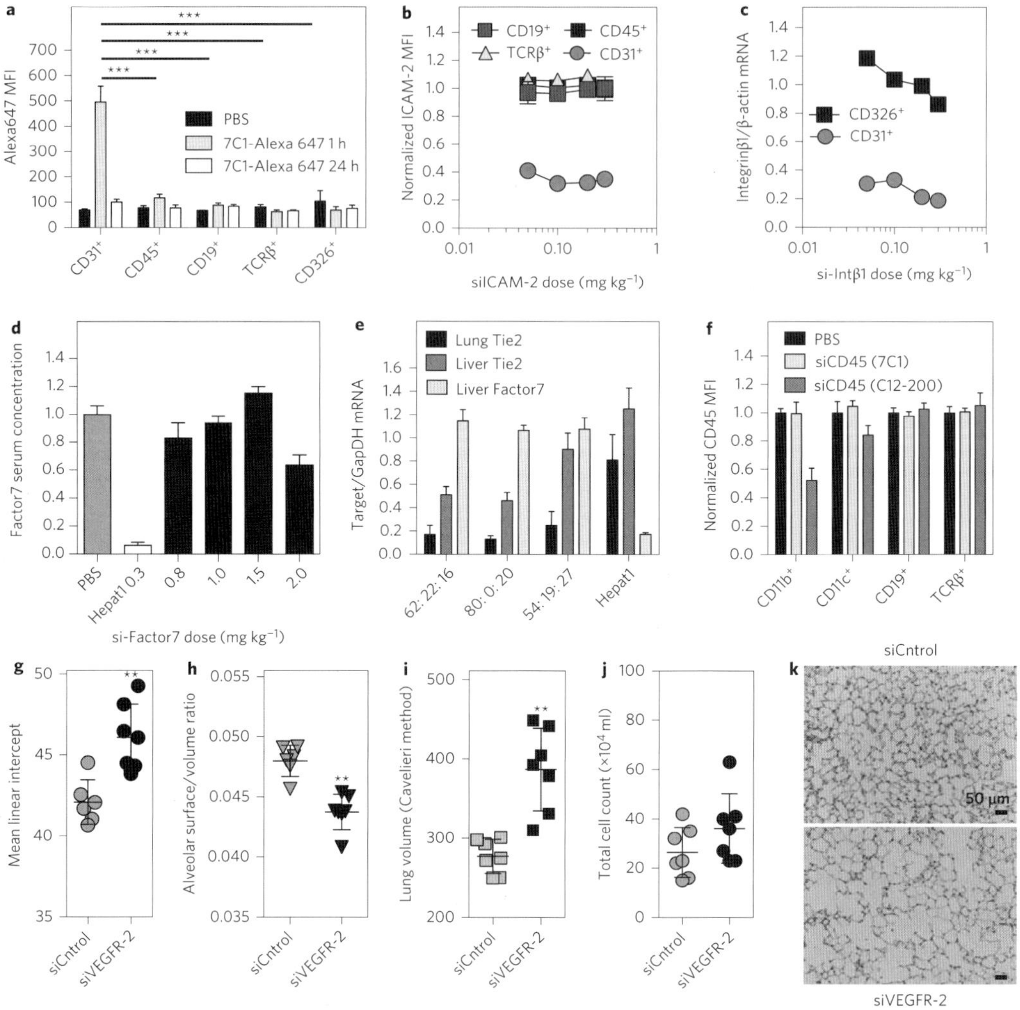

Figure 3 | 7C1 preferentially delivers siRNA to pulmonary endothelial cells *in vivo*. **a**, Alexa647 median fluorescent intensity in pulmonary endothelial ($CD31^+$), haematopoietic ($CD45^+$), epithelial ($CD326^+$), B ($CD19^+$) or T ($TCR\beta^+$) cells isolated from mice after treatment with 7C1 formulated with Alexa647-tagged siRNA. Statistical significance calculated between endothelial cells and other pulmonary cell types 1 h after injection. **b**, ICAM-2 median fluorescent intensity in pulmonary cells (normalized to siCntrol-treated mice) isolated from mice three days following treatment with 7C1-siICAM-2. **c**, Integrinβ1/β-actin mRNA ratios (normalized to siCntrol-treated mice) in pulmonary endothelial and epithelial cells isolated from mice two days after treatment with siIntegrinβ1. **d**, Factor 7 serum concentration (normalized to PBS-treated animals) two days following treatment with liver-targeting molecule Hepat01-siFactor7 or 7C1-siFactor7. **e**, Tie2 and Factor7/GapDH mRNA expression following a 0.15 mg kg^{-1} injection of 7C1 concurrently formulated with siTie2 and siFactor7. Particles were formulated with different 7C1:cholesterol:$C_{14}PEG_{2000}$ molar ratios. 7C1 decreased Tie2 mRNA expression in pulmonary, renal and hepatic endothelium without reducing F7 mRNA expression. **f**, CD45 median fluorescent intensity following treatment with 7C1-siCD45 or positive control C12-200-siCD45. **g–k**, Mean linear intercept (MLI) between alveoli, pulmonary surface/volume ratio, total volume and pulmonary histology following two 0.5 mg kg^{-1} injections of siCntrol or siVEGFR2. Increased MLI, alveolar volume, decreased surface/volume ratios and constant infiltrating myeloid cells are consistent with an induced emphysema-like phenotype ($N = 6$–7 animals per group; data shown as average ± s.d.). $^{**}P < 0.002$, $^{***}P < 0.001$.

would have similar effects as therapeutic antibodies, 7C1 was complexed with siCntrol, siVEGFR-1 or siDll4. Compared with siCntrol, siVEGFR-1 and siDll4 formulations had a significant therapeutic effect, reducing primary tumour growth by 40% and 70%, respectively, and increasing tumour necrosis (Fig. 4a–c). While some tumours treated with siVEGFR-1 exhibited high levels of cell death, others showed low levels[30]. These data suggest that targeted deletion of both the intracellular and extracellular portion of VEGFR-1 or Dll4 may reduce primary tumour growth.

The roles of VEGFR-1 and Dll4 in primary tumour growth have been studied extensively, but the role of these genes, particularly Dll4, in metastasis is less clearly understood[31,32]. Consequently, the effect of VEGFR-1 and Dll4 deletion on lung tumour metastases was studied in a metastatic Lewis lung carcinoma model (Fig. 4d,e, Supplementary Fig. 3). siVEGFR-1 reduced surface metastases by 52%, while siDll4 reduced surface metastases by 63% compared with siCntrol-treated mice (Fig. 4d,e). Lung weight, which correlates to the growth of lung metastases, decreased

ARTICLES NATURE NANOTECHNOLOGY DOI: 10.1038/NNANO.2014.84

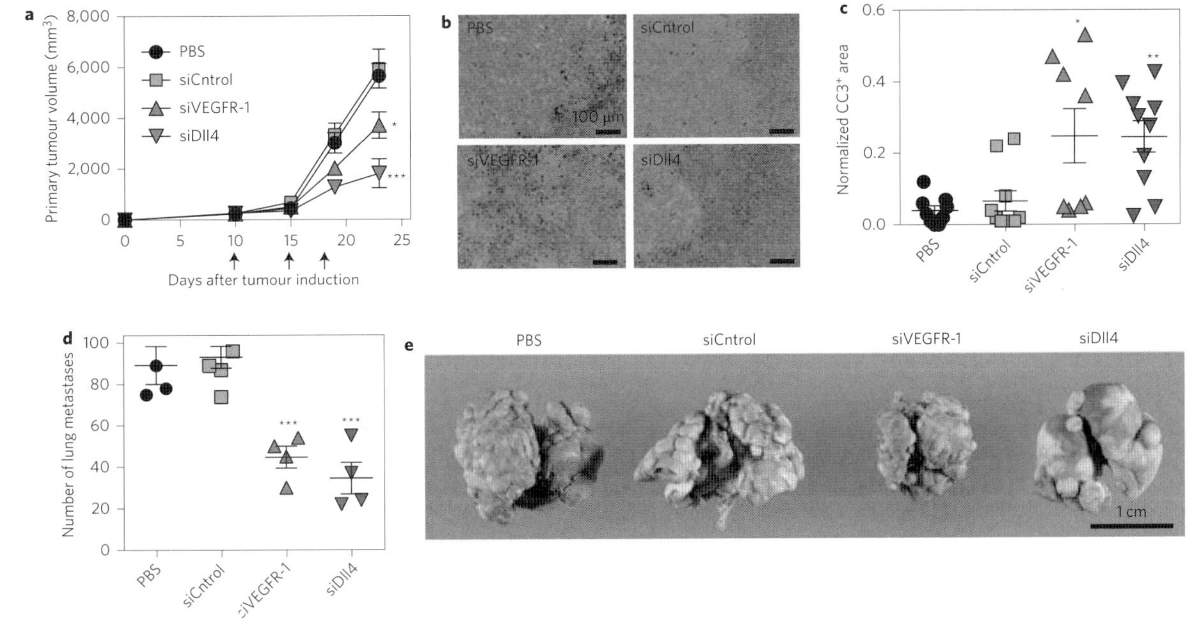

Figure 4 | 7C1-mediated mRNA silencing modifies endothelial function *in vivo*. **a**, Primary Lewis lung carcinoma (LLC) growth following three 1.0 mg kg^{-1} treatments with PBS, siCntrol, siVEGFR-1 or siDll4 ($N = 7$–10 animals per group; data shown as average $\pm$ s.e.m.). **b**,**c**, Representative images (**b**) and quantification of cleaved caspase 3 (CC3) staining (**c**), a marker for apoptosis, following treatment with PBS, siCntrol, siVEGFR-1, siDll4. Normalized CC3$^+$ area defined as the total CC3$^+$ surface area divided by the tumour surface area. **d**, Number of pulmonary surface metastases following four 1.0 mg kg^{-1} injections with PBS, siCntrol, siVEGFR-1 or siDll4 ($N = 4$–6 per group; data shown as average $\pm$ s.e.m.). To measure effects independent of primary tumour growth, animals were not treated until after primary tumour resection. **e**, Murine lungs with metastatic lesion removed after treatment with PBS, siLuc, siVEGFR-1 or siDll4. $^*P < 0.05$, $^{**}P < 0.002$, $^{***}P < 0.001$.

by 50% after siVEGFR-1 therapy and 60% after siDll4 therapy (Supplementary Fig. 3).

7C1 *in vivo* tolerability

7C1 nanoformulations were well tolerated in animal models of toxicity following acute and chronic high-dose treatment (Supplementary Fig. 4). We measured the serum concentrations of markers associated with toxicity after four 0.6 mg kg^{-1} intravenous injections in highly immunoactive CD1$^+$ mice, a mouse model used for preclinical toxicology studies. Mice were injected with a 0.6 mg kg^{-1} dose of siCntrol or siICAM-2 once a week for four weeks. Forty-eight hours after the last injection, the serum concentrations of markers for hepatic, cardiovascular and renal injury were quantified. Importantly, we observed no evidence of kidney damage (Supplementary Fig. 4a).

7C1 tolerability was also investigated in BL/6 mice at doses much higher than those required for functional gene silencing (2 mg kg^{-1}) in acute and chronic models. Over 28 days, mice were injected eight times with PBS solution or 2 mg kg^{-1} 7C1. Murine weight gain equalled that of mice injected with PBS (Supplementary Fig. 4b). Four hours after the final injection, mice were killed and lungs were removed before mRNA expression of cytokines (IL-6, TNF-α), markers of endothelial dysfunction (ICAM-2, E-selectin) and immune cell infiltration (CD45) were quantified (Supplementary Fig. 4c). Serum concentrations of 30 cytokines were also quantified. mRNA expression and cytokine concentration did not increase significantly when compared to PBS-treated mice (Supplementary Fig. 4d).

We then measured serum cytokine concentration 2, 4, 6 and 24 h following a 2.0 mg kg^{-1} injection. Although the serum concentrations of five factors did increase between 2 and 6 h, only one factor (CXCL2) equalled the concentration in mice treated with a low-dose lipopolysaccharide (LPS) control and all five returned to baseline 24 h after injection (Supplementary Fig. 4e). These data suggest that, while 7C1 probably interacts with cells to produce a transient response at doses much higher than those required for gene silencing, the formulation appears to be well tolerated in multiple mouse models *in vivo*.

Conclusions

We have identified a nanoparticle (7C1) that efficiently delivers siRNA to endothelial cells. Unlike previously reported lipid and lipidoid-based nanoparticles, 7C1 transfected endothelial cells *in vivo* at low doses, without significantly reducing gene expression in hepatocytes, peritoneal immune cells, pulmonary epithelial cells or pulmonary immune cells. The exact molecular mechanism governing this effect remains to be determined, but it seems to involve the interaction of 7C1 with serum proteins, which can promote delivery to certain cell types[33]. As a result, 7C1 may be an interesting system to study how physiochemical interactions between nanomaterials and serum proteins direct nanoparticles to endothelial cells *in vivo*[34–37]. This nanoformulation enables the simultaneous silencing of multiple endothelial genes. Although some immune cells are known to express Tie1, Tie2 and ICAM-2, the lack of significant pulmonary immune cell gene silencing (mediated by 7C1) indicates that such multigene silencing occurs primarily in endothelial cells. We therefore anticipate that 7C1 can be useful in the study of gene combinations in complex biological pathways *in vivo*, a strategy called *in vivo* genomics.

Moreover, we have found that 7C1 reduces primary tumour growth and lung metastases in a model of lung cancer. As a result, although this study examined targets whose extracellular

NATURE NANOTECHNOLOGY DOI: 10.1038/NNANO.2014.84 ARTICLES

activity can also be inhibited by antibodies, future therapies may be designed to target combinations of proteins currently considered 'undruggable'. Similarly, future RNA therapies may enhance the effects of non-RNA drugs (for example, modifying the expression of the gene involved in the exocytosis of a small molecule might enhance its delivery).

Finally, we have shown that 7C1 durably reduces target gene expression in multiple animal models. This extended therapeutic effect may increase the utility of *in vivo* endothelial RNAi. Because 7C1 is well tolerated at doses far higher than those required for gene silencing, we believe this technology will be used to manipulate gene expression *in vivo*. In summary, 7C1 nanoformulations should provide biologists and engineers with a new tool to efficiently deliver siRNA to endothelium.

Received 1 August 2013; accepted 25 March 2014; published online 11 May 2014; corrected online 20 June 2014

References

1. Pober, J. S. & Sessa, W. C. Evolving functions of endothelial cells in inflammation. *Nature Rev. Immunol.* **7,** 803–815 (2007).
2. Hagberg, C. E. *et al.* Targeting VEGF-B as a novel treatment for insulin resistance and type 2 diabetes. *Nature* **490,** 426–430 (2012).
3. Kumar, V., Abbas, A., Fausto, N. & Aster, J. *Robbins and Cotran Pathologic Basis of Disease* 8th edn (Elsevier, 2009).
4. Kanasty, R., Dorkin, J. R., Vegas, A. & Anderson, D. Delivery materials for siRNA therapeutics. *Nature Mater.* **12,** 967–977 (2013).
5. Love, K. T. *et al.* Lipid-like materials for low-dose, *in vivo* gene silencing. *Proc. Natl Acad. Sci. USA* **107,** 1864–1869 (2010).
6. Semple, S. *et al.* Rational design of cationic lipids for siRNA delivery. *Nature Biotechnol.* **28,** 172–176 (2010).
7. Whitehead, K. A., Langer, R. & Anderson, D. G. Knocking down barriers: advances in siRNA delivery. *Nature Rev. Drug Discov.* **8,** 129–138 (2009).
8. Aleku, M. *et al.* Atu027, a liposomal small interfering RNA formulation targeting protein kinase N3, inhibits cancer progression. *Cancer Res.* **68,** 9788–9798 (2008).
9. Aleku, M. *et al.* Intracellular localization of lipoplexed siRNA in vascular endothelial cells of different mouse tissues. *Microvasc. Res.* **76,** 31–41 (2008).
10. Santel, A. *et al.* RNA interference in the mouse vascular endothelium by systemic administration of siRNA-lipoplexes for cancer therapy. *Gene Ther.* **13,** 1360–1370 (2006).
11. Santel, A. *et al.* A novel siRNA-lipoplex technology for RNA interference in the mouse vascular endothelium. *Gene Ther.* **13,** 1222–1234 (2006).
12. Polach, K. J. *et al.* Delivery of siRNA to the mouse lung via a functionalized lipopolyamine. *Mol. Ther.* **20,** 91–100 (2012).
13. Kaufmann, J., Ahrens, K. & Santel, A. RNA interference for therapy in the vascular endothelium. *Microvasc. Res.* **80,** 286–293 (2010).
14. Davis, M. E. *et al.* Evidence of RNAi in humans from systemically administered siRNA via targeted nanoparticles. *Nature* **484,** 1067–1070 (2010).
15. Rozema, D. B. *et al.* Dynamic polyconjugates for targeted *in vivo* delivery of siRNA to hepatocytes. *Proc. Natl Acad. Sci. USA* **104,** 12982–12987 (2007).
16. Godbey, W. T., Wu, K. K. & Mikos, A. G. Poly(ethylenimine) and its role in gene delivery. *J. Control. Rel.* **60,** 149–160 (1999).
17. Breunig, M., Lungwitz, U., Liebl, R. & Goepferich, A. Breaking up the correlation between efficacy and toxicity for nonviral gene delivery. *Proc. Natl Acad. Sci. USA* **104,** 14454–14459 (2007).
18. Richards Grayson, A. C., Doody, A. M. & Putnam, D. Biophysical and structural characterization of polyethylenimine-mediated siRNA delivery *in vitro*. *Pharm. Res.* **8,** 1868–1876 (2006).
19. Crawford, R. *et al.* Analysis of lipid nanoparticles by cryo-EM for characterizing siRNA delivery vehicles. *Int. J. Pharm.* **403,** 237–244 (2011).
20. Jayaraman, M. *et al.* Maximizing the potency of siRNA lipid nanoparticles for hepatic gene silencing *in vivo*. *Angew. Chem. Int. Ed.* **51,** 8529–8533 (2012).
21. Huang, H., Bhat, A., Woodnutt, G. & Lappe, R. Targeting the ANGPT-TIE2 pathway in malignancy. *Nature Rev. Cancer* **10,** 575–585 (2010).
22. Novobrantseva, T. I. *et al.* Systemic RNAi-mediated gene silencing in nonhuman primate and rodent myeloid cells. *Mol. Ther. Nucleic Acids* **1,** e4 (2012).
23. Kasahara, Y. *et al.* Inhibition of VEGF receptors causes lung cell apoptosis and emphysema. *J. Clin. Invest.* **106,** 1311–1319 (2000).
24. Tuder, R. M. & Yun, J. H. Vascular endothelial growth factor of the lung: friend or foe. *Curr. Opin. Pharmacol.* **8,** 255–260 (2008).
25. Thurston, G., Noguera-Troise, I. & Yancopoulos, G. D. The Delta paradox: DLL4 blockade leads to more tumour vessels but less tumour growth. *Nature Rev. Cancer* **7,** 327–331 (2007).
26. Fischer, C., Mazzone, M., Jonckx, B. & Carmeliet, P. FLT1 and its ligands VEGFB and PlGF: drug targets for anti-angiogenic therapy? *Nature Rev. Cancer* **8,** 942–956 (2008).
27. Lyden, D. *et al.* Impaired recruitment of bone-marrow-derived endothelial and hematopoietic precursor cells blocks tumor angiogenesis and growth. *Nature Med.* **7,** 1194–1201 (2001).
28. Kaplan, R. N. *et al.* VEGFR1-positive haematopoietic bone marrow progenitors initiate the pre-metastatic niche. *Nature* **438,** 820–827 (2005).
29. Tammela, T. *et al.* VEGFR-3 controls tip to stalk conversion at vessel fusion sites by reinforcing Notch signalling. *Nature Cell Biol.* **13,** 1202–1213 (2011).
30. Stevens, J. B. *et al.* Heterogeneity of cell death. *Cytogenet. Genome Res.* **139,** 164–173 (2013).
31. Kuramoto, T. *et al.* Dll4-Fc, an inhibitor of Dll4-notch signaling, suppresses liver metastasis of small cell lung cancer cells through the downregulation of the NF-kappaB activity. *Mol. Cancer Ther.* **11,** 2578–2587 (2012).
32. Garcia, A. & Kandel, J. J. Notch: a key regulator of tumor angiogenesis and metastasis. *Histol. Histopathol.* **27,** 151–156 (2012).
33. Akinc, A. *et al.* Targeted delivery of RNAi therapeutics with endogenous and exogenous ligand-based mechanisms. *Mol. Ther.* **18,** 1357–1364 (2010).
34. Monopoli, M. P., Aberg, C., Salvati, A. & Dawson, K. A. Biomolecular coronas provide the biological identity of nanosized materials. *Nature Nanotech.* **7,** 779–786 (2012).
35. Chen, D. *et al.* Rapid discovery of potent siRNA-containing lipid nanoparticles enabled by controlled microfluidic formulation. *J. Am. Chem. Soc.* **134,** 6948–6951 (2012).
36. Whitehead, K., Dahlman, J. E., Langer, R. S. & Anderson, D. G. Silencing or stimulation? siRNA delivery and the immune system. *Ann. Rev. Chem. Biomol. Eng.* **2,** 77–96 (2011).
37. Panigrahy, D. *et al.* Epoxyeicosanoids stimulate multiorgan metastasis and tumor dormancy escape in mice. *J. Clin. Invest.* **122,** 178–191 (2012).

Acknowledgements

The authors thank J. Cattie, T. O'Shea and T. Tammela. J.E.D. was supported by National Defense Science and Engineering (NDSEG), the National Science Foundation (NSF) and MIT Presidential Fellowships. D.P. was supported by R01 CA148663. M.W.K. was supported by the Stop and Shop Pediatric Brain Tumour Fund, as well as the Pediatric Brain Tumour Fund. H.S. was supported by the Deutsche Forschungsgemeinschaft (SA1668/2-1). Research was also supported by Alnylam and the Center for RNA Therapeutics and Biology.

Author contributions

J.E.D., C.B., V.K., R.L. and D.G.A. conceived the experiments. J.E.D., C.B., O.F.K., A.T., S.J., T.E.S., Y.X., H.B.S., G.S., L.S., A.B., R.L.B., H.Y., T.R., Y.D., S.J., D.S., A.D., K.S.S., M.J.W., T.N., V.M.R., A.K.R.L.J., C.G.L., B.K., D.K.M., M.P., L.A., P.D., L.S., K.C., M.W.K., K.F., M.N., D.D., R.M.T., U.H.V.A., A.A., A.S. and D.P. performed experiments. J.E.D., C.B., V.K., R.L. and D.G.A. co-wrote the paper. All authors discussed the results and commented on the manuscript.

Additional information

Supplementary information is available in the online version of the paper. Reprints and permissions information is available online at www.nature.com/reprints. Correspondence and requests for materials should be addressed to D.G.A.

Competing financial interests

R.L. is a shareholder and member of the Scientific Advisory Board of Alnylam. R.L and D.G.A have sponsored research grants from Alnylam. Alnylam also has a licence to certain intellectual property that was invented at the Massachusetts Institute of Technology by D.G.A. and R.L.

ARTICLES NATURE NANOTECHNOLOGY DOI: 10.1038/NNANO.2014.84

James E. Dahlman[1,2,3†], Carmen Barnes[4†], Omar F. Khan[2,5], Aude Thiriot[6], Siddharth Jhunjunwala[2], Taylor E. Shaw[2], Yiping Xing[2], Hendrik B. Sager[7], Gaurav Sahay[2], Lauren Speciner[4], Andrew Bader[2], Roman L. Bogorad[2], Hao Yin[2], Tim Racie[4], Yizhou Dong[2], Shan Jiang[2], Danielle Seedorf[2], Apeksha Dave[2], Kamaljeet Singh Sandhu[2], Matthew J. Webber[2], Tatiana Novobrantseva[4], Vera M. Ruda[2], Abigail K. R. Lytton-Jean[2], Christopher G. Levins[2], Brian Kalish[8], Dayna K. Mudge[8], Mario Perez[9], Ludmila Abezgauz[10], Partha Dutta[7], Lynelle Smith[9], Klaus Charisse[4], Mark W. Kieran[6], Kevin Fitzgerald[4], Matthias Nahrendorf[7], Dganit Danino[10], Rubin M. Tuder[9], Ulrich H. von Andrian[6], Akin Akinc[4], Dipak Panigrahy[6], Avi Schroeder[11], Victor Koteliansky[12], Robert Langer[1,2,3,5] and Daniel G. Anderson[1,2,3,5]*

[1]Harvard-MIT Division of Health Sciences and Technology, Massachusetts Institute of Technology, Cambridge, Massachusetts 02139, USA. [2]David H. Koch Institute for Integrative Cancer Research, Massachusetts Institute of Technology, Cambridge, Massachusetts 02139, USA. [3]Institute for Medical Engineering and Science, Massachusetts Institute of Technology, Cambridge, Massachusetts 02139, USA. [4]Alnylam Pharmaceuticals, Cambridge, Massachusetts 02139, USA. [5]Department of Chemical Engineering, Massachusetts Institute of Technology, Cambridge, Massachusetts 02139, USA. [6]Department of Microbiology and Immunology, Harvard Medical School, Boston 02115, USA. [7]Center for Systems Biology, Massachusetts General Hospital and Harvard Medical School, Boston 02114, USA. [8]Vascular Biology Program, Children's Hospital Boston, and Division of Pediatric Oncology, Dana Farber Cancer Institute, Harvard Medical School, Boston 02115, USA. [9]Program in Translational Lung Research, Division of Pulmonary Sciences and Critical Care Program, Department of Medicine, University of Colorado School of Medicine, USA. [10]Department of Biotechnology and Food Engineering and The Russell Berrie Nanotechnology Institute, Technion Israel Institute of Technology, Haifa 3200, Israel. [11]Department of Chemical Engineering, Technion Israel Institute of Technology, Haifa 32000, Israel. [12]Skolkovo Institute of Science and Technology, Skolkovo 143025, Russian Federation. †These authors contributed equally to this work. *e-mail: dgander@mit.edu

RESEARCH ARTICLE

NANOMEDICINE

Preclinical Development and Clinical Translation of a PSMA-Targeted Docetaxel Nanoparticle with a Differentiated Pharmacological Profile

Jeffrey Hrkach,[1] Daniel Von Hoff,[2] Mir Mukkaram Ali,[1] Elizaveta Andrianova,[1] Jason Auer,[1] Tarikh Campbell,[1] David De Witt,[1] Michael Figa,[1] Maria Figueiredo,[1] Allen Horhota,[1] Susan Low,[1] Kevin McDonnell,[1] Erick Peeke,[1] Beadle Retnarajan,[1] Abhimanyu Sabnis,[1] Edward Schnipper,[1] Jeffrey J. Song,[1] Young Ho Song,[1] Jason Summa,[1] Douglas Tompsett,[1] Greg Troiano,[1] Tina Van Geen Hoven,[1] Jim Wright,[1] Patricia LoRusso,[3] Philip W. Kantoff,[4] Neil H. Bander,[5] Christopher Sweeney,[4] Omid C. Farokhzad,[6]* Robert Langer,[7]* Stephen Zale[1]*

We describe the development and clinical translation of a targeted polymeric nanoparticle (TNP) containing the chemotherapeutic docetaxel (DTXL) for the treatment of patients with solid tumors. DTXL-TNP is targeted to prostate-specific membrane antigen, a clinically validated tumor antigen expressed on prostate cancer cells and on the neovasculature of most nonprostate solid tumors. DTXL-TNP was developed from a combinatorial library of more than 100 TNP formulations varying with respect to particle size, targeting ligand density, surface hydrophilicity, drug loading, and drug release properties. Pharmacokinetic and tissue distribution studies in rats showed that the NPs had a blood circulation half-life of about 20 hours and minimal liver accumulation. In tumor-bearing mice, DTXL-TNP exhibited markedly enhanced tumor accumulation at 12 hours and prolonged tumor growth suppression compared to a solvent-based DTXL formulation (sb-DTXL). In tumor-bearing mice, rats, and nonhuman primates, DTXL-TNP displayed pharmacokinetic characteristics consistent with prolonged circulation of NPs in the vascular compartment and controlled release of DTXL, with total DTXL plasma concentrations remaining at least 100-fold higher than sb-DTXL for more than 24 hours. Finally, initial clinical data in patients with advanced solid tumors indicated that DTXL-TNP displays a pharmacological profile differentiated from sb-DTXL, including pharmacokinetics characteristics consistent with preclinical data and cases of tumor shrinkage at doses below the sb-DTXL dose typically used in the clinic.

INTRODUCTION

Targeting of therapeutic agents to diseased cells and tissues was first suggested more than 100 years ago (*1*). Since then, disease targeting has been pursued as a strategy to develop safer and more effective drugs for myriad diseases, most notably cancer, where the mainstay of therapy often involves cytotoxic agents (*2*, *3*). The past decade has seen the advent of molecularly targeted agents capable of inhibiting specific pathways that are required for survival, proliferation, and metastasis of diseased cells (*4*, *5*). However, like cytotoxic drugs, these agents are subject to dose-limiting toxicities that affect their overall clinical effectiveness. Targeted nanoparticles (TNPs) have the potential to overcome the toxicity and efficacy limitations associated with traditional cytotoxic agents and newer-generation molecularly targeted drugs by trafficking a greater fraction of the administered drug directly to cancer cells in a controllable and tunable manner (*6*, *7*).

[1]BIND Biosciences Inc., Cambridge, MA 02139, USA. [2]TGen Clinical Research Services at Scottsdale Healthcare, Phoenix, AZ 85004, USA. [3]Karmanos Cancer Institute, Detroit, MI 48201, USA. [4]Lank Center for Genitourinary Oncology, Dana-Farber Cancer Institute, Harvard Medical School, Boston, MA 02215, USA. [5]Weill Cornell Medical College, New York, NY 10065, USA. [6]Laboratory of Nanomedicine and Biomaterials and Department of Anesthesiology, Brigham and Women's Hospital, Harvard Medical School, Boston, MA 02115, USA. [7]Department of Chemical Engineering, Massachusetts Institute of Technology, Cambridge, MA 02139, USA.

*To whom correspondence should be addressed. E-mail: ofarokhzad@zeus.bwh.harvard.edu (O.C.F.); rlanger@mit.edu (R.L.); szale@bindbio.com (S.Z.)

Even though TNPs were first described more than 30 years ago (*8*), only a handful have entered clinical trials, and none have advanced beyond early-phase testing in humans (*7*). A key challenge to clinical translation of TNPs has been defining the optimal physicochemical parameters that simultaneously confer molecular targeting, immune evasion, and controlled drug release (*9*). This is mainly due to the complex interdependence of NP properties (composition, size, shape, rigidity, surface charge, hydrophilicity, and ligand type and density), payload properties (drug type, solubility, loading, and release kinetics), and in vivo physiological barriers to NP trafficking (immune surveillance, particle extravasation, tissue penetration, and cellular uptake) (*10*). Although a considerable amount of information is available regarding individual factors that affect the in vivo fate of polymeric NPs (*11*–*15*), a system that achieves optimal characteristics has remained elusive (*16*).

Recently, combinatorial and high-throughput technologies have been explored as means to address the multifactorial challenge of NP optimization (*11*, *17*–*19*). For example, combinatorial biomaterial libraries have been created by introducing wide chemical diversity in the polymeric components and then screened for a specific readout, such as small interfering RNA silencing in vitro, to identify NPs with the desired in vivo activity (*17*, *20*, *21*). Here, we investigated a new approach to NP development and optimization, by which we introduced physicochemical diversity in the NP design, while restricting the particle makeup to a clinically validated set of biomaterials. This approach consisted of evaluating a combinatorial library of TNPs that varied systematically with respect to critical parameters (size, surface hydrophilicity, targeting ligand density, drug loading, and drug release) that affect pharmacokinetics (PK), biodistribution, and efficacy of the encapsulated therapeutic agent.

Here, we present a platform to optimize the design of TNPs and its use in the development of a TNP containing the chemotherapeutic docetaxel (DTXL), which was selected from a combinatorial library

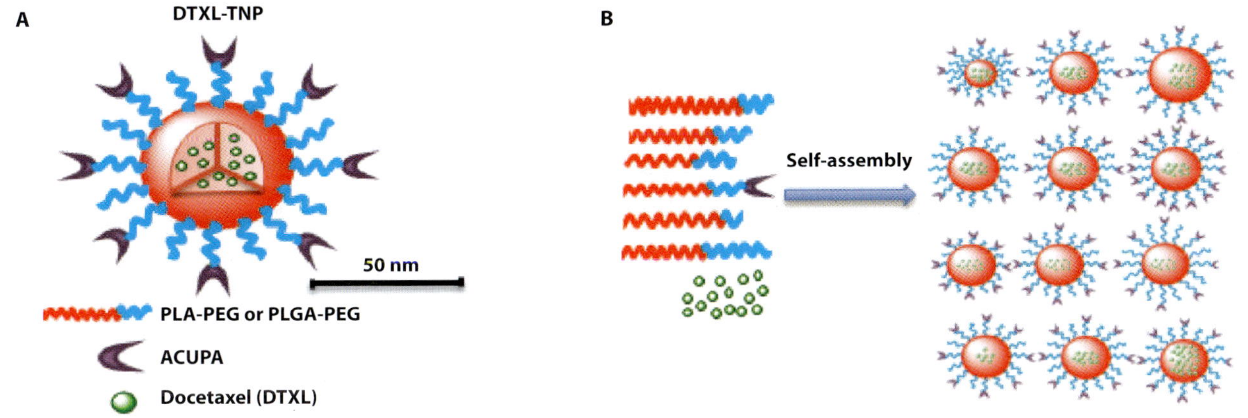

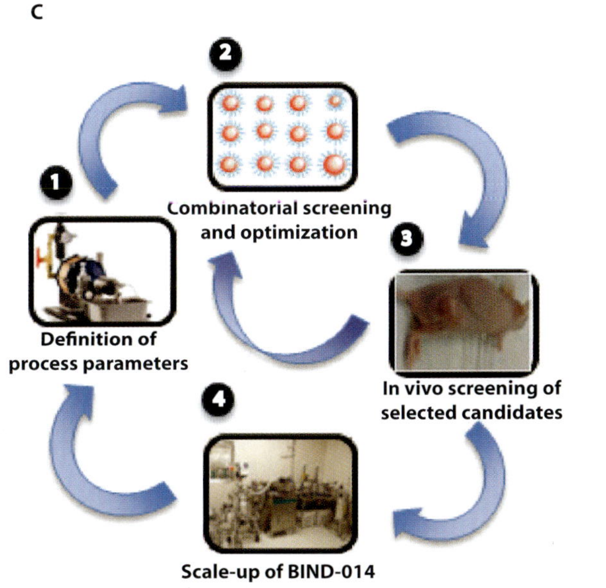

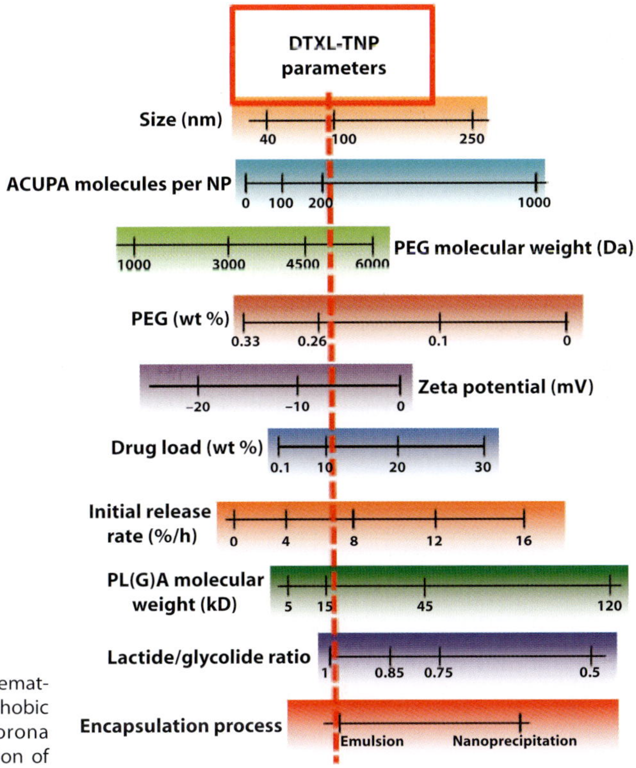

Fig. 1. Combinatorial screening and optimization of DTXL-TNPs. (**A**) Schematic of DTXL-TNP, a PSMA-targeted polymeric NP composed of a hydrophobic PLA polymeric core encapsulating DTXL and a hydrophilic PEG corona decorated with small-molecule (ACUPA) targeting ligands. (**B**) Generation of a library of DTXL-TNPs prepared by self-assembly of particles from mixtures of DTXL, PLA, PLGA, PLA- or PLGA-PEG (with varying PLA, PLGA, and PEG block lengths), and PLA-PEG-ACUPA. (**C**) Development and clinical translation of PSMA-targeted DTXL-TNPs. (1) A nanoemulsion process for efficiently encapsulating DTXL in NPs was developed. (2) Small-scale batches of DTXL-TNPs were prepared and evaluated with respect to drug load and encapsulation efficiency, particle size distribution and reproducibility, and in vitro release kinetics. (3) DTXL-TNPs with promising physicochemical properties were evaluated with respect to PK in rats, and tolerability, tumor accumulation, and efficacy in tumor-bearing mouse models. In vivo results informed additional formulation optimization and led to selection of DTXL-TNP composition and process. (4) The DTXL-TNP manufacturing process was scaled up and used to manufacture sterile clinical supplies under cGMPs. (**D**) Range of formulation parameters and NP physicochemical properties evaluated during development of DTXL-TNPs, with optimized DTXL-TNP parameters and target properties indicated by the red dotted line.

RESEARCH ARTICLE

spanning more than 100 distinct NP compositions. DTXL-TNP was designed to target the extracellular domain of prostate-specific membrane antigen (PSMA) using a stable small-molecule ligand displayed on the NP surface. PSMA is a clinically validated transmembrane receptor that is overexpressed on the surface of prostate cancer cells and in the neovasculature of nearly all nonprostate solid tumors (*22*). DTXL is a semisynthetic taxane approved for treatment of several major solid tumor cancers, including breast, prostate, lung, gastric, and head and neck (*23*). The in vivo performance of DTXL-TNP, including PK, biodistribution, and tolerability, was tested in multiple animal species (rat, mouse, and cynomolgus monkey). In addition, efficacy in mouse xenograft models of prostate, lung, and breast cancer was examined. Finally, we initiated a clinical study to evaluate the tolerability and PK of DTXL-TNP in humans and to obtain a preliminary assessment of its efficacy in patients with advanced and metastatic solid tumor cancers.

RESULTS

Combinatorial synthesis and optimization of targeted DTXL-TNPs

The TNPs were composed of (i) a hydrophobic biodegradable polymeric core that allowed for the encapsulation and controlled release of drugs, (ii) a hydrophilic corona that protects the NP from immune surveillance, and (iii) a targeting ligand that mediated molecular interactions between NPs and PSMA expressed on prostate cancer cells and in nonprostate solid tumor neovasculature (Fig. 1A). The particles consisted primarily of block copolymers of either poly(D,L-lactide) (PLA) or poly(D,L-lactide-*co*-glycolide) (PLGA) and poly(ethylene glycol) (PEG), in which DTXL was physically entrapped without chemical modification. PLA, PLGA, and PEG are widely used in pharmaceutical products and medical devices and have a long history of safe human use (*24*). The NP size fell within the range reported to be retained in the vascular compartment after intravenous administration and to access tumors via their leaky vasculature by means of the enhanced permeability and retention (EPR) effect (<400 nm) (*25*). The NPs were targeted to the extracellular domain of PSMA using *S,S*-2-[3-[5-amino-1-carboxypentyl]-ureido]-pentanedioic acid (ACUPA), a PSMA substrate analog inhibitor (*26*). The ACUPA moiety is a constituent of a prostate cancer–imaging agent currently in clinical development (*27*). In addition, targeting of DTXL-TNPs with ACUPA has been shown to enhance the cytotoxicity of DTXL in a PSMA-expressing prostate cancer cell line (LNCaP) in vitro (*28*).

Here, we prepared a combinatorial library of ACUPA-targeted DTXL TNPs with diverse physicochemical properties using single-step self-assembly processes wherein organic solutions containing DTXL and polymeric components were contacted with an external aqueous phase, resulting in encapsulation of DTXL in solid particles. The library was prepared by combining systematically varied proportions of PLA-PEG polymer conjugated to ACUPA (PLA-PEG-ACUPA), DTXL, and PLA, PLGA, PLA-PEG, and PLGA-PEG copolymers of varying PLA, PLGA, and PEG block lengths and the PLGA ratio of glycolic to lactic acid units (Fig. 1B). These mixtures self-assembled into a core-shell structure with PEG and PEG-ACUPA chains oriented toward the NP surface, as demonstrated previously for similar NPs (*29*, *30*). From enzyme-linked immunosorbent assay (ELISA) and surface plasmon resonance (SPR)–based binding studies, we found that NPs decorated with ACUPA bound tightly to PSMA even at mole fractions of PLA-PEG-ACUPA as low as 0.5% relative to nonfunctionalized polymeric components (Fig. 2).

Our approach for development, optimization, and clinical manufacturing of DTXL-TNP is depicted in Fig. 1C. Initial small-scale studies were carried out to identify a process capable of encapsulating at least 5 weight percent (wt %) DTXL and producing NPs with consistent particle size distribution across batches. Specifically, we evaluated two general processes: nanoprecipitation (in which the organic polymer-drug solution comprised water-miscible solvents, such as acetone or acetonitrile) and nanoemulsion (in which the organic phase consisted of water-immiscible solvents, such as ethyl acetate or dichloromethane). Our studies revealed that a nanoemulsion process using organic phase mixtures of ethyl acetate and benzyl alcohol in ratios between 90:10 and 70:30 (w/w) yielded stable NPs with higher DTXL content compared to nanoprecipitation or nanoemulsion processes using other solvent systems. For example, the DTXL content in NPs produced by nanoemulsion ranged from 2 to 20 wt % depending

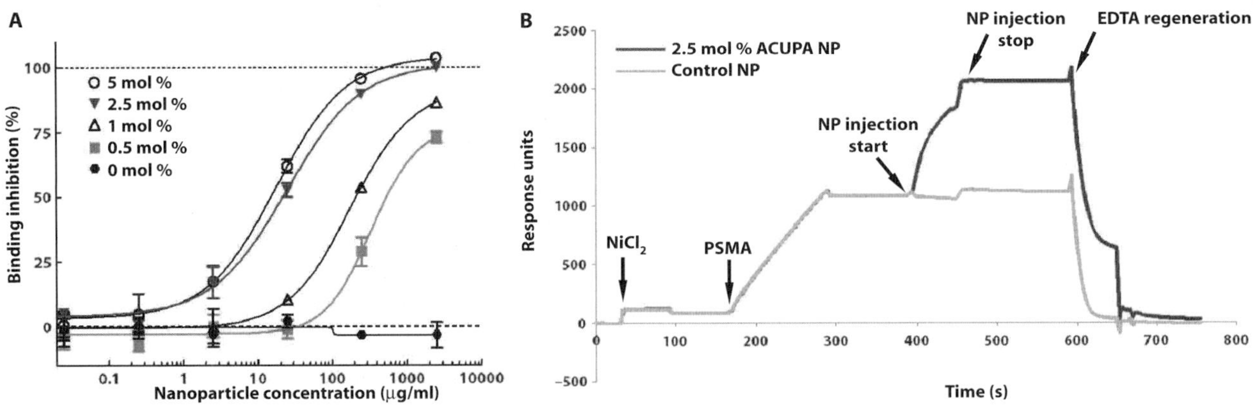

Fig. 2. (**A**) ELISA of TNP binding to PSMA. NPs containing PLA-PEG-ACUPA mole fractions of 0 to 5% were assessed for binding to recombinant human PSMA in a competitive ELISA using a biotinylated ACUPA analog. Data are means ± SD. (**B**) SPR of TNP binding to PSMA. NP samples with or without 2.5 mole percent (mol %) PLA-PEG-ACUPA relative to nonfunctionalized polymeric components were compared with respect to PSMA binding.

on formulation and process parameters, but did not exceed 1 wt % in NPs produced by nanoprecipitation.

We then produced a combinatorial library consisting of more than 100 NP compositions varying with respect to formulation and process parameters (Fig. 1D), including the molecular weight and composition of the polymeric components, the theoretical DTXL content, the targeting ligand content, the concentrations of NP components, and processing times, temperatures, flow rates, and mixing speeds. NPs were characterized with respect to particle size, ligand density, DTXL loading, and DTXL release kinetics. Data for 63 of these compositions, which span the range of formulation and process parameters evaluated, are compiled in table S1.

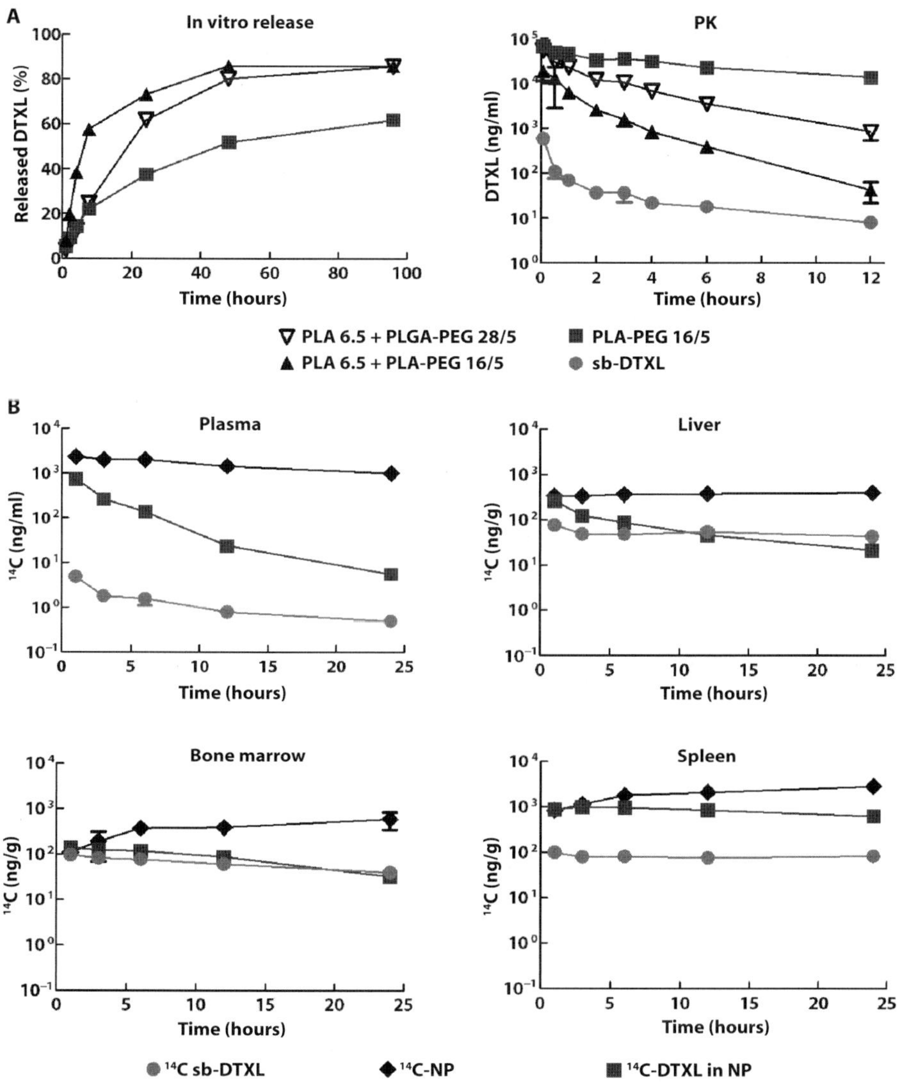

Fig. 3. In vitro release, PK, and biodistribution of DTXL-TNPs. (**A**) In vitro release and PK of DTXL-TNP formulations. For PK, rats (n = 6 per group) were administered a single bolus dose of DTXL-TNPs or sb-DTXL (5 mg/kg) through a lateral tail vein. Serial blood samples were collected from each animal at various times, and total DTXL plasma concentrations were analyzed. NPs contained a mixture of equal amounts of 6.5-kD DL-PLA and a PLGA-PEG copolymer comprising 28-kD PLGA (50:50 lactide/glycolide monomer ratio) and 5-kD PEG (28/5 PLGA-PEG), a mixture of equal amounts of 6.5-kD DL-PLA and a PLA-PEG copolymer comprising 16- and 5-kD PEG (16/5 PLA-PEG), or 16/5 PLA-PEG. (**B**) Biodistribution in rats (n = 3 per group) of sb-DTXL and NPs in plasma, liver, bone marrow, and spleen after a single intravenous dose of ^{14}C-labeled sb-DTXL, ^{14}C-labeled NPs (labeled on the targeting polymer), or ^{14}C-DTXL encapsulated in NPs. In vitro release data are means (n = 2); PK and biodistribution data are means ± SD.

Drug release and PK of DTXL-loaded NPs

DTXL was released in vitro under physiological conditions in a sustained fashion over about 48 hours, with kinetics that depended on the NP composition and physicochemical properties. For example, Fig. 3A shows the in vitro release kinetics of three candidate NPs that varied with respect to their polymeric components and DTXL release rate. The NP composition with the fastest release rate (50% release at 6 hours) contained a mixture of equal amounts of PLA (6.5 kD) and a 28/5 PLGA-PEG copolymer (Table 1). The other two slower-releasing NPs contained either a mixture of equal amounts of PLA (6.5 kD) and a 16/5 PLA-PEG, or 16/5 PLA-PEG only. Among these formulations, the time at which 50% of the encapsulated DTXL was released varied 7.5-fold, ranging from about 6 to 45 hours.

DTXL-TNPs with promising physicochemical and release properties were characterized with respect to PK. Figure 3A shows a typical study in which the same three particle compositions were evaluated in healthy Sprague-Dawley rats (n = 6). A bolus dose of DTXL-TNPs (5 mg/kg) or the commercially available solvent-based DTXL formulation (sb-DTXL, or Taxotere) was injected intravenously, and blood samples were collected at different time points over the course of 12 hours to measure the plasma concentration of DTXL (Fig. 3A). After administration of sb-DTXL, plasma DTXL concentrations rapidly declined as the drug distributed to the peripheral compartment, which was consistent with the reported 0.02-hour distribution half-life ($t_{1/2\alpha}$) of DTXL in rats (*31*). In contrast, DTXL plasma concentrations in animals receiving the same dose of DTXL-TNP remained up to three orders of magnitude higher for the duration of the experiment, yielding areas under the concentration-time curve (AUCs) between 46- and 670-fold higher than sb-DTXL (Table 1). We observed that the rate of decay of DTXL plasma concentration in vivo depended on the NP formulation and correlated with the rate of DTXL release in vitro (for instance, slower DTXL release from NPs in vitro corresponded to slower DTXL clearance in vivo). This indicated that the DTXL PK profile could be controlled by manipulating DTXL-TNP formulation and process parameters.

Overall, DTXL release kinetics and PK were dependent on the type and molecular weight of the polymeric components, with PLA-PEG NPs displaying slower release kinetics than PLGA-PEG, as exemplified in Fig. 3. Inclusion of lower–molecular weight polymers, such as 4.5- or 6.5-kD PLA, resulted in faster release kinetics, and higher PLA/PEG ratios resulted in more rapid clearance of DTXL in vivo, presumably owing to faster particle clearance rates (Fig. 3). Finally, levels of PLA-PEG-ACUPA within the range of 0.5 to 2.5 wt % relative to nonfunctionalized polymeric components imparted increasing PSMA targeting with no additional benefit observed at higher levels (Fig. 2A).

On the basis of these findings, we selected and manufactured for clinical evaluation a DTXL-TNP composition containing 10 wt % DTXL encapsulated in 100-nm NPs (polydispersity index <0.2), with a surface charge (ζ potential) of about −6 mV (Fig. 1D). The optimal DTXL-TNP was composed of a PLA-PEG copolymer comprising 16-kD PLA and 5-kD PEG (16/5 PLA-PEG) and a PLA-PEG-ACUPA targeting polymer also comprising 16-kD PLA and 5-kD PEG, with the PLA-PEG and PLA-PEG-ACUPA copolymers representing 97.5 and 2.5% of the polymer mass, respectively (Fig. 1D).

Tissue distribution of DTXL-TNPs

To characterize the biodistribution of DTXL TNPs in vivo in rats, we examined the plasma and tissue concentrations in the major clearing organs—the liver, spleen, and bone marrow—after intravenous administration of (i) PSMA-targeted NPs with ^{14}C-labeled polymer, (ii) PSMA-targeted NPs with ^{14}C-labeled DTXL, or (iii) ^{14}C-labeled sb-DTXL (Fig. 3B). Circulating NPs with the ^{14}C-labeled polymer exhibited a decrease in plasma concentration of about twofold over the 24-hour period studied. This prolonged circulation time is indicative of limited or delayed particle clearance via the mononuclear phagocyte system, which mediates the rapid clearance of most intravenously administered polymeric NPs (*25*). The plasma profiles for ^{14}C-DTXL encapsulated in NPs and ^{14}C-sb-DTXL were consistent with the PK data presented in Fig. 3A, indicating that encapsulation of DTXL prevents its rapid distribution from the plasma compartment and results in substantially higher plasma DTXL concentrations relative to sb-DTXL. With respect to tissue distribution, lower levels of ^{14}C-labeled NPs were detected in liver and bone marrow relative to plasma, whereas in the spleen, NP concentrations were slightly higher than plasma at 12 and 24 hours (Fig. 3B).

Tolerability of TNPs

To obtain a preliminary characterization of the toxicological properties of NPs exclusive of the known toxicities of the DTXL active ingredient, we performed a single-dose tolerability study in rats of PSMA-targeted and identical nontargeted NPs, both containing no DTXL. Rats received single intravenous doses of NPs at 1000, 1500, or 2000 mg/kg [about 6000, 9000, and 12,000 mg/m^2, respectively, corresponding to up to 20 g of NP in an adult human of average body surface area (*32*)] and were monitored for clinical observations, changes in body weight, and blood chemistry. No abnormal clinical observations or appreciable body weight loss were observed (fig. S1), and blood chemistry parameters determined at study termination (tables S2 and S3) were within normal ranges (*33*). These results indicated that NPs on which the DTXL-TNP formulation was based, including the ACUPA targeting ligand, were well tolerated without clinical observations suggestive of hypersensitivity or anaphylactic reactions, which have been reported after infusion of high doses of particles (*34*).

Drug efficacy and tumor accumulation

To evaluate the antitumor activity of DTXL-TNPs, we performed efficacy studies in mice bearing PSMA-expressing human LNCaP prostate cancer xenografts. Twenty-nine days after inoculation with LNCaP cells, mice were treated every 4 days for a total of four doses with equivalent doses of DTXL-TNP, nontargeted DTXL-NPs (lacking the ACUPA targeting ligand), sb-DTXL, or a 10% sucrose vehicle. Treatment with DTXL-TNPs resulted in tumor regression over the 50-day period of study, outperforming nontargeted DTXL-NPs and sb-DTXL (Fig. 4A). For DTXL-TNP, tumor mass decreased by an average of 26% compared to that of nontargeted DTXL-NPs, which saw a 75% average increase. sb-DTXL–treated tumors increased by 100% over the 50-day study period. DTXL-TNPs appeared to be better tolerated on the basis of minimal body weight loss compared to

Table 1. In vitro release and in vivo PK parameters of DTXL-TNPs. Time to 50% in vitro release was determined on the basis of measurements performed in duplicate. PK parameters are average values.

Major components	sb-DTXL	DTXL-TNP formulation		
	EtOH:polysorbate 80 (1:2)	PLA 6.5/PLGA-PEG 28/5 (1:1)	PLA 6.5/PLA-PEG 16/5 (1:1)	PLA-PEG 16/5
In vitro at 37°C				
Time to 50% release (hours)	—	6	19	45
In vivo (n = 6)				
C_{max} (ng/ml)	595	19,602	54,767	67,267
AUC (hours × ng/ml)	489	22,752	96,844	327,680
Volume of distribution (V_z) (ml/kg)	70,417	665	187	103
Clearance (ml hour^{-1} kg^{-1})	9,289	237	51	11
Terminal half-life ($t_{1/2}$) (hours)	5.2	1.9	2.6	6.6

nontargeted DTXL-NPs and sb-DTXL (Fig. 4B), but this was not statistically significant.

These results suggest that, for long-circulating NPs, active targeting improves tumor response compared to passively targeted NPs. This hypothesis was further tested in tumor models that do not express PSMA. Specifically, we compared the efficacy of DTXL-TNP, nontargeted DTXL-NPs, and sb-DTXL in mouse xenograft models of breast cancer (MX-1) and non–small cell lung cancer (NSCLC) (NCI-H460). Mice were implanted with fragments of MX-1 or NCI-H460 tumors from an in vivo passage. Fourteen days after inoculation, mice bearing MX-1 tumor xenografts were administered 10% sucrose vehicle or DTXL-TNP, sb-DTXL, or DTXL (10 mg/kg) encapsulated in nontargeted NPs, every 4 days for a total of three doses. Mice bearing NCI-H460 tumors received two doses (30 mg/kg) of the same test articles 7 days apart beginning 8 days after inoculation. (These treatments varied across different tumor models owing to differences in sensitivity to DTXL.) DTXL-TNP and the corresponding nontargeted DTXL NPs were equally effective to one another in both models and, in the MX-1 breast cancer model, exhibited significantly enhanced efficacy over sb-DTXL by drastically and durably shrinking tumors, whereas sb-DTXL resulted in only temporary cessation of tumor growth (Fig. 4, C and D). Although not statistically significant, a similar trend was observed in the NCI-H460 lung cancer model. The similar effects of targeted and nontargeted NPs in non–PSMA-expressing tumors were presumably due to increased tumor drug accumulation by means of the EPR effect.

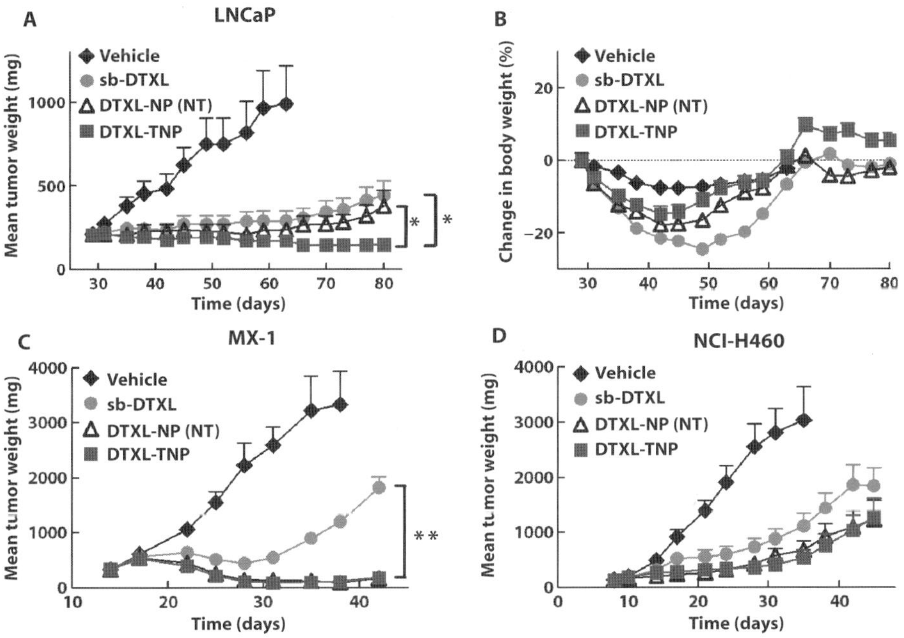

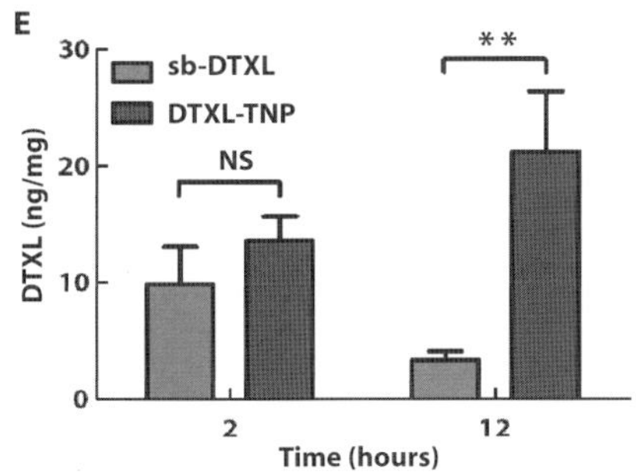

Fig. 4. Therapeutic effect of DTXL-NPs and accumulation of DTXL in mouse tumor xenograft models. (**A**) Efficacy of DTXL-TNP, nontargeted DTXL-NPs [DTXL-NP (NT)], and sb-DTXL in athymic nude mice with LNCaP tumor xenografts. Beginning on day 29 after inoculation, mice were administered 5 mg/kg every 4 days for a total of four doses. Data are mean tumor weights ± SEM ($n = 6$ per treatment group). *$P < 0.05$ (DTXL-TNP versus sb-DTXL and DTXL-TNP versus nontargeted DTXL-NPs at day 80), Wilcoxon test. (**B**) Body weight in the LNCaP efficacy study was monitored as an indicator of tolerability and is presented as mean body weight change ± SD. (**C**) Efficacy of DTXL NPs in MX-1 breast cancer xenografts. Beginning on day 14 after inoculation, mice ($n = 10$ per treatment group) were administered 10 mg/kg every 4 days for a total of three doses. Treatment groups were the same as in (A). **$P < 0.01$ (DTXL-TNP versus sb-DTXL on day 46; DTXL-TNP versus nontargeted DTXL-NPs; $P > 0.05$, not significant), Wilcoxon test. (**D**) Efficacy of DTXL NPs in NCI-H460 NSCLC xenografts. Beginning on day 8 after inoculation, mice ($n = 10$ per treatment group) were administered two doses (30 mg/kg) 7 days apart. Treatment groups were the same as in (A). Comparison of tumor weights for all DTXL treatment groups at day 46 was not significant ($P > 0.05$). (**E**) Accumulation of DTXL-TNP or sb-DTXL in LNCaP prostate xenografts. Mice ($n = 6$ to 10 per treatment group) were administered a single intravenous bolus dose of sb-DTXL or DTXL-TNP (50 mg/kg), tumors were excised 2 or 12 hours after dose administration, and DTXL content was quantified. Data are means ± SEM. **$P < 0.01$. NS, not significant.

To verify the tumor-targeting capabilities of DTXL-TNPs, we measured intratumoral DTXL concentrations in mice bearing LNCaP prostate cancer xenografts at different time points after intravenous administration of DTXL-TNPs and compared to the same dose of sb-DTXL (Fig. 4E). Tumors were excised 2 and 12 hours after dosing, and the concentration of DTXL was determined by liquid chromatography–mass spectrometry (LC-MS). At 2 hours, similar levels of DTXL were found for both DTXL-TNP and sb-DTXL. However, at 12 hours, the DTXL concentration in tumors had decreased appreciably in the sb-DTXL group, whereas the DTXL concentration continued to increase in the DTXL-TNP group such that the concentration of DTXL in the tumor was about sevenfold higher than that in the sb-DTXL group (21.2 ± 5.2 versus 3.30 ± 0.76 ng of DTXL per milligram of tumor mass ± SEM). Overall, these results demonstrate that a combination of long NP circulation time, NP targeting, and controlled release of DTXL markedly increases the amount of DTXL delivered to the tumor.

PK across different animal species

A key challenge commonly encountered in drug discovery is that results obtained with one animal species do not necessarily translate across different species (*32*). Therefore, we compared the PK of DTXL-TNP and sb-DTXL in tumor-bearing mice and in healthy rats and nonhuman primates (Fig. 5A). The PK of DTXL in the TNPs was consistent across species, maintaining substantially higher plasma concentrations compared to sb-DTXL for at least 48 hours. Plasma concentration profiles in all species were consistent with retention of DTXL-TNPs in the plasma compartment and the controlled release of DTXL.

In addition to assessing interspecies PK profiles, we evaluated dose proportionality and the effect of repeat dosing on the PK of DTXL-TNP in cynomolgus monkeys. Both the maximum DTXL plasma concentration (C_{max}) and AUC of DTXL-TNP showed excellent dose proportionality over the range of doses tested ($r^2 > 0.99$) (Fig. 5B). With respect to repeat dosing, DTXL-TNP was administered to cynomolgus monkeys on a weekly schedule and the PK profiles after administration of the first (day 1) and third (day 15) doses were evaluated (Fig. 5C). There was no appreciable change in the PK profile from repeat exposure to DTXL-TNP, indicating that at the dose tested and within the time frame of the study, the PK of DTXL-TNP was not affected by saturation of particle clearance mechanisms and that repeated exposure to DTXL-TNP did not result in any detectable acceleration or inhibition of particle clearance.

The measurement of the plasma DTXL levels in these studies reflects the total DTXL concentration (released DTXL plus DTXL remaining encapsulated in NPs). To characterize the PK of DTXL-TNP, we developed an analytical method to measure the concentration of encapsulated DTXL and then analyzed plasma samples collected from cynomolgus monkeys during the PK study described above. The concentration profiles of total and encapsulated DTXL were nearly superimposable (Fig. 5D), indicating that essentially all of the measured DTXL was encapsulated in NPs. This is consistent with the low toxicity of DTXL-TNP observed in animal models, despite plasma exposure levels of total DTXL two to three orders of magnitude higher than sb-DTXL (Fig. 5A).

PK and initial efficacy in humans

Given the promising and favorable preclinical results obtained with DTXL-TNP, we initiated a phase 1 human clinical trial (NCT01300533) where DTXL-TNP was given by intravenous infusion every 3 weeks to patients with advanced or metastatic cancer. This study is currently in progress and full results will be reported in a subsequent publication. The primary objective of the phase 1 study is to assess the dose-limiting toxicities and to determine the maximum tolerated dose of DTXL-TNP. Secondary goals of the study are to characterize the PK of DTXL-TNP administered on day 1 and to assess preliminary evidence of antitumor activity through imaging evaluation based on Response Evaluation Criteria in Solid Tumors (RECIST) (*35*). In view of the widespread expression of PSMA in solid tumor neovasculature, the activity of sb-DTXL in multiple important cancer indications, and the differentiated efficacy of DTXL-TNP in preclinical models of both prostate and non-prostate solid tumors, we chose to conduct the trial in all solid tumors instead of restricting the population to prostate cancer patients. The clinical trial uses a standard dose-escalation study design in which patients are assigned to cohorts receiving progressively higher doses until a dose is reached at which dose-limiting toxicities are observed. For these studies, DTXL-TNPs were prepared under current Good Manufacturing Practices (cGMPs) (*36*) using a multi-kilogram manufacturing process that was capable of producing clinical and commercial quantities of sterile drug product while maintaining the same physicochemical properties as the NPs produced at the bench scale (fig. S2).

Interim data for patients receiving doses up to 75 mg/m^2 indicate that DTXL-TNP displayed pharmacological properties differentiated from sb-DTXL and consistent with the preclinical data we described

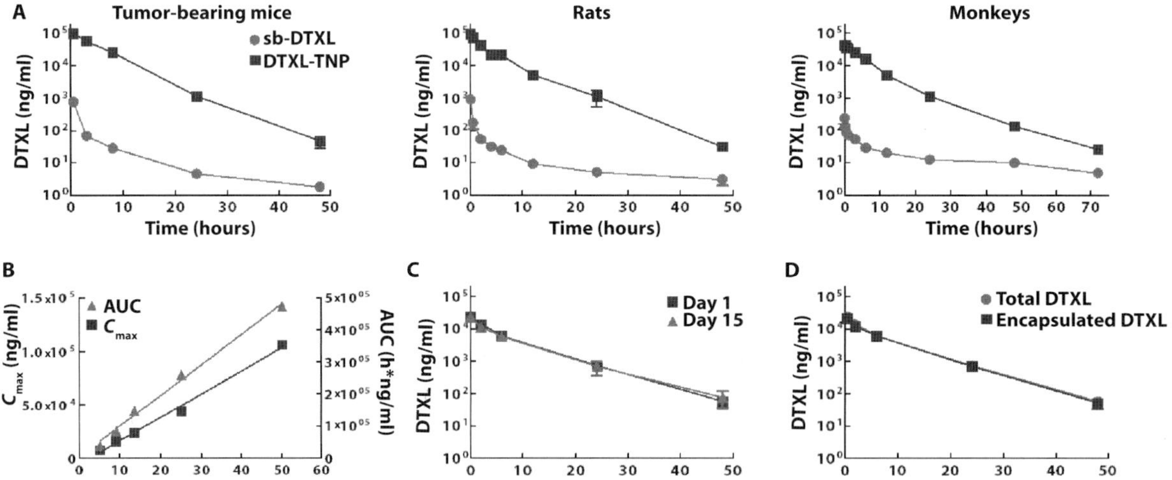

Fig. 5. PK of DTXL-TNP and sb-DTXL in tumor-bearing mice, rats, and cynomolgus monkeys. (**A**) Single-dose intravenous PK studies were performed in tumor-bearing mice (n = 4 per group) administered DTXL or DTXL-TNP (10 mg/kg), Sprague-Dawley rats (n = 4 per group) administered sb-DTXL or DTXL-TNP (5 mg/kg), and cynomolgus monkeys (n = 2 per group) administered sb-DTXL or DTXL-TNP (2.1 mg/kg). (**B**) Dose proportionality of DTXL-TNP PK in cynomolgus monkeys (n = 2 to 4 per dose level) administered doses ranging from 5 to 50 mg/m^2. $r^2 > 0.99$ for C_{max} and AUC versus dose. (**C**) PK of DTXL-TNP in cynomolgus monkeys after repeat dosing (n = 4 per treatment group). Animals were administered DTXL-TNP (13.5 mg/m^2) once a week for 3 weeks. Total DTXL plasma concentrations were quantified after the first and third doses. (**D**) Total and encapsulated DTXL in cynomolgus monkeys receiving DTXL-TNP. Animals (n = 4) were administered a single intravenous dose (13.5 mg/m^2) of DTXL-TNP, and total and encapsulated DTXL concentrations were quantified. For all graphs, data are means ± SD.

for the animal models. Figure 6A shows a comparison of the mean PK profile for DTXL-TNP in three patients versus published data for sb-DTXL in solid tumor patients at a dose of 30 mg/m^2 (*37*). The PK of DTXL-TNP in humans closely resembled the PK observed in preclinical species, in that the plasma levels of DTXL administered as DTXL-TNP were at least two orders of magnitude higher than those of an equivalent dose of sb-DTXL for all time points >1 hour after administration. Furthermore, the high plasma concentrations of total DTXL persisted for at least 48 hours, potentially enabling a larger fraction of the administered DTXL to traffic to tumor sites. Comparison of the DTXL plasma concentration profiles at DTXL-TNP doses ranging from 3.5 to 75 mg/m^2 indicated that PK was essentially dose proportional (correlation coefficients for C_{max} and AUC versus dose were 0.87 and 0.79, respectively; Fig. 6B).

Finally, early signs of DTXL-TNP activity have been observed in patients receiving doses below the dose of 75 mg/m^2 at which sb-DTXL is typically administered (*23*), with two patients exhibiting stable disease at doses of ≤30 mg/m^2. For example, computed tomography (CT) images of a 51-year-old male patient with metastatic cholangiocarcinoma receiving DTXL-TNP (15 mg/m^2) revealed disappearance or shrinkage of multiple lung metastases after two treatment cycles (Fig. 6C). This result is unexpected because sb-DTXL has only minimal activity in cholangiocarcinoma (*38*), and the dose of DTXL-TNP was only 20% of the typical dose for sb-DTXL (75 mg/m^2). Similarly, after two dosing cycles at 30 mg/m^2, we observed 25% shrinkage of a target tonsillar lesion in a 63-year-old patient with tonsillar cancer (Fig. 6D).

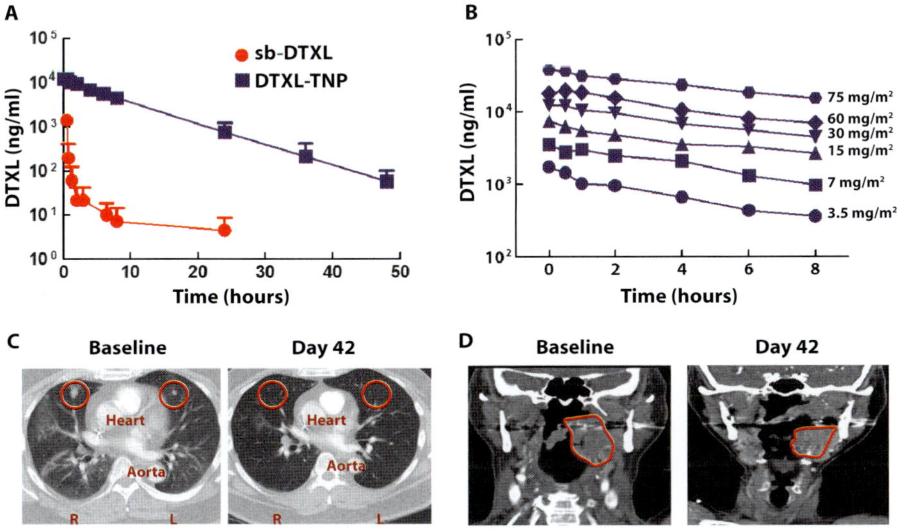

Fig. 6. PK and efficacy of DTXL-TNP in humans. (**A**) PK in patients with advanced solid tumors of DTXL-TNP at a dose of 30 mg/m^2 (n = 3) compared to published sb-DTXL data at the same dose (n = 3). Data are means ± SD. (**B**) PK profiles over the first 8 hours after single-dose administration of DTXL-TNP (3.5 to 75 mg/m^2). n = 1 patient per dose level at 3.5 and 7 mg/m^2, n = 2 at 15 and 75 mg/m^2, and n = 3 at 30 and 60 mg/m^2. C_{max} versus dose r^2 = 0.87; AUC versus dose r^2 = 0.79. (**C**) Axial images from contrast-enhanced CT scans obtained from a 51-year-old male cholangiocarcinoma patient with lung metastases at baseline and at day 42 after two treatment cycles of DTXL-TNP (15 mg/m^2). Red circles indicate locations of metastatic lesions observed in the baseline scan. (**D**) Coronal images from contrast-enhanced CT scans obtained from a 63-year-old male patient with tonsillar cancer at baseline and at day 42 after two treatment cycles of DTXL-TNP (30 mg/m^2). Target tonsillar lesion is outlined in red.

DISCUSSION

It is widely believed that nanotechnology will have a revolutionary impact on the treatment of cancer (*39*). A particularly promising application of nanotechnology in cancer therapy is the use of TNPs to enhance the accumulation and uptake of anticancer agents at specific sites, thus limiting exposure to off-target tissues and improving the therapeutic index of conventional chemotherapeutic drugs and molecularly targeted therapeutics (*40*). Despite this promise, the translation of these technologies from basic research into improving human health has remained elusive (*41*). A major obstacle preventing this translation is the inability to achieve the optimal interplay of physicochemical parameters that confer tumor targeting, evasion of particle clearance mechanisms, and controlled drug release (*42*). Our approach has been to first select a biomaterial with favorable pharmaceutical properties and a long history of safe use in humans and then introduce physicochemical diversity in the formulation by producing NPs of distinct size, targeting ligand density, surface hydrophilicity, drug loading, and drug release kinetics. To form the NPs, we used a reproducible, scalable, one-step self-assembly process that favors the orientation of hydrophilic PEG chains and targeting ligands toward the particle surface (*11*).

The optimal DTXL-TNP composition displayed properties previously unobserved in polymeric NPs, such as a blood circulation half-life approaching 20 hours. Although long-circulating liposomes have been described, they generally lack the particle stability, high drug loading, and controlled-release properties of polymeric NPs (*43*). Moreover, minimal liver accumulation was observed, which was unexpected considering the predominant role of the liver in clearance of previously described polymeric NPs (*25*, *44*–*47*). The in vitro release and in vivo PK results suggest that DTXL-TNPs largely remained in circulation and gradually released DTXL at a rate determined by the NP composition. These characteristics resulted in high initial levels and a gradual decay in the DTXL plasma concentration over time, yielding an increase of up to 1000-fold in the concentration of DTXL in the plasma that was available to access tumor sites compared to a conventional solution formulation of DTXL. These differentiated PK properties are the product of systematic parallel screening of multiple NP parameters to arrive at the optimal combination of precisely controllable physicochemical characteristics.

Among nanotechnology therapeutic products currently approved for clinical use, liposomal drugs and polymer-drug conjugates are two dominant classes (*48*). The first-generation NP therapeutic products for cancer therapy function by accumulating in tumor tissue through the EPR effect and releasing their payload in the extravascular tumor tissue. Tumor tissue accumulation is a passive process that requires a long circulating half-life to facilitate time-dependent extravasation through the leaky tumor microvasculature. This process is largely mediated by the physicochemical properties of the NPs and not by active targeting. However, once particles enter the tumor tissue, their retention and specific uptake by cancer cells could potentially be facilitated by active targeting and receptor-mediated endocytosis. This process can result in higher intracellular drug concentration and increased cellular cytotoxicity. Accordingly, our studies showed that long-circulating DTXL NPs targeting PSMA significantly enhanced the antitumor efficacy when compared to equivalent nontargeted NPs in the PSMA-expressing LNCaP prostate cancer xenograft model.

Here, we show in mice, rats, and nonhuman primates that DTXL administered as DTXL-TNP remains in the plasma at concentrations at least an order of magnitude higher than equal doses of sb-DTXL over periods of at least 24 hours, and that in mouse xenograft models, DTXL-TNP enhances the trafficking of DTXL to solid tumors, thereby improving its efficacy. The interim clinical data for DTXL-TNP are consistent with these observations. The preclinical results suggest that DTXL-TNP may potentiate the current treatment of solid tumor cancers by enhancing the therapeutic index of DTXL through PSMA targeting and controlled drug release of DTXL in the tumor microenvironment. Because PSMA is expressed on prostate tumor cells and on the neovasculature of nonprostate solid tumors, the potential exists for DTXL-TNP to have utility in the treatment of many solid tumors in which DTXL has activity but confers modest clinical benefit.

The key factors that contributed to our ability to translate this targeted nanotherapeutic platform from the bench to human clinical trials included the use of polymeric materials already used in pharmaceutical products and the development of a robust, scalable process for NP formation that enabled optimization to yield a clinical candidate NP with differentiated preclinical pharmacological properties and subsequent cGMP manufacturing in kilogram quantities. To maximize the potential of DTXL-TNP and other similar therapies, a number of areas are worthy of further exploration, including optimization of the dosing schedule, which in the phase 1 trial is based on the 3-week treatment cycle widely used for sb-DTXL, identification of patient populations most likely to benefit from treatment with DTXL-TNP based on, for example, PSMA expression levels, or assessment of the impact of PSMA targeting on activity in DTXL-resistant tumors or in androgen-resistant prostate cancer cells. Looking beyond DTXL-TNP, the presented approach of screening and optimizing physicochemically diverse, targeted NPs comprising clinically validated biomaterials are potentially applicable to other agents, including cytotoxic drugs and molecularly targeted therapeutics, for which preferential tumor accumulation can increase efficacy and decrease toxicity.

MATERIALS AND METHODS

Targeting polymer preparation

The PLA-PEG-ACUPA targeting polymer was synthesized from the allyl-protected ACUPA precursor (*S*)-diallyl 2-(3-((*S*)-1-(allyloxy)-6-amino-1-oxohexan-2-yl)-ureido) pentanedioate, trifluoroacetate salt (Organix Inc.), as described in the Supplementary Materials.

NP preparation

NPs were prepared by a single-emulsion method (otherwise called nanoemulsification) (*24*). Briefly, polymers and drugs were dissolved in an organic phase (typically a mixture of ethyl acetate and benzyl alcohol) and combined with a water phase undergoing high-energy emulsification with a microfluidizer (Microfluidics). The resulting NP suspension was diluted with an aqueous polysorbate 80 solution, purified and concentrated with tangential flow ultrafiltration/diafiltration (Millipore), and then stored as a frozen suspension in a 10% aqueous sucrose solution. NP characterization methods are described in the Supplementary Materials.

In vivo studies

All procedures were approved by the appropriate Institutional Animal Care and Use Committee before initiation. All animals used in PK, biodistribution, and efficacy studies were allowed to acclimate for at least 72 hours in the respective animal facilities before experimentation. Animals were exposed to a 12-hour light/dark cycle and received food and water ad libitum throughout the studies. Dosages of DTXL NPs are expressed as the quantity of DTXL administered. For quantitation of DTXL plasma concentration, blood samples were collected into lithium heparin tubes and plasma was generated. Total DTXL was extracted from plasma with a supported liquid extraction with methyl *tert*-butyl ether and analyzed by LC-MS.

PK studies

Mice. Female athymic nude mice (6 to 8 weeks of age) were implanted with 1×10^6 MX-1 breast cancer cells [National Cancer Institute (NCI)] in 50% growth medium and 50% Matrigel (BD Biosciences) by subcutaneous injection in the right flank. When tumors were about 200 mg in weight, mice were randomized ($n = 4$ per group) such that the mean and median tumor weights were similar between groups. Mice were then administered a single intravenous dose of sb-DTXL or DTXL-TNP (10 mg/kg) diluted in 0.9% sterile saline. At various times, a group of four mice was euthanized and blood collected by cardiac puncture was analyzed for DTXL plasma concentration.

Rats. Male Sprague-Dawley rats (about 6 weeks; $n = 4$ to 6 per group) with indwelling jugular vein cannulae were administered DTXL-TNP or sb-DTXL (5 mg/kg) diluted as appropriate in 0.9% saline. At various times, serial bleeds were collected from the jugular vein cannulae and analyzed for DTXL plasma concentration. NP tissue distribution studies in rats are described in the Supplementary Materials.

Nonhuman primates. Cynomolgus monkeys were administered DTXL-TNP or sb-DTXL by 30-min infusion through a saphenous vein. Doses ranged from 5 to 50 mg/m^2. At various times after dosing, serial bleeds were collected from a femoral vein and analyzed for total DTXL plasma concentration. To evaluate the PK after multiple-dose administrations, animals were administered three weekly doses of DTXL-TNP (13.5 mg/m^2) and serial blood samples were collected after the first and third doses and analyzed for total DTXL concentration. To evaluate the amount of DTXL remaining encapsulated in NPs, plasma samples collected from cynomolgus monkeys administered a single dose (13.5 mg/m^2) were used. Encapsulated DTXL was separated from released DTXL with solid-phase extraction. Plasma samples were applied to hydrophilic-lipophilic balance (HLB) 96-well plates (Waters Corp.) to remove free DTXL before LC-MS analysis.

Efficacy and DTXL tumor accumulation studies

LNCaP prostate tumor xenografts. Male athymic nude mice (about 6 weeks) were subcutaneously inoculated in the right flank with 2×10^7 human LNCaP prostate cancer cells (American Type Culture Collection) resuspended in 50% growth medium and 50% Matrigel. Twenty-nine days after inoculation, the mean tumor weight was about 200 mg. Animals were randomized into groups of six mice such that the mean tumor weights were similar between groups. Mice were administered 10% sucrose (vehicle control) or DTXL-TNP, sb-DTXL, or DTXL (5 mg/kg) encapsulated in nontargeted NPs every 4 days for a total of four doses. Tumor measurements and body weights were monitored twice weekly beginning on the first day of treatment. Tumor volume was determined by caliper measurements with the formula for an ellipsoid sphere ($L \times W^2/2 = \text{mm}^3$), where L and W refer to the larger and smaller perpendicular dimensions collected at each measurement. Tumor volume was converted to tumor weight assuming unit density ($1\ \text{mm}^3 = 1$ mg).

To measure tumor accumulation of DTXL in prostate cancer xenografts, male severe combined immunodeficient (SCID) mice bearing LNCaP prostate tumors ($n = 6$ to 10 per treatment group) were administered a single intravenous dose of DTXL (50 mg/kg) as either DTXL-TNP or sb-DTXL. Mice were euthanized 2 or 12 hours later. The tumors from each group were excised and homogenized. DTXL was extracted with supported liquid extraction and quantified by LC-MS, as described for the PK studies.

MX-1 breast and NCI-H460 NSCLC tumor xenografts. Female athymic nude mice (about 6 weeks) were implanted subcutaneously near the right flank with a 30- to 40-mg fragment of MX-1 human mammary or NCI-H460 NSCLC (American Type Culture Collection) tumors from an in vivo passage. Mean tumor weights were about 300 mg 14 days after MX-1 inoculation and 130 mg 8 days after NCI-H460 inoculation. Animals were randomized into groups of 10 mice such that the mean tumor weight was similar between groups. Mice were administered 10% sucrose (vehicle control) or DTXL-TNP, sb-DTXL, or DTXL encapsulated in nontargeted NPs. Mice bearing MX-1 tumors received doses of 10 mg/kg every 4 days for a total of three doses. NCI-H460 mice received two doses (30 mg/kg) 7 days apart. Tumor measurements and body weights were monitored twice weekly beginning on the first day of treatment. Tumor weight was determined as described for LNCaP xenograft studies.

Human clinical study

DTXL-TNP (designated BIND-014) is currently undergoing evaluation in a phase 1 clinical trial (NCT01300533), where DTXL-TNP is given by intravenous infusion every 3 weeks to cancer patients. The main eligibility criteria are ≥18 years old; advanced or metastatic cancer for which no standard or curative therapy exists; measurable or evaluable disease per RECIST version 1.1 (*35*); Eastern Cooperative Oncology Group performance status of 0 or 1; and life expectancy of >12 weeks. The clinical trial uses a standard dose-escalation design in which patients are assigned to cohorts receiving progressively higher doses until a dose is reached at which dose-limiting toxicities are observed. Single-patient cohorts are enrolled at low-dose levels; subsequent dose levels are enrolling three-patient cohorts. Blood samples for PK analysis are analyzed for total DTXL concentration with LC-MS. All patients provide written informed consent before participation in the study.

PK and statistical analyses

PK parameters were assessed with Phoenix WinNonlin (version 6.0). Statistical comparison of efficacy and tumor accumulation data used the Wilcoxon/Kruskal-Wallis test (JMP 9, SAS Institute Inc.).

SUPPLEMENTARY MATERIALS

www.sciencetranslationalmedicine.org/cgi/content/full/4/128/128ra39/DC1
Materials and Methods
Fig. S1. Tolerability of targeted and nontargeted NPs in rats.
Fig. S2. Clinical-scale manufacturing of DTXL-TNPs.
Table S1. Combinatorial synthesis of DTXL-TNPs.
Table S2. Clinical chemistry measurements in rats ($n = 3$ per group) treated with nontargeted placebo NPs.
Table S3. Clinical chemistry measurements in rats ($n = 2$ to 3 per group) treated with PSMA-targeted placebo NPs.

REFERENCES AND NOTES

1. K. Strebhardt, A. Ullrich, Paul Ehrlich's magic bullet concept: 100 years of progress. *Nat. Rev. Cancer* **8**, 473–480 (2008).
2. B. A. Chabner, T. G. Roberts Jr., Timeline: Chemotherapy and the war on cancer. *Nat. Rev. Cancer* **5**, 65–72 (2005).
3. D. Peer, J. M. Karp, S. Hong, O. C. Farokhzad, R. Margalit, R. Langer, Nanocarriers as an emerging platform for cancer therapy. *Nat. Nanotechnol.* **2**, 751–760 (2007).
4. C. Sawyers, Targeted cancer therapy. *Nature* **432**, 294–297 (2004).
5. M. R. Green, Targeting targeted therapy. *N. Engl. J. Med.* **350**, 2191–2193 (2004).
6. O. C. Farokhzad, R. Langer, Impact of nanotechnology on drug delivery. *ACS Nano* **3**, 16–20 (2009).
7. J. Shi, Z. Xiao, N. Kamaly, O. C. Farokhzad, Self-assembled targeted nanoparticles: Evolution of technologies and bench to bedside translation. *Acc. Chem. Res.* **44**, 1123–1134 (2011).
8. T. D. Heath, R. T. Fraley, D. Papahdjopoulos, Antibody targeting of liposomes: Cell specificity obtained by conjugation of F(ab')2 to vesicle surface. *Science* **210**, 539–541 (1980).
9. J. S. Murday, R. W. Siegel, J. Stein, J. F. Wright, Translational nanomedicine: Status assessment and opportunities. *Nanomedicine* **5**, 251–273 (2009).
10. L. Y. Chou, K. Ming, W. C. Chan, Strategies for the intracellular delivery of nanoparticles. *Chem. Soc. Rev.* **40**, 233–245 (2011).
11. F. Gu, L. Zhang, B. A. Teply, N. Mann, A. Wang, A. F. Radovic-Moreno, R. Langer, O. C. Farokhzad, Precise engineering of targeted nanoparticles by using self-assembled biointegrated block copolymers. *Proc. Natl. Acad. Sci. U.S.A.* **105**, 2586–2591 (2008).
12. O. C. Farokhzad, J. Cheng, B. A. Teply, I. Sherifi, S. Jon, P. W. Kantoff, J. P. Richie, R. Langer, Targeted nanoparticle-aptamer bioconjugates for cancer chemotherapy in vivo. *Proc. Natl. Acad. Sci. U.S.A.* **103**, 6315–6320 (2006).
13. S. D. Perrault, C. Walkey, T. Jennings, H. C. Fischer, W. C. Chan, Mediating tumor targeting efficiency of nanoparticles through design. *Nano Lett.* **9**, 1909–1915 (2009).
14. C. Salvador-Morales, L. Zhang, R. Langer, O. C. Farokhzad, Immunocompatibility properties of lipid–polymer hybrid nanoparticles with heterogeneous surface functional groups. *Biomaterials* **30**, 2231–2240 (2009).
15. J. Cheng, B. A. Teply, I. Sherifi, J. Sung, G. Luther, F. X. Gu, E. Levy-Nissenbaum, A. F. Radovic-Moreno, R. Langer, O. C. Farokhzad, Formulation of functionalized PLGA–PEG nanoparticles for in vivo targeted drug delivery. *Biomaterials* **28**, 869–876 (2007).
16. A. E. Nel, L. Mädler, D. Velegol, T. Xia, E. M. Hoek, P. Somasundaran, F. Klaessig, V. Castranova, M. Thompson, Understanding biophysicochemical interactions at the nano-bio interface. *Nat. Mater.* **8**, 543–557 (2009).
17. D. J. Siegwart, K. A. Whitehead, L. Nuhn, G. Sahay, H. Cheng, S. Jiang, M. Ma, A. Lytton-Jean, A. Vegas, P. Fenton, C. G. Levins, K. T. Love, H. Lee, C. Cortez, S. P. Collins, Y. F. Li, J. Jang, W. Querbes, C. Zurenko, T. Novobrantseva, R. Langer, D. G. Anderson, Combinatorial synthesis of chemically diverse core-shell nanoparticles for intracellular delivery. *Proc. Natl. Acad. Sci. U.S.A.* **108**, 12996–13001 (2011).
18. R. Weissleder, K. Kelly, E. Y. Sun, T. Shtatland, L. Josephson, Cell-specific targeting of nanoparticles by multivalent attachment of small molecules. *Nat. Biotechnol.* **23**, 1418–1423 (2005).
19. H. Wang, K. Liu, K. J. Chen, Y. Lu, S. Wang, W. Y. Lin, F. Guo, K. Kamei, Y. C. Chen, M. Ohashi, M. Wang, M. A. Garcia, X. Z. Zhao, C. K. Shen, H. R. Tseng, A rapid pathway toward a superb gene delivery system: Programming structural and functional diversity into a supramolecular nanoparticle library. *ACS Nano* **4**, 6235–6243 (2010).
20. K. T. Love, K. P. Mahon, C. G. Levins, K. A. Whitehead, W. Querbes, J. R. Dorkin, J. Qin, W. Cantley, L. L. Qin, T. Racie, M. Frank-Kamenetsky, K. N. Yip, R. Alvarez, D. W. Sah, A. de Fougerolles, K. Fitzgerald, V. Koteliansky, A. Akinc, R. Langer, D. G. Anderson, Lipid-like materials for low-dose, in vivo gene silencing. *Proc. Natl. Acad. Sci. U.S.A.* **107**, 1864–1869 (2010).
21. A. Akinc, A. Zumbuehl, M. Goldberg, E. S. Leshchiner, V. Busini, N. Hossain, S. A. Bacallado, D. N. Nguyen, J. Fuller, R. Alvarez, A. Borodovsky, T. Borland, R. Constien, A. de Fougerolles, J. R. Dorkin, K. Narayanannair Jayaprakash, M. Jayaraman, M. John, V. Koteliansky, M. Manoharan,

RESEARCH ARTICLE

L. Nechev, J. Qin, T. Racie, D. Raitcheva, K. G. Rajeev, D. W. Sah, J. Soutschek, I. Toudjarska, H. P. Vornlocher, T. S. Zimmermann, R. Langer, D. G. Anderson, A combinatorial library of lipid-like materials for delivery of RNAi therapeutics. *Nat. Biotechnol.* **26**, 561–569 (2008).
22. A. K. Rajasekaran, G. Anilkumar, J. J. Christiansen, Is prostate-specific membrane antigen a multifunctional protein? *Am. J. Physiol. Cell Physiol.* **288**, C975–C981 (2005).
23. *Taxotere Prescribing Information* (sanofi-aventis U.S. LLC, Bridgewater, NJ, 2010).
24. K. Avgoustakis, Pegylated poly(lactide) and poly(lactide-co-glycolide) nanoparticles: Preparation, properties and possible applications in drug delivery. *Curr. Drug Deliv.* **1**, 321–333 (2004).
25. F. Alexis, E. Pridgen, L. K. Molnar, O. C. Farokhzad, Factors affecting the clearance and biodistribution of polymeric nanoparticles. *Mol. Pharm.* **5**, 505–515 (2008).
26. K. P. Maresca, S. M. Hillier, F. J. Femia, D. Keith, C. Barone, J. L. Joyal, C. N. Zimmerman, A. P. Kozikowski, J. A. Barrett, W. C. Eckelman, J. W. Babich, A series of halogenated heterodimeric inhibitors of prostate specific membrane antigen (PSMA) as radiolabeled probes for targeting prostate cancer. *J. Med. Chem.* **52**, 347–357 (2009).
27. S. M. Hillier, K. P. Maresca, F. J. Femia, J. C. Marquis, C. A. Foss, N. Nguyen, C. N. Zimmerman, J. A. Barrett, W. C. Eckelman, M. G. Pomper, J. L. Joyal, J. W. Babich, Preclinical evaluation of novel glutamate-urea-lysine analogues that target prostate-specific membrane antigen as molecular imaging pharmaceuticals for prostate cancer. *Cancer Res.* **69**, 6932–6940 (2009).
28. S. S. Chandran, S. R. Banerjee, R. C. Mease, M. G. Pomper, S. R. Denmeade, Characterization of a targeted nanoparticle functionalized with a urea-based inhibitor of prostate-specific membrane antigen (PSMA). *Cancer Biol. Ther.* **7**, 974–982 (2008).
29. T. Riley, C. R. Heald, S. Stolnik, M. C. Garnett, L. Illum, S. S. Davis, S. M. King, R. K. Heenan, S. C. Purkiss, R. J. Barlow, P. R. Gellert, C. Washington, Core–shell structure of PLA–PEG nanoparticles used for drug delivery. *Langmuir* **19**, 8428–8435 (2003).
30. P. M. Valencia, M. H. Hanewich-Hollatz, W. Gao, F. Karim, R. Langer, R. Karnik, O. C. Farokhzad, Effects of ligands with different water solubilities on self-assembly and properties of targeted nanoparticles. *Biomaterials* **32**, 6226–6233 (2011).
31. M. C. Bissery, G. Nohynek, G. J. Sanderink, F. Lavelle, Docetaxel (Taxotere): A review of preclinical and clinical experience. Part I: Preclinical experience. *Anticancer Drugs* **6**, 339–355, 363–368 (1995).
32. S. Reagan-Shaw, M. Nihal, N. Ahmad, Dose translation from animal to human studies revisited. *FASEB J.* **22**, 659–661 (2008).
33. L. E. Lillie, N. J. Temple, L. Z. Florence, Reference values for young normal Sprague-Dawley rats: Weight gain, hematology and clinical chemistry. *Hum. Exp. Toxicol.* **15**, 612–616 (1996).
34. L. Baranyi, J. Szebeni, S. Savay, M. Bodo, B. Milan, T. B. Bentley, R. Bunger, C. R. Alving, Complement-dependent shock and tissue damage induced by intravenous injection of cholesterol-enriched liposomes in rats. *J. Appl. Res.* **3**, 221–231 (2003).
35. E. A. Eisenhauer, P. Therasse, J. Bogaerts, L. H. Schwartz, D. Sargent, R. Ford, J. Dancey, S. Arbuck, S. Gwyther, M. Mooney, L. Rubinstein, L. Shankar, L. Dodd, R. Kaplan, D. Lacombe, J. Verweij, New response evaluation criteria in solid tumours: Revised RECIST guideline (version 1.1). *Eur. J. Cancer* **45**, 228–247 (2009).
36. Food and Drug Administration, HHS, Current good manufacturing practice and investigational new drugs intended for use in clinical trials. Final rule. *Fed. Regist.* **73**, 40453–40463 (2008).
37. E. G. Chiorean, J. M. Porter, A. E. Foster, A. S. Al Omari, C. A. Yoder, K. L. Fife, R. M. Strother, D. J. Murry, M. Yu, D. R. Jones, C. J. Sweeney, A phase I and pharmacokinetic trial of erlotinib in combination with weekly docetaxel in patients with taxane-naive malignancies. *Clin. Cancer Res.* **14**, 1131–1137 (2008).
38. R. Pazdur, M. E. Royce, G. I. Rodriguez, D. A. Rinaldi, Y. Z. Patt, P. M. Hoff, H. A. Burris, Phase II trial of docetaxel for cholangiocarcinoma. *Am. J. Clin. Oncol.* **22**, 78–81 (1999).
39. M. Ferrari, Cancer nanotechnology: Opportunities and challenges. *Nat. Rev. Cancer* **5**, 161–171 (2005).
40. R. A. Petros, J. M. DeSimone, Strategies in the design of nanoparticles for therapeutic applications. *Nat. Rev. Drug Discov.* **9**, 615–627 (2010).
41. R. K. Jain, T. Stylianopoulos, Delivering nanomedicine to solid tumors. *Nat. Rev. Clin. Oncol.* **7**, 653–664 (2010).
42. D. E. Owens III, N. A. Peppas, Opsonization, biodistribution, and pharmacokinetics of polymeric nanoparticles. *Int. J. Pharm.* **307**, 93–102 (2006).
43. L. Zhang, J. M. Chan, F. X. Gu, J. W. Rhee, A. Z. Wang, A. F. Radovic-Moreno, F. Alexis, R. Langer, O. C. Farokhzad, Self-assembled lipid–polymer hybrid nanoparticles: A robust drug delivery platform. *ACS Nano* **2**, 1696–1702 (2008).
44. Y. Yamamoto, Y. Nagasaki, Y. Kato, Y. Sugiyama, K. Kataoka, Long-circulating poly(ethylene glycol)–poly(D,L-lactide) block copolymer micelles with modulated surface charge. *J. Control Release* **77**, 27–38 (2001).
45. H. Cabral, Y. Matsumoto, K. Mizuno, Q. Chen, M. Murakami, M. Kimura, Y. Terada, M. R. Kano, K. Miyazono, M. Uesaka, N. Nishiyama, K. Kataoka, Accumulation of sub-100 nm polymeric micelles in poorly permeable tumours depends on size. *Nat. Nanotechnol.* **6**, 815–823 (2011).
46. E. Blanco, E. A. Bey, C. Khemtong, S. G. Yang, J. Setti-Guthi, H. Chen, C. W. Kessinger, K. A. Carnevale, W. G. Bornmann, D. A. Boothman, J. Gao, b-Lapachone micellar nanotherapeutics for non–small cell lung cancer therapy. *Cancer Res.* **70**, 3896–3904 (2010).
47. N. Bertrand, J. C. Leroux, The journey of a drug-carrier in the body: An anatomo-physiological perspective. *J. Control Release* 10.1016/j.jconrel.2011.09.098 (2011).
48. K. K. Jain, Advances in the field of nanooncology. *BMC Med.* **8**, 83 (2010).

Acknowledgments: We thank P. M. Valencia for assistance with drafting and critical review of this manuscript, and R. Korn for assistance in reviewing CT scans. **Funding:** BIND received funding from NCI Small Business Innovation Research Contract No. HHSN261200700060C and National Institute of Standards and Technology Advanced Technology Program Award No. 70NANB7H7021. **Author contributions:** J.H. led the DTXL-TNP development program and performed study designs and interpretation of the data; D.V.H. is the principal investigator on the DTXL-TNP clinical study; J. Summa was the project leader and oversaw design and conduct of mouse xenograft studies; E.S. designed the clinical study; M.M.A., J.A., and T.C. synthesized the polymers; B.R., D.D.W., M. Figueiredo, and Y.H.S. performed the combinatorial optimization of DTXL-TNPs; A.S., G.T., M. Figa, and T.V.G.H. developed and manufactured the DTXL-TNPs; K.M. and A.H. developed and performed SPR analyses; J.J.S. and E.A. developed and performed PSMA binding ELISA analyses; E.P. and D.T. performed analytical characterization of DTXL-TNPs; S.L. designed and oversaw animal studies and development of the encapsulated DTXL bioanalytical method, and performed PK analyses and writing of the manuscript; J.W. performed interpretation of data and statistical analysis; P.L. is an investigator on the clinical study; P.W.K., N.H.B., and C.S. were scientific advisors on the preclinical animal models and clinical development strategy; O.C.F. and R.L. developed the combinatorial NP optimization platform and performed data interpretation and writing of the manuscript; S.Z. performed study design, data interpretation, and writing of the manuscript. **Competing interests:** E.S., M.M.A., J.A., B.R., D.D.W., M. Figueiredo, Y.H.S., J. Summa, A.S., G.T., M. Figa, T.V.G.H., K.M., A.H., J.J.S., E.A., E.P., D.T., S.L., J.W., S.Z., and J.H. are employees of BIND Biosciences, a biotechnology company developing nanoparticle therapeutics. P.W.K., N.H.B., C.S., O.C.F., and R.L. disclose financial interest in BIND Biosciences: P.W.K., N.H.B., and C.S. serve as members of its Scientific and Clinical Advisory Boards. O.C.F. and R.L. cofounded BIND and serve as members of its Board of Directors and Scientific Advisory Board. No aspect of this research was carried out in the laboratories of O.C.F. and R.L., and no academic research grants were used to support this research and development effort. **Data and materials availability:** All reasonable requests for collaboration involving materials used in the research will be fulfilled provided that a written agreement is executed in advance between BIND Biosciences Inc. and the requester (and his or her affiliated institution). Such inquiries or requests for additional data should be directed to the corresponding authors.

Submitted 27 December 2011
Accepted 16 March 2012
Published 4 April 2012
10.1126/scitranslmed.3003651

Citation: J. Hrkach, D. Von Hoff, M. M. Ali, E. Andrianova, J. Auer, T. Campbell, D. De Witt, M. Figa, M. Figueiredo, A. Horhota, S. Low, K. McDonnell, E. Peeke, B. Retnarajan, A. Sabnis, E. Schnipper, J. J. Song, Y. H. Song, J. Summa, D. Tompsett, G. Troiano, T. Van Geen Hoven, J. Wright, P. LoRusso, P. W. Kantoff, N. H. Bander, C. Sweeney, O. C. Farokhzad, R. Langer, S. Zale, Preclinical development and clinical translation of a PSMA-targeted docetaxel nanoparticle with a differentiated pharmacological profile. *Sci. Transl. Med.* **4**, 128ra39 (2012).

Novel Biomaterials

Poly(iminocarbonates) as potential biomaterials

Joachim Kohn and Robert Langer*,†

*Department of Applied Biological Sciences and Whitaker College of Health Sciences, Massachusetts Institute of Technology, Cambridge, MA 02139, USA and *Department of Surgery, Boston Children's Hospital, Boston, MA 02115, USA*

(Received 5 August 1985; accepted 28 September 1985)

Poly(iminocarbonates), based on diphenols such as hydroquinone or Bisphenol A were synthesized. Poly(iminocarbonates) degraded under physiological conditions. Compression moulded discs of poly-(hydroquinone–iminocarbonate) showed near zero-order release for low loadings (1%, w/w) of Eosin Y. Poly(Bisphenol A–iminocarbonate) formed transparent films and strong fibres. It is a tough polymer with high tensile strength. Under physiological conditions complete erosion of a thin film of poly(Bisphenol A–iminocarbonate) required about 200 d. The release profile for *p*-nitroaniline from solvent cast films of poly(Bisphenol A–iminocarbonate) was characterized by a lag period (20 d), followed by near zero-order release for 90 d. Initial toxicological studies revealed no inflammatory cell response after implantation in the corneas of rabbits or subcutaneously in mice.

Keywords: Polymers, biodegradable polymer, Bisphenol A, controlled release, drug delivery, hydroquinone, poly(iminocarbonate)

During the last decade, major research efforts have been directed towards the development of biodegradable polymers for medical applications[1]. Such polymers must not only be biocompatible and nontoxic, but many practical applications require a variety of good mechanical properties, such as tensile strength and toughness, as well as solubility in organic solvents, thermal stability in the molten state, and preferably low crystallinity and a low glass transition temperature (*Tg*). Unfortunately it has been difficult to find nontoxic, biodegradable polymers which possess the required mechanical properties, and therefore the clinical implementation of such devices as biodegradable controlled release systems or temporary vascular grafts has been slow as compared to the rapidly increasing use of biostable polymers in medicine. Hence there is a need to identify new biodegradable polymers and examine them for their suitability for medical applications. In addition to the obvious need to find a nontoxic material we placed special emphasis in our search for possible candidates on high tensile strength, toughness, low crystallinity and easy film and fibre formation, since these properties would ultimately determine the range of potential applications.

In this paper we report initial data on the chemical, mechanical, and toxicological properties of poly(iminocarbonates) in general and on poly(Bisphenol A‡-iminocarbonate) in particular. The poly(iminocarbonates) are a relatively new and scarcely known class of hydrolytically labile polymers which have so far not found any industrial application.

†To whom all correspondence should be addressed.

‡Bisphenol A (abbreviated BPA) is the common name for 2,2-di-(4′-hydroxyphenyl) propane.

As shown in *Figure 1*, the iminocarbonate linkage is structurally derived from the carbonate bond by replacing the carbonyl oxygen by an imino-group. Iminocarbonates were first synthesized by Sandmeyer in 1886[2], and were later investigated by other researchers[3-6]. These early studies indicated that simple iminocarbonates (such as diethyliminocarbonate) are hydrolysed rapidly both by acid (*Figure 2, Equation 1*) and by base (*Figure 2, Equation 2*). Although the mechanism of hydrolysis is different in acid or base, the final degradation products are in both cases ammonia, carbon dioxide, and the alcohol or phenol from which the iminocarbonate was originally derived[3,5]. Poly(iminocarbonates)[7,8]

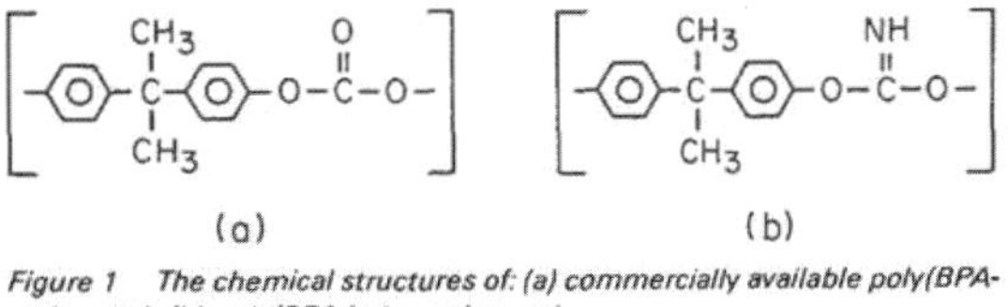

(a) (b)

Figure 1 The chemical structures of: (a) commercially available poly(BPA-carbonate); (b) poly(BPA-iminocarbonate).

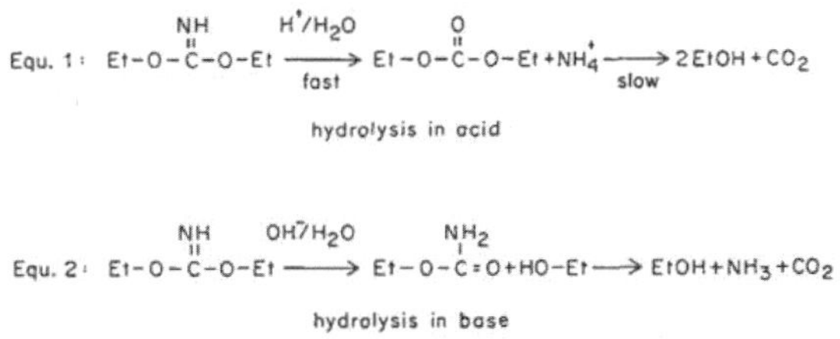

Figure 2 Hydrolysis of diethyliminocarbonate in acidic medium (Equation 1) and in basic medium (Equation 2).

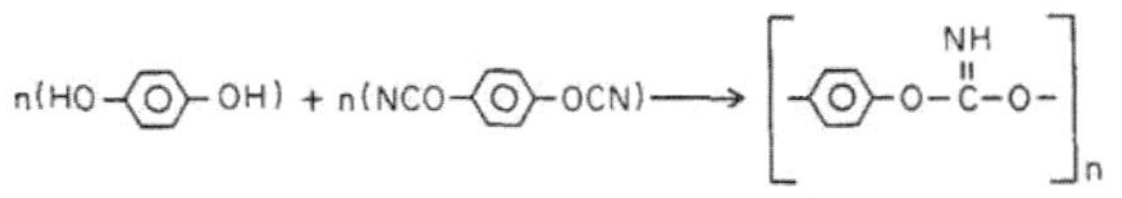

hydroquinone 1,4-dicyanatobenzene poly(hydroquinone-iminocarbonate)

Figure 3 Preparation of poly(hydroquinone-iminocarbonate). In the first step hydroquinone is cyanylated to yield 1,4-dicyanatobenzene. In the second step equimolar quantities of hydroquinone and 1,4-dicyanatobenzene are mixed resulting in rapid polymerization. This basic reaction sequence can be used for the synthesis of a variety of poly(iminocarbonates)[9, 10].

are best synthesized by the reaction sequence shown in *Figure 3:* in the first step cyanate esters are prepared and highly purified, thereby removing all traces of cyanogen halide. In the second step equimolar quantities of dicyanates** and di-alcohols are mixed, resulting in a clean, well-defined polymerization reaction, which was reported by Schminke *et al.*[9, 10] to yield high molecular weight poly(iminocarbonates). However, Schminke's work did not result in the development of commercial products based on poly(iminocarbonates[11], presumably because of the pronounced hydrolytic instability of the iminocarbonate bond. This feature, in conjunction with the excellent mechanical properties of the structurally closely related poly(carbonates)[12, 13], prompted us to investigate the suitability of poly(iminocarbonates) as biodegradable polymers for medical applications.

MATERIALS AND METHODS

All solvents were analytical grade. Hydroquinone (99% purity), potassium-*t*-butoxide, Bisphenol A (> 99% purity, Gold Label) and *p*-nitroaniline (99% purity) were obtained from Aldrich, triethylamine (reagent grade) from Mallinckrodt and cyanogen bromide (reagent grade) from Fisher Scientific. Poly(Bisphenol A-carbonate) [*MW* = 32 000] was obtained from Polysciences Inc., 400 Valley Rd, Warrington, PA. Eosin Y (Certification no. CE-D71) and charcoal (activated Darco G60) were purchased from MCB, Norwood, Ohio.

I.r.-spectroscopy was performed on a Perkin-Elmer Series 1420 dispersive spectrophotometer. For transmission spectra KBr pellets were made, or thin polymer films were directly cast onto NaCl plates from diluted solutions of the polymer in methylene chloride. For 'multiple internal reflectance' spectra the MIR accessory from Perkin-Elmer was used, employing a KRS-5 crystal and a reflectance angle of 45°.

Thermal analysis of polymers was done by 'differential scanning calorimetry' (d.s.c.), employing a Perkin-Elmer DSC 2 Differential scanning calorimeter, equipped with a Perkin-Elmer System 4 Thermal analysis microprocessor controller and the Perkin-Elmer Graphics plotter 2. Average sample size was 10 mg and a heating rate of 20°C/min was employed.

Molecular weight of polymers was determined by vapour pressure osmometry in a Knauer apparatus. For poly(hydroquinone-iminocarbonate) the solvent was dimethylformamide, and the measuring temperature was 90°C. For all other polymers the measurement was performed in chloroform at 45°C.

**Organic cyanate function R-OCN is only minimally toxic.

Mechanical properties of poly(BPA-iminocarbonate) and poly(BPA-carbonate) were determined on an Instron Model 1122 stress-strain tester at an initial strain rate of 0.1 mm mm^{-1} min^{-1}. Polymer films of 0.1 mm thickness were cast in glass dishes from methylene chloride solutions as described below. The films were cut into dog-bone shape in accordance with the ASTM-designation D882-81 (Standard test methods for tensile properties of thin plastic sheeting). Prior to testing, the polymer films were dried for 5 d in vacuum and preconditioned by storage over phosphorous pentoxide (0% humidity) at room temp. for 48 h.

Compression moulding. Compression moulded discs (diam. 1.5 cm; thickness 0.1 cm; wt 250 mg) were prepared on a 'Carver' laboratory press in a stainless steel mould. The appropriate quantities of powdered polymer and sieved particles (mesh 270, < 53 μm) of model drug were intimately mixed by grinding with a mortar and pestle. The mixture was then transferred into the mould and pressed at 9 metric tons. The conditions were 20 min and 100°C for poly(BPA-iminocarbonate), 30 min and 150°C for poly(BPA-carbonate) and 10 min and 70°C for poly(hydroquinone-iminocarbonate).

Solvent casting. Solutions of poly(BPA-iminocarbonate) (20%, w/w) and poly(BPA-carbonate) (10%, w/w) were prepared in methylene chloride and cast at 25 and 37°C respectively in petri dishes made of pyrex glass. Solvent cast release devices were prepared by adding the appropriate amount of model drug to the polymer solutions. Addition of sieved Eosin Y (mesh 270, < 53 μm) and vortexing resulted in a uniform suspension of Eosin Y. Addition of *p*-nitroaniline resulted in a homogeneous solution. After casting, circular discs (diam. 1.0 cm; thickness 0.05–0.1 cm) were excised from the dried films with a cork borer.

Polymer characterization. Determinations of solubility melting range, and thermal stability were performed as described in Reference 14.

Release studies. Solvent cast release devices were initially placed for 24 h into 1 l of distilled water at room temp. to remove dye which had accumulated on the film surface during the casting process. All dye released during this time was recorded as 'initial burst' and was not included in the release profile. Thereafter the devices were transferred into 10 ml of phosphate buffer (0.1 M, pH 7.4) and kept at 37°C. Periodically the phosphate buffer was analysed by spectroscopy for the amount of released dye (at pH 7.4 Eosin Y has a λ_{max} at 518 nm, ε = 79 400 l mol^{-1} cm^{-1}, and *p*-nitroaniline has a λ_{max} at 382 nm, ε = 12 900 l mol^{-1} cm^{-1}). After each such analysis the device was transferred into fresh, sterile phosphate buffer, and occasionally the supernatant was examined microscopically for the presence of microorganisms. Compression moulded devices were treated identically, except for the omission of the initial 24 h washing step.

Polymer degradation was followed by vapour pressure osmometry and by measuring spectrophotometrically the amount of BPA released into the hydrolysis medium (at pH 14, BPA has a λ_{max} at 244 nm, ε = 20 800 l mol^{-1} cm^{-1}).

Synthesis

Purification of hydroquinone. As obtained from the manufacturer hydroquinone (99% purity) is lightly coloured. This

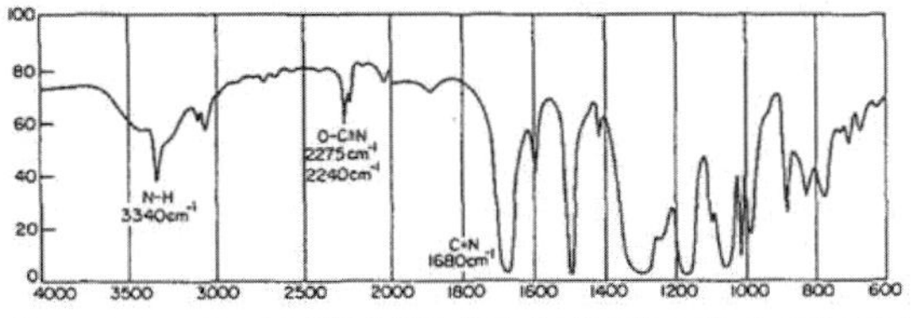

Figure 4 *I.r. spectrum (KBr pellet) of poly(hydroquinone-iminocarbonate).*

colouration was removed as follows: under nitrogen 20 g hydroquinone were dissolved in 200 ml ethyl acetate. The coloured solution was repeatedly treated with 1 g activated carbon until a clear, colourless solution was obtained upon filtration. Colourless crystals of high purity hydroquinone were precipitated from this solution by the slow addition of 800 ml hexane. The purified material was dried in vacuum and used immediately.

Preparation of 1,4-dicyanatobenzene. The general procedure, described in Reference 15, for preparation of cyanate esters was followed. However, the reaction was performed in a closed reaction vessel which was purged with nitrogen in order to exclude moisture and oxygen. Crude 1,4-dicyanatobenzene (84% yield) was obtained as white crystals. The crude material was recrystallized 3 times from boiling carbon tetrachloride (30 g crude/l). Sometimes an insoluble oily residue remained at the bottom of the hot solution of 1,4-dicyanatobenzene in carbon tetrachloride. This residue was removed by decantation. All recrystallizations were done as quickly as possible in order to minimize exposure of the product to high temperatures. The recrystallized material consisted of fine, long, transparent needles, (M.P. = 117–119°C; in Ref. 24: 115–116°C). I.r. absorptions were at 2275, 2237 (CN stretching) and 1169 cm^{-1} (C-O-C stretching)[25]. The elemental analysis was: C = 60.13%, H = 2.65%, N = 17.32% (calculated: C = 60.00%, H = 2.52%, N = 17.49%).

Preparation of poly(hydroquinone-iminocarbonate). The procedure given by Schminke *et al.*[9] was modified as follows: 2.753 (± 0.001) g of hydroquinone (freshly purified) and 4.003 (± 0.001) g of 1,4-dicyanatobenzene were dissolved under nitrogen in 50 ml acetone in a sealed reaction vessel. *Via* a syringe, 250 µl of catalyst solution (56 mg/ml of potassium-*t*-butoxide in *t*-butanol) was added. Within 1 min the solution became noticeably viscous and within 3–5 min a dense, white precipitate formed. After standing for 1 h at 37°C, the precipitate was extensively extracted with acetone (3 times with 250 ml), collected on a Buchner funnel and dried to constant weight *in vacuo*. Poly(hydroquinone-iminocarbonate) was obtained as a white powder (6.2 g, 92% yield). The i.r. spectrum is shown in *Figure 4*.

Preparation of BPA-dicyanate (2,2-di-(4'-cyanatophenyl) propane). BPA-dicyanate was prepared in the same way as 1,4-dicyanatobenzene but no efforts were made to exclude oxygen from the reaction vessel. Crude BPA-dicyanate was obtained as a white, crystalline solid (86% yield). The crude material was recrystallized 3 times from hexane (16 g crude/l) in the same way as already described for 1,4-dicyanatobenzene. The recrystallized material consisted of long needles, (M.P. = 83–83°C; in Ref. 26: 81.5–82°C). The elemental analysis was: C = 73.28%, H = 5.20%, N = 9.93% (calculated: C = 73.37%, H = 5.07%, N = 10.07%).

Preparation of poly(BPA-iminocarbonate). 5.753 (± 0.001) g BPA and 7.014 (± 0.001) g BPA-dicyanate were dissolved at room temp. in 15 ml acetone. With stirring, 250 µl catalyst solution (56 mg/ml of potassium-*t*-butoxide in *t*-butanol) was added. Within 5 min the reaction vessel warmed noticeably and the solution became viscous. After 15 min the reaction mixture completely solidified. It was kept at 37°C for 2 h, thereafter the polymer was extracted 3 times with 250 ml acetone and dried *in vacuo*. Crude poly(BPA-iminocarbonate) was obtained as a tough, transparent solid (12.6 g, 98% yield). The i.r. spectrum is shown in *Figure 6 a*.

Toxicological screening of poly(BPA-iminocarbonate)

Preparation of the polymer for toxicological testing was done as follows: a 10% (w/w) solution of crude poly(BPA-iminocarbonate) in methylene chloride was filtered and

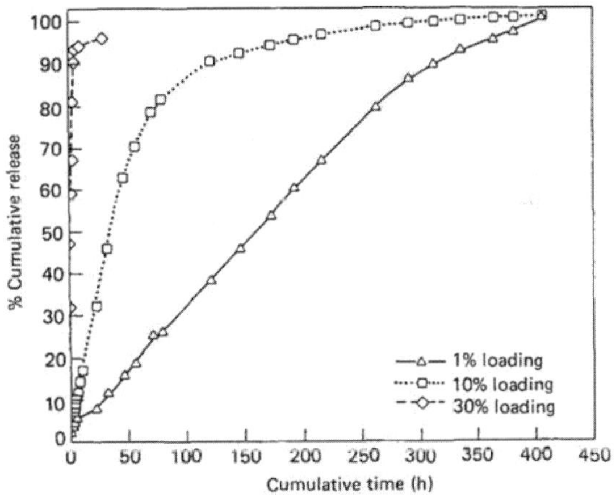

Figure 5 *Release of Eosin Y from compression moulded discs of poly(hydroquinone-iminocarbonate) at 37°C in 0.1 M phosphate buffer (pH 7.4). Eosin Y loadings were 1, 10 and 30% (w/w) as indicated.*

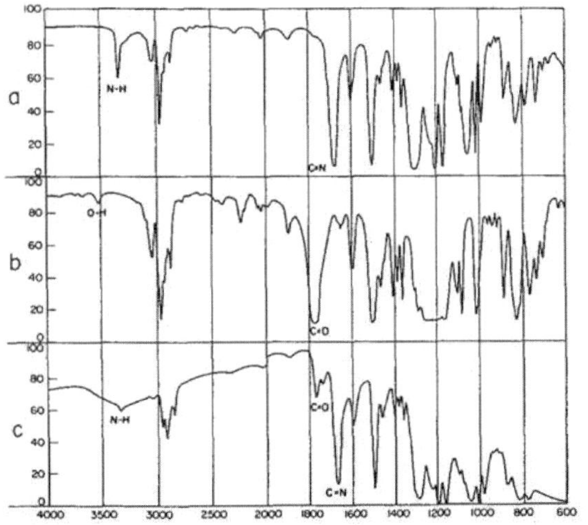

Figure 6 *I.r. spectra of (a) poly(BPA-iminocarbonate) and (b) commercial poly(BPA-carbonate). These spectra were obtained from thin films cast onto NaCl plates; (c) MIR-IR spectrum of the surface of a film of poly(BPA-iminocarbonate) after hydrolysis in 0.1 M phosphate buffer (pH 7.4) for 60 min. The spectrum shows the carbonate band at 1770 cm^{-1}.*

slowly added into a 20-fold excess of isopropanol. Poly(BPA-iminocarbonate) precipitated in large lumps. The precipitated material was collected under sterile conditions and dried in high vacuum to constant weight.

Rabbit eye tests were performed as described previously[16] using solvent cast films (approx. size: 0.1 × 0.1 × 0.05 cm) of poly(BPA-iminocarbonate), which were implanted into the corneas of male New Zealand White rabbits.

For subcutaneous implantation circular discs (diam. 0.3 cm; thickness 0.05 cm; wt 2.5 mg) were cut from solvent cast films of purified poly(BPA-iminocarbonate). The discs were sterilized by short immersion in 70% aqueous isopropanol and implanted subcutaneously in the back of male CD-1 mice (CRBL-Kingston). At specific intervals animals were sacrificed, the residual polymer retrieved, the implantation site was inspected for any pathological changes and the histology of tissues surrounding the implant was examined.

RESULTS AND DISCUSSION

Since hydroquinone is the structurally simplest diphenol we chose poly(hydroquinone-iminocarbonate) (*Figure 3*) as a starting point for our investigations. Upon solution polymerization of hydroquinone and 1,4-dicyanatobenzene, rapid precipitation of a white material was observed. This precipitate was determined to be poly(hydroquinone-iminocarbonate) by i.r. spectroscopy (*Figure 4*) and its molecular weight (number average) was 2000 (± 200) as measured by vapour pressure osmometry, indicating an approx. degree of polymerization of 16. It is possible that the premature precipitation of oligomeric hydroquinone-iminocarbonate prevented the formation of a truly polymeric material. However, no attempt was made to increase the molecular weight by optimizing the synthetic procedure.

A number of properties of this material are summarized in *Table 1*. Due to insolubility in all tested volatile solvents, poly(hydroquinone-iminocarbonate) could not be solvent cast. Likewise its marked thermal degradation in the molten state precluded high temperature fabrication methods such as injection moulding. However, compression moulding proved satisfactory and opaque, slightly brittle discs could be prepared. By blending the sieved polymer powder with up to 30% (w/w) of Eosin Y (which served as the model compound for hydrophilic, low molecular weight drugs), Eosin Y containing discs were prepared. The release characteristics of these discs (*Figure 5*) were strongly dependent on the loading: at a loading of 30% (w/w), the device started to crumble within 3 h and Eosin Y was released rapidly and without correlation to polymer erosion. Interestingly, at low loadings (1%), Eosin Y was released over 400 h at a rate which remained nearly constant with time throughout the entire release process, suggesting that Eosin Y release is not due to a simple diffusional mechanism.

Table 1 Some physical properties of poly(iminocarbonates)

Property	Poly(hydroquinone-iminocarbonate)	Poly(BPA-iminocarbonate)
Appearance	White powder	Tough, transparent solid
Glass transition temperature (Tg)	n/a	75°C
Melting range	Decomposes before melting	Decomposes before melting
Molecular weight (number average)	2000	12 000
Thermal stability	Degrades above 130°C	Degrades above 160°C
Solubility	Soluble in: dimethylsulphoxide, dimethylformamide Insoluble in: water, alcohol, acetone ether, benzene, chlorinated hydrocarbons, hexane	Soluble in: chlorinated hydrocarbons, dimethylformamide, dimethylsulphoxide Partially soluble in: acetone, benzene Insoluble in: water, alcohol, ether, hexane

Next we synthesized poly(BPA-iminocarbonate) (*Figure 1 b*). Some physical and chemical properties of poly(BPA-iminocarbonate) are summarized in *Table 1*. Contrary to poly(hydroquinone-iminocarbonate), poly(BPA-iminocarbonate) is soluble in many organic solvents. Poly(BPA-iminocarbonate) could be solvent cast, yielding clear, transparent films. From concentrated solutions in methylene chloride (30% w/w) fibres could be drawn. The molecular structure of poly(BPA-iminocarbonate) was confirmed by elemental analysis and i.r. spectroscopy. The i.r.-spectrum of a thin film of poly(BPA-iminocarbonate) is shown in *Figure 6 a*. This spectrum is similar to the spectrum of commercially available poly(BPA-carbonate) (*Figure 6 b*) except for the iminocarbonate absorptions at 3340 cm^{-1} (N-H) and 1680 cm^{-1} (C=N).

The stress-strain behaviour of poly(BPA-iminocarbonate) was investigated and compared to the behaviour of a low molecular weight commercial sample of poly(BPA-carbonate) (*MW* = 32 000). Since the stress-strain behaviour of polymers depends to a large extent on the way the samples are prepared, cut and preconditioned prior to measurement, care was taken to prepare samples of both polymers in an identical fashion and to test them under the same conditions. The results of these experiments are summarized in *Table 2*. These data indicate that poly(BPA-iminocarbonate) is a tough, strong polymer with exceptionally high tensile strength. As a general rule, both toughness and tensile strength increase with increasing molecular weight. Therefore it is encouraging that poly(BPA-iminocarbonate) with a molecular weight of only 12 000 (degree of polymerization: 45) would have mechanical properties which compare favourably to those of commercial poly(BPA-carbonate) resins of much higher molecular

Table 2 Mechanical properties of poly(BPA-iminocarbonate)

Property	Poly(BPA-iminocarbonate)[a] (*MW* = 12 000)	Poly(BPA-carbonate) (*MW* = 32 000[a])	(*MW* = 70 000–80 000[b])
Modulus of elasticity	7875 kg/cm^2	9421 kg/cm^2	22 500 kg/cm^2
Yield stress	492 kg/cm^2	406 kg/cm^2	-
Tensile strength (nominal)	529 kg/cm^2	486 kg/cm^2	738 kg/cm^2
Elongation at break	38%	109%	60–100%
Tensile energy to break	16×10^6 J/m^3	34×10^6 J/m^3	-

[a]Solvent cast films, prepared and tested as described in Materials and Methods.
[b]Literature values[12, 13] reported for commercial grade resins, having a molecular weight of about 70 000–80 000.

weight. In this context it is interesting to note that for poly(lactic acid) (MW = 13 000) a tensile strength of 220 kg/cm^2 has been reported[17]. Although not directly comparable to our measured value, it seems that poly(BPA-iminocarbonate) has a higher tensile strength than poly (lactic acid), which is used in biodegradable sutures.

Monomeric iminocarbonates were previously shown[2-5] to have two different degradation pathways, depending on the pH of the surrounding medium (Equations 1 and 2 in *Figure 2*). However, no information was available pertaining to the mode of hydrolysis of poly(iminocarbonates) at neutral pH. Therefore an initial study of the hydrolysis of poly(BPA-iminocarbonate) was performed, employing multiple internal reflectance-i.r.-spectroscopy (MIR-IR). Freshly cast films of poly(BPA-iminocarbonate) had MIR-spectra indistinguishable from regular transmission i.r.-spectra. However, after exposure of these films for only 60 min to 0.1 M phosphate buffer (pH 7.4) at 70°C, a strong band at 1770 cm^{-1} appeared (*Figure 6 c*). This absorption band is characteristic for carbonates (see *Figure 6 b*)[18] and provides evidence for the formation of carbonate bonds during the degradation of poly(BPA-iminocarbonate) at neutral pH.

In order to obtain quantitative data on the extent of carbonate formation during hydrolysis, an accelerated degradation study was performed at 70°C. Films of poly-(BPA-iminocarbonate) (dimensions: 2 × 1× 0.1 cm) were extensively hydrolysed in either 0.1 N hydrochloric acid (pH 1), 0.1 M phosphate buffer (pH 7.4) or 0.1 N sodium hydroxide (pH 14) as hydrolysis medium. During the course of hydrolysis the amount of BPA accumulating in the release medium was monitored by u.v.-spectroscopy.

The most rapid hydrolysis was observed in sodium hydroxide: films of poly(BPA-iminocarbonate) dissolved completely within 5 d. Of the theoretical amount of BPA, 96% could be recovered in the hydrolysis medium. This finding is consistent with hydrolysis according to *Figure 2* (Equation 2). (Interestingly, the films became noticeably thinner while maintaining their structural integrity up to the very last stage of degradation. This behaviour seemed to indicate 'surface erosion'. However, since erosion in hot 0.1 N NaOH is of no practical interest, this phenomenon was not further investigated). The mechanism of hydrolysis was different in hydrochloric acid and in phosphate buffer (pH 7.4). In those media the initially transparent films became opaque within 24 h and started to crumble within 2 d. By day 3 only white, flaky precipitates were left. Surprisingly, at this point only about 7-14% of the theoretical amount of BPA could be recovered from the hydrolysis medium. The release of BPA continued slowly for about 25 d, finally levelling off at a total accumulated recovery of 51% (in HCl) and 72% (in phosphate buffer) of the theoretical amount of BPA. At this point the white, flaky residues were collected, dried, weighed and analysed by i.r. spectroscopy (KBr-pellet). Their i.r. spectra closely resembled the spectrum of poly(BPA-carbonate). By vapour pressure osmometry the molecular weight of these residues was 475 (± 100), suggesting that they consisted mostly of dimers or trimers of BPA-carbonate. The results of this study are summarized in *Table 3*.

These observations suggest that at neutral pH poly (BPA-iminocarbonate) hydrolyses by a combination of two different pathways; backbone cleavage and release of BPA (*Figure 2*, Equation 2) occur concomitantly with the formation of the more stable carbonate bond (*Figure 2*, Equation 1). This process results in the formation of the observed white, flaky precipitates. Since carbonate formation is catalysed by acid, the extent of carbonate formation should

Table 3 Hydrolysis of poly(BPA-iminocarbonate)

Hydrolysis Temp. (°C)	Hydrolysis medium	BPA recovered (mol %)	Carbonate recovered (mol %)	Complete hydrolysis (d)
37	0.1 N NaOH (pH 14)	> 90	0	82
	0.1 N HCl (pH 1)	20	72	200
	Phosphate (pH 7.4)	35	63	> 200
70	0.1 N NaOH (pH 14)	> 96	0	5
	0.1 N HCl (pH 1)	51	45	18
	Phosphate (pH 7.4)	72	28	31

increase as the acidity of the hydrolysis medium increases. This expectation was confirmed by experimental results i.e. the overall proportion of carbonate formation was higher in HCl (45%) than in phosphate buffer (28%), (see *Table 3*). The way carbonate formation affects the *in vivo* degradation of poly(BPA-iminocarbonate) must still be determined. It is known that high molecular weight poly(BPA-carbonate) does not degrade at a measurable rate under physiological conditions; on the other hand the fate *in vivo* of dimers and trimers of BPA-carbonate [as they are formed during poly(BPA-iminocarbonate) degradation] has yet to be established.

In order to obtain an estimate for the rate of hydrolysis at 37°C, the hydrolysis experiments were repeated at that temperature: the hydrolysis of poly(BPA-iminocarbonate) showed the same general pattern as at 70°C (*Table 3*), but was markedly slower: for instance, in phosphate buffer (physiological conditions) a solvent cast film (dimensions: 2 × 1 × 0.1 cm) of poly(BPA-iminocarbonate) became opaque by day 30 (no BPA released), holes and cracks were visible by day 61 (5 mol% BPA released), the polymer crumbled on day 109 (11 mol% BPA released), and by day 200 only flaky precipitates of BPA-carbonate were left (35 mol% BPA released).

In a parallel experiment the changes in the molecular weight (number average) of solvent cast films (dimensions: 2 × 1 × 0.1 cm) of poly(BPA-iminocarbonate) were followed during hydrolysis under physiological conditions. After various hydrolysis times, samples of film were removed, dried, weighed, and their residual molecular weight determined by vapour pressure osmometry. As shown in *Figure 7*, the molecular weight decreased rapidly, indicating that extensive backbone cleavage occurred within 30 d of exposure to buffer. During this period the films became opaque and brittle and lost 99% of their initial tensile strength. However, no weight loss occurred — in correspondence to our observation that no BPA is released from the films during this initial stage of the degradation. This phenomenon is probably due to the virtual insolubility of the BPA oligomers in phosphate buffer.

The use of poly(BPA-iminocarbonate) for drug delivery applications was investigated using *p*-nitroaniline and Eosin Y as model compounds for low molecular weight, hydrophobic and hydrophilic drugs respectively. First compression moulded discs of poly(BPA-iminocarbonate) were prepared, containing loadings of 1, 10 or 30% (w/w) of Eosin Y. The release profiles (*Figure 8*) were dependent on the degree of loading: at high loading (30%, w/w), all Eosin Y was released within 8 h, whereas at low loadings (1 and

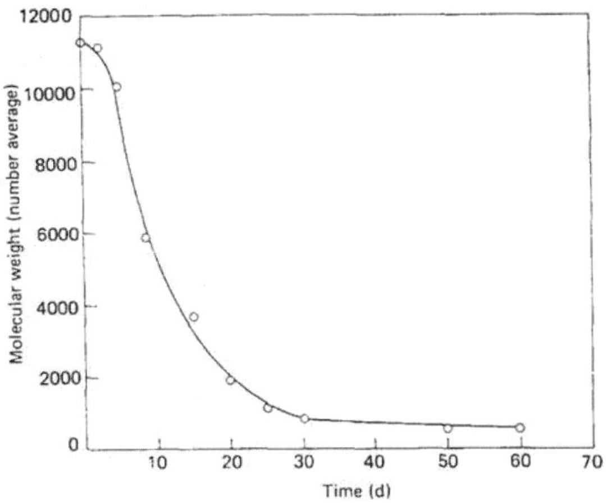

Figure 7 Decrease of molecular weight (number average) during hydrolysis in 0.1 M phosphate buffer, pH 7.4, 37°C of thin films of poly(BPA-iminocarbonate). See text for experimental details.

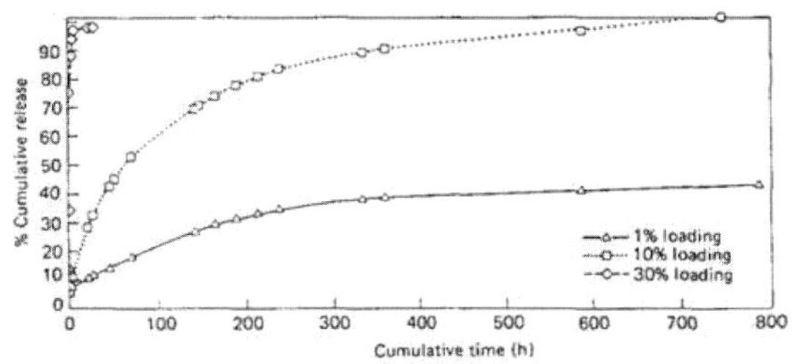

Figure 8 Release of Eosin Y from compression moulded discs of poly(BPA-iminocarbonate) at 37°C in 0.1 M phosphate buffer, pH 7.4. Eosin Y loadings were 1, 10 and 30% (w/w) as indicated.

10%, w/w) more sustained release was obtained.

Solvent cast films of poly(BPA-iminocarbonate) containing 10% (w/w) of Eosin Y were also tested. Since Eosin Y is insoluble in methylene chloride, these films were heterogeneous two-phase systems, consisting of sieved Eosin Y particles (mesh 270, < 53 μm) dispersed within the polymeric phase. Surprisingly, no Eosin Y could be released from these films beyond a short, initial burst during the first 24 h. This phenomenon is presently under further investigation. In order to study the behaviour of more hydrophobic model drugs, solvent cast films were prepared containing 10% (w/w) of *p*-nitroaniline. Since *p*-nitroaniline is soluble in methylene chloride the resulting films were homogeneous. These films were transparent, clear, and pliable. After an initial washing step during which about 15% of the total dye load was removed from the surface of the films, the release profile of *p*-nitroaniline (*Figure 9*) was characterized by a lag period of about 15 d. Thereafter a prolonged period of close to zero-order release was observed (90 d), which is desirable for many controlled release applications[19].

For the assessment of the possible toxicity of poly-(BPA-iminocarbonate) the individual toxicities of the monomers (BPA and BPA-dicyanate) must be considered in addition to the individual toxicities of the known degradation products (carbon dioxide, ammonia, BPA-carbonate and BPA).

Poly(BPA-carbonate) has been studied extensively and is regarded as a nontoxic, biocompatible polymer. Large subcutaneous implants (25 g/kg) in rats had no toxic effect[13]. BPA was also extensively investigated: the LD_{50} for oral administration is well above 2 g/kg in all tested species[20]. Furthermore continuous exposure of mice to a total dose of 865 g/kg over a period of 2 yr (equivalent to approx. 1.18 g/kg per d) failed to show carcinogenicity (NCI bioassay)[20]. For comparison, use of a polymeric drug delivery system with a volume of about 1 cm^3 could expose the patient to about 1 g of BPA over a period of 200 d resulting in an average daily exposure of about 0.00007 g/kg.

The cyanate function (OCN) seems to be devoid of the acute toxicity associated with derivatives of cyanide (CN^-). For example, sodium or potassium cyanate (KOCN) are only moderately toxic [LD_{50} for potassium cyanate, i.p. (mouse) = 320 mg/kg[21]]. Sodium cyanate has been used as an experimental drug in humans for treatment of sickle-cell anaemia[22], and oestrogen cyanate was adminstered to women as a contraceptive agent[23]. In general, the non-volatile cyanates have no irritating effect on skin or mucous membranes and very low systemic toxicity in humans[23]. In particular BPA-dicyanate was extensively tested. These studies (*Table 4*) indicate that the cyanate function in BPA-dicyanate does not increase the acute toxicity of this compound as compared to BPA itself. Likewise, chronic respiratory exposure of guinea pigs, mice and rats to BPA-dicyanate at a concentration of 400 mg/m^3 (4 h per day for a total of 4 wk) did not produce any symptoms of toxicity, and no changes in blood chemistry and kidney or liver function were noted[23].

Based on these data we concluded that neither the monomers nor the known degradation products of poly-(BPA-iminocarbonate) were sufficiently toxic to preclude — *a priori* — the possible use of poly(BPA-iminocarbonate) in some medical applications. Therefore we performed some initial *'in vivo'* tests of the tissue compatibility of poly(BPA-iminocarbonate). In the rabbit cornea test[16], no signs of

Table 4 Toxicity of BPA-dicyanate[a]

Mode of administration	LD_{50} (rat)
Oral	> 2500 mg/kg
Subcutaneous	> 2500 mg/kg
Inhalation[b]	> 440 mg/m^3

[a]according to data of Farbenwerke Bayer, Wuppertal-Elberfeld, W. Germany[23].
[b]particles in air.

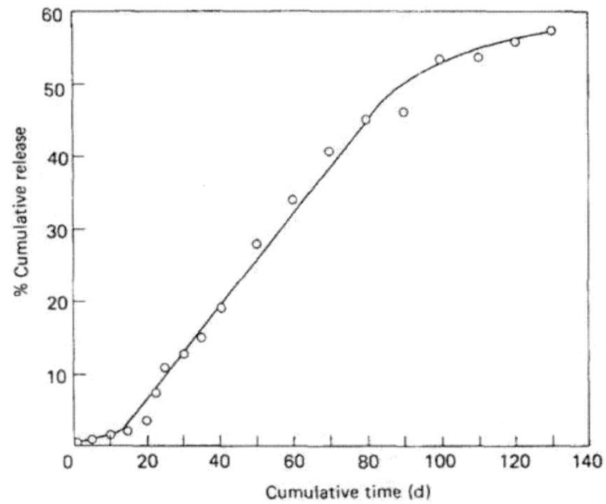

Figure 9 Release of p-nitroaniline from solvent cast films of poly(BPA-iminocarbonate) at 37°C in 0.1 M phosphate buffer, pH 7.4. p-Nitroaniline loading was 10% (w/w). The films were pre-released as described in Materials and Methods. During this washing step, 15% of the total loading of p-nitroaniline was released.

inflammation (4 animals) were detected over a period of 4 mnth post-implantation (*Figure 10*). In a second test, poly(BPA-iminocarbonate) was implanted subcutaneously into male CD-1 mice (10 animals). These animals were sacrificed at times ranging from 8 to 24 wk post-implantation. Gross postmortem observation revealed no inflammation or other lesions at the implantation site in any of the animals. Histological examination showed a very thin layer of connective tissue around the implant, however there was a complete lack of inflammatory cells (e.g. polymorphonuclear leucocytes, macrophages) at the implantation site (*Figure 11*). While the above tests do not constitute a complete toxicological assessment, they suggest that poly-(BPA-iminocarbonate) implants have a high degree of tissue compatibility.

CONCLUSION

Poly(iminocarbonates) such as poly(hydroquinone-iminocarbonate) and poly(Bisphenol A-iminocarbonate) were synthesized, characterized, and their potential use as biomaterials was examined. Poly(iminocarbonates) hydrolyse under physiological conditions. *In vitro* drug release studies showed that near zero-order release is possible with these polymers for up to several months. Poly(Bisphenol A-iminocarbonate) has many of the excellent mechanical properties of poly(BPA-carbonate) (e.g. facile film and fibre formation, high tensile strength). In two tested animal models (corneas of rabbits and subcutaneous tissue of mice) poly (BPA-iminocarbonate) was found to have a high degree of tissue compatibility. These properties suggest the possibility of using poly(iminocarbonates) in various medical applications.

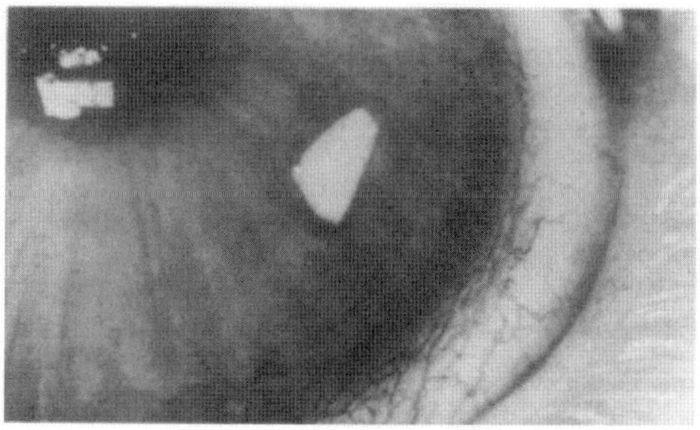

Figure 10 Rabbit cornea 30 d after implantation of a solvent cast film of poly(BPA-iminocarbonate) (size: 0.15 × 0.1 × 0.05 cm). The implant appears as the white object near the centre of the photograph. The cornea is free of signs of irritation or inflammation as indicated by the clarity of the cornea and the complete absence of ingrowth of new blood vessels.

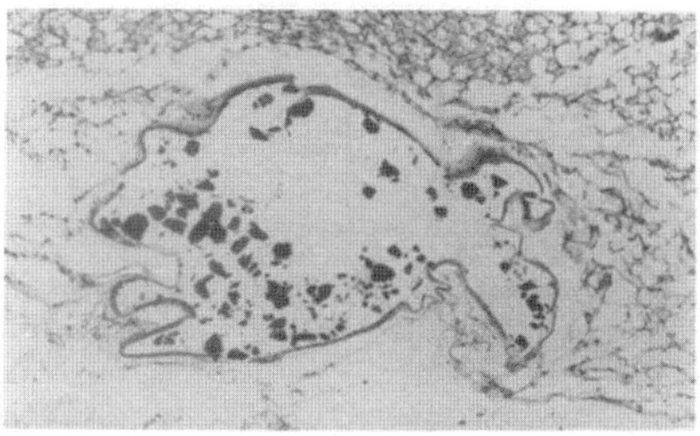

Figure 11 Histological section stained with haematoxylin and eosin of mouse subcutaneous tissue, 24 wk post-implantation. The section shows particulate remnants of the implanted polymer, surrounded by a very thin layer of connective tissue. No inflammatory cells are evident in the tissue adjacent to the implant.

ACKNOWLEDGEMENTS

We thank E.C. Albert, R. Satcher, A. Agrawal, and D. Kline for their excellent technical assistance. The authors are also grateful to Prof. F. McGarry for helpful discussions. We are indebted to Dr A. Braun for his help in reviewing the toxicity data, and we express our special thanks to Dr James C. Murphy for conducting the histological examinations. This work was supported by NIH Grant GM26698. J.K. is a recipient of the Weizmann Fellowship.

REFERENCES

1 Pitt, C.G. and Schindler, A., Biodegradation of polymers, in *Controlled Drug Delivery*, (Ed. S.D. Bruck,) Vol. 1, CRC Press, Boca Raton, FL, 1983, p 53

2 Sandmeyer, T., Ueber Derivate der Kohlensäure, *Chem. Ber.* 1886, **19**, 862-867

3 Hantzsch, A. and Mai, L., Ueber Imidokohlensäureäther und die sogen. normalen Cyansäureäther, *Chem. Ber.* 1895, **28**, 2460-2472

4 Nef, J.U., Die Chemie des Cyans und des Isocyans, *Ann. Chemie* 1895, **287**, 265-359

5 Houben, J. and Schmidt, E., Ueber Imido- und Oximido-Kohlensäureester, *Chem. Ber.* 1913, **46**, 2447-2461

6 Houben, J. and Zivadinovitsch R., Ueber Iminokohlensäure-ester, *Chem. Ber.* 1936, **69**, 2352-2360

7 Grigat, E. and Pütter, R., Umsetzung von Cyansäureestern mit hydroxylgruppenhaltigen Verbindungen, *Chem. Ber. 1964*, **97**, 3018-3021

8 Hedayatullah, M., Cyanates et iminocarbonates d'aryle, *Bull. Soc. Chim. France* 1967, 416-430

9 Schminke, H.D., Gobel, W., Grigat, E. and Pütter, R., US Patent no. 3491060, 1970 (issued to Farbenwerke Bayer)

10 Schminke, H.D., Gobel, W., Grigat, E. and Pütter, R., DB Patent no. 1220133, 1964 (Federal Republic of Germany, Patent issued to Farbenwerke Bayer)

11 Official notification, Farbenwerke Bayer, February 29, 1984

12 Goodman, I. and Rhys, J.A., *Polyesters*, Vol. 1, American Elsevier Publ. Co., New York, 1965, p 147

13 Vieweg, R. and Goerdan, L., *Polyester*, (Kunststoff-Handbuch, Band 8), Carl Hanser Verlag, Munich, West Germany, 1973

14 Collins, E.A., Bares, J. and Billmeyer, F.W., *Experiments in Polymer Science*, Wiley, New York, 1973, Ch. 6, pp 99-115

15 Grigat, E. and Pütter, R., Cyansäureester aus Hydroxylverbindungen und Halogencyan, *Chem. Ber.* 1964, **97**, 3012-3017

16 Langer, R., Brem, H. and Tapper, D., Biocompatibility of polymeric delivery systems for macromolecules, *J. Biomed. Mater. Res.* 1981, **15**, 267-277

17 Feng, X.D. (Voong, S.T.), Song, C.X. and Chen, W.Y., Synthesis and evaluation of biodegradable block polymers of ε-caprolactone and D,L-lactide, *J. Polymer Sci. (Polymer Lett. Ed.)* 1983, **21**, 593-600

18 *The Aldrich Library of Infrared Spectra, (3rd Edn), p 1589, (Spectra A-E)*

19 Bruck, S.D., *Controlled Drug Delivery*, (S.D. Bruck Ed.) Vol. 1, CRC Press, Boca Raton, Fl, 1983, pp 2-14

20 *NIOSH, Registry of Toxic Effects of Chemical Substances*, 1981-82, p 70 (entry SL 6300 000)

21 *Merck Index*, (10th Edn), p 1100, (entry 7500)

22 *Peterson, C.M., Cerami, A. and Gillette, P.N., Studies with intravenous sodium cyanate in patients with sickle cell anemia, J. Pharmacol. Exp. Ther.* 1974, **189**, 577-584

23 Martin, D., Bacaloglu, R. *Organische Synthesen mit Cyansäureestern*, Academie-Verlag, Berlin (east), GDR, 1980, p 25

24 Martin, D., Bacaloglu, R. *Organische Synthesen mit Cyansäureestern*, Academie-Verlag, Berlin (east), GDR, 1980, p 155 (compound No. 1)

25 Martin, D., Bacaloglu, R. *Organische Synthesen mit Cyansäureestern*, Academie-Verlag, Berlin (east), GDR, 1980 pp 33-35

26 Martin, D., Bacaloglu, R. *Organische Synthesen mit Cyansäureestern*, Academie-Verlag, Berlin (east), GDR, 1980, p 156 (compound No.14)

The Synthesis of Poly(hydroxamic Acid) from Poly(acrylamide)

A. J. DOMB, *Department of Chemical Engineering, Massachusetts Institute of Technology, Cambridge, Massachusetts 02139,* E. G. CRAVALHO, *Harvard–MIT Division of Health Sciences and Technology, Cambridge, Massachusetts 02139* and R. LANGER,* *Department of Chemical Engineering, Massachusetts Institute of Technology, and Harvard–MIT Division of Health Sciences and Technology, Cambridge, Massachusetts 02139*

Synopsis

Poly(hydroxamic acid) in gel or water soluble form was prepared from the reaction of poly(acrylamide) and hydroxylamine in basic aqueous solution (pH > 12) at room temperature. The polymers were composed of 70% hydroxamic acid groups, less than 5% carboxylic acid groups, and 25% unreacted amide groups. The polymers exhibited high affinity to iron(III) and copper(II) in the pH range of 1 to 5 with a high binding rate. A binding of 3 mmol/g for both metals was achieved. Preliminary tests demonstrated the urease inhibitory activity of both linear and crosslinked poly(hydroxamic acids).

INTRODUCTION

Hydroxamic acids have long been known for their special ability to form complexes with heavy metals, particularly iron(III).[1–7] A number of hydroxamic acids show biological activity, such as urease inhibition[8,9] and anticoagulant activity.[10] Reports of the use of hydroxamic acid polymers as ion exchangers have appeared sporadically over the past 40 years. Polymers were synthesized by either polymerization of vinyl monomers bearing hydroxamic acid groups[11–16] or by converting functional groups in a polymer into hydroxamic acids.[17–26] Our interest in poly(hydroxamic acids) (PHA) stems from our involvement in biopolymer research. It occurred to us that polymers bearing hydroxamic acid groups might possess anticoagulant or urease inhibitory activity like their low molecular weight analogs. If this were the case they may have a unique application as a protective coating for implantable medical devices[27] or as a polymeric drug for the treatment of urinary stones associated with urea-splitting bacteria.[8] In addition, such polymers could be useful as ion exchangers. For these applications and for a systematic bioactivity study we searched for a method of synthesizing PHA with a high content of hydroxamic acids and in variable physical forms (e.g., crosslinking, solubility, molecular weight) which may affect the biological activity.[28]

*To whom reprints should be addressed at MIT.

Journal of Polymer Science: Part A: Polymer Chemistry, Vol. 26, 2623–2630 (1988)
 CCC 0360-6376/88/102623-08$04.00

PHA are most commonly prepared from the reaction of hydroxylamine with acrylic polymers-poly(acrylonitrile)[17-22,25,26] or poly(acrylic esters).[23,24] The use of these hydrophobic polymers [e.g., poly(acrylonitrile), poly(acrylic esters)] for the formation of PHA has several disadvantages which are a result of the heterogenic reaction in hydrophilic medium: (1) high surface area (macroporous polymers) is required for high conversion yield,[26] (2) the hydrophilicity of the final product depends on the conversion yield; most of the reported polymers have a water regain of less than 2,[17-26] (3) water soluble PHA especially with high molecular weight are difficult to synthesize, and (4) the PHA obtained contained a large number of acid groups which may interfere with the polymer selectivity to metal ions[25c] and biological activity.[28] We considered that these disadvantages could possibly be overcome by using hydrophilic polymers compatible with the reaction medium as starting materials. We report here a method using poly(acrylamides), as starting materials for the synthesis of hydroxamic acid polymers where the polymers are reacted with hydroxylamines at room temperature under basic conditions. The possibility of converting amides into hydroxamic acids under basic conditions, was brought to our attention from a report on the formation of low molecular weight hydroxamic acids for the colorimetric determination of amides.[29] In this report, reactions at 26°C for a few hours gave the highest conversion yields with minimal hydrolysis. The method is as follows [eq. (1)]:

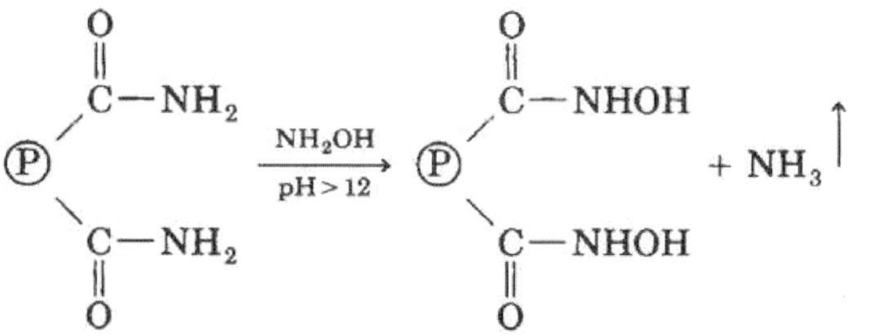

(1)

where P = polymer backbone.

The hydroxamic acid content of the formed polymers was estimated from the elemental analysis and their affinity to metal ions—iron(III) and copper(II).[29]

EXPERIMENTAL

Materials

The following chemicals: Acrylamide, *N,N'*-methylenebisacrylamide (Biorad, Richmond, CA) ammonium persulfate (Mallinckrodt, Paris, KY) *N,N,N',N'*-tetramethyl-ethylenediamine (International Biotechnologies Inc., New Haven, CT), hydroxylamine hydrochloride (Aldrich, Milwaukee, WI), cupric chloride dihydrate (J. T. Baker Chemical Co., Phillipsburg, NJ), and ferric chloride (MCB reagents, Gibbstown, NJ), were analytical-reagent grade and used as received. Linear poly(acrylamides), molecular weight 5–6 $\times 10^6$ (Aldrich, Milwaukee, WI), and molecular weight 25 $\times 10^4$ (Pfaltz & Bauer Inc., Waterbury, CT) were washed thoroughly with methanol and dried before use.

Preparation of Poly(acrylamide) (PAA)

Crosslinked PAA was prepared by polymerizing a solution of acrylamide and *N,N'*-methylenebisacrylamide (95 : 5 molar ratio, 30 g in 100 mL of water) with ammonium persulfate (2 mL, 2% w/v in water) and *N,N,N',N'*-tetramethylethylenediamine (0.2 mL, 1% w/v in water) at 25°C. A gel was formed after 20 min of reaction. After 24 h the gel was crumpled and washed with 2 L of water.

Reaction of PAA with Hydroxylamine

A solution of hydroxylamine-hydrochloride (3.3*M*, 100 mL) was added to a suspension of 20 g dry crosslinked PAA in 300 mL water. The resulting mixture was stirred for 30 min at room temperature, followed by addition of NaOH (3.9*M*, 50 mL). The reaction was continued for 24 h at room temperature. Ammonia was liberated during the reaction. The resulting white polymer was washed with water, acidified with 3*N* hydrochloric acid solution and then washed with water until chloride-free. When soluble polymers were synthesized, solutions of 20 g PAA in 300 mL of water were used. The reaction solution was acidified to pH 1 with 3*N* hydrochloric acid and the polymer was isolated from the reaction mixture by precipitation in methanol. When reacted under various basic conditions, the alkalinity of the reaction medium was adjusted with 2*N* NaOH or 20% w/v sodium carbonate or bicarbonate.

Physical and Chemical Characterization

The metal capacities were obtained from the reaction of the polymer with 50 mL 0.1*M* $FeCl_3$, pH 1 or 0.1*M* $CuCl_2$, pH 5 for 24 h. The sodium capacity of the moist polymer (5.0 g) was determined after equilibration with 10 mL of 1*M* sodium bicarbonate for 48 h, filtration and washing with distilled water. The total filtrate was acidified with 10 mL of 1*N* hydrochloric acid, boiled to remove CO_2 and back-titrated with 1*N* sodium hydroxide. The fully swollen resin was dried at 95°C overnight to determine the weight of water associated with 1 g of resin.

The metal capacities of 5 g of fully hydrated resin or 0.50 g PHA in 50 mL water solution was measured after equilibration for 24 h with 50 mL of 0.1*M* metal solutions at various pHs. The metal capacities were determined from the decrease in metal ion concentration in solution after equilibration or gravimetrically from the increase in weight of the polymer-metal complex after thoroughly washing water till metal free (determined spectrophotometrically) and air drying. Poly(acrylamide) was used as a control. 0.1*M* metal solutions of various pH's were prepared by dissolving $CuCl_2 \cdot 2H_2O$ or $FeCl_3$ in 1*M* Na_2SO_4 and adjusting the pH by adding 0.1*N* H_2SO_4. Metal capacities were measured at pHs less than or equal to 3 for $FeCl_3$ and pHs less than or equal to 5 for $CuCl_2$. Higher pH caused precipitation in metal solutions. The concentrations of iron and copper were determined spectrophotometrically[30] using a Perkin-Elmer 553 UV/VIS spectrophotometer. The time to 50%

saturation ($t_{1/2}$) was determined by equilibrating 5 g of the hydrated polymer (0.5 g dry polymer) with 50 mL of 0.1*M* $CuCl_2$ at pH 5. The reaction was followed spectrophotometrically by the decrease in the concentration of copper in the solution.

Urease Inhibition Testing

PHA was tested for urease inhibition activity using the procedure of Griffith.[31] Briefly, 1 cm^2 of an agar slab containing a layer of urease forming bacteria, *Proteus*, was dissolved in 25 mL of artificial urine made of 25 g urea dissolved in one liter of water. 30 mg of the tested material dissolved or suspended in 1 mL of artificial urine was added and the mixture was incubated overnight at 37°C. The control experiment was conducted without any additional material. The production of ammonia, indicative of urease activity, was measured using an amino acid analyzer (Beckman).

RESULTS AND DISCUSSION

Synthesis and Characterization of Poly(hydroxamic Acid)

Crosslinked and soluble PHA were synthesized from the reaction of the corresponding PAA and hydroxylamine. The reaction is strongly pH dependent (Table I). The reaction is most rapid at high pH, yielding the highest metal affinities. At pH 8.1 (run 11) no reaction occurred, while at pH 11.5 longer reaction times improved the metal binding capacities (runs 12, 13). Based on the mechanism suggested by Jencks[32] for the reaction of carbonyls and hydroxylamines under basic conditions (pH > 10), the following mechanism is proposed [eq. (2)]:

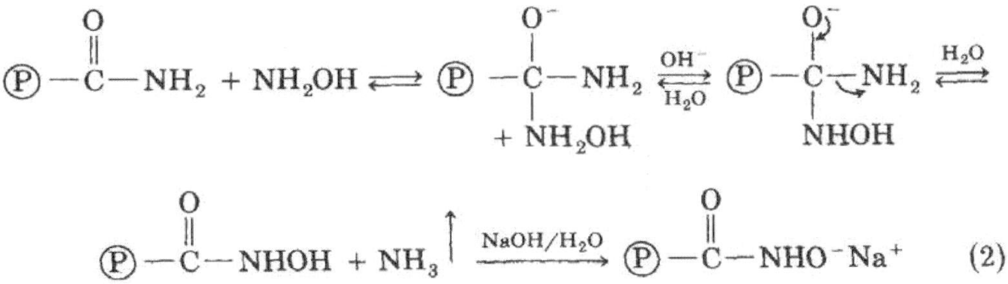

(2)

where P = polymer backbone.

The base catalysis permits the formation of product by removing a proton from the unstable dipolar intermediate, which would otherwise break down rapidly to the starting materials,[32] in addition to the product stabilization by salt formation.

The properties of the crosslinked and soluble poly(hydroxamic acid) are summarized in Table II. The sodium-hydrogen exchange shows that the polymer contains only 5% (0.8 mmol/g) carboxylic acid group. The water regain of the PHA was 10.5. This is much higher than the value reported for

TABLE I
Reaction of Poly(acrylamide) and Hydroxylamine at Various pHs[a]

Run no.	pH	NH_2—OH · HCl/N_aOH	Complexation[b] (mmol/g dry polymer) Fe^{+3}	Cu^{+2}
1	6.8	—	0	0
2	9.0	—	0	0
3	11.0	—	0.3	0
4	14	1 : 1.1	1.2	0.3
5	14	1 : 1.2	2.0	1.05
6	14	1 : 1.3	3.1	2.0
7	14	1 : 2.0	3.2	3.0
8	14	1 : 3.0	3.0	3.2
9	14	NaOH only	0	0.5
10[c]	6.8	—	1.5	0.6
11[d]	8.1	1 : 5 $NaHCO_3$	0	0
12	11.5	1 : 5 Na_2CO_3	0.8	0.2
13[d]	11.5	1 : 5 Na_2CO_3	2.2	1.1

[a] Poly(acrylamide) crosslinked with 5% *N,N'*-methylenebisacrylamide (1 g) was reacted with 1.5*M* aqueous solution of NH_2—OH · HCl (1 : 1.2 molar ratio, acrylamide units : hydroxylamine) at 25°C for 24 h. pH was adjusted by adding 2*N* NaOH.

[b] Polymer was reacted with 50 mL 0.1*M* Fe(III), pH1 or 0.1*M* Cu(II), pH5 for 24 h.

[c] Reaction was performed at 70°C for 8 h.

[d] Reacted for 7 days.

most PHA previously described in the literature, which show water's regain of less than 2.[17-26] High water regain may allow better interaction with metal ions.[25f] The higher water regain is presumably a result of the high hydrophilicity of the amide and hydroxamic acid groups. The degree of crosslinking affected the water regain of the polymer but did not affect the activity of the amide groups toward hydroxylamine. However, increasing the amount of crosslinker in the polymer decreases the concentration of free amide groups available for the conversion to hydroxamic acid groups. Thus, PHA crosslinked with 1, 10, and 20% *N,N'*-methylenebisacrylamide prepared from the

TABLE II
Physical and Chemical Characteristics of PHA

	Crosslinked	Linear[b]
Percent of crosslinking	5	—
Water regain	10.5 gg^{-1}	soluble
Sodium-hydrogen exchange capacity	0.8 mmol/g	0.7 mmol g^{-1}
Copper(II) capacity at pH 5	3.0 mmol g^{-1}	3.0 mmol g^{-1}
Iron(III) capacity at pH 1	3.2 mmol g^{-1}	3.0 mmol g^{-1}
Equilibration rate ($t_{1/2}$)[a]	5 min	5 min

[a] Determined for Cu^{+2} at pH 5.0.

[b] Prepared from poly(acrylamide) of 5–6 × 10^6 or 2.5 × 10^5 molecular weight.

TABLE III
Elemental Analysis of Poly(hydroxamic acids)

	%C	%H	%N	·%O
Poly(hydroxamic acid)	44.69	6.68	16.05	31.46
Poly(acrylamide)	52.10	6.92	18.74	22.86
Poly(hydroxamic acid)[a]	42.50	5.74	15.74	35.99
Poly(acrylamide)[b]	51.29	6.92	18.99	22.80

[a] Calculated on the basis of 100% conversion of poly(acrylamide) crosslinked with 5 molar % *N,N'*-methylenebisacrylamide.

[b] Calculated for poly(acrylamide) crosslinked with 5% *N,N'*-methylenebisacrylamide.

corresponding PAA had water regains of 20.2, 7.2, and 4.6, respectively. These polymers showed the following iron binding: 3.3, 2.6, and 2.3 mmol/g polymer, respectively. Both linear PHA showed high metal binding similar to the crosslinked polymer (Table II).

The elemental analysis of the PHA is summarized in Table III. The nitrogen and oxygen analysis of the PHA shows a high conversion to hydroxamic acid groups. Based on the nitrogen and oxygen analysis of the PHA and the sodium bicarbonate titration, the polymer contains: 9.4 mmol hydroxamic groups/g (70 molar %), 0.8 mmol carboxylic acid groups/g (5 molar %), and 3.5 mmol unreacted amide groups/g (25 molar %). The low acid content is because the mild temperature used on the polymer during the hydroxamic acid formation minimizes hydrolysis. Low acid content may be advantageous since acid groups in the polymer have antitumor activity[28] and some affinity to metal ions[25c] and thus may interfere in the polymer selectivity.

Binding and Activity of Hydroxamic Acid Polymers

On addition of a solution of Fe(III) or Cu(II) to the fully hydrated polymer, the gel turns dark brown or green respectively, with the color deepening as the metal ion is consumed. After a few minutes of ion uptake, the polymer gel contracts to a powder, the polymer no longer possessing the character of a polymer gel. This behaviour is evidence for a significant concentration of hydroxamic acid in the polymer backbone. Upon addition of a metal solution (iron or copper) to the polymer solution the polymer precipitates as a dark brown or green solid, respectively. The insolubility of the iron and copper complexes is a result of crosslinking through the metal complex.[15]

As a control, poly(acrylamide) was reacted with copper and iron solutions under the same conditions. No adsorption of metals to the polymer was observed; the polymer gel remained uncolored. The copper adsorbed by the polymer was removed in a $2M$ H_2SO_4 solution, while complete removal of the iron required a solution of $4M$ H_2SO_4.

The iron(III) and copper(II) capacities of the crosslinked PHA at various pHs are described in Table IV. Results of both determinations are close and show bindings of 3 mmol/g of polymer of both metals. Similar high capacities were reported by Vernon[26] for a macroporous PHA prepared from macroporous poly(acrylonitrile). The metal binding is pH dependent. No copper is bound in $1N$ H_2SO_4 while iron(III) shows some binding. Increasing the pH

TABLE IV
Iron(III) and Copper(II) Binding to Poly(hydroxamic Acid)[a]

Acidity of metal solution	Iron(III) (mmol/g)		Copper (II) (mmol/g)	
	Spectro photo.	Gravimetric	Spectro photo.	Gravimetric
$1N$ H_2SO_4	1.5	0.4	0	0
pH 1	3.1	2.4	0.2	0.1
pH 3	3.2	2.8	2.7	2.5
pH 5	—	—	3.0	2.8

[a] Five grams of fully hydrated crosslinked PHA (0.5 g dry polymer) were reacted with 50 mL of $0.1M$ $FeCl_3$ or $CuCl_2 \cdot 2H_2O$ at room temperature for 24 h.

increases the capacities. The differences in metal complexation between iron and copper at low pH may be useful in the separation of iron ions from copper ions in solutions at a pH of less than 1. Under our reaction conditions [high HA content polymers reacted with excess Fe(III)], we may assume a 2 : 1 ratio for hydroxamic acid: iron complexes in the polymer.[14] This is evident by the dark-brown color of the PHA–Fe(III) complexes.[14, 15]

The PHA were tested for urease inhibitory activity using acetyl hydroxamic acid (AHA) as a reference for activity. AHA is the clinical drug of choice for the treatment of infection-induced urinary stones caused by urease producing bacteria.[8] The results are summarized in Table V. Both polymers exhibited urease inhibition. PAA, the starting material in the synthesis of PHA, exhibited no urease inhibition. These initial results demonstrate the potential of these polymers as urease inhibitors. Furthermore, the synthetic procedure reported here provides a new means for the simple formulation of PHA with a high content of hydroxamic acid groups and variable physical forms. Future studies can now comprehensively examine the bioactivity and complexing capability of these polymers.

This work was supported by NIH grant GM 26698.

TABLE V
Urease Inhibition Activity of Hydroxamic Polymers[a]

Active agent	Ammonia released[c]	%Hydroxamic acid in active agent
AHA	1165–1331	100
Soluble PHA[b]	1467–2111	70
Crosslinked PHA	1625–2326	70
Crosslinked PAA	3702–4630	0
Control	3535–3768	0

[a] Determined by the Griffith procedure,[31] using 30 mg of the tested material.
[b] Molecular weight of 5–6 × 10^6.
[c] Area under the curve.

References

1. H. L. Yale, *Chem. Rev.*, **33**, 209 (1943).
2. R. E. Buckiles and J. C. Thelen, *Anal. Chem.*, **22**, 676 (1950).
3. J. B. Neilands, *Struct. Bonding*, **1**, 59 (1966).
4. H. Maehr, *Pure Appl. Chem.*, **28**, 603 (1971).
5. V. C. Bass and J. H. Yoe, *Talanta*, **13**, 735 (1966).
6. G. Askenes, *Acta Chem. Scand.*, **11**, 710 (1957).
7. S. Soloway and A. Lipschitz, *Anal. Chem.*, **24**, 898 (1952).
8. J. W. Munson, *Chem. Biol. Hydroxamic Acids* (*Proc. Int. Symp.*), 1982, pp. 1–3, 94–103.
9. I. Ines and K. Horst, *IRC5 Med. Sci.*, **13**, 579 (1985).
10. A. Bruins, N. M. Hermanus, and G. Vollema, Ger. offen **2**, 756–216 (1978); *Chem. Abstr.*, **89**, 108737x (1978).
11. F. Schouteden, *Makromol. Chem.*, **27**, 246 (1958).
12. F. Schouteden, *Chim. Ind.* (*Paris*), **79**, 749 (1958); *Chem. Abstr.*, **52**, 21116e (1958).
13. A. Winston and E. T. Mazza, *J. Polym. Sci. Polym. Chem. Ed.*, **13**, 2019 (1975).
14. A. Winston and D. Kirchner, *Macromolecules*, **11**, 597 (1978).
15. J. W. Rosthouser and A. Winston, *Macromolecules*, **14**, 538 (1981).
16. A. Winston, D. V. P. R. Varaprasad, J. J. Metterville, and H. J. Rosenkrantz, *Pharmacol. Exp. Ther.*, 232, 644 (1985).
17. F. Schouteden and J. A. Herbots, Belg. Pat. 560, 782, (1958); *Chem. Abstr.*, **53**, 5739f (1959).
18. F. Schouteden, *J. Soc. Dyers Colour.*, **75**, 309 (1959).
19. C. A. Fetscher, U.S. Pat. 3,154,499 (1964); *Chem. Abstr.*, **62**, 4882d (1965).
20. C. A. Fetscher and S. A. Lipowski, U.S. Pat. 3,345,344 (1967); *Chem. Abstr.*, **67**, 109273a (1967).
21. H. Smith, Brit. Pat. 852,176 (1960); *Chem. Abstr.*, **55**, 9284i (1961).
22. A. Peden, H. Smith, and H. Vickers, Brit. Pat. 887,175; (1962) *Chem. Abstr.*, **47**, 3630a (1962).
23. W. Kern and R. C. Schulz, *Angew Chem.*, **69**, 153 (1957).
24. M. Hatano, Y. Nose, T. Nozawa, and S. Kambara, *Kogyo Kagakeu Zasshi*, **69**, 571 (1966); *Chem. Abstr.* **65**, 15532g (1966).
25. (a) F. Vernon and H. Eccles, *Anal. Chim Acta*, **72**, 331 (1974); (b) **77**, 145 (1975); (c) **79**, 229 (1975); (d) **82**, 369 (1976); (e) **83**, 187 (1976); (f) **85**, 317 (1977).
26. F. Vernon, *Pure Appl. Chem.*, **54**, 2151 (1982).
27. J. P. Cozenave, J. A. Davies, M. D. Kazatchkine, and W. G. van Aken. Eds., *Blood-Surface Interaction*, Elsevier, New York, 1986, chapter 1.3.
28. C. E. Carraher and C. G. Geblein, Eds., *Biological Activity of Polymers*, American Chemical Society, Washington, DC, 1982, chapters 15 and 16.
29. F. Bergman, *Anal. Chem.*, **24**, 1367 (1952).
30. L. Meites, Ed., *Handbook of Analytical Chemistry*, 1st ed., McGraw-Hill, New York, 1962, pp. 12–14.
31. D. P. Griffith, D. M. Musher, and C. Itins, *Investigative Urol.*, **13**, 346 (1976).
32. (a) J. M. Sayer and W. P. Jencks, *J. Am. Chem. Soc.*, **94**, 3262 (1972); (b) J. March, Ed., *Advanced Organic Chemistry*, 3rd ed., Wiley, New York, 1985.

J. Am. Chem. Soc. **2000,** *122,* 10761–10768 10761

Degradable Poly(β-amino esters): Synthesis, Characterization, and Self-Assembly with Plasmid DNA

David M. Lynn and Robert Langer*

Contribution from the Department of Chemical Engineering, Massachusetts Institute of Technology, Cambridge, Massachusetts 02139

Received May 3, 2000

Abstract: Poly(β-aminoesters) **1–3** were synthesized *via* the addition of N,N′-dimethylethylenediamine, piperazine, and 4,4′-trimethylenedipiperidine to 1,4-butanediol diacrylate. Polymerization proceeded exclusively *via* the conjugate addition of the secondary amines to the bis(acrylate ester). Polymers were isolated in up to 86% yields with molecular weights ranging up to 31 200 relative to polystyrene standards. The polymers degraded hydrolytically in acidic and alkaline media to yield 1,4-butanediol and β-amino acids **4a–6a** and the degradation kinetics were investigated at pH 5.1 and 7.4. In general, the polymers degraded more rapidly at pH 7.4 than at pH 5.1. In initial screening assays, both the polymers and their degradation products were determined to be noncytotoxic relative to poly(ethylene imine), a polymer conventionally employed as a synthetic transfection vector. Polymers **1–3** interacted electrostatically with polyanionic plasmid DNA in water and buffer at physiological pH, as determined by agarose gel electrophoresis, quasi-elastic dynamic light scattering (QELS), and ζ-potential measurements. All three polymers condensed DNA into soluble DNA/polymer particles on the order of 50–200 nm. Particles formed from polymers **1** and **2** aggregated extensively, while particles formed from polymer **3** exhibited positive ζ-potentials (e.g., +10 to +15 mV) and did not aggregate for up to 18 h. The nanometer-sized dimensions and reduced cytotoxicities of these DNA/polymer complexes suggest that these types of polymers may be useful as degradable polymeric gene transfer vectors.

Introduction

The treatment of human diseases through the application of nucleotide-based drugs such as DNA has the potential to revolutionize the medical field.[1] Thus far, the use of modified viruses as gene transfer vectors has generally represented the most clinically successful approach to gene therapy. While viral vectors are currently the most efficient gene transfer agents, concerns surrounding the overall safety of viral vectors, which include the potential for unsolicited immune responses, have resulted in parallel efforts to develop nonviral alternatives.[2] Current alternatives to viral vectors include polymeric delivery systems,[3,4] liposomal formulations,[5] and "naked" DNA injection protocols.[6] While these strategies have yet to achieve the clinical effectiveness of viral vectors, the potential safety, processing, and economic benefits offered by these methods[1a] have ignited interest in the continued development of nonviral approaches to gene therapy.[7–9]

Cationic polymers have been widely used as transfection vectors due to the facility with which they condense and protect negatively charged strands of DNA. Amine-containing polymers such as poly(lysine),[3,4] poly(ethylene imine) (PEI),[7] and poly(amidoamine) dendrimers[9] are positively charged at physiological pH, form ion pairs with DNA, and mediate transfection in a variety of cell lines. Despite their common use, however, cationic polymers such as poly(lysine) and PEI can be significantly cytotoxic.[3,5c,10] As a result, the choice of cationic polymer for a gene transfer application generally requires a tradeoff between transfection efficiency and short- or long-term cytotoxicity. Additionally, the long-term biocompatibility of these polymers remains an important issue for use in therapeutic applications *in vivo*, since several of these polymers are not readily biodegradable.[11] For use *in vivo*, new cationic polymers should be designed incorporating hydrolyzable moieties such that the polymers readily degrade into nontoxic byproducts.

To develop safe alternatives to existing polymeric vectors and other functionalized biomaterials, we[8a,12] and others[8b–c,13] have pursued the synthesis of degradable polyesters bearing

(1) (a) Anderson, W. F. *Nature* **1996**, *392 Suppl.* 25–30. (b) Friedman, T. *Nature Med. 2* **1996**, 144–147. (c) Crystal, R. G. *Science* **1995**, *270*, 404–410. (d) Mulligan, R. C. *Science* **1993**, *260*, 926–932.

(2) For leading references, see: (a) Luo, D.; Saltzman, W. M. *Nat. Biotechnol.* **2000**, *18*, 33–37. (b) Behr, J.-P. *Acc. Chem. Res.* **1993**, *26*, 274–278.

(3) Zauner, W.; Ogris, M.; Wagner, E. *Adv. Drug Del. Rev.* **1998**, *30*, 97–113.

(4) Kabanov, A. V.; Kabanov, V. A. *Bioconjugate Chem.* **1995**, *6*, 7–20.

(5) (a) Miller, A. D. *Angew. Chem., Int. Ed. Engl.* **1998**, *37*, 1768–1785. (b) Hope, M. J.; Mui, S.; Ansell, S.; Ahkong, Q. F. *Molec. Membr. Technol.* **1998**, *15*, 1–14. (c) Deshmukh, H. M.; Huang, L. *New J. Chem.* **1997**, *21*, 113–124.

(6) Sanford, J. C. *Trends Biotechnol.* **1988**, *6*, 288–302.

(7) Boussif, O.; Lezoualc'H., F.; Zanta, M. A.; Mergny, M. D.; Scherman, D.; Demeneix, B.; Behr, J. P. *Proc. Natl. Acad. Sci. U.S.A.* **1995**, *92*, 7297–7301.

(8) (a) Putnam, D.; Langer, R. *Macromolecules* **1999**, *32*, 3658–3662. (b)Lim, Y.; Choi, Y. H.; Park, J.-S. *J. Am. Chem. Soc.* **1999**, *121*, 5633–5639. (c) Lim, Y.-B.; Kim, C.-H.; Kim, K.; Kim, S. W.; Park, J.-S. *J. Am. Chem. Soc.* **2000**, *122*, 6524–6525. (d) Gonzalez, H.; Hwang, S. J.; Davis, M. E. *Bioconjugate Chem.* **1999**, *10*, 1068–1074.

(9) (a) Kukowska-Latallo, J. F.; Bielinska, A. U.; Johnson, J.; Spindler, R.; Tomalia, D. A.; Baker, J. R., Jr. *Proc. Natl. Acad. Sci. U.S.A.* **1996**, *93*, 4897–4902. (b) Tang, M. X.; Redemann, C. T.; Szoka, F. C., Jr. *Bioconjugate Chem.* **1996**, *7*, 703–714. (c) Haensler, J.; Szoka, F. C., Jr. *Bioconjugate Chem.* **1993**, *4*, 372–379.

(10) (a) Choksakulnimitr, S.; Masuda, S.; Tokuda, H.; Takakura, Y.; Hashida, M. J. *Controlled Release* **1995**, *34*, 233–241. (b) Brazeau, G. A.; Attia, S.; Poxon, S.; Hughes, J. A. *Pharm. Res.* **1998**, *15*, 680–684.

(11) (a) Uhrich, K. *Trends Polym. Sci.* **1997**, *5*, 388–393. (b) Roberts, J. C.; Bhalgat, M. K.; Zera, R. T. *J. Biomed. Mater. Res.* **1996**, *30*, 53–65.

10.1021/ja0015388 CCC: $19.00

cationic side chains. Examples of such polymers include poly(L-lactide-*co*-L-lysine),[12a] poly(serine ester),[13] poly(4-hydroxy-L-proline ester),[8a,b] and, more recently, poly[α-(4-aminobutyl)-L-glycolic acid].[8c] Poly(4-hydroxy-L-proline ester) and poly[α-(4-aminobutyl)-L-glycolic acid] were recently demonstrated to condense plasmid DNA through electrostatic interactions and mediate gene transfer.[8a−c] Generally, these new polymers are less-toxic than poly(lysine) and PEI and they degrade into nontoxic metabolites.[8] It is clear from these investigations that the rational design of amine-containing polyesters can be a productive route to the development of safe, effective transfection vectors. Unfortunately, however, present syntheses of these polymers require either the independent preparation of specialized monomers[12a,8c] or the use of stoichiometric amounts of expensive coupling reagents.[8a] Additionally, the amine functionalities in the monomers must be protected prior to polymerization,[8,12] necessitating additional postpolymerization deprotection steps before the polymers can be used as transfection agents.

We sought to develop a complementary strategy for the synthesis of amine-containing polyesters that would address several of these synthetic drawbacks. While degradability, reduced cytotoxicity, and an ability to complex DNA remained fundamental design criteria, we also desired an approach that would yield a class of polymers that were structurally different from previously studied polymeric vectors so that structure/activity relationships could be further explored. To date, most cationic polymers synthesized and investigated for transfection efficacy have contained pendant amines in the polymer side chains.[3,4,14] Herein, we report a strategy for the preparation of poly(*β*-aminoesters) containing tertiary amines in their backbones based on the conjugate addition of bis(secondary amine) monomers to diacrylate esters (eq 1). These polymers interact electrostatically with plasmid DNA at physiological pH and assemble it into nanometer-scale polymer/DNA complexes. The polymers themselves are noncytotoxic and they degrade into nontoxic small molecule byproducts.

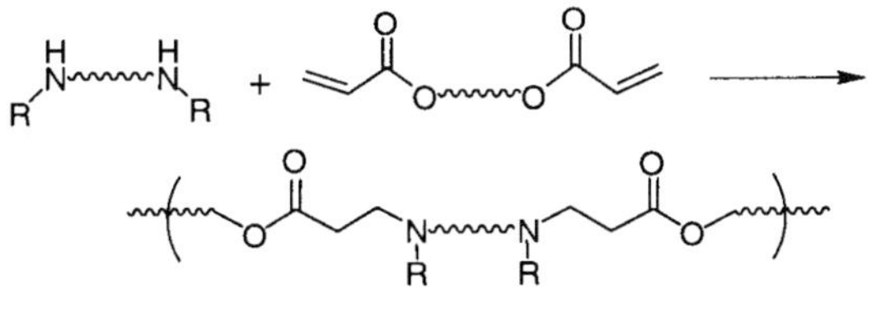

(1)

Results and Discussion

Polymer Synthesis and Characterization. The synthesis of linear poly(amido amines) containing tertiary amines in their backbones was reported by Ferruti *et al.* in 1970 by the addition of bifunctional amines to bisacrylamides.[15] Linear poly(amido amines) were initially investigated as heparin and ion complexing biomaterials.[15c−f] Dendritic poly(amido amines) (PAMAMs) have seen increasing use in gene transfer applications due to their ability to complex DNA,[9] and a recent report describes the application of a family of linear poly(amido amines) to cell transfection and cytotoxicity studies.[16] In contrast, analogous poly(ester amines) formed from the Michael-type addition of bifunctional amines to diacrylate esters have received less attention.[15,17]

The poly(ester amine) approach presents a particularly attractive basis for the development of new polymeric transfection vectors for several reasons: (1) the polymers contain the requisite amines and readily degradable linkages, (2) multiple analogues could potentially be synthesized directly from commercially available starting materials, and (3) if the resulting polymers were useful as DNA condensing agents, future generations of polymer could easily be engineered to possess amine pK_a values spanning the range of physiologically relevant pH. This last point was particularly intriguing, because the buffering capacity of polyamines has recently been implicated as a factor influencing the escape of DNA from cell endosomes following endocytosis.[7,9c,18] While complexation of DNA with cationic polymers is required to compact and protect DNA during early events in the transfection process, DNA and polymer must ultimately decomplex to allow efficient transcription.[2a] In view of this requirement, degradable polycations could play an important role in "vector unpackaging" events in the nucleus.[2a,19] Finally, we hypothesized that polymers of this general structure, and the *β*-amino acid derivatives into which they would presumably degrade, would be significantly less toxic than poly(lysine) and PEI. As outlined above, degradable polycations[8a−c] and linear polymers containing relatively hindered amines located close to the polymer backbone[8d] are less toxic than poly(lysine) and PEI.

We initially investigated the synthesis of polymers **1**−**3** *via* the addition of the bis(secondary amines) N,N′-dimethylethylenediamine, piperazine, and 4,4′-trimethylenedipiperidine to 1,4-butanediol diacrylate.[20] The polymerization of these monomers proceeded in THF and CH_2Cl_2 at 50 °C to yield the corresponding polymers in up to 86% yields (Table 1). Polymers were purified through repeated precipitation into diethyl ether or hexane. Polymer **1** was isolated as a clear viscous liquid; polymers **2** and **3** were obtained as white solids after drying under high vacuum. Alternatively, polymers **1**−**3** could be isolated as solid hydrochloride salts upon addition of diethyl ether/HCl to a solution of polymer in THF or CH_2Cl_2. All three polymers were soluble in organic solvents such as THF, CH_2Cl_2, $CHCl_3$, and MeOH and were also soluble in water at

(12) (a) Barrera, D. A.; Zylstra, E.; Lansbury, P. T., Jr.; Langer, R. *J. Am. Chem. Soc.* **1993**, *115*, 11010−11011. (b) Kwon, H. Y.; Langer, R. *Macromolecules* **1989**, *22*, 3250−3255.

(13) (a) Trollsås, M.; Lee, V. Y.; Mecerreyes, D.; Löwenhielm, P.; Möller, M.; Miller, R. D.; Hedrick, J. L. *Macromolecules* **2000**, *33*, 4619−4627. (b) Zhou, Q.; Kohn, J. *Macromolecules* **1990**, *23*, 3399−3406.

(14) van de Wetering, P.; Moret, E. E.; Schuurmans-Nieuwenbroek, N. M. E.; van Steenbergen, M. J.; Hennink, W. E. *Bioconjugate Chem.* **1999**, *10*, 589−597.

(15) (a) Danusso, F.; Ferruti, P. *Polymer* **1970**, *11*, 88−113. (b) Ferruti, P.; Marchisio, M. A.; Barbucci, R. *Polymer* **1985**, *26*, 1336. (c) Ferruti, P.; Barbucci, R. *Adv. Polym. Sci.* **1984**, *58*, 55−92. (d) Ferruti, P.; Ranucci, E.; Sartore, L.; Bignotti, F.; Marchisio, M. A.; Bianciardi, P.; Veronese, F. M. *Biomaterials* **1994**, *15*, 1235−1241. (e) Ferruti, P.; Ranucci, E.; Trotta, F.; Gianasi, E.; Evagorou, E. G.; Wasil, M.; Wilson, G.; Duncan, R. *Macromol. Chem. Phys.* **1999**, *200*, 1644−1654. (f) Ferruti, P.; Ranucci, E.; Sartore, L.; Bignotti, F.; Marchisio, M. A.; Bianciardi, P.; Veronese, F. M. *Biomaterials* **1994**, *15*, 1235−1241.

(16) Hill, I. R. C.; Garnett, M. C.; Bignotti, F.; Davis, S. S. *Biochim. Biophys. Acta* **1999**, *1427*, 161−174.

(17) (a) Kargina, O. V.; Mishustina, L. A.; Kiselev, V. Y.; Kabanov, V. A. *Vysokomol. Soedin. Ser. A* **1986**, *28*, 1139−1144. (b) Rao, W,; Smith, D. J. *J. Bioact. Compat. Polym.* **1999**, *14*, 54−63.

(18) (a) Behr, J.-P. *Chimia* **1997**, *51*, 34−36. (b) Demeneix, B. A.; Behr, J. P. In *Artificial Self-Assembling Systems for Gene Delivery*; Felgner, P. L., Heller, M. J.; Lehn, P.; Behr, J. P.; Szoka, F. C., Jr., Eds.; American Chemical Society: Washington, D. C., 1996; pp 146−151. (c) Kabanov, A. V.; Felgner, P. L.; Seymour, L. W. In *Self-Assembling Complexes for Gene Delivery: From Laboratory to Clinical Trial*; John Wiley and Sons: New York, 1998.

(19) (a) Schaffer, D. V.; Fidelman, N. A.; Dan, N.; Lauffenburger, D. A. *Biotechnol. Bioeng.* **2000**, *67*, 598−606. (b) Kabanov, A. V. *Pharm. Sci. Technol. Today* **1999**, *2*, 365−372.

(20) The preparation of polymers **1** and **2** has been briefly described; however, the synthetic details and structural characterization were not reported (see refs 15a and 17a). A general investigation into the properties of these polymers does not appear in the literature to the best of our knowledge.

Table 1. Representative Molecular Weight Data for Polymers **1**−**3**

polymer	solvent	M_n[c]	PDI	yield, %
1[a]	THF			[d]
1[a]	CH_2Cl_2			82%[21]
2[a]	THF	10 000	1.77	64%
2[a]	CH_2Cl_2	17 500	2.15	75%
3[a]	THF	24 400	1.55	58%
3[a]	CH_2Cl_2	30 800	2.02	70%
1[b]	THF	5800	2.83	55%
2[b]	CH_2Cl_2	16 500	2.37	80%[e]
3[b]	CH_2Cl_2	31 200	2.55	86%[e]

[a] Conditions: [diamine] = [1,4-butanediol diacrylate] = 0.38 M, 50 °C, 48 h. [b] Conditions: [diamine] = [1,4-butanediol diacrylate] = 1.08 M, 50 °C, 48 h. [c] GPC analysis was performed in THF/0.1 M piperidine and molecular weights are reported versus polystyrene standards. [d] No polymer was isolated under these conditions. [e] The reaction solution became very viscous and eventually solidified under these conditions.

reduced pH. Polymer **1** and the hydrochloride salts of polymers **1**−**3** were freely soluble in water.

The molecular weights of polymers **1**−**3** and their corresponding hydrochloride salts were determined by both organic and aqueous phase gel permeation chromatography (GPC). Polymer molecular weights (M_n) ranged from up to 5800 for polymer **1** to up to 32 000 for polymer **3**, relative to polystyrene standards. Molecular weight distributions for these polymers were monomodal with polydispersity indices (PDIs) ranging

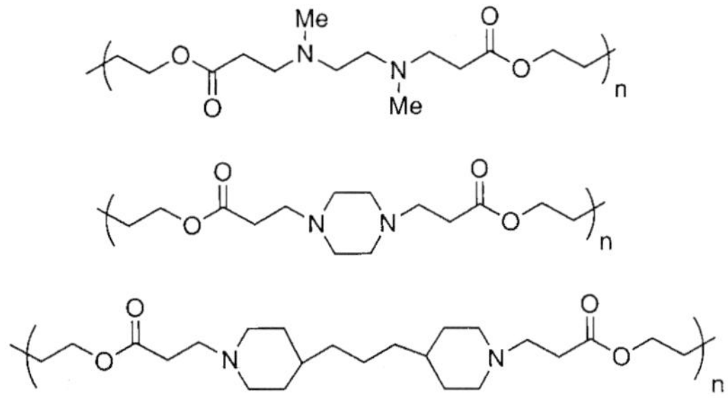

Poly-**1**

Poly-**2**

Poly-**3**

from 1.55 to 2.55. Representative molecular weight data are presented in Table 1. While the synthesis of linear poly(amido amines) is generally performed using alcohols or water as solvents,[15] we employed anhydrous THF and CH_2Cl_2 to minimize hydrolysis reactions during synthesis. The yields and molecular weights of polymers synthesized employing CH_2Cl_2 as solvent were generally higher than those of polymers synthesized in THF (Table 1).[21]

The structures of polymers **1**−**3** were confirmed by ^{1}H and ^{13}C NMR spectroscopy. These data indicate that the polymers were formed through the conjugate addition of the secondary amines to the acrylate moieties of 1,4-butanediol diacrylate and not through the formation of amide linkages under our reaction conditions. Additionally, the newly formed tertiary amines in the polymer backbones do not participate in subsequent addition reactions with diacrylate monomer, which would lead to branching or polymer cross-linking. This fortunate result appears to be unique to polymers of this type produced from bis-(secondary amine) monomers. In our hands, the synthesis of analogous polymers employing difunctional *primary* amines as monomers (such as 1,4-diaminobutane) generally leads to polymer branching and the formation of insoluble cross-linked polymer networks if conditions are not explicitly controlled.[22]

In view of the juxtaposition of amines and esters within the backbones of polymers **1**−**3**, we were initially concerned that hydrolysis might occur too rapidly for the polymers to be of practical use. For example, poly(4-hydroxy-L-proline ester) and poly[α-(4-aminobutyl)-L-glycolic acid] degrade quite rapidly near neutral pH, having half-lives of roughly 2 h[8b] and 30 min,[8c] respectively.[23] Analysis of polymers **1** and **2** by aqueous GPC using 1% acetic acid/water as eluent, however, revealed that degradation was sufficiently slow in acidic media. For example, the GPC traces of polymers **1** and **2** sampled under these conditions over a period of 4−5 h revealed no changes in molecular weights or polydispersities.[24] We were also concerned that significant backbone hydrolysis might occur during the isolation of the hydrochloride salts of polymers **1**−**3**. To prevent hydrolysis during the protonation and isolation of these polymers, anhydrous solvents were employed and reactions were performed under an argon atmosphere. Analysis of the polymers before and after protonation revealed no observable hydrolysis. For example, the GPC trace of a sample of polymer **3** after precipitation from CH_2Cl_2 with 1.0 M diethyl ether/HCl (M_n = 15 300; PDI = 1.90) was virtually identical to the molecular weight of the polymer prior to protonation (M_n = 15 700; PDI = 1.92) and no lower molecular weight species were evident.[25] Solid samples of polymers **1**−**3** could be stored for several months without detectable decreases in molecular weight.

Polymers **1**−**3** were specifically designed to degrade by hydrolysis of the ester bonds in the polymer backbones. However, an additional concern surrounding the overall stability and biocompatibility of these polymers is the potential for unwanted degradation to occur through retro-Michael reaction under physiological conditions. Because these polymers were synthesized via the Michael-type reaction of a secondary amine to an acrylate ester, it is possible that the polymers could undergo retro-Michael reaction to regenerate free acrylate groups, particularly under acidic conditions. Acrylate esters are potential DNA-alkylating agents and are therefore suspected carcinogens.[26] Because these polymers are expected to encounter the reduced pH environment within the endosomal vesicles of cells (pH = 5.0−5.5) during transfection, we addressed the potential for the degradation of these polymers to occur through a retro-Michael pathway.

Under extreme acidic (pH < 3) or basic (pH > 12) conditions, polymers **1**−**3** degraded rapidly and exclusively to 1,4-butanediol and the anticipated bis(β-amino acid) byproducts **4a**−**6a** as determined by ^{1}H NMR spectroscopy. We found no spectroscopic evidence for retro-Michael addition under these

(21) We were unable to synthesize polymer **1** in CH_2Cl_2. The color of the reaction solution progressed from colorless to an intense pink color almost immediately after the introduction of a solution of *N,N′*-dimethylethylenediamine in CH_2Cl_2 to a solution of 1,4-butanediol diacrylate in CH_2Cl_2 (see Experimental Section). The color progressed to light orange over the course of the reaction, and an orange polymer was isolated after precipitation in hexane. The isolated polymer was insoluble in CH_2Cl_2, THF, and water at reduced pH and was not structurally characterized. This problem was not encountered for the analogous reaction in THF.

(22) A recent report describes the synthesis of polyesters containing secondary amines in their backbones *via* the addition of difunctional primary amines to 1,4-butanediol diacrylate. The authors report the characterization of soluble, linear polymer. In our hands, however, we observe considerable cross-linking and the subsequent formation of insoluble polymer networks. For synthetic details and a description of the application of these polymers to the delivery of NO, see ref 17b.

(23) Such rapid degradation times did not preclude the application of these polymers to gene delivery (see refs 8b and 8c). However, extremely rapid degradation rates generally introduce additional concerns surrounding the manipulation, storage, and application of degradable polymers.

(24) Polymer **3** could not be analyzed by aqueous GPC.

(25) Comparative GPC data were collected employing THF/0.1 M piperidine as eluent (see Experimental Section). The HCl salts of the polymers were insoluble in THF, but were soluble in THF/0.1 M piperidine concomitant with the production of a fine white precipitate which was filtered prior to injection.

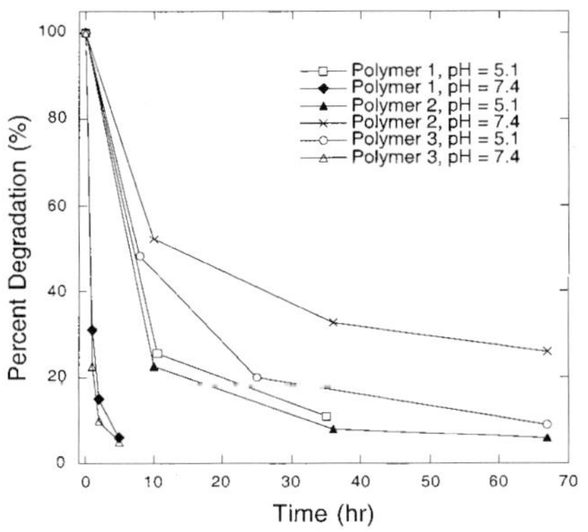

Figure 1. Degradation of polymers **1**−**3** at 37 °C at pH 5.1 and 7.4. Degradation is expressed as percent degradation over time based on GPC data.

conditions. It is worth noting that bis(β-amino acid) degradation products **4a**−**6a** would be less likely to undergo a retro-Michael reaction, as acrylic acids are generally less activated Michael addition partners.[27] We did not observe further degradation of compounds **4a**−**6a** under these conditions.

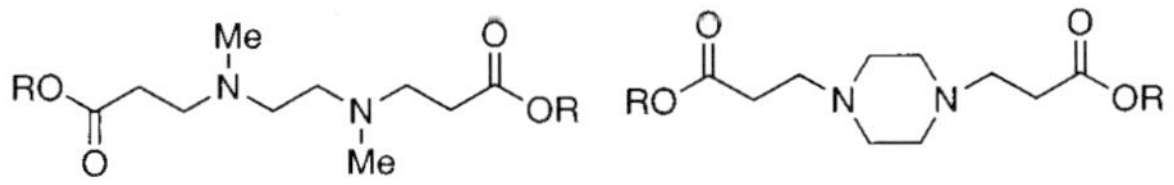

4a, R = H
4b, R = Me

5a, R = H
5b, R = Me

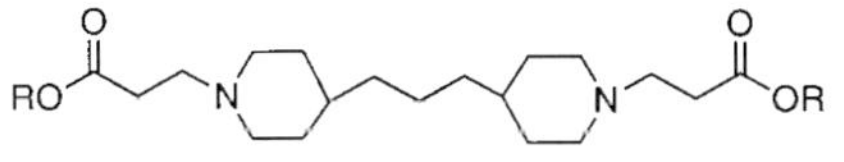

6a, R = H
6b, R = Me

The kinetics of polymer degradation were investigated under the range of conditions likely to be encountered by these polymers during transfection. Degradation was monitored at 37 °C at buffered pH values of 5.1 and 7.4 in order to approximate the pH of the environments within endosomal vesicles and the cytoplasm, respectively. The hydrochloride salts of polymers **1**−**3** were added to the appropriate buffer, incubated at 37 °C, and aliquots were removed at appropriate times. Aliquots were frozen immediately and lyophilized and polymer was extracted into THF/0.1M piperidine for analysis by GPC. Figure 1 shows the degradation profiles of polymers **1**−**3** as a function of time. In general, the polymers degraded more slowly at pH 5.1 than at pH 7.4. Polymers **1**−**3** displayed similar degradation profiles at pH 5.1, each polymer having a half-life of approximately 7−8 h. In contrast, polymers **1** and **3** were completely degraded in less than 5 h at pH 7.4. These results are consistent with the pH−degradation profiles of other amine-containing polyesters, such as poly(4-hydroxy-L-proline ester), in which pendant amine functionalities are hypothesized to act as intramolecular nucleophilic catalysts and contribute to more rapid degradation at higher pH.[8b,c] While we cannot rule out the possibility of intramolecular assistance, it is less likely for polymers **1**−**3** because the tertiary amines in these polymers should be less nucleophilic. The degradation of polymer **2** occurred more slowly at pH 7.4 than at pH 5.1 (Figure 1). This anomalous behavior is most likely due to the incomplete solubility of polymer **2** at pH 7.4 and the resulting heterogeneous nature of the degradation mileu.[28]

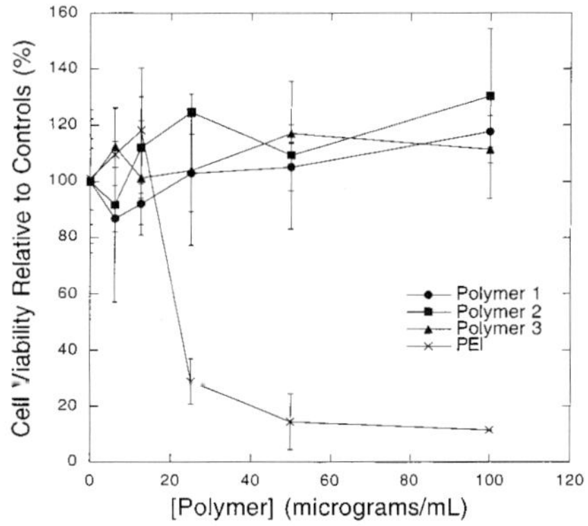

Figure 2. Cytotoxicity profiles of polymers **1**−**3** and PEI. Viability of NIH 3T3 cells is expressed as a function of polymer concentration. The molecular weights of polymers **1**, **2**, and **3** were 5800, 11 300, and 22 500, respectively. The molecular weight of the PEI employed was 25 000.

Cytotoxicity Assays. Poly(lysine) and PEI have been widely studied as DNA condensing agents and transfection vectors[2−4,7,18] and are the standards to which new polymeric vectors are often compared.[8] Unfortunately, as outlined above, these polymers are also associated with significant levels of cytotoxicity and high levels of gene expression are usually realized only at a substantial cost to cell viability. To determine the toxicity profile of polymers **1**−**3**, we conducted a MTT/thiazolyl blue dye reduction assay using the NIH 3T3 cell line and the hydrochloride salts of polymers **1**−**3** as initial indicators. The 3T3 cell line is commonly employed as a first level screening population for new transfection vectors, and the MTT assay is generally used as an initial indicator of cytotoxicity, as it determines the influences of added substances on cell growth and metabolism (Figure 2).[29]

Cells were incubated with polymer **1** (M_n = 5800), polymer **2** (M_n = 11 300), and polymer **3** (M_n = 22 500) as described in the Experimental Section. As shown in Figure 2, cells incubated with these polymers remained 100% viable relative to controls at concentrations of polymer up to 100 μg/mL. These results compare impressively to data obtained for cell populations treated with PEI ($M_n \approx$ 25 000), included as a positive control for our assay as well as to facilitate comparison to this well-known transfection agent. Fewer than 30% of cells treated with PEI remained viable at a polymer concentration of 25 μg/mL, and cell viability was as low as 10% at higher concentrations of PEI under otherwise identical conditions. We performed an analogous MTT assay using independently synthesized bis(β-amino acid)s **4a**−**6a** to screen the cytotoxicity of the hydrolytic degradation products of these polymers.[30] Compounds **4a**−**6a** and 1,4-butanediol did not perturb cell growth or metabolism

(26) For recent examples, see: (a) Schweikl, H.; Schmalz, G. *Mutat. Res.* **1999**, *438*, 71−78. (b) Yang, J.; Duerksen-Hughes, P. *Carcinogenesis* **1998**, *19*, 1117−1125.

(27) Perlmutter, P., in *Conjugate Addition Reactions in Organic Synthesis* Pergamon Press: New York, 1992.

(28) Polymers **2** and **3** are not completely soluble in water at pH 7.4. While polymer **3** dissolved relatively rapidly during the degradation experiment, solid particles of polymer **2** were visible for several days. We continue to investigate the degradation profiles of solid samples formed from polymers **2** and **3**.

(29) Hansen, M. B.; Neilson, S. E.; Berg, K. J. *Immunol. Methods* **1989**, *119*, 203-210.

(30) Bis(β-amino acid)s **4a**−**6a** should either be biologically inert or possess mild or acute toxicities which are lower than traditional polycationic transfection vectors. In either case, the degradation of these materials should facilitate rapid metabolic clearance.

in this initial screening assay (data not shown).[31] A more direct structure/function-based comparison between polymers **1**−**3** and PEI cannot be made due to differences in polymer structure and molecular weight, both of which contribute to polycation toxicity. Nonetheless, the excellent cytotoxicity profiles of polymers **1**−**3** alone suggested that they were interesting candidates for further study as DNA condensing agents.

Self-Assembly of Polymers 1−3 with Plasmid DNA. The tendency of cationic polyamines to interact electrostatically with the polyanionic backbone of DNA in aqueous solution is well-known. Provided that the polymers are sufficiently protonated at physiological pH, and that the amines are sterically accessible, such interactions can result in a self-assembly process in which the positively and negatively charged polymers form well-defined conjugates.[18c] The majority of polyamines investigated as DNA-complexing agents and transfection vectors have incorporated amines at the terminal ends of short, conformationally flexible side chains [e.g., poly(lysine) and methacrylate/methacrylamide polymers],[3,4,14] or accessible amines on the surfaces of spherical or globular polyamines (e.g., PEI and PAMAM dendrimers).[7,9] Because polymers **1**−**3** contain tertiary amines, and those tertiary amines are located in a sterically crowded environment (flanked on two sides by the polymer backbones), we were initially concerned that the protonated amines might not be sufficiently able to interact intimately with DNA.

The ability of polymers **1**−**3** to complex plasmid DNA was demonstrated through an agarose gel shift assay. Agarose gel electrophoresis separates macromolecules on the basis of both charge and size. Therefore, the immobilization of DNA on an agarose gel in the presence of increasing concentrations of a polycation has been widely used as an assay to determine the point at which complete DNA charge neutralization is achieved.[8] As mentioned above, the hydrochloride salts of polymers **1**−**3** are soluble in water. However, polymers **2** and **3** are not completely soluble at pH 7.2 over the full range of desired polymer concentrations. Therefore, DNA/polymer complexes were prepared in MES buffer (25 mM, pH = 6.0). DNA/polymer complexes were prepared by adding an aqueous solution of DNA to vortexing solutions of polymer in MES at desired DNA/polymer concentrations (see Experimental Section). The resulting DNA/polymer complexes remained soluble upon dilution in the electrophoresis running buffer (20 mM HEPES, pH = 7.2) and data were obtained at physiological pH. As a representative example, Figure 3 depicts the migration of plasmid DNA (pCMV−Luc) on an agarose gel in the presence of increasing concentrations of polymer **1**.

1 2 3 4 5 6 7 8 9 10

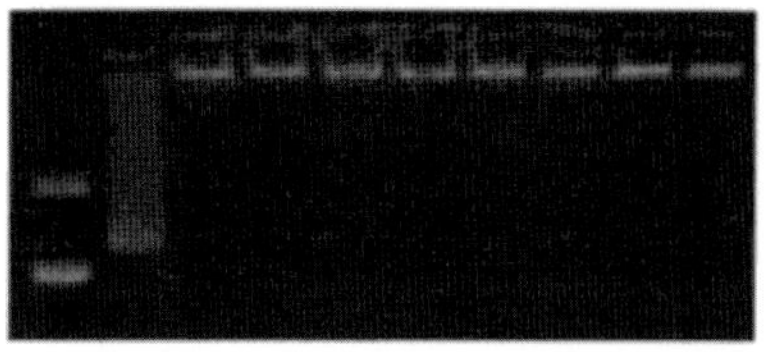

Figure 3. Agarose gel electrophoresis retardation of pCMV-Luc DNA by polymer **1**. Lane numbers correspond to different DNA/polymer weight ratios as follows: (1) 1:0 (DNA only), (2) 1:0.5, (3) 1:1, (4) 1:2, (5) 1:3, (6) 1:4, (7) 1:5. (8) 1:6, (9) 1:7, (10) 1:8.

As shown in Figure 3, retardation of DNA migration begins at DNA/**1** ratios as low as 1:0.5 (w/w) and migration is completely retarded at DNA/polymer ratios above 1:1.0 (w/w).[32] Polymers **2** and **3** completely inhibit the migration of plasmid DNA at DNA/polymer ratios (w/w) above 1:10 and 1:1.5, respectively (data not shown). These results vary markedly from gel retardation experiments conducted using model "monomers". Since the true monomers and the degradation products of polymers **1**−**3** do not adequately represent the repeat units of the polymers, we used bis(methyl ester)s **4b**−**6b** to examine the extent to which the polyvalency and cooperative binding of *poly*cations **1**−**3** is necessary to achieve DNA immobilization. "Monomers" **4b**−**6b** did not inhibit the migration of DNA at DNA/"monomer" ratios (w/w) of up to 1:30 (data not shown).

The reasons for the less-efficient complexation employing polymer **2** in the above gel electrophoresis assays most likely results from differences in the pK_a values of the amines in these polymers. The direct measurement of the pK_a values of polymers **1**−**3** is complicated by their degradability and solubility profiles. However, we predict the range of pK_a values of the amines in polymers **1** and **2** to extend from approximately 4.5 and 8.0 for polymer **1** and to 3.0 and 7.0 for polymer **2**, based on comparisons to structurally related poly(β-amino amides).[33] As a result, polymer **2** should be protonated to a lesser extent than polymer **1** at physiological or near-neutral pH and would therefore be a less effective DNA condensing agent. The range of pK_a values for polymer **3** should be higher than the range for polymers **1** and **2** due to the increased distance between the nitrogen atoms. Accordingly, polymer **3** forms complexes with DNA at substantially reduced concentrations relative to polymer **2**.

Agarose gel retardation assays are useful in determining the extent to which polycations interact with DNA. To be useful transfection agents, however, polycations must also be able to self-assemble plasmid DNA into polymer/DNA complexes small enough to enter a cell through endocytosis. For most cell types, this size requirement is on the order of 200 nm or less,[3] although larger particles can also be accommodated.[18b,c] The ability of polymers **1**−**3** to compact plasmid DNA into nanometer-sized structures was determined by quasi-elastic laser light scattering (QELS), and the relative surface charges of the resulting complexes were quantified through ζ-potential measurements. DNA/polymer particles used for particle sizing and ζ-potential measurements were formed as described above for agarose gel electrophoresis assays and diluted in 20 mM HEPES buffer (pH = 7.0) for analysis, as described in the Experimental Section.

Polymer **1** formed complexes with diameters ranging from 90 to 150 nm at DNA/polymer ratios above 1:2 (w/w), and polymer **2** condensed DNA into particles on the order of 60−125 nm at DNA/polymer ratios above 1:10. These results are consistent with the data obtained from agarose gel electrophoresis experiments above. However, the particles in these experiments aggregated over a period of hours to yield larger complexes with diameters in the range of 1−2 μm. The tendency of these particles to aggregate can be rationalized by the low

(31) It should be noted that the MTT assay is only a preliminary indicator of biocompatibility, and additional work must be done to more firmly establish the safety of these polymers.

(32) We report here DNA/polymer weight ratios rather than DNA/polymer charge ratios. Although both conventions are used in the literature, we find weight ratios to be more practical and universal, since the overall charge on a polyamine is subject to environmental variations in pH and temperature. While DNA/polymer charge ratios are descriptive for polymers such as poly(lysine), they are less meaningful for polymers such as PEI and **1**−**3**, which incorporate less basic amines.

(33) The pK_a values of structurally related poly(β-amino amides) containing piperazine and dimethylethylenediamine units in their backbones have been reported. (a) Barbucci, R.; Ferruti, P.; Micheloni, M.; Delfini, M.; Segre, A. L.; Conti, F. *Polymer* **1980**, *21*, 81−85. (b) Barbucci, R.; Ferruti, P.; Improta, C.; Delfini, M.; Segre, A. L.; Conti, F. *Polymer* **1978**, *19*, 1329−1334. (c) Barbucci, R.; Casolaro, M.; Ferruti, P.; Barone, V.; Lelj, F.; Oliva, L. *Macromolecules* **1981**, *14*, 1203−1209.

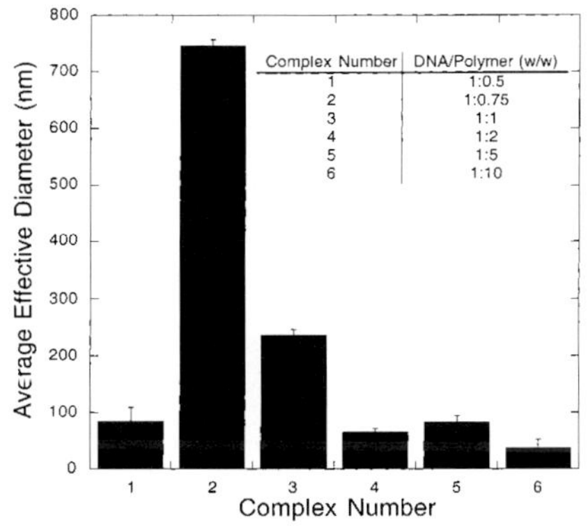

Figure 4. Average effective diameters of DNA/polymer complexes formed from pCMV-Luc plasmid and polymer **3** (M_n = 31 000) as a function of polymer concentration.

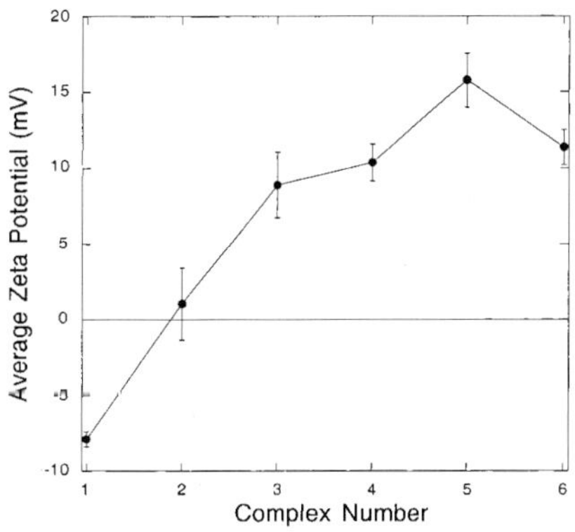

Figure 5. Average ζ-potentials of DNA/polymer complexes formed from pCMV-Luc plasmid and polymer **3** (M_n = 31 000) as a function of polymer concentration. The numbers for each complex correspond to the complex numbers in Figure 4.

ζ-potentials of the DNA/polymer particles observed under these conditions. For example, complexes formed from polymer **1** at DNA/polymer ratios above 1:10 had average ζ-potentials of +4.51 (±0.50) mV. The ζ-potentials of complexes formed from polymer **2** at DNA/polymer ratios above 1:20 were lower, reaching a limiting value of +1.04 (±0.57) mV. These differences correlate with the estimated pK_a values for these polymers, as the surfaces of particles formed from polymer **1** would be expected to be slightly more protonated than particles formed from polymer **2** at pH = 7.0.

Polymer **3** formed complexes with diameters in the range of 50−150 nm at DNA/polymer ratios above 1:2. As a representative example, Figure 4 shows the average effective diameters of particles formed with polymer **3** as a function of polymer concentration. The diameters of the particles varied within the above range from experiment to experiment under otherwise identical conditions, possibly due to subtle differences during the stirring or addition of DNA solutions during complex formation.[34] The ζ-potentials for complexes formed from polymer **3** were on the order of +10 to +15 mV at DNA/polymer ratios above 1:1, and the complexes did not aggregate extensively over an 18-h period (pH = 7, 25 °C). The positive ζ-potentials of these complexes may be significant beyond the context of particle stability, as net positive charges on particle surfaces may play a role in triggering endocytosis.[4,8c,18]

Particles formed from polymer **3** were also relatively stable at 37 °C. For example, a sample of DNA/**3** (DNA/**3** = 1:5, average diameter = 83 nm) was incubated at 37 °C for 4 h. After 4 h, a bimodal distribution was observed consisting of a fraction averaging 78 nm (>98% by number, 70 vol %) and a fraction of larger aggregates with average diameters of approximately 2.6 μm. These results suggest that the degradation of complexes formed from polymer **3** occurred more slowly than the degradation of polymer in solution or that partial degradation did not significantly affect the stability of the particles. The apparently increased stability of DNA/polymer complexes formed from degradable polycations relative to the degradation of the polymers in solution has also been observed for DNA/polymer complexes formed from poly(4-hydroxy-L-proline ester).[8b]

(34) The order of addition of polymer and DNA solutions had considerable impact on the nature of the resulting DNA/polymer complexes. To form small particles, for example, it was necessary to add the DNA solution to a vortexing solution of polymer. For cases in which polymer solutions were added to DNA, only large micrometer-sized aggregates were observed. Thus, it is possible that subtle differences in stirring or rate of addition could be responsible for variation in particle sizes.

The particle size and ζ-potential data in Figures 4 and 5 are consistent with models of DNA condensation observed with other polycations.[4,8] DNA is compacted into small negatively charged particles at very low polymer concentrations and particle sizes increase with increasing polymer concentration.[35] Complexes reach a maximum diameter as charge neutrality is achieved and aggregation occurs. Particle sizes decrease sharply at DNA/polymer concentrations above charge neutrality up to ratios at which additional polymer does not contribute to a reduction in particle diameter. This model is confirmed by ζ-potential measurements made on complexes formed from these polymers. As shown in Figure 5, the ζ-potentials of polymer/DNA particles formed from polymer **3** were negative at low polymer concentrations and charge neutrality was achieved near DNA/polymer ratios of 1:0.75, resulting in extensive aggregation. The ζ-potentials of the particles approached a limiting value ranging from +10 to +15 mV at DNA/polymer ratios above 1:2.

The average diameters of the complexes described above fall within the general size requirements for cellular endocytosis. We have initiated transfection experiments employing the NIH 3T3 cell line and the luciferase reporter gene (pCMV-Luc). Thus far, polymers **1** and **2** have shown no transfection activity in initial screening assays. By contrast, polymer **3** has demonstrated transfection efficiencies exceeding those of PEI under certain conditions.[36] These data suggest that polymers of this general

(35) Accurate light scattering data could not be obtained for DNA alone or for DNA/polymer associated species at DNA/polymer ratios lower than 1:0.5, since flexible, uncondensed DNA does not scatter light as extensively as compacted DNA (ref 18C).

(36) Transfection experiments were performed according to the following general protocol: Cells were grown in 6-well plates at an initial seeding density of 100 000 cells/well in 2 mL of growth medium. Cells were grown for 24 h after which the growth medium was removed and replaced with 2 mL of serum-free medium. DNA/polymer complexes were formed as described in the Experimental Section (2 μg of DNA, DNA/**3** = 1: 2 (w/w), 100 μL in MES (pH = 6.0)] and added to each well. DNA/PEI complexes were formed at a weight ratio of 1: 0.75, a ratio generally found in our laboratory to be optimal for PEI transfections. Transfections were carried out in serum-free medium for 4 h, after which medium was replaced with growth medium for 20 additional hours. Relative transfection efficiencies were determined by using luciferase (Promega) and cell protein assay (Pierce) kits. Results are expressed as relative light units (RLU) per milligram of total cell protein: for complexes of polymer **3**, 1.07 (±0.43) × 10^6 RLU/mg; for PEI complexes, 8.07 (±0.16) × 10^5 RLU/mg). No luciferase expression was detected for control experiments employing naked DNA or performed in the absence of DNA. These transfection data are the results of initial screening experiments, and a full report on the structure/activity relationships of polymers **1**−**3** under optimized conditions will be reported in due course.

structure hold promise as synthetic vectors for gene delivery and are interesting candidates for further study. The relative efficacy of polymer **3** relative to PEI is interesting, as our initial screening experiments were performed in the absence of chloroquine and polymer **3** does not currently incorporate an obvious means of facilitating endosomal escape. It should be noted, however, that the pK_a values of the amines in these polymers can be "tuned" to fall more directly within the range of physiologically relevant pH using this modular synthetic approach. Therefore, it will be possible to further engineer the "proton sponge" character[18] of these polymers, and thus enhance their transfection efficacies, directly through the incorporation of or copolymerization with different diamine monomers.

Summary

A general strategy for the preparation of new degradable polymeric DNA transfection vectors is reported. Poly(β-amino esters) **1−3** were synthesized *via* the conjugate addition of *N,N'*-dimethylethylenediamine, piperazine, and 4,4'-trimethylenedipiperidine to 1,4-butanediol diacrylate. The amines in the bis(secondary amine) monomers actively participate in bond-forming processes during polymerization, obviating the need for amine protection/deprotection processes which characterize the synthesis of other poly(amino esters). Accordingly, this approach should enable the generation of a variety of structurally diverse polyesters containing tertiary amines in their backbones in a single step from commercially available staring materials. Polymers **1−3** degraded hydrolytically in acidic and alkaline media to yield 1,4-butanediol and β-amino acids **4a−6a** and the degradation kinetics were investigated at pH 5.1 and 7.4. The polymers degraded more rapidly at pH 7.4 than at pH 5.1, consistent with the pH/degradation profiles reported for other poly(amino esters). An initial screening assay designed to determine the effects of polymers **1−3** on cell growth and metabolism suggested that these polymers and their hydrolytic degradation products were noncytotoxic relative to PEI, a nondegradable cationic polymer conventionally employed as a transfection vector.

Polymers **1−3** interacted electrostatically with plasmid DNA at physiological pH, initiating self-assembly processes that resulted in nanometer-scale DNA/polymer complexes. Agarose gel electrophoresis, quasi-elastic dynamic light scattering (QELS), and ζ-potential measurements were used to determine the extent of the interactions between the oppositely charged polyelectrolytes. All three polymers were found to condense DNA into soluble DNA/polymer particles on the order of 50−200 nm. Particles formed from polymers **1** and **2** aggregated extensively, while particles formed from polymer **3** exhibited positive ζ-potentials (e.g., +10 to +15 mV) and did not aggregate for up to 18 h. The nanometer-sized dimensions and reduced cytotoxicities of these DNA/polymer complexes suggest that polymers **1−3** may be useful as degradable polymeric gene transfection vectors. We are currently evaluating and optimizing the transfection activities of DNA/polymer complexes formed from these polymers and structurally related derivatives. A thorough understanding of structure/activity relationships existing for this class of polymer will expedite the design of safer polymer-based alternatives to viral transfection vectors for gene therapy.

Experimental Section

General Considerations. All manipulations involving live cells or sterile materials were performed in a laminar flow using standard sterile technique. ^1H NMR (300.100 MHz) and ^{13}C NMR (75.467 MHz) spectra were recorded on a Varian mercury spectrometer. All chemical shift values are given in parts per million and are referenced with respect to residual proton or carbon signal from solvent. Organic phase gel permeation chromatography (GPC) was performed using a Hewlett-Packard 1100 series isocratic pump, a Rheodyne model 7125 injector with a 100-μL injection loop, and two PL-Gel mixed-D columns in series (5 μm, 300 × 7.5 mm, Polymer Laboratories, Amherst, MA). THF/0.1M piperidine was used as the eluent at a flow rate of 1.0 mL/min. Data were collected using an Optilab DSP interferometric refractometer (Wyatt Technology, Santa Barbara, CA) and processed using the TriSEC GPC software package (Viscotek Corporation, Houston, TX). The molecular weights and polydispersities of the polymers are reported relative to monodisperse polystyrene standards. Aqueous phase GPC was performed by American Polymer Standards (Mentor, OH) using Ultrahydrogel L and 120A columns in series (Waters Corporation, Milford, MA). Water (1% acetic acid, 0.3 M NaCl) was used as the eluent at a flow rate of 1.0 mL/min. Data were collected using a Knauer differential refractometer and processed using an IBM/PC GPC-PRO 3.13 software package (Viscotek Corporation, Houston, TX). The molecular weights and polydispersities of the polymers are reported relative to poly(2-vinylpyridine) standards. For cytotoxicity assays, absorbance was measured using a Dynatech Laboratories MR5000 microplate reader at 560 nm.

Materials. *N,N'*-Dimethylethylenediamine, piperazine, and 4,4'-trimethylenedipiperidine were purchased from Aldrich Chemical Co. (Milwaukee, WI). 1,4-Butanediol diacrylate was purchased from Alfa Aesar Organics (Ward Hill, MA). 3-[4,5-Dimethylthiazol-2-yl]-2,5-diphenyltetrazolium bromide (MTT) was purchased from Sigma Chemical Co. (St. Louis, MO). Plasmid DNA (pCMV-Luc) was produced in *E. coli* (DH5α, a kind gift from Zycos, Inc., Cambridge, MA), isolated with a Qiagen kit, and purified by ethanol precipitation. NIH 3T3 cells were purchased from American Type Culture Collection (Manassas, VA) and grown at 37 °C, 5% CO_2 in Dulbecco's modified Eagle's medium, 90%; fetal bovine serum, 10%; penicillin, 100 units/mL; streptomycin, 100 μg/mL. All other materials and solvents were used as received without further purification.

General Polymerization Procedure. In a typical experiment, 1,4-butanediol diacrylate (0.750 g, 0.714 mL, 3.78 mmol) and diamine (3.78 mmol) were weighed into two separate vials and dissolved in THF (5 mL). The solution containing the diamine was added to the diacrylate solution *via* pipet. A Teflon-coated stirbar was added, the vial was sealed with a Teflon-lined screw-cap, and the reaction was heated at 50 °C. After 48 h, the reaction was cooled to room temperature and dripped slowly into vigorously stirring diethyl ether or hexanes. Polymer was collected and dried under vacuum prior to analysis.

Synthesis of Polymer 1. Polymer **1** was prepared according to the general procedure outlined above. ^1H NMR δ ($CDCl_3$, 300 MHz) 4.11 (br t, 4H), 2.75 (br t, J = 7.05 Hz, 4 H), 2.53 (br s, 4H), 2.50 (br t, (obsc), J = 7.20 Hz, 4H), 2.28 (br s, 6 H), 1.71, (br m, 4 H). ^{13}C NMR δ ($CDCl_3$, 75.47 MHz) 172.55, 64.14, 55.31, 53.39, 42.47, 32.54, 25.53.

Synthesis of Polymer 2. Polymer **2** was prepared according to the general procedure outlined above. ^1H NMR δ ($CDCl_3$, 300 MHz) 4.11 (br t, 4 H), 2.74 (br t, J = 7.35, 4 H), 2.56 (br m, 12 H), 1.71 (br t, 4 H). ^{13}C NMR δ ($CDCl_3$, 75.47 MHz) 172.24, 64.19, 53.55, 52.75, 32.27, 25.52.

Synthesis of Polymer 3. Polymer **3** was prepared according to the general procedure outlined above. ^1H NMR δ ($CDCl_3$, 300 MHz) 4.11 (br t, 4 H), 3.00 (br m, 4 H), 2.79 (br m, 4 H), 2.65 (br m, 4 H), 2.11 (br m, 4 H), 1.70 (br m, 8 H), 1.25 (br m, 12 H). ^{13}C NMR δ ($CDCl_3$, 75.47 MHz) 172.37, 64.13, 53.89 (br), 36.74, 35.58, 32.11 (br), 25.45, 24.05.

Polymer Degradation Studies. The hydrochloride salts of polymers **1−3** were dissolved in acetate buffer (1 M, pH = 5.1) or HEPES buffer (1 M, pH = 7.4) at a concentration of 5 mg/mL (the use of millimolar concentrations of buffer resulted in substantial reduction of pH during degradation due to the production of acidic degradation products). Samples were incubated at 37 °C on a mechanical rotator and aliquots (1 mL) were removed at appropriate time intervals. Aliquots were frozen immediately in liquid nitrogen and lyophilized. Polymer was extracted from dried buffer salts using THF/0.1 M piperidine (1 mL) and samples were analyzed directly by GPC.

Formation of DNA/Polymer Complexes and Agarose Gel Retardation Assays. DNA/polymer complexes were formed by adding 50 μL of a plasmid DNA solution (pCMV Luc, 2 μg/50 μL in water) to a gently vortexing solution of the hydrochloride salt of polymers **1**–**3** (50 μL in 25 mM MES, pH = 6.0, concentrations adjusted to yield desired DNA/polymer weight ratios). The samples were incubated at room temperature for 30 min, after which 20 μL was run on a 1% agarose gel (HEPES, 20 mM, pH = 7.2, 65 V, 30 min). Samples were loaded on the gel with a loading buffer consisting of 10% Ficoll 400 (Amersham Pharmacia Biotech, Uppsala, Sweden) in HEPES (25 mM, pH = 7.2). Bromphenol blue was not included as a visual indicator in the loading buffer, since this charged dye appeared to interfere with the complexation of polymer and DNA. DNA bands were visualized by ethidium bromide staining.

Quasi-elastic Laser Light Scattering (QELS) and Measurement of ζ-Potentials. QELS experiments and ζ-potential measurements were made using a ZetaPALS dynamic light scattering detector (Brookhaven Instruments Corporation, Holtsville, NY, 15 mW laser, incident beam = 676 nm). DNA/polymer complexes were formed as described above for agarose gel retardation assays. Samples were diluted with 900 μL of HEPES (20 mM, pH = 7.0), added to a gently vortexing sample of DNA/polymer complex (total volume = 1 mL, pH = 7.0). Average particle sizes and ζ-potentials were determined at 25 °C. Correlation functions were collected at a scattering angle of 90°, and particle sizes were calculated using the MAS option of BIC's particle sizing software (version 2.30), using the viscosity and refractive index of pure water at 25 °C. Particle sizes are expressed as effective diameters assuming a log-normal distribution. Average electrophoretic mobilities were measured at 25 °C using BIC PALS zeta potential analysis software and zeta potentials were calculated using the Smoluchowsky model for aqueous suspensions. Three measurements were made on each sample and results are reported as average diameters and zeta potentials.

Cytotoxicity Assays. Immortalized NIH 3T3 cells were grown in 96-well plates at an initial seeding density of 10 000 cells/well in 200 μL growth medium (90% Dulbecco's modified Eagle's medium, 10% fetal bovine serum, penicillin 100 units/mL, streptomycin 100 μg/mL). Cells were grown for 24 h, after which the growth medium was removed and replaced with 180 μL of serum-free medium. Appropriate amounts of polymer were added in 20 μL aliquots. Samples were incubated at 37 °C for 5 h, and the metabolic activity of each well was determined by using a MTT/thiazolyl blue assay: to each well was added 25 μL of a 5 mg/mL solution of MTT stock solution in sterile PBS buffer. The samples were incubated at 37 °C for 2 h, and 100 μL of extraction buffer (20% w/v SDS in DMF/water (1:1), pH = 4.7) was added to each well. Samples were incubated at 37 °C for 24 h. Optical absorbance was measured at 560 nm with a microplate reader and expressed as a percent relative to control cells.

Acknowledgment. We thank Drs. David Putnam, Christine Gentry, Daniel Anderson, and Sachiko Hirosue for insightful discussions and their significant contributions to this work. Christine Gentry is thanked especially for the generous donation of purified plasmid, cells, and her patience in teaching cell culture techniques and assays. Aqueous GPC analysis was performed by American Polymer Standards (Mentor, OH). This work was supported in part by the NSF (cooperative agreement ECC9843342 to the MIT Biotechnology Process Engineering Center), the Department of the Army (cooperative agreement DAMD 17-99-2-9-001 to the Center for Innovative Minimally Invasive Therapy), and the NIH (grant number GM26698). D.M.L. wishes to thank the NIH for a Postdoctoral Fellowship (NRSA Fellowship 1 F32 GM20227-01).

JA0015388

J. Am. Chem. Soc. **2001,** *123,* 8155−8156 8155

Accelerated Discovery of Synthetic Transfection Vectors: Parallel Synthesis and Screening of a Degradable Polymer Library

David M. Lynn, Daniel G. Anderson, David Putnam, and Robert Langer*

Department of Chemical Engineering
Massachusetts Institute of Technology
Cambridge, Massachusetts 02139

Received May 27, 2001
Revised Manuscript Received June 25, 2001

The safe and efficient delivery of therapeutic DNA to cells represents a fundamental obstacle to the clinical success of gene therapy.[1] The challenges facing synthetic delivery vectors are particularly clear, as both cationic polymers and liposomes are less effective at mediating gene transfer than viral vectors. The incorporation of new design criteria has led to recent advances toward functional delivery systems.[2,3] However, the paradigm for the development of polymeric gene delivery vectors remains the incorporation of these design elements into materials as part of an iterative, linear process—an effective, albeit slow, approach to the discovery of new vectors. Herein, we report a parallel approach suitable for the synthesis of large libraries of degradable cationic polymers and oligomers and the discovery of new synthetic vector families through rapid cell-based screening assays.[4]

Poly(β-amino ester)s are hydrolytically degradable, condense plasmid DNA at physiological pH, and are readily synthesized via the conjugate addition of primary or secondary amines to diacrylates (eqs 1 and 2).[5] An initial screen of model polymers identified these materials as potential gene carriers and demonstrated that structural variations could have a significant impact on DNA binding and transfection efficacies.[5] We reasoned that this approach provided an attractive framework for the elaboration of large libraries of structurally unique polymers for several reasons: (1) diamine and diacrylate monomers are inexpensive, commercially available starting materials, (2) polymerization can be accomplished directly in a single synthetic step, and (3) purification steps are generally unnecessary as no byproducts are generated during polymerization.

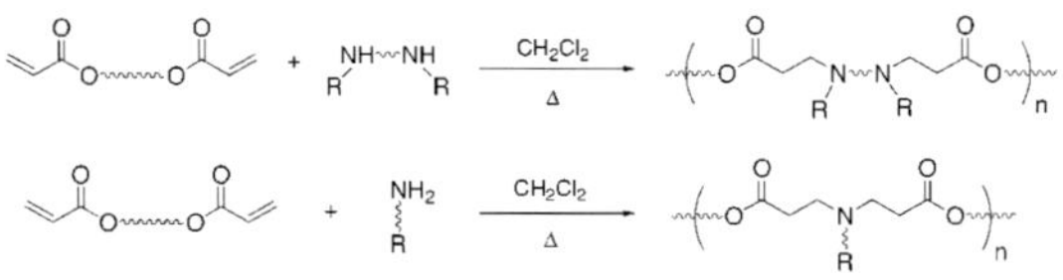

(Eq 1)

(Eq 2)

The paucity of commercially available bis(secondary amines) limits the degree of structural diversity that can be achieved using the above synthetic approach. However, the pool of useful, commercially available monomers is significantly expanded when primary amines are considered as potential library building blocks. Because the conjugate addition of amines to acrylate groups is generally tolerant of functionalities such as alcohols, ethers, and tertiary amines,[6] we believed that the incorporation of functionalized primary amine monomers into our synthetic strategy would serve to broaden structural diversity. Diacrylate monomers **A**−**G** and amine monomers **1**−**20** were selected for the synthesis of an initial screening library (Figure 1).

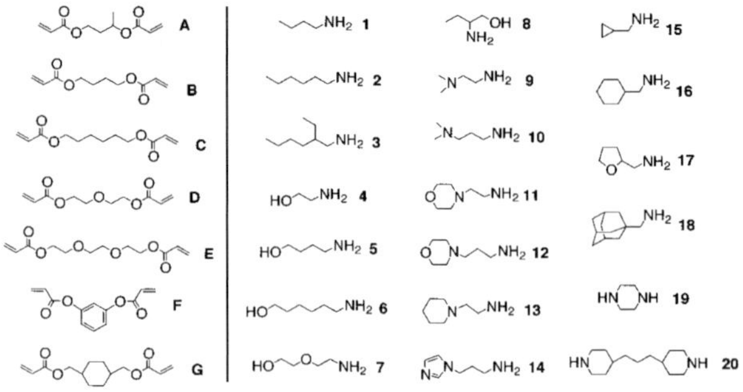

Figure 1. Diacrylate (**A**−**G**) and amine (**1**−**20**) monomers chosen for the synthesis of an initial screening library.

The size of the library constructed from this set of monomers (7 diacrylates × 20 amines = 140 structurally unique polymers) was chosen to be large enough to incorporate sufficient diversity, yet small enough to be practical without the need for automation in our initial studies. It was unclear at the outset whether a polymer such as **G16** (formed from hydrophobic and sterically bulky monomers **G** and **16**) would be water-soluble at physiological pH or be able to condense DNA sufficiently. However, monomers of this type were deliberately incorporated to fully explore diversity space, and in anticipation that this library may ultimately be useful as a screening population for the discovery of materials for applications other than gene delivery.[4,7]

Polymerization reactions were conducted simultaneously as an array of individually labeled vials. Reactions were performed in methylene chloride at 45 °C for 5 days, and polymers were isolated by removal of solvent to yield 600−800 mg of each material. Reactions performed on this scale provided amounts of each material sufficient for routine analysis by GPC and all subsequent DNA-binding, toxicity, and transfection assays. A survey of 55% of the library by GPC indicated molecular weights ranging from 2000 to 50000 (relative to polystyrene standards). As high molecular weights are not required for DNA-complexation and transfection (as shown below),[8] this library provided a collection of polymers and oligomers suitable for subsequent screening assays.

Of the 140 members of the screening library, 70 samples were sufficiently water-soluble (2 mg/mL, 25 mM acetate buffer, pH = 5.0) to be included in an electrophoretic DNA-binding assay (Figure 2). To perform this assay as rapidly and efficiently as possible, samples were mixed with plasmid DNA at ratios of 1:5 and 1:20 (DNA/polymer, w/w) in 96-well plates and loaded into an agarose gel slab capable of assaying up to 500 samples using a multichannel pipettor. All 70 water-soluble polymer samples were assayed simultaneously at two different DNA/polymer ratios in less than 30 min. As shown in Figure 2, 56 of the 70 water-

* To whom correspondence should be addressed.

(1) (a) Luo, D.; Saltzman, W. M. *Nat. Biotechnol.* **2000**, *18*, 33−37. (b) Anderson, W. F. *Nature* **1996**, *392 Suppl.* 25−30.

(2) (a) Lim, Y.; Kim, S.; Lee, Y.; Lee, W.; Yang, T.; Lee, M.; Suh, H.; Park, J. *J. Am. Chem. Soc.* **2001**, *123*, 2460−2461. (b) Lim, Y.; Kim, C.; Kim, K.; Kim, S. W.; Park, J. *J. Am. Chem. Soc.* **2000**, *122*, 6524−6525. (c) Hwang, S. J.; Bellocq, N. C.; Davis, M. E. *Bioconjugate Chem.* **2001**, *12*, 280−290.

(3) (a) Putnam, D.; Gentry, C. A.; Pack, D. W.; Langer, R. *Proc. Natl. Acad. Sci. U.S.A.* **2001**, *98*, 1200−1205. (b) Benns, J. M.; Choi, J.; Mahato, R. I.; Park, J.; Kim, S. W. *Bioconjugate Chem.* **2000**, *11*, 637−645. (c) Midoux, P.; Monsigny, M. *Bioconjugate Chem.* **1999**, *10*, 406−411.

(4) For a report on the parallel synthesis and screening of degradable polymers for tissue engineering, see: Brocchini, S.; James, K.; Tangpasuthadol, V.; Kohn, J. *J. Am. Chem. Soc.* **1997**, *119*, 4553−4554.

(5) Lynn, D. M.; Langer, R. *J. Am. Chem. Soc.* **2000**, *122*, 10761−10768.

(6) Ferruti, P.; Barbucci, R. *Adv Polym. Sci.* **1984**, *58*, 55−92.

(7) Lynn, D. M.; Amiji, M. M.; Langer, R. *Angew. Chem., Int. Ed.* **2001**, *40*, 1707−1710.

(8) Kabanov, A. V.; Felgner, P. L.; Seymour, L. W. *Self-Assembling Complexes for Gene Delivery: From Laboratory to Clinical Trial*, John Wiley and Sons: New York, 1998.

10.1021/ja016288p CCC: $20.00

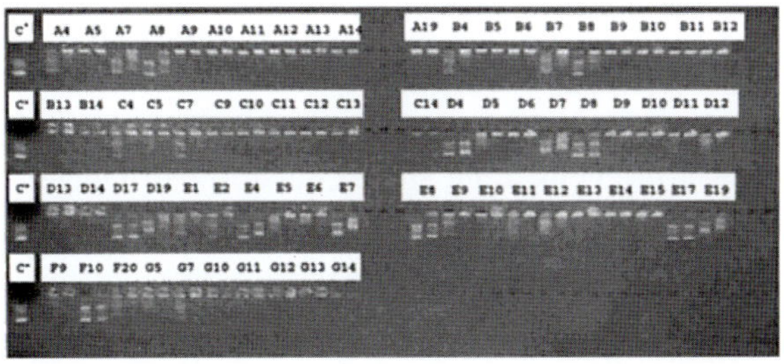

Figure 2. Gel electrophoresis assay used to identify DNA-complexing polymers. Lane annotations correspond to the 70 water-soluble members of the screening library. For each polymer, assays were performed at DNA/polymer ratios of 1:5 (left well) and 1:20 (right well). Lanes marked C* contain DNA alone (no polymer) and were used as a control.

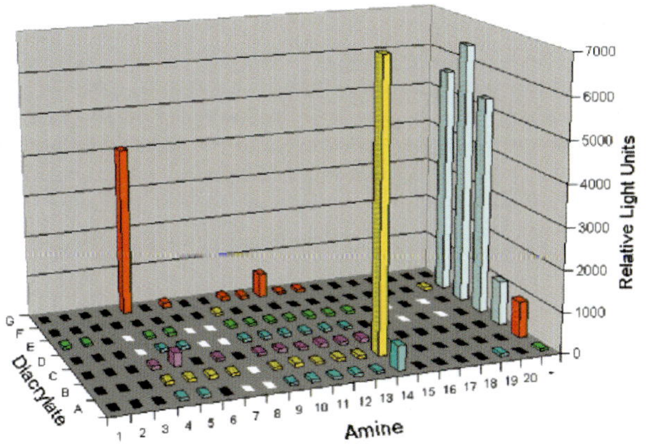

Figure 3. Transfection data as a function of structure for an assay employing pCMV-Luc (600 ng/well, DNA/polymer = 1:20). Light units are arbitrary and not normalized to total cell protein; experiments were performed in triplicate (error bars not shown). Black squares represent water-insoluble polymers, white squares represent water-soluble polymers that did not complex DNA in Figure 2. The right column (marked *) displays values for the following control experiments: no polymer (green), PEI (red), and Lipofectamine 2000 (light blue).

soluble polymer samples interacted sufficiently with DNA to retard migration through the gel matrix (e.g., **A4** or **A5**), employing the 1:20 DNA/polymer ratio as an exclusionary criterion. Fourteen polymers were discarded from further consideration (e.g., **A7** and **A8**), as these polymers did not complex DNA sufficiently.

The DNA-complexing materials identified in the above assay were further investigated in transfection assays employing plasmid DNA and the COS-7 cell line. All assays were performed simultaneously with the firefly luciferase reporter gene (pCMV−Luc), and levels of expressed protein were determined using a commercially available assay kit and a 96-well luminescence plate reader. Figure 3 displays data generated from an assay employing pCMV-Luc (600 ng/well) at DNA/poly ratios of 1:20 (w/w). The majority of the polymers screened did not mediate transfection above a level typical of "naked" DNA (no polymer) controls under these conditions. However, several wells expressed higher levels of protein and the corresponding polymers were identified as "hits" in this assay. Polymers **B14** (MW = 3180) and **G5** (MW = 9170), for example, yielded transfection levels 4−8 times higher than control experiments employing poly(ethylene imine) (PEI), a polymer conventionally employed as a synthetic transfection vector,[9] and transfection levels within or exceeding the range of expressed protein using Lipofectamine 2000,[9b] a leading commercially available lipid-based transfection vector system. Polymers **A14**, **C5**, **G7**, **G10**, and **G12** were also identified as positive "hits" in the above experiment, but levels of gene expression were lower than those for **B14** and **G5**.

Structural differences among synthetic polymers typically preclude a general set of optimal transfection conditions. For example, polymers **C5**, **C14**, and **G14** were toxic at the higher concentrations employed above,[10] but mediated transfection at lower DNA and polymer concentrations (data not shown). The assay system described above can easily be modified to evaluate polymers as a function of DNA concentration, DNA/polymer ratio, cell seeding densities, or incubation times. Additional investigation will be required to more fully evaluate the potential of this screening library, and experiments to this end are currently underway.

The assays above were performed in the absence of chloroquine, a weak base commonly added to enhance in vitro transfection,[3,8] and the results therefore reflect the intrinsic abilities of those materials to mediate transfection. The polymers containing monomer **14** are structurally similar to other histidine containing "proton sponge" polymers,[3] and could enhance transfection by buffering acidic intracellular compartments and mediating endosomal escape.[3,9] The efficacy of polymers containing monomer **5** is surprising in this context, as these materials do not incorporate an obvious means of facilitating endosomal escape. While the efficacy of these latter polymers is not yet understood, their discovery helps validate our parallel approach and highlights the value of incorporating structural diversity, as these polymers may not have been discovered on an ad hoc basis. Polymers incorporating hydrophilic diacrylates **D** and **E** have not produced "hits" under any conditions thus far, providing a possible basis for the development of more focused libraries useful for the elucidation of structure/activity relationships.

We have generated a library of 140 degradable polymers and oligomers useful for the discovery of new DNA-complexing materials and gene delivery vectors. Several of these materials are capable of condensing DNA into structures small enough to be internalized by cells and releasing the DNA in a transcriptionally active form. The total time currently required for library design, synthesis, and initial screening assays is approximately 2 weeks. However, the incorporation of robotics and additional monomers could significantly accelerate the pace at which new DNA-complexing materials and competent transfection vectors are identified. We continue to evaluate the potential of this approach through the elucidation of structure/function relationships and the development of second-generation screening libraries.

Acknowledgment. This work was supported by the NSF (via the MIT Biotechnology Process and Engineering Center) and the Department of the Army (through the Center for Innovative Minimally Invasive Therapy). D.M.L and D.G.A thank the NIH for Postdoctoral Fellowships.

Supporting Information Available: Synthetic details and experimental protocols (PDF). This material is available free of charge via the Internet at http://pubs.acs.org.

JA016288P

(9) (a) Boussif, O.; Lezoualc'H, F.; Zanta, M. A.; Mergny, M. D.; Scherman, D.; Demeneix, B.; Behr, J. P. *Proc. Natl. Acad. Sci. U.S.A.* **1995**, *92*, 7297−7301. (b) Available from Invitrogen Corp. (Carlsbad, CA).

(10) Determined by the absence of cells in wells containing these polymers as observed upon visual inspection. These polymers were less toxic and mediated transfection at lower concentrations.

COMMUNICATIONS

Erodible Conducting Polymers for Potential Biomedical Applications**

Alexander N. Zelikin, David M. Lynn, Jian Farhadi, Ivan Martin, Venkatram Shastri,* and Robert Langer*

Electrically conducting polymers have been investigated for numerous applications, including organic substitutes for metals in electronic circuits,[1] coatings for electromagnetic shielding,[2] analytical and biological sensing devices,[3–5] and as substrates for the manipulation of mammalian cell growth and function.[6] From a separate standpoint, the development and application of biodegradable polymers has had a profound impact in numerous medical and surgical applications.[7] It occurred to us that the creation of biocompatible, degradable conducting polymers could open the door to a number of new biomedical applications.[8] Several groups have reported the synthesis and characterization of conducting poly(thiophene) derivatives containing hydrolyzable ester groups in the polymer backbone.[9] However, to our knowledge, the degradability and biocompatibility of these polymers has not been established, and the reduced environmental stability of oxidized polythiophenes could limit their application under physiological conditions.

Of the various electrically conducting polymers, polypyrrole (Ppy) has been the most widely studied material for potential biomedical applications because it is relatively stable to air and water and can be readily synthesized through chemical and electrochemical routes.[10] Additionally, Ppy is a suitable substrate for cell attachment and proliferation[6] and possesses excellent biocompatibility in vivo.[8, 11] Our attempts to synthesize biodegradable Ppy derivatives through the incorporation of backbone ester moieties (in analogy to the polythiophenes discussed above) have thus far been unsuccessful because of extensive side reactions accompanying pyrrole coupling chemistry and the generally poor oxidative stability of the requisite oligopyrrole intermediates. Herein, we report an alternative strategy for the design of *erodible* Ppy materials based on the chemical and electrochemical polymerization of β-substituted pyrrole monomers containing ionizable and/or hydrolyzable side groups [Eq. 1]. These polymers can be fabricated into conductive materials that erode slowly under physiological conditions and support the growth, proliferation, and differentiation of primary human cells in in vitro cell culture assays.

[*] V. Shastri, R. Langer, A. N. Zelikin, D. M. Lynn
Department of Chemical Engineering, Room E25-342
Massachusetts Institute of Technology
Cambridge, Massachusetts 02139 (USA)
Fax: (+1) 617-258-8827
E-mail: shastriv@seas.upenn.edu, rlanger@mit.edu
J. Farhadi, I. Martin
Department of Surgery, Research Division
Kantonsspital Basel, Basel, (Switzerland)

[**] Financial support was provided by the Department of the Army (Cooperative Agreement No. DAMD 17-99-2-9001 to the Center for Innovative Minimally Invasive Therapy) and the NIH (GM26698). DML wishes to thank the NIH for a Postdoctoral Fellowship (NRSA Fellowship No. 1 F32 GM20227-01).

COMMUNICATIONS

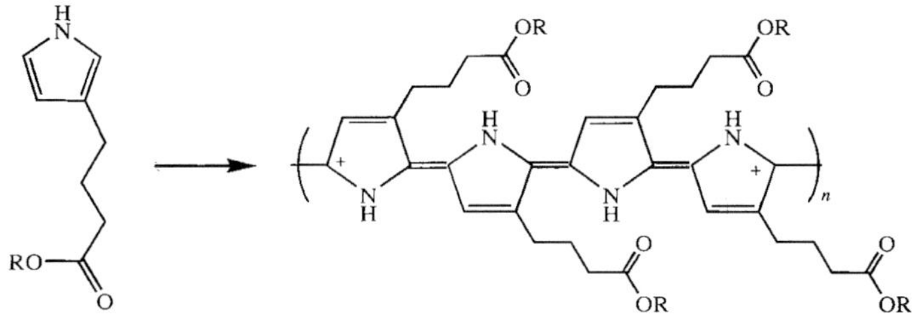

(1)

1: R = H
2: R = CH_3

poly-1
poly-2

We recently reported the synthesis of carboxylic acid functionalized pyrrole monomer **1**.[12] Direct esterification of compound **1** by reflux in methanol in the presence of phenylsulfonic acid afforded methyl ester functionalized monomer **2** in 80% yield. Both monomers underwent oxidative electrochemical polymerization and ferric chloride mediated chemical polymerization to yield conductive thin films and colloidal dispersions [poly-**1** and poly-**2**, Eq. 1]. In contrast to the insoluble cross-linked polymer deposits typically obtained from the oxidative polymerization of pyrrole, poly-**2** was completely soluble in common organic solvents such as THF and chloroform. Poly-**1** was insoluble in organic solvents, but was sparingly soluble in water. The solubility of these polymers suggested that they were not extensively cross-linked, presumably because of the presence of the β-substituents that prevent the cross-linking at this position. The sheet resistance of thin films prepared from poly-**1** was 300 ($\pm$100) Ω, which is consistent with literature values for β-substituted polypyrroles.[10, 13]

In view of the low solubility of poly-**1** in water, we hypothesized that thin films formed from poly-**1** might undergo a gradual erosion process under physiological conditions in a manner approximating the behavior of conventional biodegradable polymer systems. To test this hypothesis, we incubated thin films of poly-**1** in acetate (pH 5.0), HEPES (pH 7.2 HEPES = 2-[4-(2-hydroxyethyl)-1-piperazinyl]ethanesulfonic acid) and TAE (pH 8.2; TAE = tris(hydroxymethyl)aminomethane acetate, 10 mM ethylendiaminetetraacetate) buffers and investigated their erosion/dissolution profile at 37 °C as a function of pH value. Thin films deposited on iridium tin oxide (ITO) glass dissolved within 24 hours at pH 8.2, but dissolved more slowly at pH 5. Intermediate dissolution rates were observed at pH 7.2. Unfortunately, the deterioration of film quality over time, coupled with the tendency of the films to detach from the ITO substrates and float, prevented a quantitative analysis of the erosion process over an extended period of time.

To facilitate a more quantitative analysis, we prepared pressed pellets of chemically polymerized poly-**1** (20 mg, diameter = 8 mm, thickness = 0.5 mm) and followed the erosion/dissolution of polymer by UV/Vis spectroscopy in acetate (pH 5.0), HEPES (pH 7.2), and TAE (pH 8.2) buffers (see Experimental Section). As shown in Figure 1, the erosion rate of pellets at pH 7.2 was twice the rate of erosion at pH 5.0, and the rate was dramatically enhanced under alkaline conditions. Pressed pellets formed from hydrophobic methyl ester functionalized poly-**2** eroded much more slowly than their acid-functionalized counterparts (50 mg, diameter = 8 mm, thickness = 1 mm) at pH 7.2. As shown in Figure 2, the mass loss from pellets of poly-**1** was 27% after 80 days of incubation at 37 °C, while the mass loss from pellets of poly-**2** was 6%. These results parallel an increase in pyrrole oligomer concentration in solution over time (as determined by UV/Vis spectroscopy, Figure 2). The disparity in the erosion/dissolution rates between poly-**1** and poly-**2** suggests a means of tailoring erosion rates through judicious copolymerization of both

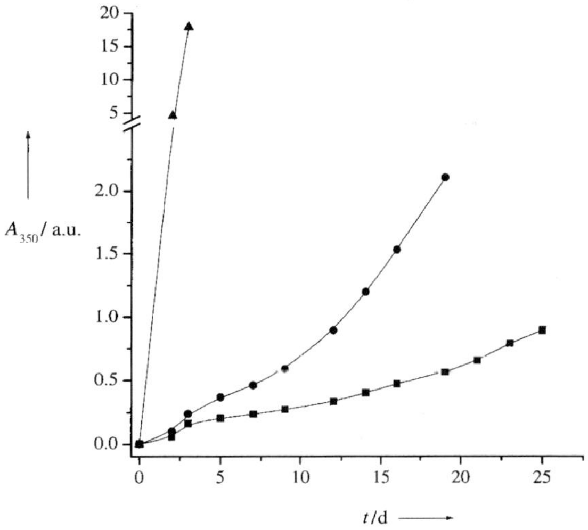

Figure 1. Dissolution of pressed pellets of poly-**1**, monitored by UV/Vis spectroscopy, at different pH: 8.2 (▲), 7.2 (●), and 5.0 (■)

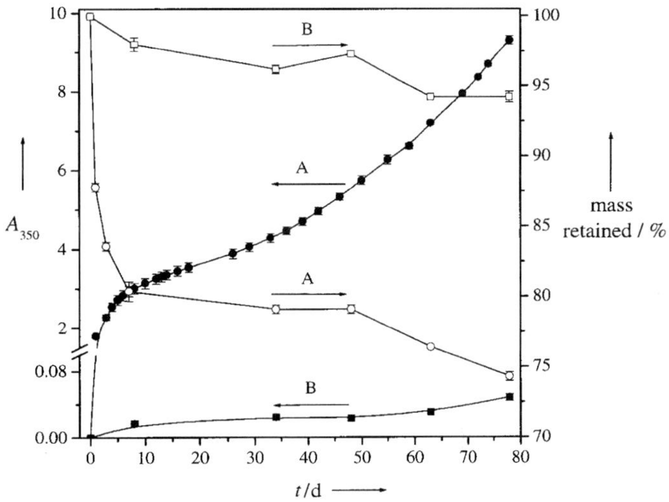

Figure 2. Dissolution of pressed pellets of poly-**1** (A) and poly-**2** (B), as monitored by solution UV/Vis absorbance (●, ■) and mass loss (○, □).

  1433-7851/02/4101-0142 $ 17.50+.50/0 Angew. Chem. Int. Ed. 2002, 41, No. 1

hydrophilic and hydrophobic monomers.[14] Unfortunately, the surface roughness arising from erosion (scanning electron microscopy, data not shown), prevented a detailed study of the conductivity of the pellets as a function of degradation time; the design of a system to address this issue is currently underway. The sheet resistance of thin films prepared from poly-**1** rises slightly upon incubation in water, and reaches a plateau at approximately 700 ($\pm$200) Ω after two days. These data indicate that the films remain conductive under aqueous or water-equilibrated conditions. The resistance of these equilibrated films is similar to that of films formed from Ppy previously used for the stimulation of cells and tissue.[8]

We also carried out cytocompatibility studies on thin films of poly-**1** using primary human bone-marrow derived mesenchymal progenitor cells (MPCs) as a model cell system. All seeded cells attached to the substrate and proliferated at an average rate of 0.63 doublings per day, comparable to that measured for cells growing on tissue-culture-treated plastic.[15] MPCs cultured for nine days on the erodible Ppy substrate in medium promoting osteogenic differentiation progressed towards the osteoblast lineage, as assessed by the increased mRNA expression of collagen Type I (1.8-fold), bone sialoprotein (2.4-fold), and osteopontin (19.2-fold; Figure 3).

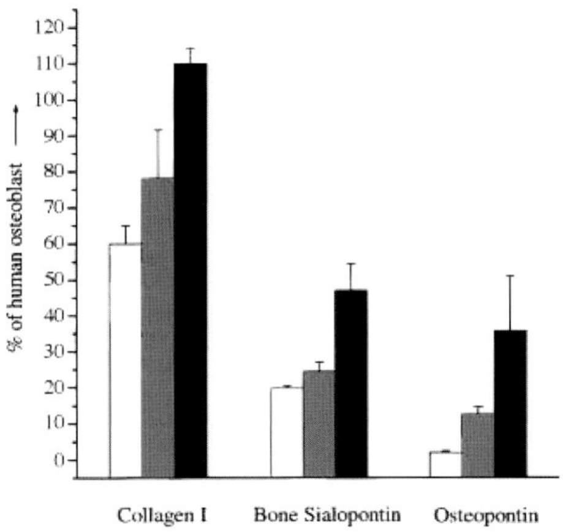

Figure 3. mRNA expression levels of osteoblast-related markers (collagen Type I, bone sialoprotein, and osteopontin) in mesenchymal progenitor cells grown on thin films of poly-**1**, before (white) and after exposure to soluble osteogenic stimuli for five (gray) or nine (black) days. The values, quantified and normalized to the amounts of 18-S rRNA (housekeeping gene) by using real-time reverse transcriptase polymerase chain reaction (RT-PCR), are expressed as a percentage of those measured in human osteoblasts.

Although additional experiments in vivo will be required to more accurately determine the biocompatibility of the developed polymers, these results demonstrate the potential for erodible Ppy-based materials as cell substrates in biomedical applications. The conductive nature of Ppy has been shown to allow interactive control over cell shape, differentiation, and function.[8, 11, 16] Ongoing studies are aimed towards influencing the behavior of cells cultured on erodible Ppy substrates through the application of magnetic fields and/or electrical currents.

Experimental Section

4-β-pyrrolylbutyric acid (**1**) was prepared as previously described.[12] ^{1}H NMR (300 MHz, $CDCl_3$) $\delta = 1.9$ (m, 2H), 2.4 (t, 2H), 2.55 (t, 2H), 6.1 (s, 1H), 6.4 (s, 1H), 6.72 (s, 1H).

Methyl-4-β-pyrrolylbutyrate (**2**): **1** (2.0 g, 13 mmole) in methanol (25 mL) was heated under reflux overnight in the presence of catalytic amounts of phenylsulfonic acid. The reaction mixture was poured onto ice cold water (200 mL) and extracted with diethyl ether (5×25 mL), and the combined organic layers were dried over Na_2SO_4. Removal of the solvent followed by a silica column chromatography (eluent ethyl acetate:hexane = 1:3) and removal of solvents gave 1.6 g of a white solid. ^{1}H NMR (300 MHz, $CDCl_3$) $\delta = 1.9$ (m, 2H), 2.4 (t, 2H), 2.55 (t, 2H), 3,64 (s, 3H), 6.1 (s, 1H), 6.4 (s, 1H), 6.72 (s, 1H).

Chemical polymerization: Monomer **1** (2.52 g, 16.5 mmol) and tetrabutylammonium *p*-toluenesulfonate (6.8 g, 16.5 mmol) were dissolved in acetonitrile (25 mL) and mixed with an equal volume of acetonitrile charged with ferric chloride (5.36 g, 33 mmol). The mixture was stirred vigorously for 20 min, then poured onto ice-cold water (500 mL). The precipitate was collected, washed with water (5×50 mL), and dried under vacuum. Polymer **2** was synthesized in an analogous manner. Gel permeation chromatography (GPC) analysis of polymer **2** (THF, 0.1 M piperidine) showed that the sample primarily consists of oligomers ($M_n = 1700$, $M_w = 3200$ relative to monodisperse polystyrene standards). ^{1}H NMR (300 MHz, $CDCl_3$) $\delta = 1.9$ (m, 2H), 2.4 (t, 2H), 2.55 (t, 2H), 6.1–7.9. Elemental analysis of the polymers showed trace amounts of Cl (0.14%) and Fe (0.15%).

Electrochemical polymerization: **1** was electrochemically deposited onto indium tin oxide (ITO) coated glass slides from a 1 M solution in acetonitrile containing 0.5 M of tetrabutylammonium tetrafluoroborate by passing 1 Coulomb of electrical current at a constant potential (1.0 V vs Ag/AgCl reference electrode).

pH Dependent degradation: Pressed pellets formed from poly-**1** were incubated at 37 °C in 2 mL of the media with the desired pH value (0.01 M acetate buffer, pH 5.0; 0.01 M HEPES buffer, pH 7.2; $1 \times$ TAE buffer, pH 8.2). Media was changed and subsequently analyzed with UV/Vis spectroscopy every 2–4 days.

Degradation of pressed pellets: Pressed pellets were placed into HEPES buffer (10 mL, 0.1 M; pH 7.2) and incubated at 37 °C on an orbital shaker. For experiments employing poly-**1**, buffer media was removed for absorbance measurements and replaced with fresh media every 2–5 days. As pellets formed from poly-**2** degraded more slowly than poly-**1**, buffer media was collected and analyzed every 7–14 days. The mass loss of the pellets was analyzed at selected time points by removing the incubation media, washing the pellets with de-ionized water (3×5 mL), and subsequently freeze-drying each pellet. All experiments were performed in triplicate.

Cell culture: Culture medium (CM) was Dulbecco's modified Eagle medium with 10% fetal bovine serum (LOT40G2470J, Gibco BRL Life Technology, Basel, CH), D-glucose (4.5 mg/ml), nonessential amino acids (0.1 mM), sodium pyruvate (1 mM), HEPES buffer (100 mM), penicillin (100 U/ml), streptomycin (100 µg/ml), and L-glutamine (0.29 mg/ml). Bone marrow aspirates were obtained during routine orthopedic surgical procedures involving exposure of the iliac crest, after informed consent. Nucleated cells were isolated using a density gradient solution (Histopaque, Sigma Chemical, Buchs, CH), plated at a density of 3,000 cells cm^{-2} in CM supplemented with fibroblast growth factor-2 (5 ng/ml; R&D Systems, Wiesbaden, D) and dexamethasone (10 nM; Sigma) and cultured in a humidified 37 °C/5% CO_2 incubator.[15] MPC were selected on the basis of adhesion and proliferation on the plastic substrate. After about 10 days, upon reaching subconfluency, MPC were detached using 0.25% trypsin/1 mM EDTA (EDTA = ethylendiaminetetraacetate) and plated on glass slides coated with the polymer **1**, at a density of 6,000 cells cm^{-2}. Cells were cultured in medium known to promote MPC osteogenic differentiation, which consisted of CM supplemented with dexamethasone (10 nM), L-ascorbic acid-2-phosphate (0.1 mM), and β-glycerophosphate (10mM).[17] Cultures were harvested at timed intervals (5 and 9 days) and assessed for DNA content and mRNA analysis as described below.

Cell proliferation: The DNA content of MPC cultures was assessed using the CyQuant kit (Molecular Probes, Eugene, Oregon). MPC proliferation

  1433-7851/02/4101-0143 $ 17.50+.50/0

COMMUNICATIONS

rate was defined as the number of doublings between day 5 and day 9 (determined as the logarithm in base 2 of the increase in the DNA amount) divided by the elapsed time (4 days).

Real-Time quantitative RT-PCR: RNA was extracted from cell layers using Trizol (Life Technologies, Basel, CH) according to the single step acid-phenol guanidinium method,[18] treated with DNAse I using the DNA-free kit (AMS Biotechnology Ltd, CH) and transcribed into cDNA by using random hexamers (Catalys AG, CH) and Stratrascript reverse transcriptase (Stratagene, NL). Real-time RT-PCR reactions were performed and monitored using an ABI Prism 7700 sequence detection system (Applied Biosystems, Rotkreuz, Switzerland). In the same reaction, cDNA samples were analyzed both for the gene of interest and the reference housekeeping gene (18-S rRNA), by using a multiplex approach (Perkin Elmer User Bulletin No. 2). The probe for 18-S rRNA was fluorescently labeled with VIC and TAMRA (Applied Biosystems), whereas probes for the genes of interest were labeled with 6-carboxy-fluorescein (FAM) and TAMRA. The efficiency of each set of primers and probes, assessed as previously described,[19] was always higher than 90%. Primers were purchased from Microsynth (Balgach, CH) and probes were from Perkin-Elmer or Eurogentech (Seraing, B).

Received: July 16, 2001
Revised: October 24, 2001 [Z17518]

[1] a) A. Berlin in *Electrical and Optical Polymer Systems: Fundamentals, Methods and Applications* (Eds.: D. L. Wise, G. E. Wnek, D. J. Trantolo, T. M. Cooper, J. D. Gresser), Marcel Dekker Inc., New York, **1998**, pp. 47–95; b) H. Shirakawa, *Angew. Chem.* **2001**, *113*, 2642–2648; *Angew. Chem. Int. Ed.* **2001**, *40*, 2574–2580; c) A. G. MacDiarmid, *Angew. Chem.* **2001**, *113*, 2649–2659; *Angew. Chem. Int. Ed.* **2001**, *40*, 2581–2590; d) A. J. Heeger, *Angew. Chem.* **2001**, *113*, 2660–2682; *Angew. Chem. Int. Ed.* **2001**, *40*, 2591–2611.

[2] R. V. Gregory, W. C. Kimbrell, H. H. Kuhn, *Synth. Met.* **1989**, *28*, C823–C835.

[3] A. Guiseppi-Elie, J. M. Tour, D. L. Allara, N. F. Sheppard, *Mater. Res. Soc. Symp. Proc.* **1996**, *413*, 439–444.

[4] T. Livache, A. Roget, E. Dejean, C. Barthet, G. Bidan, R. Téoule, *Nucl. Acid Res.* **1994**, *22*, 2915–2921.

[5] S. Ibrahim, C. Pickett, C. Sudbrake, *J. Electroanal. Chem.* **1995**, *387*, 139–142.

[6] V. Shastri, PhD Dissertation, Rensselaer Polytechnic Institute **1995**.

[7] a) R. Langer, *Nature* **1998**, *392 (suppl.)*, 5–10; b) R. Langer, J. B. Vacanti, *Science* **1993**, *260*, 920–926.

[8] C. Schmidt, V. Shastri, J. Vacanti, R. Langer, *Proc. Natl. Acad. Sci. USA* **1997**, *94*, 8948–8956.

[9] a) Y. Hong, L. L. Miller, *Chem. Mater.* **1995**, *7*, 1999–2000; b) A. Donat-Bouillud, L. Mazerolle, P. Gagnon, L. Goldenberg, M. C. Petty, M. Leclerc, *Chem. Mater.* **1997**, *9*, 2815–2821.

[10] T. V. Vernitskaya, O. N. Efimov, *Russ. Chem. Rev.* **1997**, *66*, 443–457.

[11] V. Shastri, N. Rahman, I. Martin, R. Langer, *Mater. Res. Soc. Symp. Proc.* **1999**, *550*, 215–219.

[12] A. N. Zelikin, P. Shastri, R. Langer, *J. Org. Chem.* **1999**, *64*, 3379–3380.

[13] J. R. Reynolds, P. A. Poropatic, R. L. Toyooka, *Macromolecules* **1987**, *20*, 958–961.

[14] K. Leong, P. D'Amore, R. Langer, *J. Biomed. Mater. Res.* **1986**, *20*, 51–64.

[15] A. Muraglia, I. Martin, R. Cancedda, R. Quarto, *Bone* **1998**, *22*, 131S–134S.

[16] J. Wong, D. Ingber, R. Langer, *Proc. Natl. Acad. Sci. USA* **1994**, *91*, 3201–3204.

[17] C. Maniatopoulos, J. Sodek, A. H. Melcher, *Cell Tissue Res.* **1988**, *254*, 317–330.

[18] P. Chomczynski, N. Sacchi, *Anal. Biochem.* **1987**, *162*, 156–159.

[19] I. Martin, M. Jakob, D. Schafer, W. Dick, G. Spagnoli, M. Heberer, *Osteoarthritis Cart.* **2001**, *9*, 112–118.

Biodegradable, Elastic Shape-Memory Polymers for Potential Biomedical Applications

Andreas Lendlein[1]* and Robert Langer[2]

The introduction of biodegradable implant materials as well as minimally invasive surgical procedures in medicine has substantially improved health care within the past few decades. This report describes a group of degradable thermoplastic polymers that are able to change their shape after an increase in temperature. Their shape-memory capability enables bulky implants to be placed in the body through small incisions or to perform complex mechanical deformations automatically. A smart degradable suture was created to illustrate the potential of these shape-memory thermoplastics in biomedical applications.

Current approaches for implanting medical devices, many of which are polymeric in nature, often require complex surgery followed by device implantation. With the advent of minimally invasive surgery (*1*), it is possible to place small devices with laprascopes. Such advances create new opportunities but also new challenges. How does one implant a bulky device or knot a suture in a confined space? It occurred to us that the creation of biocompatible (and ideally in many cases degradable) shape-memory polymers with the appropriate mechanical properties might enable the development of novel types of medical devices.

Shape-memory polymers possess the ability to memorize a permanent shape that can substantially differ from their initial temporary shape. Large bulky devices could thus potential-

[1]mnemoScience GmbH, Pauwelsstraße 19, D-52074 Aachen, and Institute for Technical and Macromolecular Chemistry, RWTH Aachen, Germany. [2]Department of Chemical Engineering, Massachusetts Institute of Technology, 45 Carleton Street, Cambridge, MA 02139, USA.

*To whom correspondence should be addressed. E-mail: a.lendlein@mnemoscience.de

ly be introduced into the body in a compressed temporary shape by means of minimally invasive surgery and then be expanded on demand to their permanent shape to fit as required. In the same way, a complex mechanical deformation could be performed automatically instead of manually by the surgeon. The transition from the temporary to the permanent shape could be initiated by an external stimulus such as a temperature increase above the switching transition temperature T_{trans} of the polymer [movie S1 (*19*)].

The thermally induced shape-memory effect has been described for different material classes: polymers (*2*, *3*), such as polyurethanes (*4–7*), poly(styrene-*block*-butadiene) (*8*) and polynorbornene (*9*, *10*); hydrogels (*11*, *12*); metallic alloys (*13*); and ceramics (*14*). All of these materials are nondegradable in physiological environments and many lack either biocompatibility or compliance in mechanical properties.

In metallic alloys, the shape-memory effect is due to a martensitic phase transition (*13*). In contrast, the polymers designed to exhibit a thermally induced shape-memory effect require two components on the molecular level: cross-links to determine the permanent shape and switching segments with T_{trans} to fix the temporary shape. Above T_{trans}, the permanent shape can be deformed by application of an external stress. After cooling below T_{trans} and subsequent release of the external stress, the temporary shape is obtained. The sample recovers its permanent shape upon heating to $T > T_{trans}$.

Cross-links can be either covalent bonds or physical interactions. Recently, we have reported on shape-memory polymers (*15*), which are covalently cross-linked polymer networks containing hydrolyzable switching segments. Emphasis in the present work was put on the development of a group of polymers that contain physical cross-links. These thermoplastics are easily processed from solution or melt and are substantially tougher than polymer networks. In particular, they are degradable, showing linear mass loss during hydrolytic degradation.

We selected linear, phase-segregated multiblockcopolymers as the structural concept for our polymer system, because this polymer architecture allows tailoring of macroscopic properties by variation of molecular parameters.

In the first step of the polymer synthesis, macrodiols with different thermal characteristics are synthesized through ring-opening polymerization of cyclic diesters or lactones, with a low-molecular-weight diol as initiator, and purified (*16*). In the current study, oligo(ε-caprolactone)diol (OCL) was chosen as the precursor for the switching segments having a melting transition temperature (T_m). Crystallizable oligo(*p*-dioxanone)diol (ODX), with a higher T_m than OCL was chosen as a hard segment to provide the physical cross-links (*17*). The melting transition of the latter macrodiols is determined by the average chain length, which can be tailored by the monomer/initiator ratio (*16*, *17*).

In the second step, the two macrodiols are coupled with 2,2(4),4-trimethylhexanediisocya-

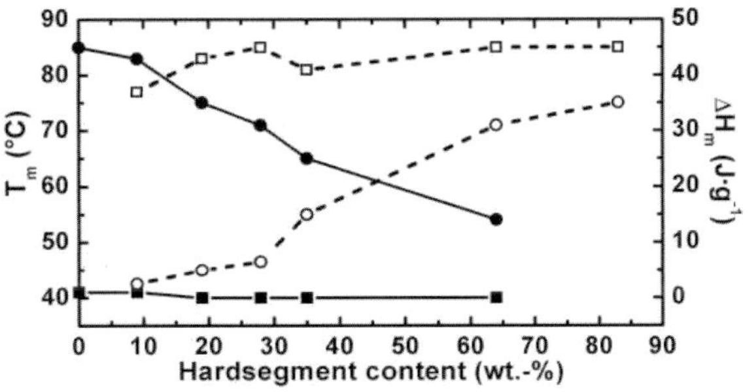

Fig. 1. T_m and enthalpies ΔH_m of multiblockcopolymers (*36*). T_m (OCL), solid squares; ΔH_m (OCL), solid circles; T_m (ODX), open squares; ΔH_m (ODX), open circles.

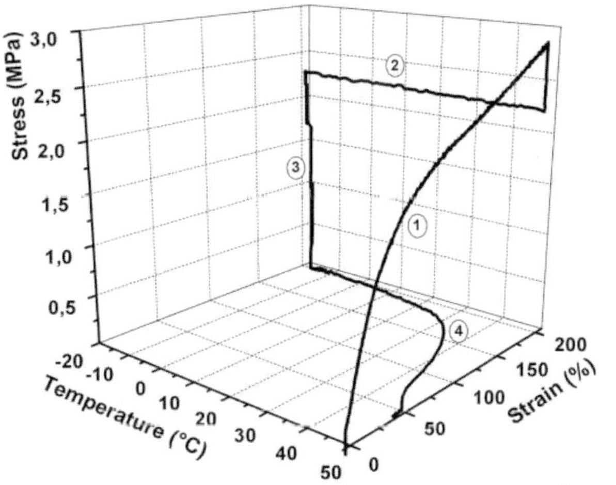

Fig. 2. Cyclic thermomechanical experiment of PDC35 (*37*) with T_{trans} = 40°C. Results of the first cycle are shown. Step 1 of the experiment was strain-controlled; steps 2 through 4 to the beginning of next cycle were stress-controlled.

Fig. 3. Hydrolytic degradation of thermoplastic shape-memory elastomers in aqueous buffer solution (pH 7) at 37°C. The relative mass loss for multiblockcopolymers differing in their hard segment content is shown (PDC10, circles; PDC17, squares; PDC31, upward-pointing triangles; PDC42, downward-pointing triangles). m(*t*), Sample mass after a degradation period *t*; m(t_0), original sample mass.

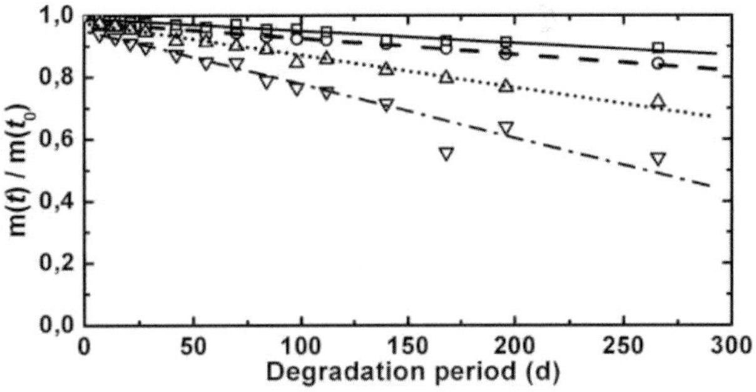

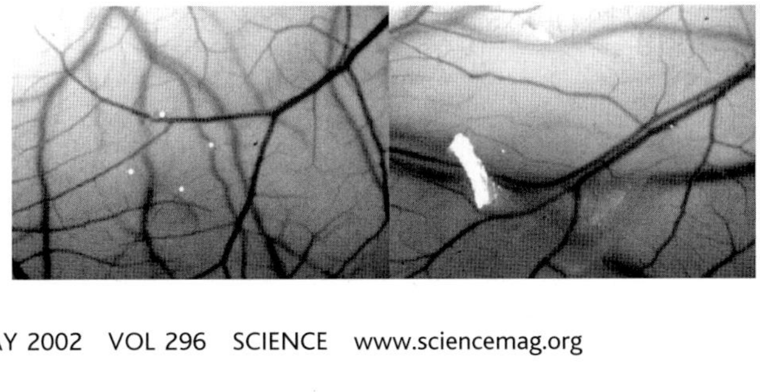

Fig. 4. Results of CAM tests of PDC38 (sample length: left, ~0.3 cm; right, ~0.5 cm). For a positive control sample, see (*25*).

nate (*18*). Hard segment contents of the synthesized polymers range from 0 to 83 weight % (wt %); and number-average molecular weights (M_n), which were determined by means of gel permeation chromatography relative to polystyrene standards, are between 35,000 and 77,000, with polydispersities around 2. Figure 1 shows melting properties of multiblockcopolymers differing in their hard segment contents. Glass transition temperatures are between –51° and 0°C [table S2 (*19*)].

The multiblockcopolymers can be elongated up to 1000% [table S1 (*19*)] before they break. This allows deformations between permanent and temporary shape up to 400%, whereas the maximum deformation for Ni-Ti alloys is 8% (*20*). The mechanical properties strongly depend on the hard segment content. Increasing the amount of ODX in the reaction mixture leads to a stiffer polymer and a decrease of the corresponding elongations at break. This can be observed at all three investigated temperatures and is due to increased crystallinity [table S1 (*19*)].

To quantify shape-memory properties, programming and recovery were investigated by cyclic thermomechanical tests (*21*, *22*). This simple test describes shape memory in one dimension; however, the effect takes place in all three dimensions. The effect is commonly described with two important parameters. The strain fixity rate R_f describes the ability of the switching segment to fix the mechanical deformation, which is applied during the programming process. For our polymers, R_f lies between 98 and 99.5%. The strain recovery rate R_r quantifies the ability of the material to recover its permanent shape. R_r depends on the cycle number and gradually approaches 100% because of reorientation of the polymer chains in the unoriented pressed films during the early cycles, because of inelastic behavior. In the first cycle, R_r has values between 76 and 80% for our multiblockcopolymers and reaches 98 to 99% in the third cycle. Ni-Ti alloys show stresses in the range of 200 to 400 MPa during shape-memory transition, whereas the shape-memory thermoplastics produce stresses in the range between 1 and 3 MPa, depending on the hard segment content (*23*). The lower value for shape-memory polymers resembles the mechanical stresses in soft tissue (*24*).

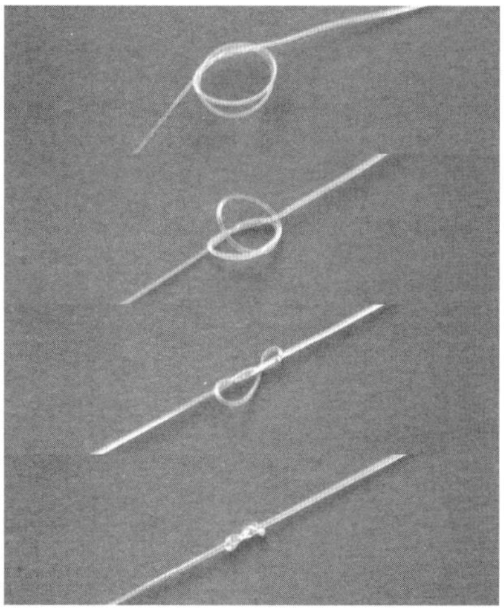

Fig. 5. A fiber of a thermoplastic shape-memory polymer was programmed by stretching about 200%. After forming a loose knot, both ends of the suture were fixed. The photo series shows, from top to bottom, how the knot tightened in 20 s when heated to 40°C. This experiment is also available as movie S2 (*19*).

To record the change in elongation during the shape-memory effect, another cyclic thermomechanical experiment was performed (Fig. 2). Step 1 is the deformation of the permanent shape and corresponds to a standard stress-strain test. After maintaining this strain for 5 min to allow relaxation for chains, the stress is then held constant while the sample is cooled (step 2), whereby the temporary shape is fixed. Then stress is completely removed after waiting for 10 min (step 3), and the sample is now in its temporary shape. Heating in step 4 (2 K min^{-1}) actuates the shape-memory effect. The contraction of the sample can be observed on the strain axis, and the fastest shape change is recorded at T_{trans} = 40°C.

We introduced hydrolyzable ester bonds in our polymers so that they would cleave under physiological conditions. The degradation kinetics can be controlled through the composition and relative mass content of the precursor macrodiols. An increase in the ODX content leads to a faster loss in mass (Fig. 3), because the concentration of rapidly hydrolyzable ODX-ester bonds in the amorphous phase is increased.

Established synthetic degradable suture materials are mainly aliphatic polyhydroxy acids showing bulk degradation. This degradation process can be split into several stages (*25*), the first three of which are swelling, loss of molecular weight, and loss of sample mass.

The degradation of L-lactide–based polyesters shows a nonlinear mass loss leading to a sudden release of potentially acidic degradation products from the bulk material, which may cause a strong inflammatory response (*26*). The high crystallinity of oligomer particles slows down degradation at the end of the process and leads to the formation of fibrous capsules in vivo (*27*). In contrast, the multiblockcopolymers presented here show linear mass loss in vitro (Fig. 3), resulting in a continuous release of degradation products.

The tissue compatibility of our polymer was investigated with chorioallantoic membrane (CAM) tests, which are a sensitive method of evaluating toxicity (*28*). Nine separate experiments were carried out. All tests showed good tissue compatibility when graded according to Folkman (*29*). There was no detectable change in the number or shape of blood vessels or damage under or in the vicinity of the polymer film (Fig. 4).

A challenge in endoscopic surgery is the tying of a knot with instruments and sutures to close an incision or open lumen. It is especially difficult to manipulate the suture so that the wound lips are pressed together under the right stress. When the knot is fixed with a force that is too strong, necrosis of the surrounding tissue can occur (*30*). If the force is too weak, scar tissue, which has poorer mechanical properties, forms and may lead to the formation of hernias (*31*). A possible solution is the design of a smart surgical suture, whose temporary shape would be obtained by elongating the fiber with controlled stress. This suture could be applied loosely in its temporary shape; when the temperature was raised above T_{trans}, the suture would shrink and tighten the knot, applying the optimum force (*32*) (Fig. 5).

An additional set of experiments to test the feasibility of this concept was performed. The highly elastic shape-memory thermoplastics were extruded into monofilaments (*33*). A sterilized suture (*34*) was programmed under sterile conditions by exerting a controlled stress on the extruded fiber and subsequent thermal quenching. This smart suture was tested in the following animal model: A rat (WAG; weight, 250 g; albino) was killed and shaved. An incision was made through the belly tissue and the abdominal muscle. The wound was loosely sutured with a standard surgical needle (Hermann Butsch, size 15, ⅜ circle). When the temperature was increased to 41°C, the shape-memory effect was actuated (Fig. 6). This test was carried out four times using two different animals. For these tests, the fibers were elongated by 200% during programming and were able to generate a force of 1.6 N upon actuation of the shape-memory effect in vitro. During the animal experiment, 0.1 N could be detected in the surrounding tissue (*35*).

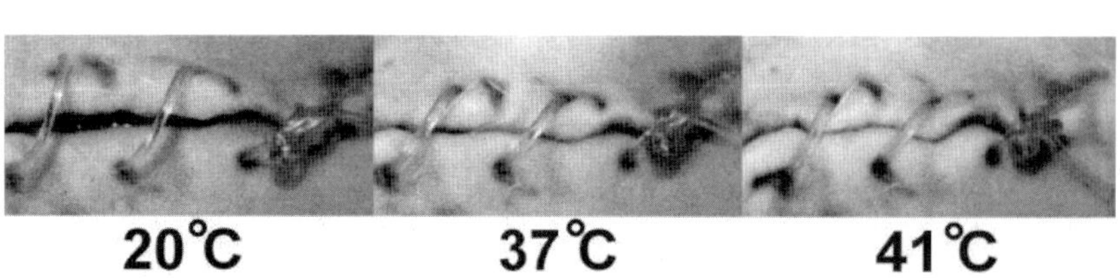

Fig. 6. Degradable shape-memory suture for wound closure. The photo series from the animal experiment shows (left to right) the shrinkage of the fiber while temperature increases.

REPORTS

This feasibility study suggests that this type of material has the potential to influence how implants are designed and could enable new surgical devices in the future.

References and Notes

1. J. G. Hunter, Ed., *Minimally Invasive Surgery* (McGraw-Hill, New York, 1993).
2. A. Charlesby, *Atomic Radiation and Polymers* (Pergamon, Oxford, 1960), pp. 198–257.
3. Y. Kagami, J. P. Gong, Y. Osada, *Macromol. Rapid Commun.* **17**, 539 (1996).
4. B. K. Kim, S. Y. Lee, M. Xu, *Polymer* **37**, 5781 (1996).
5. J. R. Lin, L. W. Chen, *J. Appl. Polym. Sci.* **69**, 1563 (1998).
6. ———, *J. Appl. Polym. Sci.* **69**, 1575 (1998).
7. T. Takahashi, N. Hayashi, S. Hayashi, *J. Appl. Polym. Sci.* **60**, 1061 (1996).
8. K. Sakurai, Y. Shirakawa, T. Kahiwagi, T. Takahashi, *Polymer* **35**, 4238 (1994).
9. K. Sakurai, T. Takahashi, *J. Appl. Polym. Sci.* **38**, 1191 (1989).
10. K. Sakurai, T. Kashiwagi, T. Takahashi, *J. Appl. Polym. Sci.* **47**, 937 (1993).
11. Y. Osada, A. Matsuda, *Nature* **376**, 219 (1995).
12. Z. Hu, X. Zhang, Y. Li, *Science* **269**, 525 (1995).
13. L. M. Schetky, *Sci. Am.* **241**, 68 (November 1979).
14. M. V. Swain, *Nature* **322**, 234 (1986).
15. A. Lendlein, A. Schmidt, R. Langer, *Proc. Natl. Acad. Sci. U.S.A.* **98**, 842 (2001).
16. A. Lendlein, P. Neuenschwander, U. W. Suter, *Macromol. Chem. Phys.* **201**, 1067 (2000).
17. H. G. Grablowitz, A. Lendlein, in preparation.
18. Both macrodiols were dissolved in 1,2-dichloroethane and heated to 80°C. An equimolar amount of 2,2(4),4-trimethylhexanediisocyanate was added. The synthesis was carried out under exclusion of water, and solvents and monomers were dried by standard techniques. The crude product was precipitated in hexane.
19. Data and video are available as supporting material on *Science* Online.
20. J. Van Humbeeck, *Mater. Sci. Eng.* **A273-275**, 134 (1999).
21. The material was pressed into films having a thickness of 300 to 500 μm. Dog bone–shaped samples (length between clamps, 6 mm; width, 3 mm) were punched out of the films and mounted in a tensile tester equipped with a thermo chamber (*8*, *21*). The tests were carried out at 200% strain at a strain rate of 10 mm min^{-1} with a low temperature (T_{low}) of –20°C and a high temperature (T_{high}) of 50°C. The samples were held at T_{low} for 10 min before the load was removed.
22. H. Tobushi, H. Hara, E. Yamada, S. Hayashi, *Soc. Photo-Opt. Instrum. Eng.* **2716**, 46 (1996).
23. H. Tobushi, S. Hayashi, A. Ikai, H. Hara, *J. Physique IV* **6**, C1-377 (1996).
24. A. F. T. Mak, M. Zhang, in *Handbook of Biomaterial Properties*, J. Black, G. Hastings, Eds. (Chapman and Hall, New York, ed. 1, 1998), pp. 66–69.
25. A. Lendlein, *Chem. Unserer Zeit* **33**, 279 (1999).
26. K. Fu, D. W. Pack, A. M. Klibanov, R. S. Langer, *Pharm. Res.* **17:1**, 100 (2000).
27. K. A. Hooper, N. D. Macon, J. Kohn, *J. Biomed. Mater. Res.* **32**, 443 (1998).
28. K. Spanel-Borowski, *Res. Exp. Med.* (*Berlin*) **189**, 69 (1989).
29. R. Crum, S. Szabo, J. Folkman, *Science* **230**, 1375 (1985).
30. J. Hoer, U. Klinge, A. Schachtrupp, Ch. Töns, V. Schumpelick, *Langenb. Arch. Surg.* **386**, 218 (2001).
31. N. C. F. Hodgson, R. A. Malthaner, T. Østbye, *Ann. Surg.* **231**, 436 (2000).
32. Control of stress was achieved on three levels: in the material itself through hard segment content, by programming, and during application through the looseness of the loops of the suture.
33. Extrusion was at 90°C through a 1-mm rod die on a Haake Polylab single-screw extruder.
34. Sterilization was done with ethylene oxide at 45°C.
35. The force of the fiber was determined with a tensile tester equipped with a thermo chamber. The force on the surrounding tissue was estimated by mounting a spring of known stiffness close to the wound and measuring the length change.
36. T_m and enthalpies ΔH_m of multiblockcopolymers were measured on a Perkin-Elmer DSC 7 at a heating rate of 10 K min^{-1}. The results were taken from the second heating run.
37. The weight content of ODX in the polymer is given by the two-digit number in the sample ID.

38. We thank H. Grablowitz for degradation experiments, J. Schulte for mechanical tests, W. Grasser for graphics, D. Rickert and M. Moses (Children's Hospital, Boston) for CAM tests, and R.-P. Franke (Zentralinstitut für biomedizinische Technik, University of Ulm) for the animal experiment. A.L. is grateful to Fonds der Chemischen Industrie for a Liebig fellowship. Partially funded by Bundesministerium für Bildung und Forschung BioFuture award no. 0311867.

Supporting Online Material

www.sciencemag.org/cgi/content/full/1066102/DC1
Tables S1 to S3
Movies S1 and S2

10 September 2001; accepted 11 April 2002
Published online 25 April 2002;
10.1126/science.1066102
Include this information when citing this paper.

A tough biodegradable elastomer

Yadong Wang[1], Guillermo A. Ameer[1,2], Barbara J. Sheppard[3], and Robert Langer[1*]

Biodegradable polymers have significant potential in biotechnology and bioengineering. However, for some applications, they are limited by their inferior mechanical properties and unsatisfactory compatibility with cells and tissues. A strong, biodegradable, and biocompatible elastomer could be useful for fields such as tissue engineering, drug delivery, and *in vivo* sensing. We designed, synthesized, and characterized a tough biodegradable elastomer from biocompatible monomers. This elastomer forms a covalently crosslinked, three-dimensional network of random coils with hydroxyl groups attached to its backbone. Both crosslinking and the hydrogen-bonding interactions between the hydroxyl groups likely contribute to the unique properties of the elastomer. *In vitro* and *in vivo* studies show that the polymer has good biocompatibility. Polymer implants under animal skin are absorbed completely within 60 days with restoration of the implantation sites to their normal architecture.

Biodegradable polymers have significant potential in various fields of bioengineering, such as tissue engineering, drug delivery, and *in vivo* sensing[1,2]. Because many biomedical devices are implanted in a mechanically dynamic environment in the body, the implants must sustain and recover from various deformations without mechanical irritations to the surrounding tissues. In many cases, the matrices and scaffolds of these implants would ideally be made of biodegradable polymers whose properties resemble those of the extracellular matrix (ECM), a soft, tough, and elastomeric proteinaceous network that provides mechanical stability and structural integrity to tissues and organs. Hence a soft biodegradable elastomer that readily recovers from relatively large deformations is advantageous for maintaining the implant's proper function without mechanical irritation to the host. Three classes of biodegradable elastomers have been reported: hydrogels[3,4], elastin-like peptides[5–7], and polyhydroxyalkanoates (PHAs)[8,9]. Here we report a tough and inexpensive biodegradable elastomer with excellent biocompatibility. This elastomer is much tougher than most hydrogels[3,4]. Compared with elastin-like peptides made by bacterial fermentation[10–12], it is inexpensive, nonimmunogenic, and endotoxin-free. The elastomer is capable of much larger reversible deformations than a reportedly elastomeric PHA (see Results and discussion), poly-4-hydroxybutyrate (P4HB).

This biodegradable elastomer is analogous to vulcanized rubber in that it forms a crosslinked, three-dimensional network of random coils (it is this characteristic that is considered to give vulcanized rubber its elasticity)[13]. Such a strategy to achieve tough and elastomeric materials is also found in nature: collagen and elastin, the major fibrous protein components of ECM, are both crosslinked[14]. In addition to covalent crosslinking, hydrogen-bonding interactions through hydroxyproline hydroxyl groups also contribute to the mechanical strength of collagen[14,15]. This relationship is manifested in various diseases in which the strength of collagen fibers decreases markedly when crosslink density decreases considerably, or when the production of hydroxyproline is interrupted. Collagen fibers can usually recover from deformations of ~20% (refs 16, 17), far greater than the most prevalent biodegradable polymers: poly(lactide), poly(glycolide), and their copolymers (PLGA)[18,19]. The deformation is also greater than that of P4HB (~10%). The design of our polymer is based on two hypotheses: (i) Good mechanical properties could be obtained through covalent crosslinking and hydrogen-bonding interactions (comparison of the *in vivo* degradation profiles of this polymer and PLGA will be published elsewhere); (ii) rubberlike elasticity could be obtained by building a three-dimensional network of random coils through copolymerization in which at least one monomer is trifunctional.

To realize this design, we considered five criteria: (i) as a degradation mechanism, we chose hydrolysis to minimize individual differences in degradation characteristics caused by enzymes[2]; (ii) to provide a hydrolyzable chemical bond, we chose ester for its established and versatile synthetic methods[20]; (iii) because a high degree of crosslinking usually leads to rigid and brittle polymers, we wanted a low density of crosslinking; (iv) we chose crosslink chemical bonds to be hydrolyzable and identical to those in the backbone to minimize the possibility of heterogeneous degradation; (v) we required that specific monomers be nontoxic, at least one be trifunctional, and at least one provide hydroxyl groups for hydrogen bonding. We chose glycerol ($CH_2(OH)CH(OH)CH_2OH$), the basic building block for lipids, as the alcohol monomer because it satisfies all three requirements. From the same toxicological and polymer chemistry standpoints, we chose sebacic acid ($HOOC(CH_2)_8COOH$) as the acid monomer. Sebacic acid is the natural metabolic intermediate in ω-oxidation of medium- to long-chain fatty acids[21–24]. It has been shown to be safe *in vivo*[25]. The US Food and Drug Administration has approved glycerol and polymers containing sebacic acid for medical applications. Furthermore, sebacic acid has the appropriate chain length. Short-chain dicarboxylic acids are more acidic and more likely to cyclize during polymerization reactions, whereas long-chain dicarboxylic acids are more hydrophobic and mix poorly with glycerol. Both monomers are inexpensive, an advantage for large-scale applications. We anticipated that this approach would yield biodegradable

[1]Department of Chemical Engineering, 77 Massachusetts Avenue, Massachusetts Institute of Technology, Cambridge, MA 02139. [2]Current address: Biomedical Engineering Department, Northwestern University, Evanston, IL 60208. [3]Division of Comparative Medicine, 77 Massachusetts Avenue, Massachusetts Institute of Technology, Cambridge, MA 02139. *Corresponding author (rlanger@mit.edu).

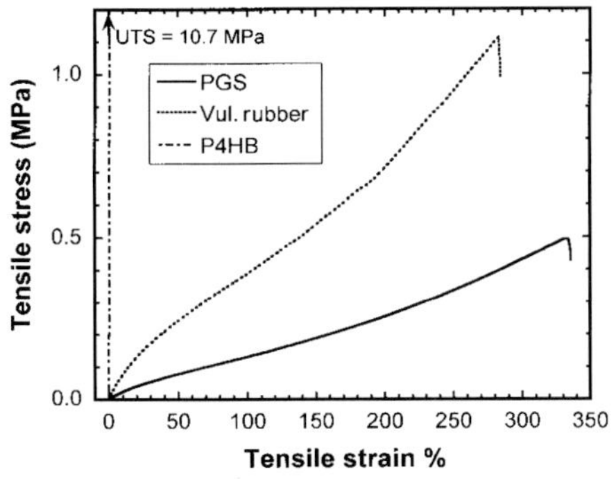

Figure 1. Stress–strain curves of PGS, vulcanized rubber, and P4HB. Both PGS and vulcanized rubber are marked by low modulus and large elongation ratio, indicating elastomeric and tough materials. The high modulus and low yield strain (~10%) of P4HB indicate a stiffer material. UTS, ultimate tensile strength.

polymers with improved mechanical properties and biocompatibility. The polymer is referred to as poly(glycerol-sebacate) (PGS).

Results and discussion

Polycondensation of glycerol and sebacic acid yields a transparent, almost colorless elastomer (note that under different conditions, a rigid, totally crosslinked polymer has been synthesized from glycerol and sebacic acid in the molar ratio 2:3)[26]. The resulting polymer features a small number of crosslinks and hydroxyl groups directly attached to the backbone. The intense C=O stretch at 1,740 cm^{-1} in Fourier-transformed infrared (FTIR) spectrum confirms the formation of ester bonds (Supplementary Fig. 1 online).

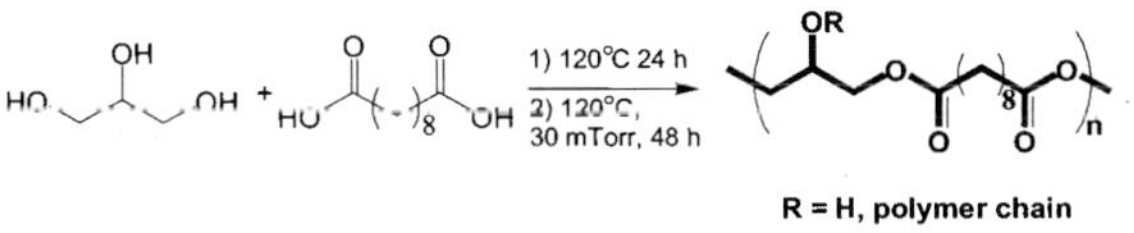

FTIR also shows a broad, intense OH stretch at 3,448 cm^{-1}, indicating that the hydroxyl groups are hydrogen bonded. The polymer surface is very hydrophilic because of the hydroxyl groups attached to its backbone. Its water-in-air contact angle is 32.0°, almost identical to that of a flat 2.7-nm-thick type I collagen film (31.9°)[27]. The polymer is insoluble in water, and swells 2.1 ± 0.33% after soaking

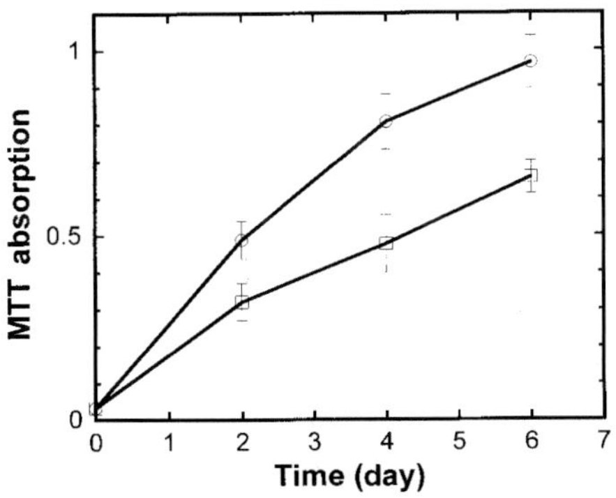

Figure 3. Comparison of growth rate of NIH 3T3 fibroblast cells in PGS (○) wells and PLGA (□) wells. MTT absorption measured at 570 nm, normalized value shown.

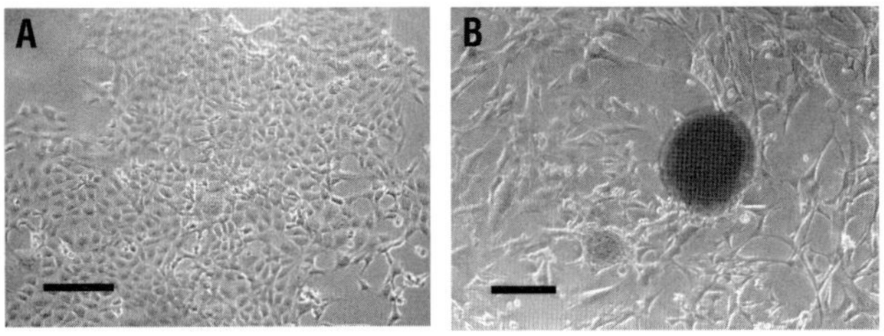

Figure 2. Comparison of NIH 3T3 fibroblast cell morphology and number. PGS sample wells (A) and PLGA control wells (B), six days after seeding. The PGS wells had more adherent cells and the cell morphology appeared normal, while those in the control well adopted a long, thin, threadlike shape. Scale bar = 200 μm.

in water for 24 h. Elemental analysis confirms the composition of the polymer as approximately 1:1 glycerol/sebacic acid (calculated for $C_{13}H_{22}O_5$: C, calculated 60.47—found 60.46; H, calculated 8.53—found 8.36). The crosslinking density is expressed by n (moles of active network chains per unit volume), which is 38.3 ± 3.40 mol/m^3, and M_c, the relative molecular mass between crosslinks, which is 18,300 ± 1,620, calculated from the following equation[28]:

$$n = E_0/3RT = \rho/M_c$$

where E_0 is Young's modulus, R is the universal gas constant, T is the temperature, and ρ is the density. Differential scanning calorimetery (DSC) measurement showed two crystallization temperatures at −52.14°C and −18.50°C, and two melting temperatures at 5.23°C and 37.62°C, respectively. No glass transition temperature was observed above −80°C, which is the lower detection limit of the instrument. The DSC results indicate that the polymer is totally amorphous at 37°C. Similar to vulcanized rubber, this elastomer is a thermoset polymer. However, the uncrosslinked prepolymer can be processed into various shapes, because it can be melted into liquid and is soluble in common organic solvents, such as 1,3-dioxolane, tetrahydrofuran, ethanol, isopropanol, and N,N-dimethylformamide. We have prepared PGS sheets and foams with these methods. Briefly, a mixture of NaCl particles of appropriate size and an anhydrous 1,3-dioxolane solution of the prepolymer is poured into a PTFE mold. The polymer is cured in the mold

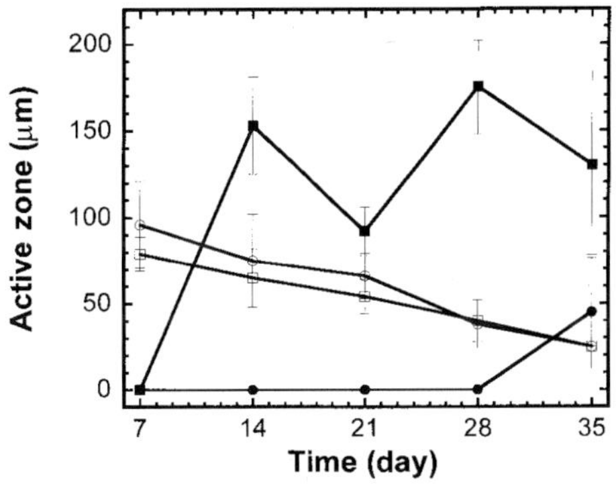

Figure 4. Change of thickness of the immune responses with time for PGS and PLGA. The inflammatory response decreased with time for both polymers, which had similar inflammatory-zone thickness. However, the thickness of fibrous capsules surrounding PLGA was consistently and significantly larger than that of PGS. Inflammatory zone: PGS (○); PLGA (□). Fibrous capsule: PGS (●); PLGA (■).

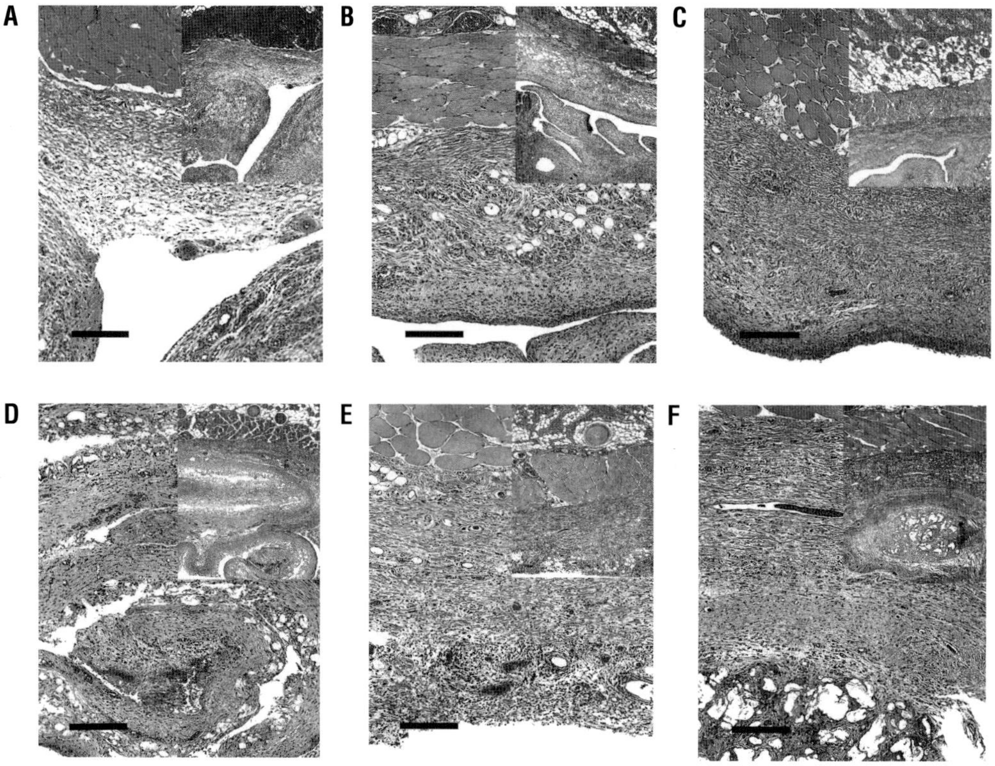

Figure 5. Photomicrographs of rat skin. Comparisons of lumen wall characteristics (H&E, 10×) and fibrous capsule thickness (insets, MTS, 5×) at implantation sites across time. (A, C, E) PGS; (B, D, F) PLGA. (A, B) At 7 days postimplantation (p.i.), the lumenal wall was markedly thickened by a zone of dense vascular proliferation and mild inflammation without detectable collagen deposition. (C, D) At 21 days p.i., the lumenal wall was significantly thinner with a modest degenerative inflammatory infiltrate immediately adjacent to the polymer. The PLGA implantation site (D) was marked by a significant collagen fibrous capsule, which was absent in PGS. (E, F) At 35 days p.i., the lumenal wall was reduced to a thin zone of cell debris with no vascular proliferation. Collagen deposition in PGS implantation site was much thinner than that surrounding the fragmented PLGA implant. Top, skin; blank area, implantation site; scale bar = 200 μm.

in a vacuum oven at 120°C and 100 mtorr. A porous scaffold is obtained after salt leaching in deionized water.

Tensile tests on thin strips of PGS reveal a stress–strain curve characteristic of an elastomeric and tough material (Fig. 1). The nonlinear shape of the tensile stress–strain curve is typical for elastomers, and resembles those of ligament[29–31] and vulcanized rubber[32]. Compared with hard and brittle materials, which have high modulus (the initial slope of the stress–strain curve) and low strain (relative deformation), PGS can be elongated repeatedly to at least three times its original length without rupture. The total elongation is unknown, because grip breaks occurred at 267 ± 59.4% strain. The tensile Young's modulus of the polymer is 0.282 ± 0.0250 MPa, indicating a soft material. The ultimate tensile strength is >0.5 MPa, the point at which the PGS strips broke from the grip of the mechanical tester. P4HB, a reportedly elastomeric degradable PHA (refs 8,9), has a strain to failure value of 11.1 ± 0.491% and a Young's modulus of 253 ± 5.29 MPa, similar to that of low-density polyethylene. The ultimate tensile strength is 10.4 ± 0.554 MPa. Overall, P4HB has a much higher modulus (stiffer) and much lower strain to failure compared with either PGS or vulcanized rubber. The value of the Young's modulus of PGS is between those of ligaments (kPa scale)[29–31], which contain a large amount of elastin in addition to collagen, and tendon (GPa scale)[16,17,33], which is mainly made of collagen. The strain to failure of PGS is similar to that of arteries and veins (up to 260%)[34], and much larger than that of tendons (up to 18%)[35]. After soaking for 24 h in water, the weight of PGS barely changes, and the mechanical properties are virtually the same as dry polymer.

PGS appears to be biocompatible both *in vitro* and *in vivo*. NIH 3T3 fibroblast cells were seeded homogeneously on PGS-coated glass Petri dishes with PLGA-coated dishes as controls. The cells in PGS sample wells are viable and showed normal morphology with higher growth rate than the control, as tested by MTT assay[36] (Figs 2A, 3). Cells in PLGA wells tend to form clusters, and there is a greater number of floating cells; furthermore, most of the attached cells adopted a long, thin, threadlike morphology (Fig. 2B). These experiments suggested that PGS is at least as biocompatible as PLGA *in vitro*.

Subcutaneous (s.c.) implantation in Sprague–Dawley rats was used to compare the *in vivo* biocompatibility of PGS and PLGA. These PGS and PLGA implants have the same surface area/volume ratio (1.33 ± 0.04). Both PGS and PLGA samples were implanted symmetrically into the back of the same animal. The inflammatory responses subsided with time for both polymer implants. In the first three weeks, the inflammatory response of PLGA implantation sites was ~16% thinner than that of PGS sites (Figs 4, 5). The thickness of the inflammatory zone in both implantation sites was approximately the same at weeks 4 and 5. Fibrous capsules (thick avascular collagen layer) surrounding PLGA implants developed within 14 days, and their thickness remained ~140 μm. Collagen deposition did not appear around PGS implants until 35 days post implantation. The collagen layer was highly vascularized and was only ~45 μm thick. The inflammatory response and fibrous capsule formation observed for PLGA are similar to those reported in the literature[37,38]. Thick fibrous capsules block mass transfer between the implants and surrounding tissues, which can impair implant functions. In an *in vivo* study with PGS alone, the s.c. implantation sites were undetectable despite repeated sectioning of the specimens at multiple levels in 60 days (two implantation sites each in three animals). The implants were completely absorbed without granulation or formation of scar tissues, and the implantation site was restored to its normal histologic architecture. Overall, the inflammatory response of PGS is similar to that of PLGA. However, unlike PLGA, PGS induces little, if any, fibrous capsule formation.

We examined the degradation characteristics of PGS both *in vitro* and *in vivo*. Agitation for 60 days in PBS solution at 37°C caused the polymer to degrade 17 ± 6%, as measured by change of dry sample weight. In contrast, s.c. implants in rats were totally absorbed in 60 days. For the *in vivo* experiment, enzymes, and perhaps macrophages as well, may have caused differences in degradation rate. *In vivo* degradation thinned the polymer implants, with the explants maintaining their square shape and relatively sharp edges up to at least 35 days. Preliminary data indicate that mechanical strength likely decreases linearly with mass loss (comparison of the *in vivo* degradation profiles of this polymer and PLGA will be published elsewhere). Both results suggest that PGS disks degrade predominantly through surface erosion. The preservation of integrity during the degradation process can be important for certain types of tissue-engineered implants, drug delivery devices, and *in vivo* sensors.

Compared with existing biodegradable elastomers, PGS appears to be tougher, inexpensive, and more flexible. In the models tested, the material is biocompatible both *in vitro* and *in vivo*. The polymer's properties, such as hydrophilicity, and degradation rate and pattern can potentially be tailored by grafting hydrophobic moieties to the hydroxyl groups[39,40]. To provide further control or regulation of polymer interaction with cells, biomolecules could be coupled to the hydroxyl groups or integrated into the polymer backbone[41–43].

Experimental protocol

Synthesis and characterization of the polymer. The polymer was synthesized by polycondensation of 0.1 mol each of glycerol (Aldrich, Milwaukee, WI) and sebacic acid (Aldrich) at 120°C under argon for 24 h before the pressure was reduced from 1 torr to 40 mtorr over 5 h. The reaction mixture was kept at 40 mtorr and 120°C for 48 h. A KBr pellet of newly prepared polymer was used for FTIR analysis on a Nicolet Magna-IR 550 spectrometer. A Perkin-Elmer DSC differential scanning calorimeter was used for DSC measurement. Elemental analysis on vacuum-dried samples was performed by QTI (Whitehouse, NJ). The water-in-air contact angle was measured at room temperature using the sessile-drop method and an image analysis of the drop profile with VCA2000 video contact angle system on slabs of polymer fixed on glass slides.

Mechanical properties. Tensile tests were conducted on six 25 × 5 × 0.7 mm polymer strips cut from polymer sheets according to ASTM standard D 412-98a on an Instron 5542 mechanical tester equipped with a 50 N load cell. Vulcanized rubber and P4HB (Metabolix, Cambridge, MA) strips (25 × 5 × 0.5 mm) were cut from polymer sheets. Deflection rate was kept at 50 mm/min. The samples were elongated to failure.

***In vitro* degradation.** Slabs of dry polymer (5 × 5 × 2 mm) were weighed and transferred to 15 ml centrifuge tubes (Falcon, Bedford, MA) filled with PBS (pH 7.4; Gibco, Carlsbad, CA). After 60 days, the samples were removed and washed with deionized water. The surface water was removed with Kimwipe, and the samples were weighed after drying at 40°C in an oven for seven days. The degree of degradation was determined by dry-weight change.

***In vitro* biocompatibility.** Nine glass Petri dishes (60 mm diameter) were coated with 1,3-dioxolane solution of the prepolymer (1%). The coated dishes were transferred into a vacuum oven after evaporation of the solvent in air. The prepolymer was crosslinked into the elastomer after 24 h at 120°C and 120 mtorr. Nine control dishes were coated with 1% CH_2Cl_2 solution of PLGA (50:50, carboxyl ended, relative molecular mass 15,000 (M_r 15K), Boehringer Ingelheim, Ingelheim, Germany), and the solvent was evaporated for 24 h in air. The coated dishes were sterilized by UV radiation for 15 min. Each dish was soaked in growth medium for 4 h, replaced with fresh medium, and soaked for 4 h before cell seeding to remove any unreacted monomers or residual solvents. Each dish was seeded with 100,000 NIH 3T3 fibroblast cells and 8 ml of growth medium. The cells were incubated at 37°C with 5% CO_2. Cell density was measured by MTT assay[44]. Medium exchange was carried out every 48 h. At day 6, phase contrast images were taken for both the polymer wells and the control wells on a Zeiss Axiovert 200 microscope equipped with a Dage 240 digital camera.

***In vivo* biocompatibility.** Autoclaved PGS slabs of approximately 6 × 6 × 3 mm, and ethylene oxide–sterilized PLGA (50:50, carboxyl ended, M_r 15K, Boehringer Ingelheim) disks (2 mm thick, 12.5 mm diameter) were implanted s.c. in 15 seven-week-old female Sprague–Dawley rats (Charles River Laboratories, Wilmington, MA) by blunt dissection under deep isoflurane–O_2 general anesthesia. The animals were cared for in compliance with the regulations of MIT and the NIH[45]. The surface area/volume ratio was kept the same for both PGS and PLGA implants. Two implants each of PGS and PLGA were implanted symmetrically on the upper and lower back of the same animal. Every implantation site was marked by two tattoo marks 2 cm away from the implantation center. The animals were randomly divided into five groups. At each predetermined time point (7, 14, 21, 28, and 35 days), one group of rats was killed and tissue samples (~15 × 15 mm) surrounding the implants were harvested with the intact implant. The samples were fixed in 10% formalin for 24 h and embedded in paraffin after a series of dehydration steps in ethanol and xylenes. The slides were stained with hematoxylin and eosin (H&E) and Masson's trichrome stain (MTS). At each time point, 12 slides for each polymer were obtained. All histologic preparations were assessed by a pathologist (B.J.S.), who was not informed of the identity of the polymer implant in each slide. The thickness of the inflammatory zone (H&E) and collagen deposition (MTS) for each polymer implant is expressed as the average value of three readings per slide of six slides at each time point.

Note: Supplementary information is available on the Nature Biotechnology *website.*

Acknowledgments

The authors thank Dr. David LaVan and Dr. Daniel Anderson for advice and discussions. We appreciate comments from Dr. David LaVan, Prof. Hiroyuki Ijima, and Ms. Sheryl Villa on the manuscript. This work was supported by NIH grant 5-R01-HL60435.

Received 21 September 2001; accepted 25 February 2002

1. Peppas, N.A. & Langer, R. New challenges in biomaterials. *Science* **263**, 1715–1720 (1994).
2. Langer, R. Biomaterials: status, challenges, and perspectives. *AIChE J.* **46**, 1286–1289 (2000).
3. Lee, K.Y. *et al.* Controlling mechanical and swelling properties of alginate hydrogels independently by cross-linker type and cross-linking density. *Macromolecules* **33**, 4291–4294 (2000).
4. Temenoff, J.S., Athanasiou, K.A., LeBaron, R.G. & Mikos, A.G. Effect of poly(ethylene glycol) molecular weight on tensile and swelling properties of oligo(poly(ethylene glycol)fumarate) hydrogels for cartilage tissue engineering. *J. Biomed. Mater. Res.* **59**, 429–437 (2002).
5. van Hest, J.C.M. & Tirrell, D.A. Protein-based materials, toward a new level of structural control. *Chem. Comm.* **19**, 1897–1904 (2001).
6. Welsh, E.R. & Tirrell, D.A. Engineering the extracellular matrix: a novel approach to polymeric biomaterials. I. Control of the physical properties of artificial protein matrices designed to support adhesion of vascular endothelial cells. *Biomacromolecules* **1**, 23–30 (2000).
7. Urry, D.W. *et al.* Elastic protein-based polymers in soft tissue augmentation and generation. *J. Biomater. Sci., Polym. Ed.* **9**, 1015–1048 (1998).
8. Poirier, Y., Nawrath, C. & Somerville, C. Production of polyhydroxyalkanoates, a family of biodegradable plastics and elastomers, in bacteria and plants. *Bio/Technology* **13**, 142–150 (1995).
9. Sodian, R. *et al.* Fabrication of a trileaflet heart valve scaffold from a polyhydroxyalkanoate biopolyester for use in tissue engineering. *Tissue Eng.* **6**, 183–187 (2000).
10. Katsarava, R. *et al.* Amino acid–based bioanalogous polymers. Synthesis and study of regular poly(ester amide)s based on bis(α-amino acid) α,ω-alkylene diesters, and aliphatic dicarboxylic acids. *J. Polym. Sci. Part A: Polym. Chem.* **37**, 391–407 (1999).
11. Kemnitzer, J. & Kohn, J. In *Handbook of biodegradable polymers* (eds Domb, A.J., Kost, J. & Wiseman, D.M.) 251–272 (Harwood Academic Publishers, Amsterdam; 1998).
12. Petsch, D. & Anspach, F.B. Endotoxin removal from protein solutions. *J. Biotechnol.* **76**, 97–119 (2000).
13. Erman, B. & Mark, J.E. In *Science and technology of rubber* (eds Mark, J.E., Burak, E. & Eirich, F.R.) 189–210 (Academic Press, San Diego, CA; 1994).
14. Voet, D. & Voet, J.G. *Biochemistry* (John Wiley & Sons, New York, 1995).
15. Stryer, L. *Biochemistry* (W.H. Freeman and Company, New York; 1995).
16. Fratzl, P. *et al.* Fibrillar structure and mechanical properties of collagen. *J. Struct. Biol.* **122**, 119–122 (1998).
17. Wang, J.L., Parnianpour, M., Shirazi-Adl, A. & Engin, A.E. Failure criterion of collagen fiber: viscoelastic behavior simulated by using load control data. *Theor. Appl. Fract. Mech.* **27**, 1–12 (1997).
18. Storey, R.F., Warren, S.C., Allison, C.J. & Puckett, A.D. Methacrylate-endcapped

poly(d,l-lactide-co-trimethylene carbonate) oligomers. Network formation by thermal free-radical curing. *Polymer* **38**, 6295–6301 (1997).

19. Helminen, A., Korhonen, H. & Seppala, J.V. Biodegradable crosslinked polymers based on triethoxysilane terminated polylactide oligomers. *Polymer* **42**, 3345–3353 (2001).
20. March, J. *Advanced organic chemistry* (John Wiley & Sons, New York; 1992).
21. Liu, G., Hinch, B. & Beavis, A.D. Mechanisms for the transport of α,ω-dicarboxylates through the mitochondrial inner membrane. *J. Biol. Chem.* **271**, 25338–25344 (1996).
22. Grego, A.V. & Mingrone, G. Dicarboxylic acids, an alternate fuel substrate in parenteral nutrition: an update. *Clin. Nutr.* **14**, 143–148 (1995).
23. Mortensen, P.B. & Gregersen, N. The biological origin of ketotic dicarboxylic aciduria. *In vivo* and *in vitro* investigations of the omega-oxidation of C6-C16-monocarboxylic acids in unstarved, starved and diabetic rats. *Biochim. Biophys. Acta* **666**, 394–404 (1981).
24. Mortensen, P.B. C6–C10-dicarboxylic aciduria in starved, fat-fed and diabetic rats receiving decanoic acid or medium-chain triacylglycerol. An *in vivo* measure of the rate of β-oxidation of fatty acids. *Biochim. Biophys. Acta* **664**, 349–355 (1981).
25. Tamada, J. & Langer, R. The development of polyanhydrides for drug delivery applications. *J. Biomater. Sci. Polym. Ed.* **3**, 315–353 (1992).
26. Nagata, M. *et al.* Synthesis, characterization, and enzymatic degradation of network aliphatic copolyesters. *J. Polym. Sci. Part A: Polym. Chem.* **37**, 2005–2011 (1999).
27. Dupont-Gillain, C.C., Nysten, B. & Rouxhet, P.G. Collagen adsorption on poly(methyl methacrylate): net-like structure formation upon drying. *Polym. Int.* **48**, 271–276 (1999).
28. Sperling, L.H. *Introduction to physical polymer science* (John Wiley & Sons, New York; 1992).
29. Yamaguchi, S. Analysis of stress–strain curves at fast and slow velocities of loading *in vitro* in the transverse section of the rat incisor periodontal ligament following the administration of β-aminopropionitrile. *Arch. Oral Biol.* **37**, 439–444 (1992).
30. Komatsu, K. & Chiba, M. The effect of velocity of loading on the biomechanical responses of the periodontal ligament in transverse sections of the rat molar *in vitro*. *Arch. Oral Biol.* **38**, 369–375 (1993).
31. Chiba, M. & Komatsu, K. Mechanical responses of the periodontal ligament in the transverse section of the rat mandibular incisor at various velocities of loading *in vitro*. *J. Biomech.* **26**, 561–570 (1993).
32. Nagdi, K. *Rubber as an engineering material: guideline for users* (Hanser, Munich;1993).
33. Misof, K., Rapp, G. & Fratzl, P. A new molecular model for collagen elasticity based on synchrotron X- ray scattering evidence. *Biophys. J.* **72**, 1376–1381 (1997).
34. Lee, M.C. & Haut, R.C. Strain rate effects on tensile failure properties of the common carotid artery and jugular veins of ferrets. *J. Biomech.* **25**, 925–927 (1992).
35. Haut, R.C. The effect of a lathyritic diet on the sensitivity of tendon to strain rate. *J. Biomech. Eng.* **107**, 166–174 (1985).
36. Northup, S.J. & Cammack, J.N. In *Handbook of biomaterials evaluation* (ed. von Recum, A.F.) 325–339 (Taylor & Francis, Philadelphia; 1999).
37. Cadee, J.A., Brouwer, L.A., den Otter, W., Hennink, W.E. & Van Luyn, M.J.A. A comparative biocompatibility study of microspheres based on crosslinked dextran or poly(lactic-co-glycolic)acid after subcutaneous injection in rats. *J. Biomed. Mater. Res.* **56**, 600–609 (2001).
38. van der Elst, M., Klein, C.P.A.T., de Blieck-Hogervorst, J.M., Patka, P. & Haarman, H.J.T.M. Bone tissue response to biodegradable polymers used for intramedullary fracture fixation: a long-term *in vivo* study in sheep femora. *Biomaterials* **20**, 121–128 (1999).
39. Jayachandran, K.N. & Chatterji, P.R. Synthesis of dense brush polymers with cleavable grafts. *Eur. Polym. J.* **36**, 743–749 (2000).
40. Laschewsky, A., Rekai, E.D. & Wischerhoff, E. Tailoring of stimuli-responsive water- soluble acrylamide and methacrylamide polymers. *Macromol. Chem. Phys.* **202**, 276–286 (2001).
41. Barrera, D.A., Zylstra, E., Lansbury, P.T., Jr. & Langer, R. Synthesis and RGD peptide modification of a new biodegradable copolymer: poly(lactic acid-co-lysine). *J. Am. Chem. Soc.* **115**, 11010–11011 (1993).
42. West, J.L. & Hubbell, J.A. Polymeric biomaterials with degradation sites for proteases involved in cell migration. *Macromolecules* **32**, 241–244 (1999).
43. Mann, B.K., Gobin, A.S., Tsai, A.T., Schmedlen, R.H. & West, J.L. Smooth muscle cell growth in photopolymerized hydrogels with cell adhesive and proteolytically degradable domains: synthetic ECM analogs for tissue engineering. *Biomaterials* **22**, 3045–3051 (2001).
44. Hansen, M.B., Nielsen, S.E. & Berg, K. Re-examination and further development of a precise and rapid dye method for measuring cell growth/cell kill. *J. Immunol. Methods* **119**, 203–210 (1989).
45. National Institutes of Health. *Principles of Laboratory Animal Care* (NIH pub. no. 85-23, rev. 1985).

A Reversibly Switching Surface

Joerg Lahann,[1] Samir Mitragotri,[2] Thanh-Nga Tran,[1] Hiroki Kaido,[1] Jagannathan Sundaram,[2] Insung S. Choi,[1]* Saskia Hoffer,[3] Gabor A. Somorjai,[3] Robert Langer[1]†

We report the design of surfaces that exhibit dynamic changes in interfacial properties, such as wettability, in response to an electrical potential. The change in wetting behavior was caused by surface-confined, single-layered molecules undergoing conformational transitions between a hydrophilic and a moderately hydrophobic state. Reversible conformational transitions were confirmed at a molecular level with the use of sum-frequency generation spectroscopy and at a macroscopic level with the use of contact angle measurements. This type of surface design enables amplification of molecular-level conformational transitions to macroscopic changes in surface properties without altering the chemical identity of the surface. Such reversibly switching surfaces may open previously unknown opportunities in interfacial engineering.

Interfacial properties, such as wetting behavior, are defined by the molecular-level structure of the surface (*1*). Diverse modification procedures have been used to permanently alter wettability (*2–4*). Control of wettability has been recently demonstrated by elegant methods including light-induced (*5–6*) and electrochemical surface modifications (*7–10*). These systems require chemical reactions in order to control wettability.

We demonstrate an alternative approach for dynamically controlling interfacial properties that uses conformational transitions (switching) of surface-confined molecules. Polymers have been shown to undergo conformational reorientations when changed from one solvent to another (*11*) or from one temperature to another (*12*, *13*) because of phase transitions between a well solvated and a poorly solvated state (*14*). In contrast, our approach maintains the system's environment unaltered (including solvent, electrolyte content, pH, temperature, and pressure) while using an active stimulus, such as an electrical potential, to trigger specific conformational transitions (e.g., switching from an all-trans to a partially gauche oriented conformation; see Fig. 1). Amplification of conformational transitions to macroscopically measurable changes requires synergistic molecular reorientations of ordered molecules. In principle, this is attainable with a single-molecular layer, such as a self-assembled monolayer (SAM) of alkanethiolates on gold (*15*). However, the dense molecular packing in SAMs and the strong interactions between the methylene groups restrict dynamic molecular motions to the outermost atoms (*16*, *17*). All in situ evidence so far indicates that applied electrical potentials have no effect on long-chain alkanethiolate monolayers on gold within the range of chemical stability of the SAM (*18*). In other words, conventional SAMs are too dense to allow conformational transitions and consequently do not allow for switching. To explore SAMs as a model system for switching, we must establish sufficient spatial freedom for each molecule. Once a low-density SAM is created, preferential exposure of either hydrophilic or hydrophobic moieties of the SAM to the surrounding medium could be exploited for the switching of macroscopic surface properties.

(16-Mercapto)hexadecanoic acid (MHA) was chosen as a model molecule because it (i) self-assembles on Au(111) into a monolayer and (ii) has a hydrophobic chain capped by a hydrophilic carboxylate group, thus potentially facilitating changes in the overall surface properties. To create a monolayer with sufficient spacing between the individual MHA molecules, we used a strategy that exploits synthesis and self-assembly of a MHA derivative with a globular end group, which results in a SAM that is densely packed with respect to the space-filling end groups but shows low-density packing with respect to the hydrophobic chains. Subsequent cleavage of the space-filling end groups establishes a low-density SAM of MHA. The spatial dimensions of the precursor molecule to be used were adapted to match the optimum alkanethiolate density for conformational rearrangements.

The equilibrium low-energy conformational state of each of the sparsely packed MHA

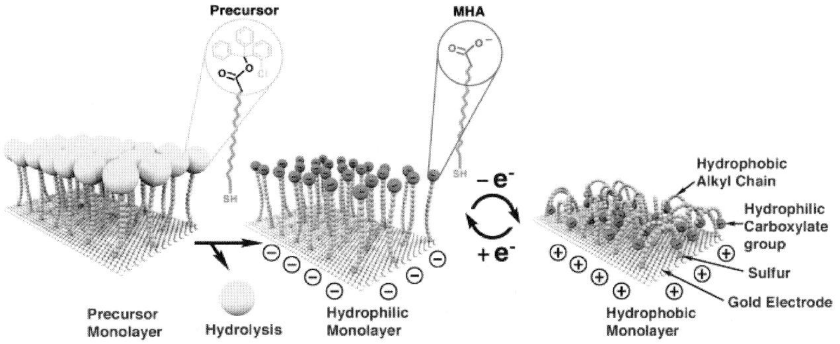

Fig. 1. Idealized representation of the transition between straight (hydrophilic) and bent (hydrophobic) molecular conformations (ions and solvent molecules are not shown). The precursor molecule MHAE, characterized by a bulky end group and a thiol head group, was synthesized from MHA by introducing the (2-chlorophenyl)diphenylmethyl ester group.

[1]Department of Chemical Engineering, Massachusetts Institute of Technology (MIT), 45 Carleton Street, Cambridge, MA 02139, USA. [2]Department of Chemical Engineering, University of California at Santa Barbara, Santa Barbara, CA 93106, USA. [3]Department of Chemistry, University of California at Berkeley, Material Science Division, Lawrence Berkeley Laboratory, Berkeley, CA 94720, USA.

*Present address: Department of Chemistry and School of Molecular Science (BK21), Korean Advanced Institute of Science and Technology, Daejeon 305-701, Korea.

†To whom correspondence should be addressed. E-mail: rlanger@mit.edu

REPORTS

molecules is all trans (all C-C-C-C torsion angles of the hydrophobic core are 180 degrees) (15). Upon applying an electrical potential, the negatively charged carboxylate groups experience an attractive force to the gold surface, causing the hydrophobic chains to undergo conformational changes. This chain bending disrupts the all-trans conformational state and causes the aliphatic MHA chains to adopt a mixture of trans and gauche conformations. Thus, the "bent" state of the MHA chains is characterized by an ensemble of molecules in mixed conformations, which maximizes the intermolecular van der Waals contact, and exposes the hydrophobic chains of the MHA molecules to the surrounding medium. To obtain a theoretical estimate of the packing density that would provide sufficient conformational freedom for optimal arrangement of the bent states of the MHA molecules, we performed molecular simulations of MHA monolayers, computing intramolecular bonded interactions and inter- and intramolecular nonbonded interactions. Simulations were conducted starting from assemblies of MHA molecules in bent states. Six assemblies with area-per-molecule values in the range between 0.29 and 2.59 nm^2 were studied (Materials and Methods). For small area-per-molecule values, relaxation of the highly constrained assemblies was dominant resulting in a steep decline of the potential energy as S-S spacings were widened. The plot of potential energy versus area per molecule (Fig. 2) shows that steric constraints are reasonably low for area-per-molecule values of 0.65 nm^2 or higher. On the other hand, the spacing between the MHA molecules for an area-per-molecule value of 0.65 nm^2 still permits a substantial overlap of MHA molecules, leading to favorable hydrophobic interactions between alkyl chains. Thus, even for the system with the widest S-S spacing of 1.73 nm, sufficient overlap of neighboring chains occurred and prevented the overall energy of the system from approaching zero. We conclude from these simulations that an area-per-molecule of 0.65 nm^2 is optimum because it supports steric relaxation and allows for substantial chain overlap.

On the basis of these theoretical considerations, we synthesized a MHA derivative with a globular end group [(16-mercapto) hexadecanoic acid (2-chlorophenyl)diphenylmethyl ester, MHAE]. Its space-filling end group (about 0.67 nm^2) matched closely with the previously determined area of 0.65 nm^2. Self-assembly of MHAE on gold and subsequent removal of the acid-labile end groups resulted in a low-density SAM of MHA (19). The quantitative removal was verified by the absence of the signal of chlorine as detected by x-ray photoelectron spectroscopy (XPS). In addition, the characteristic signal of the Cl-triphenyl ester group at 1743 cm^{-1} in the infrared (IR) spectrum was no longer observed, and the characteristic C=O peak appeared (20). We then exposed a low-density SAM of MHA to a solution of *n*-butanethiol immediately after cleavage of the bulky end group and conducted electrochemical desorption. For this mixed SAM, a single desorption peak was detected (orange line in fig. S1). In contrast, two desorption peaks were found for phase-separated SAMs composed of MHA and *n*-butanethiol formed via co-adsorption (blue line in fig. S1), corresponding to the coexistence of two distinct phases (21). We conclude from this experiment that preparation of a SAM following the previously mentioned strategy resulted in a rather homogeneously distributed monolayer and not a phase-segregated system. These results support the assumption that the low-density SAM consists of molecules that show increased spacing between individual molecules; the formation of MHA clusters is less likely.

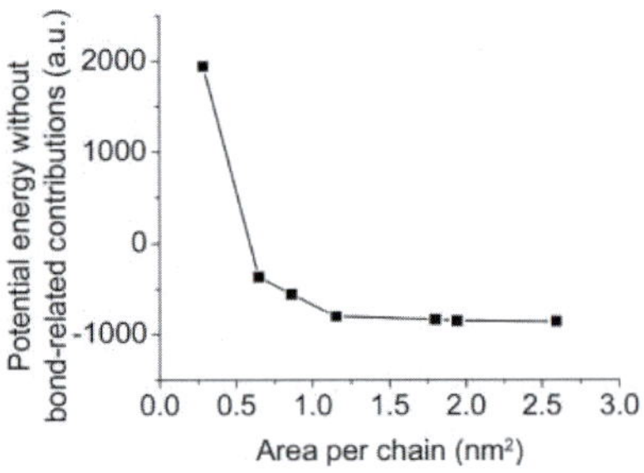

Fig. 2. Nonbonded interaction energy of MHA molecules versus area per molecule. a.u., arbitrary units. To evaluate the effect of the packing density on the conformational transitions and steric interactions of the MHA molecules, we chose as a starting point an assembly of 64 alkanethiolates that were forced into a bent molecular state (distance between sulfur and carbonyl carbon below 0.4 nm). Electrostatics were ignored because the goal of this simulation was to estimate the optimal geometrical spacing between the MHA molecules to allow for conformational transitions that provide switching capabilities to the monolayer. Nearest neighbor distances were set to be between 0.58 and 1.73 nm (35). A distance of 0.58 nm between two sulfur atoms corresponds to the densest packing we could establish with a bent assembly, and the distance of 1.73 nm was chosen as an upper limit to guarantee chain-chain overlap (an all-trans oriented chain of MHA has an approximate length of 2.24 nm). All alkanethiolates were confined on a hexagonal gold lattice (Au-Au distance of 0.29 nm), and sulfur and gold atoms were fixed in space. The intermolecular potential was adopted from (36, 37), and the potential describing the interactions between carbon and sulfur atoms with the gold surface was selected to mimic the values reported by Hautman and Klein (38).

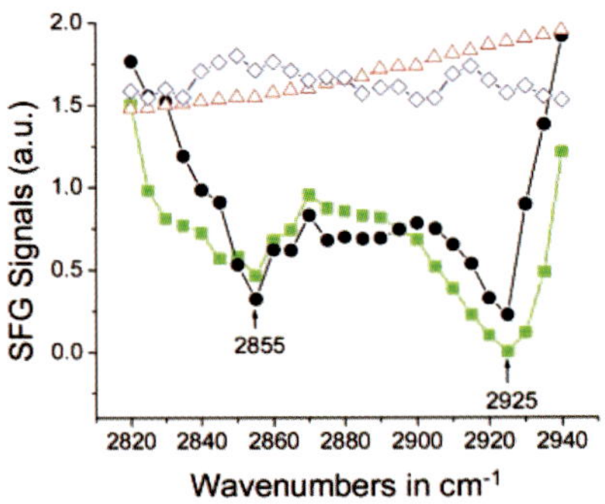

Fig. 3. In situ SFG spectra of the CH-stretch region (2820 to 2940 cm^{-1}) for the low-density SAM exposed to air (green squares), d^3-acetonitrile (red triangles), d^2-water (blue diamonds), and d^8-toluene (black circles). Tunable mid-IR light (1400 to 4000 cm^{-1}) was generated by difference frequency mixing of near-IR light with the fundamental beam of the neodymium: yttrium-aluminum-garnet (Nd:YAG) pump laser in a $LiNbO_3$ or $AgGaS_2$ crystal. The near-IR light was produced through optical parametric generation and amplification of 532-nm light in angle-tuned barium borate crystals. The IR and visible beams were incident on the liquid-solid interface at 40° and 35° and have energy densities of 4 mJ cm^{-2} and 15 mJ cm^{-2}, respectively. SFG spectra were an average of 10 scans with 5 cm^{-1} resolution. SFG signals were collected for 1 to 5 s at every 5 cm^{-1} interval. SFG data were normalized with respect to the reflected IR beam from the electrode to ensure that no false SFG peaks were observed due to strong IR absorption by the thin film of electrolyte above the electrode surface. The lines are drawn as a guide to the eye.

Before examining whether the low-density SAM of MHA undergoes conformational transitions in response to electric potentials, we confirmed the conformational freedom of the alkanethiolates by studying their response to changes in the chemical environment. Sum-frequency generation (SFG) spectroscopy was used for this purpose because it exploits a highly surface-sensitive nonlinear optical process (22). The intensity of a peak in the SFG spectrum is affected by the orientation of the adsorbed molecules. An ordered monolayer of all-trans oriented molecules is locally centrosymmetric and hence, by rule of mutual exclusion, its methylene modes are sum-frequency inactive. Gauche conformations break the local symmetry and give rise to SFG signals of the methylene groups (23). Figure 3 shows SFG spectra of the low-density SAMs of a MHA that was deprotonated. When exposing the low-density SAM to an apolar medium such as air or toluene, the molecules were found to be in disordered conformations. This state was characterized by the presence of gauche conformations as indicated by SFG signals of the methylene groups at wavelengths of 2925 and 2855 cm^{-1} (Fig. 3, green and black lines). When the same surface was brought in contact with a polar environment (acetonitrile or water), molecules straightened up, presumably by exposing their polar end groups to the solvent. SFG signals associated with the methylene groups were no longer detected. For the dense SAM of MHA, structural reorganizations were far less pronounced and were comparable to reported studies (*24*). On the basis of the detected conformational transitions in response to changes in the polarity of the surrounding medium, we conclude that only the low-density SAM provided sufficient chain mobility to allow conformational reorientations from a hydrophilic (straight chains) to a hydrophobic (bent chains) state.

We next assessed switching induced by an active stimulus, such as application of an electrical potential. Information regarding the prospective range of the electrical potential for switching may be obtained from basic energy considerations. First, alkanethiolates adsorbed on gold and exposed to a surrounding solution show electrochemical stability only in a relatively narrow range of electrical potential [–1.046 V to +654 mV with respect to (w.r.t.) the standard calomel electrode (SCE)] (*25*). Applied electrical potentials must be within this range of stability. Second, the change in Gibbs free energy of the system must be negative for a change in the monolayer structure to occur. With the use of a basic model, we estimated a potential above about +150 mV (–118 mV w.r.t. SCE) to be required to satisfy the conditions of molecular reorientation in a low-density SAM (*26*).

A SFG spectrometer equipped with an electrochemical quartz cell (*27*) was used to assess the switching of the surface in response to electrical potentials (Fig. 4, A and B). Without applied electrical potential, the SFG spectra recorded in acetonitrile [0.1 M cesium trifluoromethanesulfonate (CT)] were featureless in the range between 2820 and 2940 cm^{-1}, signifying straight molecular conformations of an all-trans orientation. Slightly positive polarization of the gold surfaces (+25 mV w.r.t. SCE), however, initiated simultaneous switching of the molecules indicated by characteristic methylene modes at wavelengths of 2855 and 2925 cm^{-1} (Fig. 4A). The presence of gauche conformations in the molecules implies that the molecules bend their negatively charged end groups toward the positively charged gold surface (Fig. 1). After the positive potential was turned off, the low-density SAM returned to an assembly of straight molecules with all-trans orientation, because SFG signals of the methylene groups were no longer detected. Switching was reversible, with intensities of the SFG signals being nearly constant as the electrical potential was repeatedly applied. In contrast, the dense SAM of MHA did not show reorientations induced by an applied electrical potential, because the SFG signals remained unaltered when potential was applied (Fig. 4B).

We then addressed the question of whether the observed rearrangements could be amplified into macroscopically detectable changes in surface properties. Advancing (open symbols) and receding (solid symbols) contact angles for the low-density (Fig. 4C) and the dense (Fig. 4D) SAMs were determined while applying either +80 or –300 mV w.r.t. SCE between the underlying gold electrode and a platinum-made microelectrode. Four subsequent switch cycles were examined, and contact angles with an aqueous solution were measured. Figure 4C indicates switching of the receding contact angles as the surface polarization was alternately changed. Although the advancing contact angle was independent from the applied potential, the receding contact angle showed a sharp step whenever the polarization of the surface was changed (Fig. 4C, solid squares). The large hysteresis in contact angle can be caused by surface roughness or chemical heterogeneity of the surface (*28*). Because scanning force microscopy did not reveal notable differences in surface roughness between systems configured of molecules in bent and straight states (*29*), the assumption of a chemical discontinuity along the solid-air-liquid contact line best explains the large hysteresis in contact angles and the pronounced sensitivity of the receding contact angle. Following Good and Neumann's wettability model (*30*), the system reflects the behavior of a smooth but chemically heterogeneous system that is composed of a low-energy area (exposed to air) and a high-energy area (exposed to solution). The observed changes in receding contact angle then signify molecular transitions at the high-energy area. SFG results support this assumption (Fig. 3). The drop in contact angle was a reversible phenomenon, because the assembly was switched several times between its hydrophilic (straight molecules) and hydrophobic (bent molecules) state (four times for the study shown in Fig. 4C). In contrast, switching was not observed for the dense SAM (Fig. 4D). We excluded the possibility of electrochemically induced protonation, because it would affect both dense and low-density SAMs similarly (in addition, experiments were conducted at pH = 11.5 to avoid protonation). The applied electrical potentials were well above the estimated lower limit of electrical potentials that permits conformational transitions (*26*) but low enough to be within the potential window of greatest stability for SAMs of alkanethiolates on gold (*25*, *31*). Thus, electrochemical reactions can be excluded as driving forces for the observed changes in surface properties. We conclude that the observed switching in surface properties is microscopically driven by conformational transitions.

This study demonstrates reversible control of surface switching for a low-density monolayer. Because of synergistic molecular reorientations, amplification into macroscopic changes in surface properties is observed. Future research might be directed toward enhancement of macroscopic effects and development of alternative stimuli. The fact that controlled conformational reorientations of single-layered molecules induced observable changes in wettability suggest that these findings may, with further study, have implications in dynamic regulation of macroscopic properties, such as wettability, adhesion, friction, or biocompatibility. Potential applications might include microfluidics, microengineering of smart templates for bioseparation or data storage, or microfabrication of controlled-release devices.

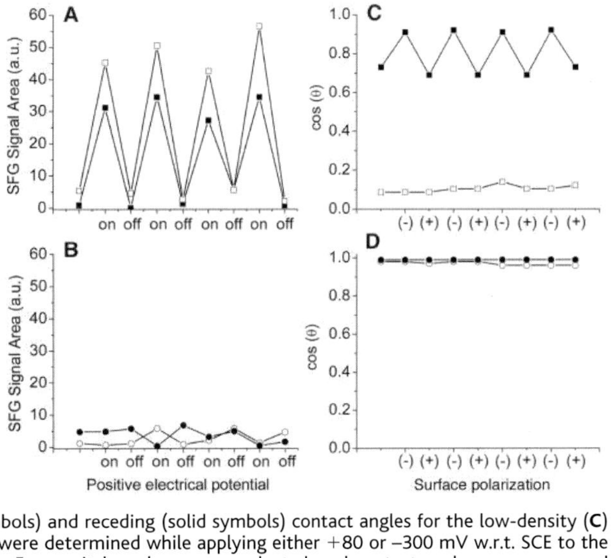

Fig. 4. Microscopic and macroscopic responses of the low-density SAM to an electrical potential as monitored by SFG spectroscopy and contact angle measurements. Relative SFG intensities (peak areas) of the methylene modes at wavelengths of 2855 cm^{-1} (solid symbols) and 2925 cm^{-1} (open symbols) are shown for the low-density (**A**) and the dense (**B**) SAMs measured in d^3-acetonitrile (0.1 M CT) when a potential of +25 mV w.r.t. SCE was repeatedly applied to the system. Cosine of the advancing (open symbols) and receding (solid symbols) contact angles for the low-density (**C**) and the dense (**D**) SAMs were determined while applying either +80 or –300 mV w.r.t. SCE to the underlying gold electrode. Four switch cycles were conducted, and contact angles were measured with an aqueous solution (0.1 M CT, pH = 11.5) at air using a goniometer (VCA-2500XE, Advanced Surface Technology) equipped with an electrometer (6517A, Keithley Instruments) and platinum and carbon fiber microelectrodes (Kation Scientific). Contact angles averaged at least 100 data points from nine samples with maximum errors of ±3°. The SAMs were examined for chemical integrity and deprotonation by IR spectroscopy after an electrical potential was applied. The lines are drawn as a guide to the eye.

REPORTS

References and Notes

1. P. G. deGennes, *Rev. Mod. Phys.* **57**, 827 (1985).
2. G. M. Whitesides, P. E. Laibinis, *Langmuir* **6**, 87 (1990).
3. P. E. Laibinis, C. D. Bain, R. G. Nuzzo, G. M. Whitesides, *J. Phys. Chem.* **99**, 7663 (1995).
4. M. K. Chaudhury, G. M. Whitesides, *Science* **256**, 1539 (1992).
5. S. Abbott, J. Ralston, G. Reynolds, R. Hayes, *Langmuir* **15**, 8923 (1999).
6. K. Ichimura, S.-K. Oh, M. Nakagawa, *Science* **288**, 1624 (2000).
7. N. L. Abbott, C. B. Gorman, G. M. Whitesides, *Langmuir* **11**, 16 (1995).
8. M. Byloos, H. Al-Maznai, M. Morin, *J. Phys. Chem. B* **105**, 5900 (2001).
9. A. Iannelli, J. Merza, J. Lipkowski, *J. Electroanal. Chem.* **376**, 49 (1994).
10. B. S. Gallardo *et al.*, *Science* **283**, 57 (1999).
11. M. D. Wilson, G. M. Whitesides, *J. Am. Chem. Soc.* **110**, 8718 (1988).
12. G. De Crevoisier, P. Favre, J. Corpart, L. Leibler, *Science* **285**, 1246 (1999).
13. Y. G. Takei *et al.*, *Macromolecules* **27**, 6163 (1994).
14. T. P. Russell, *Science* **297**, 964 (2002).
15. Y. Xia, G. M. Whitesides, *Angew. Chem. Int. Ed.* **37**, 550 (1998).
16. M. H. Schoenfisch, J. E. Pemberton, *Langmuir* **15**, 509 (1999).
17. R. Anderson, M. Gatin, *Langmuir* **10**, 1638 (1994).
18. M. A. Hines, J. A. Todd, P. Guyot-Sionnest, *Langmuir* **11**, 493 (1995).
19. A low-density monolayer was prepared by (i) immersing Au(111) in a solution of MHAE (1mM, in ethanol; refer to Materials and Methods for details regarding the synthesis of MHAE) and (ii) removing the acid-labile end group by incubation in trifluoroacetic acid (40% in ethanol). The thickness of the SAM of MHAE was 2.05 (±0.1) nm as examined by ellipsometry. Formation of islands of adsorbed thiolates was ruled out by cyclic voltammetry (Gamry PC4-300 potentiostat; polycrystalline Au counter and Ag-AgCl reference electrodes; scan rate, 50 mV s^{-1}; 10 cycles; scan range, −0.2 to 0.5 V w.r.t. Ag-AgCl), revealing a shielding of the gold surface by the SAM of MHAE that was comparable to dense SAMs (*32*).
20. In air, the low-density monolayer collapsed into a disordered configuration with a thickness of 1.1 (±0.1) nm as confirmed by ellipsometry. This value was significantly lower than the thickness of a dense SAM [2.03 (±0.1) nm] and indicated the loss of two-dimensional order due to increased spacing between the adsorbed molecules. The collapsed structure was also characterized by IR spectroscopy, which revealed general features of a disordered monolayer, such as the red shift of the methylene bands from 2921 and 2852 cm^{-1} (dense SAMs of MHA) to 2929 and 2858 cm^{-1}. Furthermore, we detected an increased static contact angle with water of 62° as compared to 10° measured for the dense SAMs of MHA. The carboxylic acid group was deprotonated under argon atmosphere by incubation with a degassed aqueous cesium hydroxide solution (0.1 M), and completeness was verified by IR spectroscopy (BioRad FTS 175, grazing angle of 80°), revealing the absence of the carbonyl signals at 1742 and 1716 cm^{-1} and the presence of signals at 1550 and 1440 cm^{-1}. The low-density SAM showed chemical stability because it remained unaltered after exposure to ambient conditions for 4 weeks (confirmed by XPS and IR spectroscopy).
21. K. Shimazu, T. Kawaguchi, T. Isomura, *J. Am. Chem. Soc.* **124**, 652 (2002).
22. Y. R. Shen, *Nature* **337**, 519 (1989).
23. T. H. Ong, P. B. Davies, C. D. Bain, *Langmuir* **9**, 1836 (1993).
24. S. M. Stole, M. D. Porter, *Langmuir* **6**, 1199 (1990).
25. W. R. Everett, I. Fritsch-Faules, *Anal. Chim. Acta* **307**, 253 (1995).
26. In a first-order approximation, we neglected interactions between the gold surface and solvent or electrolyte molecules, implying that their net energetic contributions are comparably small. We then assumed the negatively charged end groups act as dynamically mobile acceptors dictating the molecular reorientations and assumed the change in Gibbs free energy upon transition of the system to an electrified state ($N \rightarrow E$) as follows: $\Delta g_{total}|_{N\rightarrow E} = \Delta g_{int}|_{N\rightarrow E} + \Delta g_{sur}|_{N\rightarrow E} + g_{ele}$ (1) Δg_{int} is the change in the Gibbs free energy associated with internal reorganization of the molecules, Δg_{sur} reflects the intermolecular interactions of the hydrocarbon chains and their interactions with the surrounding solvent molecules, and g_{ele} describes the electrostatic component associated with attractive forces between the gold surface and the carboxylate groups. Because steric hindrance between the chains is minimal in the low-density SAM, gauche conformations associated with chain bending mainly account for Δg_{int}. Δg_{int} can be approximated as $0.9nkT$, with n being the number of gauche-oriented bonds, k being Boltzmann constant, and T being temperature (*33*). Δg_{sur} primarily originates from exposure of methylene groups to the surrounding aqueous environment that were embedded in the hydrophobic environment before application of an electrical potential. This term is approximated by $1.5mkT$, where m is the number of methylene groups exposed to the aqueous surrounding because of application of electric field (*34*). Assuming further that at least two bonds need to be gauche-oriented in order to observe a detectable change in contact angle, the required contribution from the electrostatic field based on Eq. 1 is $|g_{ele}| > 4.8kT$. The electrostatic contribution of the free energy can be approximately described by $e\psi_o$, where e is the electronic charge and ψ_o is the surface potential (*33*).
27. S. Baldelli, N. Markovic, P. Ross, Y. R. Shen, G. Somorjai, *J. Phys. Chem.* **103**, 8920 (1999).
28. Y. L. Chen, A. Helm, J. N. Israelachvili, *J. Phys. Chem.* **95**, 10736 (1991).
29. Scanning force microscopy did not reveal notable differences in surface roughness between systems configured of molecules in bent [root mean square (RMS) roughness = 1.9 nm] and straight (RMS roughness = 1.5 nm) states. The different states were realized by conducting scanning force microscopy of a low-density SAM in toluene and water, respectively. Our experiments indicated that surface roughness is predominately caused by the intrinsic surface roughness of the gold surface (RMS roughness < 2 nm). All RMS values refer to an area of 500 × 500 nm^2. Refer to fig. S2 for detailed information.
30. A. W. Neumann, R. J. Good, *J. Colloid Interface Sci.* **38**, 341 (1972).
31. H. O. Finklea, S. Avery, M. Lynch, T. Furtsch, *Langmuir* **3**, 409 (1987).
32. F. P. Zamborini, R. M. Crooks, *Langmuir* **14**, 3279 (1998).
33. N. Aydogan, B. S. Gallardo, N. L. Abbott, *Langmuir* **15**, 722 (1999).
34. M. H. Abraham, *J. Chem. Soc. Faraday Trans. 1* **80**, 153 (1984).
35. T. Li, I. Chao, Y. Tao, *J. Phys. Chem. B* **102**, 2935 (1998).
36. W. D. Cornell *et al.*, *J. Am. Chem. Soc.* **117**, 5179 (1995).
37. S. J. Weiner *et al.*, *J. Am. Chem. Soc.* **106**, 765 (1984).
38. J. Hautman, M. L. Klein, *J. Chem. Phys.* **91**, 4995 (1989).

39. We thank S. Baldelli, N. Flynn, R. Raman, M. Stevens, and G. Venkataraman for valuable discussions; G. M. Whitesides and J. Israelachvili for insightful comments on the manuscript; P. E. Laibinis for the use of the IR; and the Materials Research Science and Engineering Center shared facilities (MIT) supported by NSF under award DMR-0213282. J.L. thanks the Fonds der Chemischen Industrie, Germany, for a Justus-Liebig fellowship, and T.-N.T. acknowledges support from the Whitaker Foundation. This work was sponsored in part by the U.S. Army Research Office through the Institute for Soldier Nanotechnologies at MIT under contract DAAD-19-02-002 and NIH. The content of the information does not necessarily reflect the position or the policy of the Government, and no official endorsement should be inferred.

Supporting Online Material

www.sciencemag.org/cgi/content/full/299/5605/371/DC1

Materials and Methods

Figs. S1 and S2

1 October 2002; accepted 27 November 2002

review article

Designing materials for biology and medicine

Robert Langer[1] **& David A. Tirrell**[2]

[1]*Department of Chemical Engineering, Massachusetts Institute of Technology, Building E25-342, Cambridge, Massachusetts 02139, USA*
[2]*Division of Chemistry and Chemical Engineering, California Institute of Technology, Mail Code 164-30, Pasadena, California 91125, USA*

Biomaterials have played an enormous role in the success of medical devices and drug delivery systems. We discuss here new challenges and directions in biomaterials research. These include synthetic replacements for biological tissues, designing materials for specific medical applications, and materials for new applications such as diagnostics and array technologies.

Biomaterials have been defined as substances other than foods or drugs contained in therapeutic or diagnostic systems[1] and, in some cases, have been described as materials composed of biologically derived components (for example, amino acids) irrespective of their application. Throughout history, biomaterials have played an important role in the treatment of disease and the improvement of health care. Early biomaterials include metals such as gold that were used in dentistry over 2,000 years ago. Other early examples of biomaterials include wooden teeth and glass eyes[2]. However, with the advent of synthetic polymers at the end of the nineteenth century, these materials became increasingly used in health care. For example, polymethylmethacrylate, PMMA, was used in dentistry in the 1930s[1] and cellulose acetate was used in dialysis tubing[2] in the 1940s. Dacron was used to make vascular grafts; polyetherurethanes, the materials used in ladies' girdles, were used in artificial hearts; and PMMA and stainless steel were used in total hip replacements[1,2]. Naturally occurring materials such as collagen have also been used as biomaterials[3]. However, in nearly every case, these materials were adopted from other areas of science and technology without substantial redesign for medical use. Although these materials helped usher in new medical treatments, critical problems in biocompatibility, mechanical properties, degradation and numerous other areas remain. To this end, scientists are creating new materials including those with improved biocompatibility, stealth properties, responsiveness (smart materials), specificity and other critical properties. Modern biomaterials science is characterized by a growing emphasis on identification of specific design parameters that are critical to performance, and by a growing appreciation of the need to integrate biomaterials design with new insights emerging from studies of cell–matrix interactions, cellular signalling processes, and developmental and systems biology.

Biomaterials are already having an enormous effect on medicine. Controlled drug delivery systems that largely involve polymers[4] are used by tens of millions of people annually[5]. Recent examples are polymer-coated stents, which have recently been approved both in Europe and the United States. Hundreds of thousands of lives are expected to be saved each year[6]. In addition, various controlled release systems for proteins, such as human growth hormone, as well as molecules decorated with polyethylene glycol (PEG), such as pegylated interferon[4–7], have recently been approved by regulatory authorities, and are showing how biomaterials can be used to positively affect the safety, pharmacokinetics and duration of release of important new drugs. Another area where biomaterials have recently had an impact is in tissue engineering. By combining polymers with mammalian cells, it is now possible to make skin for patients who have burns or skin ulcers, and various other polymer/cell combinations are in clinical trials, including corneas, cartilage, bone and liver[8]. Biomaterials have also had a major impact as the central components of dental implants, sutures, and numerous medical devices[2].

Here we describe novel material concepts that are shaping future directions in biomaterials science. In particular, we discuss (1) creating synthetic replacements for biological tissues using naturally occurring building blocks, (2) synthesizing materials using man-made building blocks for specific medical and biological applications, and (3) design concepts for new *in vitro* applications such as diagnostics and array technologies.

Synthetic replacements for biological tissues

Materials composed of naturally occurring (biologically derived) building blocks, including extracellular matrix (ECM) components, are being studied for applications such as direct tissue replacement and tissue engineering. The ECM, a complex composite of proteins, glycoproteins and proteoglycans, provides an important model for biomaterials design[9]. ECM-derived macromolecules (for example, collagen) have been used for many years in biomaterials applications[3], and it is now possible to create artificial analogues of ECM proteins using recombinant DNA technology[10]. Through the design and expression of artificial genes, it is possible to prepare artificial ECM proteins with controlled mechanical properties and with domains chosen to modulate cellular behaviour. This approach avoids several important limitations encountered in the use of natural ECM proteins, including batch-to-batch (or source-to-source) variation in materials isolated from tissues, restricted flexibility in the range of accessible materials properties, and concerns about disease transmission associated with materials isolated from mammalian sources.

Elastin-based systems have been of special interest in this regard. Urry and co-workers have shown over many years that simple repeating polypeptides related to elastin can be engineered to exhibit mechanical behaviour reminiscent of the intact protein[11]. Crosslinking can be accomplished via radiative[12] or chemical[13] means, and electrospinning has been used to prepare fibrous forms of engineered elastins[14] (Fig. 1). Incorporation of cell-adhesion ligands allows attachment and spreading of cultured cells, and in the specific case of materials for vascular grafts, retention of endothelial cell adhesion in the face of shear stresses characteristic of the normal circulation[15].

The promise of biosynthetic approaches to biomaterials design must be weighed against the fact that very little is known about the *in vivo* performance of systems prepared in this way. Animal experiments on elastin-like polypeptides prepared by chemical synthesis have shown these materials to evoke relatively mild tissue reactions, but the range of materials investigated to date has been small[16]. As more is learned about the mechanical properties of such materials and about their interactions with cells in culture, the groundwork will be laid for more extensive evaluation in animal models. Recent progress in the development of methods for incorporation of non-natural amino acids into recombinant proteins points the way to an alternative strategy for preparing artificial ECM proteins with diverse chemical, physical and biological properties[17].

Substantially more experience has been gained in evaluating the *in vivo* performance of engineered biomaterials based on polysaccharides. Alginate hydrogels bearing cell-adhesion ligands have been used as scaffolds for cell encapsulation and transplantation, and have yielded promising results in experiments directed towards the engineering of bone tissue capable of growth from small numbers of implanted cells[18]. The prospect of growing tissues from small numbers of precursor cells is an attractive alternative to harvesting and encapsulating large cell masses before transplantation.

Molecular self assembly of peptides or peptide-amphiphiles may also lead to unique biomaterials. A number of self assembled peptide systems have been developed, including systems that can potentially be used in tissue engineering and nanotechnology[19,20].

An alternative to synthesizing polymers composed of natural components is the synthesis of biomimetic polymers, which combine the information content and multifunctional character of natural materials (such as a particular amino acid sequence that might be desirable for cell attachment) with the tailorability of a synthetic polymer, such as control of molecular mass or polymer degradation, and the ability to impart appropriate mechanical properties. An example of this concept has been the synthesis of polymers composed of lactic acid and lysine. Like polylactic-glycolic acid, these polymers can be made to degrade at desired times. However, by adding lysine as a co-monomer with lactic acid, a free amino acid is provided, which allows coupling reactions to take place and does not affect the overall biocompatibility of the polymer[21]. By coupling specific amino acids (such as the tripeptide sequence RGD) to this polymer, cell adhesion can be regulated[22].

Another strategy that has been used in modifying a variety of natural and synthetic polymers has been the inclusion of PEG into the material to reduce non-specific effects of protein adsorption and colloidal aggregation. The molecular origins of these phenomena are not yet thoroughly understood. One example of the effects of the PEG can be seen in the creation of polylactic-glycolic acid (PLGA) PEG diblock polymers that, when formed into nanospheres, can, like cells, circulate in the body for long time periods. For example, in mice, only 30% of PLGA-PEG nanoparticles were cleared after 5 h whereas 66% of non-PEG-containing PLGA nanoparticles were cleared in only 5 min (ref. 23). Halstenberg *et al.*[24] have adopted a related approach in engineering protein-based biomaterials, by grafting poly(ethylene glycol) diacrylate onto an artificial protein that contained multiple cysteine residues. The protein was designed to serve several functions, including cell adhesion, heparin binding, and degradation by plasmin to facilitate penetration by invading cells. The mechanical properties of the material were controlled by photopolymerization of the pendant acrylate units.

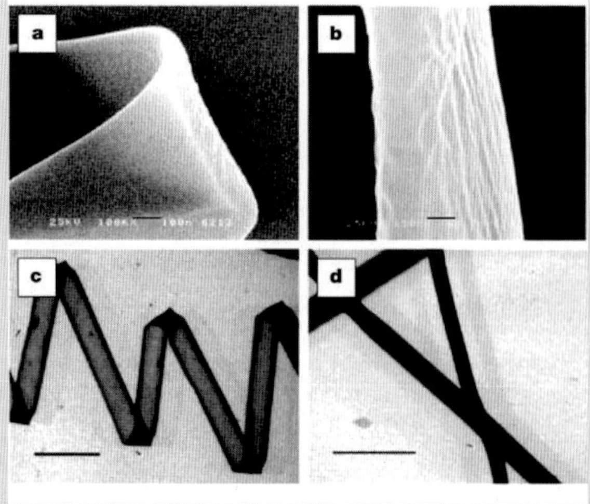

Figure 1 Electrospun fibres of elastin-like artificial proteins made by expression of artificial genes in bacterial cells. **a**, **b**, Scanning electron micrographs. **c**, **d**, Transmission electron micrographs. Scale bars: **a** and **b**, 100 nm; **c**, 3.3 μm; **d**, 2.0 μm. Images from ref. 14.

Another approach to creating a biomimetic reversible system is the creation of an antigen responsive hydrogel. Corresponding antibody pairs are used to form reversible non-covalent crosslinks in a polyacrylamide system. In the presence of excess free antigen, the hydrogel swells, but in its absence, the gel collapses back to a crosslinked network. Swelling does not occur when foreign antigens are added, showing that the system is antigen specific. Release of a model protein such as haemoglobin has been demonstrated in response to specific antigens[25].

Materials for specific medical and biological applications

A variety of new materials are being synthesized from man-made building blocks, and being used to create devices for specific medical applications. One area of increasing attention has been the development of shape-memory materials that have one shape at one temperature and another shape at a different temperature[26]. Such materials might permit new medical procedures. For example, current approaches for implanting medical devices often require complex surgery followed by device implantation. However, with the development of minimally invasive surgery, it is possible to place small devices inside the body using laproscopes. These types of surgical advances may create new opportunities to enable a bulky device to be implanted into the human body in a convenient way.

Shape-memory materials might provide such an opportunity, because they have the ability to memorize a permanent shape that can be substantially different from an initial temporary shape. Thus bulky devices could potentially be introduced into the body in a temporary shape, like a string, that could go through a small laproscopic hole, but then be expanded on demand into a permanent shape (for example, a stent, a sheet, and so on) at body temperature (Fig. 2). New polymers have been synthesized with this concept in mind, including phase segregated multiblock copolymers whose starting materials are known biocompatible monomers such as ϵ-caprolactone and *p*-dioxanone. Generally, these materials have at least two separated phases, each with thermal transition (glass or melting) temperatures. The phase with the higher transition temperature is responsible for the permanent shape, whereas the second phase can act as a molecular switch and enable fixation of a temporary shape. By regulating temperatures above and below the second phase's transition temperature, shape can be shifted from one form to another. In addition to permitting new opportunities for implanting devices, these polymers have been developed into sutures that are able to tie themselves on demand, triggered by a temperature change[27] (Supplementary video 1).

Materials that are liquids at room temperature but that harden in response to a change in temperature or an external stimulus, such as light, are also being studied[28–30]. Such systems offer an opportunity to inject materials containing drugs, for example, into the body through a small needle, but still enable the formation of an implant.

A variety of smart gels have also been developed. These gels can respond (and swell, for example) to triggers such as temperature or pH, or even specific molecules in the body such as glucose. Such systems may, with further study, be valuable in the treatment of disease: in the case of diabetes, for example, smart gels could provide direct feedback control, allowing more insulin to be delivered in response to excess glucose[31].

It may be possible to change not only the bulk properties of materials but also their surface properties, using a simple 'switch'

review article

such as temperature or electric charge. For example, alkanethioates such as 16-mercaptohexadecanoic acid, which has a hydrophobic chain capped by a hydrophilic carboxyl group, forms self-assembled monolayers on gold surfaces. These chains have an upright equilibrium conformation presenting the carboxylic groups to the surrounding medium. But when these alkanethioates are placed on gold at the correct density, application of an electric potential causes the carboxyl groups to be attracted to the gold surface electrostatically, causing the molecules to reversibly rearrange and expose the hydrophobic chain (Supplementary video 2). Such surface switches might offer new opportunities in such areas as drug delivery, microfluidics and biosensors[32].

The development of high-throughput approaches to create novel biopolymers and screen them for various applications is garnering increased attention. For example, Kohn and co-workers have created polymer libraries and then screened them for different applications[33,34]. This type of high-throughput approach has also been used in the creation of gene therapy agents. For example, poly-β-amino esters can be synthesized in a high-throughput manner, and a number of these new polymers have been shown to have higher DNA transfection activities in cell-based assays than existing materials such as lipofectamine[35,36].

Gene therapy represents an area where appropriate molecular design is critical to achieving a successful outcome. Although viral vectors are highly effective, their use has raised serious safety concerns. This has motivated research on synthetic gene therapy vectors, which, although safer, have thus far been much less effective than viral vectors. To be effective, there are a number of attributes that the material must possess, including the ability to condense DNA to sizes less than 150 nm so that it can be taken up by receptor-mediated endocytosis, the ability to be taken up by endosomes in the cell and to allow DNA to be released in active form, and to enable it to travel to the cell's nucleus[37]. An interesting example of the design of such new materials is provided by the cationic-cyclodextrin polymers developed by Davis and co-workers. Cyclodextrins are relatively non-toxic and do not elicit an immune response. When Davis and co-workers initially used cyclodextrins for packaging DNA they found the resulting complexes to be relatively unstable. To address this, they added adamantane-conjugated polyethylene glycol to the surface of the cyclodextrin particles. This enabled the development of uniformly sized nanoparticles that resisted aggregation. By modifying the surface of the particles in this way, chemical groups could be exposed that could attach other molecules and allow the particles to be targeted to, and deliver genes to, specific cells[38]. Another novel approach for gene therapy involves creating a triplex where low-density lipoprotein is used for targeting and stearyl polylysine is used for DNA complexation. This approach has been used to deliver vascular endothelial cell growth factor to heart muscle to aid in treating blockage of blood vessels[39]. With the advent of new gene therapy agents such as RNA interference[40], the ability to design improved materials to deliver these agents will have increasing importance.

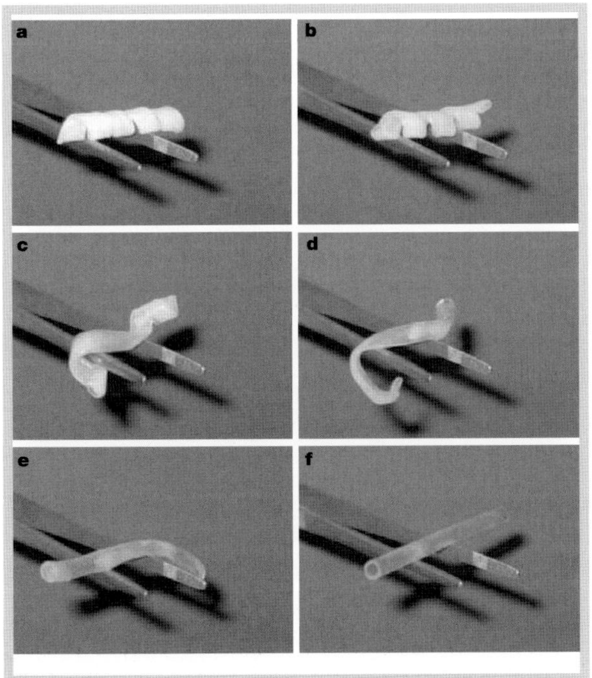

Figure 2 Time series of photographs showing recovery of a shape-memory tube. **a–f**, Start to finish of the process; total time, 10 s at 50 °C. The tube was made of a poly(ε-caprolactone)dimethacrylate polymer network (the M_n of the network's switching segments was 10^4) that had been programmed to form a flat helix. Images courtesy of A. Lendlein.

Biomaterials are also being used to affect bioadhesion in novel ways, even enabling materials to be used to potentially deliver complex molecules orally. It appears that polymers with high concentrations of hydroxyl groups bind with the intestinal mucosa. For example, Mathiowitz and co-workers[41] have designed polyanhydride nanoparticles (which can encapsulate DNA or other molecules, and which expose carboxyl groups on their exterior as the polymer erodes), and have shown that they can attach to mucous membranes and bind to the intestinal wall. Poly(fumaric-co-sebacic) anhydride showed greater adhesive forces than other materials tested, had a longer contact time with cells in the intestine, and was able to pass through the intestinal wall with a protein inside it. Peppas and co-workers have developed polymers that are not only bioadhesive, but also swell in response to a pH change. These polymers are able to protect proteins from the acidic pH of the stomach, and release it in the more basic pH of the intestine. These materials also appear to temporarily open connections between intestinal cells, allowing the proteins to pass through[42].

Microfabrication-based devices may also provide a novel approach for creating a variety of new biomaterials and delivery systems. For example, silicon microchips have been engineered to contain over 100 nanometre-sized drug-containing wells covered with gold on a chip 1 cm × 1 cm. By applying approximately 1 V to individually addressable wells, the drug in any of these wells can be released[43]. These types of systems have also been recently made with degradable materials (Fig. 3)[44].

Microfabrication has also been used in the creation of sensors. For example, small sensors are being used to measure intraocular pressure for glaucoma patients[45], and silicone microstimulators have been developed that can be controlled by telemetry[46] and are being used for retinal stimulation to aid in photoreceptor generation to treat diseases of the back of the eye that cause blindness. Prausnitz and co-workers have developed microneedles that are able to penetrate into the skin to depths far enough for drugs to enter the circulation, but shallow enough not to reach skin layers that contain nerves: therefore they do not cause pain[47]. Microfabrication has also been used to create polymer scaffolds containing an intricate vascular network[48]. All of these represent important ways of using micro- and nanotechnology to create potential new medical devices.

New applications: diagnostics and array technologies

Much of the preceding discussion has focused on the development of materials that display biological information, often in the form of peptide or protein domains, for the purpose of controlling cell and tissue behaviour. A related challenge arises in the engineering of

review article

materials for diagnostics and array technologies, in which large numbers (typically hundreds or thousands) of nucleic acids[49] or proteins[50] are presented in a format that allows rapid and highly parallel read-out of information concerning gene expression or protein function. Analogous technologies are being pursued with peptides[51], carbohydrates[52] and other small molecules, as well as with cells and tissues[53].

The role of materials science in array technologies was demonstrated in striking fashion by Fodor and co-workers[54], who used photolithographic techniques to prepare an array of 1,024 peptides in just ten sequential operations. Similar methods have been used to create oligonucleotide arrays[55]. Brown and co-workers introduced an alternative approach to DNA arrays by high-speed robotic spotting of complementary DNAs on treated glass surfaces[56]. Both techniques yield 'DNA chips' characterized by high densities of biological information.

Although these technologies are now highly developed and widely used, they are far from optimized. Hybridization of DNA arrays typically requires many hours[56], precluding their use in circumstances that demand rapid diagnosis or detection of pathogens, and limiting the possibilities for high-throughput screening of large numbers of samples. Array technologies that allow local application of electric fields (to enhance DNA migration)[57], ultrasonic mixing, or acoustic microstreaming[58], have all been shown to reduce substantially the time required for read-out of microarray data. Thermal-gradient DNA chips have been developed to allow local (site-to-site) control of the hybridization temperature[59], which can be important in distinguishing perfect sequence matches from single-site mismatches, and microporous arrays have been used in flow systems to speed up hybridization[60].

Additional challenges arise in the fabrication of protein arrays. Unlike nucleic acids, which share a common physical chemistry that is largely independent of sequence, proteins are highly diverse in terms of charge, hydrophobic character, and so on. So whereas robotic printing of DNAs on a common substrate can be done reliably, similar treatment of a complex protein library is unlikely to result in uniform retention of protein structure and function. The potential importance of protein arrays in the direct determination of protein concentrations in cells and tissue fluids, in the identification of protein–protein interactions, and in the study of drug-binding behaviour, has prompted substantial effort directed towards the development of reliable methods for array fabrication[50].

A successful fabrication method must be applicable to diverse protein libraries, and must achieve robust attachment of each protein to the surface of the array device. The attached proteins must not denature (unfold), and at least some fraction of each protein must be presented in an orientation that is consistent with retention of function; the site(s) of interest cannot be blocked by the array surface or by the linking groups used to achieve surface attachment. The substrate should not be prone to non-specific protein adsorption, because non-specific adsorption can give rise to high background noise and false positive signals.

Brown and co-workers examined the simple approach of printing proteins (either antigens or antibodies) on poly(L-lysine)-coated glass microscope slides[61]. They examined 115 antibody/antigen pairs, and found that approximately 50% of the printed antigens yielded signals that were quantitatively consistent with the known concentrations of the corresponding antibodies in the analyte solutions. Antibody arrays performed at a significantly lower level. This study is especially instructive because it provides a careful analysis of the quantitative reliability of the technology, and because it illustrates the challenges involved in fabricating protein arrays. The fact that fewer than 50% of arrayed antibodies yielded reliable signals is particularly striking given the structural uniformity of antibodies. More diverse protein libraries are likely to be substantially more difficult to array.

Alternative approaches to the fabrication of protein microarrays have used aldehydic surfaces to capture protein-bound lysine side chains[62], polyacrylamide gel pads to entrap proteins and to immobilize antibodies via reaction with oxidized carbohydrate substituents[63], and 'capture ligands' that react covalently and specifically with a single 'capture protein' that can be fused to each of the diverse proteins in the library to be arrayed[64]. The most ambitious experi-

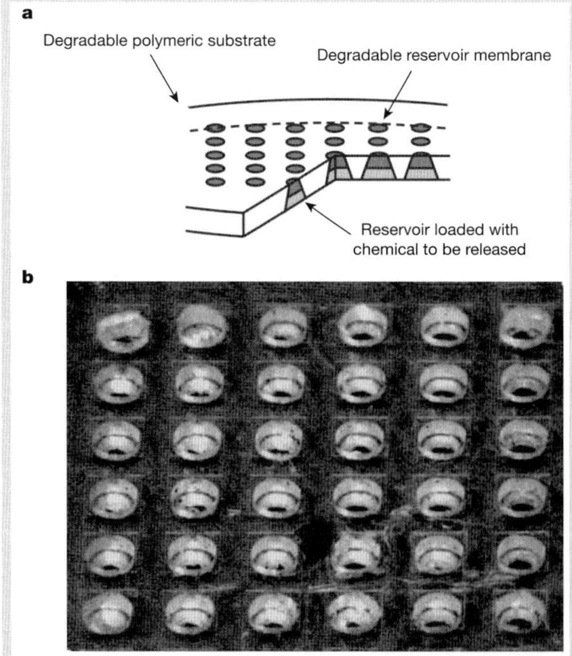

Figure 3 Degradable polymeric microchip. **a**, Cut-away diagram. **b**, Close-up photograph of reservoirs on a degradable microchip composed of polylactic-glycolic acid. Image courtesy of A. Richards-Grayson.

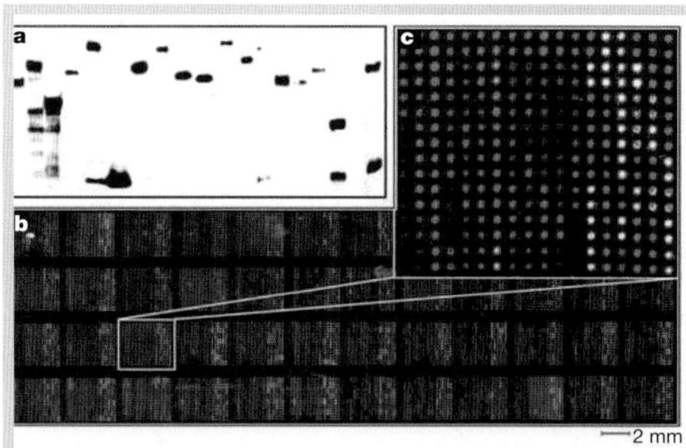

Figure 4 Miroarray of 5,800 yeast proteins, each fused to a hexahistidine sequence that facilitates immobilization on a nickel-coated glass slide. A second fused sequence, that of glutathione *S*-transferase (GST), allows visualization with anti-GST antibodies. Arrays of this kind have been developed by Snyder and co-workers, and used in global analyses of the yeast proteome. **a**, Immunoblot analysis of 19 representative fusion proteins purified in 96-well format; proteins were visualized with anti-GST. **b**, Protein samples spotted on nickel-coated glass slide and probed with anti-GST. **c**, Enlarged image of a portion of the protein array. Image from ref. 65.

ment to date[65] has used a polyhistidine tag to immobilize more than 5,000 yeast proteins to the surface of a glass slide functionalized with Ni^{2+} (Fig. 4)

Arraying of membrane proteins presents special challenges. Membrane proteins are especially difficult to isolate and to maintain in active form. A promising approach to membrane protein arrays has emerged from studies of supported lipid bilayers, and from the discovery that supported bilayers can be partitioned into 'corrals' that do not allow lipid intermixing via lateral diffusion[66]. Partitioning can be accomplished by microcontact printing, by deposition of molecular barriers, or by a variety of photochemical methods. The supported bilayer stabilizes membrane proteins with respect to denaturation, and the lateral mobility of the system facilitates protein–protein interactions that are critical to many biological processes.

There is as yet no consensus regarding the preferred method(s) of array fabrication, and further development will be required before protein arrays become widely available for use in research and clinical practice. Materials that reduce non-specific adsorption and protein denaturation, that allow further reduction in feature size, that facilitate address of individual array features, and that enable signal amplification and enhanced sensitivity, will constitute an important part of such development.

Concluding remarks

Biomaterials have already had an enormous impact on health care, and are already widely used in prosthetic and drug delivery devices. Much of this success has been achieved through judicious selection of existing materials, with little real design for biomedical use. Current work is laying the foundations for a much richer application of biomaterials through elucidation of the fundamental design considerations that determine the success or failure of biomaterials systems.

Numerous challenges remain in biomaterials development. These challenges include targeting materials (containing drugs), for example, to specific cells; designing materials that can sense biochemical signals in the body; and developing materials with improved biocompatibility. The ability to address these challenges will be facilitated by advances in biology and materials science. Understanding more about extracellular matrix biology, cell receptors, and immunology will help, for example, in understanding how the body responds to specific materials. Analogously, advances in nanomaterials will create new opportunities to mimic entities in the body (such as cells), and advances in materials characterization will aid in understanding how materials interface with cells and tissues. Finally, we note that many significant recent advances in biomaterials occurred at the interface of clinical medicine and materials science and engineering. Creating opportunities and training programs for cross-disciplinary research for individuals engaged in these areas could significantly accelerate the advancement of biomaterials and create new applications for these materials in medicine. □

doi:10.1038/nature02388.

1. Peppas, N. A. & Langer, R. New challenges in biomaterials. *Science* **263,** 1715–1720 (1994).
2. Ratner, B. D., Hoffman, A. S., Schoen, J. F. & Lemons, J. E. *Biomaterials Science, an Introduction to Materials in Medicine* 1–8 (Academic, San Diego, 1996).
3. Bell, E., Ivarsson, B. & Merrill, C. Production of a tissue-like structure by contraction of collagen lattices by human fibroblasts of different proliferative potential *in vitro*. *Proc. Natl Acad. Sci. USA* **76,** 1274–1278 (1979).
4. Langer, R. Drug delivery and targeting. *Nature* **392** (Suppl.), 5–10 (1998).
5. Langer, R. Where a pill won't reach. *Sci. Am.* **288,** 50–57 (2003).
6. Morice, M. *et al.* A randomized comparison of a sirolimus-eluting stent with a standard stent for coronary revascularization. *N. Engl. J. Med.* **346,** 1773–1780 (2002).
7. Langer, R. Perspectives: Drug delivery—Drugs on target. *Science* **293,** 58–59 (2001).
8. Vacanti, J. P. & Langer, R. Tissue engineering: the design and fabrication of living replacement devices for surgical reconstruction and transplantation. *Lancet* **354,** 32–34 (1999).
9. Yurchenco, P. D., Birk, D. E. & Mecham, R. P. (eds) *Extracellular Matrix Assembly and Structure* (Academic, San Diego, 1994).
10. van Hest, J. C. M. & Tirrell, D. A. Protein-based materials: Toward a new level of structural control. *Chem. Commun.* **19,** 1897–1904 (2001).
11. Lee, J., Macosko, C. W. & Urry, D. W. Mechanical properties of cross-linked synthetic elastomeric polypentapeptides. *Macromolecules* **34,** 5968–5974 (2001).
12. Nagapudi, K. *et al.* Photomediated solid-state cross-linking of an elastin-mimetic recombinant protein polymer. *Macromolecules* **35,** 1730–1737 (2002).
13. McMillan, R. A. & Conticello, V. P. Synthesis and characterization of elastin-mimetic protein gels derived from a well-defined polypeptide precursor. *Macromolecules* **33,** 4809–4821 (2000).
14. Huang, L. *et al.* Generation of synthetic elastin-mimetic small diameter fibers and fiber networks. *Macromolecules* **33,** 2989–2997 (2000).
15. Heilshorn, S. C., DiZio, K. A., Welsh, E. R. & Tirrell, D. A. Endothelial cell adhesion to the fibronectin CS5 domain in artificial extracellular matrix proteins. *Biomaterials* **24,** 4245–4252 (2003).
16. Urry, D. W., Parker, T. M., Reid, M. C. & Gowda, D. C. Biocompatibility of the bioelastic material poly(GVGVP) and its γ-irradiation crosslinked matrix. *J. Bioact. Compat. Polym.* **3,** 263–282 (1991).
17. Kwon, I., Kirshenbaum, K. & Tirrell, D. A. Breaking the degeneracy of the genetic code. *J. Am. Chem. Soc.* **125,** 7512–7513 (2003).
18. Alsberg, E., Anderson, K. W., Albeiruti, A., Rowley, J. A. & Mooney, D. J. Engineering growing tissues. *Proc. Natl Acad. Sci. USA* **99,** 12025–12030 (2002).
19. Zhang, S. Emerging biological materials through molecular self-assembly. *Biotechnol. Adv.* **20,** 321–339 (2002).
20. Hartgerink, J. D., Beniash, E. & Stupp, S. Peptide-amphiphile nanofibers: A versatile scaffold for the preparation of self-assembling materials. *Proc. Natl Acad. Sci. USA* **99,** 5133–5138 (2002).
21. Barrera, D. A., Zylstra, E., Lansbury, P. T. & Langer, R. Synthesis and RGD peptide modification of a new biodegradable copolymer system: Poly(lactic acid-co-lysine). *J. Am. Chem. Soc.* **115,** 11010–11011 (1993).
22. Cook, A. D. *et al.* Characterization and development of RGD-peptide-modified poly(lactic acid-co-lysine) as an interactive, resorbable biomaterial. *J. Biomed. Mater. Res.* **35,** 513–523 (1997).
23. Gref, R. *et al.* Biodegradable long-circulating polymeric nanospheres. *Science* **263,** 1600–1603 (1994).
24. Halstenberg, S., Panitch, A., Rizzi, S., Hall, H. & Hubbell, J. A. Biologically engineered protein-graft-poly(ethylene glycol) hydrogels: A cell adhesive and plasmin-degradable biosynthetic material for tissue repair. *Biomacromolecules* **3,** 710–723 (2002).
25. Miyata, T., Asami, N. & Uragami, T. A reversibly antigen-responsive hydrogel. *Nature* **399,** 766–769 (1999).
26. Kelch, S. & Lendlein, A. Shape memory polymers. *Angew. Chem. Int. Edn Engl.* **41,** 2034–2057 (2002).
27. Lendlein, A. & Langer, R. Biodegradable, elastic shape-memory polymers for potential biomedical applications. *Science* **296,** 1673–1676 (2002).
28. Pathak, C. P., Swahney, A. S. & Hubbell, J. A. Rapid photopolymerization of immunoprotective gelins in contact with cells and tissue. *J. Am. Chem. Soc.* **114,** 8311–8312 (1992).
29. Anseth, K., Shastri, V. & Langer, R. Photopolymerizable degradable polyanhydrides with osteocompatibility. *Nature Biotechnol.* **17,** 156–159 (1999).
30. Elisseff, J. *et al.* Transdermal photopolymerization for minimally invasive implantation. *Proc. Natl Acad. Sci. USA* **96,** 3104–3107 (1999).
31. Peppas, N. A. Hydrogels and drug delivery. *Curr. Opin. Colloid Interf. Sci.* **2,** 531–537 (1997).
32. Lahann, J. *et al.* A reversible switching of surfaces. *Science* **299,** 371–374 (2003).
33. Brocchini, S., James, K., Tangpasuthadol, V. & Kohn, J. Structure-property correlations in a combinatorial library of degradable biomaterials. *J. Biomed. Mater. Res.* **42,** 66–75 (1998).
34. Belu, A. M., Brocchini, S., Kohn, J. & Ratner, B. D. Characterization of combinatorially designed polyarylates by time-of-flight secondary ion mass spectrometry. *Rapid Commun. Mass Spectrom.* **14,** 564–571 (2000).
35. Lynn, D. M., Anderson, D. G., Putnam, D. & Langer, R. Accelerated discovery of synthetic transfection vectors: Parallel synthesis and screening of degradable polymer library. *J. Am. Chem. Soc.* **123,** 8155–8156 (2001).
36. Anderson, D., Lynn, D. & Langer, R. Semi-automated synthesis and screening of a large library of degradable cationic polymers for gene delivery. *Angew. Chem.* **42,** 3153–3158 (2003).
37. Luo, D. & Saltzman, W. M. Synthetic DNA delivery systems. *Nature Biotechnol.* **18,** 33–37 (2000).
38. Pun, S. H. & Davis, M. E. Development of a nonviral gene delivery vehicle for systemic application. *Bioconjugate Chem.* **13,** 630–639 (2002).
39. Affleck, D. G., Yu, L., Bull, D. A., Bailey, S. H. & Kim, S. W. Augmentation of myocardial transfection using TerplexDNA: a novel gene delivery system. *Gene Ther.* **8,** 349–353 (2001).
40. McManus, M. T. & Sharp, P. A. Gene silencing in mammals by small interfering RNAs. *Nature Rev. Genet.* **3,** 737–747 (2002).
41. Mathiowitz, E. *et al.* Biologically erodable microspheres as potential oral drug delivery systems. *Nature* **386,** 410–414 (1997).
42. Torres-Lugo, M., Garcia, M., Record, R. & Peppas, N. A. pH-sensitive hydrogels as gastrointestinal tract absorption enhancers: Transport mechanisms of salmon calcitonin and other model molecules using the Caco-2 cell model. *Biotechnol. Progr.* **18,** 612–616 (2002).
43. Santini, J. T., Cima, M. J. & Langer, R. A controlled-release microchip. *Nature* **397,** 335–338 (1999).
44. Grayson, A. *et al.* Multi-pulse drug delivery from a resorbable polymeric microchip device. *Nature Mater.* **2,** 767–772 (2003).
45. Stangel, K. *et al.* A programmable intraocular CMOS pressure sensor system implant. *IEEE J. Solid-State Circuits* **36,** 1094–1100 (2001).
46. Schwartz, M. *et al.* Single chip CMOS imagers and flexible microelectronic stimulators for a retina implant system. *Sensors Actuators* **83,** 40–46 (2000).
47. Kaushik, S. *et al.* Lack of pain associated with microfabricated microneedles. *Anesth. Analg.* **92,** 502–504 (2001).
48. Borenstein, J. T. *et al.* Microfabrication technology for vascularized tissue engineering. *Biomed. Microdevices* **4,** 671–680 (1999).
49. Eisen, M. B. & Brown, P. O. DNA arrays for analysis of gene expression. *Methods Enzymol.* **303,** 179–205 (1999).
50. Zhu, H. & Snyder, M. Protein chip technology. *Curr. Opin. Chem. Biol.* **7,** 55–63 (2003).
51. Houseman, B. T., Huh, J. H., Kron, S. J. & Mrksich, M. Peptide chips for the quantitative evaluation of protein kinase activity. *Nature Biotechnol.* **20,** 270–274 (2002).
52. Houseman, B. T. & Mrksich, M. Carbohydrate arrays for the evaluation of protein binding and enzymatic modification. *Chem. Biol.* **9,** 443–454 (2002).

review article

53. Kononen, J. *et al.* Tissue microarrays for high-throughput molecular profiling of tumor specimens. *Nature Med.* **4,** 844–847 (1998).
54. Fodor, S. P. *et al.* Light-directed, spatially addressable parallel chemical synthesis. *Science* **251,** 767–773 (1991).
55. McGall, G. *et al.* Light-directed synthesis of high-density oligonucleotide arrays using semiconductor photoresists. *Proc. Natl Acad. Sci. USA* **93,** 13555–13560 (1996).
56. Schena, M., Shalon, D., Davis, R. W. & Brown, P. O. Quantitative monitoring of gene expression patterns with a complementary DNA microarray. *Science* **270,** 467–470 (1995).
57. Sosnowski, R., Tu, E., Butler, W., O'Connell, J. & Heller, M. Rapid determination of single base mismatch mutations in DNA hybrids by direct electric field control. *Proc. Natl Acad. Sci. USA* **94,** 1119–1123 (1997).
58. Hui Liu, R., Lenigk, R., Druyor-Sanchez, R. L., Yang, J. & Grodzinski, P. Hybridization enhancement using cavitation microstreaming. *Anal. Chem.* **75,** 1911–1917 (2003).
59. Kajiyama, T. *et al.* Genotyping on a thermal gradient DNA chip. *Genome Res.* **13,** 467–475 (2003).
60. Cheek, B. J., Steel, A. B., Torres, M. P., Yu, Y. & Yang, H. Chemiluminescence detection for hybridization assays on the flow-thru chip, a three-dimensional microchannel biochip. *Anal. Chem.* **73,** 5777–5783 (2001).
61. Haab, B. B., Dunham, M. J. & Brown, P. O. Protein microarrays for highly parallel detection and quantitation of specific proteins and antibodies in complex solutions. *Genome Biol.* **2,** 1–13 (2001).
62. MacBeath, G. & Schreiber, S. L. Printing proteins as microarrays for high-throughput function determination. *Science* **289,** 1760–1763 (2000).
63. Arenkov, P. *et al.* Protein microchips: Use for immunoassay and enzymatic reactions. *Anal. Biochem.* **28,** 123–131 (2000).
64. Hodneland, C. D., Lee, Y.-S., Min, D.-H. & Mrksich, M. Selective immobilization of proteins to self-assembled monolayers presenting active site-directed capture ligands. *Proc. Natl Acad. Sci. USA* **99,** 5048–5052 (2002).
65. Zhu, H. *et al.* Global analysis of protein activities using proteome chips. *Science* **293,** 2101–2105 (2001).
66. Groves, J. T. & Boxer, S. G. Micropattern formation in supported lipid membranes. *Acc. Chem. Res.* **35,** 149–157 (2002).

Supplementary Information accompanies the paper on **www.nature.com/nature**.

Acknowledgements This work was supported in part by the National Institutes of Health.

Competing interests statement The authors declare that they have no competing financial interests.

Correspondence and requests for materials should be addressed to R.L. (rlanger@mit.edu).

Light-induced shape-memory polymers

Andreas Lendlein[1,2], **Hongyan Jiang**[2,3], **Oliver Jünger**[2*] & **Robert Langer**[4]

[1]*GKSS Research Center Geesthacht GmbH, Institute of Chemistry, Kantstraße 55, D-14513 Teltow, Germany*
[2]*Institute for Technology and Development of Medical Devices, RWTH Aachen, Pauwelsstraße 30, and* [3]*mnemoScience GmbH, Pauwelsstraße 19, D-52074 Aachen, Germany*
[4]*Department of Chemical Engineering, Massachusetts Institute of Technology, 45 Carleton Street, Cambridge, Massachusetts 02139, USA*

* Present address: Prof.-Staudinger-Strasse, D-65451 Kelsterbach, Germany

Materials are said to show a shape-memory effect if they can be deformed and fixed into a temporary shape, and recover their original, permanent shape only on exposure to an external stimulus[1–3]. Shape-memory polymers have received increasing attention because of their scientific and technological significance[4,5]. In principle, a thermally induced shape-memory effect can be activated by an increase in temperature (also obtained by heating on exposure to an electrical current[6] or light illumination[6,7]). Several papers have described light-induced changes in

letters to nature

the shape of polymers[8–12] and gels[13–15], such as contraction[8–10], bending[11–13] or volume changes[14,15]. Here we report that polymers containing cinnamic groups can be deformed and fixed into pre-determined shapes—such as (but not exclusively) elongated films and tubes, arches or spirals—by ultraviolet light illumination. These new shapes are stable for long time periods, even when heated to 50 °C, and they can recover their original shape at ambient temperatures when exposed to ultraviolet light of a different wavelength. The ability of polymers to form different pre-determined temporary shapes and subsequently recover their original shape at ambient temperatures by remote light activation could lead to a variety of potential medical and other applications.

As an example of the shape-memory functionality of such photoresponsive materials, results of programming and recovering induced by light are shown in Fig. 1. First, a photoresponsive polymer is conventionally processed into its permanent shape, a plain film (Fig. 1A, a). Afterwards, the polymer film is stretched to 20% elongation by external stress, and both sides of the film were evenly irradiated with ultraviolet (UV) light, wavelength $\lambda > 260$ nm, for 60 min. When the external stress is released, the sample stayed in an elongated form (Fig. 1A, b). Irradiating the elongated shape with UV light of $\lambda < 260$ nm for 60 min induces cleaving of newly formed photosensitive crosslinks. As a consequence, the recovery of the memorized, permanent shape, that is, the original length (Fig. 1A, c), is observed. Other temporary shapes can also be produced. For example, when only the top side of a polymer film in a stretched state (100% elongation) is irradiated with UV light of $\lambda > 260$ nm for 60 min, a corkscrew spiral shape is obtained after release of external stress (Fig. 1B). The formation of such a spiral shape is caused by the formation of two layers. While the deformation is well-fixed for the top layer, the bottom layer keeps its elasticity. As a result, the latter contracts much more than the former when the external stress is released, forming an arch or corkscrew spiral shape.

Thermoresponsive shape-memory polymers consist of two components on the molecular level: first, molecular switches, which are segments having a thermal transition at temperature T_{trans} and that fix the temporary shape by forming physical crosslinks; and second, 'netpoints' that link these segments and determine the permanent shape[4]. In photoresponsive shape-memory polymers (Fig. 1C), we want the netpoints to be part of

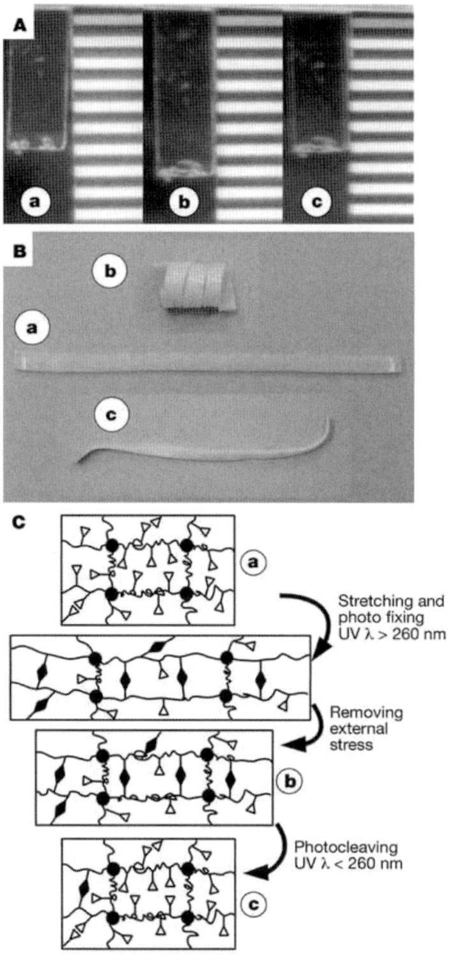

Figure 1 Shape-memory effect of photoresponsive polymers. **A**, A film of grafted polymer BHCA(10,2,1). **a**, Permanent shape (6 cm × 1.2 cm × 0.05 cm); **b**, temporary shape; **c**, recovered permanent shape. **B**, An IPN polymer film. **a**, Permanent shape (8 cm × 0.4 cm × 0.05 cm); **b**, corkscrew spiral temporary shape; **c**, recovered shape obtained by irradiation with UV light of $\lambda < 260$ nm for 60 min. **C**, Molecular mechanism of shape-memory effect of the grafted polymer network: the chromophores (open triangles) are covalently grafted onto the permanent polymer network (filled circles, permanent crosslinks), forming photoreversible crosslinks (filled diamonds); fixation and recovery of the temporary shape are realized by UV light irradiation of suitable wavelengths.

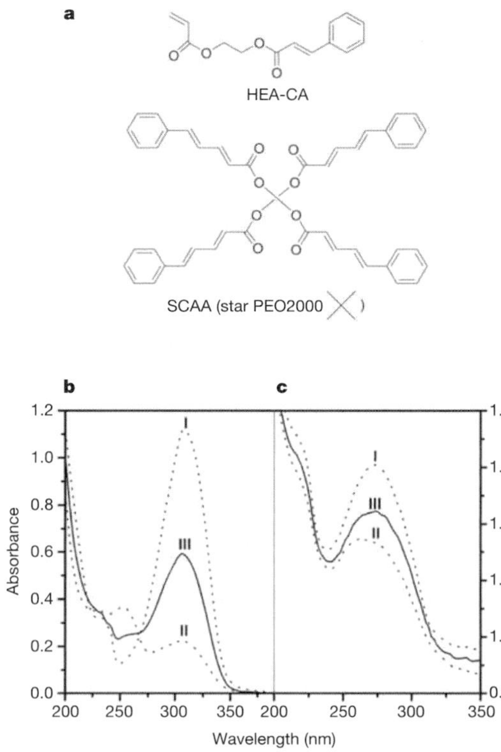

Figure 2 UV absorption spectra. **a**, Chemical structure of molecular switches HEA-CA and SCAA. **b**, A thin film of SCAA. Shown are spectra taken 0 min (spectrum I) and 6 min (II) after exposure to UV light of $\lambda > 260$ nm (crosslinking), and after another 6 min exposure to UV light of $\lambda < 260$ nm (III; de-crosslinking). **c**, A thin film of grafted polymer BHCA(10,2,1): Shown are spectra taken 0 min (I) and 60 min (II) after exposure to UV light of $\lambda > 260$ nm, and after 60 min exposure to UV light of $\lambda < 260$ nm (III).

a covalently crosslinked, amorphous permanent polymer network. The molecular switches are photoresponsive cinnamic acid type molecules such as cinnamic acid (CA)[16] and cinnamylidene acetic acid (CAA)[17], which are able to undergo efficient photoreversible [2 + 2] cycloaddition reactions when exposed to alternating wavelengths (here: $\lambda > 260$ nm or $\lambda < 260$ nm). When such a photoresponsive polymer film (Fig. 1C, a, permanent shape) is stretched, the coiled segments of the amorphous polymer chains between two netpoints are elongated. Upon exposure to UV light of $\lambda > 260$ nm, the elongated segments of the chains are partially fixed owing to formation of new photoresponsive crosslinks, resulting in an elongated new shape (Fig. 1C, b, temporary shape). The photoresponsive crosslinks can be reversibly cleaved by irradiation with UV light of $\lambda < 260$ nm. As a result, the fixed elongated film can recover to the original permanent shape (Fig. 1C, c, recovered permanent shape).

We created two photoresponsive shape-memory polymers. In the first polymer ('grafted polymer'), CA molecules are grafted onto the permanent polymer network, forming a polymer architecture, as schematically illustrated in Fig. 1C. Grafted polymers are obtained by copolymerization of *n*-butylacrylate (BA), hydroxyethyl methacrylate (HEMA) and ethyleneglycol-1-acrylate-2-CA (HEA-CA, in Fig. 2a) with poly(propylene glycol)-dimethacrylate (number-average molecular mass $M_n = 560\,\mathrm{g\,mol^{-1}}$, PPG2M560) as crosslinker. Here the molecular switches are photoreversible crosslinks, fixing the deformed temporary shape when irradiated with UV light of $\lambda > 260$ nm, and recovering the shape when exposed to UV light of $\lambda < 260$ nm, as schematically illustrated in Fig. 1C. Only the CA chromophore is selected as a molecular switch, because CAA-containing (meth)acrylates cannot be incorporated into the permanent network via free-radical polymerization. CAA groups may possibly work as free-radical scavengers during the reaction. We synthesized a series of polymer networks with CA groups (Supplementary Table 1). Increasing the amount of HEA-CA increases both the elastic moduli and the glass-transition temperatures (T_g) of the grafted polymer networks, probably owing to the bulky side chains of the CA groups and/or π–π interaction of CA molecules. All polymer networks are amorphous, and have glass-transition temperatures lower than 20 °C.

In the second polymer ('IPN polymer'), the permanent network is formed from BA with 3.0 wt% poly(propylene glycol)-dimethacrylate ($M_n = 1{,}000\,\mathrm{g\,mol^{-1}}$, PPG2M1000) as crosslinker. The permanent network was loaded with about 20 wt% star-poly(ethylene glycol) containing CAA terminal groups (SCAA, in Fig. 2a) by swelling it in a 10 wt% SCAA chloroform solution, forming an opaque yellow film (BP-SCAA) after the solvent is removed. Under the influence of light, these SCAA molecules can be reversibly crosslinked, fixing the temporary shape by interpenetrating the permanent network.

Figure 2b shows changes in the absorption spectra of SCAA when exposed to UV light of $\lambda > 260$ nm and UV light of $\lambda < 260$ nm. The absorption maximum at 310 nm (Fig. 2b, spectrum I) relating to conjugated diene groups decreased from 1.1 to 0.2 after 6 min exposure to UV light of $\lambda > 260$ nm (Fig. 2b, spectrum II). At the same time, the absorption peak at 250 nm relating to the formed styryl groups increased. The spectral changes are indicative of the disappearance of the double bonds adjacent to carbonyl groups of the CAA moiety, and the formation of a cyclobutane ring bearing styryl groups. On exposure to UV light of $\lambda < 260$ nm for another 6 min, the formed cyclobutane ring is reversibly cleaved and the absorption peak at 310 nm increases again (Fig. 2b, spectrum III). The extent of photoreversibility is around 40%, which is determined from the intensity changes of the absorption peak at 310 nm. The grafted polymer networks having CA groups show a maximum absorption at 275 nm, and a similar photoreversibility (Fig. 2c).

Unlike azo-containing liquid crystal elastomers[9–13] but similar to polymers having a thermally induced shape-memory effect[2–4], our photosensitive polymers require pre-determined programming—including deformation into a new shape by application of an external force. Irradiating free-standing, non-deformed polymer films did not change their shape. For potential specific applications, different original and deformed temporary shapes can be made in a pre-determined manner.

The light-induced shape-memory functionality of the polymers was quantified by cyclic photomechanical experiments at 25 or 35 °C under stress-controlled and/or strain-controlled conditions. These cyclic experiments have been designed in analogy to those characterizing a thermally induced effect[4,18]. A typical elongation–time diagram of the grafted polymer is shown in Fig. 3, which was obtained during a cyclic photomechanical test under stress-controlled conditions. Initially, the sample is elongated to ε_{max} and the stress is kept constant. During 1.5 h of photo-crosslinking in the first cycle, the elongation stayed constant in the stretched state, indicating no relaxation of the polymer. After switching off the UV light ($\lambda > 260$ nm) and lowering the controlled stress to zero, a temporarily fixed elongation ε_u was observed. The fixed temporary shape is maintained for 3 h until UV light of $\lambda < 260$ nm is switched on for 1.5 h, activating the shape recovery of the temporarily fixed shape to nearly its original length characterized by a remaining elongation ε_p. This cycle is then repeated twice. The experiment allows the determination of the strain-fixity rate of the *N*th cycle $R_f(N) = \varepsilon_u(N)/\varepsilon_{max}$, as well as the strain-recovery rate $R_r(N) = (\varepsilon_{max} - \varepsilon_p(N))/\varepsilon_{max}$[18,19]. The strain fixity rate $R_f(N)$ describes the ability to fix the mechanical deformation that has been applied during the photo-crosslinking process. The strain recovery rate $R_r(N)$ quantifies the ability of the material to recover its permanent shape in the *N*th cycle. Similar shape-memory properties were obtained under strain-controlled conditions (Table 1). Increasing the amount of photo-crosslinker HEA-CA improves both R_f and R_r. However, increasing HEA-CA to more

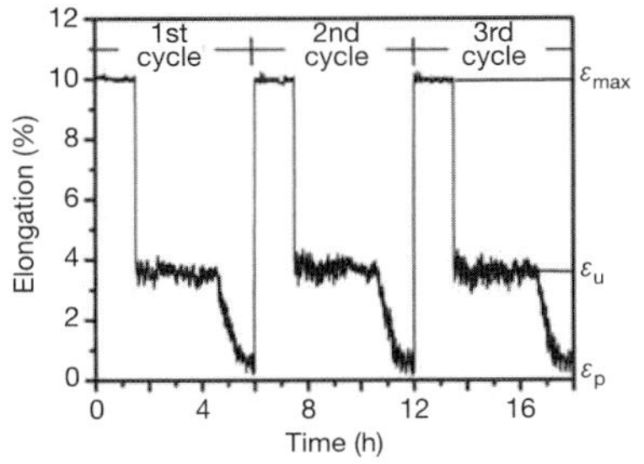

Figure 3 Cyclic photomechanical experiment on a grafted polymer BHCA (10,2,1) under stress-controlled conditions. Here $\varepsilon_{max} = 10\%$, $T = 25$ °C). ε_u is the temporarily fixed elongation, and ε_p is the remaining elongation after shape recovery.

Table 1 **Shape-memory properties of photoresponsive polymer networks at 25 °C**

Polymer type	Sample ID	ε_{max} (%)	Stress control		Strain control	
			R_f (%)	R_r (%)	R_f (%)	R_r (%)
Grafted polymer	BHCA(10,2,0.1)*	10	11	78	14	54§
	BHCA(10,2,0.25)	10	17	82	18	79§
	BHCA(10,2,0.5)	10	25	85	28	82§
	BHCA(10,2,1)	10	36	94	38	97§
	BHCA(10,2,2)	10	36	98	38	98§
	BHCA(10,2,3)	20	n.d.	n.d.	52	95§
IPN polymer	BP-SCAA†	20	20	90	33 (30)‡	98 (95)‡
	BP-SCAA†	50	15	94	20 (20)‡	95 (95)‡
	BP-SCAA†	100	n.d.	n.d.	20 (18)‡	88 (90)‡

Data from third cycle (IPN); fifth cycle (Grafted). n.d., not determined.
*Data in brackets indicate the molar co-monomer ratio (BHCA), with B = BA, H = HEMA, and CA = HEA-CA.
†IPN from BA (B) and 3.0 wt% PPG2M1000 (P) as crosslinker, loaded with 20 wt% SCAA.
‡Data in brackets were obtained at 35 °C.
§$R_r(N) = (\varepsilon_{max} - \varepsilon_p(N))/(\varepsilon_{max} - \varepsilon_p(N-1))$

than 10 mol% has no further influence on the shape-memory properties. After two cycles of UV irradiation, the clear film turned pale yellow, possibly owing to photo-Fries rearrangement[20], but no changes of shape memory properties were observed (Fig. 3). Although the extent of photoreversibility of the photoresponsive chromophores is around 40%, it is likely that partial cleavage of the photoreversible crosslinks on the microscopic level can lead to the changes of macroscopic mechanical properties and to the shape recovery.

Polymers from the IPN polymer system show an R_f of 20–33% and an R_r of more than 88%. Data obtained at 35 °C (Table 1) are not much different from those obtained at 25 °C. During the irradiation process, the temperature of the polymers does not change, indicating that heating is not the trigger of shape-memory behaviour. When the cyclic photomechanical experiment is performed at lower ε_{max} (such as 20%), higher values for R_f are obtained. Large deformation will broadly distribute the photoresponsive groups within the specimen under test and make aggregation of these functional groups difficult, thereby influencing the photo-fixing reactions. The relaxation behaviour of IPN polymer networks was investigated in control tests, in which an IPN polymer film was examined in the cyclic mechanical test under the same conditions but without applying UV light. The extent of relaxation determined from these experiments is about 2% in a test with $\varepsilon_{max} = 20\%$, and 5% for $\varepsilon_{max} = 50\%$.

Relative to thermoresponsive shape-memory polymers, the strain-recovery rate of photoresponsive shape-memory polymers is comparable, but the strain-fixity rate is much lower, because the mechanism is different. In photoresponsive polymers, the amorphous permanent network itself has long and coiled segments between two netpoints. During the programming process, the coiled segments of the chains are stretched and elongated (Fig. 1C). The strain-fixation via light irradiation is due to formation of new chemical netpoints, rather than freezing the stretched chains as in the case of thermoresponsive shape-memory polymers. The elastic contraction of the stretched chain segments accounts for the lower strain-fixity rate of light-induced shape-memory polymers. However, the unique characteristics of the photoresponsive shape-memory polymers enable shape-recovery at ambient temperatures by using remote activation, and could eliminate the temperature constraints of thermoresponsive shape-memory polymers for medical and other applications arising from external sample heating. □

Methods

Synthesis

As an example of the grafted polymer, the synthesis of BHCA(10,2,1) (see Table 1 for nomenclature) is described: a solution of 12.80 g (100 mmol) BA, 2.60 g (20 mmol) HEMA, 2.46 g (10 mmol) HEA-CA, 0.22 g (0.4 mmol) PPG2M560 and 0.05 g azoisobutyronitril (AIBN) was polymerized between two glass sheets with a Teflon spacer of ~0.5 mm at 80 °C for 12 h. The resulting films were swollen in chloroform to remove the residual monomers. SCAA was synthesized by reacting 4-arm star-poly(ethylene glycol) ($M_n = 2{,}000$, Shearwater Polymers) with an excess of cinnamylidene acetyl chloride in dry THF solution with triethylene amine as the acid receptor.

Cyclic photomechanical and mechanical tests

These were carried out at 25 or 35 °C (selected to be $T > T_g$ and $T > T_{m,SCAA}$, the melting point of SCAA) on a Zwick-Z005, a tensile tester equipped with a 10 N load cell and a thermochamber (model 091250, Climatic Systems Ltd) controlled by a Eurotherm 902-904 unit (Eurotherm Controls Ltd). Films are cut into standard samples (ISO 527-2/1BB, length 32 mm, width 2 ± 0.2 mm). The elongation rate is 10 mm min^{-1}. For photomechanical tests, an optical system consisting of two UV lamps (30 W, Cathodeon) and an optical filter having the ability to split a light beam reaching the filter at an angle of 45° were added to this set-up. While light having wavelength $\lambda > 260$ nm is passing through the filter, light with wavelength $\lambda < 260$ nm is being reflected.

The influence of UV irradiation on the temperature of the polymer film in the Zwick thermochamber was investigated by inserting a digital thermocouple (RS Components) into the polymer film. During a 1.5 h irradiation period with UV light ($\lambda > 260$ nm or $\lambda < 260$ nm) at 25 or 35 °C, the average fluctuation of the polymer temperature is ±0.1 °C.

Received 6 December 2004; accepted 2 February 2005; doi:10.1038/nature03496.

1. Osada, Y. & Matsuda, A. Shape memory in hydrogels. *Nature* **376,** 219 (1995).
2. Liu, C. & Mather, P. T. Thermomechanical characterization of a tailored series of shape memory polymers. *J. Appl. Med. Polym.* **6,** 47–52 (2002).
3. Rousseau, I. A. & Mather, P. T. Shape memory effect exhibited by smectic-C liquid crystalline elastomers. *J. Am. Chem. Soc.* **125,** 15300–15301 (2003).
4. Lendlein, A. & Kelch, S. Shape-memory polymers. *Angew. Chem. Int. Edn Engl.* **41,** 2034–2057 (2002).
5. Wei, Z. G., Sandström, R. & Miyazaki, S. Shape-memory materials and hybrid composites for smart systems: Part I. Shape-memory materials. *J. Mater. Sci.* **33,** 3743–3762 (1998).
6. Koerner, H. *et al.* Remotely actuated polymer nanocomposites—stress-recovery of carbon-nanotube-filled thermoplastic elastomers. *Nature Mater.* **3,** 115–120 (2004).
7. Maitland, D. J. *et al.* Photothermal properties of shape memory polymer micro-actuators for treating stroke. *Lasers Surg. Med.* **30,** 1–11 (2002).
8. Eisenbach, C. D. Isomerization of aromatic azo chromophores in poly(ethyl acrylate) networks and photomechanical effect. *Polymer* **21,** 1175–1179 (1980).
9. Li, M. H. *et al.* Light-driven side-on nematic elastomer actuators. *Adv. Mater.* **15,** 569–572 (2003).
10. Finkelmann, H., Nishikawa, E., Pereira, G. G. & Warner, M. A new opto-mechanical effect in solids. *Phys. Rev. Lett.* **87,** 015501 (2001).
11. Camacho-Lopez, M., Finkelmann, H., Palffy-Muhoray, P. & Shelley, M. Fast liquid-crystal elastomer swims into the dark. *Nature Mater.* **3,** 307–310 (2004).
12. Yu, Y. L., Nakano, M. & Ikeda, T. Directed bending of a polymer film by light. *Nature* **425,** 145 (2003).
13. Ikeda, T. *et al.* Anisotropic bending and unbending behavior of azobenzene liquid-crystalline gels by light exposure. *Adv. Mater.* **15,** 201–205 (2003).
14. Suzuki, A. & Tanaka, T. Phase transition in polymer gels induced by visible light. *Nature* **346,** 345–347 (1990).
15. Juodkazis, S. *et al.* Reversible phase transitions in polymer gels induced by radiation forces. *Nature* **408,** 178–181 (2000).
16. Akabori, S. *et al.* The novel synthesis of photoreversible cyclobutanocrown ethers by the intramolecular photoaddition of α,ω-dicinnamoyl polyethylene glycol derivatives. *Bull. Chem. Soc. Jpn* **60,** 3453–3455 (1987).
17. Tanaka, H. & Honda, K. Photoreversible reactions of polymers containing cinnamylideneacetate derivatives and the model compounds. *J. Polym. Sci. Polym. Chem. Edn* **15,** 2685–2689 (1977).
18. Kim, B. K., Lee, S. Y. & Xu, M. Polyurethanes having shape memory effects. *Polymer* **37,** 5781–5793 (1996).
19. Tobushi, H., Haray, H., Yamaday, E. & Hayashi, S. Thermomechanical properties in a thin film of shape memory polymer of polyurethane series. *Smart Mater. Struct.* **5,** 483–491 (1996).
20. Haddleton, D. M., Creed, D., Griffin, A. C., Hoyle, C. E. & Venkataram, K. Photochemical crosslinking of main-chain liquid-crystalline polymers containing cinnamoyl groups. *Makromol. Chem. Rapid Commun.* **10,** 391–396 (1989).

Supplementary Information accompanies the paper on **www.nature.com/nature**.

Acknowledgements We thank the Bundesministerium für Bildung und Forschung for a BioFuture Award; H.J. is grateful for an AvH fellowship in 1999/2000.

Competing interests statement The authors declare competing financial interests: details accompany the paper on **www.nature.com/nature**.

Correspondence and requests for materials should be addressed to A.L. (Lendlein@gkss.de).

Tissue Engineering and Stem Cells

Selective Cell Transplantation Using Bioabsorbable Artificial Polymers as Matrices

By Joseph P. Vacanti, Martin A. Morse, W. Mark Saltzman, Abraham J. Domb, Antonio Perez-Atayde, and Robert Langer
Boston and Cambridge, Massachusetts

● To date, selective cell transplantation has involved injecting cell suspensions into tissues or the vascular system. This study describes attaching cell preparations to bioerodable artificial polymers in cell culture and then implanting this polymer-cell scaffold into animals. Using standard techniques of cell harvest, single cells and clusters of fetal and adult rat and mouse hepatocytes, pancreatic islet cells, and small intestinal cells have been seeded onto biodegradable polymers of polyglactin 910, polyanhydrides, and polyorthoester. Sixty-five fetuses and 14 adult animals served as donors. One hundred fifteen polymer scaffolds were implanted into 70 recipient animals: 66 seeded with hepatocytes; 23 with intestinal cells and clusters; and 26 with pancreatic islet preparations. The cells remained viable in culture, and in the case of fetal intestine and fetal hepatocytes, appeared to proliferate while on the polymer. After four days in culture, the cell-polymer scaffolds were implanted into host animals, either in the omentum, the interscapular fat pad, or the mesentery. In three cases of fetal intestinal implantation coupled with partial hepatectomy, successful engraftment occurred in the omentum, one forming a visible 6.0 mm cyst. Three cases of hepatocyte implantation, one using adult cells and two using fetal cells, have also engrafted, showing viability of hepatocytes, mitotic figures, and vascularization of the cell mass. To date, no pancreatic islets have survived implantation. This method of cell transplantation, which we have termed "chimeric neomorphogenesis," is an alternative to current methods and requires further study.

INDEX WORDS: Transplantation; cell culture; fetal cell transplantation.

SELECTIVE cell transplantation of only those parenchymal elements necessary to replace lost function has been studied as an alternative to whole or partial organ transplantation.[1] In the past, cells have been harvested, dispersed into a suspension, and then innoculated into various tissues. We report attaching parenchymal cells from liver, intestine, and pancreas onto bioerodable artificial polymers and then implanting these polymer-cell scaffolds into animals as a novel method of cell transplantation.

MATERIALS AND METHODS

Polymers

Three synthetic absorbable polymers have been used to fabricate filaments and discs as matrices for cell attachment, growth, and implantation (Fig 1).

Polyglactin. This polymer was developed as absorbable synthetic suture material. It is a 90-10 copolymer of glycolide and lactide and is produced as Vicryl braided absorbable suture (Ethicon Co, Somerville, NJ).[2]

Polyorthoesters. The specific polymer used was 3,9-bis(ethylidene-2,4,8,10-tetraoxaspiro[5.5]) undecane copolymer with *trans*-1,4-cyclohexanedimethanol and 1,6-hexandiol in a molar ratio 2:1:1, respectively (SRI, CA).[3]

Polyanhydride. The specific polymer used was a polyanhydride of bis(1,4-carboxyphenoxy) propane and sebacic acid. It is biocompatible and has been used extensively in drug delivery applications.[3-6]

Polymer Configuration

Small wafer discs or filaments of polyanhydrides and polyorthoesters were fabricated using one of the following methods.

Solvent casting. A solution of the polymer (10% in methylene chloride) was cast on a branching pattern relief structure as a disc 10 mm in diameter for 10 minutes at 25°C using a Carver press. After solvent evaporation, a film 0.5 mm in thickness with an engraved branching pattern on its surface was obtained.

Compression molding. One hundred mg of the polymer was pressed (30,000 psi) into a branching pattern relief structure 10 mm in diameter. 0.5 mm discs were obtained.

Filament drawing. Filaments were drawn from the molten polymer (30 μm in diameter). Small flattened 1.0 cm tufts were used for the experiments.

Polyglactin 910. Multiple fibers of 90-10 copolymer of glycolide and lactide converging to a common base were fashioned from suture material of 0-Vicryl by fraying the braided end of the polymer (Fig 2). These branching fiber clusters were approximately 1.0 cm in height. The individual fibrils were 30 μm in diameter.

Animals

Young adult and fetal Sprague Dawley rats and C57 B1/6 mice (Charles River Labs, Wilmington, MA) were used as cell donors for all experiments. The animals were housed individually, allowed access to food and water ad lib, and maintained at 12 hour light and dark intervals. Anesthesia was obtained with an intraperitoneal injection of pentobarbital (Abbott Labs, North Chicago, IL) at a dose of 0.05 mg/g and supplemented with methoxyflurane (Pitman-

From the Departments of Surgery and Pathology, The Children's Hospital and the Harvard Medical School, Boston, and the Department of Applied Biological Sciences, Massachusetts Institute of Technology, Cambridge.
Supported by the March of Dimes Grant No. 5-543, Basil O'Connor Starter Research Grant, and NIH Grant No. 6M 26698.
Presented at the 18th Annual Meeting of the American Pediatric Surgical Association, Hilton Head Island, South Carolina, May 6-9, 1987.
Address reprint requests to Joseph P. Vacanti, MD, The Children's Hospital, 300 Longwood Ave, Fegan Building, Fifth Floor, Boston, MA 02115.

0022-3468/88/2301-0002$03.00/0

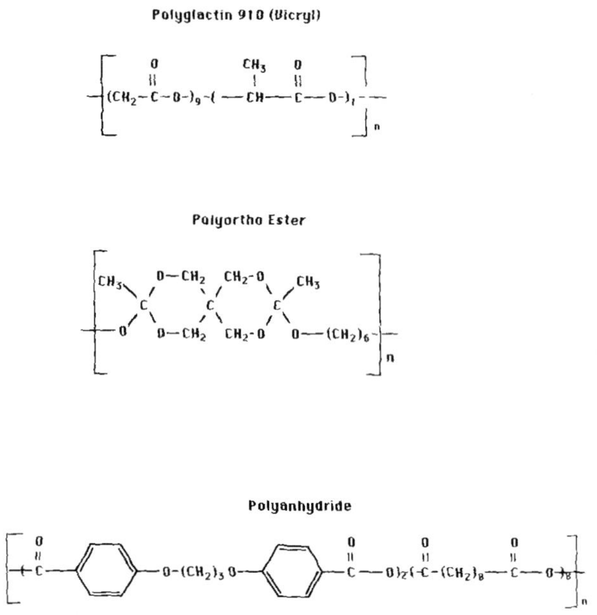

Fig 1. Chemical structure of polymers used for biodegradable cellular matrices. Preparation and configurations are described in text.

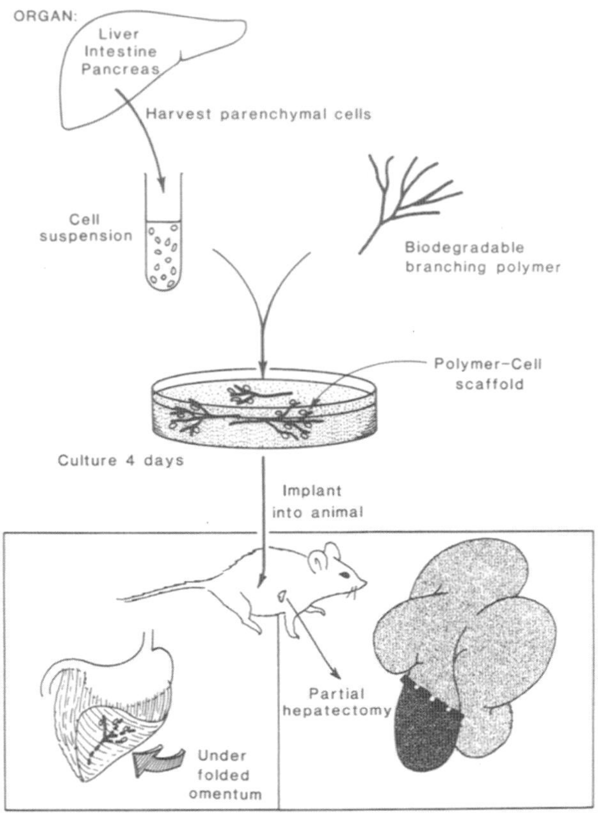

Fig 2. Technique of chimeric neomorphogenesis. Appropriate parenchymal cells are harvested and dispersed. They are seeded onto the polymer matrix in cell culture, where attachment and growth occur. Partial hepatectomy is performed to stimulate growth of the transplant. The polymer-cell scaffold is then implanted into the recipient animal where neovascularization, cell growth, and reabsorption of the polymer matrix occurs.

Moore, Inc, Washington Crossing, NJ) by cone administration. Fetal animals were harvested at 13, 17, and 20 days gestation and were used as liver, pancreas, and intestinal donors. Young adult animals were used as liver and pancreas donors and as recipients of the cell-scaffold matrices.

Cell Harvest and Cell Culture

Liver. After the induction of anesthesia, the abdomen of young adult animals was shaved, prepped with betadine, and opened using sterile technique. The liver was isolated and after heparinization with 100 U of heparin (Elkins-Sinn, Inc, Cherry Hill, NJ), the portal vein was cannulated with a 23 gauge plastic intravenous cannula (Critikon, Inc, Tampa, FL). The inferior vena cava was transected, and the liver was flushed with 2 to 3 mL of sterile saline. It was then removed from its bed and transferred to a sterile dish where it was perfused with an oxygenated solution of 0.025% collagenase Type II (BCA/Cappel Products, West Chester, PA) by a technique modified from Selgen.[7] After a 20-minute perfusion, the liver was transferred to a sterile hood for cell dispersion and culture.

Fetal animals were harvested by isolating and removing the gravid uterine horns from pregnant animals of the appropriate gestation. The intact uterus with multiple fetuses was transferred in saline to a sterile room, equipped with a dissection microscope. Individual fetuses were opened and the liver, intestine, and pancreas were harvested and pooled. Organs were then transferred to a sterile hood for cell isolation. The tissues were minced, treated with 0.025% Type II collagenase, and dispersed into cell suspensions (Fig 2).

Pancreas. After the induction of anesthesia, the abdomen of young adult animals was shaved, prepped with betadine, and opened in the midline using sterile technique. The common bile duct was isolated, and the pancreas visualized. Two and a half milliliters of 2.0% type II collagenase was infused into the pancreas by injection into the common bile duct using the technique described by Gotoh et al.[8] After five minutes, the pancreas was transferred to a sterile hood for islet cell isolation. Briefly, the tissue was placed into a 25% Ficoll solution and layered under a discontinuous Ficoll gradient (21%, 11%) and centrifuged at 800 × g for ten minutes. Islets that aggregated at the 21% to 11% interface were washed with cold Hank's solution and centrifuged at 320 × g three times. The islets were resuspended in RPMI 1640 (Gibco, Grand Island, NY) media supplemented with 10% fetal calf serum, and overlaid onto polymer scaffolds. Fetal animals were harvested as donors as described previously.

Intestine. Fetal intestine was obtained as described previously.

Polymer-cell scaffolds and implantation. Cells in suspension were plated onto polymer matrices at 1×10^5 or 1×10^6 cells/mL. They were maintained in Chee's media supplemented with 10% fetal calf serum for 3 to 4 days in a 10% CO_2 environment. Viability of cells on the scaffold immediately preimplantation was assessed by the trypan blue exclusion method. Young adult Sprague Dawley rats were anesthetized, shaved over the operative site, and prepped with betadine.

The polymer-cell scaffold was implanted in one of three sites: (1) the interscapular fat pad; (2) the omentum; and (3) the bowel mesentery (Fig 3).

Most animals underwent a partial hepatectomy to stimulate cell growth. Animals were sacrificed at day 3, 7, or 14 and the implants were examined histologically with hematoxylin and eosin. Polymers without cells served as controls. Polymer-cell scaffolds were exam-

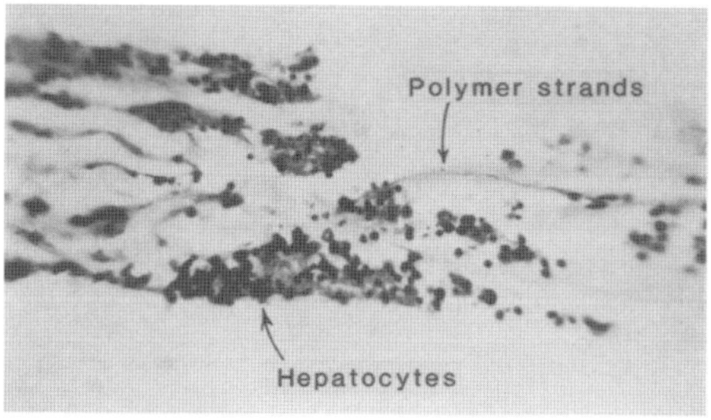

Fig 3. Hepatocytes attached to fibers of polyglactin 910 after four days in culture. Cells vary from healthy to degenerated and necrotic (Hematoxylin and eosin, original magnification ×172).

ined histologically after four days in culture and before implantation to assess cell attachment and viability.

RESULTS

Seventy-nine animals, including 14 adults and 65 fetuses, were used as donors for cell harvest; 115 polymer scaffolds were prepared for implantation. Sixty-six of these scaffolds were seeded with hepatocytes, 23 with intestinal cells and clusters, and 26 with pancreatic islets and cell preparations. Implantation was performed in 70 recipient animals. Fifty-eight were killed at seven days for histologic examination of the implant while three were examined at three days, and nine at 14 days after implantation.

Cell viability on the polymer scaffold at 3 to 4 days in culture varied with the type of polymer material used. Less than 10% of the cells were viable on the polyanhydride discs, whereas 80% of cells cultured on polyorthoester discs and filaments remained viable, and over 90% survived on polyglactin 910 (Fig 3).

Blood vessel ingrowth was noted three days after implantation with all of the polymer types and configurations. In the implanted fiber networks, new blood vessels formed in the interstices between the polymer filaments. The polymer discs showed capillary formation immediately adjacent to the polymer material. This angiogenic response accompanied an inflammatory infiltrate that displayed both an acute phase and a chronic foreign body reaction to the implanted polymers. The intensity of inflammation varied with the polymer type tested: polyanhydride elicited the most severe acute and chronic response although the inflammation surrounding branching fibers of either polyorthoester or polyglactin appeared proportionately greater than the disc configuration because of the greater surface area of exposed foreign material to host.

Histologic examination of liver cell implants in three animals showed evidence of successful engraftment of hepatocytes at seven days. Small clusters of healthy appearing hepatocytes were seen with bile canaliculi between adjacent cell membranes and some areas demonstrated mitotic figures. The cells were surrounded by an inflammatory response and blood vessels coursed around and through the cell clusters. Polymer material was seen immediately adjacent to the cells (Fig 4).

Successful engraftment of intestinal cells and clusters were observed in three animals. Histologic findings were similar to the hepatocyte implants with one exception (Figs 5A-C). On gross examination of the implant at seven days, a cystic structure approximately 6.0 mm in length was found at the implant site (Fig 5A) with polymer fibers splayed within its wall. Microscopic examination revealed well-differentiated intestinal epithelium lining the cavity with mucous and cellular debris within the lumen. One wall of the cyst contained polymer fibers, blood vessels, and inflammatory cells immediately adjacent to the intestinal epithelium (Fig 5B). The other wall included a muscular coating that suggested that the polymer held a small minced piece of fetal intestine as the origin of the cyst that eventually developed. The cyst displayed well-differentiated intestinal epithelium with mucous secreting cells (Fig 5C). Other clusters of intestinal epithelium demonstrated active mitoses.

No viable pancreatic islets or other cell types were found in any of the pancreas implants tested. However, all of the implants were accompanied by an inflammatory and angiogenic response similar to those noted with liver and intestine.

Control polymers implanted without prior cell seed-

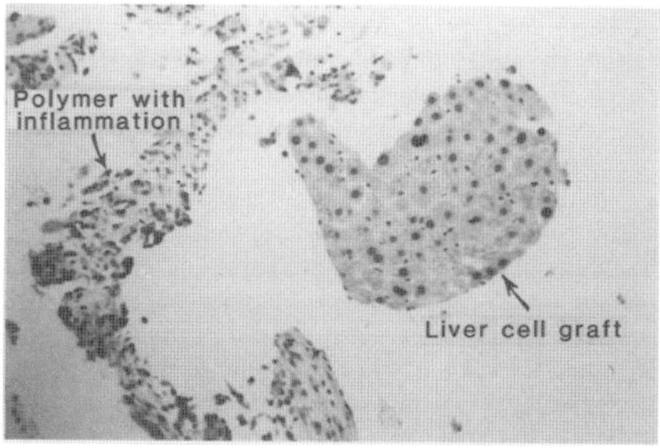

Fig 4. Implant of hepatocytes from adult rat donor into omentum. The polymer-cell implant has been in place for seven days before death. Hepatocytes are healthy and several mitotic figures can be seen. Blood vessels are present in the mass. To the left, an inflammatory infiltrate in the area of the polymer is observed (Hematoxylin and eosin, original magnification, ×172).

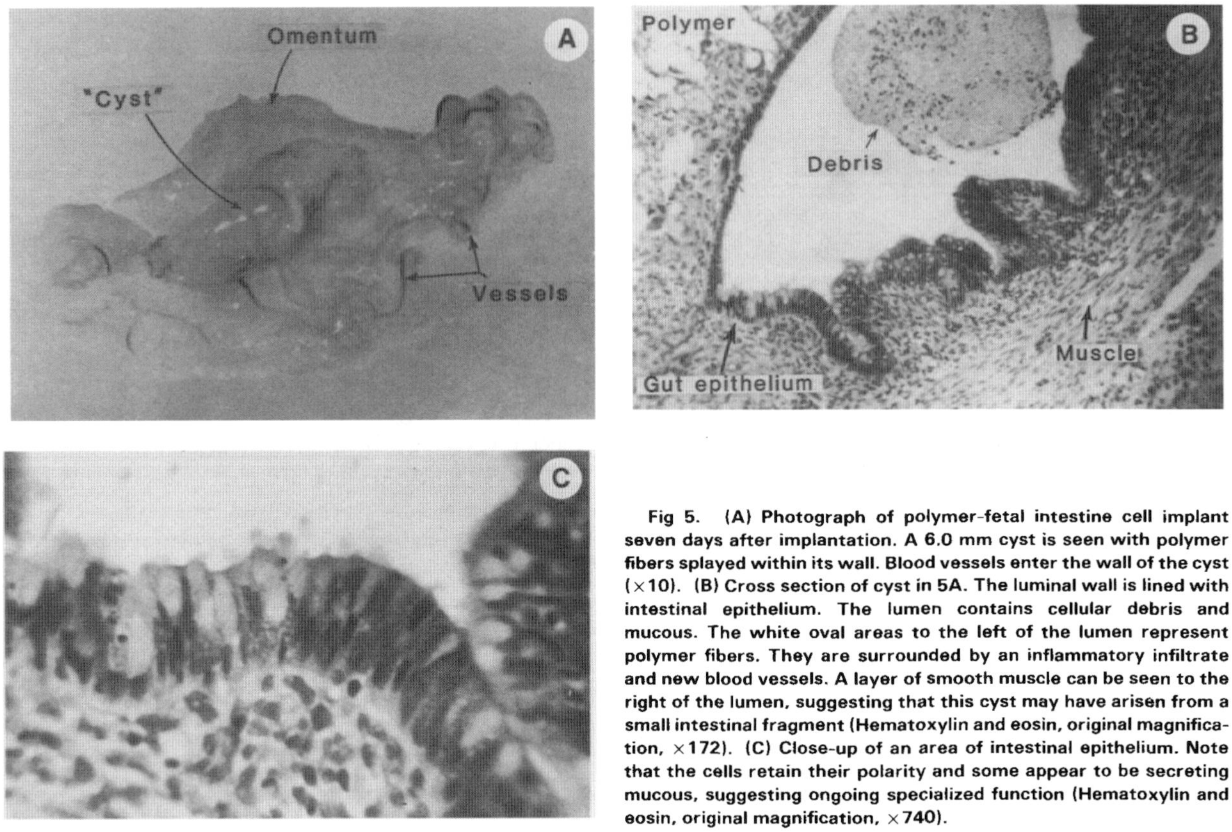

Fig 5. (A) Photograph of polymer-fetal intestine cell implant seven days after implantation. A 6.0 mm cyst is seen with polymer fibers splayed within its wall. Blood vessels enter the wall of the cyst (×10). (B) Cross section of cyst in 5A. The luminal wall is lined with intestinal epithelium. The lumen contains cellular debris and mucous. The white oval areas to the left of the lumen represent polymer fibers. They are surrounded by an inflammatory infiltrate and new blood vessels. A layer of smooth muscle can be seen to the right of the lumen, suggesting that this cyst may have arisen from a small intestinal fragment (Hematoxylin and eosin, original magnification, ×172). (C) Close-up of an area of intestinal epithelium. Note that the cells retain their polarity and some appear to be secreting mucous, suggesting ongoing specialized function (Hematoxylin and eosin, original magnification, ×740).

ing elicited an angiogenic and inflammatory response similar to their counterparts that had been seeded with parenchymal cells and maintained in culture. This suggested that the cells themselves did not play a major role in the inflammation and neovascularization seen.

DISCUSSION

Many diseases of the liver, intestine, and pancreas result in organ failure. This century has seen the development of specific pharmacologic therapy to replace lost function. Insulin replacement for diabetes mellitus is an example.

The emergence of organ transplantation and the science of immunobiology has allowed replacement of the kidney, heart, liver, and other organs. However, as our ability to perform these complex operations has improved, the limitations of the technology have become more evident. For example, in pediatric liver transplantation, donor scarcity has increased as more programs have opened. Only a small number of donors are available in the United States for 800 to 1,000 children in liver failure and those children that undergo transplantation are often so ill by the time a liver is found that the likelihood of success is diminished. The surgery is complex and usually associated with major blood loss. The preservation time is short and, therefore, results in major logistical problems in matching a distant donor with a recipient. For these reasons, the undertaking is expensive and labor intensive, requiring a major investment of resources available only in tertiary care facilities.

Selective cell transplantation of only those parenchymal elements necessary to replace lost function has been proposed as an alternative to whole or partial organ transplantation.[1] It has several attractive features. It avoids major surgery with its attendant blood loss, anesthetic difficulties and complications. It replaces only those cells that supply the needed function and, therefore, problems with passenger leukocytes, antigen presenting cells, and other cell types which may promote the rejection process are avoided. Adding the techniques of cell culture provides another set of tools to aid in the transplantation process. The ability to expand cell numbers with proliferation of cells in culture allows autotransplantation of one's own tissue. Skin equivalents using contracted collagen lattices and epidermal cells have been described.[9,10] Insertion of gene segments, and deletion of antigenic components while the cells are in culture is also possible with current technology.

Islet cell transplantation as an experimental treatment of diabetes mellitus is an area of current research. Although there is evidence of short-term function, long-term results have been less satisfactory.[11,12] Currently, whole organ pancreatic transplantation is the preferred replacement. Hepatocyte injections into the portal circulation have been attempted to support hepatic function. A recent novel approach in which hepatocytes were attached to collagen-coated microcarrier beads prior to injection into the peritoneal cavity demonstrated successful implantation, viability of the implanted hepatocytes, and function. The authors suggested that cell attachment to a maxtrix prior to implantation was an important component of successful engraftment and function.[13]

Our studies of isolating parenchymal cells, attaching them to biodegradable polymer scaffolds in cell culture, and implanting them into hosts are based on several biologic observations:

(1) Every structure in living organisms is in a dynamic state of equilibrium. It undergoes constant renewal, remodeling, and replacement of functional tissue. The degree of change varies from organ to organ and structure to structure.

(2) Structural cells, if placed in a dissociated state, tend toward reforming structure. Their ability to do so depends on the environment in which they are placed and the degree of alteration they have undergone. Examples include capillary endothelial cells that form tubes in vitro under certain conditions[14] and bile duct cells that form tubes under the proper conditions.[15]

(3) Tissue cannot be implanted in volumes >1.0 to 3.0 μL because nutrition is limited by the maximum diffusion distance until angiogenesis occurs.[16]

(4) Cell shape determined by cytoskeletal components and attachment to matrix plays an important role in cell division and differentiated function.[17,18] If dissociated cells are placed into mature tissue as a suspension without cell attachment, they may have a difficult time finding attachment sites, achieving polarity, and functioning because they begin with intrinsic organization. This may limit the total number of implanted cells that can remain viable to organize, proliferate, and function.

We reasoned that if we provided an organized scaffolding to which the cells were already attached, we could increase the total number of implanted cells. Artificial biodegradable polymers were chosen for several reasons. We could engineer configuration, manageability, tensile strength, and rate of degradation to a great degree with a man-made plastic and also might be able to modify the inflammatory response by modifying the material. The polymer material could be coated with other cell types or with attachment factors to increase cell attachment. In addition, we have considerable experience in placing biologically active molecules such as growth factors directly into the polymer, and allowing slow release of these agents in a controlled and predictable way.[19] Finally, by the use of biodegradable matrices, we provide only a temporary scaffold, which eventually is reabsorbed, leaving structural support of the mass to mesenchymal elements supplied by the host, and modified by the implanted cells. Although some of the polymers studied caused somewhat of an inflammatory response, it is likely that further purification will reduce or eliminate this problem. This has already been shown to be the case with ethylene-vinyl acetate[19] and polyanhydrides.[5]

This technology allowed us to design the polymers to meet the biologic needs of the system we wished to create. The configuration of the polymer scaffold must have enough surface area for the cells to be nourished by diffusion while neovascularization occurs. The new blood vessels must interdigitate with the implanted parenchymal elements to continue to support their growth, organization, and function. Polymer discs seeded with a monolayer of cells and branching fiber networks both satisfy these needs. The branching fibers are based on the same principles that nature has used to solve the problem of increasing surface area proportionate to volume increases. All multicellular organisms utilize this repeating branching structure. Branching systems represent communication networks between organs as well as the functional units of individual organs.[20] Seeding this configuration with cells and implanting the structure as fibers allows us to implant large numbers of cells, each of which is exposed to the environment of the host. Therefore, free exchange of nutrients and waste can occur while neovascularization is achieved. If functional, the new three-dimensional mass would be a true chimera of parenchymal elements of the donor, and mesenchymal elements of the recipient. The term "chimeric neomorphogenesis" describes the process that has occurred.

The results of this study demonstrate that cells from liver, intestine, and pancreas can be successfully harvested, and will attach to artificial biodegradable polymers. They will survive in culture in this configuration and can then be implanted into a host in a variety of locations. An inflammatory response that is mediated by both the wound and the nature of the polymer will occur. Successful engraftment of small clusters of hepatocytes and intestinal cells has been demonstrated. However, we do not yet have evidence of cell function in the new environment. Further studies to define the optimal characteristics of the polymer, attachment parameters, growth criteria, and function, need to be performed. Clinical application in diabetes with beta cells of the pancreas, in hepatic failure with hepatocytes and biliary cells, and in intestinal insufficiency with intestinal epithelium remains the long-term goal.

ACKNOWLEDGMENT

We wish to express our sincere gratitude to Judah Folkman for advice, Dr Jorge Heller for his generous gift of polyorthoester, Deborah Stark for her technical assistance, and Susann Antczak for preparation of this manuscript.

REFERENCES

1. Russell PS: Selective transplantation. Ann Surg 201:255-262, 1985

2. Craig PH, Williams JA, Davis KW, et al: A biological comparison of polyglactin 910 and polyglycolic acid synthetic absorbable sutures. Surgery 141:1-10, 1975

3. Heller J, Penhale WH, Helwing RF, et al: Release of norethindrone from polyacetals and polyorthoesters. AIChE Symposium Series, 206:28-36, 1981

4. Leong KW, D'Amore P, Marletta M, et al: Bioerodable polyanhydrides as drug carrier matrices. II. Biocompatability and chemical reactivity. J Biomed Mat Res 20:51, 1986

5. Domb AJ, Langer R: Polyanhydrides I. Preparation of high molecular weight polyanhydrides. J Poly Sci (in press)

6. Kopecek J, Ulbrich K: Biodegradation of biomedical polymers. Prog Poly Sci 9:1, 1983

7. Selgen PO: Preparation of rat liver cells. III. Enzymatic requirements of tissue dispersion. Exp Cell Res 82:391-398, 1973

8. Gotoh M, Maki T, Kiyozumi T, et al: An improved method of isolation of mouse pancreatic islets. Trans 40:437-438, 1985

9. Bell E, Ehrlich HP, Buttle DJ, et al: Living tissue formed in vitro and accepted as skin-equivalent tissue of full thickness. Science 211:1052-1054, 1981

10. Gallico GG, O'Connor NE, Compton CC, et al: Permanent coverage of large burn wounds with autologous cultured human epithelium. N Engl J Med 311:448-451, 1984

11. Sutherland DER: Pancreas and islet transplantation. I. Experimental studies. Diabetologia 20:161-185, 1981

12. Sutherland DER: Pancreas and Islet Transplantation. II. Clinical Trials. Diabetologia 20:435-450, 1981

13. Demetriou AA, Whiting JF, Feldman D, et al: Replacement of liver function in rats by transplantation of microcarrier-attached hepatocytes. Science 233:1190-1192, 1986

14. Folkman J, Haudenschild C: Angiogenesis in vitro. Nature 288:551-556, 1980

15. Vacanti JP, Morse MA, Haudenschild C, et al: The isolation and growth of biliary epithelial cells in long-term monolayer cell culture. (Submitted)

16. Folkman J, Hochberg MM: Self-regulation of growth in three dimensions. J Exp Med 138:745-753, 1973

17. Folkman J, Moscona A: Role of cell shape in growth control. Nature 273:345-349, 1978

18. Ben-Ze'ev A, Farmer SR, Penman S: Protein synthesis requires cell-surface contact while nuclear events respond to cell shape in anchorage-dependent fibroblasts. Cell 2:365-372, 1980

19. Langer R, Folkman J: Polymers for the sustained release of proteins and other macromolecules. Nature 263:797-800, 1976

20. MacDonald N: Trees and Networks in Biological Models. New York, Wiley, 1983

Discussion

P. Donahoe (Boston): This interesting approach, which uses biodegradable materials as an artificial matrix, has been successfully done in two-dimensional systems by Drs John Burke and Howard Green for the replacement of skin for major burns. Dr Vacanti is attempting this technique in a three-dimensional system. I look forward to further screening of matrix materials in the future. After maintaining dispersed cells from liver, pancreas, and intestine in culture for three days before implantation, small viable implants were found in three of 66 hepatic implants, three of 26 small intestinal implants with a 6 mm cystic structure, and 0 of 23 pancreatic implants reflecting also our own nonsuccess with that particular organ. Though falling far short of providing needed organs for transplantation, which is the author's dream, it does provide us with some promising techniques. I would make a number of suggestions for improving the rate of graft survival. First of all, allogeneic grafts were used and rejection can expect to be high. I would suggest using syngeneic mice to test the hypothesis more strongly. Ontogeny studies should be done in order to determine at which fetal age growth would be allowed. The position of implantation is also vital. We too lost a high percentage of grafts in omentum and in mesentery, as well as the hepatic capsule and in the retroperitoneum, so we turned to the subrenal capsule where structures can grow without being ejected. Can the authors tell us why the polyorthoester was so unsuccessful? Were implantation toxicity studies done before undertaking the present studies? What will be your next experimental designs? We will follow subsequent studies with great interest.

J.P. Vacanti (closing): Syngeneic mice eliminate rejection as a potential mediator of the inflammatory response, which has been significant in some of the polymers we have tested. The time-line studies with different aged animals are very important and I appreciate that point. Our hope is that adult cells can be used with success as can fetal cells. We have evolved the idea that terminal differentiation of cells is probably not very common and that in the appropriate circumstance with the appropriate communication with other cells and matrix, cells can be manipulated and can express genes that they do not usually express. We have come to the conclusion that the matrix component of different tissues probably contributes to the successful engraftment of different cell types. So which particular location for which particular cell type needs to be worked out. We wonder whether we can take different polymers for which we have a long experience as slow release reservoirs for bioactive molecules and then use them to precondition different tissues with various growth factors and differentiation factors before we do the implant. We might be able to fool the tissue into thinking it is fetal tissue as opposed to adult tissue. The polyanhydrides at the configuration that we used were the least successful and the most toxic and they are very acidic, and it is very difficult to keep up with the acid release in culture and probably in in vivo. More work needs to be done with this particular polymer family. The polyorthoester is reasonably noninflammatory and provided good cell attachment.

Neocartilage formation *in vitro* and *in vivo* using cells cultured on synthetic biodegradable polymers

L. E. Freed,[1] J. C. Marquis,[1] A. Nohria,[1] J. Emmanual,[2] A. G. Mikos,[3] and R. Langer[1,3,*]
[1]*Division of Health Sciences and Technology, Massachusetts Institute of Technology, Cambridge, Massachusetts 02139;* [2]*Harrington Arthritis Research Center, Phoenix, Arizona 85006; and* [3]*Department of Chemical Engineering, Massachusetts Institute of Technology, Cambridge, Massachusetts 02139*

Cartilaginous implants for potential use in reconstructive or orthopedic surgery were created using chondrocytes grown on synthetic, biodegradable polymer scaffolds. Chondrocytes isolated from bovine or human articular or costal cartilage were cultured on fibrous polyglycolic acid (PGA) and porous poly(L)lactic acid (PLLA) and used in parallel *in vitro* and *in vivo* studies. Samples were taken at timed intervals for assessment of cell number and cartilage matrix (sulfated glycosaminoglycan [S-GAG], collagen). The chondrocytes secreted cartilage matrix to fill the void spaces in the polymer scaffolds that were simultaneously biodegrading. *In vitro,* chondrocytes grown on PGA for 6 weeks reached a cell density of 5.2×10^7 cells/g, which was 8.3-fold higher than at day 1, and equalled the cellularity of normal bovine articular cartilage. *In vitro,* the cell growth rate was approximately twice as high on PGA as it was on PLLA; cells grown on PGA produced S-GAG at a high steady rate, while cells grown on PLLA produced only minimal amounts of S-GAG. These differences could be attributed to polymer geometry and biodegradation rate. *In vivo,* chondrocytes grown on both PGA and PLLA for 1–6 months maintained the three-dimensional (3-D) shapes of the original polymer scaffolds, appeared glistening white macroscopically, contained S-GAG and type II collagen, and closely resembled cartilage histologically. These studies demonstrate the feasibility of culturing isolated chondrocytes on biodegradable polymer scaffolds to regenerate 3-D neocartilage.

INTRODUCTION

Reconstructive plastic and orthopedic surgeries are done on a daily basis at major medical centers to treat many patients with damaged cartilage due to congenital abnormalities or trauma.[1,2] Current therapies include: (a) transplantation (i.e., excising healthy host cartilage, carving it into the desired shape, and reimplanting it where needed into the same patient); and (b) implantation of artificial prosthetic devices. Problems arise however, in that the amount of donor tissue available for transplantation is limited, and it is often not possible to form delicate three-dimensional (3-D) implants from the harvested tissue. Artificial prostheses are complicated by infections and a high incidence of adhesive breakdown at the host/prosthesis interface. Also, a prosthesis cannot adapt in response to environmental stresses as does living cartilaginous tissue.

The clinical need for improved treatment options in reconstructive surgery has motivated research aimed at creating new cartilage *in vitro* and *in vivo*. Neocartilage formation has been reported using systems based on isolated chondrocytes,[3,4] organic support matrices,[5,6] and 2-D synthetic polymers.[7–10] In the present work, neocartilage with specific, reproducible 3-D shapes was formed from chondrocytes cultured on synthetic, biodegradable polymeric scaffolds, and assessed with respect to cell adhesion, proliferation, and cartilage matrix secretion both *in vitro* and *in vivo* (Fig. 1).

MATERIALS

Calf knee joint or rib specimens were obtained from a local abattoir within 4 h of slaughter. Discarded human costal cartilage was obtained following repair of congenital defects from Boston's Children's Hospital. Athymic "nude" mice were from Taconic (Germantown, NY).

*To whom correspondence should be addressed.

Journal of Biomedical Materials Research, Vol. 27, 11–23 (1993)

CCC 0021-9304/93/010011-13

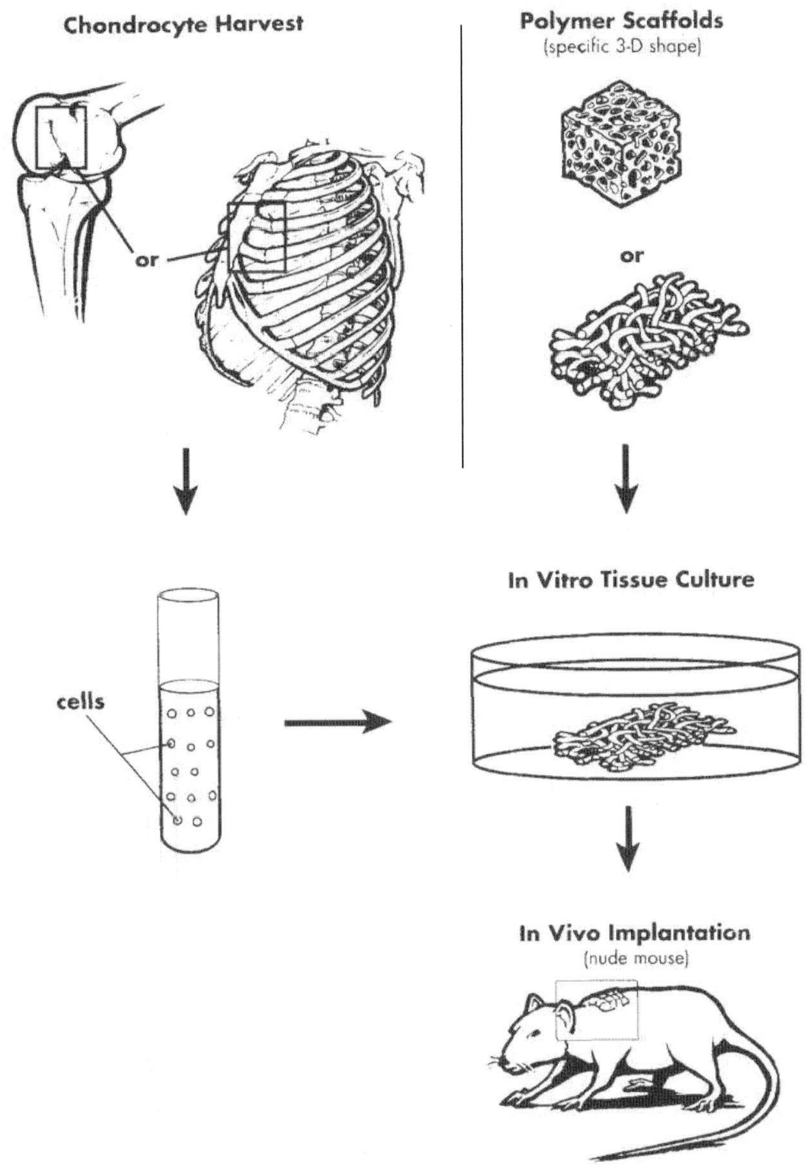

Figure 1. (A) Chondrocytes were isolated from bovine or human articular or rib cartilage; (B) scaffolds were formed of synthetic, biodegradable polymers; (C) cells were seeded on polymers (10^6 cells each) and cultured *in vitro* for up to 2 months; (D) in parallel studies, some cell-polymer samples were implanted *in vivo* (subcutaneously in athymic mice) for up to 6 months.

Polyglycolic acid (PGA) mesh was generously donated by American Cyanamid (Danbury, CT). Poly(L)lactic acid (PLLA) was from Polysciences (Warrington, PA). Sodium chloride (NaCl) and chloroform were from Mallinckrodt (Paris, KY). Gypsum plaster was from Miles (St. Louis, MO); separating film was from Lincoln Dental (Cherry Hill, NJ). Ethylene oxide was from Andersen (Chapel Hill, NC). Triple-distilled mercury (Hg) used in the porosimetry studies was from Bethlehem Apparatus (Hellertown, PA). Tribromoethanol and tertiary amyl alcohol were from Aldrich (Milwaukee, WI). Ethilon sutures were from Johnson & Johnson (Somerville, NJ).

Type II collagenase was from Worthington (Freehold, NJ). Dulbecco's modified Eagle medium (DMEM), fetal bovine serum (FBS), phosphate buffered saline (PBS), 0.05% trypsin/0.53 mmol/L Na_4 ethylene dinitrilo-tetraacetic acid (EDTA), and N-2-hydroxy-ethyl piperazine N′-2-ethane sulfonic acid (HEPES) were from GIBCO (Grand Island, NY). Type III papain, bovine dermal type I collagen, penicillin-streptomycin, nonessential amino acids,

TABLE I
List of Experiments

Study No.	Figure No.	Study Type	Donor Tissue	Polymer	Time Points	Total *n*
1	3(A), 4(A)–(C)	*In vitro*	Bov FPG	PGA	1 day, 1 week 2, 3, 4, 6 weeks	18
2	3(C)–(D)	*In vitro*	Bov rib	PGA	1, 3, 5, 7 weeks	12
3	3(B), 4(A)–(C)	*In vitro*	Bov FPG	PLLA or PLLA + collagen	1 day, 1, 2, 3, 4, 5, 6, 7, 8 weeks	72
4	5(A), (B)	*In vivo*	Hum rib	PLLA	4, 8 weeks	4
5	6(A), 7(A)	*In vivo*	Bov rib	PGA	2, 5, 8, 12, 16, 24 weeks	6
6	6(B), 7(B)	*In vivo*	Bov rib	PLLA	2, 5, 8, 12, 16, 24 weeks	6
7	7(A)	*In vivo*	Bov FPG	PGA	2, 5, 8, 12 weeks	4
8	7(A)	*In vivo*	Bov FPG (passaged)	PGA	2, 5, 8, 12 weeks	4
9	7(B)	*In vivo*	Bov FPG	PLLA	2, 5, 8, 12, 16, 24 weeks	6
10	5(C)	*In vivo*	No cells	PGA or PLLA	2, 5, 8, 12, 16, 24 weeks	10
11	—	*In vivo*	Bov FPG	No polymer	2, 5, 8, 12 weeks	10

proline, ascorbic acid, and neutral buffered formalin were from Sigma (St. Louis, MO). Glutaraldehyde was from Electron Microscopy Sciences (Fort Washington, PA). Histology cassettes were from Baxter (Beford, MA). Hoechst 33258 dye was from Polysciences. Dimethyl-methylene blue dye was from Aldrich. Poly(methylmethacrylate) embedding resin and *n*-butyl phthalate plasticizer were from Fischer (Pittsburgh, PA), while the catalyst, benzoyl peroxide, was from Aldrich. Epon 812 embedding medium was from Ted Pella (Redding, CA). Polyclonal rabbit anti-bovine type I collagen (AB 749) and anti-bovine type II collagen (AB 746) were from Chemicon (Temecula, CA). The Labeled StreptAvidinBiotin (LSAB) alkaline phosphatase staining kit (K682) was from DAKO (Carpinteria, CA).

METHODS

Chondrocytes

Bovine (Bov) articular or costal (rib) cartilage was aseptically excised from the FemoroPatellar Groove (FPG) or anterior ribs respectively, of 1–2-week-old calves (Table I). Human (Hum) costal cartilage was obtained following repair of a congenital defect in an 11-year-old male, whose parents signed informed consent forms according to protocols approved by the M.I.T. Committee on Human Research. Chondrocytes were isolated and cultured using sterile technique as follows. Cartilage samples were diced into 1-mm^3 cubes and placed in test tubes containing 10–15 cm^3/g tissue of DMEM with 5% FBS and 4–5 U/cm^3 of type II collagenase. The tubes were placed horizontally on an orbital shaker (Bellco, Vineland, NJ) and stirred at 40 rpm for 15 h at 37°C. The digestate was repipetted with a 10-cm^3 pipette, centrifuged at 200*g* for 10 min at 4°C, and the cell pellet resuspended in 40 volumes of PBS without calcium or magnesium, and containing 0.02% EDTA. The suspension was filtered through a 150-μm nylon screen (Tetko, Elmsford, NY) and recentrifuged. The cell pellet was washed twice with PBS and resuspended in culture medium (DMEM containing 4500 mg/L glucose and 584 mg/L glutamine, 10% FBS, 100 U/cm^3 penicillin, 100 $\mu g/cm^3$ streptomycin, 12.5 mmol/L HEPES, 0.1 mmol/L nonessential amino acids, 0.4 mmol/L proline and 50 mg/L ascorbic acid).[11] Cell number and viability were quantitated via trypan blue exclusion using a hemocytometer and an inverted phase-contrast light microscope (Nikon, Tokyo, Japan).

Polymers

Poly-glycolic acid (PGA) had an average molecular weight of 60,000 daltons, as measured by viscometry.[20] PGA fibers, 14 μm diameter, were formed as a non-woven mesh with a fiber density of 8.5 mg/cm^2.

Poly(L)lactic acid (PLLA) had a number average molecular weight of 104,810 daltons, as measured by gel permeation chromatography.[12] PLLA scaffolds were prepared using a solvent-casting particulate-leaching technique. NaCl was ground with an analytical mill (Tekmar, Cincinnati, OH) and sieved using a sieve shaker (CSC Scientific, Fairfax, VA). Salt particles (100–500 μm in diameter) were added to a solution of PLLA in chloroform. The salt:polymer ratio was 9:1. The vortexed dispersion was cast in a

5-cm Petridish, loosely covered, and the chloroform was allowed to evaporate for 48 h. Residual amounts of chloroform were removed by vacuum-drying at 100 μm Hg for 24 h. The salt was leached out by immersing the scaffold in 250 cm^3 distilled water, and changing the water every 6 h for a total of 72 h. The scaffold was then air-dried for 24 h, vacuum-dried at 100 μm Hg for 48 h and stored in a desiccator under vacuum until use. PLLA hemispheres were made by casting the PLLA–NaCl–chloroform mixture into 1 cm diameter × 0.5 cm deep hemispheric wells made of gypsum plaster coated with separating film, followed by the same solvent evaporation and salt leaching procedures described above. Collagen-coated PLLA was prepared by prewetting PLLA with 70% ethanol for 5 min, then replacing the ethanol with 100 μL of 0.1% bovine dermal type I collagen and vacuum-drying the samples for 15 h at 37°C.

The porosity, ϵ, and total internal pore area, α, of PGA and PLLA were measured using a mercury (Hg) porosimeter (Poresizer 9320, Micromeritics, Norcross, GA). The polymer sample weighing approximately 0.05 g was placed into a solid penetrometer with a bulb volume of 5 cm^3 (Micromeritics). The Hg intrusion volume was measured as a function of the applied pressure until a plateau was reached at an applied pressure of approximately 30 psi. Porosity was calculated from the maximal Hg intrusion volume and the polymer density; α was calculated from the work required to fill the pores, which in turn depended on the advanced contact angle, θ, between the Hg and the pore wall.[13] Contact angles were found to be 160° for PLLA and 132° for PGA.[12] The interstitial distance, d, was approximated by the diameter of an equivalent cylindrical pore, which in turn was calculated according to[13]:

$$d = (-4\gamma \cos\theta)/P \qquad (1)$$

where γ = the surface tension of Hg (485 dyn/cm at 25°C) and P = the Hg filling pressure of the penetrometer. For PGA scaffolds, the minimal Hg filling pressure (P_{min}) of the penetrometer was 0.61 psi and the corresponding maximal interstitial distance (d_{max}) measurable was 308 μm. For PLLA scaffolds, P_{min} was 0.51 psi and d_{max} was 515 μm. Polymer samples were assessed using a scanning electron microscope (SEM) (JSM-T300, JEOL, Tokyo, Japan). Prior to cell culture, all polymers were sterilized with ethylene oxide and aerated for at least 24 h.

In vitro studies

Bovine FPG or rib chondrocytes were cultured on polymer scaffolds in 35 mm tissue-culture treated polystyrene dishes (Table I, studies 1–3, and Fig. 1). Each PGA square was seeded with 100 μL of a suspension of culture medium containing 1 × 10^7 cells/cm^3 via capillary action. PLLA and collagen-coated PLLA cubes were prewetted with 70% ethanol for 5 min, rinsed twice with PBS, and seeded with 50 μL of a suspension containing 2 × 10^7 cells/cm^3 dropwise using a 1-cm^3 syringe fitted with a 25-g needle; PLLA hemispheres were similarly seeded with 100 μL of 3 × 10^7 cells/cm^3. PLLA prewetting with ethanol was necessary to overcome the negative spreading coefficient of water on PLLA.[14] Samples were incubated at 37°C in a humidified atmosphere containing 5% CO_2 for 8 h to permit cell adhesion to and entrapment within the polymer scaffold; 1.5 cm^3 of culture medium was then carefully added. Another 1.0 cm^3 of medium was added after 24 h. Cell growth was monitored using an inverted phase-contrast light microscope. Medium was replaced every 2–3 days for up to 8 weeks of tissue culture. Triplicate samples were taken at intervals for analysis as described below.

In vivo studies

Human and bovine FPG and rib cell–polymer constructs were cultured *in vitro* and then implanted subcutaneously in 4–6 athymic "nude" mice (Table I, studies 4–11, and Fig. 1). Mice (Swiss CD1 nu/nu males weighing 22–25 g) were handled aseptically according to standard procedures and within the N.I.H. guidelines for the care and use of laboratory animals.[15] Mice were anesthetized with avertin (2.5% v/v solution of 1 g/cm^3 tribromoethanol in tertiary amyl alcohol and PBS) administered intraperitoneally at a dosage of 0.3–0.5 mg/gm body weight. Mice were scrubbed dorsally with betadine, and subcutaneous pouches (2 cm × 2 cm) were created on the left and right sides using sterile technique. Cell–polymer constructs grown for 2–3 weeks *in vitro* were inserted into one pouch; polymer squares without cultured cells were inserted into the opposite pouch. Incisions were closed using 5–0 Ethilon sutures, which were removed 10 days postoperatively. In study 8, chondrocytes which had been serially passaged five times in monolayer culture were cultured on polymers and implanted. In study 11, 1 × 10^6 isolated chondrocytes in 250 μL of phosphate buffered saline were injected subcutaneously. After *in vivo* intervals of 2, 5, 8, 12, 16, and 24 weeks, animals were sacrificed via CO_2 inhalation, and explants removed for analysis as described below.

Analytical assays

Cell–polymer samples and explants from *in vitro* and *in vivo* studies were assessed as follows. Samples for biochemical analysis were frozen and lyophilized.

Samples were then digested with 1 cm^3 of 2.4 U/cm^3 type III papain for 15 h at 60°C.[16] The number of chondrocytes was determined for *in vitro* samples by measuring the amount of DNA in papain-digests using Hoechst 33258 dye and a spectroflorometer (Aminco, Urbana IL).[16] The amount of sulfated glycosaminoglycan (S-GAG) was determined for papain digests of *in vitro* and *in vivo* samples using dimethylmethylene blue dye and a UV-VIS spectrophotometer (Perkin-Elmer, Oak Bridge, IL).[17] Histologic samples were fixed in 2% glutaraldehyde for 15 min, transferred to tissue cassettes, and stored in neutral buffered formalin. Samples were then processed for embedding in poly(methylmethacrylate) as previously described.[18] Sections 8 μm thick were then cut using a Polycut-S microtome (Reichert-Jung, Cambridge Instruments, Deerfield, IL). Sections were stained with both (a) hematoxylin and eosin and (b) safranin-O/fast green to demonstrate presence of proteoglycans. Additional sections were stained with antibodies to type I or type II collagen using a commercially available Labeled StreptAvidinBiotin (LSAB) Alkaline Phosphatase kit. Nonspecific binding of the type I collagen antibody to type II collagen or of the type II collagen antibody to type I collagen were < 0.5% and < 3%, respectively.[19]

RESULTS

Cells and polymers

Harvested bovine FPG or rib cartilage typically yielded 4–5 × 10^7 and 1 × 10^7 viable cells per gram, respectively (75–95% viability). Harvested human rib cartilage yielded 3 × 10^6 viable cells/g (74% viability).

Fibrous polyglycolic acid (PGA) and porous poly(L-lactic acid) (PLLA) scaffolds are shown in Figures 2(A) and 2(B), respectively:

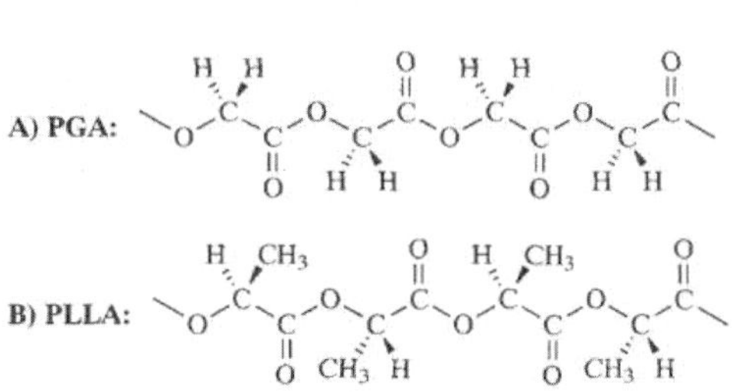

PGA scaffolds were 1 cm × 1 cm × 0.15 cm "patches" which weighed 7 ± 0.5 mg. A porosity of 97% was calculated for PGA based on sample size, weight, and a fiber density of 1.6 g/cm^3.[20] A slightly lower porosity of 91% (6.7 μL/mg) was measured for PGA using Hg porosimetry, which detects only pores with diameters less than 308 μm (see Methods, Eq. [1]). For PGA, a total pore area of 3.10 cm^2/mg was measured using Hg porosimetry.

PLLA scaffolds were 1 cm × 0.5 cm × 0.3 cm "sponges" which weighed 17 ± 3 mg. A porosity of 91% was calculated for PLLA based on sample size, weight, and a polymer density of 1.26 g/cm^3.[21] A slightly lower porosity of 87% (5.1 μL/mg) was measured for PLLA using Hg porosimetry. For PLLA, a total pore area of 2.15 cm^2/mg was measured using Hg porosimetry. PLLA hemispheres weighed 41 ± 4 mg.

In vitro culture

The combination of adhesion to and entrapment of bovine FPG cells within PGA and PLLA scaffolds was assessed using 1-day samples, since we found that the cells did not divide during the first 3 days of *in vitro* culture on polymers (Table I, studies 1, 3). PGA absorbed 1 × 10^6 cells suspended in a volume of 100 μL; PLLA absorbed 1 × 10^6 cells in a 50-μL volume if prewet with ethanol (see Methods). One-day cell–PGA or cell–PLLA constructs yielded 5.4 ± 0.3 × 10^5 cells or 4.7 ± 0.4 × 10^5 cells, respectively; initial cell number thus did not differ significantly between the experimental groups (two-tailed Student *t*-test). The distributions of cells throughout both PGA and PLLA scaffolds 1 day after seeding were spatially uniform, as assessed using histologic sections stained with hematoxylin and eosin.

Over 6–8 weeks of *in vitro* culture, bovine FPG chondrocytes cultured on PGA, PLLA, and collagen-coated PLLA proliferated and regenerated cartilaginous matrix to fill the void spaces in the polymeric scaffolds which were simultaneously biodegrading (Table I, studies 1, 3). Four-week PGA-based samples consisted mostly of cells and cartilage matrix, since PGA is 99% degraded over this time period.[22] In contrast, the scaffold was still present in 8-week PLLA-based samples, since PLLA does not degrade significantly over 6 months[23] (Figs. 3[A], [B]).

The histologic appearance of the samples depended on which polymer scaffold (PGA or PLLA), but not on which cell type (bovine FPG or rib) was used (Table I, studies 1, 3). On PGA, both FPG and rib chondrocytes appeared rounded and within lacunae surrounded by cartilaginous matrix (Figs. 3[A], [C], [D]). The extracellular matrix stained strongly positive for glycosaminoglycan (GAG) using safranin-O (Fig. 3[A]). Pericellular staining was positive for both type I and type II collagens using anti-collagen antibodies (Figs. 3[C], [D]). If either the primary or the link antibody was omitted from the immunochemical (LSAB) assay, no significant collagen staining occurred, indicating that nonspecific antibody binding was minimal. On PLLA, in contrast, the cells appeared spindle-shaped and were evenly distributed within the pores of the scaffold; minimal matrix staining for GAG or collagen was observed (Fig. 3[B]).

Chondrocyte proliferation and matrix regeneration also depended on which polymer scaffold was used. Bovine FPG chondrocytes grown on PGA for 6 weeks *in vitro* reached a density of $5.2 \pm 0.3 \times 10^7$ cells/g (Fig. 4[A], triangles; Table I, studies 1, 3), which was 8.3-fold higher than at day one. The final cellularity of these *in vitro* constructs equalled that of the original cartilage specimen, as assessed after collagenase digestion. Bovine FPG chondrocytes cultured on PLLA for 8 weeks *in vitro* reached a density of $2.4 \pm 0.2 \times 10^7$ cells/g, and this did not change significantly even if the PLLA was coated with type I collagen prior to cell seeding (Fig. 4[A], *closed and open circles*). The final cell density per unit scaffold void volume was also higher on PGA ($2.8 \pm 0.05 \times 10^7$ cells/cm^3) than on PLLA ($2.1 \pm 0.2 \times 10^7$ cells/cm^3) (Fig. 4[B]). The initial growth rate (0–2 weeks) was 2.3 times as high for cells grown on PGA (1.2×10^7 cells/cm^3 void/week) than for those grown on PLLA (0.50×10^7 cells/cm^3 void/week).

The sulfated glycosaminoglycan (S-GAG) content of normal bovine FPG cartilage, as quantitated with dimethylmethylene blue, was 42.0 ± 5.3 mg S-GAG/g ($n = 6$). The S-GAG content of *in vitro* samples grown for 6 weeks on PGA or PLLA was 10.5 ± 0.01 mg S-GAG/g or 1.0 ± 0.1 mg S-GAG/g, respectively. This

(a)

(b)

Figure 2. Scanning electron micrographs of: (a) polyglycolic acid (PGA) 14 μm diameter fibers, 97% void volume; and (b) poly(L)lactic acid (PLLA) 200–500 μm pore diameter, 91% void volume.

corresponded to 25 or 2.4% of the amount of S-GAG in normal cartilage, respectively. The S-GAG content of *in vitro* samples, normalized per cell, (S-GAG/cell) increased rapidly and linearly with time in PGA-based samples, but was minimal in PLLA- and collagen-coated PLLA-based samples (Fig. 4[C]).

In Vivo studies

A nude mouse with a subcutaneous implant of human rib chondrocytes cultured on PLLA is shown in Figure 5(A) (Table I, study 4). The implant maintained the exact dimensions of the original polymer scaffold, and appeared glistening white macroscopically (Fig. 5[B]). Similar results were obtained in 6-month *in vivo* studies using bovine FPG or rib chondrocytes cultured on PGA or PLLA (Table I, studies 5–9).

In the absence of cultured chondrocytes, implantation of polymer scaffolds did not result in neocartilage formation (Table I, study 10). PGA scaffolds not seeded with cells had completely degraded by 5 weeks; PLLA scaffolds not seeded with cells remained

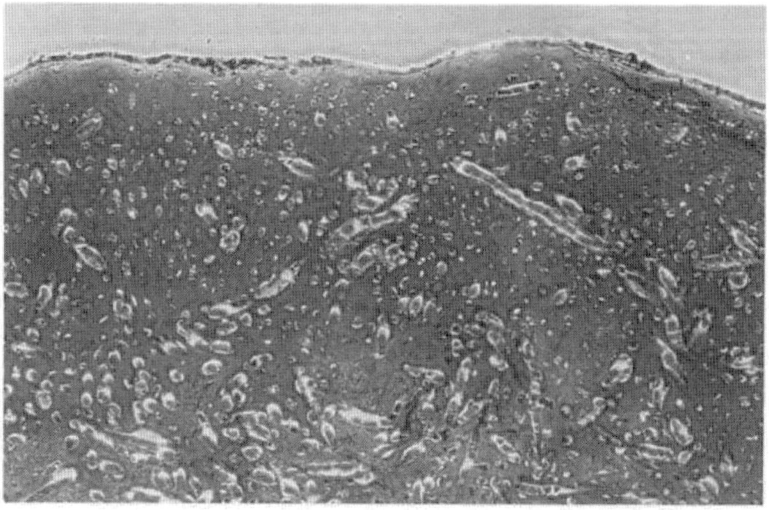

(a)

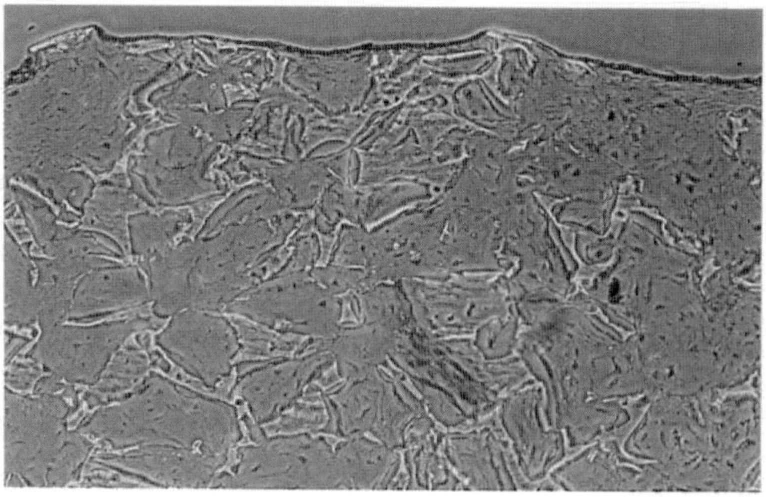

(b)

Figure 3. *In vitro* histologies: (A) safranin-O stain of bovine FPG chondrocytes cultured on PGA for 4 weeks (×100), (B) safranin-O stain of bovine FPG chondrocytes on PLLA for 8 weeks (×100), (C, D) anti-type I collagen and anti-type II collagen staining, respectively, of bovine rib chondrocytes cultured on PGA for 7 weeks (×1000) (Table I, studies 1 and 3).

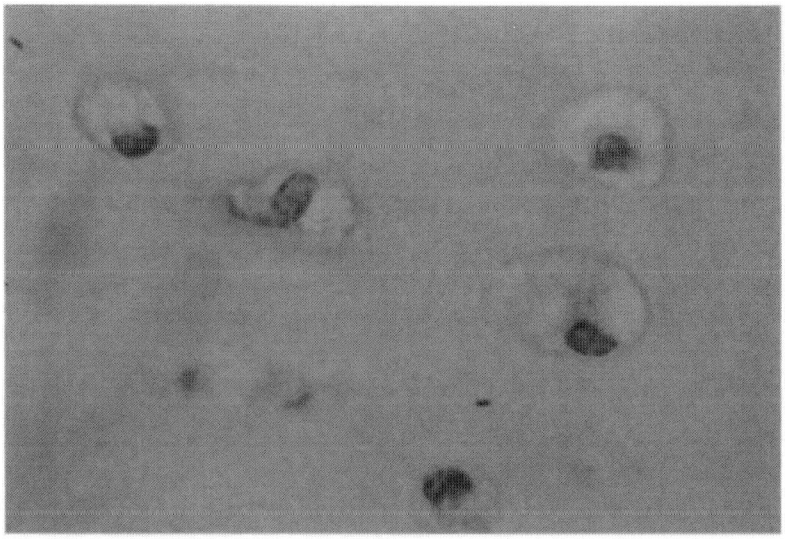

(c)

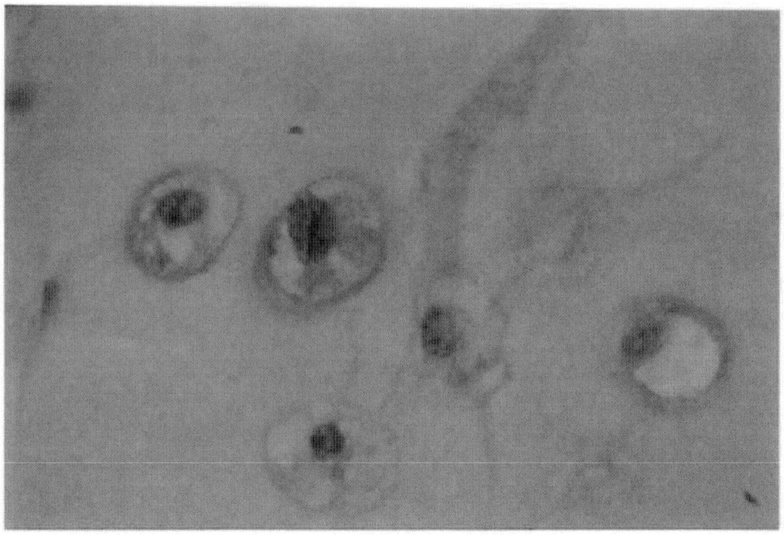

(d)

intact for up to 6 months. Explants of the latter appeared macroscopically red (Fig. 5[C]); histologically the polymers appeared infiltrated by host cells including red blood cells in new blood vessels, lymphocytes, multinucleated giant cells, and fibroblasts. In the absence of a polymer scaffold, subcutaneous injection of isolated bovine FPG chondrocytes resulted in irregular cartilaginous nodules less than 0.2 cm in diameter, weighing 2–3 mg (Table I, study 11).

Histologic analysis of chondrocyte-polymer explants (rib–PGA, rib–PLLA, FPG–PGA, or FPG–PLLA) revealed rounded cells in lacunae in all cases (Table I, studies 4–9). Explants of bovine rib chondrocytes cultured on PGA or PLLA for 2 weeks *in vitro* and then 5–8 weeks *in vivo* are shown in Figure 6 (Table I, studies 5, 6). Both PGA- and PLLA-based *in vivo* samples stained strongly positive for GAG with safranin-O (Figs. 6[A], [B]) and also for both type I and type II collagen with anticollagen antibodies (similar to Figs. 3[C], [D]). Control implants of polymers without cultured chondrocytes (Table I, study 10) did not stain positively with either safranin-O or anticollagen type I or anticollagen type II antibodies. A 3–8-cell-thick fibrous capsule was observed surrounding all subcutaneous cell–polymer implants (Fig. 6).

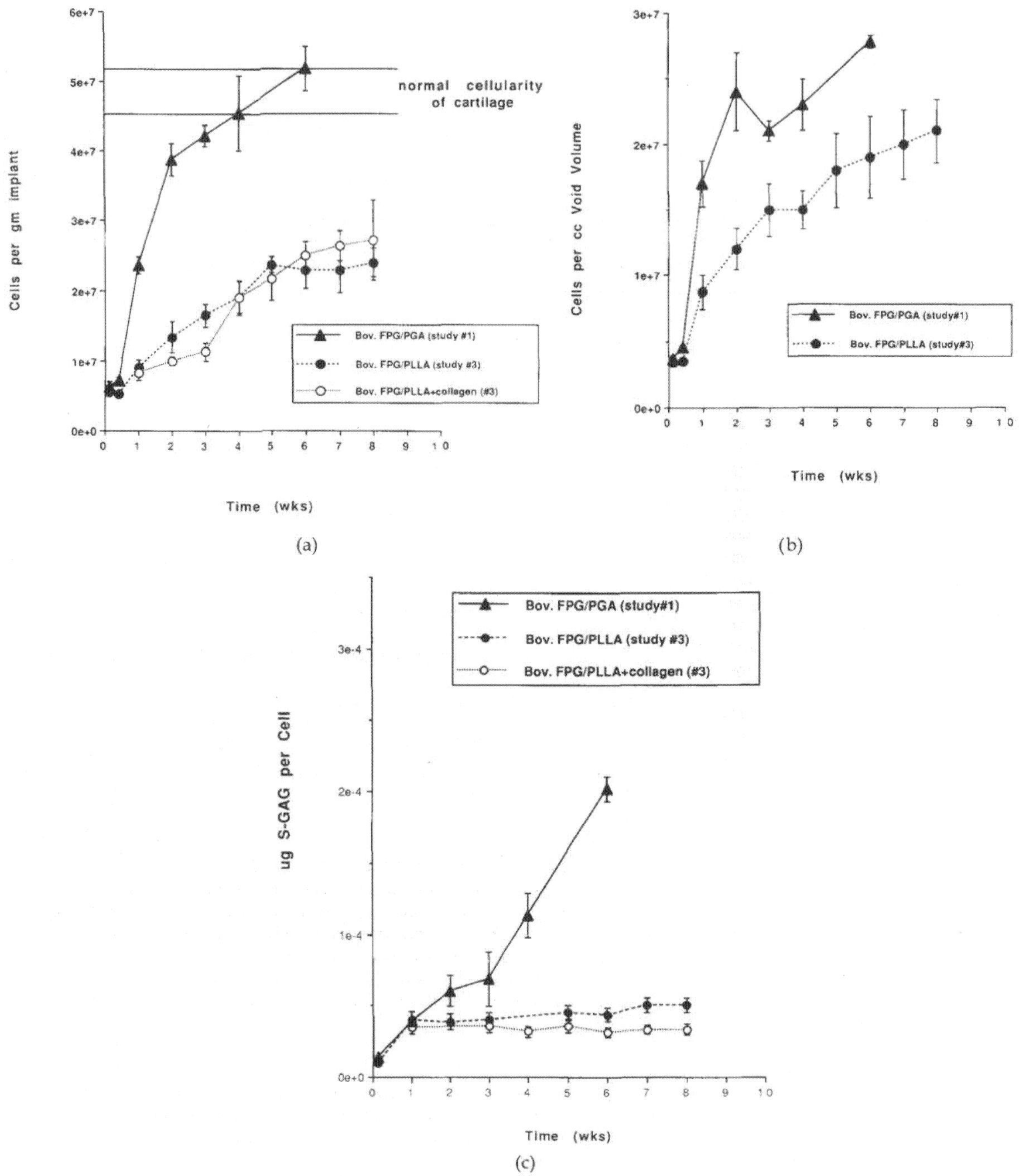

Figure 4. *In vitro* studies of bovine FPG chondrocytes on polymers: (A) cell number per gm wet weight, (B) cell number per cm^3 void volume, and (C) sulfated glycosaminoglycan (S-GAG) per cell (Table I, studies 1 and 3).

The sulfated glycosaminoglycan content of all *in vivo* explants was normalized per gram of sample instead of per cell, due to the combined presence of graft and host cells. No significant S-GAG/g was observed over 6 months in control explants (study 10, Table I, Fig. 7[B]). The amount of S-GAG/g in PGA-based explants increased and then either plateaued or decreased over 4–6 months (studies 5, 7, 8, Table I, Fig. 7[A]). The amount of S-GAG/g in all PLLA-based explants increased steadily over 4–6 months (studies 4, 6, 9, Table I, Fig. 7[B]). The S-GAG contents of rib/PGA or rib/PLLA explants grown for 2 weeks *in vitro* and then 6 months *in vivo* were 22.9 or 35.8 mg/g, respectively. These values corresponded to 55 or 85%

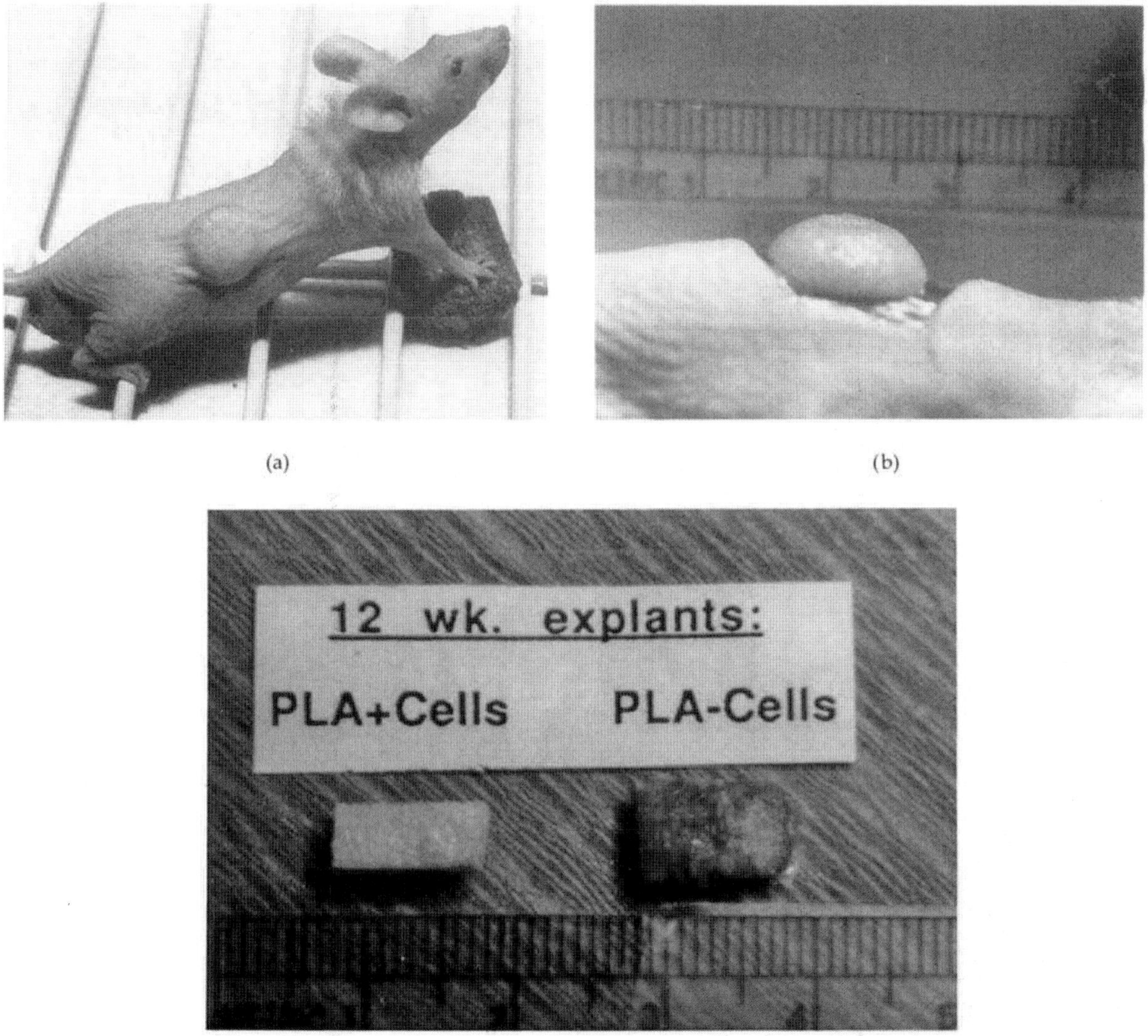

(a) (b)

(c)

Figure 5. (A, B) Nude mouse with a subcutaneous implant of human rib chondrocytes cultured on a PLLA hemisphere, (C) explants with or without cultured bovine FPG chondrocytes (Table I, studies 4, 9, 10).

of the amount of S-GAG in normal cartilage for PGA- or PLLA-based explants, respectively.

DISCUSSION

In short-term (2 month) *in vitro* studies, PGA proved a better scaffold than porous PLLA with respect to bovine FPG or rib cell proliferation and expression of differentiated function (Figs. 3, 4). However, in longer-term (6 month) *in vivo* studies, both PGA and PLLA proved excellent scaffolds for the regeneration of "neocartilage" from both FPG and rib chondrocytes (Figs. 5–7).

In vitro, it is likely that chondrocyte proliferation on PLLA was limited by diffusional constraints, as indicated by the characteristic shapes of the growth curves (Figs. 4[A], [B]). The initial cell growth rate on PGA was twice as high as that on PLLA; final cell densities per gram implant and per cm^3 void volume were also higher for PGA than for PLLA. PGA scaffolds had a higher porosity (97%) and were only half as thick (1.5 mm) as PLLA scaffolds (91% porous, 3 mm thick). It is also likely that faster degradation of the PGA increased the space available for the cells to multiply and regenerate matrix. At the onset of this study, the desired rate of scaffold degradation was unknown. PGA which degrades rapidly (99%

within 4 weeks)[22] was thus compared to PLLA which degrades slowly (no change in scaffold morphology over 6 months).[23] We found that *in vitro* growth and functional differentiation of bovine FPG chondrocytes grown on PLLA were limited under the conditions studied by (a) diffusional constraints and (b) slow biodegradation of the scaffold.

The S-GAG contents of samples of bovine FPG chondrocytes cultured on PGA *in vitro* for 6 weeks were up to 25% of normal for articular cartilage. Similar values (26% of normal) were reported for chondrocytes cultured for 7 weeks *in vitro* in agarose gels.[25] Potential advantages of using a synthetic, biodegradable polymeric scaffold (e.g., PGA) instead of a naturally occurring material (e.g., agarose) include: (a) fabrication of scaffolds with specific 3-D shapes for clinical use, (b) selection of polymers which can biodegrade in parallel with cell secretion of cartilaginous matrix, (c) reduction of the host inflammatory response by implanting cells cultured on biocompatible scaffolds, and (d) no inflammatory response once the scaffold has biodegraded.

In vivo, both PGA and PLLA-based explants maintained the exact 3-D shapes of the original scaffolds for up to 6 months, while in the absence of a polymer scaffold, injection of isolated bovine FPG chondrocytes

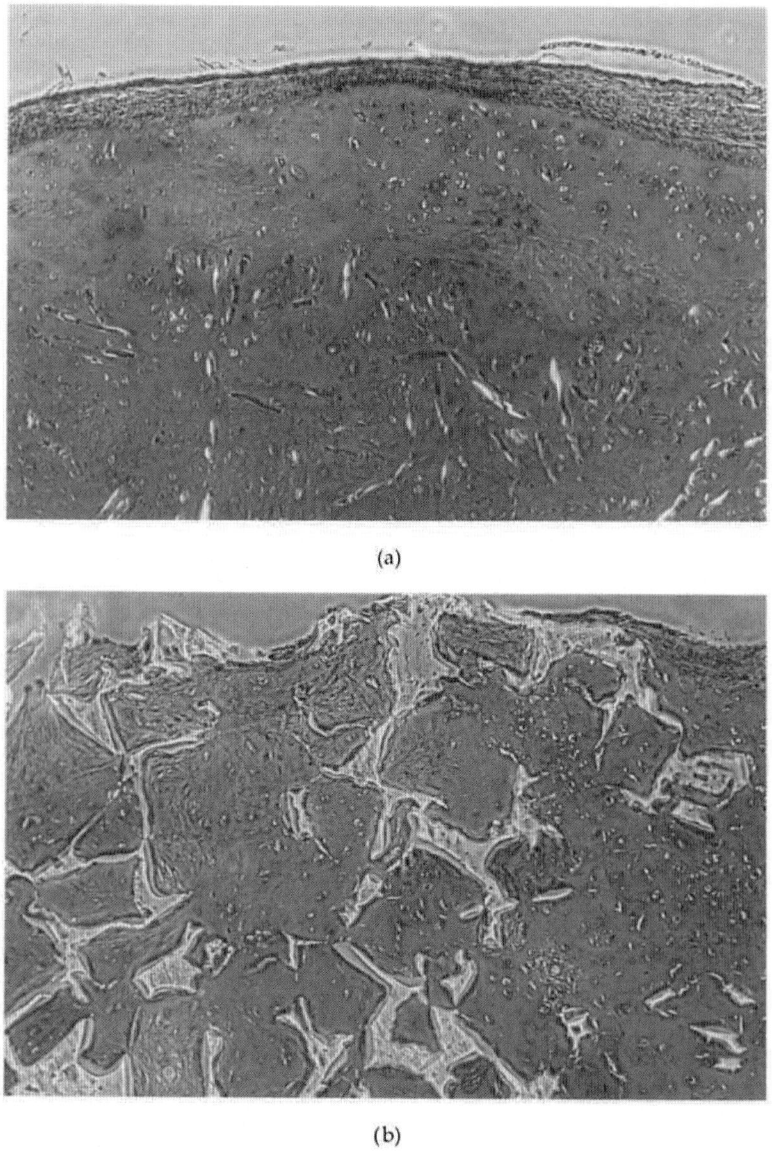

(a)

(b)

Figure 6. Explant histologies: safranin-O stains of bovine rib chondrocytes grown on: (A) PGA for 2 weeks *in vitro* and then 5 weeks *in vivo*, and (B) PLLA for 2 weeks *in vitro* and then 8 weeks *in vivo* (×100) (Table I, studies 5 and 6).

formed only small (< 0.2 cm) cartilaginous nodules. Bovine FPG chondrocytes cultured on PLLA which had appeared spindle-shaped *in vitro* became round and began secreting safranin-O positive matrix upon *in vivo* implantation (Figs. 3[B] vs. Fig. 6[B]). This finding is consistent with previous reports that apparent chondrocyte "dedifferentiation" which occurs *in vitro*

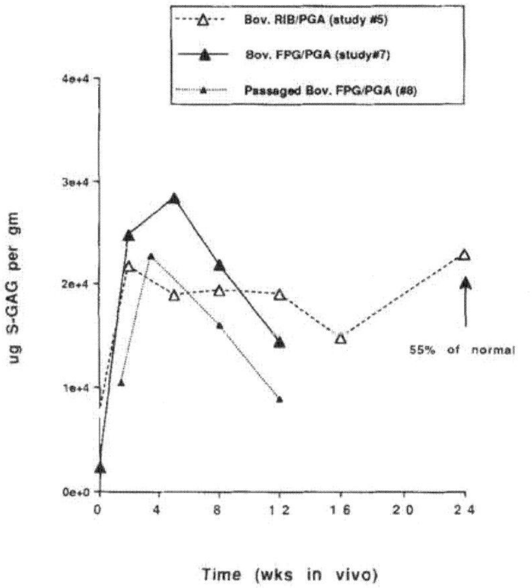

(a)

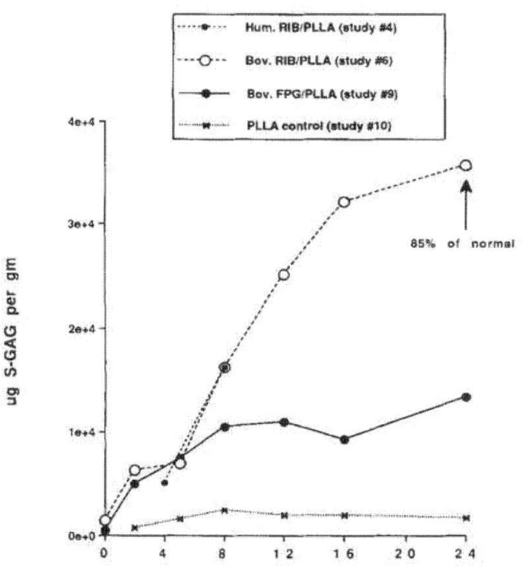

(b)

Figure 7. *In vivo* studies of bovine (Bov) or human (Hum) chondrocytes on: (A) PGA or (B) PLLA: sulfated glycosaminoglycan content (μg S-GAG/g) of chondrocyte–polymer and control explants (Table I, studies 5–10).

can be reversed given the appropriate environmental conditions, including unspecified minimal cell density and *in vivo* growth factors supplied by the host.[4,24]

In vivo, we found increasing amounts of S-GAG per gram over 6 months when polymers with cultured bovine or human chondrocytes were implanted, but insignificant amounts of S-GAG per gram over 6 months when polymers without cultured chondrocytes were implanted (Fig. 7B). All S-GAG in chondrocyte–polymer explants could thus be attributed to the presence of the grafted cells. The S-GAG contents of 6-month explants based on bovine rib chondrocytes were up to 55–85% of normal for articular cartilage. This study measured the bulk mass of S-GAG and not the structure of these molecules, which might be important in cartilage function.

The finding that S-GAG per gram decreased over time in some samples was attributed to the presence of a capsule, composed of host fibroblasts. Capsule formation depended on the implant's environment; it was present in subcutaneous implants (in nude mice) but not in our separate ongoing studies of articular implants (in dogs). Explants of polymers with cultured chondrocytes appeared nonvascular and noninflammatory, as compared to explants of polymers without cultured chondrocytes. This effect might be attributed to chondrocyte secretion of an angiogenesis inhibitory factor.[26] Further studies are needed to assess the very-long-term (years) survival and function of chondrocyte-based implants.

In summary, neocartilage was formed both *in vitro* and *in vivo* (subcutaneously in nude mice) using chondrocytes and synthetic, biodegradable polymers. The feasibility of using human rib cartilage to create implants in specific 3-D configurations was demonstrated. The long-term goals of this work are to create neocartilage implants with clinically appropriate form and function for use in reconstructive or orthopedic surgery, and to provide basic information for the development of other cell–polymer systems which could be applied to a variety of clinical disorders.

This work was supported by the National Institutes of Health (N.I.H. fellowship F32GM14191 to L. Freed) and Neomorphics, Inc. (Lexington, MA). The authors thank E. Gluzband, A. J. Grodzinsky, J. Sudhalter, J. Upton, and S. B. Weinstock for their contributions.

References

1. G. Bently, "Grafts and implants for cartilage repair and replacement," *Crit. Rev. Biocompatibility,* **5,** 245–267 (1989).
2. D. J. Zukor, R. D. Oakeshott, P. J. Brooks, J. P. McAuley, F. Langer, and A. E. Gross, "Fresh, small fragment osteochondral allografts for posttraumatic defects of the knee," *Articular Cartilage and knee joint function: Basic science and arthroscopy,* J. W. Ewing (ed.), Raven Press, New York, 1990.

3. D. A. Grande, M. I. Pitman, L. Peterson, D. Menche, and M. Klein, "The repair of experimentally produced defects in rabbit articular cartilage by autologous chondrocyte transplantation," *J. Orthop. Res.*, **7**, 208–218 (1989).
4. M. Takigawa, E. Shirai, K. Fukuo, K. Tajima, Y. Mori, and F. Suzuki, "Chondrocytes dedifferentiated by serial monolayer culture form cartilage nodules in nude mice," *Bone & Mineral*, **2**, 449–462 (1987).
5. S. Italy, A. Abramovici, and Z. Nevo, "Use of cultured embryonal chick epiphyseal chondrocytes as grafts for defects in chick articular cartilage," *Clin. Orthop. Rel. Res.*, **220**, 284–303 (1987).
6. S. Wakitani, T. Kimura, A. Hirooka, T. Ochi, M. Yoneda, N. Yasui, H. Owaki, and K. Onno, "Repair of rabbit articular surfaces with allograft chondrocytes embedded in collagen gel," *J. Bone Joint Surg.*, 71B: 74–80 (1989).
7. L. E. Freed, S. B. Weinstock, A. G. Mikos, J. C. Marquis, A. Nohria, A. J. Grodzinsky, and R. Langer, "Chondrocytes cultured on synthetic biodegradable polymers grow and secrete cartilage-like matrix,," presented at the 1991 Annual meeting of the American Institute of Chemical Engineers, Los Angeles, CA, November 18–22, 1991 (Abstract #3j, p. 5).
8. B. Hemmen, C. W. Archer, and G. Bently, "Repair of articular cartilage defects by carbon fibre plugs loaded with chondrocytes," 37th Annual Meeting of the Orthopedic Research Society, Aneheim CA, March 4–7, 1991 (p. 278).
9. C. Vacanti, R. Langer, B. Schloo, and J. P. Vacanti, "Synthetic biodegradable polymers seeded with chondrocytes provide a template for new cartilage formation in vivo," *Plast. Reconstruct. Surg.*, **88**, 753–759 (1991).
10. H. P. von Schroeder, M. Kwan, D. Amiel, and R. D. Coutts, "The use of polylactic acid matrix and periosteal grafts for the reconstruction of rabbit knee articular defects," *J. Biomed. Mater. Res.*, **25**, 329–339 (1991).
11. R. L. Y. Sah, Y. J. Kim, J. Y. H. Doong, A. J. Grodzinsky, A. H. K. Plaas, and J. D. Sandy, "Biosynthetic response of cartilage explants to dynamic compression," *J. Orth. Res.*, **7**, 619–636 (1989).
12. A. G. Mikos, G. Sarakinos, S. M. Leite, J. P. Vacanti, and R. Langer, "Laminated three-dimensional biodegradable foams for use in tissue engineering," *Biomaterials*, (1993), in press.
13. D. N. Winslow, "Advances in experimental techniques for mercury intrusion porosimetry," in *Surface and colloid science*, E. Matijevic and R. J. Good (eds.), Plenum Press, New York, 1984, pp. 259–282.
14. C. A. Miller, "Stability of interfaces," in *Surface and colloid science, Vol. 10*, E. Matijevic (ed.), Plenum Press, New York, 1978, pp. 227–277.
15. "Guide for the Care and Use of the Nude (Thymus-Deficient) Mouse in Biomedical Research," *ILAR News*, **19**, M1–M20 (1976).
16. Y. J. Kim, R. L. Sah, J. Y. H. Doong, and A. J. Grodzinsky, "Fluorometric assay of DNA in cartilage explants using Hoechst 33258," *Anal. Biochem.*, **174**, 168–176 (1988).
17. R. W. Farndale, D. J. Buttle, and A. J. Barrett, "Improved quantitation and discrimination of sulphated glycosaminoglycans by use of dimethylene blue," *Biochim. Biophys. Acta.*, **883**, 173–177 (1986).
18. J. Emmanual, C. Hornbeck, and R. D. Bloebaum, "A polymethyl methacrylate method for large specimens of mineralized bone with implants," *Stain Technol.*, **62**, 401–410 (1987).
19. Chemicon company literature.
20. American Cyanamid company literature.
21. E. W. Fischer, H. J. Sterzel, and G. Wegner, "Investigation of the structure of solution grown crystals of lactide copolymers by means of chemical reactions," *Kolloid-Zeitschrift und Zeitschrift fur Polymere*, **251**, 980–990 (1973).
22. C. C. Chu and A. Browning, "The study of thermal and gross morphologic properties of polyglycolic acid upon annealing and degradation treatments," *J. Biomed. Mater. Res.*, **22**, 699–712 (1978).
23. A. G. Mikos, H.-L. Lai, S. M. Leite, M. D. Lyman, J. A. Tamada, and R. Langer, "Characterization of degradation of poly (L-lactic acid) foams," unpublished results.
24. F. Mallein-Gerin, F. Ruggiero, and R. Garrone, "Proteoglycan core protein and type II collagen gene expressions are not correlated with cell shape changes during low density chondrocyte cultures," *Differentiation*, **43**, 204–211 (1990).
25. M. D. Buschmann, Y. A. Gluzband, A. J. Grodzinsky, K. B. King, J. H. Kimura, and K. E. Kuettner, "Chondrocytes synthesize a mechanically functional matrix in agarose gel," *Proc. Orth. Res. Soc.*, **15**, 86 (1990).
26. M. A. Moses, J. Sudhalter, and R. Langer, "Identification of an inhibitor of neovascularization from cartilage," *Science*, **248**, 1408–1410 (1990).

Received December 17, 1991
Accepted May 8, 1992

Functional Arteries Grown in Vitro

L. E. Niklason,[1*] J. Gao,[2] W. M. Abbott,[3] K. K. Hirschi,[5] S. Houser,[4] R. Marini,[6] R. Langer[7]

A tissue engineering approach was developed to produce arbitrary lengths of vascular graft material from smooth muscle and endothelial cells that were derived from a biopsy of vascular tissue. Bovine vessels cultured under pulsatile conditions had rupture strengths greater than 2000 millimeters of mercury, suture retention strengths of up to 90 grams, and collagen contents of up to 50 percent. Cultured vessels also showed contractile responses to pharmacological agents and contained smooth muscle cells that displayed markers of differentiation such as calponin and myosin heavy chains. Tissue-engineered arteries were implanted in miniature swine, with patency documented up to 24 days by digital angiography.

Atherosclerotic vascular disease, in the form of coronary artery and peripheral vascular disease, is the largest cause of mortality in the United States (*1*). Surgical mainstays of therapy for affected vessels less than 6 mm in diameter include bypass grafting with autologous veins or arteries (*2*); however, adequate tissue for bypass conduits is lacking in many patients. Artificial materials, when used to bypass arteries that are less than 6 mm in diameter, have thrombosis rates greater than 40% after 6 months (*3*). Although novel approaches for producing small-caliber arterial grafts have been developed, problems with mechanical properties (*4*) or the utilization of neonatal cells (*5*) have heretofore prevented clinical implementation.

We report here the development of tech-

[1]Departments of Anesthesia and Biomedical Engineering, Duke University, Durham, NC 27710, USA. [2]Department of Biomedical Engineering, Case Western Reserve University, Cleveland, OH 44106, USA. [3]Department of Surgery, [4]Department of Pathology, Massachusetts General Hospital, Boston, MA 02114, USA. [5]Children's Nutrition Research Center, Baylor College of Medicine, Houston, TX 77030, USA. [6]Department of Comparative Medicine, [7]Department of Chemical Engineering, Massachusetts Institute of Technology, Cambridge, MA 02139, USA.

*To whom correspondence should be addressed. E-mail: nikla001@mc.duke.edu

niques to produce small-caliber autologous arteries in vitro from vascular cells grown on a biodegradable polymer matrix, by means of a pulsatile perfusion system for vessel culture. Ideal biological grafts should possess a confluent endothelium and differentiated, quiescent, smooth muscle cells (SMCs), as well as sufficient mechanical integrity and elastic moduli to allow suture retention and tolerance of systemic arterial pressures. Because vascular cells are exposed to pulsatile physical forces during most of vasculogenesis (*6*) and throughout life, we hypothesized that the development of arteries in the laboratory would be facilitated by the application of pulsatile physical stress to cultured vascular cells (*7, 8*).

The biomimetic system we used for vessel culture is composed of bioreactors containing engineered vessels assembled in a parallel flow system (Fig. 1). In initial experiments, a suspension of cultured SMCs isolated from the medial layer of bovine aorta (*9*) was pipeted onto tubular biodegradable polyglycolic acid (PGA) scaffolds that were secured in bioreactors (*10*). The surface of the PGA scaffolds was chemically modified with sodium hydroxide (*11*), which caused ester hydrolysis on the surface of the fibers, leading to increased hydrophilicity, increased adsorption of serum proteins, and improved SMC attachment. After an initial SMC seeding period of 30 min, the bioreactors were filled with medium and the SMCs were cultured under conditions of pulsatile radial stress for 8 weeks (Fig. 1). Control vessels were cultured without pulsatile radial stress under otherwise identical conditions.

After 8 weeks of culture, the gross appearance of the vessels was identical to that of native arteries. Histologic examination of pulsed vessels revealed that SMCs migrated inward to envelop PGA fragments in the vessel wall, resulting in a smooth luminal surface onto which bovine aortic endothelial cells (ECs) (*12*) could easily be seeded (Fig. 2, A and B). In contrast, nonpulsed controls exhibited no such inward SMC migration through the polymer scaffold (Fig. 2, C and D) and possessed an uneven layer of polymer fragments in the vessel lumen. Thus, vessels cultured under pulsatile conditions had a histologic appearance more similar to that of native arteries.

On the basis of these observations of vessel lumen morphology, we applied bovine aortic EC layers to vessels that had been cultured for 8 weeks under pulsatile conditions. After EC seeding, continuous perfusion of the vessel lumens was instituted for the final 3 days of culture. After 3 days, the presence of an endothelial layer on the luminal surface was confirmed by scanning electron microscopy (Fig. 2G) and by staining for von Willebrand factor and platelet endothelial cell adhesion molecule (PECAM, Fig. 2H).

The mechanical properties of native arteries rely on contractile SMCs, collagen, and elastin. Models of the wall mechanics of muscular arteries have shown that recruited bundles of collagen fibers contribute most of the incremen-

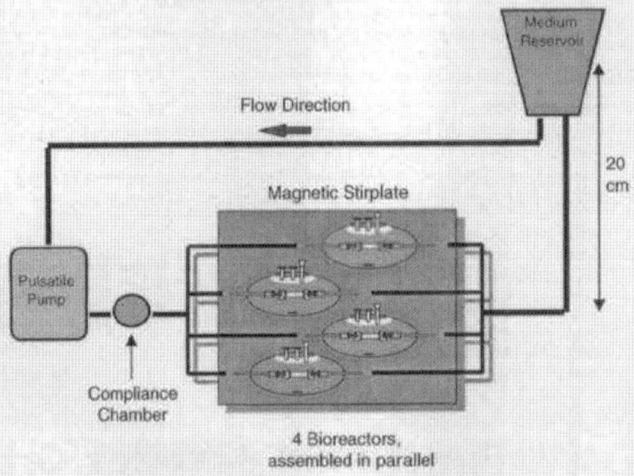

Fig 1. Biomimetic system for vessel culture. Non-cross-linked PGA mesh scaffolds were sewn into tubular form with 6-0 uncoated Dexon suture and were connected in bioreactors and sterilized by ethylene oxide. Scaffolds were seeded with 1 to 2 ml of a suspension of aortic SMCs at 5 × 10^6 cells/ml. After seeding, bioreactors were filled with Dulbecco's modified Eagle's medium (DMEM) (200 ml per vessel), supplemented with 20% fetal bovine serum, penicillin G (100 U/ml), 5 mM Hepes, ascorbic acid (0.05 mg/ml), $CuSO_4$ (3 ng/ml), proline, alanine, and glycine. The bioreactors and fluid reservoir were fitted to provide gas exchange, and buffer flowed through highly distensible silicone tubing inserted through the vessel lumen. The compliance chamber consisted of a 300-ml plastic reservoir that minimized the transmission of high-frequency vibrations to the bioreactors. Pulsatile radial stress was applied to the vessels at 165 beats per minute and 5% radial distention (strain). These conditions were chosen to approximate fetal development in large mammals. After 8 weeks, the silicone tubing was removed, and the flow of the medium was applied directly through the cultured vessel. To produce an endothelial layer, an EC suspension of 3 × 10^6 cells/ml in DMEM was injected into the lumen, and the cells were allowed to adhere for 90 min. Luminal flow rate was then gradually increased from 0.033 to 0.1 ml/s over 3 days of culture, with corresponding shear stresses at the vessel wall of 1×10^{-2} N/m^2 to 3×10^{-2} N/m^2. SMCs and ECs used for vessel culture were all below passage five at the time of seeding.

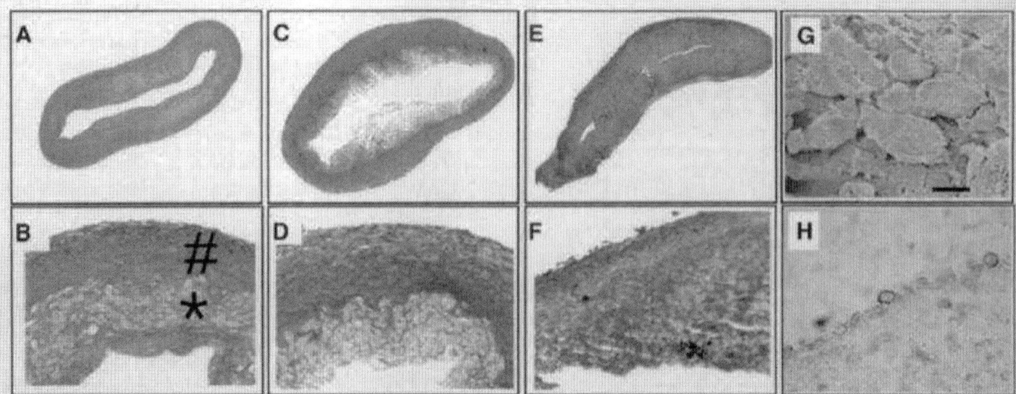

Fig 2. Histology of engineered vessels. (**A** and **B**) Pulsed vessel cultured for 8 weeks. (A) Verhoff's elastin stain [original magnification (orig. mag.) ×20]. (**B**) Masson's trichrome stain; collagen stains blue (orig. mag. ×100). Number sign indicates the dense cellular region; asterisk indicates the polymer region. (**C** and **D**) Nonpulsed vessel cultured for 8 weeks. (C) Verhoff's stain (orig. mag. ×20). (D) Masson's stain (orig. mag. 100×). (**E** and **F**) Pulsed vessel without medium supplementation. (E) Verhoff's stain (orig. mag. ×20). (F) Masson's stain (orig. mag. ×100). (**G**) Scanning electron microscopy of the endothelial cell layer in an engineered vessel; cells are less confluent and more rounded than those of the arterial endothelium. Scale bar, 10 μm. (**H**) Immunoperoxidase staining for PECAM antigen reveals an EC monolayer on the vessel lumen. Nonspecific staining of polymer fragments is apparent deeper in the vessel wall (orig. mag. ×1000).

tal elastic modulus at intraluminal pressures above 100 to 200 millimeters of mercury (mm Hg) (*13, 14*). The mechanical properties of the engineered vessels ultimately derive from the SMCs and the extracellular matrix proteins they produce, because the PGA scaffold becomes fragmented and degrades to less than 15% of its initial mass after 5 weeks in culture (*15*). To optimize extracellular matrix production and therefore mechanical properties at high pressures, we supplemented the culture medium with ascorbic acid, copper ion, and amino acids to support SMC collagen synthesis and cross-link formation (*16, 17*).

The vessels in Fig. 2, A through D, were grown in supplemented medium, whereas the vessel in Fig. 2, E and F was grown under pulsatile stress without culture medium supplementation. All vessels grown without supplementation displayed rupture strengths below 300 mm Hg. The specimen in Fig. 2E ruptured spontaneously at 270 mm Hg after 7 weeks of culture. In contrast, pulsed vessels cultured with medium supplementation ruptured at 570 ± 100 mm Hg ($n = 4$) after 5 weeks and at 2150 ± 709 mm Hg ($n = 3$) after 8 weeks. This latter rupture strength is greater than that reported for native human saphenous veins [1680 ± 307 mm Hg, (*5*)] and is well within the range suitable for arterial grafting. Thus, supplemented medium was judged essential for vessel growth.

Table 1 summarizes the effects of pulsatile stress and increasing culture time on the characteristics of vessels grown in supplemented medium. Vessel wall thickness was increased both by increasing culture time and by pulsatile culture conditions and was comparable to native vessel wall thickness after 8 weeks in culture (*18*). The vessel collagen content (*19*) observed after 8 weeks of pulsatile culture conditions was 50% and was not significantly different ($P < 0.1$) from that of native arteries (Table 1), whereas vessels grown without pulsatile stress possessed significantly ($P < 0.005$) less collagen. Applied pulsatile stress significantly increased the observed suture retention strengths (*20*) after 5 ($P < 0.001$) and 8 ($P < 0.005$) weeks of culture to values as high as 90 g, which was still less than those observed for native arteries (Table 1).

Pulsatile culture conditions did not significantly affect SMC density (*21*) ($P < 0.1$), although these SMC densities are higher than those reported in systems where collagen gels are used as the matrix for SMC growth (*22*). The difference in calculated cell density between engineered and native vessels may be partly due to the volume occupied by polymer fragments in the engineered vessel wall.

Engineered vessels displayed measurable contractions in response to serotonin, endothelin-1, and prostaglandin $F_{2\alpha}$, the magnitudes of which were not affected by pulsatile culture conditions. Representative contractile forces in response to prostaglandin $F_{2\alpha}$ ($PGF_{2\alpha}$) are shown for five vessels (Fig. 3A). The measurable forces generated by these vessels (0.17 ± 0.07 g, $n = 8$, $P < 0.001$) after 8 weeks of cell culture demonstrate retained differentiated function, which is in contrast to the frequently observed loss of contractile phenotype by SMCs in vitro (*23*).

The mechanical behavior of the intact engineered vessels was further characterized by quantifying their stress-strain relationships. Vessels were subjected to a range of static intraluminal pressures, and magnified digital images of the vessels were obtained (*24*). The wall stress (σ) and the midwall strain (ε) were calculated and plotted for six vessels (*25*) (Fig. 3, B and C). There was a linear relationship between the observed maximal incremental

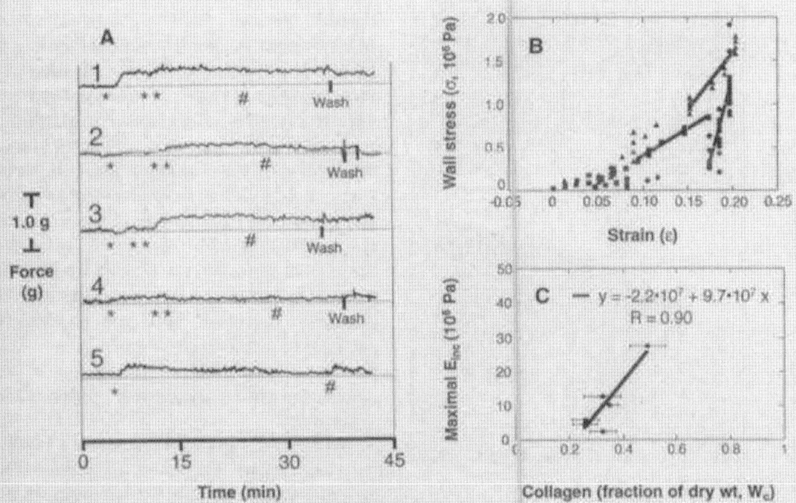

Fig 3. **(A)** Pharmacological reactivity of engineered vessels (*30*). Segments of engineered vessel 2 mm in length were mounted on tungsten wires in conventional myographs. Rabbit abdominal aorta served as a control. Vessels were maintained at 4 g of tension for 30 min before testing and were treated with indomethecin (10^{-5} M) before constriction with prostaglandin $F_{2\alpha}$ ($PGF_{2\alpha}$). Absolute magnitudes of responses were approximately 5% of control vessels and 15% of reported values for excised vein grafts (*31*). 1 and 2, nonpulsed vessels cultured for 8 weeks; 3 and 4, pulsed vessels cultured for 8 weeks; 5, explanted xenograft in which contractile function is preserved. Single asterisk indicates $PGF_{2\alpha}$, 10^{-5} M; double asterisk indicates $PGF_{2\alpha}$, 10^{-4} M; number sign indicates papaverine, 10^{-5} M. **(B)** Representative stress-strain plots for cultured vessels. External vessel diameters were extracted from the digital images, and vessel cross-sectional areas were obtained from analysis of fixed histologic sections. Circles, pulsed vessel cultured for 8 weeks; triangles, nonpulsed vessel cultured for 8 weeks; squares, pulsed vessel cultured for 5 weeks. Each data point represents a separate intraluminal pressure. At σ less than 0.5×10^6 Pa, there was incomplete recruitment of collagen fibers (that is, $f_c < 1.0$), and the value of E_{inc} was submaximal. At high stress levels, complete collagen fiber recruitment yields a maximal slope (solid lines) of the stress-strain relationship for each vessel and a maximal E_{inc}. **(C)** Maximal E_{inc} versus collagen content yields a linear relationship. The equation for the regression is shown. The value of the abscissa intercept, at $W_c = 0.23$, represents the minimum collagen content below which the mechanical properties of the engineered vessels would become negligible.

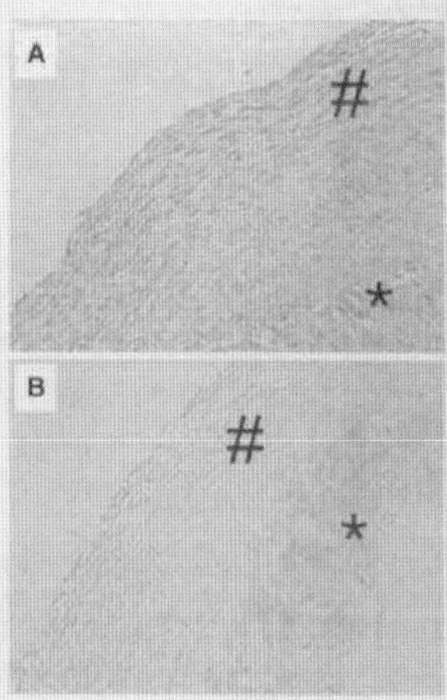

Fig. 4. Immunoperoxidase staining for myosin heavy chains. Number sign indicates the dense cellular region; asterisk indicates the polymer region. **(A)** Vessel cultured for 8 weeks with pulsatile stress. **(B)** Vessel cultured for 8 weeks without pulsatile stress. Myosin heavy chain staining was greater in the pulsed vessels and was observed in three sets of experiments. Nonspecific staining of polymer remnants is shown in (A) (orig. mag. ×200).

modulus (E_{inc}) at high wall stress and the collagen content [W_C (Fig. 3C)], with a correlation coefficient of $R = 0.9$. A linear relationship is expected and predicted from previous modeling of native vessels (*13, 14, 25*). From this relationship, we constructed a model that describes the incremental modulus of the vessel wall: $E_{inc} = (9.7 \times 10^7 \cdot W_C - 2.2 \times 10^7) \cdot f_C$, where f_C is the recruitment function for collagen fibers (*13*).

An ideal arterial conduit contains SMCs that are both contractile and quiescent with respect to proliferation, decreasing the likelihood of SMC hyperplasia and consequent luminal occlusion after implantation (*1*). Immunoperoxidase staining for markers of SMC differentiation revealed that the SMCs of vessels grown under pulsatile or static conditions express significant amounts of smooth muscle α-actin and calponin. However, the SMCs of vessels cultured under pulsatile conditions stained more intensely for myosin heavy chains, a late marker in ontological SMC development (*23*), as compared with nonpulsed vessels (Fig. 4, A and B). In all vessels, the expression of these SMC markers was very low in regions containing polymer remnants.

Assessment of mitotic rate by immunohistochemical staining for proliferating cell nuclear antigen in the vessel wall revealed 2.3 ± 1.4% positively staining SMC nuclei in the dense cellular region (Fig. 2B), as compared with 19.4 ± 9.6% in the polymer scaffold region ($P < 0.01$). Staining in the dense cellular region resembled that observed in the medial layers of native bovine and porcine vessels excised from young animals (3.7 ± 0.6%). The application of pulsatile stress did not increase the mitotic rate in the dense cellular areas. However, the percent of stained SMCs in the polymer scaffold regions of endothelialized vessels was decreased to 0.5 ± 0.7% ($n = 2$, $P < 0.01$), as compared with vessels that did not contain ECs. This result is consistent with the expected inhibition of replication of SMCs by confluent ECs (*12*) and would decrease the likelihood of SMC hyperplasia and vessel stenosis after implantation.

To assess the practical utility of these cultured arteries in vivo, we undertook initial implantation studies in four Yucatan miniature swine. Autologous arterial SMCs and ECs were cultured for each 6-month-old animal from small biopsies of the common carotid artery. For the first animal, a xenograft vessel that was cultured under pulsatile conditions from bovine arterial SMCs was seeded with autologous porcine arterial ECs 2 days before implantation. For the latter three animals, completely autologous grafts were cultured from carotid SMCs and ECs under conditions identical to those described for bovine cells, with the addition of platelet-derived growth factor BB (5 ng/ml) to the culture medium to promote SMC growth. Of the completely autologous grafts, one was grown under pulsatile conditions, and the remaining two were grown under static conditions.

All animals underwent implantation into the right saphenous artery, which is a branch of the femoral artery and is the largest artery in the distal half of the hindlimb. All animals were treated with daily aspirin (325 mg by mouth). Animals were followed for up to 4 weeks, after which time the grafts were explanted for assessment of histology and contractile function. All implanted vessels demonstrated good flow at implantation, and all grafts remained open for 2 weeks postoperatively as assessed by Doppler. The pulsed xenograft vessel was examined by digital subtraction angiography at 24 days after implantation and was patent, without evidence of stenosis or dilatation (Fig. 5A). The xenograft also had unchanged contractile responses to prostaglandin $F_{2\alpha}$ (Fig. 3A) when explanted at 4 weeks (Fig. 5B). The pulsed autologous graft remained open for 4 weeks as assessed by Doppler, although decreased flow was noted at

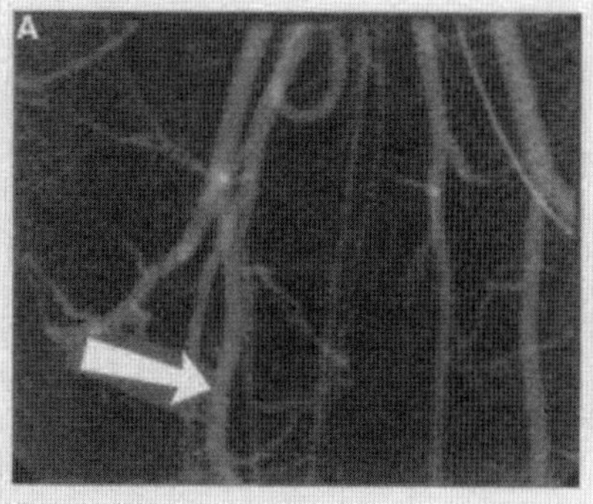

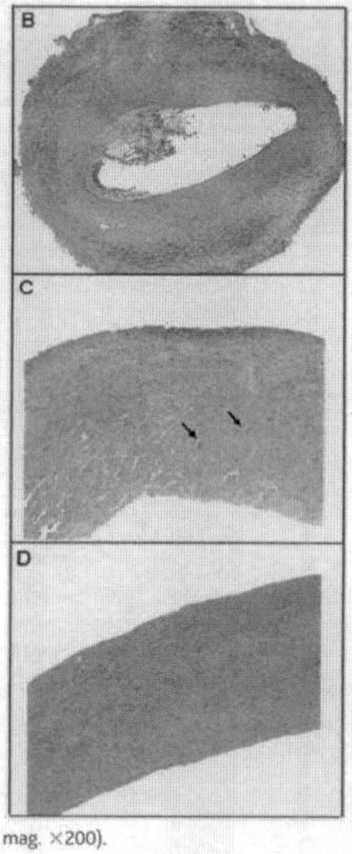

Fig. 5. **(A)** Digital subtraction angiogram of the hindlimb vessels of a Yucatan miniature pig 24 days after implantation of an engineered xenograft vessel into the right saphenous artery. The location of the graft on the angiogram was determined from skin markers over the incision. The animal with the xenograft also received cyclosporin A (12 mg per kilogram of body weight per day). The grafted vessel had a length of 3.5 cm and is indicated by the arrow. Patency of the implanted vessel was assessed weekly by Doppler ultrasound and daily by clinical examination of the hindlimb. All animal procedures were conducted in accordance with institutional guidelines for animal welfare. **(B** through **D)** show the histology of the implanted vessels. **(B)** Hematoxylin and eosin stain of explanted xenograft vessel; an inflammatory response is visible in the vessel wall (orig. mag. ×20). **(C)** Autologous porcine vessel, nonpulsed, preimplantation; arrows indicate polymer remnants (orig. mag. ×200). **(D)** Explanted autologous vessel. The wall structure is intact with minimal inflammatory response, loosely organized SMCs are present in the midsection of the vessel wall, and polymer remnants are no longer visible (orig. mag. ×200).

Table 1. Characteristics of engineered vessels.

Vessel type*	Wall thickness (cm)	Collagen (% dry weight)	Suture retention (g)	SMC density (10^8 cells/ml)
5 P	0.019 ± 0.006 ($n = 4$)	29 ± 6 ($n = 4$)	40 ± 16 ($n = 4$) ($P < 0.001$)	0.40 ± 0.16 ($n = 4$)
5 NP	0.009 ($n = 1$)	†	2.7 ± 1.2 ($n = 3$)	†
8 P	0.038 ± 0.004 ($n = 5$)	50 ± 5 ($n = 6$) ($P < 0.005$)	91 ± 26 ($n = 6$) ($P < 0.005$)	0.93 ± 0.37 ($n = 5$)
8 NP	0.023 ± 0.004 ($n = 3$)	35 ± 3 ($n = 4$)	22 ± 8 ($n = 6$)	1.19 ± 0.14 ($n = 3$)
Native	0.029 (*18*)	45 ± 9‡	273 ± 31‡	2.87 ± 0.14§

*P, pulsed; NP, nonpulsed; number represents weeks. †Data insufficient. ‡Measured from bovine muscular arteries stripped of adventitia. §Measured from bovine arteries by fluorometric assay.

the time of explantation. The mechanical strength of the explanted grafts could not be assessed accurately because of postsurgical fibroblast migration and collagen deposition at the outer surface of the engineered vessel.

In contrast, the two nonpulsed autologous grafts remained open for 3 weeks and then developed thrombosis, which may have been caused by gradual shearing loss of the luminal polymer region and of the endothelial layer due to arterial flows (Fig. 5, C and D). Histologically, the walls of the autologous explanted vessels showed highly organized structure and minimal inflammation as compared to the xenograft. For all vessels, there was no evidence of bleeding at the anastomoses or mechanical breakdown at explantation.

Important areas of future work include the effects of culture conditions on graft longevity, the stimulation of elastin in the vessel wall (*26, 27*), and the minimization of residual polymer fragments (*28*) in the engineered tissues. Clinically useful engineered vessels should approximate the patency rate of 90% at 30 days that is observed with autologous vein grafts (*29*). Although further studies are required to assess the biological function of these vessels during short-term and long-term implantation, the feasibility of culturing autologous implantable arteries and the important effects of pulsatile culture conditions have been demonstrated.

References and Notes

1. R. Ross, *Nature* **362**, 801 (1993).
2. J. V. Tu *et al.*, *N. Engl. J. Med.* **336**, 1500 (1997).
3. R. D. Sayers, S. Raptis, M. Berce, J. H. Miller, *Br. J. Surg.* **85**, 934 (1998).
4. C. B. Weinberg and E. Bell, *Science* **231**, 397 (1986).
5. N. L'Heureux, S. Paquet, R. Labbe, L. Germain, F. A. Auger, *FASEB J.* **12**, 47 (1998).
6. W. Risau and I. Flamme, *Annu. Rev. Cell Dev. Biol.* **11**, 73 (1995).
7. G. K. Owens, *Circ. Res.* **79**, 1054 (1996).
8. P. F. Davies, *Physiol. Rev.* **75**, 519 (1995).
9. R. Ross, *J. Cell. Biol.* **50**, 172 (1971).
10. L. E. Niklason and R. S. Langer, *Transplant. Immunol.* **5**, 303 (1997).
11. J. Gao, L. E. Niklason, R. S. Langer, *J. Biomed. Mater. Res.* **42**, 417 (1998).
12. A. B. Dodge, X. Lu, P. A. D'Amore, *J. Cell. Biochem.* **53**, 21 (1993).
13. R. L. Armentano *et al.*, *Am. J. Physiol.* **260**, H1870 (1991).
14. J. G. Barra *et al.*, *Circ. Res.* **73**, 1040 (1993).
15. L. E. Freed *et al.*, *Bio/Technology* **12**, 689 (1994).
16. J. M. Davidson, P. A. LuValle, O. Zoia, D. Quaglino, M. G. Giro, *J. Biol. Chem.* **272**, 345 (1997).
17. D. Tinker and R. B. Rucker, *Physiol. Rev.* **65**, 607 (1985).
18. C. G. Caro, T. J. Pedley, R. C. Schroter, W. A. Seed, *The Mechanics of the Circulation* (Oxford Univ. Press, Oxford, 1978).
19. J. F. Woessner, *Arch. Biochem. Biophys.* **93**, 440 (1961).
20. Association for the Advancement of Medical Instrumentation (AAMI), *American National Standard for Cardiovascular Implants-Vascular Prostheses* (American National Standards Institute (AAMI, Arlington, VA, 1994).
21. Y. J. Kim, R. L. Sah, J. Y. H. Doong, A. J. Grodzinsky, *Anal. Biochem.* **174**, 168 (1988).
22. T. Ziegler and R. M. Nerem, *J. Cell. Biochem.* **56**, 204 (1994).
23. G. K. Owens, *Physiol. Rev.* **75**, 487 (1995).
24. G. J. L'Italien *et al.*, *Am. J. Physiol.* **267**, H574 (1994).
25. The wall stress (σ) was calculated from (*13, 14*) as follows:

$$\sigma = 8 \cdot P \cdot (r_e r_i)^2 / [(r_e^2 - r_i^2) \cdot (r_e + r_i)^2] \quad (1)$$

where r_e is the measured external radius, r_i is the internal radius calculated from the measured external radius and cross-sectional area, and P is the intraluminal pressure. Wall strain (ε) was calculated at the midwall radius as follows:

$$\varepsilon = [(r_e + r_i)/2]/[(R_e + R_i)/2] - 1 \quad (2)$$

where R_e and R_i are the external and internal vessel radii under unstressed conditions (at $P = 25$ mm Hg) (*13, 14*). The formulation for E_{inc} is based on models previously reported to describe native vessels (elastin staining in engineered vessels was negative, so elastin is neglected) (*13, 14*):

$$E_{inc} = E_C \cdot W_C \cdot f_C + E_{SM} \cdot W_{SM} \quad (3)$$

where E_{inc} is the incremental modulus, calculated from the slope of the stress-strain curves (Fig. 3B): $E_{inc} = 0.75(d\sigma/d\varepsilon)$; E_C and E_{SM} are elastic moduli for collagen and SMCs, respectively; W_C and W_{SM} are amounts of collagen and SMCs in the vessel wall; and f_C is the recruitment function for collagen. The recruitment function describes the fraction of total collagen fibers that support wall stress at a given strain. Previously reported values of maximal $E_C \cdot W_C$ fall in the range of $1.3 \pm 0.6 \times 10^8$ Pa. From Fig. 3A, we calculate the maximal SMC-supported wall stress for any vessel to be approximately 6×10^3 Pa. This negligible SMC contribution effectively reduces the model to a function of the collagen content alone.
26. D. Y. Li *et al.*, *Nature* **393**, 276 (1998).
27. S. K. Rao, K. V. Reddy, J. R. Cohen, *Ann. N.Y. Acad. Sci.* **800**, 131 (1996).
28. H. P. Greisler, C. Gosselin, D. Ren, S. S. Kang, D. U. Kim, *Biomaterials* **17**, 329 (1996).
29. J. L. Cox, D. A. Chiasson, A. I. Gotlieb, *Progr. Cardiovasc. Dis.* **34**, 45 (1991).
30. W. Steudel *et al.*, *Circ. Res.* **81**, 34 (1997).
31. K. S. Cross *et al.*, *Br. J. Sur.* **81**, 699 (1994).

32. We thank J. Bevan for assistance with the contractile function testing and P. D'Amore for helpful discussions. This work was supported by National Heart, Lung and Blood Institute grant K08HL03492, NIH grant HL60435, and the Foundation for Anesthesia Education and Research.

11 January 1999; accepted 18 March 1999

Endothelial cells derived from human embryonic stem cells

Shulamit Levenberg*, Justin S. Golub*, Michal Amit†, Joseph Itskovitz-Eldor†, and Robert Langer*‡

*Department of Chemical Engineering, Massachusetts Institute of Technology, Cambridge, MA 02139; and †Department of Obstetrics and Gynecology, Rambam Medical Center, Faculty of Medicine, The Technion, 35245 Haifa, Israel

Contributed by Robert Langer, February 7, 2002

Human embryonic stem cells have the potential to differentiate into various cell types and, thus, may be useful as a source of cells for transplantation or tissue engineering. We describe here the differentiation steps of human embryonic stem cells into endothelial cells forming vascular-like structures. The human embryonic-derived endothelial cells were isolated by using platelet endothelial cell-adhesion molecule-1 (PECAM1) antibodies, their behavior was characterized *in vitro* and *in vivo*, and their potential in tissue engineering was examined. We show that the isolated embryonic PECAM1+ cells, grown in culture, display characteristics similar to vessel endothelium. The cells express endothelial cell markers in a pattern similar to human umbilical vein endothelial cells, their junctions are correctly organized, and they have high metabolism of acetylated low-density lipoprotein. In addition, the cells are able to differentiate and form tube-like structures when cultured on matrigel. *In vivo*, when transplanted into SCID mice, the cells appeared to form microvessels containing mouse blood cells. With further studies, these cells could provide a source of human endothelial cells that could be beneficial for potential applications such as engineering new blood vessels, endothelial cell transplantation into the heart for myocardial regeneration, and induction of angiogenesis for treatment of regional ischemia.

Human vascular endothelial cells are important for developing engineered vessels for treatment of vascular disease (1) and also may be useful for augmenting vessel growth to areas of ischemic tissue or after implantation (2). A potential source of cells for these applications are embryonic stem cells which, in murine systems, were shown to differentiate into endothelial cells that form vascular structures in a process called vasculogenesis (3). Vasculogenesis is defined as the *in situ* assembly of capillaries from undifferentiated endothelial cells, as opposed to angiogenesis, the sprouting of capillaries from preexisting blood vessels (4). The vasculogenic potential of the embryonic cells could be of use specifically in tissue engineering for the induction of tissue vascularization. Early endothelial progenitor cells isolated from differentiating mouse embryonic stem cells were shown to give rise to three blood vessel cell components, hematopoietic, endothelial, and smooth muscle cells (5). In addition, it was recently shown that endothelial progenitors and embryonic endothelial cells could differentiate into beating cardiomyocytes when cocultured with neonatal cardiomyocytes or when injected near a damaged heart area (6). It also has been shown that embryonic endothelial cells are critical for the earliest stages of liver and pancreas organogenesis (7, 8). Therefore, in addition to potential clinical applications, purified human embryonic endothelial cells could be important for studying early human development and the differentiation of embryonic stem cells into various tissues.

Differentiation of embryonic stem cells into endothelial cells and the formation of vessel structure have been studied extensively in murine embryogenesis, including maturation steps, molecular events, and growth factor involvement (9–11). However, lack of experimental cell systems has made it difficult to study these developmental processes in the human until now. Human embryonic stem cell lines (hES) recently established from the inner cell mass of human blastocytes (12) provide a unique system for studying these events in human embryonic development. hES cells have the potential to generate all embryonic cell lineages when they undergo differentiation (13). Differentiation of hES can be induced by removing the cells from their feeder layer and growing them in suspension. This differentiation in suspension results in aggregation of the cells and formation of embryoid bodies (EBs), in which successive differentiation steps occur (14).

In the present study, we show an increase in the expression of several endothelial cell-specific genes during EB differentiation, reaching a maximum between days 13–15, and development of extensive vasculature-resembling structures within the EB. We isolated human embryonic endothelial cells from day 13–15 EBs by using platelet endothelial cell-adhesion molecule-1 (PECAM1) antibodies and characterized their behavior *in vitro* and *in vivo*.

Materials and Methods

Cell Culture. hES cells (H9 clone) were grown on mouse embryonic fibroblasts (Cell Essential, Boston, MA) in knockout medium, as described (14). Tissue culture plates were coated with 0.1% gelatin (Sigma). Cultures were grown in 5% CO_2 and were routinely passaged every 5–6 days after disaggregating with 1 mg/ml collagenase type IV (GIBCO/BRL). To induce formation of EBs, hES colonies were digested by using either 1 mg/ml collagenase type IV or trypsin/EDTA (0.1%/1 mM) and transferred to Petri dishes to allow their aggregation and prevent adherence to the plate. Human EBs were grown in the same culture medium without lymphocyte inhibitory factor (LIF) and basic fibroblast growth factor (bFGF). Isolated PECAM1+ cells were grown on plates coated with 1% gelatin in endothelial growth medium, EGM-2 (Clonetics, San Diego) and passaged by using 0.025%/0.01% trypsin/EDTA (Clonetics). HUVEC cells (Clonetics) were grown on regular tissue culture plates in EGM-2 medium. For matrigel differentiation assay, cells removed from confluent culture by trypsin treatment were seeded in matrigel-coated 35-mm plates (BD Biosciences, Bedford, MA) at a concentration of 1×10^5 cells per 300 μl of culture medium. After 30 min of incubation at 37°C, 1 ml of medium was added. Cord formation was evaluated by contrast-phase microscopy 24 h or 3 days after seeding the cells.

Reverse Transcription (RT)-PCR Analysis. Total RNAs from undifferentiating hES cells and from EBs were isolated by using RNEasy Mini Kit (Qiagen, Chatsworth, CA). RT-PCR was

Abbreviations: hES, human embryonic stem cells; EB, embryoid bodies; PECAM1, platelet endothelial cell adhesion molecule-1; HUVEC, human umbilical vain endothelial; LIF, lymphocyte inhibitory factor; RT-PCR, reverse transcription–PCR; vWF, von Willebrand factor; VE-cad, vascular endothelial-cadherin.

‡To whom reprint requests should be addressed at: Department of Chemical Engineering, E25-342 Massachusetts Institute of Technology, 77 Massachusetts Avenue, Cambridge, MA 02139. E-mail: rlanger@mit.edu.

The publication costs of this article were defrayed in part by page charge payment. This article must therefore be hereby marked "*advertisement*" in accordance with 18 U.S.C. §1734 solely to indicate this fact.

performed by using Qiagen OneStep RT-PCR kit with the addition of 10 units RNase inhibitor (GIBCO/BRL) and 40 ng RNA. To ensure semi-quantitative results of the RT-PCR assays, the number of PCR cycles for each set of primers was checked to be in the linear range of the amplification. In addition, all RNA samples were adjusted to yield equal amplification of glyceraldehyde-3-phosphate dehydrogenase (GAPDH) as an internal standard. Primer sequences, reaction conditions, and optimal cycle numbers are published as supporting information on the PNAS web site, www.pnas.org.

The amplified products were separated on 1.2% agarose gels with ethidium bromide (E-Gel, Invitrogen). For each time point, mean pixel intensities of each band were measured and normalized to mean pixel intensities of the GAPDH band. The values for three experiments then were averaged and graphed with SD.

Immunochemical Reagents and Procedures. For staining, EBs were transferred to gelatin-coated cover slips with medium containing 10% (vol/vol) FBS. After attachment to the cover slips, EBs or cells grown on gelatin-coated cover slips were fixed with methanol for 5 min at −20°C or with 3% (wt/vol) paraformaldehyde at room temperature and stained for 30 min with the relevant primary antibodies: anti-human PECAM1, anti-human vinculin (Sigma), anti-human von Willebrand factor (vWF, Dako), purified monoclonal anti-N-cadherin (15), and anti-human VE-cad (7B4; ref. 16). The secondary antibodies were Cy3-labeled goat anti-mouse IgG (Jackson ImmunoResearch) and AlexaFluor goat anti-rabbit IgG (Molecular Probes). In some cases, cells or EBs also were stained with 4′,6-diamidino-2-phenylindole (DAPI) and FITC-phalloidin (Sigma). After the indirect immunolabeling, cells were mounted in Floromount-G (Southern Biotechnology) and were examined with either a conventional fluorescence microscope (Nikon) or Zeiss LSM 510 confocal microscope.

For uptake of Dill-labeled ac-LDL, PECAM1+ cells and control PECAM− cells were incubated with 10 μg/ml Dill-labeled ac-LDL (Biomedical Technologies, Stoughton, MA) for 4 h at 37°C. After incubation, cells were washed three times with PBS, fixed with 3% (wt/vol) paraformaldehyde for 30 min, and visualized with a fluorescent microscope (Nikon).

For immunohistology, tissues sections were deparaffinized, blocked with sniper (Biocare Medical, Walnut Creek, CA) for 5 min and stained by using Vector ABC or ARK (DAB) kits with 2-h incubation with the antibodies. The antibodies used include anti-human PECAM1, anti-human vWF (DAKO), and anti-human CD34 (Lab Vision, Fremont, CA).

Flow Cytometry. For isolation of PECAM1+ cells, EBs at days 13–15 were dissociated with 0.025%/0.01% trypsin/EDTA, washed with PBS containing 5% (vol/vol) FBS, and incubated for 30 min with fluorescent-labeled PECAM1 antibodies (30884X, PharMingen) on ice. Fluorescent-labeled cells were isolated by using a FACStar flow cytometry cell sorter (Becton Dickinson) and plated on 1% gelatin-coated plates with endothelial cell growth medium (Clonetics). For analysis of endothelial cell markers, PECAM1+ cells (grown in culture for six passages) and HUVEC cells were dissociated by using cell dissociation buffer (GIBCO/BRL) and washed with PBS containing 5% (vol/vol) FBS. The cells were incubated with either isotype control (mouse IgG1κ, PharMingen) or antigen-specific antibodies: PECAM1-FITC (PharMingen), CD34-FITC (AC136, Miltenyi Biotec, Auburn, CA) and Flk-1/VEGFR-2-PE (ImClone Systems, New York). Cells were analyzed live (without fixation) by using propidium iodide to exclude dead cells on a FACScan with CELLQUEST software.

Electron Microscopy. Cells seeded in Matrigel-coated 35 mm plates were fixed for 1 hr in 2.5% (wt/vol) glutaraldehyde, 3% (wt/vol) paraformaldehyde, and 7.5% (wt/vol) sucrose in 0.1 M sodium cacodylate buffer (pH 7.4) and then post-fixed in 1% (wt/vol) OsO_4 in veronal-acetate buffer for 1 h. The cells were stained *en bloc* overnight with 0.5% uranyl acetate in veronal-acetate buffer (pH 6.0), dehydrated, and embedded in Spurrs resin. Sections were cut on a Reichert Ultracut E at a thickness of 70 nm with a diamond knife. Sections were examined with a Phillips EM410 electron microscope.

Biodegradable Polymer Matrix. Porous sponges composed of poly-(L-lactic acid) (PLLA) and polylactic-glycolic acid (PLGA) were fabricated mainly as described (17). Briefly, PLLA (Polysciences) and PLGA (Boehringer Ingelheim) 1:1 were dissolved in chloroform to yield a solution of 5% (wt/vol) polymer; 0.24 ml of this solution was loaded into molds packed with 0.4 g of sodium chloride particles. The solvent was allowed to evaporate, and the sponges were subsequently immersed for 8 h in distilled water (changed every hour) to leach the salt and create an interconnected pore structure. The sponges, which had an average pore diameter of 250 μm, were cut to $0.5 \times 4 \times 5$ mm^3. Before transplantation, sponges were soaked in 70% (vol/vol) ethyl alcohol overnight and washed three times with PBS.

Transplantation into SCID Mice. PECAM1+ cells (1×10^6) were resuspended in 50 μl of 1:1 mix of culture medium and matrigel (BD Biosciences) and allowed to absorb into the PLLA/PLGA polymer sponges. After a 30-min incubation at 37°C to allow for gelation of matrigel, the cells plus scaffolds were implanted s.c. in the dorsal region of 4-week-old SCID mice (CB.17.SCID, Taconic Farms). After transplantation (7 or 14 days), the implants were retrieved, fixed overnight in 10% (vol/vol) buffered formalin at 4°C, embedded in paraffin, and sectioned for histological examination.

Results

Endothelial Gene Expression During hEB Differentiation. To isolate endothelial cells from hES cells, we first characterized their vasculogenic potential by analyzing the expression of endothelial-specific genes and proteins during hES differentiation. Spontaneous *in vitro* differentiation of H9 hES cells (12) into endothelial cells was investigated after removing undifferentiated cells from their mouse embryonic fibroblast (MEF) feeder layer and placing them into Petri dishes with culture medium lacking LIF and bFGF for induction of EB formation. At different time points during the differentiation process, the cultured hEBs were collected, and RNA was extracted for analysis of endothelial-related gene expression by using RT-PCR. The genes analyzed included endothelial cell adhesion molecules such as PECAM1/CD31 (18), vascular endothelial-cadherin (VE-cad; ref. 16) and CD34 (19); growth factor receptors such as vascular endothelial growth factor receptor-2 (VEGFR-2/Flk-1/KDR; ref. 20) and Tie-2 (21); transcription factors GATA-2 and GATA-3 (22); and AC133/CD133 a cell surface marker of vascular/hematopoietic stem and progenitor cells (23). As shown in Fig. 1, the levels of endothelial markers PECAM1, VE-cad, and CD34 increased during EB differentiation, reaching a maximum at days 13–15 and indicating a differentiation process toward endothelial cells. GATA-2 was expressed earlier and rose dramatically toward day 18. Unlike the mice system, the VEGF receptor Flk-1 is expressed in undifferentiated cells (also reported recently by Kaufman *et al.* in H1 line, ref. 24), and increased very slightly during differentiation. The tyrosine kinase receptor Tie-2 and the transcription factors GATA-3 also are expressed in hES cells, and their expression increased during the first 6 days of EB differentiation and then decreased (Fig. 1 *A* and *B*). AC133 is expressed in undifferentiated cells as well as in differentiated EB cells in a pattern similar to that of Flk-1. The levels of Oct-4, which is known to be expressed in undifferentiated cells (25),

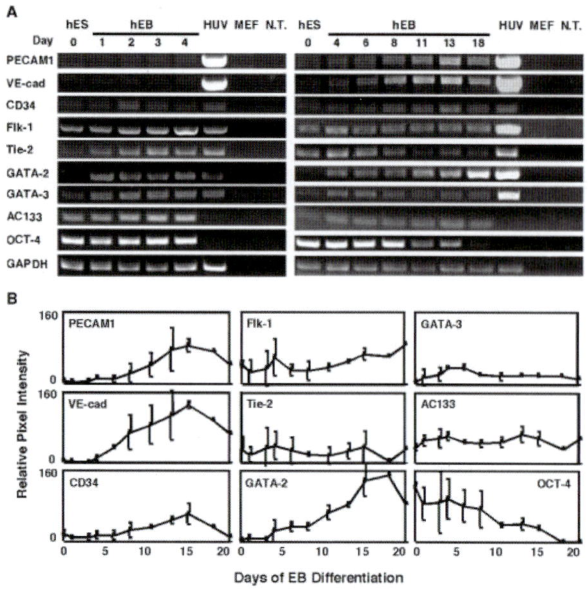

Fig. 1. Endothelial gene expression in hES-derived EBs by RT-PCR analysis. (*A*) RNA was isolated from undifferentiated hES cells and from hEBs at different time points (days) during differentiation and subjected to RT-PCR analysis. The negative controls, no template (N.T.) and MEF, and the HUVEC positive control (HUV) are shown to the right. (*B*) Quantitative analysis of gene expression. Relative pixel intensity corresponds to gene expression level; for each time point, mean pixel intensities of each band were measured and normalized to mean pixel intensities of GAPDH band. The results shown are mean values ± SD of three different experiments.

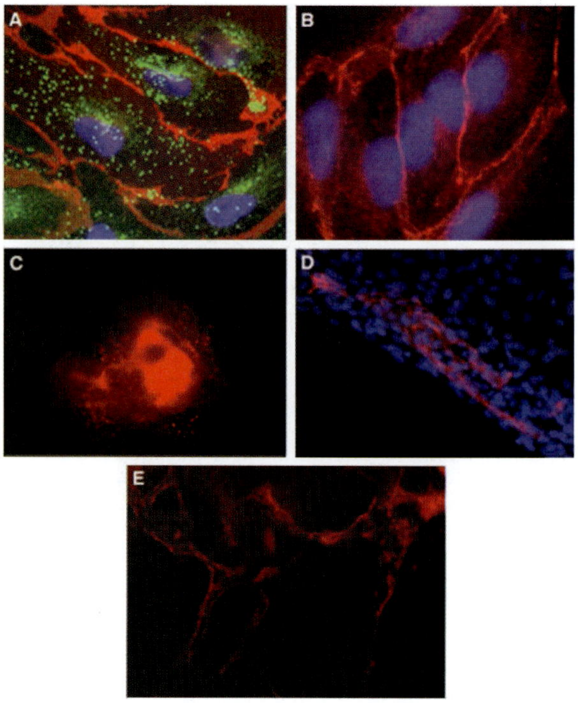

Fig. 2. Expression of endothelial cell markers in vessel-like structure within hEBs. (*A*) EBs at day 13 stained with human PECAM1 antibodies (red), vWF antibodies (green), and DAPI for nuclear staining (blue). PECAM1 is organized at cell–cell junctions, whereas vWF is found in organelles in the cytoplasm. (*B*) EB cells stained with human VE-cadherin antibodies (red) and DAPI (blue) (magnification, ×1,000). (*C*) Low magnification (×100) of EB stained with PECAM1 antibodies. (*D*) Areas of PECAM1+ cells (red) within part of an EB, organized in elongated clusters. Cell nuclei stained with DAPI (blue) (magnification, ×400). (*E*) Channels forming PECAM1+ cells within a 13-day-old EB (magnification, ×200).

served as a control. Oct-4 expression shows the undifferentiated stage of the cells at day 0 as it is expressed in the cells in high levels. Oct-4 expression subsequently goes down, indicating that the differentiation process is proceeding in the EBs. HUVEC cells were used as a positive control for the expression of the various human endothelial genes. The MEF feeder layer cells were used as a negative control and did not express any of the human-specific genes examined. These data demonstrate an increase in expression of several endothelial cell genes during EB differentiation, reaching a maximum at days 13–15 (Fig. 1 *A* and *B*). Some genes were expressed in the undifferentiated cells in either high levels (Flk-1, AC133, Tie-2) or low levels (GATA-3, CD34), and others became notable after EB formation and differentiation (PECAM1, VE-cad, GATA-2; Fig. 1 *A* and *B*).

Formation of Vessel-Like Structure in Differentiating hEBs. Analysis of endothelial-specific protein expression in day-13 EBs indicated that all EBs had defined cell areas expressing PECAM1 (Fig. 2*C*). Further analysis of PECAM1+ cells, with various endothelial specific proteins, indicated that these cells are endothelial-like-expressing PECAM1 and VE-cad adhesion molecules at cell–cell adhesion sites and vWF in large granules dispersed throughout the cytoplasm (Fig. 2 *A* and *B*). Within these EBs, the endothelial cells were not found as single cells but in groups organized in specific channel-like structures (Fig. 2 *D* and *E*), showing that hES cells cultivated as EBs spontaneously differentiate to endothelial cells and blood vessel-like structures.

To study this vascularization-like process further, EBs at different time points were stained with PECAM1 antibodies and analyzed with confocal microscopy. Fig. 3 demonstrates that the capillary area increased during subsequent maturation steps up to day 13. On day 4, PECAM1+ cells were observed in a low percentage of the EBs and concentrated in small cell clusters (Fig. 3*A*). From day 6 onward, some sprouting of endothelial structures that resembled capillaries became evident (Fig. 3*B*). From day 10 onward, 100% of EBs contained extended areas of network-like capillary structures (Fig. 3*C*). The positive area was larger at day 13, and the network structure became more complex (Fig. 3*D*). The time course of cell differentiation and the development of extensive vasculature-resembling structures within the EB correlates with the RT-PCR analysis that demonstrates the subsequent increase in RNA levels of the endothelial genes PECAM1, VECAD, CD34, reaching a maximum between day 13–15 (Fig. 1).

Endothelial Cells Derived from hEBs. Based on the analysis of endothelial gene and protein expression, we determined the method and time point at which to isolate human embryonic endothelial cells. We decided to use antibodies against PECAM1 for the isolation, as PECAM1 has been shown as the definitive marker for mice embryonic endothelial cells (26), and, in human EBs, is expressed in vessel-like structures in correlation with VE-cad and vWF expression (Figs. 2 and 3), suggesting that it could serve as a marker for human embryonic-endothelial cells as well. EBs at day 13 were dissociated, stained with fluorescent-labeled anti-PECAM1 antibodies, and the PECAM1+ cells (2%) were sorted by using flow cytometry (Fig. 4*A*). To confirm an endothelial-like phenotype of PECAM1+ cells grown in culture, we assayed them for the expression of endothelial cell markers. Isolated PECAM1+ cells (after several passages in

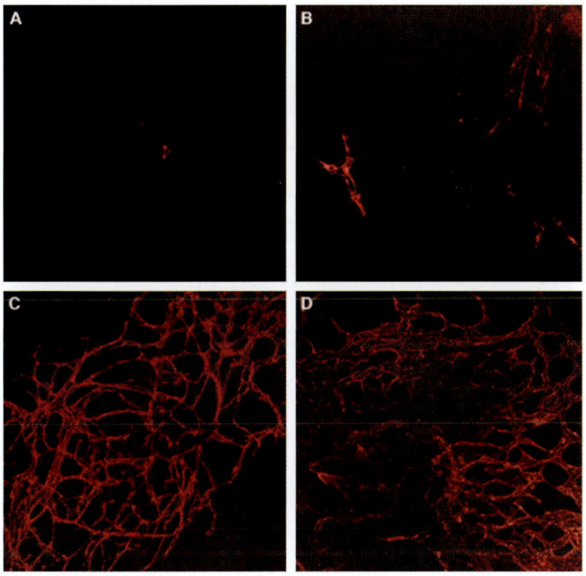

Fig. 3. Confocal microscopy of EBs stained for PECAM1 showing three-dimensional network formations, vascular-like channels. (*A*) 4-day-old EB. (*B*) 6-day-old EB. (*C*) 10-day-old EB. (*D*) 13-day-old EB. Notice the intensive and complicated vascular network developed at day 10 of 13-day-old EB. (magnification, ×100).

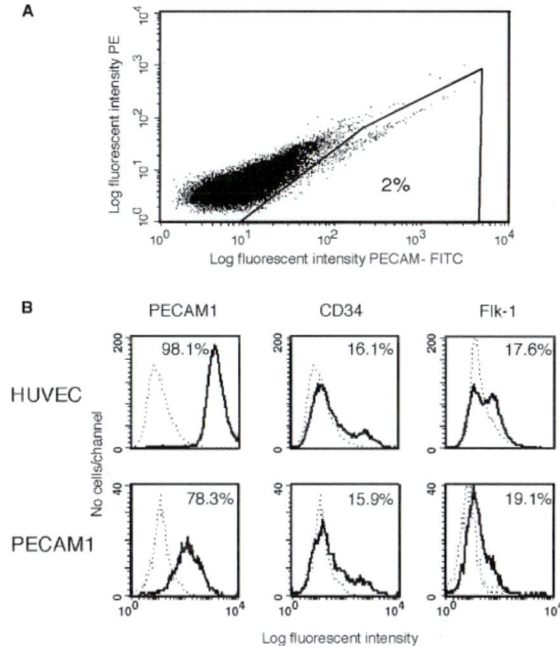

Fig. 4. Isolation of endothelial cells from human embryoid bodies by using fluorescent-labeled anti-PECAM1 antibodies and analysis of the sorted cells. (*A*) EBs at day 13 were dissociated and incubated with PECAM1 antibodies. Fluorescent-labeled cells were isolated by using a flow cytometry cell sorter. (*B*) Flow cytometric analysis of endothelial cell markers in PECAM1+ cells grown in culture for six passages and HUVEC cells. The cells were dissociated and incubated with either isotype control (dashed lines) or antigen-specific antibodies, as indicated (solid lines). Percent positive cells are shown.

culture) and HUVEC cells were incubated with fluorescent-labeled antibodies and analyzed by fluorescence-activated cell sorter (FACS, Becton Dickinson). Fig. 4*B* shows that the expression profile of CD34 and Flk-1 in isolated PECAM1+ cells is similar to the HUVEC cells. Expression of PECAM1 also is comparable but with higher expression in the HUVEC cells (98%) compared with PECAM1+ isolated cells (78%). In addition to FACS analysis, we studied the distribution of adhesion molecules by immunofluorescence microscopy. PECAM1+ cells appear to present a correct organization of endothelial junctions (27); N-cadherin and the endothelium-specific VE-cadherin are distributed at adherent-type junctions (Fig. 5 *C* and *D*), a class of cell adhesions characterized by their interaction with the actin microfilament system. Actin stress fibers are found throughout the cells and end in both the cell–cell adherence junctions and focal contacts, as seen by double staining with vinculin (Fig. 5*E*). The tight-junction component, PECAM1, is distributed at the intercellular clefts, and the endothelial marker vWF is highly expressed in the cytoplasm (Fig. 5 *A* and *B*).

Take-up of ac-LDL has been used to characterize endothelial cells (28). To evaluate whether embryonic-derived PECAM1+ cells are able to incorporate ac-LDL, cells were incubated with Dill-ac-LDL and subsequently examined by fluorescence microscopy. As shown in Fig. 5*F*, embryonic-derived PECAM1+ cells were brightly fluorescent, whereas the fluorescence intensity of PECAM1 cells was at background levels.

The characteristics of human embryonic PECAM1+ cells also were assessed by culture in matrigel, an extracellular matrix basement membrane that can be used to promote differentiation of endothelial cells (29). When PECAM1+ cells were cultured on matrigel, they were able to spontaneously reorganize in cord-like structures when maintained in culture for several days (Fig. 5 *G* and *H*). Electron microscopy analysis of the cord cross section indicated that the cords have a lumen (Fig. 5 *I* and *J*), suggesting that the cells have the capacity to differentiate and form tube-like structures under suitable conditions.

Transplantation of PECAM+ Cells into SCID Mice. To analyze the therapeutic potential of hES-derived endothelial cells, we studied their behavior *in vivo*. The cells were seeded on highly porous PLLA/PLGA biodegradable polymer scaffolds, commonly used as scaffolds for tissue engineering (30). Sponges seeded with embryonic-derived PECAM+ cells were implanted in the s.c. tissue of SCID mice. At the time of implant retrieval (up to 14 days), no signs of infection were detected, and inflammation was minimal. Implants maintained in mice for at least 7 days became encapsulated by fibrous connective tissue that was permeated by mouse blood vessels. Histological examination with antibodies that are human specific and do not react with mice microvessels show microvessels that are immunoreactive with human PECAM1 and CD34 (Fig. 6*A*–*E*). Some of these human-positive vessels had mouse blood cells in their lumen, suggesting that microvessels had formed and anastomosed with the mouse vasculature, becoming functional blood-carrying microvessels.

Discussion

This study indicates that human ES cells, when induced to form EBs, can spontaneously differentiate into the endothelial lineage, ultimately forming vascular structures. Our data demonstrate an increase in expression of several endothelial cell genes during EB differentiation, reaching a maximum at days 13–15. Some genes were expressed in undifferentiated cells in either high levels (Flk-1, AC133, Tie-2) or lower levels (GATA-3, CD34), and others (PECAM1, VE-cad, GATA-2) became notable after EB formation and differentiation. In the mouse, these genes are not expressed in ES (or expressed in very low levels that disappear by day 1 as EBs are formed, e.g., PECAM1 and Tie-2) and start to appear only around day 3 and later. (Flk-1 at

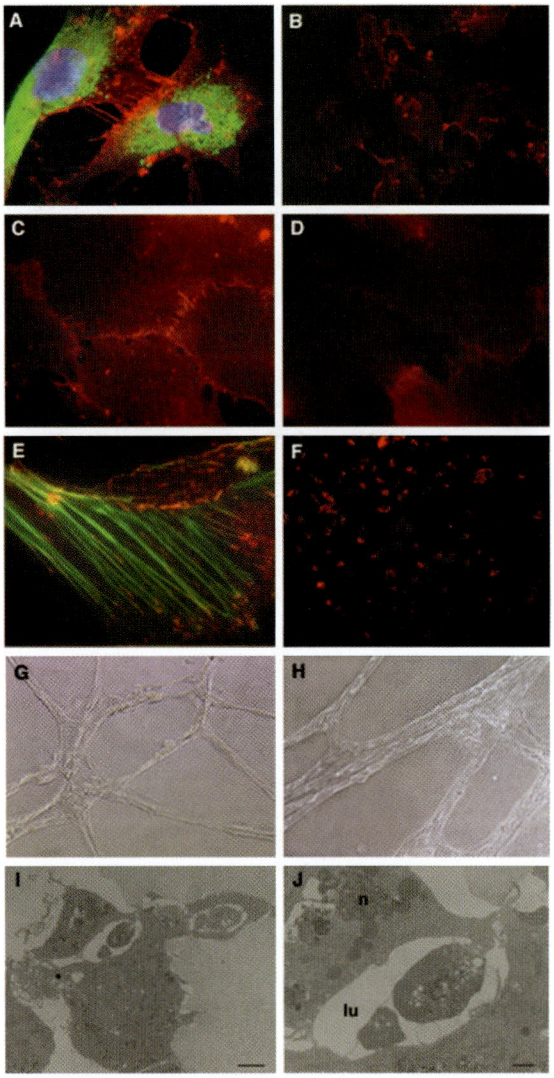

Fig. 5. Characterization of hES-derived endothelial cells grown in culture. (*A*) Immunofluorescence staining of PECAM1 (red) at cell–cell junctions and vWF (green) in the cytoplasm. The nuclei are stained with DAPI (blue). (*B*) Cells stained for PECAM1. (*C*) N-cadherin and (*D*) VE-cadherin staining in cell–cell adherent junctions. (*E*) Double staining for vinculin (red) and actin (green). Vinculin is found in both focal contacts and cell–cell adherent junctions where it associates with actin stress fiber ends (magnification: *A* and *C–E*, ×1,000; *B*, ×200). (*F*) Uptake of DilI-labeled ac-LDL by PECAM1+ cells. (*G* and *H*) Cords formation by PECAM1+ cells 24 h (*G*) or 3 days (*H*) after seeding the cells in matrigel (magnification: *G*, ×100; *H*, ×200). (*I*) Electron microscopy of the cord cross-section showing lumen formation (bar = 2 μm) and (*J*) higher magnification of the lumen (lu) area showing cell–cell interactions closing the lumen and the nucleus (n) of one cell (bar = 8 μm).

day 2–3; PECAM and Tie–2 at day 4; VE-cad and Tie-1 at day 5; refs. 3 and 31). Mouse and human ES cells differ in morphology, population doubling time, and growth factor requirements. Undifferentiated mouse cells, for example, can be maintained as undifferentiated cells independently of feeder layer if growth factors such as LIF are added to the media (32). However, human cells will differentiate if grown without feeder layer or feeder layer-conditioned medium even in the presence of LIF (12, 33). Thus, different mechanisms of response to LIF and LIF

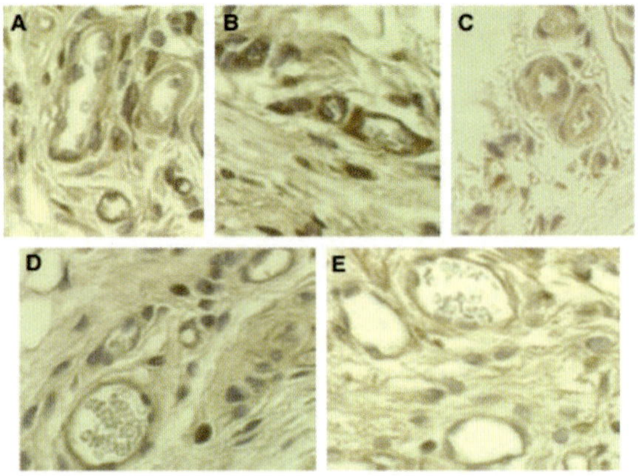

Fig. 6. Transplantation of embryonic endothelial cells (PECAM1+) in SCID mice. PECAM1+ cells were seeded onto PLLA/PLGA polymer scaffolds as described in *Materials and Methods*. The cells plus scaffolds were implanted s.c. in the dorsal region of 4-week-old SCID mice. (*A–C*) Immunoperoxidase (brown) staining of 7-day implants with anti-human PECAM1 antibodies and (*D* and *E*) of 14-day implants with anti-human CD34 antibodies, showing microvessels that are immunoreactive with these human-specific antibodies. Some of these human-positive microvessels have mouse blood cells in their lumen (magnification, ×400).

removal between mouse and human ES cells may affect differences in gene expression patterns observed in the transition from the undifferentiated to the differentiated stage of the cells. It is possible that gene expression of endothelial markers in undifferentiated hES cells can be related to the "escape" of some cells from the undifferentiated stage of hES cells or to different basic definitions (regarding gene expression) of the undifferentiated state of hES cells kept in current culture conditions. However, because of significant differences between early human and mouse development and differences in the behavior of mouse and human ES cells, the pattern of human endothelial gene expression shown here might indicate differences in the mechanism of embryonic endothelial differentiation. Our preliminary results indicate that growth factor cocktails (including bFGF and VEGF) known to induce endothelial differentiation in mice EBs do not have the same effect on hEBs (data not shown), pointing again to potential differences between the two systems in the molecular mechanism underlying this process, and emphasizing the need to analyze developmental processes by using human systems.

The assembly of developing vascular-like structures could be observed during EBs outgrowth, as soon as the cells acquired the set of endothelial markers. The data also indicate that the capillary area in the EBs increased during subsequent maturation steps up to day 13, starting from cell clusters that later sprout into capillary-like structures and eventually become organized in a network-like arrangement. The increase in RNA expression of PECAM1, CD34, VE-cad, and GATA-2 genes during EB differentiation correlates with the observed increase in the number of endothelial cells expressing PECAM1 and VE-cad proteins, as demonstrated by antibody staining of differentiating EBs (Figs. 2 and 3). Antibody staining also indicates that at different stages of maturation, most markers seem to be coexpressed by the same cells. These data demonstrate that human ES cells, similar to mice ES cells, can spontaneously differentiate and organize *in vitro* in vessel-like structures in a pattern that resembles embryonic vascularization.

In the present study, we isolated and maintained in culture endothelial cells derived from hES cells differentiated *in vitro*. PECAM1 antibodies have been used in the mouse system for isolation of endothelial cells (34). The procedure used to obtain a pure culture of endothelial cells from hES is relatively simple and allowed us to culture large numbers of human embryonic endothelial cells that can be grown in culture without losing endothelial characteristics.

The isolation of human embryonic endothelial cells has potential therapeutic implications, including cell transplantation for repair of ischemic tissues and tissue engineering of vascular grafts. Recently, several studies demonstrated the use of adult endothelial progenitor cells for such applications (2, 35). It would be important to examine the efficiency of embryonic stem cells in such applications, because ES cells can be expanded without apparent limit (36), and ES cell-derived cells could be created in virtually unlimited amounts and available for potential clinical use. Another area in which embryonic endothelial cells can potentially be beneficial is the engineering of complex tissues where vascularization of the regenerating tissue is essential. Several approaches are now being developed to vascularize engineered tissue *in vitro* before transplantation (37, 38). Vascularization *in vitro* is important to enable cell viability during tissue growth, induce structural organization, and promote integration upon implantation. Because the formation of the first capillaries takes place mostly during early stages of embryogenesis when endothelial cells are generated from precursor cells (39), isolated human embryonic endothelial cells or progenitor cells can be critical for such applications. Future studies should include further analysis of the molecular mechanism underlying endothelial differentiation and vasculogenesis during hEB differentiation and investigation of growth factor involvement in this process. In addition, efforts to identify and isolate early embryonic progenitors are important to provide tools for elucidating regulatory elements in vasculogenesis and to potentially shed light on vasculogenic and angiogenic mechanisms involved in pathological situations affecting the vascular system.

We thank Nicki Watson for excellent assistance in preparation of the confocal and electron microscopy images, Robert Marini and Massachusetts Institute of Technology Division of Comparative Medicine for help with animal procedures and histology, and the Massachusetts Institute of Technology Flow Cytometry Unit. We thank Erin Lavik for her help in scaffold preparations. S.L. wishes to thank the European Molecular Biology Organization for a postdoctoral-fellowship.

1. Niklason, L. E., Gao, J., Abbott, W. M., Hirschi, K. K., Houser, S., Marini, R. & Langer, R. (1999) *Science* **284,** 489–493.
2. Kawamoto, A., Gwon, H. C., Iwaguro, H., Yamaguchi, J. I., Uchida, S., Masuda, H., Silver, M., Ma, H., Kearney, M., Isner, J. M. & Asahara, T. (2001) *Circulation* **103,** 634–637.
3. Vittet, D., Prandini, M. H., Berthier, R., Schweitzer, A., Martin-Sisteron, H., Uzan, G. & Dejana, E. (1996) *Blood* **88,** 3424–3431.
4. Yancopoulos, G. D., Klagsbrun, M. & Folkman, J. (1998) *Cell* **93,** 661–664.
5. Yamashita, J., Itoh, H., Hirashima, M., Ogawa, M., Nishikawa, S., Yurugi, T., Naito, M. & Nakao, K. (2000) *Nature (London)* **408,** 92–96.
6. Condorelli, G., Borello, U., De Angelis, L., Latronico, M., Sirabella, D., Coletta, M., Galli, R., Balconi, G., Follenzi, A., Frati, G., *et al.* (2001) *Proc. Natl. Acad. Sci. USA* **98,** 10733–10738.
7. Matsumoto, K., Yoshitomi, H., Rossant, J. & Zaret, K. S. (2001) *Science* **294,** 559–563.
8. Lammert, E., Cleaver, O. & Melton, D. (2001) *Science* **294,** 564–567.
9. Keller, G. M. (1995) *Curr. Opin. Cell Biol.* **7,** 862–869.
10. Darland, D. C. & D'Amore, P. A. (2001) *Curr. Top. Dev. Biol.* **52,** 107–149.
11. Hirashima, M., Kataoka, H., Nishikawa, S. & Matsuyoshi, N. (1999) *Blood* **93,** 1253–1263.
12. Thomson, J. A., Itskovitz-Eldor, J., Shapiro, S. S., Waknitz, M. A., Swiergiel, J. J., Marshall, V. S. & Jones, J. M. (1998) *Science* **282,** 1145–1147.
13. Odorico, J. S., Kaufman, D. S. & Thomson, J. A. (2001) *Stem Cells* **19,** 193–204.
14. Itskovitz-Eldor, J., Schuldiner, M., Karsenti, D., Eden, A., Yanuka, O., Amit, M., Soreq, H. & Benvenisty, N. (2000) *Mol. Med.* **6,** 88–95.
15. Volk, T. & Geiger, B. (1986) *J. Cell Biol.* **103,** 1451–1464.
16. Lampugnani, M. G., Resnati, M., Raiteri, M., Pigott, R., Pisacane, A., Houen, G., Ruco, L. P. & Dejana, E. (1992) *J. Cell Biol.* **118,** 1511–1522.
17. Mooney, D. J., Sano, K., Kaufmann, P. M., Majahod, K., Schloo, B., Vacanti, J. P. & Langer, R. (1997) *J. Biomed. Mater. Res.* **37,** 413–420.
18. DeLisser, H. M., Newman, P. J. & Albelda, S. M. (1994) *Immunol. Today* **15,** 490–495.
19. Young, P. E., Baumhueter, S. & Lasky, L. A. (1995) *Blood* **85,** 96–105.
20. Yamaguchi, T. P., Dumont, D. J., Conlon, R. A., Breitman, M. L. & Rossant, J. (1993) *Development (Cambridge)* **118,** 489–498.
21. Sato, T. N., Qin, Y., Kozak, C. A. & Audus, K. L. (1993) *Proc. Natl. Acad. Sci. USA* **90,** 9355–9358.
22. Weiss, M. J. & Orkin, S. H. (1995) *Exp. Hematol. (Charlottesville, VA)* **23,** 99–107.
23. Peichev, M., Naiyer, A. J., Pereira, D., Zhu, Z., Lane, W. J., Williams, M., Oz, M. C., Hicklin, D. J., Witte, L., Moore, M. A. & Rafii, S. (2000) *Blood* **95,** 952–958.
24. Kaufman, D. S., Hanson, E. T., Lewis, R. L., Auerbach, R. & Thomson, J. A. (2001) *Proc. Natl. Acad. Sci. USA* **98,** 10716–10721.
25. Yeom, Y. I., Fuhrmann, G., Ovitt, C. E., Brehm, A., Ohbo, K., Gross, M., Hubner, K. & Scholer, H. R. (1996) *Development (Cambridge)* **122,** 881–894.
26. Vecchi, A., Garlanda, C., Lampugnani, M. G., Resnati, M., Matteucci, C., Stoppacciaro, A., Schnurch, H., Risau, W., Ruco, L., Mantovani, A., *et al.* (1994) *Eur. J. Cell Biol.* **63,** 247–254.
27. Ayalon, O., Sabanai, H., Lampugnani, M. G., Dejana, E. & Geiger, B. (1994) *J. Cell Biol.* **126,** 247–258.
28. Voyta, J. C., Via, D. P., Butterfield, C. E. & Zetter, B. R. (1984) *J. Cell Biol.* **99,** 2034–2040.
29. Grant, D. S., Lelkes, P. I., Fukuda, K. & Kleinman, H. K. (1991) *In Vitro Cell. Dev. Biol. Anim.* **27A,** 327–336.
30. Putnam, A. J. & Mooney, D. J. (1996) *Nat. Med.* **2,** 824–826.
31. Robertson, S. M., Kennedy, M., Shannon, J. M. & Keller, G. (2000) *Development (Cambridge)* **127,** 2447–2459.
32. Matsuda, T., Nakamura, T., Nakao, K., Arai, T., Katsuki, M., Heike, T. & Yokota, T. (1999) *EMBO J.* **18,** 4261–4269.
33. Xu, C., Inokuma, M. S., Denham, J., Golds, K., Kundu, P., Gold, J. D. & Carpenter, M. K. (2001) *Nat. Biotechnol.* **19,** 971–974.
34. Balconi, G., Spagnuolo, R. & Dejana, E. (2000) *Arterioscler. Thromb. Vasc. Biol.* **20,** 1443–1451.
35. Kaushal, S., Amiel, G. E., Guleserian, K. J., Shapira, O. M., Perry, T., Sutherland, F. W., Rabkin, E., Moran, A. M., Schoen, F. J., Atala, A., *et al.* (2001) *Nat. Med.* **7,** 1035–1040.
36. Amit, M., Carpenter, M. K., Inokuma, M. S., Chiu, C. P., Harris, C. P., Waknitz, M. A., Itskovitz-Eldor, J. & Thomson, J. A. (2000) *Dev. Biol.* **227,** 271–278.
37. Black, A. F., Berthod, F., L'Heureux, N., Germain, L. & Auger, F. A. (1998) *FASEB. J.* **12,** 1331–1340.
38. Kaihara, S., Borenstein, J., Koka, R., Lalan, S., Ochoa, E. R., Ravens, M., Pien, H., Cunningham, B. & Vacanti, J. P. (2000) *Tissue Eng.* **6,** 105–117.
39. Flamme, I., Frolich, T. & Risau, W. (1997) *J. Cell Physiol.* **173,** 206–210.

Functional recovery following traumatic spinal cord injury mediated by a unique polymer scaffold seeded with neural stem cells

Yang D. Teng*[†], Erin B. Lavik[†‡], Xianlu Qu*, Kook I. Park*, Jitka Ourednik*, David Zurakowski[§], Robert Langer[¶‖**], and Evan Y. Snyder*[‖**]

*Departments of Neurology, Pediatrics, and Neurosurgery, Harvard Medical School, Boston, MA 02115; ‡Harvard–MIT Division of Health Sciences and Technology, and ¶Department of Chemical Engineering, Massachusetts Institute of Technology, Cambridge, MA 02139; and §Departments of Orthopaedic Surgery and Biostatistics, Children's Hospital, Boston, MA 02115

Contributed by Robert Langer, December 17, 2001

To better direct repair following spinal cord injury (SCI), we designed an implant modeled after the intact spinal cord consisting of a multicomponent polymer scaffold seeded with neural stem cells. Implantation of the scaffold–neural stem cells unit into an adult rat hemisection model of SCI promoted long-term improvement in function (persistent for 1 year in some animals) relative to a lesion-control group. At 70 days postinjury, animals implanted with scaffold-plus-cells exhibited coordinated, weight-bearing hindlimb stepping. Histology and immunocytochemical analysis suggested that this recovery might be attributable partly to a reduction in tissue loss from secondary injury processes as well as in diminished glial scarring. Tract tracing demonstrated corticospinal tract fibers passing through the injury epicenter to the caudal cord, a phenomenon not present in untreated groups. Together with evidence of enhanced local GAP-43 expression not seen in controls, these findings suggest a possible regeneration component. These results may suggest a new approach to SCI and, more broadly, may serve as a prototype for multidisciplinary strategies against complex neurological problems.

Each year approximately 10,000 Americans sustain spinal cord injuries (SCI). Functional deficits following SCI result from damage to or severance of axons, loss of neurons and glia, and demyelination. SCI pathology is determined not only by the initial mechanical insult, but also by secondary processes including ischemia, anoxia, free-radical formation, and excitotoxicity that occur over hours and days following injury (1). Central nervous system (CNS) axonal regeneration appears to be impeded partly by myelin-associated inhibitors (1, 2), loss in adult neurons of an intrinsic ability to overcome inhibitory cues (3), and formation of a post-lesion scar barrier (4). However, if axons can traverse the injury site, there is evidence that they may regrow in unscarred regions (5, 6). Furthermore, preservation of even a small percentage of tissue significantly enhances functional recovery (7). Although there have been encouraging reports of deficit reduction (8–10) and axonal regrowth by blocking inhibitory molecules and antagonizing secondary injury mechanisms (11); myelin replacement by stem (12), Schwann (13), and olfactory ensheathing cells (14, 15); delivery of growth factors (16–18) and small molecules (19); and implantation of fetal tissue (19) and scaffolds (20–23), as yet there is no practical treatment for SCI.

We sought a conceptually new approach that simulated the architecture of the healthy spinal cord through an implant consisting of a polymer scaffold seeded with neural stem cells (NSCs) modeled after the gray and white matter of the intact cord (Fig. 1*a*). The scaffold's inner portion emulated the gray matter via a porous polymer layer designed to be seeded with NSCs (24) for cellular replacement as well as trophic support (Fig. 1 *b* and *c*).[††] The outer portion emulated the white matter with long, axially oriented pores for axonal guidance and radial porosity to allow fluid transport while inhibiting ingrowth of scar tissue (Fig. 1*d*). The scaffold is designed to be tailored to fit into a variety of cavities. In this study, the scaffold was tailored to fit into the cavity created by a midline lateral hemisection in the spinal cord of an adult rat. We hypothesized that this multicomponent, degradable, synthetic scaffold of specified architecture seeded with NSCs might mitigate secondary injury, impede glial scar formation, direct cell replacement, facilitate regeneration, and guide repair to create a more physiologically relevant structure.

Methods

Scaffold Fabrication. Both the inner and outer scaffolds were fabricated from a blend of 50:50 poly(lactic-*co*-glycolic acid) (PLGA) (75%, number average molecular weight, M_n, ≈ 40,000) and a block copolymer of poly(lactic-*co*-glycolic acid)-polylysine (25%, PLGA block $M_n \approx 30{,}000$, polylysine block $M_n \approx 2000$). The PLGA was chosen to achieve a degradation rate of about 30–60 days, and the functionalized polymer was incorporated to provide sites for possible surface modification. The inner scaffold was made using a salt-leaching process: a 5% (wt/vol) solution of the polymer blend in chloroform was cast over salt with a diameter range of 250–500 μm, and the solvent was allowed to evaporate. The salt was then leached in water. The oriented outer scaffold was fabricated using a solid–liquid phase separation technique (25) in the following way: A 5% (wt/vol) solution of the polymers was filtered and injected into silicone tubes which were lowered at a rate of 2.6×10^{-4} m/s into an ethanol/dry ice bath. Once frozen, the dioxane was sublimated using a shelf temperature-controlled freeze drier (VirTis). The scaffolds were then removed, trimmed, assembled, and stored in a vacuum desiccator until use.

Cell Maintenance and Seeding. Murine NSCs (clone C17.2) were maintained in serum-containing medium (24). Scaffolds were soaked in 70% ethanol for 24 h, rinsed three times in PBS, and seeded on an orbital shaker with 5×10^5 cells/ml at 37°C in a humidified 5% CO_2/air incubator. The medium was changed the next day, and the implants were incubated for 4 more days before implantation. For the cells-alone control, 10 μl of a 10^7 cells/ml

Abbreviations: SCI, spinal cord injuries; NSC, neural stem cell; BDA, biotinylated dextran amine; p.i., postinjury; BBB, Basso–Beattie–Bresnahan; NF, neurofilament; GFAP, glial fibrillary acidic protein.

†Y.D.T. and E.B.L. contributed equally to this work.

‖R.L. and E.Y.S. contributed equally to this work.

**To whom reprint requests may be addressed. esnyder1@caregroup.harvard.edu (for E.Y.S.) or rlanger@mit.edu (for R.L.).

††Lu, P., Jones, L. L., Park, K. I., Snyder, E. Y. & Tuszynski, M. (2000) *Soc. Neurosci. Abstr.* **26,** 332.

The publication costs of this article were defrayed in part by page charge payment. This article must therefore be hereby marked "*advertisement*" in accordance with 18 U.S.C. §1734 solely to indicate this fact.

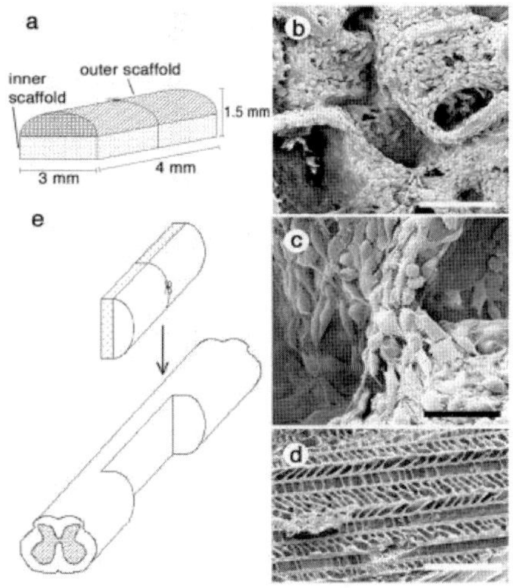

Fig. 1. (*a*) Schematic of the scaffold design showing the inner and outer scaffolds. (*b* and *c*) Inner scaffolds seeded with NSCs. (Scale bars: 200 μm and 50 μm, respectively.) The outer section of the scaffold was created by means of a solid–liquid phase separation technique that produced long, axially oriented pores for axonal guidance as well as radial pores to allow fluid transport and inhibit the ingrowth of scar tissue (*d*; scale bar, 100 μm). (*e*) Schematic of surgical insertion of the implant into the spinal cord.

suspension was injected into the cavity following the hemisection. No gelfoam or other means were used to maintain the cells in the cavity because this was thought to introduce a confounding variable.

Surgical Procedures and Animal Care. Fifty adult female Sprague–Dawley rats were used. Animals were anesthetized with a 4% chloral hydrate solution (360 mg/kg i.p.). Using a dissecting microscope, a laminectomy was made at the 9th-to-10th thoracic (T_9-T_{10}) spinal vertebrae, followed by a lateral hemisection at the T_9-T_{10} level by creating a 4-mm-long longitudinal cut along the midline of the cord with a No. 11 surgical blade, followed by lateral cuts at the rostral and caudal ends and removal of the tissue by aspiration. The surgical blade was repeatedly scraped along the ventral surface of the vertebral canal, followed by aspiration to remove any residual fibers at the lesion site. After gelfoam-triggered hemostasis occurred, an independent blinded observer confirmed the adequacy of the length and breadth of the lesion. Only at that time was the surgeon informed of the treatment (previously prepared) to be administered to the lesion. The lesion was affirmed *a priori* to be similar across all experimental groups and animals.

Either the full treatment, consisting of insertion of the NSC-seeded scaffold ("scaffold plus cells," $n = 13$; Fig. 1*e*), or one of three control treatments was performed: (*a*) polymer implant without NSCs ("scaffold alone," $n = 11$; (*b*) NSCs suspended in medium ("cells alone," $n = 12$); or (*c*) hemisection alone ("lesion control," $n = 12$).

Surgeries were performed in a randomized block design. The surgeries for the implant plus controls were performed on the same day to minimize differences between groups arising from any refinement in surgical technique during the study, and the order was varied each day to reduce surgical bias. Hemisections were alternated between the right and left sides to further reduce bias. Following either the full or control treatment, the musculature was sutured, skin closed, and the animal recovered in a clean cage on a heating pad. Ringer's lactate solution (10 ml) was given daily for 7 days post-op and bladders were evacuated twice daily until reflex bladder function was established.

Because immunosuppressive agents such as cyclosporin A have been shown to be neuroprotective on their own (26), these experiments were performed without such neuroimmunophilins to avoid this confounding variable. Donor cells were nevertheless present at the end of the study.

A separate group of scaffold plus cells animals underwent the same procedures as above and were maintained for one year.

All procedures were reviewed and approved by the Animal Care and Use Committee of Children's Hospital, Boston.

Behavioral Studies and Statistical Evaluation. One day postinjury (p.i.) and weekly thereafter, behavioral analysis was performed by two observers blinded to the treatments, using a battery of tests to rate open-field locomotion by the Basso–Beattie–Bresnahan (BBB) scale (27); the ability to maintain body position on an inclined plane; and contact righting reflex and spinal cord-mediated hindlimb withdrawal to pain (28). Behavioral tests were performed at the same time each day and graded by the same blinded observers. For the inclined plane test, the highest degree of inclination was defined as being that at which the animal could maintain its position for 5 s on two separate trials (29). Contact righting and limb withdrawal to pain were performed according to Gale *et al.* (28). One animal died following surgery because of an overdose of chloral hydrate. The remaining 49 animals survived and exhibited no signs of autophagia throughout the course of the study. One animal was excluded from the analysis because it showed incomplete paralysis at 1 day post-op. The Kolmogorov–Smirnov goodness-of-fit test was used to determine the normality (Gaussian-shaped distribution) of the data. The BBB data showed no significant departure from a normal distribution for all time points beyond day 1, so parametric methods were used including repeated-measures analysis of variance (ANOVA) and Bonferroni *post hoc* analysis for multiple group comparisons to determine the statistical significance of the results. The inclined plane data showed a significant departure from a normal distribution and was analyzed with the Kruskal–Wallis H-test, followed by the Mann–Whitney *U* test to identify specific group differences when the Kruskal–Wallis test showed significance ($P < 0.05$).

Tissue Processing for Histology. Following perfusion, spinal cords were carefully dissected, postfixed overnight in 4% paraformaldehyde, dehydrated overnight at 4°C in 30% sucrose, and frozen in isopentane. Two centimeter blocks of the thoracic region of the cords including injury epicenters were embedded in OCT and cryosectioned. Longitudinal sections for hematoxylin/eosin (H&E) were obtained from a depth of 1.6 mm into the cord (from the dorsal side). Neurofilament sections were 80 μm dorsal to H&E sections.

BDA Tracing. BDA (biotinylated dextran amine; Molecular Probes, 10% wt/vol solution) was injected at ten points in the sensory motor cortex contralateral to the side of the spinal cord hemisection at or beyond 70 days p.i. (30). Animals were maintained for 4 more weeks then killed. Following cryosectioning of the tissue, BDA was revealed histochemically on mounted sections by using a Vector Elite ABC kit (Vector Laboratories) followed by development with DAB and light counterstaining with neutral red.

Immunocytochemistry. Immunocytochemistry was performed on 20-μm mounted sections employing the following antibodies: anti-neurofilament 200 (NF) (Sigma; 1:200); anti-glial fibrillary acidic protein (GFAP; Sigma; 1:80); M2 (a mouse-specific marker; gift from Carl Lagenaur, University of Pittsburgh; 1:5); anti-GAP-43 (gift from Larry Benowitz, Children's Hospital, Boston; 1:50,000). Briefly, sections were incubated with primary antibody for 1 h at 37°C followed by the appropriate secondary antibody for 1 h at 37°C. Blocking solution consisted of 3%

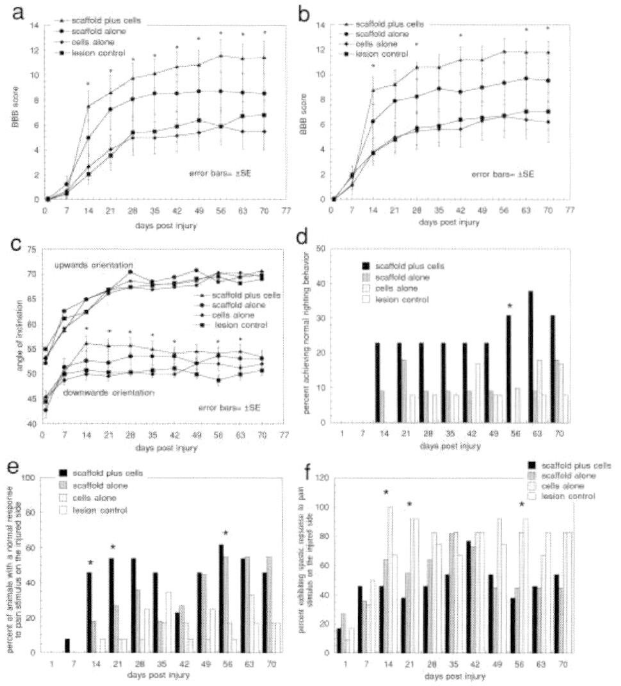

Fig. 2. (*a*) BBB open-field walking scores for the four groups on the ipsilateral, lesioned side. Hindlimbs were assessed independently to determine the degree of asymmetry. The *rate* of improvement for the scaffold plus cells group was significantly greater than the rate for the cells-alone ($P < 0.001$) and lesion-control groups ($P = 0.004$; two-way repeated measures of ANOVA). For absolute score attained, the scaffold plus cells group showed significant improvement in open-field locomotion compared with the cells-alone ($P = 0.006$) and lesion-control groups ($P = 0.007$) for all time points from 14 days p.i. on (ANOVA, Bonferroni *post hoc* analysis). (*b*) BBB open-field walking scores for the contralateral unlesioned hindlimb. There is only a 1–2 point difference between the lesioned (see *a*) and unlesioned sides, indicating that the walking behavior of the groups was relatively symmetric overall and that both sides were impacted by this unilateral lesion. Individual animals did, however, exhibit a degree of asymmetry in their stepping with a twisted trunk position leading to the rotation of both hindlimbs to one side of their bodies. (*c*) Inclined plane results. When facing upwards, animals in the four groups are statistically similar for all time points (Kruskal–Wallis test) indicating that they were generally similar in strength. When facing downwards, the scaffold plus cells group performance showed statistically significant improvement on all days from day 14 on (except days 49 and 70) compared with the cells-alone and lesion-control groups (Kruskal–Wallis test). Although the data showed significant departure from the Gaussian distribution, parametric analysis revealed similar results to nonparametric analysis, and the results are represented in the figure by their means for visual clarity. (*d*) Righting reflex results. The scaffold plus cells group exhibited a significantly higher percentage of normal righting on day 56 as compared with the lesion-control (Pearson χ^2 test of independence). (*e*) Percent of animals with a normal pain reflex in response to toe pinching was significantly higher in the plus cells group on the days marked by the asterisks (Pearson χ^2 test). (*f*) Percent of animals exhibiting a spastic response to the same toe pinching stimuli.

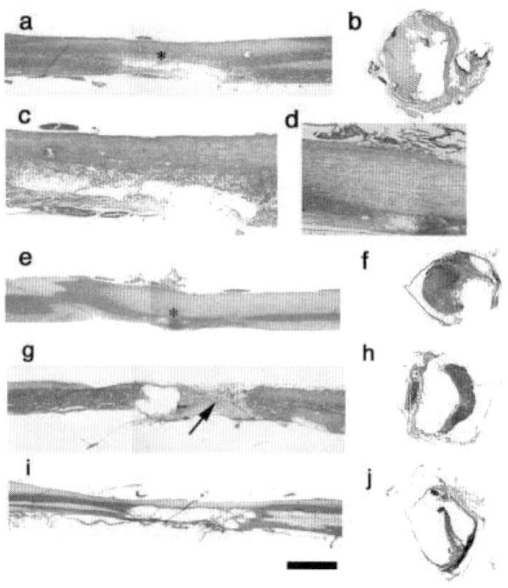

Fig. 3. Hematoxylin/eosin (H&E) staining of representative 20-μm-thick longitudinal and transverse sections from each group. (*a* and *b*) Scaffold plus cells sections in longitudinal and cross section, respectively. Area with an asterisk in *a* is shown at higher power in *c*. (*a*) A section from the cord of the scaffold plus cells animal from Movie 1, which is published as supporting information. Although there is some degeneration of gray matter, there is preserved tissue near the injury epicenter as well as rostral and caudal to the injury. There is only limited scar tissue at the injury epicenter on the lesioned side (*c*). (*d–f*) Scaffold-alone sections. Area with an asterisk in *e* is shown at higher power in *d*. (*g* and *h*) Cells-alone sections. Note the degradation of tissue both rostral and caudal to the injury epicenter as well as the extent of scar tissue at the injury epicenter on both the lesioned and unlesioned sides (arrow). (*i* and *j*) Lesion-control sections. In *i* there is a small degree of white matter on the unlesioned side, but the lesioned region has expanded almost 2-fold from the original lesion size, a 4-mm lateral hemisection. In *j* there is an incomplete rim of peripheral white matter on the unlesioned side. (Scale bar, 2.5 mm.)

serum in which the secondary antibody was raised and 5% BSA (Vector Laboratories).

Results and Discussion

Fig. 2 *a* and *b* shows the mean BBB open-field walking scores (27) for the four experimental groups on their ipsilateral, "lesioned" (Fig. 2*a*), and contralateral, "unlesioned" (Fig. 2*b*) sides. The open-field locomotion of representative animals from the scaffold plus cells and lesion-control groups who performed near the mean of their groups at day 70 is shown in Movie 1, which is published as supporting information on the PNAS web site, www.pnas.org. Open-field locomotion on both the ipsilateral and contralateral sides in the scaffold plus cells group was significantly enhanced in the subchronic phase (up to 28 days p.i.) with respect to both *rate* and *absolute level* of improvement achieved; this enhanced performance was maintained into the chronic phase of recovery (70 days p.i.). Performance of the unlesioned side mirrored the lesioned side both in absolute mean values achieved and statistical significance. Impairment of the contralateral side likely reflects the severity of the hemisection (see surgical procedures for details) in which 4 mm of tissue was removed, leading to severance of a number of blood vessels that may have resulted in ischemia- and hypoxia-related secondary injury on the unlesioned side. Similar phenomena have been reported (31). Both hindlimbs were affected, and recovery of both was improved by the scaffold plus cells intervention. The scaffold plus cells group, in the late chronic stage (i.e., 56 days p.i. and later), achieved a stabilized mean BBB score of approximately 12, which corresponds to frequent-to-consistent weight-supported plantar-stepping and occasional forelimb-hindlimb coordination. In contrast, the cells-alone and lesion-control groups only achieved stabilized mean BBB scores of $\approx$6 and 7, respectively. Encouragingly, the scaffold-alone group also achieved a mean BBB score of $\approx$9, which indicates the onset of weight-supported locomotion, suggesting that even the scaffold alone is capable of leading to improvement in functional recovery. Defining the onset of significant walking behavior as achieving a BBB score of 10 or greater at 70 days p.i., 69% of the

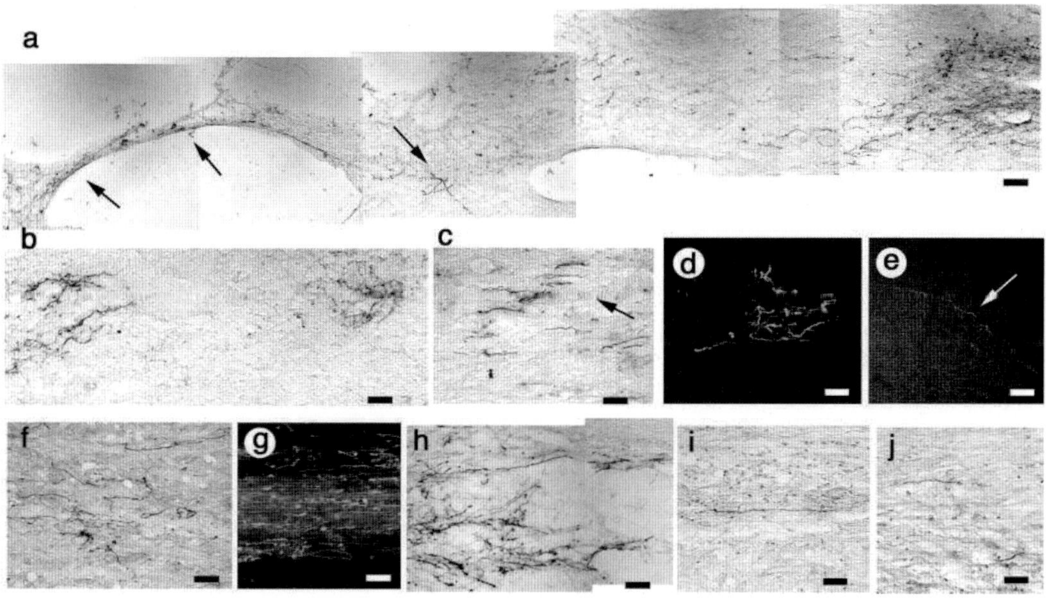

Fig. 4. (*a*) BDA tracing of the corticospinal tract in a scaffold-alone animal. BDA was injected into the sensorimotor cortex contralateral to the spinal cord injury at or beyond 70 days p.i. such that BDA-positive fibers would be visible at the site of, or caudal to, the spinal cord lesion if corticospinal tract fibers were present and passing through the injury site. Arrows indicate where the tracer passes into the lesioned side of the cord. Note the tortuous path of the fibers (arrows; see text). The rostral end is to the right, and the lesion cavity is on the bottom. (*b*) Positive BDA tracing caudal to the injury in the same cord as in *a*. (*c*) GAP-43, a marker for regenerating axons, is positive in the same cord as seen in *a* and *b* just rostral to the injury. Likewise, scaffold plus cells cords exhibit BDA tracing with the same tortuousity immediately (*d*) rostral to as well as (*e*) caudal to the injury, and (*f*) a large number of GAP-43-positive axons just rostral to the injury. The cells-alone (*g*) and lesion-control cords (*h*) exhibit BDA staining rostral to the injury but not caudal, and there are minimal fibers positive for GAP-43 (*i*, cells alone; *j*, lesion control; scale bars in *a–b*, *d–e*, and *g–h*, 10 μm; in *c*, *f*, and *i–j*, 20 μm).

scaffold plus cells groups, 54% of the scaffold alone, but only 17% of the cells-alone and 33% of lesion-control groups attained BBB score of at least 10.

For inclined plane performance, animals were tested facing both upwards and downwards (Fig. 2*c*). Whereas performance in the upward-facing orientation reflects forelimb strength, which should be unaffected by this injury in healthy animals, performance facing downwards measures coordinated hindlimb motor function (28). In the upwards orientation, all groups exhibited statistically similar performances, suggesting that the animals had similar forelimb strength. However, the scaffold plus cells group showed statistically significant improvement with respect to the cells-alone and lesion-control groups in the downward-facing orientation on all days beyond day 7 (except days 49 and 70). Overall, the inclined plane results mirror the BBB scoring, suggesting that the scaffold plus cells intervention is associated with significant long-term improvement of motor function.

Spinal cord mediated reflexes also showed a trend toward improvement following the scaffold plus cells intervention as compared with cells alone and lesion controls. In particular, a higher percentage of animals in the scaffold plus cells group exhibited normal righting behavior on all days beyond day 14 (Fig. 2*d*).

The scaffold plus cells group demonstrated more normal sensory responses. Fig. 2*e* shows the percentage of animals after SCI in each group with a normal pain withdrawal reflex induced by a controlled brief pinch of the toes (28). At 70 days p.i., approximately 50% of the animals regained a normal pain withdrawal reflex (scored as a 2 and representing an immediate and controlled withdrawal reflex in response to the stimuli) on the lesioned side in the groups treated with scaffolds plus cells and scaffold alone (Fig. 2*e*). In contrast, only approximately 15% of the cells-alone and lesion-control groups regained a normal pain withdrawal reflex on the lesioned side. Fig. 2*f* shows the percentage of animals with a spastic pain withdrawal reflex, which is defined as an immediate but hyperactive response in which the limb demonstrates spasms upon stimulation. Overall, the scaffold plus cells and scaffold-alone interventions appear to lead to a more favorable outcome of the normal, nonspastic response, suggesting more normal spinal cord-mediated reflexes.

Histology, tract tracing, and immunocytochemistry were performed on all cords to help elucidate possible mechanisms leading to the observed functional recovery. The images of the lesion area presented in Fig. 3 are representative of each experimental group at 120 days p.i., and were obtained from animals whose BBB scores placed them near the mean of their respective groups. The sections are all from the same depth at 1.6 mm into the cord. The cross sections were obtained through the injury epicenter, which was determined by the least amount of contralateral tissue sparing in each cord. The scaffold plus cells (Fig. 3 *a–c*) and scaffold-alone cords (Fig. 3 *d–f*) at 120 days p.i. had very limited residual lesions that barely crossed the midline of the cord (Fig. 3 *a* and *c*). In some cases, there appeared to be newly formed tissue within the injury epicenter (Fig. 3 *d* and *e*). The cells-alone (Figs. 3 *g* and *h*) and lesion-control cords (Fig. 3 *i* and *j*) demonstrated marked residual lesions 5–6 mm in length and encompassing the majority of the width of the cord. Furthermore, both the lesion-control and cells-alone cords exhibited atrophy both rostral and caudal to the injury. In some of the cells-alone and lesion-control cords, scar tissue and cysts were seen. Fig. 3*g*, a longitudinal section of a typical cells-alone cord, demonstrates the epidural scar tissue (marked by the arrow) as well as the cyst formation seen in some of the cells-alone and lesion-control cords, but not seen in the scaffold plus cells or scaffold alone cords.

It is hypothesized that the scaffold may impede scarring and subsequent cyst formation. The mechanism for this is not yet known but may be correlated with inhibition of cellular ingrowth by the outer scaffold. *In vitro*, the outer scaffold appears to inhibit the ingrowth of a variety of cell types, including fibroblasts, into the porous scaffold for at least 2 weeks (data not shown). The absence of scar formation is correlated with a higher degree of tissue preservation at the injury epicenter as well as preservation of tissue rostral and caudal to the epicenter. The absence of scar tissue may account for the diminished atrophy in the spinal cords of the scaffold-containing groups as compared with the cells alone and lesion controls. The expansion of the lesion has been correlated with secondary injury processes (1), and the size of the lesion appears to correlate with the degree of functional deficit (32).

Although tissue preservation appears to be the primary factor accounting for the improved functional recovery in the scaffold plus cells group, it may be augmented by possible regenerative processes. BDA tracing of the corticospinal tract (30), coupled with immunocytochemistry for GAP-43, a protein up-regulated in the growth cones of regenerating axons (33), suggests that there may be a component of regeneration occurring. Fig. 4 *a* and *b* shows BDA tracing of the corticospinal tract on the lesioned side of the cord in the scaffold-alone group. There is positive tracing of fibers through the injury epicenter (Fig. 4*a*, arrows) and caudal to the injury (Fig. 4*b*). For BDA-positive corticospinal tract axons to be detectable in regions caudal to the injury epicenter, BDA, injected in the motor cortex, had to have been transported in an antegrade manner along either spared or regenerated corticospinal fibers. The tortuous path of the traced fibers as well as the multipolar nature of the portions caudal to the injury site is most consistent with morphological profiles generally seen in regenerating axons as opposed to spared fibers, which would retain their straight, nontortuous profile. There are a large number of GAP-43-positive fibers just rostral to the injury site in this cord (Fig. 4*c*, arrows), further suggesting a possible component of regeneration at day 70 p.i. Likewise, in the scaffold plus cells cords, there was BDA-positive tracing both rostral and caudal to the injury epicenter as well as GAP-43 immunopositive staining just rostral to the injury (Fig. 4 *d–f*). In contrast, the cells-alone and lesion-control cords showed positive BDA tracing rostral but not caudal to the injury epicenter (Fig. 4 *g* and h), and very few GAP-43-positive fibers (Fig. 4 *i* and *j*). BDA tracing plus GAP-43 immunostaining, though frequently used indicators of regeneration (27), are not definitive markers.

Both the scaffold plus cells and scaffold-alone cords exhibited immunopositive staining for an antibody against neurofilament (NF) within the injury epicenter (Fig. 5 *a* and *b*). In Fig. 5*b*, the region of immunopositive staining is within the space of the primary injury, suggesting that the newly formed tissue contains neurons or neuronal elements. Fig. 5*c* shows immunohistology of a scaffold plus cells cord just rostral to the injury epicenter employing an NF-antibody (green) plus mouse-specific antibody, M2 (red), to identify donor murine NSC-derived cells within the rat host. Donor-derived cells were identified in both cells-alone and scaffold plus cells cords rostral to the injury epicenter (Fig. 5*c*, red immunostaining). Confocal microscopy of the expression pattern of these markers, however, suggests that the donor cells were not immunopositive for NF, suggesting that the neurons were actually of host origin. In comparison to the scaffold-containing groups, the lesion-control cords showed much less immunopositive staining for NF throughout the thoracic region of the cord (Fig. 5*d*). Although the donor NSCs do not appear to be immunopositive for NF, they are also not largely immunopositive for GFAP (Fig. 5*e*), suggesting that they are not contributing to the glial scar. The majority of the donor cells are, in fact, immunopositive for nestin (Fig. 5*f*), suggesting that they remained largely undifferentiated. It is possible that the major contribution of the NSCs was trophic support rather than cellular replacement.[††] Although there were substantially fewer donor-derived cells in the cords receiving NSCs alone, the NSCs appear to have behaved in a similar way, remaining largely nestin-positive.

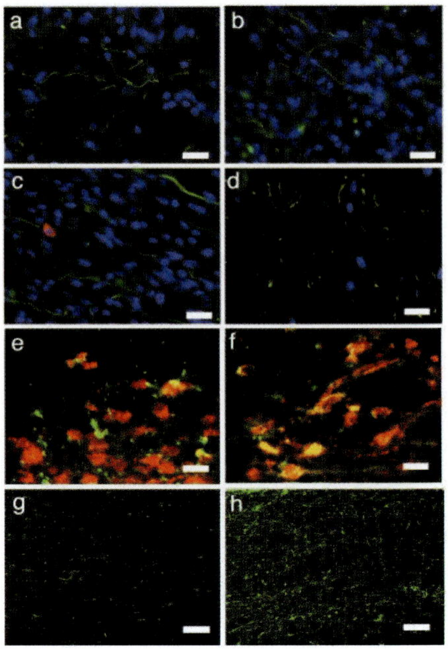

Fig. 5. All longitudinal sections. (*a*) A scaffold plus cells section at the injury epicenter immunostained for NF (green) and counterstained with DAPI (blue). (*b*) Scaffold-alone section at the injury epicenter exhibiting NF immunoreactivity (green) within the region excised by the initial lesion, suggesting that the new tissue contains neurons or neuronal elements (blue, DAPI). (*c*) Double immunostaining for NF (green) and the mouse-specific antigen M2 used to identify donor cells (red) in a scaffold plus cells cord and DAPI (blue). The cord has extensive NF immunopositivity, but there is no clear colocalization of NF and M2, suggesting that the NF+ cells were of host-origin. (*d*) There are few NF+ cells near the injury epicenter in the lesion-control cord. (*e*) Scaffold plus cells section doubled-immunostained for GFAP (green) and M2 (red). The majority of donor-derived cells (red) are not double-labeled for GFAP, suggesting that they do not contribute to the astroglial scar. (*f*) Most donor cells (stained with the M2 antibody, green in this figure) are double-labeled for nestin (red in this figure), as seen by confocal microscopy. GFAP immunostaining (green) in the spared tissue contralateral to the hemisection in a scaffold plus cells cord (*g*) compared with a lesion-control cord (*h*). There is a far greater incidence of astrocytes at the injury epicenter of the lesion-control compared with the scaffold-alone cord. (Scale bars: *a–f*, 20 μm; *g–h*, 10 μm.)

The functional results suggest that the scaffold played an important role in recovery. Immunostaining for GFAP in the scaffold-alone and lesion controls (Fig. 5 *h* and *g*) exhibited a smaller population of astrocytes in the injury epicenter of the former than of the latter. The lesion-control and cells-alone cords, in general, exhibited larger populations of GFAP-positive astrocytes at their lesion epicenters than the scaffold plus cells and scaffold alone cords. Although controversial, astrocytes have been correlated with formation of the inhibitory glial scar. Thus, a reduction in astrocyte ingrowth following injury is believed to be desirable. Although the mechanism by which the scaffold may augment recovery is not known, *in vitro* evidence, as noted above, suggests that the outer scaffold inhibits ingrowth of a variety of cell types. Because the scaffold is completely degradable, this inhibition is temporary. However, the outer scaffold may reduce inflammatory cell infiltration that occurs shortly after SCI, in turn reducing astrogliosis (34).

The mechanisms by which the scaffold and NSCs augment functional recovery need to be explored further. A detailed anatomical study at multiple time points correlated with the current behavioral results (including quantitative analysis of cell types involved and the number of myelinated and unmyelinated axons) may help to elucidate some mechanisms. In addition, because the corticospinal tract in the rat is not mainly involved in locomotor function, regenerative activity in more relevant but less accessible tracts (e.g., rubrospinal tracts) needs to be examined in future studies. Electrophysiology may allow a better understanding of the nature of the neuroprotected tissue along with any regenerated neural tissue at the injury site. The use of additional animal models may also be helpful.

In summary, implantation of the scaffold with NSCs significantly improves functional recovery as compared with cells alone and lesion controls. Indeed, scaffold plus cell animals maintained for 1 year p.i. continue to show enduring recovery with respect to BBB scoring ($n = 8$). Although the scaffold alone appears to play a significant role in functional recovery following hemisection by reducing epidural and glial scar formation, the combination of the NSCs and scaffold is critical for the greatest recovery.

Satisfactory outcomes have not been achieved to date in treating SCI by means of a single approach. Our results suggest that, with further study, the concept of polymer scaffolds seeded with NSCs may contribute to a new multifaceted approach for the treatment of SCI. More broadly, this work may provide a prototype for other multidisciplinary strategies for addressing complex neurological maladies.

We thank Umberto DeGirolami, Marion Slaney, Ali Shafarae, Karen Fu, Daniel Kohane, Larry Benowitz, and Carl Lagenaur for their assistance. This work was supported in part by Project ALS, National Institutes of Health 1-R21-NS41999-01, International Institute for Research in Paraplegia, and the A-T Children's Project, as well as a Harvard School of Dental Medicine/Massachusetts Institute of Technology National Institute of Dental and Craniofacial Research Training Grant in Biomaterials (to E.B.L.). R.L. holds equity in GMP Companies, which holds certain patents in this area.

1. Beattie, M. S., Farooqui, A. A. & Bresnahan, J. C. (2000) *J. Neurotrauma* **17,** 915–925.
2. Chen, M. S., Huber, A. B., Haar, M. E. v. d., Frank, M., Schnell, L., Spillmann, A. A., Christ, F. & Schwab, M. E. (2000) *Nature (London)* **403,** 434–439.
3. Cai, D., Qui, J., Cao, Z., McAtee, M. & Bregman, B. S. (2001) *J. Neurosci.* **21,** 4731–4739.
4. Fitch, M. T., Doller, C., Combs, C. K., Landreth, G. E. & Silver, J. (1999) *J. Neurosci.* **19,** 8182–8198.
5. Davies, S. J., Fitch, M. T., Memberg, S. P., Hall, A. K., Raisman, G. & Silver, J. (1997) *Nature (London)* **390,** 680–683.
6. Neumann, S. & Woolf, C. (1999) *Neuron* **23,** 83–91.
7. Blight, A. R. (1983) *Neuroscience* **10,** 521–543.
8. Cheng, H., Cao, Y. & Olson, L. (1996) *Science* **273,** 510–513.
9. McDonald, J. W., Liu, X.-Z., Qu, Y., Liu, S., Mickey, S. K., Turetsky, D., Gottleib, D. I. & Choi, D. W. (1999) *Nat. Med.* **5,** 1410–1412.
10. Imaizumi, T., Lankford, K. L., Burton, W. V., Fodor, W. L. & Kocsis, J. D. (2000) *Nat. Biotechnol.* **18,** 949–953.
11. Bregman, B. S., Kunkel-Bagden, E., Schnell, L., Dai, H. N., Gao, D. & Schwab, M. E. (1995) *Nature (London)* **378,** 498–501.
12. Liu, S., Qu, Y., Stewart, T. J., Howard, M. J., Chakrabortty, S., Holekamp, T. F. & McDonald, J. W. (2000) *Proc. Natl. Acad. Sci. USA* **97,** 6126–6131.
13. Xu, X., Zhang, S.-X., Li, H., Aebischer, P. & Bunge, M. (1999) *Eur. J. Neurosci.* **11,** 1723–1740.
14. Ramón-Cueto, A., Cordero, M. I., Santos-Benito, F. F. & Avila, J. (2000) *Neuron* **25,** 425–435.
15. Li, Y., Field, P. M. & Raisman, G. (1997) *Science* **277,** 2000–2002.
16. Teng, Y. D., Mocchetti, I., Taveira-DaSilva, A. M., Gillis, R. A. & Wrathall, J. R. (1999) *J. Neurosci.* **19,** 7037–7047.
17. Xu, X. M., Guénard, V., Kleitman, N., Aebischer, P. & Bunge, M. B. (1995) *Exp. Neurol.* **134,** 261–272.
18. Grill, R., Murai, K., Blesch, A., Gage, F. H. & Tuszynski, M. H. (1997) *J. Neurosci.* **17,** 5560–5572.
19. Diener, P. S. & Bregman, B. S. (1998) *J. Neurosci.* **18,** 763–778.
20. Vacanti, M., Leonard, J., Dore, B., Bonassar, L., Cao, Y., Stacheleck, S., Vacanti, J., O'Connell, F., Yu, C., Farwell, A. & Vacanti, C. (2001) *Transplant. Proc.* **33,** 592–598.
21. Woerly, S., Petrov, P., Sykova, E., Roitbak, T., Simonova, Z. & Harvey, A. (1999) *Tissue Eng.* **5,** 467–488.
22. Gautier, S. E., Oudega, M., Fragoso, M., Chapon, P., Plant, G. W., Bunge, M. B. & Parel, J.-M. (1998) *J. Biomed. Mater. Res.* **42,** 642–654.
23. Holmes, T., Lacalle, S. d., Su, X., Liu, G., Rich, A. & Zhang, S. (2000) *Proc. Natl. Acad. Sci. USA* **97,** 6728–6733.
24. Yandava, B., Billinghurst, L. & Snyder, E. (1999) *Proc. Natl. Acad. Sci. USA* **96,** 7029–7034.
25. Schurgens, C., Maquet, V., Grandfils, C., Jerome, R. & Teyssie, P. (1996) *Polymer* **37,** 1027–1038.
26. Guo, X., Dillman, J., Dawson, V. & Dawson, T. (2001) *Ann. Neurol.* **50,** 6–16.
27. Basso, D. M., Beattie, M. S. & Bresnahan, J. C. (1995) *J. Neurotrauma* **12,** 1–21.
28. Gale, K., Kerasidis, H. & Wrathall, J. (1985) *Exp. Neurol.* **88,** 123–134.
29. Rivlin, A. & Tator, C. (1977) *J. Neurosurg.* **47,** 577–581.
30. Weidner, N., Grill, R. J. & Tuszynski, M. H. (1999) *Exp. Neurol.* **160,** 40–50.
31. Dusart, I. & Schwab, M. (1994) *Eur. J. Neurosci.* **6,** 712–714.
32. Noble, L. J. & Wrathall, J. R. (1989) *Exp. Neurol.* **103,** 34–40.
33. Benowitz, L. I., Apostolides, P. J., Perrone-Bizzozero, N., Finklestein, S. P. & Zwiers, H. (1988) *J. Neurosci.* **8,** 339–352.
34. Fitch, M. T. & Silver, J. (1997) *Exp. Neurol.* **148,** 587–603.

LETTERS

nature
biotechnology

Nanoliter-scale synthesis of arrayed biomaterials and application to human embryonic stem cells

Daniel G Anderson, Shulamit Levenberg & Robert Langer

Identification of biomaterials that support appropriate cellular attachment, proliferation and gene expression patterns is critical for tissue engineering and cell therapy. Here we describe an approach for rapid, nanoliter-scale synthesis of biomaterials and characterization of their interactions with cells. We simultaneously characterize over 1,700 human embryonic stem cell–material interactions and identify a host of unexpected materials effects that offer new levels of control over human embryonic stem cell behavior.

The extracellular microenvironment plays a key role in controlling cellular behavior[1,2]. For cells growing on a surface, the properties of the surface, such as roughness and hydrophobicity, and specific cell-surface interactions can all affect cell activity[3]. The effects of the cellular substrate are also important in biomaterial-based therapies. Tissue-engineered constructs, *ex vivo* cell isolation, bioreactors and cell encapsulation require some type of interaction between cells and supporting material for growth, function and/or delivery[4]. Much research is currently focused on the development of biomaterials that provide optimal cellular substrates, including bioactive materials that incorporate ligands, DNA or growth factors[5,6]. The application of stem cells in these systems is especially difficult as it requires the additional ability to control growth and differentiation into useful cell types. The effects of biomaterials on the behavior of stem cells, and, in particular, human embryonic stem (hES) cells, have not been studied in great detail. This is due in part to the potential diversity of biomaterials, the difficulty of large-scale hES cell production and the lack of systems for rapid synthesis and testing of material-cell interactions. We have developed a high-throughput, miniaturized system to address these needs.

Design of a cell-compatible, miniaturized polymer array required that (i) a large number of materials be synthesized in nanoliter volumes, (ii) polymer spots be attached to the microarray in a manner compatible with those materials and resistant to the aqueous conditions necessary for cell-based testing, (iii) cell growth in the spaces between different polymers be inhibited to allow material effects on cells to be independent of neighboring materials and (iv) the array be in a format that allows simple, simultaneous assay of multiple cellular markers.

The first step towards microarray production required the identification of a relevant class of biomaterials. A variety of acrylate-based polymers have been used for tissue engineering, surgical glues and drug delivery[7]. There are a number of commercially available monomers, and these can be polymerized quickly using a light-activated radical initiator. Using a highly modified fluid handling system, we deposited in triplicate 576 different combinations of 25 different acrylate, diacrylate, dimethacrylate and triacrylate monomers with a radical initiator onto a layer of poly(hydroxyethyl methacrylate) (pHEMA), on top of an epoxide-coated slide. pHEMA inhibits cell growth[8], and we reasoned that a monomer deposited onto a pHEMA surface would interpenetrate and potentially become fixed in place upon polymerization. Once polymerized, the monomers formed rigid polymer spots that were firmly attached to the slide. The synthesis of polymers in arrayed form onto a conventional 25 × 75 mm glass slide allows for easy, simultaneous staining and four-color fluorescence imaging of multiple slides, each containing 1,728 individual polymer spots. Twenty such 1,728 spot polymer arrays can be synthesized in a single day (**Fig. 1**).

We tested the biomaterial microarrays for their effects on hES cell growth and differentiation. To initiate differentiation, we allowed embryoid bodies to form for 6 d before trypsinizing them and seeding them onto the arrays. The cells were incubated with retinoic acid on the arrays for 6 d. Arrays were then fixed and stained for cytokeratin 7, an intermediate filament protein found in most glandular and transitional epithelia[9]; vimentin, an intermediate filament protein common in many cells of mesenchymal origin; and DNA/nucleus with SYTO24, a fluorescent, nucleic-acid binding dye (**Fig. 2**).

In general, cell attachment and spreading to display typical cellular morphology was supported by the majority of these materials (**Fig. 2f** and **Supplementary Fig. 3**). However, certain monomers inhibited hES cell attachment or spreading, in particular, monomers 6, 18 and 21 (defined in **Fig. 1**). Under these conditions, the majority of polymers allowing for cell attachment and spreading also allowed for differentiation into cytokeratin-positive cells (**Figs. 2,3** and **Supplementary Table 1** online). This simple, one-step production of cytokeratin-positive cells might be useful for producing epithelia for tissue engineering and cell therapy. To our knowledge, a method for the production of a substantially pure population of epithelial-like cells from hES cells has not been described previously.

To examine polymer effects on other cell types, we tested the arrays with C2C12, an embryonic muscle cell line. Unlike the hES cells, C2C12 cells attached to and grew on almost all of the materials, including those containing 70% monomers 21 and 6 (**Supplementary Fig. 2** online). The mechanism underlying these cell-specific differences is unclear, but the identification of materials that selectively support the growth of

Department of Chemical Engineering, Massachusetts Institute of Technology, Cambridge, Massachusetts 02139, USA. Correspondence should be addressed to R.L. (rlanger@mit.edu).

Published online 13 June 2004; doi:10.1038/nbt981

LETTERS

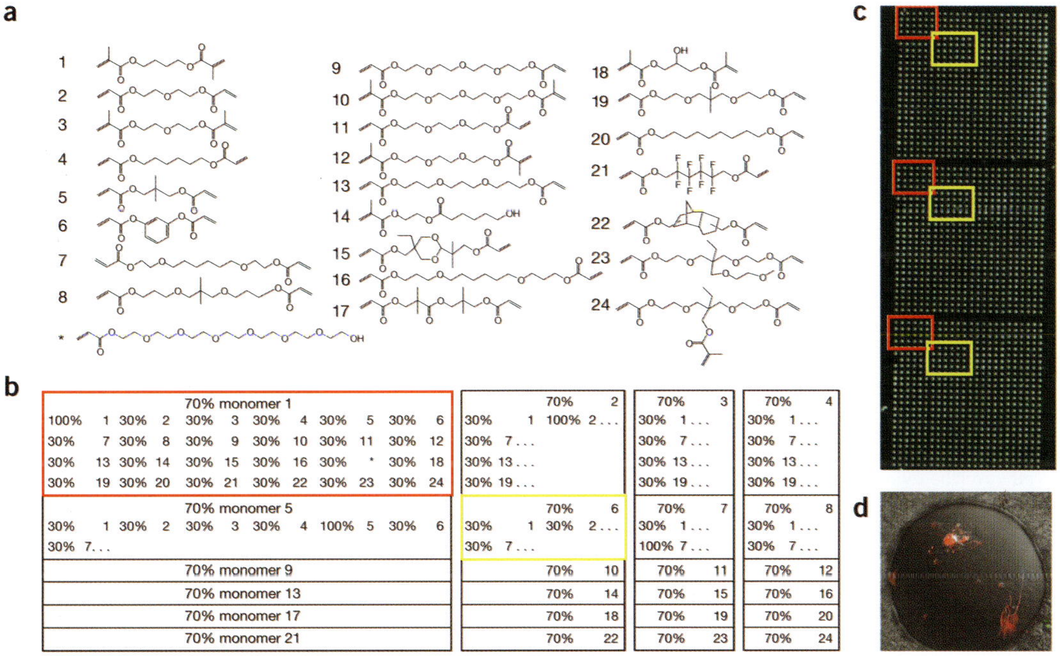

Figure 1 Biomaterial microarray design. (**a**) Monomers used for microarray synthesis. (**b**) Monomers were mixed at a 70:30 ratio pairwise in all possible combinations with the exception of monomer 17, which was substituted with * to increase polymer hydrophilicity. To facilitate analysis, all 24 polymers composed of 70% of a particular monomer were printed as a 6 × 4 group on the array, as highlighted by the red and yellow boxes. Three blocks of 576 polymers were printed on each slide, with a center to center spacing of 740 μm. (**c**) Printed polymer array imaged by Arrayworx reader. Blocks composed of 70% monomer 1 and 70% monomer 6 are highlighted in red and yellow, respectively. (**d**) Differential interference contrast light microscopy of typical polymer spot overlaid with a few fluorescent cells (red).

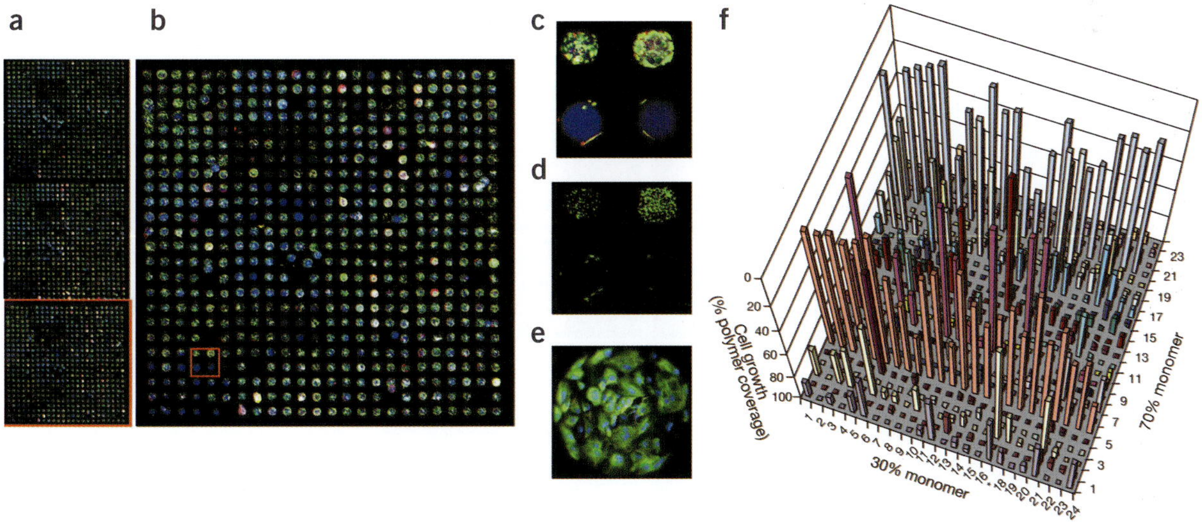

Figure 2 hES cells grown on polymer arrays. (**a–c**) Six million hES cell embryoid body day-6 cells were added on the polymer array in the presence of retinoic acid for 6 d and then stained for cytokeratin 7 (green) and vimentin (red). Polymer spots can be identified by blue fluorescence. (**d**) Nuclei were also stained (green) (not shown in other images to simplify presentation). (**e**) Typical cytokeratin 7 positive spot on polymer composed of 100% monomer 9. (**f**) Cell attachment and growth as a function of polymer composition, measured as the average percent cell coverage of a polymer spot on day 6. Although the vast majority of polymers remained attached to the matrix during analysis, certain particularly hydrophilic polymers (such as those composed of 30% monomer *) did fall off after extensive submersion.

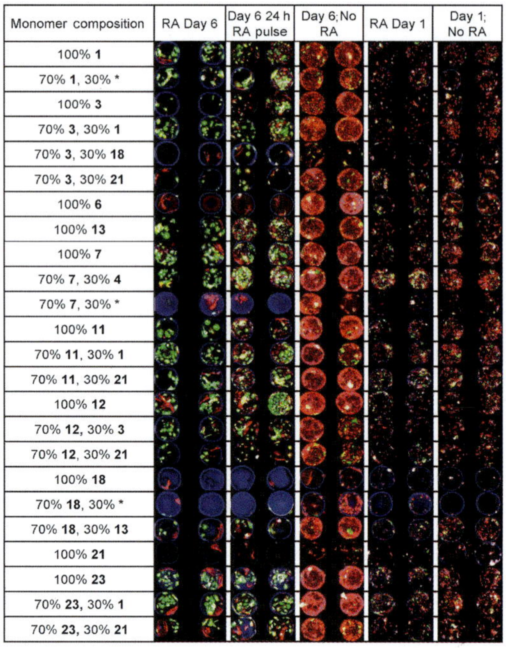

Figure 3 'Hit' polymer effects on hES cell attachment, growth and proliferation. (**a**) Four million hES cell embryoid body day-6 cells were grown on the 'hit' polymer arrays in the presence of retinoic acid, the absence of retinoic acid and with a 24-h pulse of retinoic acid for 1 or 6 d. Cells were then stained for cytokeratin 7 (green), vimentin (red) and DNA (blue). Representative images at each time point for each polymer are shown for all conditions. RA, retinoic acid.

specific cell types could be useful for the creation of complex tissue-engineered constructs in addition to fundamental studies using multiple cell types.

Polymers of interest, chosen by their ability to support cell growth in the first screen (**Fig. 2f**), were retested on 'hit' arrays for their effects on hES cell differentiation (**Supplementary Fig. 3** online). Each 'hit' array again contained 1,728 polymer spots, but only 24 polymer materials with 72 replicates per array. These arrays were seeded with fewer cells than before to more clearly identify polymer effects. Both soluble factors, such as growth factors, and the matrix on which cells grow have the potential to affect cell behavior[1,2]. To more carefully examine the interplay of polymer and soluble factors on cell behavior, we tested arrays with retinoic acid, without retinoic acid and with a 24-h pulse of retinoic acid (**Figs. 3** and **4**). Arrays were stained after 1 d and 6 d.

The absence of retinoic acid had several effects on cell behavior after 6 d: much less expression of cytokeratin was evident, the number of cells per spot was higher and the cells were smaller and more tightly packed. Analysis of arrays after 1 d (**Figs. 3** and **4i–n**) revealed that the presence of retinoic acid had, in general, little effect after 24 h. Quantitative analysis of cells grown with retinoic acid on day 1 and day 6 revealed that the total cell number per spot decreased in the presence of retinoic acid, and the percentage of cytokeratin-positive cells increased over time (**Supplementary Figs. 4,5** and **Supplementary Table 1** online). However, some polymers support growth only when retinoic acid is absent. For example, cells could attach to polymers such as 100% monomer 6 in similar quantities per spot with or without retinoic acid, as measured by cell counts after 24 h of growth (**Figs. 3**, **4j** versus **4m** and **Supplementary Fig. 4** online). However, after 6 d very few cells were present, suggesting 100% monomer 6 did not support proliferation of these cells in the presence of retinoic acid (**Fig. 4g** and **Supplementary Fig. 5** online). In contrast, some polymers supported attachment and spreading in both conditions (*e.g.*, 100% monomer 12 (**Fig. 4c,f,i,l**)), and others had very few cells present at day 1, suggesting inhibition of cell attachment (*e.g.*, 100% monomer 21 (**Fig. 4e,h,k,n**)). The discovery of polymers that support cell proliferation in a soluble factor–

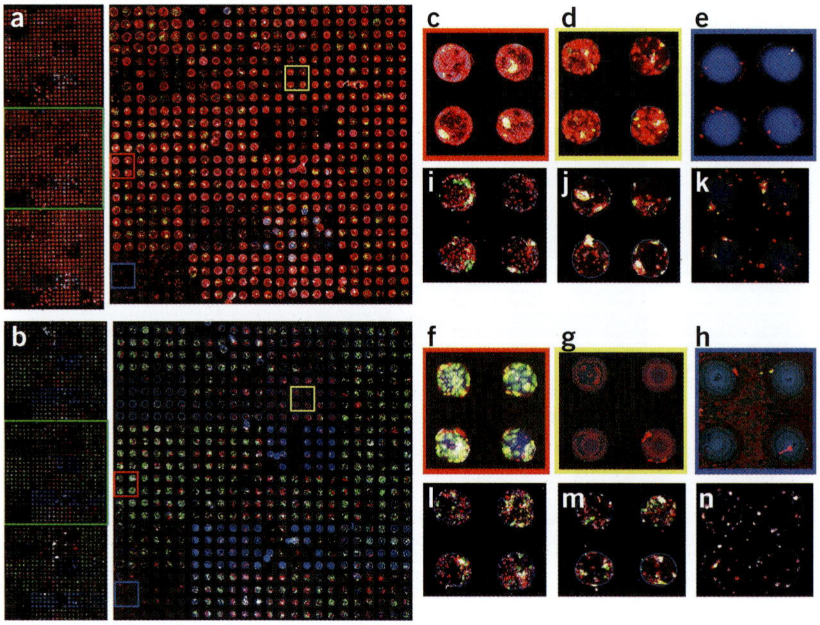

Figure 4 hES cells grown on 'hit' arrays. (**a**,**b**) Four million hES cell embryoid body day-6 cells were grown on the 'hit' polymer array in the presence (**b**) and absence (**a**) of retinoic acid for 6 d and then stained for cytokeratin 7 (green) and vimentin (red). (**c**–**n**) The blue channel can be used to identify the location of polymer spots (**b**,**e**,**h**) or cells lacking other signal by nuclear staining (blue in **a**,**c**,**d**,**f**,**g**,**i**–**n**). Using these methods we identify polymers supporting growth in both medium conditions (70% monomer 11, 30% monomer 1, (**c**,**f**)), only one (100% monomer 6 (**d**,**g**)) or growth in neither (100% monomer 21 (**e**,**h**)). Parallel experiments were done and cells were stained 24 h after loading with cells (**i**–**n**).

dependent manner could provide a new tool for controlling hES cell growth and proliferation.

The microarrays we have developed enable the rapid synthesis and cell-based screening of large libraries of biomaterials. Our proof-of-principle study with hES cells identified polymers that allow for varying levels of hES cell attachment and spreading, cell-type specific growth, and growth factor–specific proliferation. In addition, we identified a variety of materials that, in combination with proper solution conditions, allow for high levels of differentiation into cytokeratin-positive cells. The utility of these materials when used in different formats such as large coatings and gels is currently under investigation.

Quantification of these initial experiments was done manually. However, high-content image analysis is routinely applied in high-throughput cell-based screening[10] and could be readily adapted to the analysis of microarrayed cells. Future work will focus on the development and integration of automated analytical systems and chemical analysis (*e.g.*, degree of polymerization and surface properties) of materials of interest.

We believe this technology could have wide applicability to the study of polymer-cell interactions. The flexibility of acrylate polymerization allows for the easy incorporation of cell membrane components and adhesion peptides[11,12] and could be used to create targeted, rationally designed polymer libraries for specific cell types. The technology may also facilitate the identification of new materials with applications outside of biology.

METHODS

Polymer array synthesis. The use of robotic fluid handling for the production of DNA, protein and small molecule microarrays is well defined[13–15]. However, the deposition of structurally diverse acrylate monomers to produce a uniform, cell-compatible polymer microarray required substantial modification of existing robotic technology. First, some acrylate monomers are viscous, affecting all aspects of monomer printing including preprinting pin priming, fluid ejection at printing and pin washing. Another problem unique to these arrays is that radical polymerization is inhibited by oxygen, an inhibition that is particularly evident at small volumes. Consequently, we did our printing in an atmosphere of humid argon with oxygen present at <0.1%. Humidity was essential for minimizing failed printing, presumably by reducing static effects. Finally, some monomers spread soon after deposition, forming irregular polymer spots, whereas others started to evaporate if the time after deposition was longer than a few minutes. To address these issues, we modified our robot by inclusion of a long-wave UV lamp that immediately polymerized the monomers after each round of monomer deposition.

Epoxy-coated glass slides (Xenopore) were dip-coated into 4% (wt/vol) poly hydroxyethyl methacrylate (pHEMA, Aldrich) solution in ethanol and dried for 3 d before use. Monomers were purchased from Aldrich, Scientific Polymers, and Polysciences. Printing of acrylate monomers required several modifications to existing printing methods: (i) incorporation of 25% dimethyl formamide to reduce viscosity, (ii) substantially increasing washing and preprinting steps and (iii) modification of pin speed and size. Stock solutions were prepared at a ratio of (vol/vol) 75% monomer, 25% DMF and 1% (wt/vol) 2,2-dimethoxy-2-phenyl acetophenone. These were then mixed pairwise in 384-well black polypropylene plates at a ratio of 70:30 (vol/vol). Monomers were printed using the larger format CMP9B or CMP6B pins (Telechem international) with a Pixsys 5500 robot (Cartesian). After each round of printing on ten slides, the slides were polymerized by exposure to long-wave UV (UVP Blak-Ray) for ~10 s. The chips were then dried at <50 mTorr for at least 7 d. Chips were sterilized by exposure to UV for 30 min on each side, and then washed twice with PBS for 30 min and then twice with medium for 30 min before use, to remove residual monomer or solvent.

Because monomers may have different reactivity levels, the actual composition of polymers after drying and wash steps may not precisely equal the monomer ratio that was deposited.

Cell culture. H9 cells[16] were grown as described[17]. C2C12 were grown as described[18]. Embryoid bodies were formed and then trypsinized after 6 d[19]. Cells were added to the growth medium (KO DMEM, 20% heat-inactivated fetal bovine serum, L-glutamine, β-mercaptoethanol, minimal essential amino acids (Invitrogen) and 1 μM retinoic acid (Aldrich) when indicated) and then seeded onto chips in 26 × 100 mm Teflon dishes. Chips were incubated at 37 °C with 5% CO_2 and medium was changed after 1 d, and then every 2 d thereafter.

Immunohistochemistry. Chips were washed, fixed in 4% paraformaldehyde for 8 min, blocked with 10% goat serum (Zymed) and permeabilized with 0.2% Triton X-100 for 30 min. Primary antibodies, mouse anti-cytokeratin 7, mouse anti-myogenin (Dako), rabbit anti-vimentin (Biomeda) in PBS with 3% goat serum were incubated on the chips for 1 h. Chips were washed 3 times in 1% goat serum PBS. A mixture of goat anti-mouse Alexa 555, goat anti-rabbit Alexa and SYTOX24 (Molecular Probes) were diluted into 3% goat serum PBS and incubated on the chips for 1 h. Slides were washed 3 times in 1% goat serum PBS and dipped in 0.5 mM Tris Cl pH 7.5 to remove salt, and air dried immediately before scanning. Slides were then scanned using an Arrayworx autoloader scanner (API). Cell coverage and counting were done manually.

Note: Supplementary information is available on the Nature Biotechnology website.

ACKNOWLEDGMENTS
The authors would like to thank David Putnam, David Lynn, David Lavan and Daniel Kohane for helpful advice, Mara Macdonald for help with cell culture and Joseph Itskovitz-Eldor for assistance and cooperation in conducting this research. We would also like to thank Sean Milton and the BioMicro Center for help with slide analysis. This work was supported by the National Science Foundation (through the MIT Biotechnology Process and Engineering Center) and the National Institutes of Health (grant no. HL60435).

COMPETING INTERESTS STATEMENT
The authors declare that they have no competing financial interests.

Received 20 October 2003; accepted 16 April 2004
Published online at http://www.nature.com/naturebiotechnology/

1. Spradling, A., Drummond-Barbosa, D. & Kai, T. Stem cells find their niche. *Nature* **414**, 98–104 (2001).
2. Streuli, C. Extracellular matrix remodelling and cellular differentiation. *Curr. Opin. Cell Biol.* **11**, 634–640 (1999).
3. Saltzman, W.M. in *Principles of Tissue Engineering*, edn. 2 (eds. Lanza, R.P., Langer, R.S. & Vacanti, J.) 221–235 (Academic Press, San Diego, 2000).
4. Lanza, R.P., Langer, R.S. & Vacanti, J. (eds.) *Principles of Tissue Engineering*, edn. 2 (Academic Press, San Diego, 2000).
5. Chen, R.R. & Mooney, D.J. Polymeric growth factor delivery strategies for tissue engineering. *Pharm. Res.* **20**, 1103–1112 (2003).
6. Sakiyama-Elbert, S.E. & Hubbell, J.A. Functional biomaterials: Design of novel biomaterials. *Annu. Rev. Mater. Res.* **31**, 183–201 (2001).
7. Fisher, J.P., Dean, D., Engel, P.S. & Mikos, A.G. Photoinitiated polymerization of biomaterials. *Annu. Rev. Mater. Res.* **31**, 171–181 (2001).
8. Folkman, J. & Moscona, A. Role of cell shape in growth control. *Nature* **273**, 345–349 (1978).
9. Moll, R., Franke, W.W., Schiller, D.L., Geiger, B. & Krepler, R. The catalog of human cytokeratins: patterns of expression in normal epithelia, tumors and cultured cells. *Cell* **31**, 11–24 (1982).
10. Abraham, V.C., Taylor, D.L. & Haskins, J.R. High content screening applied to large-scale cell biology. *Trends Biotechnol.* **22**, 15–22 (2004).
11. Burkoth, A.K., Burdick, J. & Anseth, K.S. Surface and bulk modifications to photocrosslinked polyanhydrides to control degradation behavior. *J. Biomed. Mater. Res.* **51**, 352–359 (2000).
12. Hern, D.L. & Hubbell, J.A. Incorporation of adhesion peptides into nonadhesive hydrogels useful for tissue resurfacing. *J. Biomed. Mater. Res.* **39**, 266–276 (1998).
13. MacBeath, G., Koehler, A.N. & Schreiber, S.L. Printing small molecules as microarrays and detecting protein-ligand interactions *en masse*. *J. Am. Chem. Soc.* **121**, 7967–7968 (1999).
14. MacBeath, G. & Schreiber, S.L. Printing proteins as microarrays for high-throughput function determination. *Science* **289**, 1760–1763 (2000).
15. Schena, M., Shalon, D., Davis, R.W. & Brown, P.O. Quantitative monitoring of gene-expression patterns with a complementary-DNA microarray. *Science* **270**, 467–470 (1995).
16. Thomson, J.A. *et al.* Embryonic stem cell lines derived from human blastocysts. *Science* **282**, 1145–1147 (1998).
17. Levenberg, S., Golub, J.S., Amit, M., Itskovitz-Eldor, J. & Langer, R. Endothelial cells derived from human embryonic stem cells. *Proc. Natl. Acad. Sci. USA* **99**, 4391–4396 (2002).
18. Yaffe, D. & Saxel, O. Serial passaging and differentiation of myogenic cells isolated from dystrophic mouse muscle. *Nature* **270**, 725–727 (1977).
19. Levenberg, S. *et al.* Differentiation of human embryonic stem cells on three-dimensional polymer scaffolds. *Proc. Natl. Acad. Sci. USA* **100**, 12741–12746 (2003).

LETTERS

nature
biotechnology

Engineering vascularized skeletal muscle tissue

Shulamit Levenberg[1,2], Jeroen Rouwkema[3], Mara Macdonald[2], Evan S Garfein[4], Daniel S Kohane[5], Diane C Darland[6], Robert Marini[7], Clemens A van Blitterswijk[3], Richard C Mulligan[8], Patricia A D'Amore[6] & Robert Langer[2]

One of the major obstacles in engineering thick, complex tissues such as muscle is the need to vascularize the tissue *in vitro*. Vascularization *in vitro* could maintain cell viability during tissue growth, induce structural organization and promote vascularization upon implantation. Here we describe the induction of endothelial vessel networks in engineered skeletal muscle tissue constructs using a three-dimensional multiculture system consisting of myoblasts, embryonic fibroblasts and endothelial cells coseeded on highly porous, biodegradable polymer scaffolds. Analysis of the conditions for induction and stabilization of the vessels *in vitro* showed that addition of embryonic fibroblasts increased the levels of vascular endothelial growth factor expression in the construct and promoted formation and stabilization of the endothelial vessels. We studied the survival and vascularization of the engineered muscle implants *in vivo* in three different models. Prevascularization improved the vascularization, blood perfusion and survival of the muscle tissue constructs after transplantation.

Most approaches to engineering new tissue have relied on the host for vascularization. Although this approach has been useful in many tissues, it has not been as successful in thick, highly vascularized tissues such as muscle[1–3]. Skeletal muscle consists of individual muscle fibers arranged in parallel. Each fiber is a long, cylindrical multinucleated cell that is surrounded by connective tissue. Skeletal muscles have an abundant blood vessel supply, with branches of blood vessels following the connective tissue components of the muscle[4,5]. So far, attempts to engineer skeletal muscle tissue have involved cultivation of skeletal myoblasts only, in some cases using growth factor delivery matrices or genetically engineered myoblasts to provide vascularization factors[6–8].

We hypothesized that embryonic endothelial cells in the appropriate environment could be used to induce endothelial vessel networks in engineered skeletal muscle tissue *in vitro*. First we developed a three-dimensional (3D) coculture system in which mouse myoblasts were mixed with human embryonic endothelial cells (hES cell–derived endothelial cells)[9] or with human umbilical vein endothelial cells (HUVECs) and seeded on highly porous, 3D biodegradable polymer scaffolds. The sponge-like scaffolds were composed of 50% poly-(L-lactic acid) (PLLA) and 50% polylactic-glycolic acid (PLGA), with pore sizes of 225–500 μm[10] (**Fig. 1a**).

As indicated by desmin immunostaining of cross sections of constructs 3 d after seeding of myoblasts, the myoblasts attached to and grew on the scaffolds (**Fig. 1b**). By 14 d the myoblasts had differentiated and formed partially aligned, elongated and multinucleated myotubes. Some of the myotubes differentiated further and became myogenin positive (36% ± 3% of total nuclei) (**Fig. 1b**). When both myoblasts and endothelial cells (either hES cell–derived or HUVECs) were cultured on the scaffolds, the endothelial cells (CD31 positive) organized into tubular structures in between the myoblasts and throughout the construct, forming vessel networks within the engineered muscle tissue *in vitro* (**Fig. 1c**).

We compared the effects of two media: myoblast medium composed of DMEM with 10% FCS and 10% CS with HEPES and endothelial medium (EGM-2) with 2% FCS and endothelial growth factor supplements. Myoblast medium promoted both differentiation of the muscle cells (27% ± 6% myogenin-positive nuclei) and formation of endothelial lumens in the constructs (**Fig. 2a**), whereas endothelial medium alone did not support differentiation of the muscle cells (2% ± 1% myogenin-positive nuclei) and inhibited endothelial lumen formation (**Fig. 2a**). Why the endothelial medium inhibited vessel formation in the muscle constructs is not clear, and may be related to its inhibitory effect on muscle differentiation and cell signaling in the culture.

Because blood vessels are stabilized by association with pericytes or smooth muscle cells[11–17] and because endothelial cells can induce the differentiation of undifferentiated mesenchymal cells into smooth muscle cells[18–21], we hypothesized that the formation of vessels characterized by lumen structures in the engineered skeletal muscle tissue would be promoted by embryonic fibroblasts. Addition of embryonic fibroblasts to the cultures, together with myoblasts and endothelial cells, strongly promoted vascularization of the engineered muscle; this was evidenced by increases in the total area of endothelial cells and the number and size of endothelial lumens, compared with constructs

[1]Department of Biomedical Engineering, Technion 32000, Haifa, Israel. [2]Department of Chemical Engineering, [7]Division of Comparative Medicine, Massachusetts Institute of Technology, 77 Massachusetts Avenue, Cambridge, Massachusetts 02139, USA. [3]Institute for Biomedical Technology, Twente University, Prof. Bronkhorstlaan 10-D, Bilthoven, 3723MB, The Netherlands. [4]Department of Surgery, Brigham and Women's Hospital, 75 Francis St., 02115, [5]Department of Pediatrics, Massachusetts General Hospital, 55 Fruit St., 02114, [6]The Schepens Eye Research Institute and Department of Ophthalmology 20 Staniford St., 02114, [8]Department of Molecular Medicine, Children's Hospital 300 Longwood Avenue, 02115, Harvard Medical School, Boston, Massachusetts, USA. Correspondence should be addressed to R.L. (rlanger@mit.edu) or S.L. (shulamit@bm.technion.ac.il).

Published online 19 June 2005; doi:10.1038/nbt1109

seeded with myoblasts and endothelial cells only (**Fig. 2a**). The effect of the fibroblasts was dependent on medium conditions and the cell ratios (**Fig. 2a**, tri-cultures) with best vascularization in myoblast medium and addition of 0.2 million fibroblasts per scaffold (1.5 million cells total with 40–50% endothelial cells). The inductive effect of embryonic fibroblasts on endothelial vessels is shown even when comparing tri-culture samples that were seeded with a lower percentage (33%) of endothelial cells with cocultures of myoblasts and endothelial cells seeded with higher percentages of endothelial cells (40–50%) (**Fig. 2a**).

To analyze the stability of the *in vitro* vessel-like structures formed in the muscle constructs, we examined tri-culture constructs (a combination of myoblasts, fibroblasts and endothelial cells) at 2 weeks and 1 month. Large vessel structures ($>1{,}500\ \mu m^2$) were evident only in the 1-month-old tri-culture constructs (**Fig. 1c**). In addition, tri-cultures grown for 1 month had a twofold increase in the number of endothelial structures, a greater surface area covered by endothelial cells and a higher percentage of vessel-like structures with lumens, compared with 2-week tri-cultures (**Fig. 2b**).

These results suggest that addition of embryonic fibroblasts promoted stabilization of the vessel structures over time. Double labeling for von Willebrand factor (vWF) and smooth muscle actin on cross sections of constructs seeded with endothelial cells and embryonic fibroblasts showed that fibroblasts in the cultures became smooth-muscle-actin positive (suggesting differentiation into smooth muscle cells) and were colocalized around endothelial cells in vessel-like structures (**Fig. 2c**). Quantitative analysis of the double staining revealed that 65.7% ± 8.8% of endothelial vessel-like structures in the constructs had associated smooth muscle cells.

To study the expression of key vasculogenic and angiogenic factors in the 3D muscle constructs, we analyzed the expression of vascular endothelial growth factor (VEGF) and platelet-derived growth factor (PDGF)-B at the mRNA level in the muscle constructs. The RT-PCR results showed that addition of human endothelial cells to myoblast or fibroblast cultures resulted in an increase in mouse VEGF expression. Moreover, tri-cultures that included embryonic fibroblasts had higher levels of VEGF mRNA than myoblast-endothelial cocultures (**Supplementary Fig. 1** online). The increased VEGF expression is consistent with the increase in the endothelial network observed in the tri-cultures, and could be one of the factors affecting the induction of vascularization of the constructs. Indeed, addition of VEGF to the medium resulted in a larger area covered by endothelial cells and an increase in the number of vessel-like endothelial tubular structures in the constructs (**Fig. 2d**). Depletion of VEGF from the medium resulted in a decreased number of vessel structures, whereas addition of fibroblasts to cultures without VEGF supplementation restored vessel formation (data not shown).

To assess the therapeutic potential of our approach, we used three models to analyze the survival, differentiation, integration and vascularization of the implant *in vivo*: (i) subcutaneous implantation in the back of severe combined immunodeficient (SCID) mice, (ii) intramuscular implantation into the quadriceps muscle of nude rats, and (iii) replacement of the anterior abdominal muscle segment of nude mice with the construct. In all three models the muscle implant continued to differentiate *in vivo*. The implanted myotubes were elongated and multinucleated (6–8 nuclei), with a high percentage of myogenin-positive myotubes (67% ± 9%) (**Fig. 3a**). Control implants containing fibroblasts or no cells showed no desmin-positive myotubes or myogenin-positive nuclei within the scaffold area (**Fig. 3a**). To further ensure that the myotubes observed within the scaffold were derived from implanted cells rather than invading host

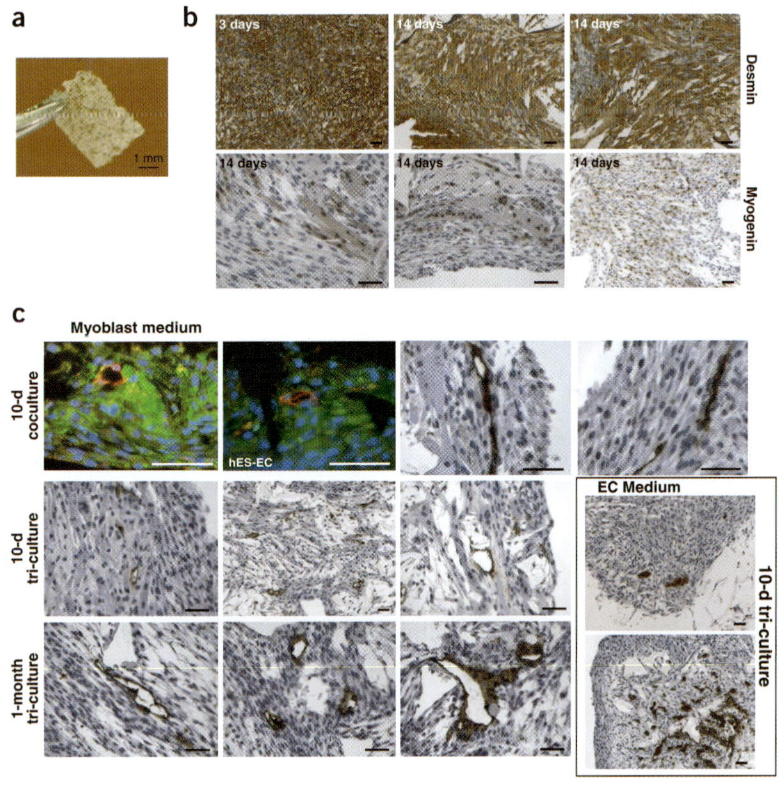

Figure 1 *In vitro* vascularization of engineered skeletal muscle tissue (**a**) PLLA/PLGA scaffold before cell seeding. Scale bar, 1 mm. (**b**) Myocyte differentiation on PLLA/PLGA polymer scaffolds. Desmin and myogein immunostaining of tissue sections taken from 3D scaffolds cocultured with skeletal myoblasts and endothelial cells (HUVEC) and grown for 3 d and 14 d. Scale bar = 50 μm. (**c**) Vessel-like network formation *in vitro* in muscle 3D constructs. Endothelial cells (HUVEC or hES cell-derived endothelial cells (hES-EC) when indicated) were coseeded with skeletal myoblasts on polymer scaffolds and grown for 10 d (cocultures). Tissue construct sections were immunofluorescently stained using anti-CD31 antibodies (red), anti-desmin antibodies (green) and DAPI for nuclear staining (blue) (left), or stained using anti-CD31 antibodies alone (brown) and counterstained with hematoxylin (blue) (right). One-month tri-cultures and 10-d, both including mouse embryonic fibroblasts, were grown either in myoblast medium or endothelial medium (EC medium) and stained using anti-CD31 antibodies. Cultured cell numbers (myoblast/endothelial/ fibroblast), $0.5/0.7/0.2 \times 10^6$; bottom picture in endothelial (EC) medium, $0.5/0.5/0.5 \times 10^6$. Scale bar, 50 μm.

cells, we incorporated 5-bromodeoxyuridine (BrdU) in the tissue-engineered constructs before implantation. BrdU labeling confirmed that the implanted myoblasts had indeed survived and differentiated to populate the scaffold (data not shown). In most cases, particularly in the abdominal muscle, the implanted myotubes were in close contact with the host muscle, with very thin and sometimes barely detectable fibrous tissue around the implant (**Fig. 3a**). The myotubes in the implanted area were relatively long and thick and in many cases appeared to have reoriented themselves to align with the fibers of the host tissue (**Fig. 3a**).

The constructs were permeated with host blood vessels (**Fig. 3b**). Quantification of the number of endothelial vessel-like structures in intramuscular implants 2 weeks after implantation indicated that there was no significant difference between constructs seeded with HUVECs or hES cell–derived endothelial cells (**Supplementary Fig. 2** online). Staining of subcutaneous implants with anti-human specific endothelial antibodies (anti-CD31) demonstrated the presence of vessels (between elongated myotubes), which were lined by implanted human endothelial cells. Moreover, construct-derived vessels contained intraluminal red blood cells, suggesting that vessels had anastomosed with the host vasculature (**Fig. 3b**).

To determine whether the vessels were functional, we injected labeled lectin into the tail vein 2 weeks after implantation and counted the perfused vessels. The results indicated that 41% ± 12% of human CD31–positive vessels (implant-derived vessels) were perfused with lectin. Quantification of the total number of

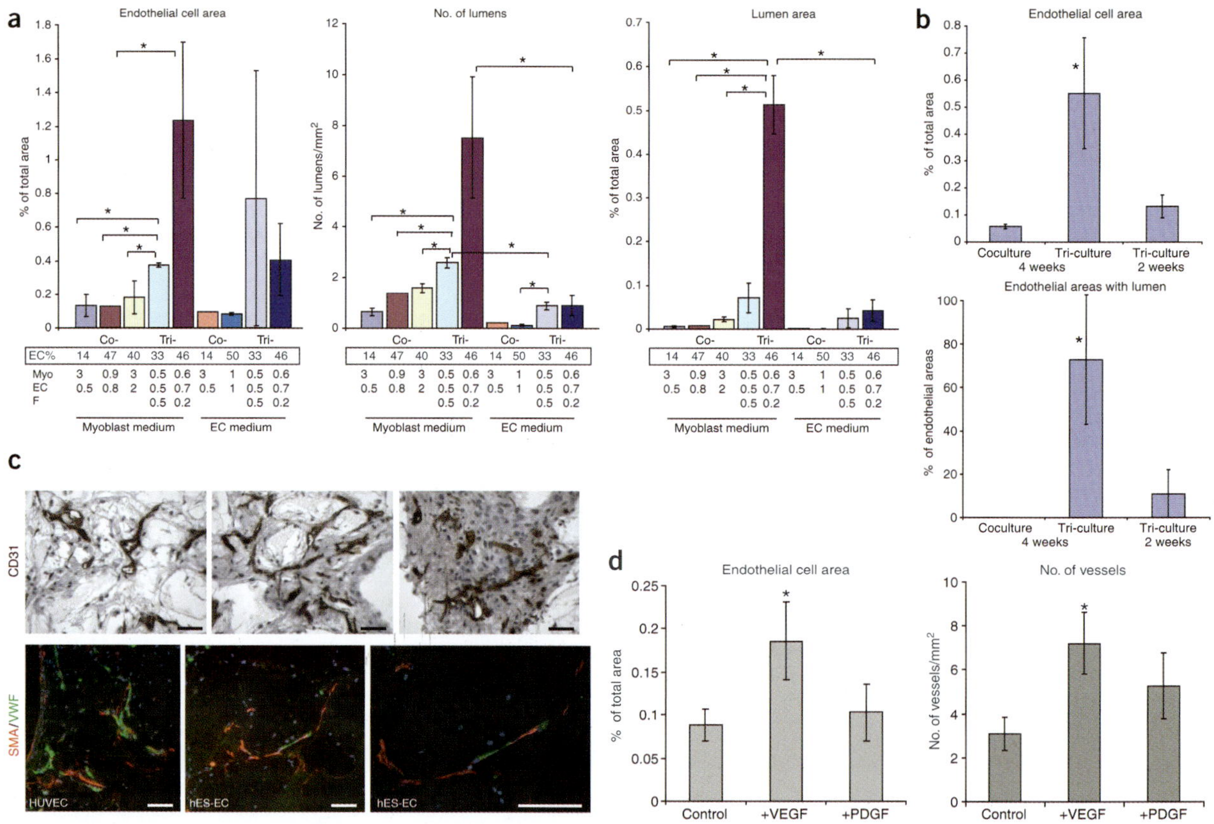

Figure 2 Quantitative analysis of endothelial vessels in muscle constructs, *in vitro*. (**a**) Comparison of vessel formation in coculture (Co) and tri-culture (Tri) constructs grown with different cell ratios (cell number $\times 10^6$) and medium conditions (myoblast medium and endothelial medium). Endothelial cell ratio (EC %) is calculated as percentage of the total cell number. Endothelial cell area corresponds to percentages of area positively stained with CD31 antibody within the tissue section. Lumen area shows the total area of all the lumens in the section as percentages of total section area. Myo, myoblasts; EC, endothelial cells (HUVEC); F, mouse embryonic fibroblasts. * denotes statistical significance ($P < 0.05$) between the indicated pairs. (**b**) Comparison of vessels in 2-week and 4-week constructs. Cocultures are myoblasts and endothelial cells (0.8 and 0.6 $\times 10^6$ cells per scaffold, respectively). Tri-cultures are myoblasts, fibroblasts and endothelial cells (0.6, 0.2 and 0.6 $\times 10^6$ cells per scaffold respectively). * denotes statistical significance ($P < 0.05$) compared with controls. (**c**) Effect of embryonic fibroblasts on vessel formation. Constructs seeded with embryonic fibroblasts and endothelial cells (0.5,1 $\times 10^6$) were grown for 2 weeks. Construct sections were immunostained with human-specific anti-CD31 antibodies (brown) showing vessel formation throughout the 3D constructs. Colocalization of smooth muscle cells and endothelial cells (HUVEC or hES-EC as indicated). Constructs were immunofluorescently double-stained using anti-VWF antibodies (green), anti-smooth muscle actin antibodies (red) and DAPI for nuclear localization. Note smooth muscle actin–positive cells around elongated endothelial structures. Scale bar, 50 μm. (**d**) Effect of VEGF or PDGF-B supplementation on endothelial vessel formation. Tri-culture constructs (myoblasts, fibroblasts and endothelial cells (0.6, 0.2 and 0.7 $\times 10^6$ cells per scaffold, respectively)) were incubated with control medium or medium supplemented with VEGF (50 ng/ml) or PDGF-B (5 ng/ml). After 2 weeks, construct sections were immunoassayed using anti-CD31 antibodies and analyzed for endothelial-positive area, and number of endothelial vessels. * denotes statistical significance ($P < 0.05$) compared with controls. The results shown are mean values ± s.d. ($n = 4$).

perfused vessels (host and implant derived) indicated that muscle constructs seeded with endothelial cells had 30 ± 2 functional vessels per square millimeter compared with 21 ± 2 vessels in constructs seeded with muscle cells only ($n = 3$, $P < 0.01$). The size distribution of functional vessels showed that including endothelial cells in the scaffolds also increased the number of larger or stabilized vessels in the muscle implants (**Fig. 3c**). These results suggest that pre-endothelialization of the construct, by promoting implant vascularization, can improve blood perfusion to the muscle implant and implant survival *in vivo*.

To further evaluate tissue-engineered muscle construct survival and integration *in vivo*, we used a luciferase-based imaging system. The *in vivo* imaging system (IVIS) works by detecting light generated by the interaction of systemically administered luciferin with locally produced luciferase. Muscle constructs seeded with or without endothelial cells or fibroblasts, and control constructs were infected with adeno-associated virus (AAV) vector, encoding luciferase, for 48 h before transplantation. Detection of luciferase expression in the constructs indicated no difference among the various muscle constructs *in vitro*. The constructs were then placed *in situ* in the anterior abdominal muscle walls of nude mice. Three weeks after surgery, the mice received luciferin (injected subcutaneously on the dorsal surface) to assess perfusion to the tissue-engineered construct. A minimal signal was detected in areas that did not receive gene transfer or

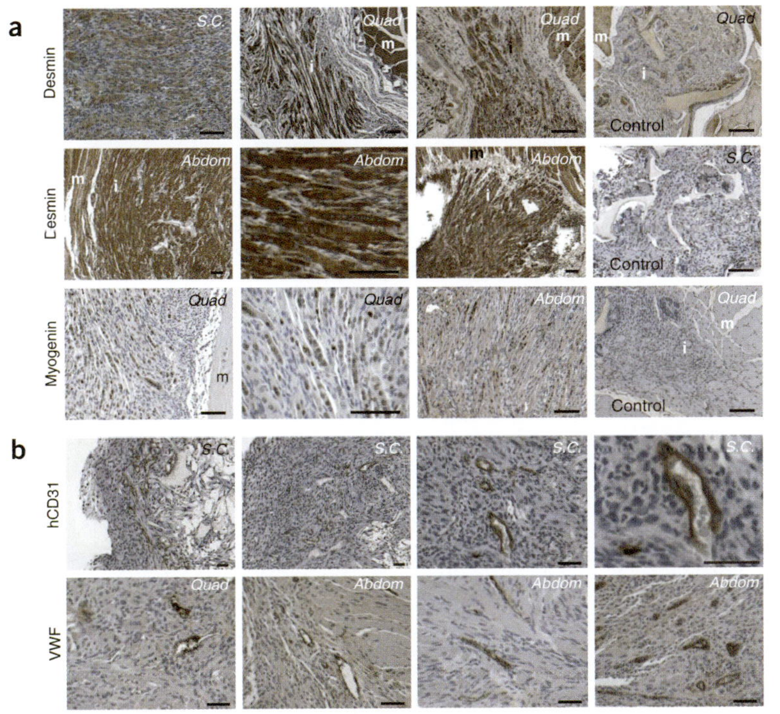

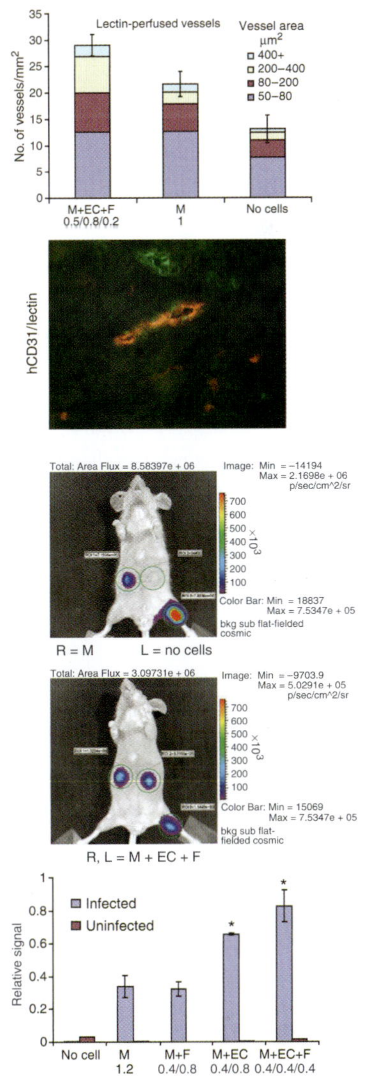

Figure 3 *In vivo* analysis of engineered muscle constructs (**a**) Differentiation of engineered muscle *in vivo*. Two-week-old engineered muscle constructs were implanted either subcutaneously into immunodeficient mice (S.C.), or intramuscularly into rat quadriceps muscle (Quad) or mouse abdominal muscle (Abdom). After 2 weeks *in vivo*, constructs were sectioned and immunostained using anti-desmin antibodies or myogenin antibodies. For control, constructs without cells were implanted and stained (Control). Note differentiation of myocytes into aligned, multinucleated elongated myotubes. Muscle area, m; implant area, i. Scale bar, 100 μm. (**b**) Vascularization of engineered muscle *in vivo*. After 2 weeks *in vivo*, constructs were sectioned and immunostained using human specific anti-CD31 (hCD31) or anti-vWF antibodies. Scale bar, 50 μm. (**c**) Functional, lectin-perfused vessels in tissue-engineered muscle implants. After 2 weeks, *in vivo* constructs were perfused with fluorescently labeled lectin (red). Bottom, implant's frozen sections were immunofluorescently stained with anti-hCD31 (green), showing implant-derived lectin-perfused vessel. Top, quantitative analysis of number and size of lectin-perfused vessels in muscle implants. Vessel formation was compared between tri-culture constructs seeded with myoblasts (M), endothelial cells (HUVEC) (EC), and embryonic fibroblasts (F), and constructs seeded with myoblasts alone or without cells (no cells). (Cell number $\times 10^6$). Standard deviation error bars relate to total number of perfused vessels ($n = 3$). (**d**) Microvascular perfusion and survival of tissue-engineered muscle implants. Cells in the constructs were infected with AAV-luciferase for 48 h before transplantation. Control constructs were not infected with virus. The constructs were then placed *in situ* in the anterior abdominal muscle walls of nude mice. AAV-luciferase was injected into the left lower extremity of each mouse at the time of surgery to serve as a positive control. Three weeks after surgery, the mice received luciferin to assess perfusion and survival of the tissue engineered implants using luciferase-based *in vivo* imaging system (IVIS). Quantification of signal detected after 3 weeks. The results are mean values ± s.d. ($n = 3$). * denotes statistical significance ($P < 0.05$) compared with myoblasts alone (M) or myoblasts + fibroblasts (M+F). (Cell number $\times 10^6$). Signal intensities are shown on a scale of purple to red ($\sim 2 \times 10^4$ to $\sim 7.7 \times 10^5$ p/s/cm^2).

in constructs that were not seeded with cells. However strong signals were detected from areas either directly transduced with AAV-luciferase (as controls) or transplanted with virally transduced cells, indicating perfused vessels. By using the luciferase system, we were able to noninvasively determine the degree to which different constructs continued to survive (and express luciferase) and maintain vascular connections with the recipient to receive systemically delivered luciferin. The relative signal detected in implants seeded with endothelial cells in cocultures (with myoblasts) and in tri-cultures (with myoblasts and fibroblasts) was higher than in myoblast-only implants (**Fig. 3d**). Coupled with similar preimplantation levels of luciferase expression and with the histological evidence of increased functional vascularity, the results suggest that the increased signal in pre-endothelialized samples is related to increased perfusion and survival of the tissue-engineered muscle constructs. Analysis of cell survival in the implant using TUNEL staining indicated a twofold increase in the number of apoptotic cells in muscle-only implants compared with pre-endothelialized implants (36 $\pm$ 9 and 19 $\pm$ 7 cells per implant, respectively ($n = 3$)).

The approach that we have developed enables formation and stabilization of endothelial vessel networks *in vitro* in 3D engineered skeletal muscle tissue. The overall *in vivo* results show that prevascularization of the implants improves implant vascularization and survival. Unlike previous studies demonstrating endothelial differentiation within fibroblast culture and fibroblast differentiation into pericytes[11,14,15], this study demonstrates engineering of 3D vascularized skeletal muscle constructs with formation of endothelial networks throughout and in between differentiating skeletal muscle fibers. This study emphasizes the importance of multicell cultures in providing appropriate signals for vascular organization in skeletal muscle tissue. Moreover, it provides evidence for the potential of endothelial cocultures in promoting *in vivo* vascularization of engineered tissues. Cocultures with endothelial cells may also be important for inducing differentiation of engineered tissues, as embryonic endothelial cells are critical for the earliest stages of organogenesis[22,23]. We believe that this approach could have potential applications in tissue engineering and may provide a tool for the *in vitro* study of multicellular processes such as tissue vascularization.

METHODS

Cell culture. Mouse skeletal myoblast cells (C_2C_{12})[24,25] were cultured in DMEM supplemented with 10% FBS, 10% calf serum and 2.5% HEPES buffer. Human umbilical vein endothelial cells (HUVEC) were cultured in endothelial cell medium (EGM-2; Cambrex Biosciences). Mouse embryonic fibroblasts (MEF) (Cell Essentials) were cultured in DMEM supplemented with 10% FBS. HES cell–derived CD31+ endothelial cells were isolated as described[9] and cultured in endothelial cell medium.

Polymer scaffolds. Porous sponges composed of 50% PLLA and 50% PLGA were fabricated as described[10] with pore sizes of 225–500 μm and 93% porosity. The PLGA was selected to degrade quickly ($\sim$3 weeks) to facilitate cellular ingrowth, whereas the PLLA was chosen to provide mechanical support to 3D structures. The degradation time of the composed sponges is $\sim$6 months. Biocompatibility of PLLA and PLGA porous scaffolds was previously shown[26]. For seeding, the desired number of cells were pooled and resuspended in 7–15 μl of a 1:1 mixture of culture medium and growth factor-reduced Matrigel (BD Biosciences). This suspension was allowed to absorb into the sponges, after which the sponges were incubated for 30 min at 37 °C to allow solidification of the gel. Culture medium was then added, the sponges were detached from the bottom, and incubated at 37 °C on a XYZ shaker. The medium was changed every other day. At the conclusion of the experiments, samples were fixed in 10% formalin and subsequently embedded in paraffin for sectioning or were transplanted into mice or rats.

Implantation of muscle constructs. Male 5- to 6-week-old SCID mice (CB.17 SCID) were anesthetized with 2.5% isoflurane in balanced oxygen, after which a construct was implanted subcutaneously on each side lateral to the dorsal midline region of each mouse. For intramuscular implantation, constructs were implanted into the outer quadriceps muscle of the right-hand side of 5- to 7-week-old male nude rats. Sutures of 6-0 Prolene in a simple interrupted pattern were used to prevent movement of the constructs from the muscle site, and the skin was closed using surgical staples. Two to eight weeks after implantation, the mice or rats were killed and the implants were retrieved. Samples were fixed in 10% natural buffered formalin, processed routinely and sectioned at 4 μm before staining. Two assays were involved in perfusion analysis. (i) Lectin perfusion. Lectin HPA (*Helix pomatia* agglutinin) conjugated to Alexa Fluor 594 (Molecular Probes) (0.5 mg/0.25 ml PBS) was injected into the tail vein of anesthetized animals (20 mg/kg body weight). Circulation was allowed for 2 min after which the animals were killed and the implants were retrieved. Samples were snap frozen (liquid nitrogen) in Cryomatrix (Thermo Shandon) and sections of 6 μm were cut with a cryotome. (ii) Luciferase assay (abdominal wall model). Tissue-engineered constructs were placed in 12-well culture dishes in 2 ml of medium and 1.0×10^9 dot blot/cc of AAV-luciferase added. After 48 h, the constructs were washed with two volumes of PBS. As a control, luciferin (150 μl of 5mg/ml) was added to each well. After incubation for 11 min, the constructs were imaged in the Xenogen IVIS device at a 3-minute exposure. Luminescence was determined by calculating the flux (photons/sec/cm^2) overlying each construct. Immediately thereafter, the constructs were implanted into isoflurane-anesthetized mice by creation of a 3 mm $\times$ 3 mm defect in the anterior abdominal wall of the mice in line with the inferior epigastric artery. The construct was then sutured in place using four 7-0 prolene sutures attached to each corner of the construct. After the ventral skin was sutured, the animal recovered. At various intervals after surgery, the animals were imaged using the Xenogen IVIS device. Mice were anesthetized using an intramuscular injection of ketamine and xylazine. Luciferin (150 μl of 5 mg/ml) was injected subcutaneously on the dorsal surface of the mouse and allowed to circulate for 11 min before the animals were imaged. Luminescence was calculated by determining the photon flux. A 1.0 cm^2 area was chosen arbitrarily as the standard. The ratio of the flux from the tissue-engineered construct relative to the hind limb was calculated. After the final imaging session, the animals were killed and the implants were retrieved and placed in 10% formalin before routine processing and histological sectioning. Unseeded scaffolds were used as controls and showed host-cell infiltration (including fibroblasts and blood vessels) as known from previous studies[26].

Tissue processing and immunohistochemical staining. Tissue constructs were fixed for 6 h in 10% neutral buffered formalin, routinely processed, and embedded in paraffin. Transverse sections (4 μm) were placed on silanized slides for immunohistochemistry or staining with hematoxylin and eosin. Immunohistochemical staining was carried out using the Biocare Medical Universal HRP-DAB kit (Biocare Medical) according to the manufacturer's instructions, with prior heat treatment at 95 °C for 20 min in ReVeal buffer (Biocare Medical) for epitope recovery. For immunofluorescent staining, the secondary antibodies used were Alexa Fluor (Molecular Probes) and Cy-3 (Jackson Laboratories) followed by DAPI (Sigma) nuclear staining. The primary antibodies were anti-human: CD31 (1:20); desmin (1:150); α-smooth muscle actin (1:50); vWF (1:200) (all from Dako); or vWF (Chemicon) (1:100). Staining with α-smooth muscle actin antibody (as well as other smooth muscle actin markers to identify fibroblasts differentiation into smooth muscle cells could not be done in the presence of C_2C_{12} cells because of expression of these markers by the C_2C_{12} myoblasts. The vWF antibody was used after deparaffinization and trypsin treatment for epitope recovery. For labeling implanted myoblasts, myoblast culture medium was supplemented with 10 μm of 5′-bromo-2′-deoxyuridine (BrdU) (Sigma) and applied to 60% confluent dish cultures for 16 h. Cells were washed and seeded on scaffolds as described. Tissue sections were stained using mouse anti-BrdU antibodies (1:1,000) as described, but with treatment with 2N HCL and 0.5% Triton X-100 for 30 min at 37 °C to denature the DNA, before addition of the antibodies. For staining apoptotic cells DeadEnd colorimetric TUNEL system (Promega) was used.

LETTERS

RT-PCR analysis. Total RNA was isolated by an RNEasy Mini Kit (Qiagen), using the isolation-from-tissue protocol. RT-PCR was carried out using a Qiagen OneStep RT-PCR kit with 10 units Rnase inhibitor (Gibco) and 40 ng RNA. To ensure semi quantitative results of the RT-PCR assays the number of PCR cycles for each set of primers was checked to be in the linear range of the amplification. Primer sequences: mouse VEGF, 5′-CCT CCG AAA CCA TGA ACT TTC TGC TC-3′ and 5′-CAG CCT GGC TCA CCG CCT TGG CTT-3′; human PDGF-B, 5′-GGA GCA TTT GGA GTG CGC CT-3′ and 5′-ACA TCC GTG TCC TGT TCC CGA-3′.

The amplified products were separated on 1.2% agarose gels with ethidium bromide (E-Gel, Invitrogen). Mean pixel intensities of each band were measured and normalized to mean pixel intensities of glyceraldehyde phosphodehydrogenase band. The values for two experiments (performed in duplicate) were then averaged and graphed with s.d.

Image analysis. Overlapping microscopic pictures were taken at a magnification of 100× so that the entire area of the sample was covered. An imaging software (AxioVision 3.1, Carl Zeiss) was used to determine the area of endothelial cells, the area of vessels or lumen and the total sample area. The number of structures with lumen was counted manually. For colocalization analysis, 3–6 randomly chosen 20× fields were analyzed using OpenLab (Density Slice Module) image analysis software (ImproVision) to quantify endothelial cell–positive areas with and without colocalization of smooth muscle cell–positive areas. An endothelial cell–positive area was identified by the presence of a vWF-positive vessel-like structure with a lumen. *P* values were calculated using Student's *t*-test.

Note: Supplementary information is available on the Nature Biotechnology website.

ACKNOWLEDGMENTS
The authors thank the MIT division of comparative medicine for excellent assistance in tissue embedding and processing, Adam Kapur for help with data analysis, and Justin S. Golub for help with RT-PCR assays. We would like to thank Joseph Itskovitz-Eldor for assistance and cooperation in conducting this research. This work was supported by National Institutes of Health grants HL60435 (R.L. and S.L.) and EY05318 (P.A.D. and D.C.D.).

COMPETING INTERESTS STATEMENT
The authors declare that they have no competing financial interests.

Received 26 January; accepted 2 May 2005
Published online at http://www.nature.com/naturebiotechnology/

1. Saxena, A.K., Marler, J., Benvenuto, M., Willital, G.H. & Vacanti, J.P. Skeletal muscle tissue engineering using isolated myoblasts on synthetic biodegradable polymers: preliminary studies. *Tissue Eng.* **5**, 525–532 (1999).
2. Neumann, T., Nicholson, B.S. & Sanders, J.E. Tissue engineering of perfused microvessels. *Microvasc. Res.* **66**, 59–67 (2003).
3. Bach, A.D., Stem-Straeter, J., Beier, J.P., Bannasch, H. & Stark, G.B. Engineering of muscle tissue. *Clin. Plast. Surg.* **30**, 589–599 (2003).
4. Wigmore, P.M. & Evans, D.J. Molecular and cellular mechanisms involved in the generation of fiber diversity during myogenesis. *Int. Rev. Cytol.* **216**, 175–232 (2002).
5. Buckingham, M. Skeletal muscle formation in vertebrates. *Curr. Opin. Genet. Dev.* **11**, 440–448 (2001).
6. Zisch, A.H., Lutolf, M.P. & Hubbell, J.A. Biopolymeric delivery matrices for angiogenic growth factors. *Cardiovasc. Pathol.* **12**, 295–310 (2003).
7. von Degenfeld, G., Banfi, A., Springer, M.L. & Blau, H.M. Myoblast-mediated gene transfer for therapeutic angiogenesis and arteriogenesis. *Br. J. Pharmacol.* **140**, 620–626 (2003).
8. Lu, Y. *et al.* Recombinant vascular endothelial growth factor secreted from tissue-engineered bioartificial muscles promotes localized angiogenesis. *Circulation* **104**, 594–599 (2001).
9. Levenberg, S., Golub, J.S., Amit, M., Itskovitz-Eldor, J. & Langer, R. Endothelial cells derived from human embryonic stem cells. *Proc. Natl. Acad. Sci. USA* **99**, 4391–4396 (2002).
10. Levenberg, S. *et al.* Differentiation of human embryonic stem cells on three-dimensional polymer scaffolds. *Proc. Natl. Acad. Sci. USA* **100**, 12741–12746 (2003).
11. Darland, D.C. & D'Amore, P.A. TGF beta is required for the formation of capillary-like structures in three-dimensional cocultures of 10T1/2 and endothelial cells. *Angiogenesis* **4**, 11–20 (2001).
12. Carmeliet, P. Angiogenesis in health and disease. *Nat. Med.* **9**, 653–660 (2003).
13. Jain, R.K. Molecular regulation of vessel maturation. *Nat. Med.* **9**, 685–693 (2003).
14. Sieminski, A.L., Padera, R.F., Blunk, T. & Gooch, K.J. Systemic Delivery of Human Growth Hormone Using Genetically Modified Tissue-Engineered Microvascular Networks: Prolonged Delivery and Endothelial Survival with Inclusion of Nonendothelial Cells. *Tissue Eng.* **8**, 1057–1069 (2002).
15. Koike, N. *et al.* Tissue engineering: creation of long-lasting blood vessels. *Nature* **428**, 138–139 (2004).
16. Black, A.F., Berthod, F., L'Heureux, N., Germain, L. & Auger, F.A. *In vitro* reconstruction of a human capillary-like network in a tissue-engineered skin equivalent. *FASEB J.* **12**, 1331–1340 (1998).
17. Shinoka, T. Tissue engineered heart valves: autologous cell seeding on biodegradable polymer scaffold. *Artif. Organs* **26**, 402–406 (2002).
18. Flamme, I., Frolich, T. & Risau, W. Molecular mechanisms of vasculogenesis and embryonic angiogenesis. *J. Cell. Physiol.* **173**, 206–210 (1997).
19. Rossant, J. & Hirashima, M. Vascular development and patterning: making the right choices. *Curr. Opin. Genet. Dev.* **13**, 408–412 (2003).
20. Hirschi, K.K., Rohovsky, S.A. & D'Amore, P.A. PDGF, TGF-beta, and heterotypic cell-cell interactions mediate endothelial cell-induced recruitment of 10T1/2 cells and their differentiation to a smooth muscle fate. *J. Cell Biol.* **141**, 805–814 (1998).
21. Darland, D.C. *et al.* Pericyte production of cell-associated VEGF is differentiation-dependent and is associated with endothelial survival. *Dev. Biol.* **264**, 275–288 (2003).
22. Lammert, E., Cleaver, O. & Melton, D. Induction of pancreatic differentiation by signals from blood vessels. *Science* **294**, 564–567 (2001).
23. Matsumoto, K., Yoshitomi, H., Rossant, J. & Zaret, K.S. Liver organogenesis promoted by endothelial cells prior to vascular function. *Science* **294**, 559–563 (2001).
24. Yaffe, D. & Saxel, O. Serial passaging and differentiation of myogenic cells isolated from dystrophic mouse muscle. *Nature* **270**, 725–727 (1977).
25. Blau, H.M. *et al.* Plasticity of the differentiated state. *Science* **230**, 758–766 (1985).
26. Holder, W.D. Jr. *et al.* Cellular ingrowth and thickness changes in poly-L-lactide and polyglycolide matrices implanted subcutaneously in the rat. *J. Biomed. Mater. Res.* **41**, 412–421 (1998).

Microscale technologies for tissue engineering and biology

Ali Khademhosseini*†, Robert Langer*‡§, Jeffrey Borenstein¶, and Joseph P. Vacanti‖**

*Harvard–Massachusetts Institute of Technology Division of Health Sciences and Technology and ‡Department of Chemical Engineering, Massachusetts Institute of Technology, Cambridge, MA 02139; †Department of Medicine, Brigham and Women's Hospital, Harvard Medical School, Boston, MA 02115; ¶The Charles Stark Draper Laboratory, Cambridge, MA 02139; and ‖Department of Surgery, Massachusetts General Hospital, Harvard Medical School, Boston, MA 02114

Contributed by Robert Langer, September 7, 2005

Microscale technologies are emerging as powerful tools for tissue engineering and biological studies. In this review, we present an overview of these technologies in various tissue engineering applications, such as for fabricating 3D microfabricated scaffolds, as templates for cell aggregate formation, or for fabricating materials in a spatially regulated manner. In addition, we give examples of the use of microscale technologies for controlling the cellular microenvironment *in vitro* and for performing high-throughput assays. The use of microfluidics, surface patterning, and patterned cocultures in regulating various aspects of cellular microenvironment is discussed, as well as the application of these technologies in directing cell fate and elucidating the underlying biology. Throughout this review, we will use specific examples where available and will provide trends and future directions in the field.

microfabrication | review | soft lithography | stem cells | microfluidics

Each year in the U.S., millions of people suffer from a variety of diseases that could be aided from therapies such as organ transplantation. However, despite the widespread need for transplantable tissues, many patients die while waiting for donor organs. It is from this need that the field of tissue engineering has emerged. Tissue engineering is an interdisciplinary field that involves the use of biological sciences and engineering to develop tissues that restore, maintain, or enhance tissue function (1). Tissue engineering has particular advantages over other therapies such as drugs because it can provide a permanent solution to the problem of organ failure. In general, there are three main approaches to tissue engineering: (*i*) to use isolated cells or cell substitutes as cellular replacement parts, (*ii*) to use acellular biomaterials capable of inducing tissue regeneration, and (*iii*) to use a combination of cells and materials (typically in the form of scaffolds) (Fig. 1) (2).

Despite significant advances in tissue engineering (3), which have resulted in successful engineering of organs such as skin and cartilage, there are a number of challenges that remain in making off-the-shelf tissue-engineered organs. These barriers include the lack of a renewable source of functional cells that are immunologically compatible with the patient; the lack of biomaterials with desired mechanical, chemical, and biological properties; and the inability to generate large, vascularized tissues that can easily integrate into the host's circulatory system with the architectural complexity of native tissues.

Microscale technologies are potentially powerful tools for addressing some of the challenges in tissue engineering (4). MEMS (microelectromechanical systems), which are an extension of the semiconductor and microelectronics industries, can be used to control features at length scales from <1 μm to >1 cm (5). These techniques are compatible with cells and are now being integrated with biomaterials to facilitate fabrication of cell–material composites that can be used for tissue engineering. In addition, microscale technologies allow for an unprecedented ability to control the cellular microenvironment in culture and miniaturize assays for high-throughput applications (Fig. 2).

In the past few years, microfabrication has been increasingly used in biomedical and biological applications, partly because of the emergence of techniques such as soft lithography to fabricate microscale devices without the use of expensive "clean rooms" and photolithographic equipment (5). Soft lithography is a set of microfabrication techniques that use elastomeric stamps fabricated from patterned silicon wafers to print or mold materials at resolutions as low as several hundred nanometers (6–9). Therefore, soft lithography can be used to control the topography and spatial distribution of molecules on a surface, as well as the subsequent deposition of cells (10, 11). Soft lithographic methods can also be used to fabricate microfluidic channels and scaffolds for tissue engineering in a convenient, rapid, and inexpensive manner (5, 12). In addition, photolithography, a technique in which microscale features are fabricated based on selective exposure of a material to light, can also be used for microfabrication of tissue engineering structures.

In this review, we will discuss the use of microfabrication technologies as they relate to tissue engineering in addressing challenges such as vascularizing engineered tissues, fabricating 3D structures, directing stem cell fate, and controlling cellular microenvironment. In addition, we will discuss the use of these technologies for providing significant insights into cellular behavior *in vitro* and for creating physiological microenvironments in culture. We will first discuss the use of microscale approaches as they apply directly to tissue engineering in fabricating scaffolds and bioreactors. We will then discuss the use of these technologies as they apply to tissue engineering indirectly through their use in controlling cellular microenvironment.

Microscale Approaches for Tissue Engineering

Cell-Seeded, Microfabricated Scaffolds. In many tissue engineering applications, scaffolds are used to provide cells with a suitable growth environment, optimal oxygen levels, and effective nutrient transport as well as mechanical integrity (13). Scaffolds aim to provide 3D environments to bring cells in close proximity so that they can assemble to form tissues. Ideally, as the scaffold is degraded, the cells deposit their own extracellular matrix (ECM) molecules and eventually form 3D structures that closely mimic the native tissue architecture. Currently, tissue engineering scaffolds are prepared by using a variety of

Conflict of interest statement: No conflicts declared.

Abbreviations: ECM, extracellular matrix; PDMS, poly(dimethylsiloxane); PLGA, poly(DL-lactide-co-glycolide); PEG, polyethylene glycol.

§To whom correspondence may be addressed. E-mail: rlanger@mit.edu.

**To whom correspondence may be addressed at: Massachusetts General Hospital, 55 Fruit Street, Warren Building, Room 1157, Boston, MA 02114. E-mail: jvacanti@partners.org.

www.pnas.org/cgi/doi/10.1073/pnas.0507681102

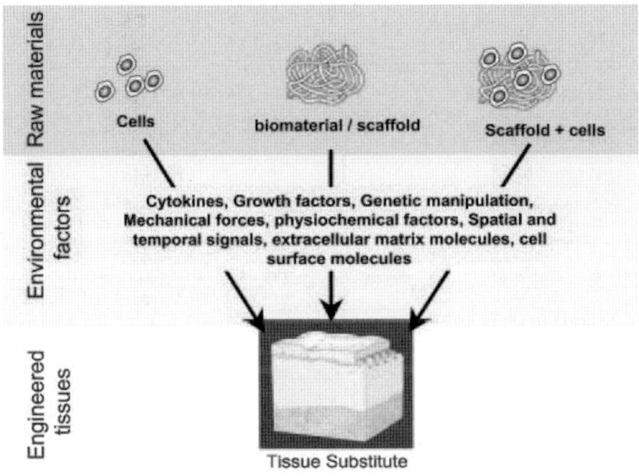

Fig. 1. Tissue engineering approaches. Tissue engineering approaches are classified into three categories: (*i*) cells alone, (*ii*) cells with scaffolds, and (*iii*) scaffolds alone. Each one of these approaches can be enhanced by *in vitro* microenvironmental factors before application as a tissue substitute.

techniques, such as solvent casting and particulate leaching (14). However, scaffold properties such as pore geometry, size, interconnectivity, and spatial distribution depend on the fabrication process rather than design. The inability to generate desired scaffolds has hindered the construction of engineered tissues that are larger than a few hundred micrometers due to oxygen diffusion limitations (15, 16).

Microfabrication approaches have been used to engineer the desired microvasculature directly into the tissue engineering scaffolds (17, 18). Initial experiments used micromachining technologies on silicon surfaces to generate vascularized systems. Subsequent work on the replica molding of biocompatible polymers from patterned silicon wafers has resulted in the fabrication of biocompatible scaffolds (Fig. 2*B*). Originally, such scaffolds were fabricated from poly(dimethylsiloxane) (PDMS) (17), a transparent, biocompatible elastomer that is widely used in replica molding (5, 19). More recently, microfabricated capillary networks have been fabricated out of biodegradable elastomers such as poly(DL-lactide-*co*-glycolide) (PLGA) (20) and poly(glycerol sebacate) (PGS) (21). PLGA has been used to generate biodegradable microfluidic systems and microfabricated scaffolds by superpositioning and stacking multiple layers of fabricated structures (20, 21). These artificial capillary networks were coated with fibronectin and seeded with endothelial cells, which grew to confluence within a few days.

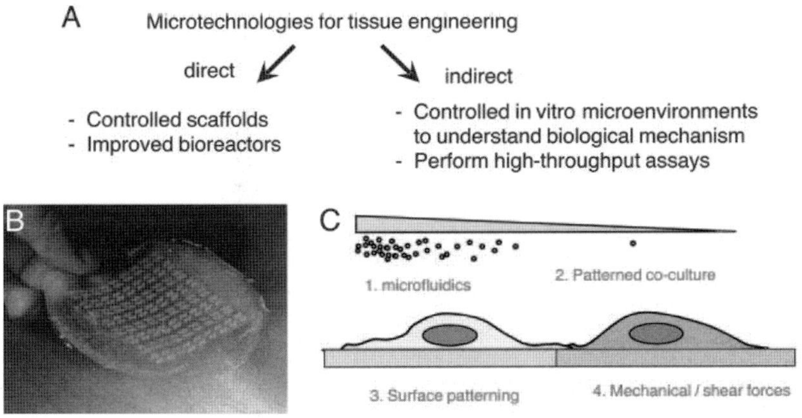

Fig. 2. Microscale technologies for tissue engineering. (*A*) Microtechnologies can be used directly to fabricate improved scaffolds and bioreactors or indirectly to study cellular behavior in controlled conditions or through the use of high-throughput experimentation. (*B*) A PDMS microfabricated tissue engineering scaffold with the vasculature directly embedded into the scaffold (17). (*C*) Various microscale techniques used to control different aspects of cell–microenvironment interactions are shown.

However, there are potential disadvantages with the use of PLGA, including its rigid mechanical properties (22) and bulk degradation kinetics (23). Thus, to improve the mechanical properties and to allow for surface erosion instead of bulk erosion of biodegradable scaffolds, other polymers such as PGS (23) have been used (24). PGS films were patterned with the desired microvasculature and bonded to a flat film to create a capillary network.

Layer-by-layer microfluidic patterning has also been used to generate scaffolds (25–27). In this method, cells and matrix biopolymers were flowed through channels with controlled flow rates. By sequential deposition of cells and matrix on particular regions within the microchannels, 3D structures were generated with controlled location of the deposited cells.

Alternative methods of fabricating scaffolds with microscale and nanoscale resolution include 3D printing, microsyringe deposition, tissue spin casting, and electrospinning of nanofibers. In 3D printing, polymer particles and salt are printed by using a bonding agent, which forms a porous scaffold once the salt is dissolved (28, 29). More recently, rapid prototyping methods such as soft lithography and microsyringe deposition have been used to fabricate PLGA (21) and polyurethane (30) scaffolds composed of multilayer structures with controlled resolution. Although a comprehensive review of these technologies is beyond the scope of this report (for detailed reviews, see refs. 31–33), they are emerging as powerful methods of fabricating tissue engineering scaffolds.

Spatially Regulated Hydrogels and Scaffolds. Many biological processes are regulated by spatially dependent signals. For example, gradients of molecules are commonly used in the body to regulate cell migration, axon extension, angiogenesis, and differentiation. Therefore, controlling the spatial location of molecules on a surface or throughout a material could be potentially beneficial for tissue engineering (34). A common approach to generate gradients is to release molecules from a source to form a concentration gradient over time as the molecule diffuses away from the source (35). However, these gradients are unstable, and it is difficult to control their shapes. Therefore, methods for conjugating biomolecules to materials have been used to increase the stability of signaling molecules in a spatially controlled manner. For example, by using lasers, specific regions within an agarose gel were tethered with RGD (Arg–Gly–Asp) peptide, which allowed for neurite extension

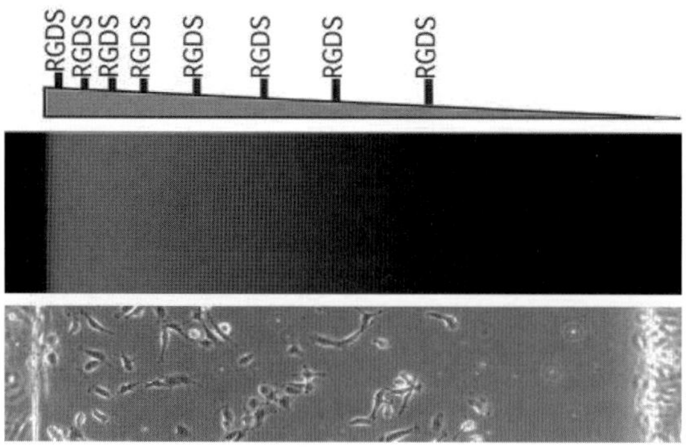

Fig. 3. Gradient hydrogels for tissue engineering. (*Top*) Hydrogels can be fabricated with control over the spatial properties of the materials by embedding a gradient of materials, such as RGD peptide, directly into the material. (*Middle* and *Bottom*) The shape of the gradient can be visualized by using fluorescent molecules (*Middle*), and its function can observed by imaging endothelial cell adhesion after a few hours on the gels (*Bottom*) (39).

within peptide-modified regions (36). By using similar techniques, multicomponent, spatially patterned, photocrosslinkable hydrogels were fabricated to localize growth factors within hydrogels (37). Microfabrication approaches such as microfluidics provide an attractive alternative to these technologies because of their availability and cost effectiveness. Recently, the ability to pattern fluids within microchannels has been merged with photopolymerization chemistry to form spatially oriented hydrogels (38, 39). As a result, gradients of the photocrosslinkable monomers were formed within microfluidic channels and subsequently gelled by exposure to UV light. Hydrogels were synthesized with gradients of signaling or adhesive molecules or with varying crosslinking density across the material (Fig. 3) with the ability to direct cell behavior such as migration, adhesion, and differentiation (39, 40).

In addition to bulk characteristics of scaffold materials, surface properties also have a significant effect on cellular behavior such as cell adhesion and migration. Microcontact printing has been used to modify surfaces of scaffold materials such as PLGA (41, 42), poly(lactic acid) (42), and chitosan (43) with variations of polyethylene glycol (PEG) or other biological ligands, such as RGD or biotin (44). In addition, other techniques such as microfluidic patterning have been used to generate patterns on various tissue engineering surfaces such as PLGA (45). In most cases, these techniques have been limited to 2D substrates, and their incorporation to porous scaffolds remains a challenge.

Microfabrication approaches have also been used to control substrate topography with resolutions as small as 50 nm to control various cell fate decisions. For example, it has been seen that surface topography of tissue culture substrates can significantly affect cardiomyocyte orientation. In these cultures, topographical features that were a few microns across were used to orient cells and enhance their function (46). Furthermore, because microtextured and nanotextured substrates have been shown to significantly influence cell adhesion, gene expression (47–49), and migration (50), such features can be incorporated into microfabricated tissue engineering scaffolds to provide functional cues to cells. For example, topographically patterned PLGA surfaces have been shown to induce alignment and elongation of smooth muscle cells (51) and to enhance the adhesion of several cell types such as endothelial cells and smooth muscle cells (52, 53). Although these approaches have not been easily incorporated into 3D PLGA scaffolds, it may be possible to stack PLGA layers with the desired topographical features to induce specific cell responses.

Controlled Microbioreactors. Bioreactors could be useful in tissue engineering either for growing cells and tissues or as extracorporeal devices for liver and kidney diseases. Microfluidic bioreactors are potentially advantageous for cellular applications because they provide a large surface-area-to-volume ratio, as well as many other biomimetic properties. Recently, microreactors have shown promise in applications where conventional bioreactors have failed to provide the cells with adequate nutrients and oxygen. For example, multilayer PDMS microfluidic networks have been used to culture hepatocytes *in vitro* (54, 55). Inside these reactors, hepatocytes were maintained for many days as they spread and grew to confluency within the channels. In addition, at least 10 layers have been stacked, indicating that such technology is scalable. In another approach, the combination of 3D architecture and fluid perfusion has been used to mimic liver sinusoids (56). Silicon microfluidic chips with holes through the plates were placed on a membrane. The medium was flowed through each hole in the membrane as the cells were retained inside each well. Inside these wells, cells formed spheroids and maintained elevated liver function. Microchannels have also been used as improved versions of flat plate bioreactors for hepatocyte culture with the ability to control parameters such as shear stress, contact with parenchymal cells, and oxygenation (57, 58).

Future generations of microfluidic reactors provide powerful means of exposing cells to various physiological stimuli. For example, a Braille system has been developed to create physiological flow conditions such as pulsatile flow. In this scheme, computer-actuated metal pins were used to deform PDMS channels to pump and regulate the flow of the fluid within the microchannels (59, 60). In addition, complex reactors have been fabricated with the aim of recapitulating the multiple organ interactions of the body. For example, compartmentalized microreactors have been generated to mimic the key functions of the lungs, liver, and fat tissues (61). These animal-on-a-chip devices could be used to test the pharmacological and toxicity effects of drugs in a controlled manner.

Cell Assembly for Tissue Engineering. Artificial microtissues can also be fabricated by inducing the reaggregation of one or more cell types (62). These microtissues may be beneficial for applications such as pancreatic, liver, vascular, and cardiac tissue engineering, as well as drug discovery. Layering of cells has been used to engineer myocardial tissues by assembling multiple sheets of cardiomyocytes (63) or to engineer blood vessels by fabricating cylindrically rolled sheets of endothelial cells (64). Although such approaches may be suitable for some tissue engineering applications, they lack the complexity associated with the architecture of more complex organs. Microscale approaches may provide a solution to this challenge as templates to generate microtissues in a reproducible manner. For example, by using a combination of microcontact printing and micromachining, hepatocyte spheroids have been formed (65). More recently, nonadhesive PEG microwells have been used as templates for formation of aggregates of various cell types, including ES cells

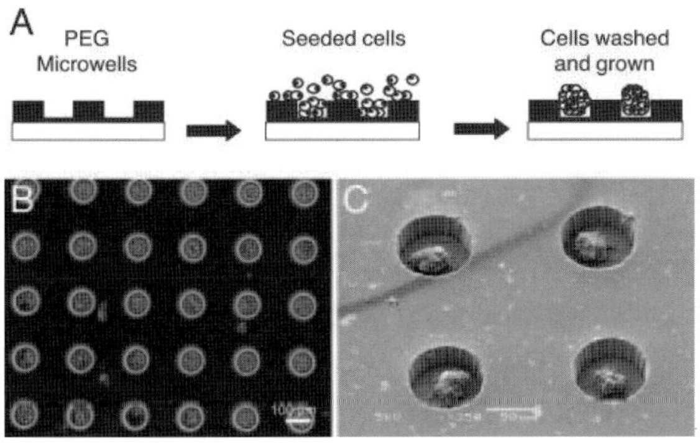

Fig. 4. Microscale tissue engineering using template-based cell assembly. (*A*) A schematic diagram of the template-based assembly method. PEG microwells were fabricated so that cells could dock within the low-shear-stress regions generated within the microstructures. Once cells had immobilized within the microwells, other cells were washed away, and the cells within the microwells formed aggregates of controlled properties. (*B*) A light microscope image of 100-μm PEG wells that were seeded with ES cells and washed. (*C*) A scanning electron micrograph of cells within PEG microwells (A.K., J. Yeh, G. Eng, J. Fukuda, O. Farokhzad, J. J. Cheng, J. Bumbling, and R.L., unpublished data).

(66) (Fig. 4). This approach aims to overcome the disadvantages associated with the hanging drop and suspension culture methods (62) by providing control over the size, shape, and other features of the cellular assembly in a scalable manner. The controlled formation of embryoid bodies may also be important in generating more homogeneous cultures that are capable of directing the differentiation of ES cells. In addition, template-based assembly of cells could be used to organize multiple cell types into specific geometries relative to each other within these aggregated tissue sections. It is envisioned that with the integration of such technologies with biomaterials such as photocrosslinkable gels and microfluidics, more complex tissue sections for therapeutic applications can be fabricated.

Controlling the Cellular Microenvironment *in Vitro*

A major challenge to the clinical feasibility of tissue engineering is a viable cell source. Cells have been traditionally derived from autografts (from the same patient), allografts (from others), or xenografts (from other species). With the emergence of adult stem cells and ES cells, potentially powerful cell sources have emerged. One of the major challenges associated with the use of stem cells is to understand the microenvironmental cues that regulate their fate. Microscale approaches can also be used to control culture conditions and perform high-throughput experimentation, hence providing a suitable tool to study cell–microenvironment interactions *in vitro*. In most tissue culture systems, the cellular microenvironment is vastly different relative to *in vivo* conditions. In the body, cells are exposed to a controlled microenvironment that is tightly regulated with respect to interactions with the surrounding cells, soluble factors, and ECM molecules. In particular, the spatial and temporal distribution of these signals is tightly controlled and unique to each organ. Furthermore, cells in the body are exposed to a 3D environment instead of the 2D environment experienced by cells in traditional culture dishes.

Extracellular cues are important in regulating adult stem cell and ES cell fate decisions. Adult stem cells are present within unique stem cell niches that regulate their self-renewal and differentiation (67–71). Engineering artificial stem cell niches, potentially through microscale techniques, is a potent bioengineering approach to regulate stem cell fates *in vitro* (72). Furthermore, ES cells differentiate based on a series of spatially and temporally regulated signals. Therefore, given the proper series of signals, it may be possible to differentiate ES cells into any desired cell type efficiently and reproducibly (73, 74). Microscale technologies provide a powerful tool to investigate the extracellular signals that regulate cell fate, because they can control cell–microenvironment interactions and be merged with high-throughput technologies to test many environmental factors simultaneously. In general, microscale approaches to control cell–microenvironment interactions can be separated into cell–cell, cell–matrix, and cell-soluble factor components (Fig. 5), which are described below.

Arrays of Cells for Clonal Analysis and Controlling Cell Shape. Cell arrays have been used to pattern stem cells on 2D substrates (Fig. 5*A*). Arrays of cells can be used to localize and track individual cells, enabling clonal analysis of stem cell fates (75, 76). For example, clonal populations of neural stem cells were immobilized within microfabricated structures, and their progeny was tracked by using real-time microscopy, yielding information about cellular kinetics and cell fate decisions in a high-throughput manner (76) (Fig. 6). By using this approach, it is possible to study the response of individual stem cells to various microenvironmental signals.

Cell patterning on geometrically defined shapes has been used to study the effects of cell shape on various cell fate decisions. As cells adhere onto micropatterned substrates, they align themselves to the shape of the underlying adhesive region. This shape change induces changes in the cytoskeletal features and has been shown to influence apoptosis and proliferation (77). More

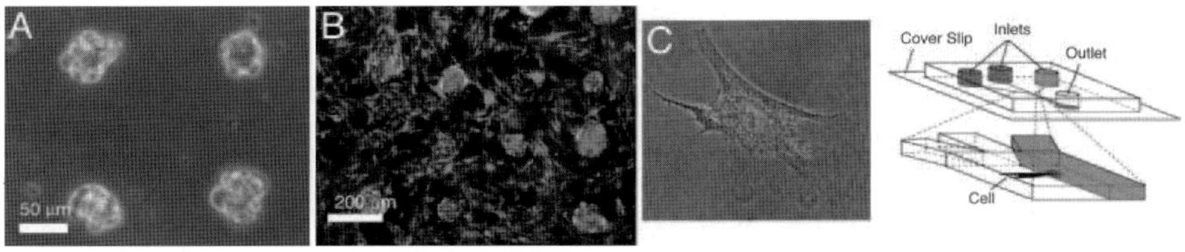

Fig. 5. Microscale approaches for controlling the *in vitro* cellular microenvironment. (*A*) Light microscope images of ES cells patterned on PEG-coated substrates as an example of surface patterning for regulating cell–ECM interactions. (*B*) A fluorescent image of patterned cocultures depicting control over the degree of heterotypic and homotypic interactions between ES cells and fibroblasts (95). (*C*) An image of a cell and a schematic diagram of microfluidic methods of regulating cell-soluble signal interactions by flowing two parallel streams of fluids on an individual cell (81).

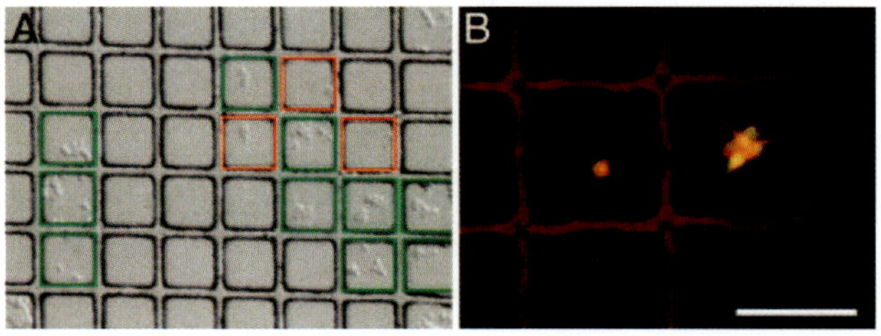

Fig. 6. Stem cell arrays for tracking cell fates in culture. (*A*) Cell microarrays used to track clonal populations of stem cells after 4 days. Green boxes had proliferating cells, whereas red boxes did not contain proliferating cells (76). (*B*) Fluorescent images of the differentiating cells inside a microfabricated array that were stained with Tuj1 (red) and glial fibrillary acidic protein (blue) (76). (Scale bar, 100 μm.)

recently, it has been determined that cell shape can also control stem cell differentiation. When human mesenchymal stem cells were patterned on fibronectin islands of various sizes, cells on large islands adhered and flattened, whereas cells on small islands generated spherically shaped cells. As these cells were stained for differentiation markers, it was observed that the spread cells generated osteoblasts, whereas spherical cells gave rise to adipocytes (78) (Fig. 7). Further elucidation of the mechanism indicated that cell shape regulated the activation of the RhoA pathway, demonstrating that mechanical stresses experienced during differentiation can be crucial for directing stem cell differentiation. Therefore, controlling the cellular microenvironment through micropatterning may be used for directing cell fate for tissue engineering applications.

Microfluidics for Regulating Cell-Soluble Factor Interactions. The ability to laminarly flow fluids within microchannels can be used to control the spatial positioning of soluble factors relative to cells. Laminar flows form within microchannels because of low convective mixing, which limits molecular transport to diffusion. Laminarly flowing fluids have been used to pattern cells and their microenvironments (79, 80). Furthermore, an individual cell can be exposed to multiple microenvironments simultaneously by being placed at the interface between two or more adjacent streams (81). One can treat distinct parts of a cell or cell aggregate to probe the influence of asymmetrically oriented signals on cell behavior or to stain specific regions of a cell for the study of intracellular molecular kinetics. The partial treatment of cell aggregates has already been used to perturb embryonic development and to study the effects of changing the temperature of part of an embryo (82). In addition, microfluidic gradient generators that are fabricated by using the same principles have been used to study the effects of concentration gradients on cells. Linear or complex soluble gradients are generated by sequential merging, mixing, and splitting of two or more inlet streams, each of which may contain a particular environmental stimulus (83, 84) (Fig. 5*C*). These stable gradients have been used to study complex biological systems. For example, cell chemotaxis in response to soluble chemoattractants has been studied to yield insights into neutrophils migration in response to various shape gradients of IL-8 (85). Microfluidic gradients have also been used to study neural axon extensions (86, 87) and neural stem cell differentiation (88). It is important to consider design parameters such as shear stress, transport phenomenon, and material–cell interactions within these devices (12); however, with the explosion of research in this area in the past few years, such parameters are continuously optimized and modified to attain optimum conditions for various cell types.

Patterned Cocultures for Controlling Cell–Cell Interactions. Inside the body, cells lie in contact or in close proximity to other cell types in a tightly controlled architecture. Tissue engineering constructs that aim to reproduce the architecture and the geometry of tissues will benefit from methods of controlling cell–cell interactions within these tissues. Patterned cocultures are a useful tool for tissue engineered constructs and for studying cell–cell interactions *in vitro*, because they can be used to control the degree of homotypic and heterotypic cell–cell contact. Some of the pioneering work in this area has been performed by studying the interaction of hepatocytes and nonparenchymal fibroblasts in cocultures (89). By using these techniques, specific information about the nature of hepatoycte and fibroblast interaction in culture has been elucidated (90–92). Although the original methods of fabricating patterned coculture required selective adhesion of each cell types to a particular material on a patterned region, subsequent methods have been developed that allow a wider array of cell types to be tested. The recently developed methods are based on thermally reversible polymers (93, 94), layer-by-layer deposition of ionic polymers (95) (Fig. 5*B*), microfluidic deposition (96), and molding of hydrogels (97). Through these approaches, other coculture techniques have been used to probe the interaction of a number of other cell types, including ES cells and other developmentally relevant cell types (95).

High-Throughput Assays for Tissue Engineering. ***Material arrays.*** As methods to develop small-molecule (98), polymeric (99), and genome-based libraries become more widespread and available, there is a great need to test these libraries by using high-throughput assays. Such libraries are useful for tissue engineering and have already yielded candidates that have been shown to induce osteogenesis (100) and cardiomyogenesis (101) from ES cells as well as the dedifferentiation of committed cells (102). Microscale technologies can miniaturize assays and facilitate high-throughput experimentation and therefore provide a

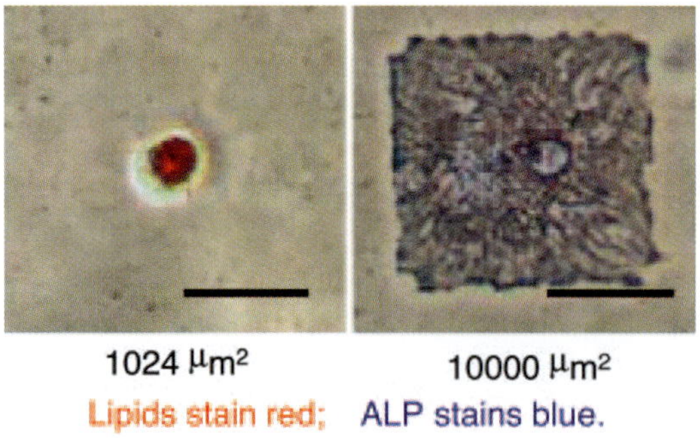

Fig. 7. Light microscope images of human mesenchymal stem cells on small and large fibronectin islands after 1 week of culture. The images indicate that cells on the small islands stained for lipids, thus differentiating into adipogenic fates, whereas cells on large islands stained for alkaline phosphatase and differentiated into osteoblasts (78). (Scale bars, 50 μm.)

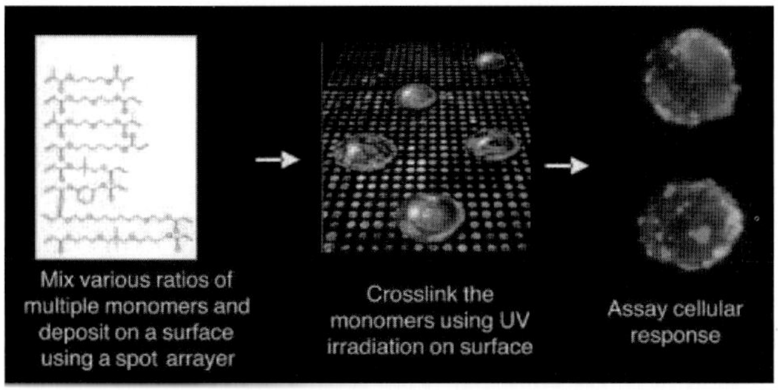

Fig. 8. Schematic diagram of the procedure used to fabricate arrays of polymeric biomaterials (103).

potential tool for screening libraries. Recently, robotic spotters capable of dispensing and immobilizing nanoliters of material have been used to fabricate microarrays, where cell–matrix interactions can be tested and optimized in a high-throughput manner. For example, synthetic biomaterial arrays have been fabricated to test the interaction of stem cells with various extracellular signals (103) (Fig. 8). In this approach, thousands of polymeric materials were synthesized, and their effect on differentiation of human ES cells (103) and human mesenchymal stem cells (104) was evaluated. These interactions have led to unexpected cell–material interactions. Although the molecular mechanisms associated with the biological responses have yet to be clarified, such technology may be widely applicable in cell microenvironment studies and in the identification of cues that induce desired cell responses. Also, the materials that yield the desired responses could be used as templates for tissue engineering scaffolds. Such an approach is a radical change from traditional methods of developing new biomaterials, where polymers have been individually developed and tested for their effect on cells.

In addition to analyzing synthetic material libraries, the effect of natural ECM molecules on cell fate can be evaluated in a high-throughput manner. In one example, combinatorial matrices of various natural ECM proteins were tested for their ability to maintain the function of differentiated hepatocytes and to induce hepatic differentiation from murine ES cells (105).

Microfluidic systems. Microfluidic systems are ideal for performing cell-based high-throughput experiments because they are cheap and can be used to miniaturize assays, reduce expensive reagent volumes, and reduce cell numbers (19). Microfluidic arrays can be used to perform screening experiments, to test drug toxicity, and to optimize culture conditions for inducing specific cell fates. To create such systems, it is important to be able to culture cells inside microfluidic devices in a reliable and long-term manner and to be able to combine these devices in such a way as to test many conditions simultaneously. Many experiments have localized cells within microfluidic channels; however, the long-term survival of cells inside these channels has not been demonstrated (75, 87). In addition, it is important to localize the cells in particular regions of microfluidic channels by using substrate patterning or topographical structures (66, 106). Also, it is important to interface the microscale channels to the outside world. Current techniques include the use of complicated valves and pumps to fabricate large arrays of microchannels that can be individually addressed (107). The ability to test various combinations and doses of drugs can be useful for performing factorial experiments, which can optimize culture conditions for inducing various biological fates (108). An alternative approach for cell-based molecular screening is to use multiphenotype cell arrays (109, 110). In these systems, the effect of a soluble factor can be tested on many cell types simultaneously. Incorporating multiphenotype cell arrays within microfluidic arrays has shown promise as a tool for fabricating high-throughput assays (110).

Future Directions

The synthesis of new materials and the incorporation of improved fabrication strategies have led to dramatic enhancements in the complexity and biomimecry of today's tissue engineering constructs. In particular, the use of microscale technologies in tissue engineering, either directly as transplantable constructs or indirectly as a tool for understanding the biology of organs, has pushed the field closer to clinical therapies. However, significant challenges need to be addressed. These barriers include the lack of suitable materials with the desired degradation rates, products, and suitable mechanical properties for the desired tissue. Another challenge that must be addressed is the optimization of scaffold architecture, including pore size, morphology, surface topography, and bioactivity. Also, new and optimized processing methods must be developed to address issues related to cell seeding, vascularization, and fabrication of 3D tissues without cumbersome methods of assembly. Furthermore, more research is required in testing and validating the *in vivo* functionality of microfabricated constructs and in assessing the performance of these constructs against competing technologies.

In addition, microscale tools have provided much insight into the fundamental biology of how cells interact with the surrounding components, such as cell–cell, cell-soluble factors, and cell–ECM molecules. This knowledge should be used to direct cell fates as a cell source or be incorporated into tissue engineering scaffolds. Thus, the continued merger of engineering, medicine, materials, and biological sciences as mediated by microscale approaches in tissue engineering and biology will enhance our ability to create *in vivo*-like physiological models that can be used for fabricating tissues or for understanding fundamental biology.

Conclusions

The widespread use and availability of lithographic technologies have made microscale engineering a powerful tool for tissue engineering and biological applications. Microfabrication techniques and engineered biomaterials are being used for tissue engineering in a variety of applications: for example, by fabricating scaffolds with control over features such as shape and pore architecture, as templates for microtissue formation, or as improved bioreactors. In addition, microscale control of cellular environments has been used to probe the influence of the spatial and temporal effects of specific cell–cell, cell–ECM, and cell-soluble factor interactions on cell fate. Finally, the ability to simultaneously test many environmental factors on cell behavior has been used to optimize culture conditions and material–cell interactions.

This work was supported by National Institutes of Health Grants HL060435, DE013023, and DE16516; the National Science Laboratory; The Charles Stark Draper Laboratory; Center for Integration of Medicine and Innovative Technology Cooperative Agreement DAMD-02-2-0006; and Institute for Soldier Nanotechnology Grant DAAD-19-02-D-002.

1. Langer, R. & Vacanti, J. P. (1993) *Science* **260,** 920–926.
2. Griffith, L. G. & Naughton, G. (2002) *Science* **295,** 1009–1014.
3. Niklason, L. E. & Langer, R. (2001) *J. Am. Med. Assoc.* **285,** 573–576.
4. Andersson, H. & van den Berg, A. (2004) *Lab Chip* **4,** 98–103.
5. Whitesides, G. M., Ostuni, E., Takayama, S., Jiang, X. Y. & Ingber, D. E. (2001) *Annu. Rev. Biomed. Eng.* **3,** 335–373.
6. Ostuni, E., Chen, C. S., Ingber, D. E. & Whitesides, G. M. (2001) *Langmuir* **17,** 2828–2834.
7. Ostuni, E., Yan, L. & Whitesides, G. M. (1999) *Colloids Surf. B Biointerfaces* **15,** 3–30.
8. Kane, R. S., Takayama, S., Ostuni, E., Ingber, D. E. & Whitesides, G. M. (1999) *Biomaterials* **20,** 2363–2376.
9. Xia, Y. N. & Whitesides, G. M. (1998) *Angew. Chem. Int. Ed.* **37,** 551–575.
10. Khademhosseini, A., Jon, S., Suh, K. Y., Tran, T. N. T., Eng, G., Yeh, J., Seong, J. & Langer, R. (2003) *Adv. Mater.* **15,** 1995–2000.
11. Suh, K. Y., Khademhosseini, A., Yang, J. M., Eng, G. & Langer, R. (2004) *Adv. Mater.* **16,** 584–588.
12. Walker, G. M., Zeringue, H. C. & Beebe, D. J. (2004) *Lab Chip* **4,** 91–97.
13. Cohen, S., Bano, M. C., Cima, L. G., Allcock, H. R., Vacanti, J. P., Vacanti, C. A. & Langer, R. (1993) *Clin. Mater.* **13,** 3–10.
14. Karp, J., Dalton, P. & Shoichet, M. (2003) *MRS Bull.* **28,** 301–306.
15. Folkman, J. & Hochberg, M. (1973) *J. Exp. Med.* **138,** 745–753.
16. Hutmacher, D. W., Sittinger, M. & Risbud, M. V. (2004) *Trends Biotechnol.* **22,** 354–362.
17. Borenstein, J. T., Terai, H., King, K. R., Weinberg, E. J., Kaazempur-Mofrad, M. R. & Vacanti, J. P. (2002) *Biomed. Microdevices* **4,** 167–175.
18. Kaihara, S., Borenstein, J., Koka, R., Lalan, S., Ochoa, E. R., Ravens, M., Pien, H., Cunningham, B. & Vacanti, J. P. (2000) *Tissue Eng.* **6,** 105–117.
19. Sia, S. K. & Whitesides, G. M. (2003) *Electrophoresis* **24,** 3563–3576.
20. King, K., Wang, C., Kaazempur-Mofrad, M., Vacanti, J. & Borenstein, J. (2004) *Adv. Mater.* **16,** 2007–2012.
21. Vozzi, G., Flaim, C., Ahluwalia, A. & Bhatia, S. (2003) *Biomaterials* **24,** 2533–2540.
22. Leatrese, D. H., Kim, B. S. & Mooney, D. J. (1998) *J. Biomed. Mater. Res.* **42,** 396–402.
23. Wang, Y., Ameer, G. A., Sheppard, B. J. & Langer, R. (2002) *Nat. Biotechnol.* **20,** 602–606.
24. Fidkowski, C., Kaazempur-Mofrad, M. R., Borenstein, J., Vacanti, J. P., Langer, R. & Wang, Y. (2005) *Tissue Eng.* **11,** 302–309.
25. Tan, W. & Desai, T. A. (2003) *Tissue Eng.* **9,** 255–267.
26. Tan, W. & Desai, T. A. (2004) *Biomaterials* **25,** 1355–1364.
27. Tan, W. & Desai, T. A. (2005) *J. Biomed. Mater. Res. A* **72,** 146–160.
28. Griffith, L. G., Wu, B., Cima, M. J., Powers, M. J., Chaignaud, B. & Vacanti, J. P. (1997) *Ann. N.Y. Acad. Sci.* **831,** 382–397.
29. Kim, S. S., Utsunomiya, H., Koski, J. A., Wu, B. M., Cima, M. J., Sohn, J., Mukai, K., Griffith, L. G. & Vacanti, J. P. (1998) *Ann. Surg.* **228,** 8–13.
30. Folch, A., Mezzour, S., During, M., Hurtado, O., Toner, M. & Muller, R. (2000) *Biomed. Microdevices* **2,** 207–214.
31. Yeong, W. Y., Chua, C. K., Leong, K. F. & Chandrasekaran, M. (2004) *Trends Biotechnol.* **22,** 643–652.
32. Bhatia, S. N. & Chen, S. C. (1999) *Biomed. Microdevices* **2,** 131–144.
33. Tsang, V. L. & Bhatia, S. N. (2004) *Adv. Drug Delivery Rev.* **56,** 1635–1647.
34. Kapur, T. A. & Shoichet, M. S. (2004) *J. Biomed. Mater. Res. A* **68,** 235–243.
35. Richardson, T. P., Peters, M. C., Ennett, A. B. & Mooney, D. J. (2001) *Nat. Biotechnol.* **19,** 1029–1034.
36. Luo, Y. & Shoichet, M. S. (2004) *Nat. Mater.* **3,** 249–253.
37. Mapili, G., Lu, Y., Chen, S. & Roy, K. (2005) *J. Biomed. Mater. Res. B* **75,** 414–424.
38. Koh, W. G., Revzin, A. & Pishko, M. V. (2002) *Langmuir* **18,** 2459–2462.
39. Burdick, J. A., Khademhosseini, A. & Langer, R. (2004) *Langmuir* **20,** 5153–5156.
40. Zaari, N., Rajagopalan, S. K., Kim, S. K., Engler, A. J. & Wong, J. Y. (2004) *Adv. Mater.* **16,** 2133–2137.
41. Lu, L., Nyalakonda, K., Kam, L., Bizios, R., Gopferich, A. & Mikos, A. G. (2001) *Biomaterials* **22,** 291–297.
42. Lin, C. C., Co, C. C. & Ho, C. C. (2005) *Biomaterials* **26,** 3655–3662.
43. Kumar, G., Wang, Y. C., Co, C. C. & Ho, C. C. (2003) *Langmuir* **19,** 10550–10556.
44. Lee, K. B., Kim, D. J., Lee, Z. W., Woo, S. I. & Choi, I. S. (2004) *Langmuir* **20,** 2531–2535.
45. Liu, V. A., Jastromb, W. E. & Bhatia, S. N. (2002) *J. Biomed. Mater. Res.* **60,** 126–134.
46. Deutsch, J., Motlagh, D., Russell, B. & Desai, T. A. (2000) *J. Biomed. Mater. Res.* **53,** 267–275.
47. den Braber, E. T., de Ruijter, J. E., Ginsel, L. A., von Recum, A. F. & Jansen, J. A. (1998) *J. Biomed. Mater. Res.* **40,** 291–300.
48. Walboomers, X. F., Croes, H. J., Ginsel, L. A. & Jansen, J. A. (1999) *J. Biomed. Mater. Res.* **47,** 204–212.
49. van Kooten, T. G., Whitesides, J. F. & von Recum, A. (1998) *J. Biomed. Mater. Res.* **43,** 1–14.
50. Teixeira, A. I., Abrams, G. A., Bertics, P. J., Murphy, C. J. & Nealey, P. F. (2003) *J. Cell Sci.* **116,** 1881–1892.
51. Thakar, R. G., Ho, F., Huang, N. F., Liepmann, D. & Li, S. (2003) *Biochem. Biophys. Res. Commun.* **307,** 883–890.
52. Miller, D. C., Thapa, A., Haberstroh, K. M. & Webster, T. J. (2004) *Biomaterials* **25,** 53–61.
53. Thapa, A., Webster, T. J. & Haberstroh, K. M. (2003) *J. Biomed. Mater. Res. A* **67,** 1374–1383.
54. Leclerc, E., Sakai, Y. & Fujii, T. (2004) *Biotechnol. Prog.* **20,** 750–755.
55. Leclerc, E., Sakai, Y. & Fujii, T. (2003) *Biomed. Microdevices* **5,** 109–114.
56. Powers, M. J., Domansky, K., Kaazempur-Mofrad, M. R., Kalezi, A., Capitano, A., Upadhyaya, A., Kurzawski, P., Wack, K. E., Stolz, D. B., Kamm, R. & Griffith, L. G. (2002) *Biotechnol. Bioeng.* **78,** 257–269.
57. Park, J., Berthiaume, F., Toner, M., Yarmush, M. L. & Tilles, A. W. (2005) *Biotechnol. Bioeng.* **90,** 632–644.
58. Tilles, A. W., Berthiaume, F., Yarmush, M. L., Tompkins, R. G. & Toner, M. (2002) *J. Hepatobiliary Pancreat. Surg.* **9,** 686–696.
59. Song, J. W., Gu, W., Futai, N., Warner, K. A., Nor, J. E. & Takayama, S. (2005) *Anal. Chem.* **77,** 3993–3999.
60. Gu, W., Zhu, X., Futai, N., Cho, B. S. & Takayama, S. (2004) *Proc. Natl. Acad. Sci. USA* **101,** 15861–15866.
61. Viravaidya, K., Sin, A. & Shuler, M. L. (2004) *Biotechnol. Prog.* **20,** 316–323.
62. Kelm, J. M. & Fussenegger, M. (2004) *Trends Biotechnol.* **22,** 195–202.
63. Shimizu, T., Yamato, M., Kikuchi, A. & Okano, T. (2003) *Biomaterials* **24,** 2309–2316.
64. L'Heureux, N., Paquet, S., Labbe, R., Germain, L. & Auger, F. A. (1998) *FASEB J.* **12,** 47–56.
65. Fukuda, J., Mizumoto, H., Nakazawa, K., Kajiwara, T. & Funatsu, K. (2004) *Int. J. Artif. Organs* **27,** 1091–1099.
66. Khademhosseini, A., Yeh, J., Jon, S., Eng, G., Suh, K. Y., Burdick, J. A. & Langer, R. (2004) *Lab Chip* **4,** 425–430.
67. Zhang, J., Niu, C., Ye, L., Huang, H., He, X., Tong, W. G., Ross, J., Haug, J., Johnson, T., Feng, J. Q., *et al.* (2003) *Nature* **425,** 836–841.
68. Calvi, L. M., Adams, G. B., Weibrecht, K. W., Weber, J. M., Olson, D. P., Knight, M. C., Martin, R. P., Schipani, E., Divieti, P., Bringhurst, F. R., *et al.* (2003) *Nature* **425,** 841–846.
69. Lemischka, I. R. (1997) *Stem Cells* **15,** 63–68.
70. Benitah, S. A., Frye, M., Glogauer, M. & Watt, F. M. (2005) *Science* **309,** 933–935.
71. Watt, F. M. & Hogan, B. L. (2000) *Science* **287,** 1427–1430.
72. Semino, C. E. (2003) *J. Biomed. Biotechnol.* **2003,** 164–169.
73. Zandstra, P. W. & Nagy, A. (2001) *Annu. Rev. Biomed. Eng.* **3,** 275–305.
74. Czyz, J. & Wobus, A. (2001) *Differentiation* **68,** 167–174.
75. Tourovskaia, A., Figueroa-Masot, X. & Folch, A. (2005) *Lab Chip* **5,** 14–19.
76. Chin, V. I., Taupin, P., Sanga, S., Scheel, J., Gage, F. H. & Bhatia, S. N. (2004) *Biotechnol. Bioeng.* **88,** 399–415.
77. Chen, C. S., Mrksich, M., Huang, S., Whitesides, G. M. & Ingber, D. E. (1997) *Science* **276,** 1425–1428.
78. McBeath, R., Pirone, D. M., Nelson, C. M., Bhadriraju, K. & Chen, C. S. (2004) *Dev. Cell* **6,** 483–495.
79. Takayama, S., Ostuni, E., Qian, X. P., McDonald, J. C., Jiang, X. Y., LeDuc, P., Wu, M. H., Ingber, D. E. & Whitesides, G. M. (2001) *Adv. Mater.* **13,** 570–574.
80. Takayama, S., McDonald, J. C., Ostuni, E., Liang, M. N., Kenis, P. J. A., Ismagilov, R. F. & Whitesides, G. M. (1999) *Proc. Natl. Acad. Sci. USA* **96,** 5545–5548.
81. Takayama, S., Ostuni, E., LeDuc, P., Naruse, K., Ingber, D. E. & Whitesides, G. M. (2001) *Nature* **411,** 1016.
82. Lucchetta, E. M., Lee, J. H., Fu, L. A., Patel, N. H. & Ismagilov, R. F. (2005) *Nature* **434,** 1134–1138.
83. Jeon, N. L., Dertinger, S. K. W., Chiu, D. T., Choi, I. S., Stroock, A. D. & Whitesides, G. M. (2000) *Langmuir* **16,** 8311–8316.
84. Dertinger, S. K. W., Chiu, D. T., Jeon, N. L. & Whitesides, G. M. (2001) *Anal. Chem.* **73,** 1240–1246.
85. Jeon, N. L., Baskaran, H., Dertinger, S. K. W., Whitesides, G. M., Van de Water, L. & Toner, M. (2002) *Nat. Biotechnol.* **20,** 826–830.
86. Dertinger, S. K. W., Jiang, X. Y., Li, Z. Y., Murthy, V. N. & Whitesides, G. M. (2002) *Proc. Natl. Acad. Sci. USA* **99,** 12542–12547.
87. Taylor, A. M., Blurton-Jones, M., Rhee, S. W., Cribbs, D. H., Cotman, C. W. & Jeon, N. L. (2005) *Nat. Methods* **2,** 599–605.
88. Chung, B. G., Flanagan, L. A., Rhee, S. W., Schwartz, P. H., Lee, A. P., Monuki, E. S. & Jeon, N. L. (2005) *Lab Chip* **5,** 401–406.
89. Bhatia, S. N., Balis, U. J., Yarmush, M. L. & Toner, M. (1999) *FASEB J.* **13,** 1883–1900.
90. Bhatia, S. N., Balis, U. J., Yarmush, M. L. & Toner, M. (1998) *Biotechnol. Prog.* **14,** 378–387.
91. Bhatia, S. N., Balis, U. J., Yarmush, M. L. & Toner, M. (1998) *J. Biomater. Sci. Polym. Ed.* **9,** 1137–1160.
92. Bhatia, S. N., Yarmush, M. L. & Toner, M. (1997) *J. Biomed. Mater. Res.* **34,** 189–199.
93. Yamato, M., Konno, C., Utsumi, M., Kikuchi, A. & Okano, T. (2002) *Biomaterials* **23,** 561–567.
94. Hirose, M., Yamato, M., Kwon, O. H., Harimoto, M., Kushida, A., Shimizu, T., Kikuchi, A. & Okano, T. (2000) *Yonsei Med. J.* **41,** 803–813.

95. Khademhosseini, A., Suh, K. Y., Yang, J. M., Eng, G., Yeh, J., Levenberg, S. & Langer, R. (2004) *Biomaterials* **25,** 3583–3592.
96. Chiu, D. T., Jeon, N. L., Huang, S., Kane, R. S., Wargo, C. J., Choi, I. S., Ingber, D. E. & Whitesides, G. M. (2000) *Proc. Natl. Acad. Sci. USA* **97,** 2408–2413.
97. Tang, M. D., Golden, A. P. & Tien, J. (2003) *J. Am. Chem. Soc.* **125,** 12988–12989.
98. Ding, S., Gray, N. S., Wu, X., Ding, Q. & Schultz, P. G. (2002) *J. Am. Chem. Soc.* **124,** 1594–1596.
99. Lynn, D. M., Anderson, D. G., Putnam, D. & Langer, R. (2001) *J. Am. Chem. Soc.* **123,** 8155–8156.
100. Wu, X., Ding, S., Ding, Q., Gray, N. S. & Schultz, P. G. (2002) *J. Am. Chem. Soc.* **124,** 14520–14521.
101. Wu, X., Ding, S., Ding, G., Gray, N. S. & Schultz, P. G. (2004) *J. Am. Chem. Soc.* **126,** 1590–1591.
102. Chen, S. B., Zhang, Q. S., Wu, X., Schultz, P. G. & Ding, S. (2004) *J. Am. Chem. Soc.* **126,** 410–411.
103. Anderson, D. G., Levenberg, S. & Langer, R. (2004) *Nat. Biotechnol.* **22,** 863–866.
104. Anderson, D. G., Putnam, D., Lavik, E. B., Mahmood, T. A. & Langer, R. (2005) *Biomaterials* **26,** 4892–4897.
105. Flaim, C. J., Chien, S. & Bhatia, S. N. (2005) *Nat. Methods* **2,** 119–125.
106. Khademhosseini, A., Suh, K. Y., Jon, S., Chen, G., Eng, G., Yeh, J. & Langer, R. (2004) *Anal. Chem.* **76,** 3675–3681.
107. Thorsen, T., Maerkl, S. J. & Quake, S. R. (2002) *Science* **298,** 580–584.
108. Prudhomme, W., Daley, G. Q., Zandstra, P. & Lauffenburger, D. A. (2004) *Proc. Natl. Acad. Sci. USA* **101,** 2900–2905.
109. Koh, W. G., Itle, L. J. & Pishko, M. V. (2003) *Anal. Chem.* **75,** 5783–5789.
110. Khademhosseini, A., Yeh, J., Eng, G., Kazi, H., Borenstein, J., Karp, J. M., Farokhzad, O. & Langer, R., *Lab Chip* **5,** 1380–1386.

ARTICLES

Accordion-like honeycombs for tissue engineering of cardiac anisotropy

GEORGE C. ENGELMAYR Jr[1], MINGYU CHENG[1], CHRISTOPHER J. BETTINGER[2,3], JEFFREY T. BORENSTEIN[3], ROBERT LANGER[1,4] AND LISA E. FREED[1*]

[1]Harvard-MIT Division of Health Sciences and Technology, Massachusetts Institute of Technology, 77 Massachusetts Avenue, E25-330 Cambridge, Massachusetts 02139, USA
[2]Department of Materials Science and Engineering, Massachusetts Institute of Technology, 77 Massachusetts Avenue, E25-330 Cambridge, Massachusetts 02139, USA
[3]Biomedical Engineering Center, Charles Stark Draper Laboratory, 555 Technology Square, Cambridge, Massachusetts 02139, USA
[4]Department of Chemical Engineering, Massachusetts Institute of Technology, 77 Massachusetts Avenue, E25-330 Cambridge, Massachusetts 02139, USA
*e-mail: Lfreed@mit.edu

Published online: 2 November 2008; doi:10.1038/nmat2316

Tissue-engineered grafts may be useful in myocardial repair; however, previous scaffolds have been structurally incompatible with recapitulating cardiac anisotropy. Here, we use microfabrication techniques to create an accordion-like honeycomb microstructure in poly(glycerol sebacate), which yields porous, elastomeric three-dimensional (3D) scaffolds with controllable stiffness and anisotropy. Accordion-like honeycomb scaffolds with cultured neonatal rat heart cells demonstrated utility through: (1) closely matched mechanical properties compared to native adult rat right ventricular myocardium, with stiffnesses controlled by polymer curing time; (2) heart cell contractility inducible by electric field stimulation with directionally dependent electrical excitation thresholds ($p < 0.05$); and (3) greater heart cell alignment ($p < 0.0001$) than isotropic control scaffolds. Prototype bilaminar scaffolds with 3D interconnected pore networks yielded electrically excitable grafts with multi-layered neonatal rat heart cells. Accordion-like honeycombs can thus overcome principal structural–mechanical limitations of previous scaffolds, promoting the formation of grafts with aligned heart cells and mechanical properties more closely resembling native myocardium.

Ventricular myocardium is a richly vascularized, quasi-lamellar tissue in which functional syncytia of cardiomyocytes (cardiac muscle fibres) are interwoven within collagen. Hierarchically, cardiac muscle fibres are surrounded and coupled by endomysial collagen sheaths[1] that are bundled within a honeycomb-like network of undulated perimysial collagen fibres[2]. These features yield directionally dependent electrical and mechanical properties[3] collectively termed cardiac anisotropy (Fig. 1). However, in developing tissue-engineered grafts for myocardial repair, it has become apparent that previous three-dimensional (3D) scaffold materials are structurally and mechanically incompatible with the formation of a fully biomimetic tissue. Isotropic collagen foams and gels yield isolated regions of cardiomyocyte alignment, requiring external stimuli such as non-homogeneous boundary constraints[4], cyclic stretch[5,6] or electrical stimulation[7] to guide parallel orientation; stiff scaffolds such as non-woven poly(glycolic acid)[8,9] tend to resist all but small-amplitude, largely isometric contractility; and hydrogels, although capable of soft-tissue-like compliance[10], can be difficult to suture[11] and are often too weak to support physiologic loads[12]. Conceptually appealing would be a scaffold microstructure borrowing more closely from nature's lessons. Indeed, Ott *et al.*[13] recently demonstrated that decellularized adult rat hearts retaining anisotropic structural and mechanical properties could provide a scaffold for cultured neonatal rat heart cells to regenerate nascent pump function. Here, we sought to microfabricate biomimetic scaffolds with anisotropic structural and mechanical properties from a synthetic bioresorbable elastomer. In particular, we reasoned accordion-like honeycomb scaffolds exhibiting distinct preferred and orthogonal, cross-preferred material directions could potentially: (1) match the anisotropic in-plane mechanical response of native myocardium within the physiologic regime, (2) provide low in-plane resistance to contraction and (3) provide an inherent structural capacity to guide cardiomyocyte orientation in the absence of external stimuli.

The accordion-like honeycomb was designed by overlapping two $200 \times 200\,\mu m$ square pores oriented at 45° (diamonds) in a pattern to leave a planar network of undulated, 50-μm-wide struts (Fig. 2a,b). To assess the utility of accordion-like honeycomb scaffolds in myocardial tissue engineering, poly(glycerol sebacate) (PGS) was chosen as the material of construction for its rubber-like elasticity[14] and structure-preserving, gradual degradation (through surface hydrolysis, with a slow, linear decrease in mechanical strength of ~8% per week *in vivo*)[15]. Initial PGS curing conditions of 16 h at 160 °C were chosen on the basis of previous PGS soft lithography studies[16,17]. Mechanical testing of the bulk 16 h/160 °C PGS yielded effective stiffness (E_{PGS}), ultimate tensile strength (UTS_{PGS}) and strain-to-failure (εf_{PGS}) values of $2,116 \pm 100$ kPa, 793 ± 117 kPa and 0.69 ± 0.15, respectively. Accordion-like honeycomb scaffolds were fabricated from PGS membranes by excimer laser microablation. A membrane thickness of ~250 μm was chosen to mitigate mass transport limitations, on the basis of our previous modelling analyses of oxygen diffusion through solid PGS (ref. 17) and tissue-engineered myocardium[18]. Excimer laser microablation yielded well-defined features comparable to the design (Fig. 2a–c). Accordion-like honeycomb scaffolds were found to be anisotropic when tested

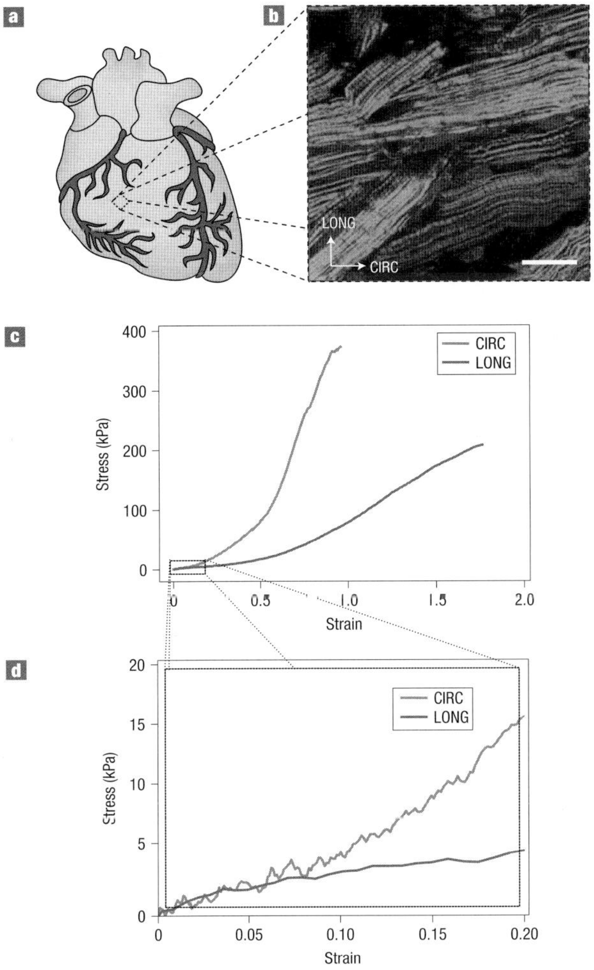

Figure 1 Structural and mechanical aspects of cardiac anisotropy. Ventricular myocardium has long been recognized as an anisotropic tissue, with tensile mechanical properties dictated by cardiac muscle fibre orientation. **a**, A schematic diagram illustrating the gross macroscopic appearance of a four-chamber mammalian heart. **b**, A full-thickness specimen of adult rat right ventricular myocardium showing preferentially oriented cardiac muscle fibres, fluorescently labelled for F-actin and cell nuclei and imaged from the epicardial surface by confocal microscopy. Scale bar: 50 μm; anatomically defined circumferential (CIRC) and longitudinal (LONG) axes are indicated. **c**,**d**, Representative uniaxial tensile stress–strain plots for circumferential and longitudinal specimens of full-thickness right ventricular myocardium demonstrated anisotropic mechanical properties consistent with observed cell orientations (**c**: full range to demonstrate failure properties; **d**: physiologic regime[19–21]). Collectively, these structural, mechanical and associated electrical properties comprise the interrelated aspects of cardiac anisotropy.

under uniaxial tension, with different effective stiffnesses in the preferred and cross-preferred directions ($E_{PD} = 195 \pm 8$ kPa versus $E_{XD} = 57 \pm 3$ kPa, $p < 0.0001$, with an associated anisotropy ratio E_{PD}/E_{XD} of 3.4 ± 0.2).

To determine if these stiffness values were appropriate, specimens of adult rat right and left ventricular myocardium were tested in the circumferential and longitudinal directions, and effective stiffnesses (E_{CIRC}, E_{LONG}) were obtained in the linear, physiologic strain regime of ± 0.15 (refs 19–21) (Fig. 1c). Of note, although organ-level cardiac muscle structures[22] and transmural variations in muscle fibre orientation[23] are complex, we defined circumferential and longitudinal axes that respectively corresponded to the epicardial surface of the heart sectioned transversely and longitudinally (apex-to-base) as in previous tissue-engineering[13], mechanical[19,21] and anatomical[2] studies. Compared to 16 h/160 °C PGS accordion-like honeycomb scaffolds, adult rat right ventricular myocardium ($E_{CIRC} = 54 \pm 8$ kPa versus $E_{LONG} = 20 \pm 4$, $p < 0.01$, and $E_{CIRC}/E_{LONG} = 2.8 \pm 0.5$) was similarly anisotropic, but less stiff. Adult rat right ventricular myocardium was more anisotropic and compliant than its left ventricular counterpart ($E_{CIRC} = 157 \pm 14$ kPa versus $E_{LONG} = 84 \pm 8$, $p < 0.01$, and $E_{CIRC}/E_{LONG} = 2.1 \pm 0.4$), leading us to focus on approximating right ventricular myocardium mechanical properties. We reasoned that right ventricular myocardial grafts with the potential to grow, regenerate and remodel could ultimately lead to new treatments for congenital heart defects, which typically affect the right ventricle[24,25], as well as for right ventricular infarctions, which frequently complicate left ventricular infarctions[26] and are associated with substantial first-year mortality[27].

We next sought a better match between the magnitudes of E_{PD} and E_{XD} of accordion-like honeycomb scaffolds and E_{CIRC} and E_{LONG} of adult right ventricular myocardium by reducing the PGS curing time[28]. Note that non-porous PGS was recently tailored for myocardial applications by varying curing temperature[29]. PGS was cured at 160 °C for times ranging from 4 to 16 h, yielding a standard curve with a linear dependence of E_{PGS} on curing time within the tested range (Fig. 2d). Using this standard curve with the ratios $E_{PD}/E_{PGS} = 0.092 \pm 0.006$ and $E_{XD}/E_{PGS} = 0.027 \pm 0.002$ calculated from the 16 h/160 °C data, we predicted 7.5 h curing would yield E_{PD} and E_{XD} values of ~77 and ~22 kPa, respectively, closely matching the corresponding values of E_{CIRC} and E_{LONG} of adult right ventricular myocardium. As predicted, mechanical testing of the 7.5 h/160 °C PGS yielded $E_{PGS} = 825 \pm 62$ kPa, and accordion-like honeycomb scaffolds fabricated and wetted for 24 h yielded $E_{PD} = 83 \pm 2$ kPa versus $E_{XD} = 31 \pm 1$ kPa ($p < 0.0002$) and $E_{PD}/E_{XD} = 2.7 \pm 0.1$ kPa (Table 1; Supplementary Information, Fig. S1). Strain-to-failure, εf, was ~1.6 for accordion-like honeycomb scaffolds made of 7.5 h/160 °C PGS (Table 1). In comparison, an isotropic PGS foam we used previously[30,31] seemed less suitable for adult rat right ventricular myocardium, with $E_{FOAM} \sim 2$–9 kPa and $\varepsilon f_{FOAM} \sim 0.4$–0.8 (ranges estimated from Fig. 9 of Crapo *et al.*[32]).

To assess longer-term mechanical stability, accordion-like honeycomb scaffolds made of 7.5 h/160 °C PGS were placed in a fatigue bath (Supplementary Information, Fig. S2) and subjected to *in vitro* cyclic stretch mimicking physiologic loading[5,6,33]. Specimens were subjected to cyclic loading at 1 Hz and peak strain of 0.1 (10%) for 1 week in buffered saline at room temperature. Fatigued scaffolds yielded: $E_{PD} = 46 \pm 4$ kPa versus $E_{XD} = 19 \pm 0.8$ kPa ($p < 0.02$), with $E_{PD}/E_{XD} = 2.4 \pm 0.2$ (Table 1; Supplementary Information, Fig. S3). Compared to adult rat right ventricular myocardium, fatigued scaffolds exhibited similar mechanical properties (Table 1) except for higher εf values. Compared to scaffolds maintained statically in water for 24 h to 3 weeks, fatigued scaffolds exhibited similar mechanical properties except for a lower E_{PD} (Table 1; Supplementary Information, Fig. S1).

To assess the mechanical properties with cultured neonatal rat heart cells, scaffolds made of 7.5 h/160 °C PGS (10 × 5 mm, $n = 5$ per group) were autoclave-sterilized, seeded with cells isolated from neonatal rat ventricles and cultured *in vitro* for 1 week. Grafts yielded an attenuated anisotropic mechanical response

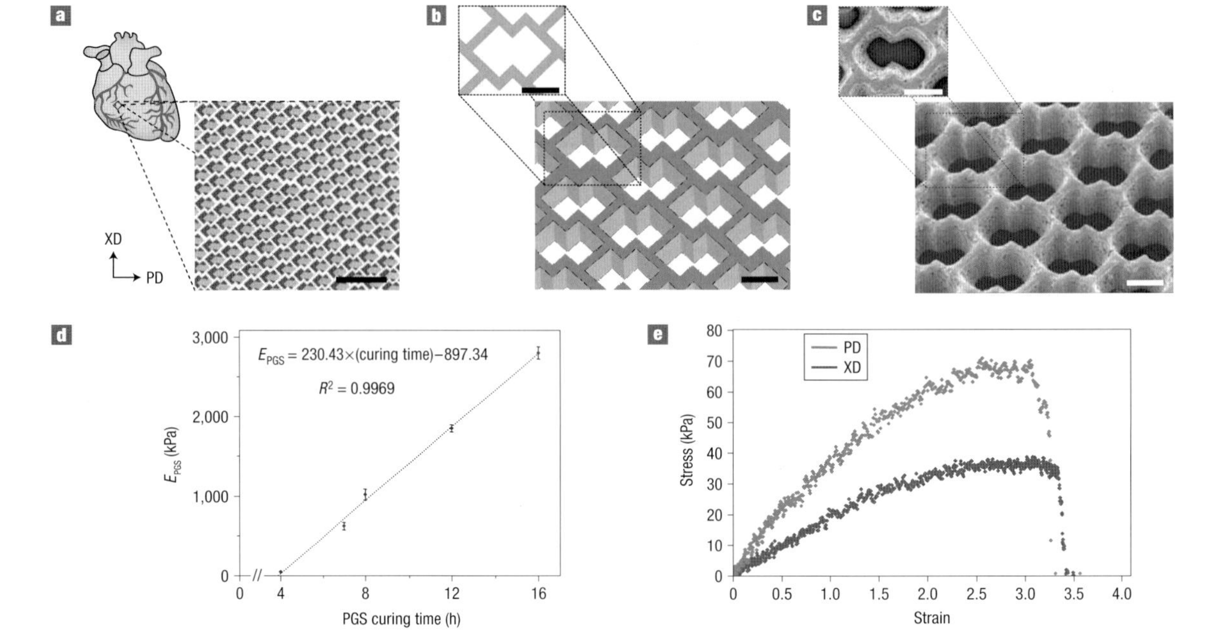

Figure 2 Accordion-like honeycomb scaffolds yield anisotropic mechanical properties similar to native myocardium. a,**b**, Schematic diagrams illustrating the accordion-like honeycomb design constructed by two overlapping 200 × 200 μm squares rotated 45° (diamonds). Preferred (PD) and orthogonal cross-preferred (XD) material directions, respectively corresponding to circumferential and longitudinal axes of the heart (Fig. 1), are indicated. Scale bars: 1 mm (**a**) and 200 μm (**b**). **c**, Scanning electron micrographs demonstrating the fidelity of excimer laser microablation in rendering an accordion-like honeycomb design in PGS. Scale bars : 200 μm. **d**, PGS curing time was systematically varied, yielding a linear dependence of PGS effective stiffness (E_{PGS}) on curing time within the tested range. **e**, Representative uniaxial stress–strain plots for accordion-like honeycomb scaffolds with cultured neonatal rat heart cells (scaffolds were fabricated from PGS membranes cured for 7.5 h at 160 °C; neonatal rat heart cells were cultured for 1 week).

Table 1 Summary of the effective stiffnesses (E_{PD}, E_{XD}), anisotropy ratio (E_{PD}/E_{XD}) and strains-to-failure (εf_{PD}, εf_{XD}) measured in uniaxial tension for accordion-like honeycomb scaffolds (cured 7.5 h/160 °C) under various conditions: 24 h wetting in water, 3 week wetting in water, 1 week fatigue by cyclic uniaxial stretch (1 Hz, 0.1 strain) in buffered saline and 1 week culture with unseparated neonatal rat heart cells. Corresponding data measured for native adult rat right ventricle (RV) is shown for comparison. Scaffold preferred and cross-preferred directions respectively correspond to the circumferential and longitudinal axes of the myocardium. (Data represent mean ± s.e.m.)

	Accordion-like honeycomb scaffolds, PGS 7.5 h/160 °C				Adult rat RV
Parameter	24 h wet ($n = 3$)	3 wks wet ($n = 3$)	1 wk fatigue ($n = 3$)	1 wk heart cells ($n = 5$)	myocardium ($n = 4$)
E_{PD} (kPa)	$83 \pm 2^{*\dagger\ddagger\S}$	$85 \pm 7^{*\dagger\ddagger\S}$	$46 \pm 4^{*}$	$32 \pm 2^{\S}$	$54 \pm 8^{*}$
E_{XD} (kPa)	31 ± 1	33 ± 2	19 ± 0.8	19 ± 4	20 ± 4
E_{PD}/E_{XD}	2.7 ± 0.1	2.6 ± 0.1	2.4 ± 0.2	1.9 ± 0.3	2.8 ± 0.5
εf_{PD}	1.6 ± 0.2	$1.9 \pm 0.3^{\S}$	$2.1 \pm 0.3^{\S}$	$2.4 \pm 0.2^{\S}$	0.9 ± 0.1
εf_{XD}	$1.6 \pm 0.02^{\ddagger}$	2.0 ± 0.1	1.3 ± 0.2	$2.7 \pm 0.2^{\S}$	1.3 ± 0.2

*Significantly different from corresponding sample tested in the orthogonal direction.

† Significantly different from corresponding fatigued sample.

‡ Significantly different from corresponding sample with cultured cells.

§ Significantly different from adult rat right ventricular myocardium.

(Fig. 2e), with $E_{PD} = 32 \pm 2$ kPa versus $E_{XD} = 19 \pm 4$ kPa ($p > 0.05$) and $E_{PD}/E_{XD} = 1.9 \pm 0.3$ (Table 1; Supplementary Information, Fig. S4). Compared with adult rat right ventricular myocardium, accordion-like honeycomb scaffolds with cultured heart cells exhibited a lower E_{PD} and higher εf values. Compared with wetting for 24 h to 3 weeks, scaffolds with heart cells exhibited a lower E_{PD} and insignificantly higher εf values (Table 1). Of note, autoclaving did not significantly change PGS mechanical response (Supplementary Information, Fig. S5). Together, these data suggested that *in vitro* culture of heart cells reduced E_{PD}, possibly owing to a combination of scaffold biodegradation and extracellular matrix deposition. These effects will need to be considered in future studies if scaffold mechanical properties are to be matched to adult rat right ventricular myocardium post-culture.

To compare grafts based on accordion-like honeycomb scaffolds of varying effective stiffness, neonatal rat heart cells were seeded on scaffolds made of PGS cured for 7.5, 12 and 16 h (5 × 5 mm, $n = 6$ per group). After 1 week *in vitro* culture, most pores were filled with neonatal rat heart cells grossly aligned along the preferred direction, as shown by confocal microscopy of grafts labelled for filamentous (F-) actin (Fig. 3a,b; Supplementary Information, Fig. S6). Note that F-actin labelling principally identified stress fibres in the cardiac fibroblast fraction of the cultured neonatal heart cells. Higher-magnification images demonstrated some elongated cells (Fig. 3c) and cross-striations (Fig. 3d) similar but significantly less developed than those characteristic of adult rat right ventricular myocardium (Fig. 3e). To assess graft cellularity, DNA contents were measured, found not to depend significantly on PGS curing time and yielded an average of 0.766 ± 0.043 mg DNA per gram wet weight comparable to our previous 1 week study of engineered myocardial grafts[9].

Electrophysiological assessment[30,31,34,35] demonstrated synchronous contractions in all scaffold groups, with all grafts capable of being paced by electrical field stimulation at up to at least 2 Hz (Supplementary Information, Video S1). Excitation thresholds were found not to depend significantly on PGS curing time. Hence, these data were pooled to yield a directional dependence of the excitation threshold for a given scaffold orientation with respect to the electric field. In particular, the excitation threshold was 11 ± 4% lower when measured with the scaffold preferred direction oriented parallel versus perpendicular to the applied electric field ($ET_{PD} = 1.05 \pm 0.02$ V versus $ET_{XD} = 1.19 \pm 0.04$ V, $p < 0.01$). This was consistent with reports in which the excitation threshold for cultured adult guinea pig[36] and canine[37] cardiomyocytes depended on the orientation of the elongated cells with respect to the electric field, with ~50% lower excitation threshold with the long cell axis parallel versus perpendicular.

Separate experiments were undertaken to compare accordion-like (Fig. 4a), square (200 × 200 μm; Fig. 4b) and rectangular (400 × 200 μm; Fig. 4c) honeycomb scaffolds fabricated from 16 h/160 °C PGS. These scaffolds exhibited different degrees of mechanical anisotropy (Fig. 4d). Accordion-like honeycombs ($E_{PD} = 195 \pm 8$ kPa, $E_{XD} = 57 \pm 3$ kPa and $E_{PD}/E_{XD} = 3.4 \pm 0.2$) and rectangular honeycombs ($E_{PD} = 206 \pm 21$ kPa versus $E_{XD} = 117 \pm 16$ kPa, $p < 0.01$, and $E_{PD}/E_{XD} = 1.8 \pm 0.3$) were anisotropic, whereas square honeycombs were isotropic ($E_{PD} = E_{XD} = 270 \pm 22$ kPa; $E_{PD}/E_{XD} = 1$). These results confirmed that of the scaffolds tested, the accordion-like honeycomb scaffold exhibited anisotropic mechanical properties most closely resembling adult rat right ventricular myocardium (Table 1).

To assess whether the electrical excitation threshold was influenced by scaffold microstructure, honeycomb scaffolds with accordion-like, square and rectangular pores (5 × 5 mm, $n = 4$ per group) were autoclave-sterilized, seeded with neonatal rat heart cells and cultured *in vitro* for 1 week. On the basis of our finding that co-cultured cardiac fibroblasts enhanced assembly of contractile grafts[30], scaffolds were sequentially seeded with cardiac fibroblasts followed by neonatal rat heart cells enriched for cardiomyocytes. As few as 4 days after seeding, spontaneous contractions were observed in all groups. After 1 week culture, most pores were filled by neonatal rat heart cells and the grafts could be synchronously paced at up to 4 Hz. The excitation threshold depended significantly on the orientation of the scaffold preferred direction with respect to the electric field (Fig. 4e). Anisotropic accordion-like and rectangular honeycomb scaffolds respectively had 12 ± 1% and 17 ± 2% lower excitation threshold values when the scaffold preferred direction was aligned parallel to the electric field ($p = 0.02$), whereas no consistent dependence was observed in isotropic square honeycombs (data not shown), similar to isotropic electrophysiologic properties observed by our group on randomly oriented non-woven poly(glycolic acid) scaffolds[38].

To assess whether heart cell alignment was influenced by scaffold microstructure, F-actin filament orientation[39] was quantified by fast Fourier Transform (FFT)-based image analysis of confocal micrographs taken at a depth of ~30 μm from the grafts (Fig. 3b; Supplementary Information, Figs S7–S9). Adult right ventricular myocardium imaged from the epicardial surface was analysed for comparison (Supplementary Information, Fig. S10). An orientation index was defined as the angular increment centred about the mean encompassing 50% of the total (F-actin filament) population (that is, a lower orientation index indicates a tighter distribution about the mean and more parallel alignment)[39–41]. Orientation index values for adult rat right ventricular myocardium were respectively 27 ± 9% and 31 ± 10% lower than for accordion-like and rectangular

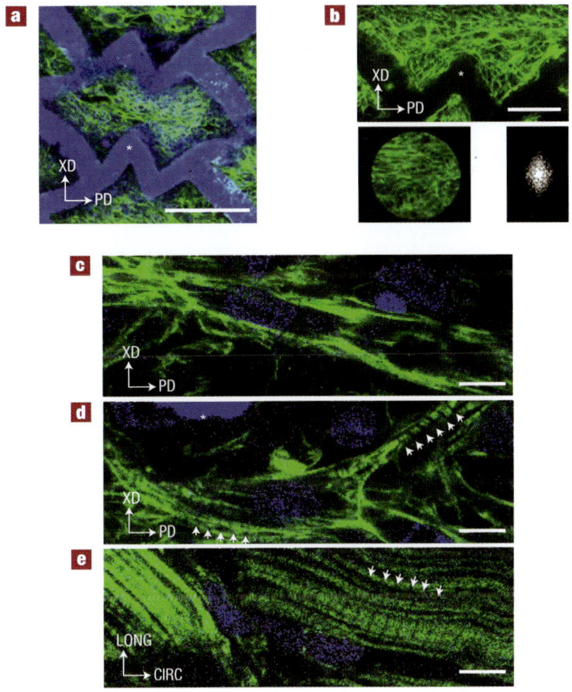

Figure 3 Accordion-like honeycomb scaffolds guide heart cell alignment. For the images in **a–d**, neonatal rat heart cells were cultured for 1 week on accordion-like honeycomb scaffolds, fluorescently labelled for F-actin (green), counterstained for nuclear DNA (blue) and imaged by confocal microscopy to assess cell morphology and alignment by FFT analysis[40,51]. Note that F-actin labelling principally identified stress fibres in the cardiac fibroblast fraction of the cultured neonatal heart cells. **a**,**b**, Low-magnification images of a representative graft demonstrated pores completely filled by neonatal rat heart cells grossly aligned in parallel to the preferred direction. Cell alignment was quantified by cropping confocal micrographs using a circular mask and determining F-actin orientation distribution in the associated FFT images (lower row in **b**; Supplementary Information, Fig. S6–S11). **c–e**, Higher-magnification images of neonatal rat heart cells cultured on scaffolds (**c**,**d**) demonstrating the presence of some elongated neonatal rat heart cells and cross-striations (**d**, white arrows) similar but significantly less developed than those present in control specimens (**e**) of adult rat right ventricular myocardium (a magnification of Fig. 1b). Scale bars: 200 μm (**a**); 100 μm (**b**); 10 μm (**c**,**d**,**e**). Scaffold is indicated by the white asterisks.

honeycombs ($p < 0.05$) (Fig. 4f,g). Orientation index values measured for accordion-like and rectangular honeycombs were respectively 37 ± 6% and 33 ± 6% lower than for isotropic square honeycombs ($p < 0.0001$) (Fig. 4f,g), indicating that each of the two anisotropic scaffolds possessed an inherent structural capacity to guide preferential heart cell alignment along the preferred material direction. Moreover, high and similar orientation index values (85 ± 5° versus 81 ± 2°, $p > 0.05$) were measured for square honeycomb scaffolds consisting of 1:1 aspect-ratio pores as well as for a previously used porogen-leached PGS foam[28,30] that we fabricated and analysed for comparison (Supplementary Information, Fig. S11). Of note, similar orientation index values (58 ± 4° versus 54 ± 1°, $p > 0.05$) were obtained for accordion-like honeycombs independent of scaffold stiffness (7.5 h/160 °C versus 16 h/160 °C) and cell seeding method (unseparated neonatal

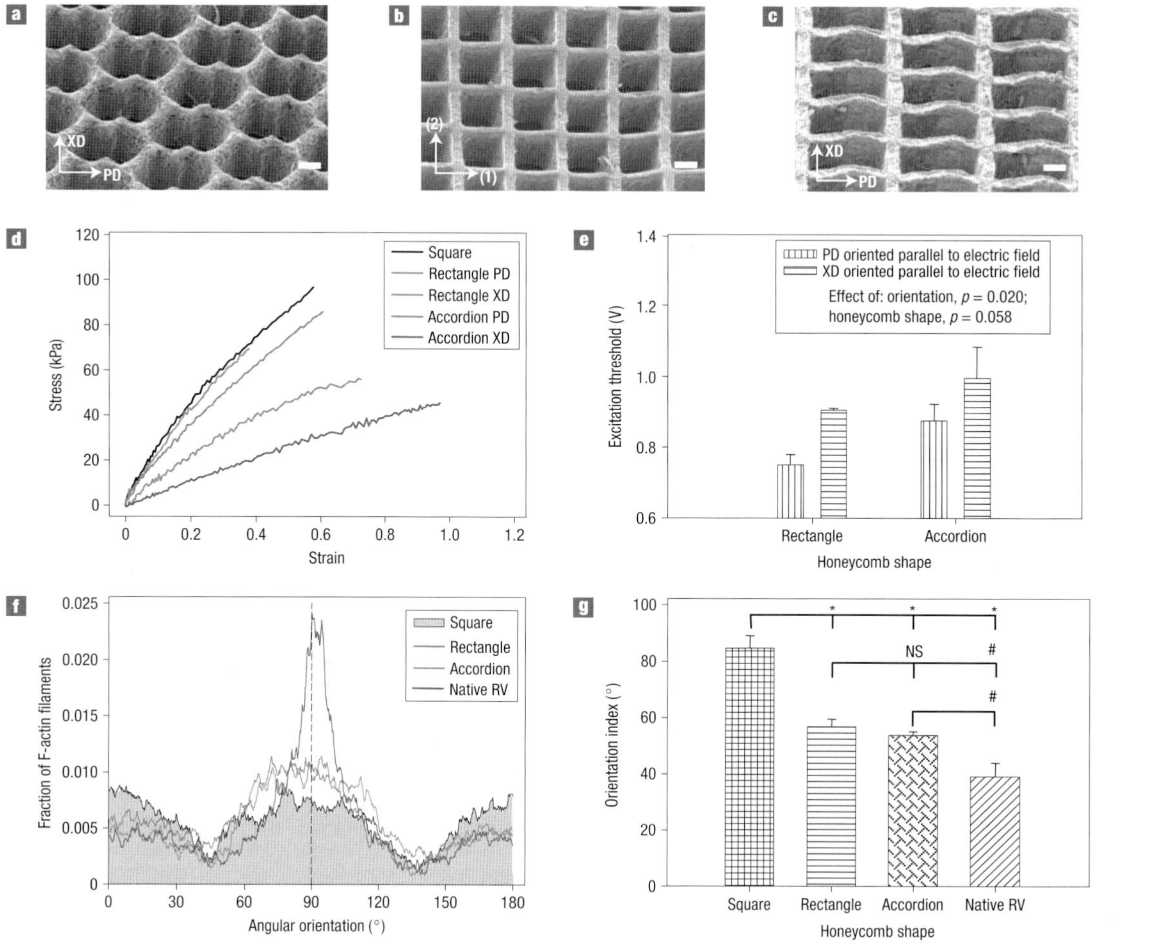

Figure 4 Anisotropic honeycomb scaffolds promote heart cell alignment and directionally dependent electrophysiologic properties. a–c, Scanning electron micrographs of accordion-like (**a**), square (**b**) and rectangular (**c**) honeycomb scaffolds. Scale bars: 100 μm. Axes show preferred (PD) and cross-preferred (XD) material directions (**a**,**c**) or are arbitrarily labelled (1) and (2) (**b**). **d**, Comparison of the uniaxial stress–strain behaviours of scaffolds with different honeycomb microstructures, demonstrating anisotropic mechanical response of the accordion-like and rectangular honeycomb scaffolds versus the isotropic mechanical response of square honeycomb scaffolds. **e**, Electrical field stimulation demonstrated directionally dependent excitation thresholds for grafts based on anisotropic scaffolds. The excitation threshold was significantly lower ($p = 0.02$) when the scaffold preferred direction was oriented parallel to the electric field, with no consistent differences observed in isotropic square honeycombs (data not shown). Data represent mean ± s.e.m. for $n = 4$ specimens. For the plots in **f** and **g**, FFT analysis of confocal images demonstrated preferential cell alignment in grafts based on anisotropic scaffolds. **f**, Plots of the mean fraction of F-actin filaments versus angular orientation. The scaffold preferred direction was arbitrarily designated as 90° (black dashed line), and an orientation index was defined as the angular increment about the distribution mean encompassing 50% of the F-actin filament population. **g**, Orientation index data demonstrated that anisotropic honeycomb scaffolds induced a significantly higher degree of heart cell alignment than isotropic honeycombs, albeit lower cell alignment than adult rat right ventricle (RV). Data are mean ± s.e.m. for $n = 6$ pores from a representative specimen from each group. $^*p < 0.0001$. # $p < 0.05$. NS indicates 'not significant.'

rat heart cells versus serially seeding with cardiac fibroblasts plus cardiomyocyte-enriched heart cells) suggesting that, under the conditions tested, pore geometry and not E_{PGS} or seeding method probably dominated cell orientation.

Although ~250-μm-thick honeycomb scaffolds provide a model for systematic *in vitro* studies, they are too thin to address reconstruction of full-thickness myocardium, except perhaps through a polysurgery strategy[42]. Moreover, with its transmural distribution of cardiac muscle fibre orientations[19,22,23], the quasi-lamellar structure of ventricular myocardium seems to demand an equivalently adaptive out-of-plane scaffold structure. As a first step towards addressing these limitations, prototype bilaminar honeycomb scaffolds of greater thickness (~400 μm) with interconnected 3D square-pore networks were fabricated by partially microablating a lateral array of troughs in one PGS lamina, overlaying a second lamina, microablating an array of top-to-bottom pores through both lamina and then stabilizing the resultant bilaminar honeycomb scaffold by thermal crosslinking[17,43] (Fig. 5a,b). In a preliminary study, unseparated neonatal rat heart cells were cultured for 1 week on bilaminar scaffolds. Cellular

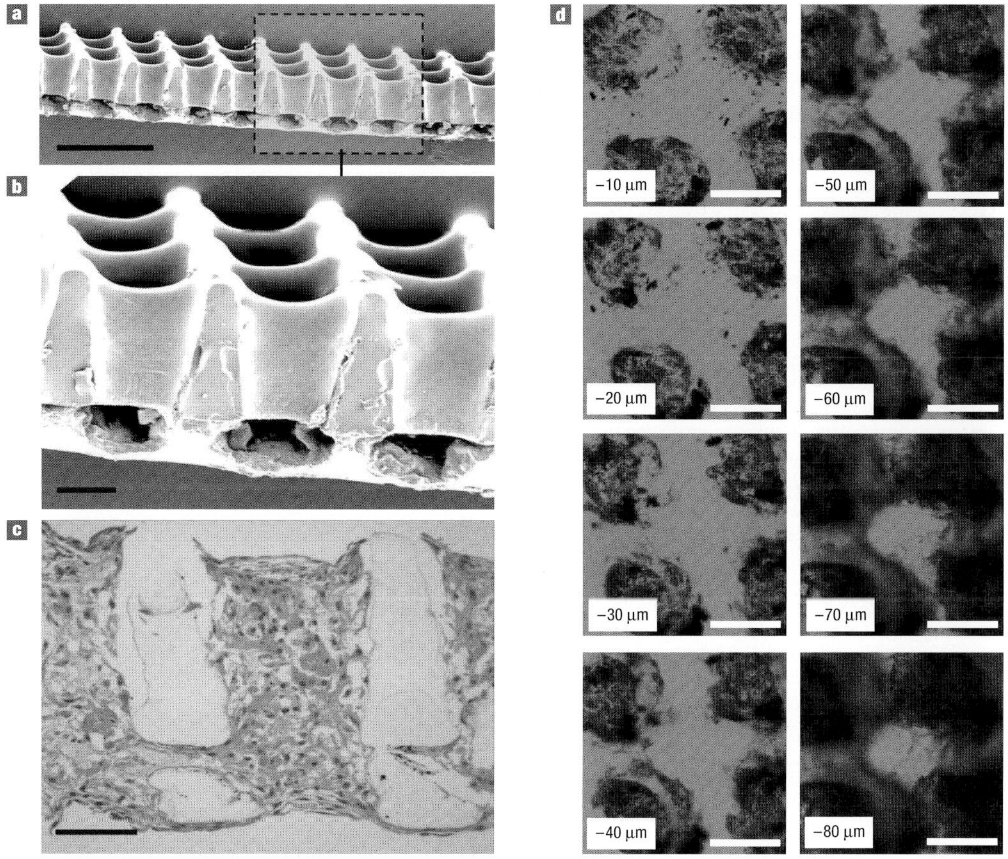

Figure 5 Prototype bilaminar honeycomb scaffolds with 3D interconnected pore networks are compatible with heart cell cultivation. Prototype bilaminar honeycomb scaffolds with 3D pore networks were fabricated from PGS through a combination of excimer laser microablation and lamination. **a**,**b**, Scanning electron micrographs of bilaminar scaffolds with fully interconnected 200 μm × 200 μm top-to-bottom and ∼200 μm × 100 μm lateral pores. **c**, Cross-section of a bilaminar honeycomb scaffold with cultured neonatal rat heart cells stained by haematoxylin and eosin. Histology demonstrated the presence of interconnected neonatal rat heart cells throughout the scaffold, spanning and penetrating its pores, and elongated near the scaffold surfaces. **d**, Serial confocal sections of an F-actin-labelled bilaminar scaffold with cultured neonatal rat heart cells provided an extra perspective of the lateral pore interconnectivity. Serial confocal sections begin 10 μm in from the bottom surface (top left panel) and end 80 μm in from the bottom surface (bottom right panel). Scale bars: 500 μm (**a**); 100 μm (**b**–**d**).

penetration of and interconnectivity between the top-to-bottom and lateral pores were observed in histological cross-section (Fig. 5c) as well as in serial, through-thickness confocal sections (Fig. 5d). These preliminary studies suggest our bilaminar scaffolds with 3D interconnected pore networks can enable the formation of multi-layered tissue structures from neonatal rat heart cells, thereby extending recent demonstrations that 2D microfabricated materials can guide the behaviour of cultured heart cells[44,45]. These findings further suggest that combined excimer laser microablation and lamination might be useful in fabricating honeycomb scaffolds of greater thickness from discrete, potentially offset lamina, with the caveat that thicker scaffolds would probably require perfusion to maintain cardiomyocyte viability[35,46–48]. Future studies will also focus on programming the excimer laser to generate the more complex arrays of troughs required for the assembly of a bilaminar accordion-like honeycomb scaffold.

To the best of our knowledge, this is the first report of a scaffold incorporating an accordion-like honeycomb microstructure. In particular, the accordion-like honeycomb scaffold demonstrated the novel ability to yield tissue-engineered grafts with closely matched anisotropic mechanical properties compared to adult rat right ventricular myocardium, while simultaneously promoting the preferential orientation of cultured neonatal rat heart cells in the absence of external stimuli. Matching of tissue mechanical properties was enabled by the materials-engineering approach we introduced, in which a standard curve (Fig. 2d) of PGS stiffness versus curing time was established at constant temperature and used in conjunction with measured ratios of scaffold-to-membrane stiffness to predict *a priori* the required PGS curing time. Also introduced was the use of excimer laser microablation in PGS microfabrication, which enabled creation of open, closely spaced (∼50 μm apart) pores with variable, precisely controlled geometries in ∼250-μm-thick PGS membranes, with the limitation that pores exhibited a slight top-to-bottom taper typical of laser drilling[49] (Figs 2c and 5a,b). Feature resolution with excimer laser microablation was higher than in our previous study wherein a carbon dioxide laser created more widely spaced (>410 μm apart), larger (>370 μm diameter) culture medium flow channels of cylindrical geometry in 2-mm-thick PGS foam[31]. PGS membrane thickness with excimer laser microablation was fivefold higher than in our previous study wherein replica moulding created more widely spaced (175 μm apart), small (50 μm) pores in thin (45 μm) PGS membranes[50]. Although electrical field stimulation did induce synchronous contractions of neonatal rat heart cells cultured on accordion-like honeycomb scaffolds, these contractions remained largely isometric under the conditions tested. Reduction in scaffold in-plane compressive resistance towards that of collagen foams[7,34,35,46] and PGS foams[30,31] (for example, by further reducing E_{PGS} or strut width of accordion-like honeycomb scaffolds) is expected to result in macroscopic contractility (fractional area change) of engineered myocardial grafts. In addition, use of soft lithography for surface micropatterning of the PGS (refs 16,17,43) before excimer laser microablation, and *in vitro* application of mechanical[5,6] or electrical[7] stimulation are expected to further increase orientation of cultured neonatal rat heart cells.

In summary, we have shown that accordion-like honeycomb scaffolds can overcome principal structural–mechanical limitations of scaffolds for myocardial tissue engineering by promoting the formation of grafts with preferentially aligned neonatal rat heart cells and mechanical properties more closely resembling adult rat right ventricular myocardium. Generalized to an enabling technology for structural biomimicry, libraries of bioresorbable honeycombs exhibiting graded microstructural features could likewise be generated, paving the way towards further integration of tissue-specialized scaffolds into advanced tissue-engineering strategies.

METHODS

EXCIMER LASER MICROABLATION

PGS membranes (~250 μm thick) on silicon wafers[16] were situated on the *x–y* stage (accuracy ±1 μm) of a Rapid X 1000 system (Resonetics) with a 248 nm krypton fluoride LPX200 excimer laser (Coherent-Lambda Physik). Pores were patterned using G-code programs. For highly crosslinked PGS (16 h/160 °C), a power of 350 mJ, burst frequency of 500 s^{-1} and burst count of 1000 were used. For partially crosslinked PGS (for example, 7.5 h/160 °C), the respective burst frequency and burst count were 25 s^{-1} and 100. Scaffolds were loosened from wafers in deionized water (24 h) and 70% ethanol (24 h). Bilaminar scaffolds (~400 μm) with 3D pore networks were fabricated by microablating a lateral array of troughs in one PGS lamina (burst spacing of 0.01 s; table speed of 0.5 mm s^{-1}), overlaying a second PGS lamina, microablating top-to-bottom pores through both lamina (burst frequency of 500 s^{-1}; burst count of 2000) and then stabilizing the resultant scaffold by thermally crosslinking[17] through autoclaving (121 °C for 30 min).

HEART CELL CULTURE ON MICROFABRICATED PGS SCAFFOLDS

Cells were obtained from 2-day-old neonatal rat hearts using an institute-approved protocol as described[8] (Supplementary Information, Methods). In studies of PGS scaffolds with varying effective stiffness and prototype bilaminar scaffolds (5 mm × 5 mm for imaging; 10 mm × 5 mm for mechanics), scaffolds were autoclave-sterilized and seeded with freshly collected, unseparated neonatal rat heart cells at $\sim 36 \times 10^6$ cells cm^{-2}. In brief, 10 ml of cell suspension was added to a vent-capped 50 ml tube (Product # 91253; TPP) containing one scaffold, and tubes were placed in a 37 °C/5% CO_2 humidified incubator and gently mixed (8 r.p.m.) using a rotisserie (Labquake, Barnstead-Thermolyne). After 4 days, culture media were replaced and culture was continued statically for 3 more days. In studies of 16 h/160 °C PGS scaffolds with varying pore structures, scaffolds (5 mm × 5 mm) were autoclave-sterilized and sequentially seeded[30] first with cardiac fibroblasts at $\sim 1 \times 10^6$ cells cm^{-2}, followed by a heart cell population enriched for cardiomyocytes at $\sim 36 \times 10^6$ cells cm^{-2}. In brief, 10 ml of cardiac fibroblast cell suspension was added to a 50 ml tissue culture tube containing a single scaffold and cultured with gentle mixing as described above. After 5 days, cardiomyocyte-enriched heart cells were added, and after a further 3 days, grafts were transferred to Petri dishes (Costar Ultra Low Attachment; Corning) (one per 35 mm well) and cultured statically for 4 more days. Culture media (6 ml/well) were replaced every 2 days.

ELECTROPHYSIOLOGIC ASSESSMENT

Specimens were assessed using our previous method, wherein an electrical pulse generator applied biphasic square waveforms at 1, 2, 3 and 4 Hz to a specimen positioned between two carbon rod electrodes within an environmentally controlled, perfused test chamber[34,35]. The excitation threshold was defined as the minimum voltage observed to initiate synchronous contractions. Video recordings made at twice the excitation threshold used a 30 frames per second video camera (Sony XCD-X710) and a Nikon Diaphot microscope at ×10 magnification.

CONFOCAL MICROSCOPY AND CELL ORIENTATION ANALYSIS

Cell orientation was evaluated using confocal laser microscopy with staining of F-actin[39] and FFT-based image analysis[40,51]. In brief, specimens were rinsed, fixed in 10% neutral buffered formalin (Sigma) for 2 h, rinsed, extracted in 0.2% (v/v) Triton X-100 (Sigma) in PBS for 2 h, rinsed, pre-incubated in 1% (w/v) bovine serum albumin (Sigma) in PBS (BSA buffer) for 2 h, incubated in Alexa Fluor 488-phalloidin (2:150 (v/v) dilution; Molecular Probes) for 3 h, rinsed, and then incubated in DRAQ5 (25 μM in PBS; Biostatus Limited) for 30 min, all at room temperature. Specimens were then rinsed, bulk mounted on glass slides in Vectashield medium (Vector Laboratories) and covered with a coverslip. Specimens were imaged with ×25 and ×100 oil-immersion objectives on a Zeiss LSM 510 laser scanning confocal microscope, with F-actin and nuclei pseudo-coloured green and blue, respectively. Micrographs of F-actin filaments (green channel) were analysed for cell orientation using a Matlab program. FFT image analysis was similar to that in our previous study[40], and adapted from Ng *et al.*[51]. See descriptions of FFT in quantifying 2D fibre network orientation and comparisons of FFT with morphometry in refs 52 and 53, respectively. Micrographs taken at ~30 μm from the graft and right ventricular epicardial surfaces were analysed ($n = 6$ pores; one representative sample per group). An orientation index was defined as the angular increment centred about the distribution mean encompassing 50% of the population[41]. For random cell orientations, the orientation index approaches 90 degrees; for highly aligned cells, the orientation index approaches zero.

STATISTICS

Data were mean ± s.e.m. unless indicated. Calculated parameter errors were determined by propagation of error. Comparisons were made by one-factor (anisotropy ratio) or two-factor (other parameters) analysis of variance with Tukey's *post-hoc* test (Statistica v7, StatSoft). A p-value < 0.05 was considered significant.

Received 16 October 2007; accepted 1 October 2008; published 2 November 2008.

References

1. Macchiarelli, G. *et al.* A micro-anatomical model of the distribution of myocardial endomysial collagen. *Histol. Histopathol.* **17**, 699–706 (2002).
2. Hanley, P. J., Young, A. A., LeGrice, I. J., Edgar, S. G. & Loiselle, D. S. Three dimensional configuration of perimysial collagen fibres in rat cardiac muscle at resting and extended sarcomere lengths. *J. Physiol.* **517**, 831–837 (1999).
3. Holmes, J. W., Borg, T. K. & Covell, J. W. Structure and mechanics of healing myocardial infarcts. *Annu. Rev. Biomed. Eng.* **7**, 223–253 (2005).
4. Costa, K. D., Lee, E. J. & Holmes, J. W. Creating alignment and anisotropy in engineered heart tissue: Role of boundary conditions in a model three-dimensional culture system. *Tissue Eng.* **9**, 567–577 (2003).
5. Akhyari, P. *et al.* Mechanical stretch regimen enhances the formation of bioengineered autologous cardiac muscle grafts. *Circulation* **106**, I137–I142 (2002).
6. Fink, C. *et al.* Chronic stretch of engineered heart tissue induces hypertrophy and functional improvement. *FASEB J.* **14**, 669–679 (2000).
7. Radisic, M. *et al.* Functional assembly of engineered myocardium by electrical stimulation of cardiac myocytes cultured on scaffolds. *Proc. Natl Acad. Sci. USA* **101**, 18129–18134 (2004).
8. Bursac, N. *et al.* Cardiac muscle tissue engineering: Toward an in vitro model for electrophysiological studies. *Am. J. Physiol. Heart Circ. Physiol.* **277**, H433–444 (1999).
9. Papadaki, M. *et al.* Tissue engineering of functional cardiac muscle: Molecular, structural, and electrophysiological studies. *Am. J. Physiol. Heart Circ Physiol.* **280**, H168–H178 (2001).
10. Yeo, Y. *et al.* Photocrosslinkable hydrogel for myocyte cell culture and injection. *J. Biomed. Mater. Res. B* **81**, 312–322 (2007).
11. Zimmermann, W. H. *et al.* Engineered heart tissue grafts improve systolic and diastolic function in infarcted rat hearts. *Nature Med.* **12**, 452–458 (2006).
12. Feng, Z., Matsumoto, T. & Nakamura, T. Measurements of the mechanical properties of contracted collagen gels populated with rat fibroblasts or cardiomyocytes. *J. Artif Organs* **6**, 192–196 (2003).
13. Ott, H. C. *et al.* Perfusion-decellularized matrix: Using nature's platform to engineer a bioartificial heart. *Nature Med.* **14**, 213–221 (2008).
14. Wang, Y., Ameer, G. A., Sheppard, B. J. & Langer, R. A tough biodegradable elastomer. *Nature Biotechnol.* **20**, 602–606 (2002).
15. Wang, Y., Kim, Y. M. & Langer, R. In vivo degradation characteristics of poly(glycerol sebacate). *J. Biomed. Mater. Res. A* **66**, 192–197 (2003).

16. Bettinger, C. J., Orrick, B., Misra, A., Langer, R. & Borenstein, J. T. Microfabrication of poly (glycerol-sebacate) for contact guidance applications. *Biomaterials* **27**, 2558–2565 (2006).
17. Bettinger, C. J. *et al.* Three-dimensional microfluidic tissue-engineering scaffolds using a flexible biodegradable polymer. *Adv. Mater.* **18**, 165 (2006).
18. Radisic, M. *et al.* Oxygen gradients correlate with cell density and cell viability in engineered cardiac tissue. *Biotechnol. Bioeng.* **93**, 332–343 (2006).
19. Chuong, C. J., Sacks, M. S., Templeton, G., Schwiep, F. & Johnson, R. L. Jr. Regional deformation and contractile function in canine right ventricular free wall. *Am. J. Physiol.* **260**, H1224–H1235 (1991).
20. Rappaport, D., Adam, D., Lysyansky, P. & Riesner, S. Assessment of myocardial regional strain and strain rate by tissue tracking in B-mode echocardiograms. *Ultrasound Med. Biol.* **32**, 1181–1192 (2006).
21. Sacks, M. S. & Chuong, C. J. Biaxial mechanical properties of passive right ventricular free wall myocardium. *J. Biomech. Eng.* **115**, 202–205 (1993).
22. Kocica, M. J. *et al.* The helical ventricular myocardial band. Global, three dimensional, functional architecture of the ventricular myocardium. *Eur. J. Cardiothorac. Surg.* **29**, S21–S40 (2006).
23. Streeter, D. D. Jr, Spotnitz, H. M., Patel, D. P., Ross, J. Jr & Sonnenblick, E. H. Fiber orientation in the canine left ventricle during diastole and systole. *Circ. Res.* **24**, 339–347 (1969).
24. Bautista-Hernandez, V. *et al.* Coarctectomy reduces neoaortic arch obstruction in hypoplastic left heart syndrome. *J. Thorac. Cardiovasc. Surg.* **133**, 1540–1546 (2007).
25. Reinhartz, O. *et al.* Homograft valved right ventricle to pulmonary artery conduit as a modification of the Norwood procedure. *Circulation* **114**, I594–I599 (2006).
26. Kinch, J. W. & Ryan, T. J. Right ventricular infarction. *N. Engl. J. Med.* **330**, 1211–1217 (1994).
27. Gumina, R. J., Murphy, J. G., Rihal, C. S., Lennon, R. J. & Wright, R. S. Long-term survival after right ventricular infarction. *Am. J. Cardiol.* **98**, 1571–1573 (2006).
28. Gao, J., Crapo, P. M. & Wang, Y. Macroporous elastomeric scaffolds with extensive micropores for soft tissue engineering. *Tissue Eng.* **12**, 917–925 (2006).
29. Chen, Q. Z. *et al.* Characterisation of a soft elastomer poly(glycerol sebacate) designed to match the mechanical properties of myocardial tissue. *Biomaterials* **29**, 47–57 (2008).
30. Radisic, M. *et al.* Pre-treatment of synthetic elastomeric scaffolds by cardiac fibroblasts improves engineered heart tissue. *J. Biomed. Mater. Res. A* **86**, 713–724 (2008).
31. Radisic, M. *et al.* Biomimetic approach to cardiac tissue engineering: Oxygen carriers and channeled scaffolds. *Tissue Eng.* **12**, 2077–2091 (2006).
32. Crapo, P. M., Gao, J & Wang, Y. Seamless tubular poly(glycerol sebacate) scaffolds: High-yield fabrication and potential applications. *J. Biomed. Mater. Res. A* **86**, 354–363 (2008).
33. Boublik, J. *et al.* Mechanical properties and remodeling of hybrid cardiac constructs made from heart cells, fibrin, and biodegradable, elastomeric knitted fabric. *Tissue Eng.* **11**, 1122–1132 (2005).
34. Cheng, M., Park, H., Engelmayr, G. C., Moretti, M. & Freed, L. E. Effects of regulatory factors on engineered cardiac tissue in vitro. *Tissue Eng.* **13**, 2709–2719 (2007).
35. Cheng, M., Moretti, M., Engelmayr, G. C. Jr & Freed, L. E. Insulin-like growth factor-I and perfusion enhance the formation of tissue engineered cardiac grafts. *Tissue Eng.* doi:10.1089/ten.tea.2008.0077 (in the press; PMID: 18759675).
36. Bardou, A. L. *et al.* Directional variability of stimulation threshold measurements in isolated guinea pig cardiomyocytes: Relationship with orthogonal sequential defibrillating pulses. *Pacing Clin. Electrophysiol.* **13**, 1590–1595 (1990).
37. Ranjan, R. & Thakor, N. V. Electrical stimulation of cardiac myocytes. *Ann. Biomed. Eng.* **23**, 812–821 (1995).
38. Bursac, N. *et al.* Cultivation in rotating bioreactors promotes maintenance of cardiac myocyte electrophysiology and molecular properties. *Tissue Eng.* **9**, 1243–1253 (2003).
39. Engelmayr, G. C. Jr, Papworth, G. D., Watkins, S. C., Mayer, J. E. Jr & Sacks, M. S. Guidance of engineered tissue collagen orientation by large-scale scaffold microstructures. *J. Biomech.* **39**, 1819–1831 (2006).
40. Nichol, J. W., Engelmayr, G. C. Jr, Cheng, M. & Freed, L. E. Co-culture induces alignment in engineered cardiac constructs via MMP-2 expression. *Biochem. Biophys. Res. Commun.* **373**, 360–365 (2008).
41. Sacks, M. S., Smith, D. B. & Hiester, E. D. A small angle light scattering device for planar connective tissue microstructural analysis. *Ann. Biomed. Eng.* **25**, 678–689 (1997).
42. Shimizu, T. *et al.* Polysurgery of cell sheet grafts overcomes diffusion limits to produce thick, vascularized myocardial tissues. *FASEB J.* **20**, 708–710 (2006).
43. Borenstein, J. T. *et al.* Microfabrication of three-dimensional engineered scaffolds. *Tissue Eng.* **13**, 1837–1844 (2007).
44. Camelliti, P., Gallagher, J. O., Kohl, P. & McCulloch, A. D. Micropatterned cell cultures on elastic membranes as an in vitro model of myocardium. *Nature Protocols* **1**, 1379–1391 (2006).
45. Feinberg, A. W. *et al.* Muscular thin films for building actuators and powering devices. *Science* **317**, 1366–1370 (2007).
46. Radisic, M. *et al.* Medium perfusion enables engineering of compact and contractile cardiac tissue. *Am. J. Physiol. Heart Circ. Physiol.* **286**, H507–H516 (2004).
47. Dvir, T., Benishti, N., Shachar, M. & Cohen, S. A novel perfusion bioreactor providing a homogenous milieu for tissue regeneration. *Tissue Eng.* **12**, 2843–2852 (2006).
48. Dvir, T., Levy, O., Shachar, M., Granot, Y. & Cohen, S. Activation of the ERK1/2 cascade via pulsatile interstitial fluid flow promotes cardiac tissue assembly. *Tissue Eng.* **13**, 2185–2193 (2007).
49. Dahotre, N. B. & Harimkar, S. P. *Laser Fabrication and Machining of Materials* (Springer Science Business Media, 2008).
50. Neeley, W. L. *et al.* A microfabricated scaffold for retinal progenitor cell grafting. *Biomaterials* **29**, 418–426 (2008).
51. Ng, C. P., Hinz, B. & Swartz, M. A. Interstitial fluid flow induces myofibroblast differentiation and collagen alignment in vitro. *J. Cell. Sci.* **118**, 4731–4739 (2005).
52. Ayres, C. E. *et al.* Measuring fiber alignment in electrospun scaffolds: A user's guide to the 2D fast Fourier transform approach. *J. Biomater. Sci. Polym. Ed.* **19**, 603–621 (2008).
53. Sander, E. A. & Barocas, V. H. Comparison of 2D fiber network orientation measurement methods. *J. Biomed. Mater. Res. A* (2008).

Supplementary Information accompanies the paper at www.nature.com/naturematerials.

Acknowledgements

Financial support for this work was provided by NIH NRSA fellowship 1 F32 HL084968-01 (to G.C.E.), a Charles Stark Draper fellowship (to C.J.B.), NASA Grant NNJ04HC72G (to L.E.F.) and NIH Grant DE013023 (to R.L.). We are indebted to Y. Wang and M. Radisic for advice on porogen-leached PGS scaffolds, S.N. Bhatia, D. Albright and D. Ward for advice on PGS membrane microfabrication, N. Watson and E. Batchelder for help with confocal and electron microscopy and S. Kangiser for help with manuscript preparation.

Author information

Reprints and permissions information is available online at http://npg.nature.com/reprintsandpermissions.

Correspondence and requests for materials should be addressed to L.E.F.

ARTICLES
PUBLISHED ONLINE: 22 AUGUST 2010 | DOI: 10.1038/NMAT2812

nature materials

Combinatorial development of biomaterials for clonal growth of human pluripotent stem cells

Ying Mei[1†], Krishanu Saha[2†], Said R. Bogatyrev[1], Jing Yang[3], Andrew L. Hook[3], Z. Ilke Kalcioglu[4], Seung-Woo Cho[5], Maisam Mitalipova[2], Neena Pyzocha[2], Fredrick Rojas[1], Krystyn J. Van Vliet[4], Martyn C. Davies[3], Morgan R. Alexander[3], Robert Langer[1,6,7]*, Rudolf Jaenisch[2,8]* and Daniel G. Anderson[1,6,7]*

Both human embryonic stem cells and induced pluripotent stem cells can self-renew indefinitely in culture; however, present methods to clonally grow them are inefficient and poorly defined for genetic manipulation and therapeutic purposes. Here we develop the first chemically defined, xeno-free, feeder-free synthetic substrates to support robust self-renewal of fully dissociated human embryonic stem and induced pluripotent stem cells. Material properties including wettability, surface topography, surface chemistry and indentation elastic modulus of all polymeric substrates were quantified using high-throughput methods to develop structure–function relationships between material properties and biological performance. These analyses show that optimal human embryonic stem cell substrates are generated from monomers with high acrylate content, have a moderate wettability and employ integrin $\alpha_v\beta_3$ and $\alpha_v\beta_5$ engagement with adsorbed vitronectin to promote colony formation. The structure–function methodology employed herein provides a general framework for the combinatorial development of synthetic substrates for stem cell culture.

Human pluripotent stem cells (both human embryonic stem (hES) and induced pluripotent stem (hiPS) cells) hold great promise for regenerative medicine[1–4] and human-disease modelling[5]. However, existing methods to grow human pluripotent stem cells are not well suited for genetic manipulation experiments and introduce animal components, increasing the risks of immune rejection. Present methods to grow hES and hiPS cells include growing them on a 'feeder' cell layer of mitotically inactivated mouse embryonic fibroblasts[1–3,6] (mEFs), and on 'feeder free' culture systems, composed of a variety of extracellular-matrix/serum proteins coated onto tissue-culture dishes[7–15] or synthetic materials[16–19] such as hyaluronic acid hydrogels. These have been reported to promote hES cell self-renewal when seeded at a suitably high cell density[9,16,17] (for example, $\sim 10^6$ cells ml^{-1} for the hydrogel) and have not been demonstrated to efficiently promote clonal growth of single hES cells (efficiencies typically <10%). However, gene targeting in pluripotent stem cells necessitates clonal outgrowth of single cells to detect rare targeting events (1 in 10^5–10^6 cells) and requires selective growth of a correctly gene-targeted cell within a population of $>10^5$ cells[20–22]. Further, current human culture methods use either animal products or undefined components, which make it problematic for the potential transplantation applications[4]. Here we employed a high-throughput approach to engineer new culture substrates that could be used to clonally expand human pluripotent stem cells in a chemically defined, xeno-free, feeder-free manner.

To facilitate rapid synthesis and analysis of synthetic substrates, we manufactured cell-compatible, biomaterial microarrays[23,24]. Microarrays were prepared from 22 acrylate monomers with diversified hydrophobicity–hydrophilicity and crosslinking densities (Fig. 1a). The arrays were prepared by copolymerization between each of 16 'major' monomers (numbered 1–16) and each of six 'minor' monomers (lettered A–F) at six different ratios (100:0, 90:10, 85:15, 80:20, 75:25, 70:30 (v/v)) (Supplementary Fig. S1). In this way, arrays with 496 $[16+(16\times5\times6)]$ different combinations were created, consisting of the major monomer (70–100%) and minor monomer (0–30%). These monomer mixtures were robotically deposited in triplicate on a non-cell adhesive layer of poly(2-hydroxyethyl methacrylate) covering conventional glass slides (75 mm × 25 mm), and then polymerized with a long-wave ultraviolet source.

We next used fluorescence-activated cell sorting of transgenic hES cells to ensure that hES cells were both dissociated from one another and undifferentiated in our assays (Fig. 1b). A transgenic green-fluorescent protein (GFP) reporter for Oct4 expression, a marker of pluripotent cells (Supplementary Fig. S2), was knocked in to the BG01 hES cell line and propagated under standard hES cell culture conditions using mEFs[25]. GFP$^+$ sorted hES

[1]Department of Chemical Engineering, Massachusetts Institute of Technology, 77 Massachusetts Avenue, Cambridge, Massachusetts 02139, USA, [2]The Whitehead Institute for Biomedical Research, 9 Cambridge Center, Cambridge, Massachusetts 02142, USA, [3]Laboratory of Biophysics and Surface Analysis, School of Pharmacy, The University of Nottingham, Nottingham, NG7 2RD, UK, [4]Department of Material Sciences and Engineering, Massachusetts Institute of Technology, 77 Massachusetts Avenue, Cambridge, Massachusetts 02139, USA, [5]Department of Biotechnology, Yonsei University, Seoul 120-749, Korea, [6]David H. Koch Institute for Integrative Cancer Research, Massachusetts Institute of Technology, 45 Carleton Street, Building E25-342, Cambridge, Massachusetts 02142, USA, [7]Harvard-MIT Division of Health Science Technology, Massachusetts Institute of Technology, 45 Carleton Street, Building E25-342, Cambridge, Massachusetts 02142, USA, [8]Department of Biology, Massachusetts Institute of Technology, 9 Cambridge Center, Cambridge, Massachusetts 02142, USA. †These authors contributed equally to this work. *e-mail: rlanger@mit.edu; jaenisch@wi.mit.edu; dgander@mit.edu.

NATURE MATERIALS DOI: 10.1038/NMAT2812 ARTICLES

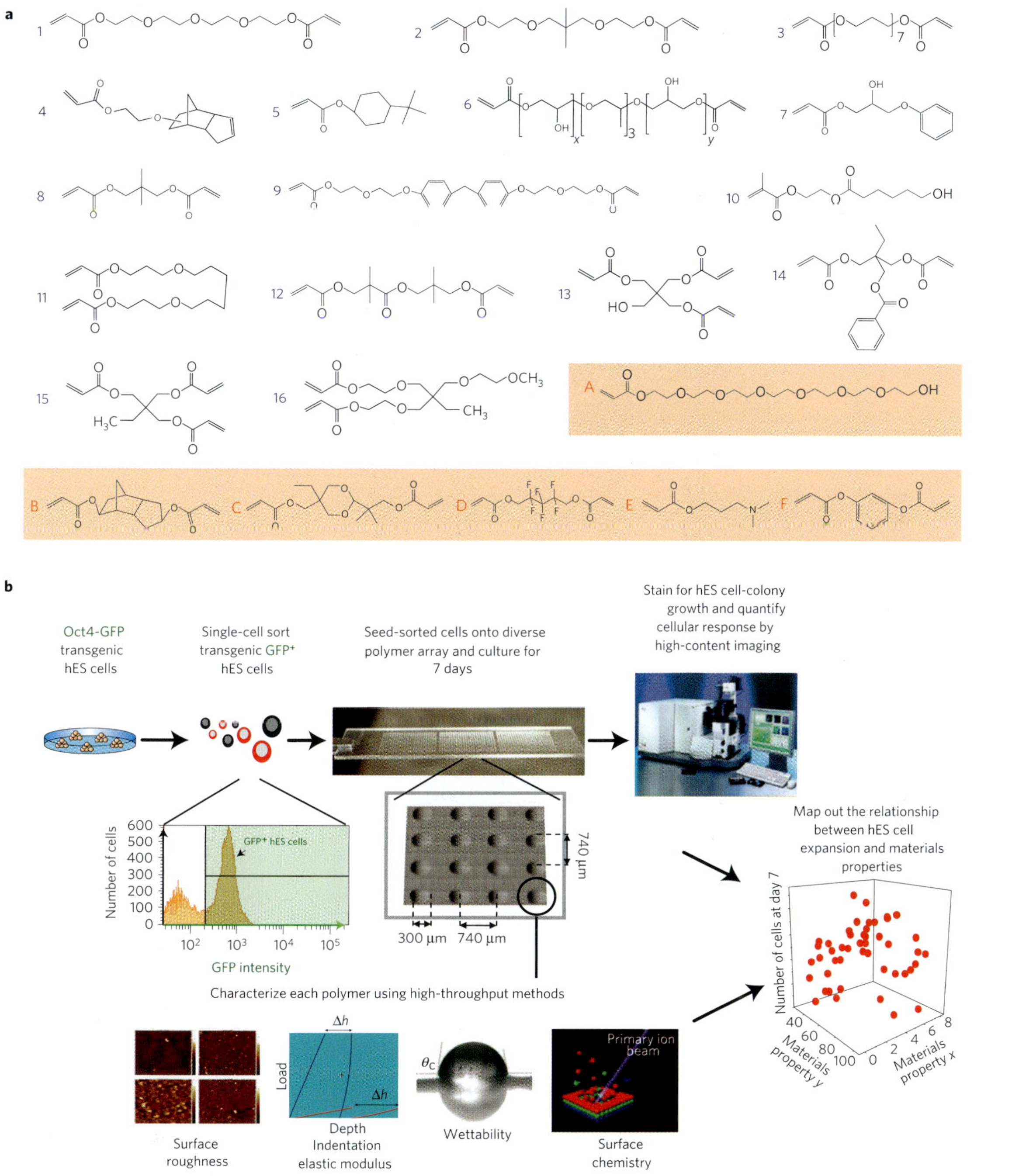

Figure 1 | High-throughput screening of biomaterials for clonal growth. a, Monomers used for array synthesis were classified into two categories: 'major' monomers that constitute >50% of the reactant mixture and 'minor' monomers that constitute <50% of the mixture. Sixteen major monomers were named numerically (blue), and six minor monomers were labelled alphabetically (orange). **b**, Schematic diagram of the screen. First, transgenic Oct4-GFP hES cells were maintained on mEFs. Then flow cytometry enabled the isolation of high-purity undifferentiated hES cells from the completely dissociated coculture of hES cells and mEFs. A flow-cytometry histogram during a representative cell sort is shown. GFP^{+} cells (right of the black gate) were seeded onto the arrays, whereas the differentiated cells and mEFs (GFP^{-}, left of the black gate) were not used. A photograph of the polymer microarray with 16 polymer spots is shown to illustrate dimensions and separation. Each polymer was also characterized using high-throughput methods to characterize its surface roughness, indentation elastic modulus, wettability (water contact angle, θ_C) and surface chemistry. Finally, the cellular response on the polymer array was quantified by using laser-scanning cytometry, and structure-function relationships were determined by numerical analysis of both the cellular response and materials characterization data.

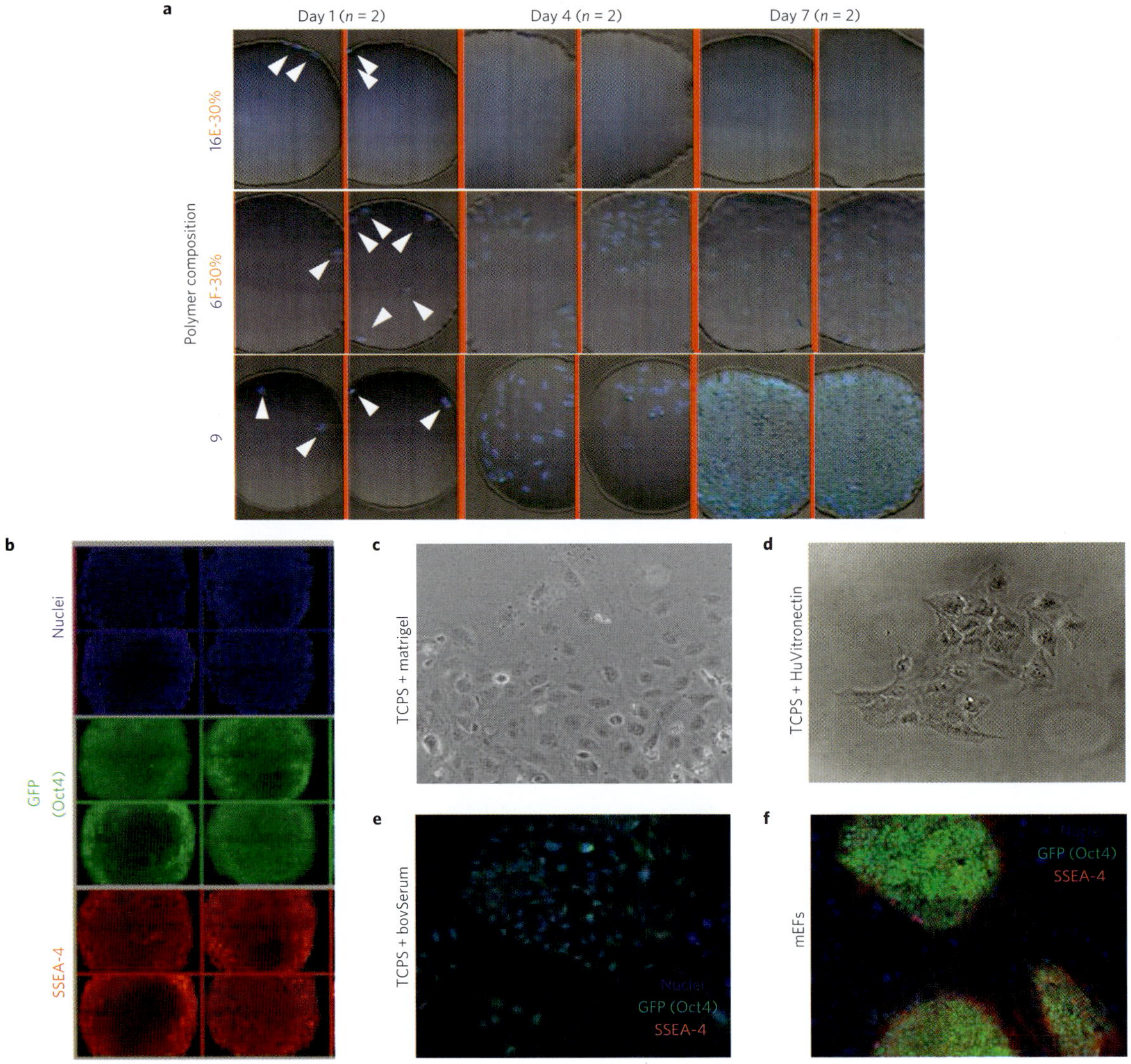

Figure 2 | Diverse hES cell behaviour on primary polymer arrays. a, Single Oct4-GFP$^+$ hES cells were seeded on the polymer arrays. White arrowheads point to cells attached after one day of culture, indicating a near-clonal seeding density for each spot. Diverse cell behaviour was seen on the array on subsequent culture in mEF-conditioned media. Representative images of cell nuclei (stained by Hoechst in blue) on three different polymers (shown are two replicates of each): the 16E-30% polymer did not support either attachment or survival of dissociated hES cells; the 6F-30% polymer supported moderate growth but also differentiation of hES cells; the 9 homopolymer (a hit polymer) supported robust growth of hES cells. **b**, Immunostaining of hES cells propagated on hit polymer spots for cell nuclei (blue) and for pluripotency markers Oct4 (green) and SSEA4 (red). Owing to the raised centre of each spot above the plane of the microscope slide, spot centres are not completely in focus, leading to lower intensity at the centre of each image. **c–e**, At the near-clonal cell densities used for the polymer experiments, hES cells spread out on matrigel-coated tissue culture polystyrene (TCPS) (**c**), human vitronectin-coated TCPS (**d**) and bovine serum-coated TCPS (**e**) substrates in mEF-conditioned media. **f**, In contrast, traditional means of culturing hES cells by using mitotically inactivated mouse embryonic feeder cells grown on gelatin-coated TCPS ('mEFs') could support colony formation at these near-clonal cell densities. In **e** and **f** immunostaining was carried out for nuclei (blue) and for pluripotency markers Oct4 (green) and SSEA-4 (red). See also Fig. 6a for colony-formation efficiencies.

single cells (Fig. 1b, Supplementary Fig. S3) were seeded onto the polymer arrays and cultured with mEF-conditioned medium, because soluble growth factors secreted by mEFs help maintain the undifferentiated hES cell state[7,17] (Supplementary Fig. S2c).

Proteins can rapidly adsorb onto the surfaces of materials used for cell culture[26–28]. The surface properties of cell-culture substrates can modulate both the amount and the conformation of adsorbed proteins, and thereby interact with cell surface receptors (for example, integrins) to initiate signal transduction and alter cell behaviour[29]. To investigate the potential of different adsorbed proteins, fibronectin, laminin, bovine serum albumin, and fetal bovine serum (FBS) were separately adsorbed onto the microarrays from solution. In general, FBS was found to most effectively support the propagation of hES cells across the entire array, whereas fibronectin and laminin coatings led to more differentiation as indicated by down-regulation of Oct4–GFP expression (Supplementary Fig. S4).

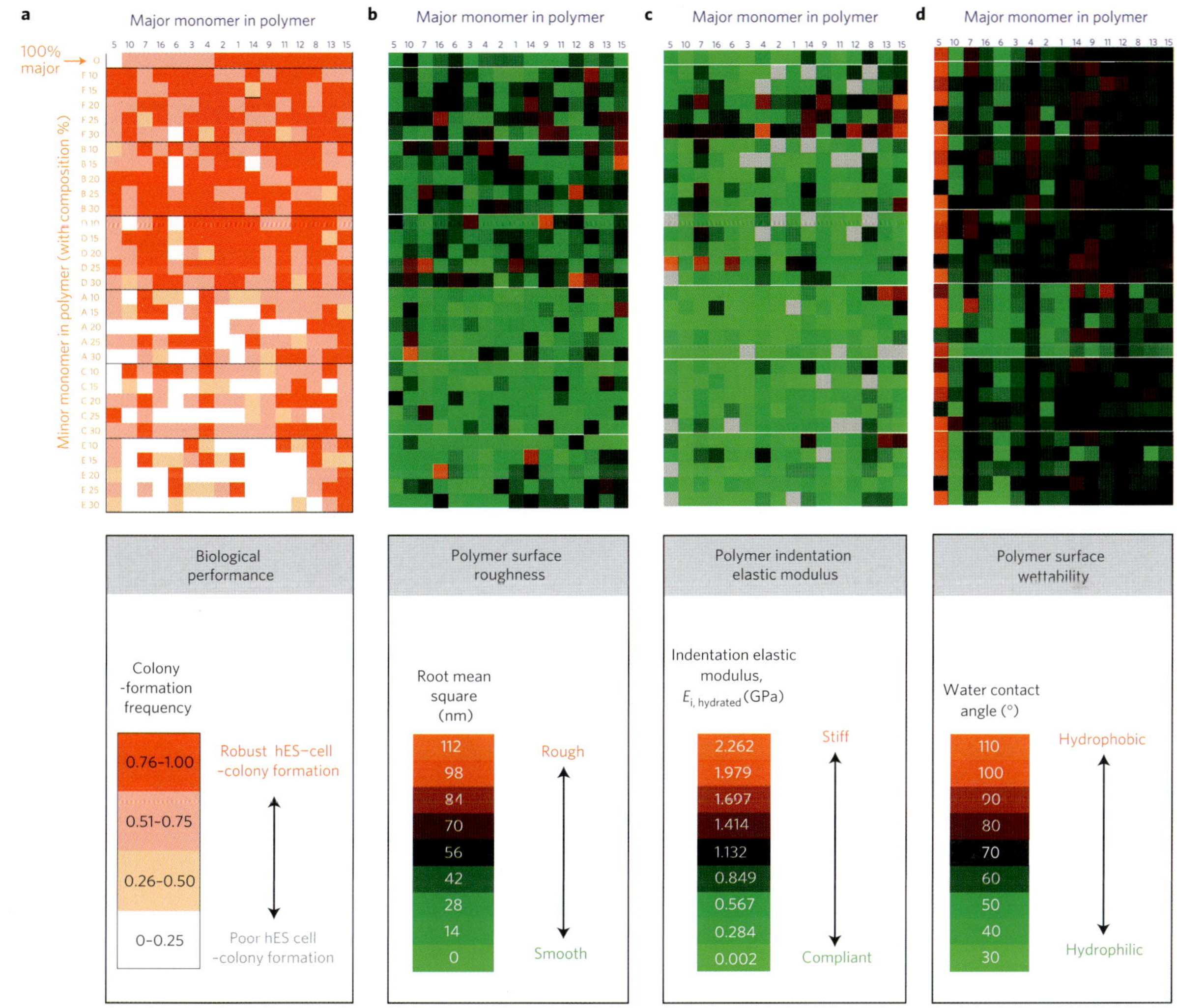

Figure 3 | Mapping hES cell behaviour to polymer properties using primary arrays. a, Map of hES cell-colony formation and polymer composition for all of the 496 monomer combinations in the primary array. For the minor and major composition axes, the numbers and letters indicate the major and minor monomer, respectively, as shown in Fig. 1a. Major monomers are listed in order, from left to right, of increasing colony formation, and minor monomers are listed in order of increasing colony formation from bottom to top. Therefore, the region of the map corresponding to highest colony formation is the top right corner, and the region with the lowest is the bottom left corner. Homopolymers are listed on the top row. The frequency of colony formation on the primary polymer array was grouped into four categories, 0–0.25, 0.25–0.50, 0.50–0.75 and 0.75–1.0 per polymeric spot, as indicated by the intensity of red. **b**, Surface roughness of primary array polymers coated with FBS in DMEM. A map indicating the root-mean-square roughness (see coloured legend below) for all of the 496 monomer combinations in the primary array. **c**, Indentation elastic modulus of primary array polymers hydrated in PBS. A map indicating the indentation elastic modulus (see the coloured legend below) for all of the 496 monomer combinations in the primary array. Grey indicates no data obtained. **d**, Wettability of primary array polymers. A map indicating the water contact angle (see coloured legend below) for all of the 496 monomer combinations in the primary array. Note that polymers in the upper right corner of **a** with higher colony-formation frequencies (dark red) have moderate water contact angles (black) in the upper right corner of **d**, whereas this region does not correlate to any specific ranges of roughness or elastic modulus in the upper right corners of **b** and **c**.

Poor cell attachment was observed when arrays were coated with bovine serum albumin. Therefore, FBS was used initially to coat the polymer array to screen for the suitable polymers ('hits') that can support hES cell growth from single cells.

The FBS-coated arrays were seeded at low cell density (40 cells mm^{-2}), to best model the ability of cells to grow in isolation. Two to ten non-contacting cells were observed on most polymer spots after 24 h cell culture (see, for example, Fig. 2a and Supplementary Fig. S5). At various time points during culture, adherent cells were fixed and stained for cell nuclei and two pluripotent stem cell markers, SSEA-4 and Oct4 (Fig. 2b). The cellular responses were quantified with laser scanning cytometry[30]. For each polymer, we defined the colony-formation frequency as the number of polymer spots on which $Oct4^+$ and $SSEA\text{-}4^+$ hES cell colonies formed divided by the total number of replicate spots of the same kind of polymer on each array ($n = 3$–18; see Methods). After 7 days of culture, a range of cellular responses was found on the polymer array (Fig. 2a): some polymers did not

ARTICLES NATURE MATERIALS DOI: 10.1038/NMAT2812

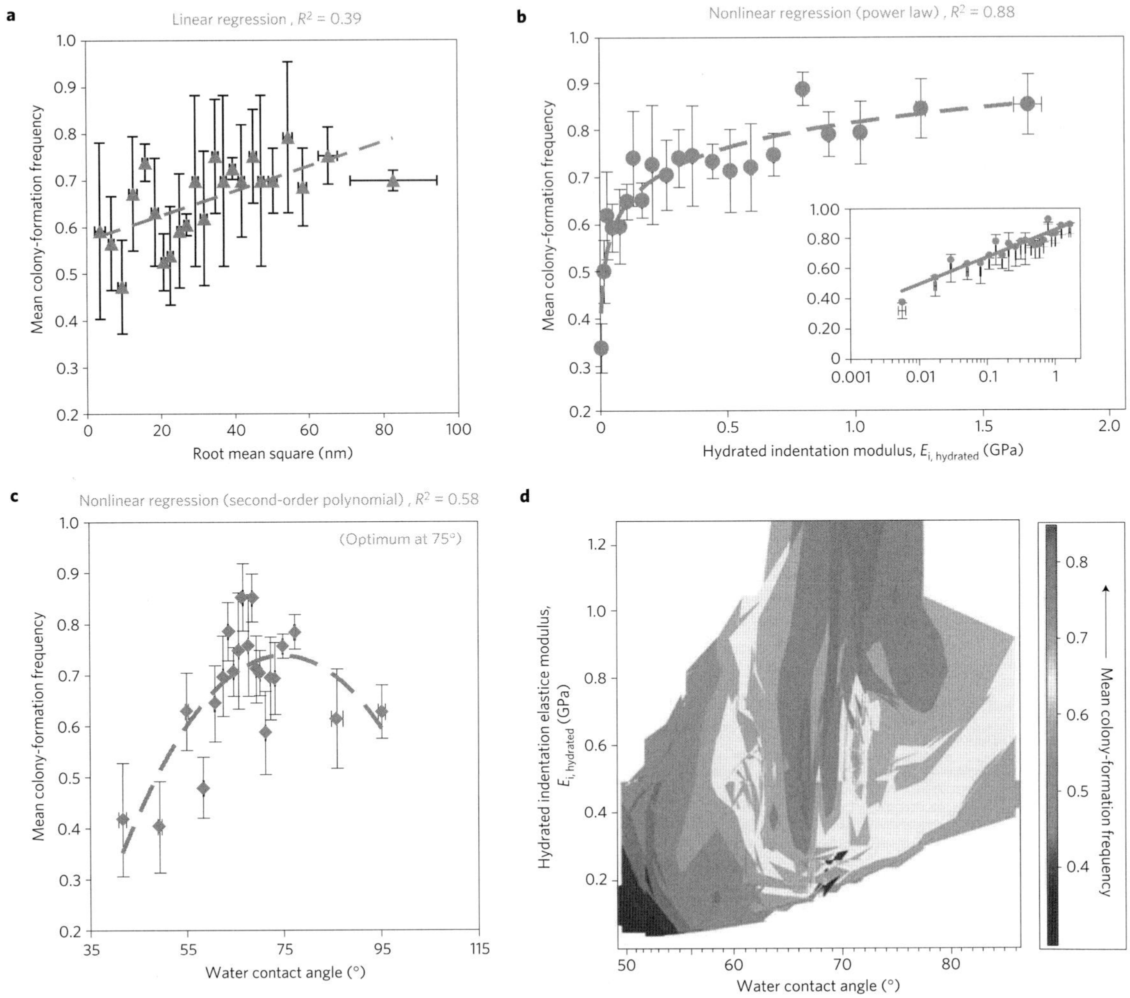

Figure 4 | Correlating hES cell behaviour to polymer properties using primary arrays. a–d, Using data in Fig. 3, regression was carried out for two properties listed at the top of each plot: colony formation versus polymer roughness (**a**), colony formation versus polymer indentation elastic modulus in the hydrated state (**b**), colony formation versus polymer wettability (**c**) and colony formation as a function of both polymer wettability and indentation elastic modulus in the hydrated state (**d**). After carrying out linear regression, second-order polynomial regression and power-law regression, only the regression with the highest R^2 is shown in each plot (dashed line). The inset in **b** is in semi-log format to indicate the behaviour at low modulus values. Data are sorted into groups of 20–25 spots as a function of increasing WCA, roughness or modulus. For the contour plot, interpolation between data points (groups of 20–25 spots) was carried out in Matlab (see Methods). In **a–c**, abscissa error bars represent the standard error of the WCA, roughness or modulus for a given group of 20–25 spots, and ordinate error bars represent the standard error of the mean of the colony-formation frequency of three replicates for a given group.

support either survival or growth of dissociated hES cells, some polymers supported the moderate growth of Oct4$^-$ differentiating cells and potential 'hit' polymers supported both robust growth and hES cell-colony formation (see also Supplementary Fig. S6). These differences in cell response demonstrate that polymers can strongly modulate hES cell behaviour between days 1 and 7 during colony growth from individual cells.

To better understand the relationship between polymer chemical composition and clonal growth of hES cells, a map of colony-formation frequency on the FBS-coated arrays against polymer monomeric composition has been generated. In Fig. 3a, major monomers are organized in order, from left to right, of increasing colony formation, and minor monomers are organized from bottom to top in order of increasing colony formation. This map indicates that the homopolymer formed from monomer 5 poorly supported clonal growth of hES cells, whereas most other homopolymers effectively supported cell growth. Tertiary amine containing minor monomer E and oligo(ethylene glycol) containing minor monomer A negatively influenced colony-formation frequency (bottom half of Fig. 3a), whereas minor monomers B, D and F could robustly support colony formation.

We characterized all polymeric substrates in the library using high-throughput techniques[31] to quantify several material properties: surface roughness (in air, PBS and culture medium after FBS adsorption), indentation elastic modulus (in air and fully hydrated in PBS) and surface wettability (Fig. 3b–d; Supplementary Fig. S7). Surface roughness and elastic properties of bulk material substrata can affect the behaviour of adult somatic and stem cells[32–34]. Surface wettability, indicating the hydrophobicity–hydrophilicity of the polymer surface (quantified as water contact angle, WCA), has been correlated previously with protein adsorption and cell adhesion[35]. To develop quantitative relationships between the colony formation and material properties, we attempted to correlate these properties with colony-formation frequency using linear and nonlinear regression (Fig. 4, where polymer spots of distinct composition are clustered as a function of the property indicated on each horizontal axis). These data indicate that polymer surface roughness in air (root mean square, RMS $\sim$ 0–25 nm; Supplementary Fig. S7a), in PBS (RMS $\sim$ 0–50 nm; Supplementary Fig. S7b) and in culture medium after FBS adsorption (RMS $\sim$ 0–110 nm; Fig. 4a) did not correlate strongly with colony-formation frequency. The standard error of measurement of material properties was low for replicate samples (for example, for WCA, $<$ 0.9–6.9% in Supplementary Fig. S8), such that error-bar heights in Fig. 4a are indicative of this weak correlation of roughness with colony formation frequency. In contrast, a positive power-law correlation was observed between the indentation elastic modulus, E_i, of hydrated polymers and colony-formation frequency (Fig. 4b, inset). However, we note that polymers exhibiting a low indentation elastic modulus (that is, high elastic compliance) also exhibited a low WCA (Supplementary Fig. S9a). Many of these highly compliant polymers contain hydrophilic major monomer 10 and hydrophilic minor monomer A (Supplementary Fig. S10a), which is consistent with our observation that the most compliant of these hydrated polymers also exhibited the greatest change in E_i between the dry and fully hydrated states (Supplementary Fig. S9b). This trend is consistent with previous studies of cell adhesion and proliferation capacity on swellable polymers, where decreased elastic stiffness correlated directly with increased absorption of aqueous solvents[36]. Thus, the power-law correlation between E_i and colony-formation frequency (Fig. 4b) probably reflects the extent of polymer hydrophobicity–hydrophilicity in the cases where a hydrophilic polymer swells to create a compliant surface (E_i < 0.2 GPa) that poorly supports colony formation. Figure 4b also demonstrates that, for the present array of hydrated polymers, colony formation is not strongly governed by polymer stiffness for E_i exceeding 0.2 GPa. In contrast, a moderate wettability (WCA $\sim$ 70°) is associated with optimal frequency of hES cell-colony formation (Fig. 4c). A contour projection of colony-formation frequency as a function of both E_i and WCA (Fig. 4d) shows clearly that the optimum wettability (65° $<$ WCA $<$ 80°) persists over a broad range of polymer stiffness, even for E_i > 0.2 GPa (see three-dimensional plot in Supplementary Fig. S10b). Thus, together these data indicate that colony-formation frequency of hES cells can be modulated more strongly by the wettability of these polymers than by variation in the elastic moduli of these polymers over the range considered.

To further refine our understanding of how this combination of material properties modulates colony formation, 48 polymers were selected to generate a 'secondary' polymer array with 36 replicates. This secondary array was designed to encompass a range of WCA similar to the range in the primary array (Supplementary Fig. S11), and the presence of 12 times more replicates significantly decreased experimental error. In good agreement with the primary-array data, a moderate wettability (WCA $\sim$ 70°) again effectively supported optimal hES cell clonal growth (Supplementary Figs S11–S13). Surfaces of all polymers in the secondary array were analysed using time-of-flight secondary-ion mass spectrometry (ToF-SIMS) in a high-throughput manner to provide molecular information on the topmost layers ($\sim$10 Å) of each polymer surface[37,38]. ToF-SIMS spectra from two homopolymers generated from similar monomers, 1 and 16 (Fig. 5a), were substantially different, suggesting that the polymer surface chemistry cannot be necessarily predicted from the monomer composition alone. Using a chemometric technique (partial least-squares (PLS) regression on ToF-SIMS spectra)[31,38], we correlated surface chemistry contained in the spectra to the colony-formation frequency observed on each polymer in the secondary array (Fig. 5b). Good agreement between measured colony-formation frequency and that predicted from the ToF-SIMS spectra was found ($R^2 = 0.78$). Each secondary ToF-SIMS ion associated with functionalities in the polymer structures could be listed with its regression coefficient, α, a quantitative measure of its contribution to colony-formation frequency (Fig. 5b and Supplementary Fig. S14a). The tertiary-amine moiety (characteristic ions $C_3H_8N^+$, $C_2H_6N^+$, CN^-) and tertiary-butyl moiety ($C_4H_9^+$) were identified by the PLS analysis to be correlated most strongly with a low colony-formation frequency, and hydrocarbon ions ($C_2H_3^+$, $C_3H_3^+$), oxygen-containing ions (CHO_2^-, $C_3H_3O^+$, $C_2H_3O^+$) from esters and ions from cyclic structures (C_6H^-, C_4H^-, C_2H^-) had the largest effect on promoting colony formation. The oxygen-containing ions and hydrocarbon ions can be attributed to the acrylate groups ($CH_2 = CHCO_2^-$) in each monomer, which form the backbone chain and the pendant ester groups after polymerization. Monomers with di- and tri-acrylates, which contain the most acrylate groups in our library, indeed showed the highest colony-formation frequencies.

The refined quantitative relationships among surface chemical structure and hES cell clonal growth generated from the secondary array provide an integrated view of all the cell responses seen on the primary array. On the primary array, the pendent functional groups in mono-acrylates (4, 5, 7 and 10) have sizeable effects on colony formation (Fig. 3a). For instance, the presence of tertiary butyl, a large non-polar functional group ($\alpha < 0$, Fig. 5b), in the major monomer 5 resulted in low colony formation. For most di- and tri-acrylate major monomers (1, 2, 8, 9, 11, 12, 13, 14 and 15) in the primary array, high acrylate content supported robust clonal formation (Fig. 3a), as expected from the large positive α (Fig. 5b). The exceptions (3, 6 and 16) can be attributed to the presence of a long chain of propylene glycol or ethylene glycol (for glycols, $\alpha < 0$, Fig. 5b). Although the ethylene glycol moiety can be found in the monomer chemistry of other di-acrylate major monomers such as 1, 2, 9 and 11, ToF-SIMS analysis indicated much higher propylene/ethylene glycol content present at the surface of homopolymers 3, 6 and 16 compared with 1, 2, 9 and 11 (Fig. 5a; Supplementary Fig. S14b). Further, the PLS model based on the secondary-array data was used to predict hES cell-colony formation of all 16 homopolymers in the primary array entirely on the basis of their ToF-SIMS spectra ($R^2 = 0.7$, Fig. 5b). This demonstrates that the model can be used to quantitatively predict hES cell clonal growth on a variety of acrylate polymers outside the training set of the model[39].

As polymers with a moderate WCA generated from multiple-acrylate-group-containing monomers performed best in these experiments, we chose the homopolymer of monomer 9, a di-acrylate with phenyl groups, and the 15A–30% copolymer, a tri-acrylate, to further validate the screening results. We fabricated 'hit arrays' where the entire polymer array is composed of one 'hit' polymer (that is, 9 or 15A-30%). The colony-formation efficiency of mEFs and hit polymer spots was quantified on the basis of the ratio of hES cell colonies formed on day 7 per attached hES cell on day 1 (Fig. 6a). About 20–25% of attached hES cells on day 1 created GFP^+, $SSEA4^+$ undifferentiated hES cell colonies after 7 days of culture on either the mEFs substrate or on the hit polymers. In contrast, cells on vitronectin- and matrigel-coated tissue-culture polystyrene (TCPS) exhibited predominantly differentiated growth, had lower Oct4 expression (similar to Fig. 2e) and did not form typical hES cell colonies with distinct borders (Fig. 2c–d). Further, nearly all (>95%) spots on hit arrays can support the expansion of $Oct4^+$, $SSEA\text{-}4^+$, $Nanog^+$ and $Tra\text{-}1\text{-}60^+$ cells after 7 day culture

ARTICLES NATURE MATERIALS DOI: 10.1038/NMAT2812

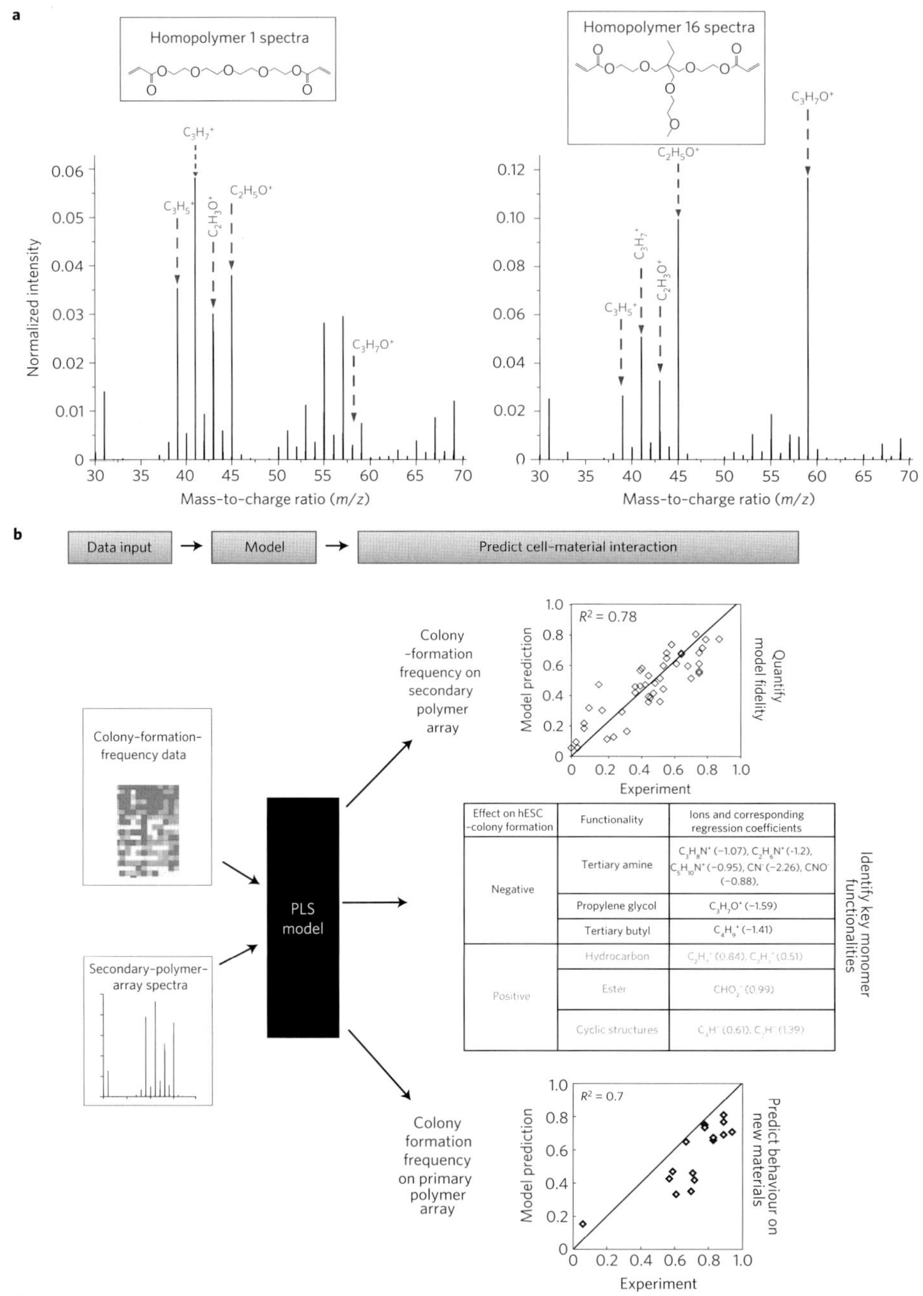

Figure 5 | Mapping cell behaviour to surface chemistry using secondary arrays. a, ToF-SIMS spectra of homopolymers 1 and 16, indicating that the surface chemistry cannot necessarily be predicted from the monomer chemistry; *m* denotes the mass and *z* denotes the charge. Arrows delineate higher intensities of hydrocarbon secondary ions ($C_3H_5^+$, $C_3H_7^+$) and ester ions ($C_2H_3O^+$) in the homopolymer 1 spectra. In contrast, higher intensities of ethylene glycol ions ($C_2H_5O^+$) and propylene glycol ions ($C_3H_7O^+$) were observed in the homopolymer 16 spectra. See the full analysis of ToF-SIMS spectra in Supplementary Fig. S14. **b**, A multivariate PLS regression method was used to quantitatively analyse and predict the cell-material interactions by correlating ToF-SIMS spectra of polymer spots to their biological performance (colony-formation frequency). The fidelity of PLS models can be quantified by a linear correlation of predicted versus measured colony-formation frequency. Top: each point in the figure represent one of 48 different polymers in the secondary array, and the inserted line represents the ideal situation when prediction match experiments completely. All ions and their regression coefficients, α, are listed in Supplementary Fig. S14a. Middle table: functionalities and their associated characteristic ions supporting or inhibiting hES colony formation. Ions were identified by correlating ToF-SIMS spectra to hES colony formation using PLS regression. Each ion was designated with a regression coefficient, α, that characterizes the relative effect on hES cell-colony formation. Bottom: as in top plot, but the PLS model was developed on ToF-SIMS spectra from the secondary array (with α values listed in the middle table) and was used to predict behaviour in the primary array.

NATURE MATERIALS DOI: 10.1038/NMAT2812 ARTICLES

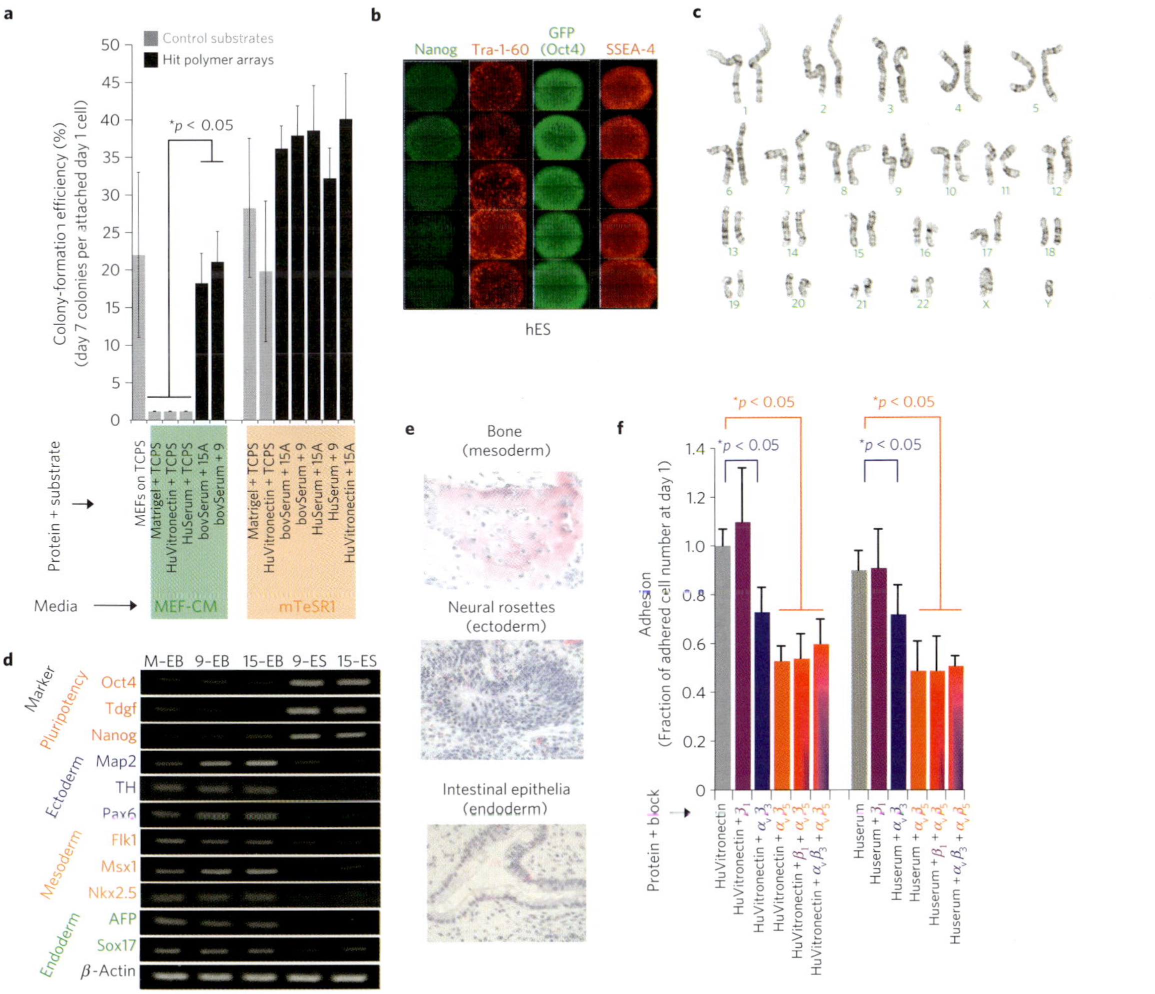

Figure 6 | Short- and long-term feeder-free culture on hit polymer arrays. a, Efficiencies of various culture systems to support undifferentiated growth of dissociated hES cells. Two medium conditions were used, labelled at the bottom: mEF-conditioned medium (MEF-CM) or chemically defined medium (mTeSR1). Several combinations of substrate and protein coating were used in conjunction with these media. Three substrates consisted of tissue culture polystyrene (TCPS), hit-polymer 9 (9; see Fig. 1a for monomer structure) and hit polymer 15A-30% (15A; see Fig. 1a for monomer structures). Four protein coatings consisted of matrigel, bovine serum, human serum and human vitronectin. Last, mEFs on gelatin-coated TCPS in regular hES medium were also used. In each condition, efficiencies were calculated as the number of SSEA-4$^+$ and Oct4$^+$ colonies seen on day 7 normalized to the number of cells attached on day 1. This metric specifically reflects the ability of substrates to promote undifferentiated clonal cell growth after correcting for any differences in initial cell attachment. **b**, Immunostaining of dissociated hES cells propagated on hit FBS-coated 15A-30% polymer for 7 days against Nanog (green) and Tra-1-60 (red), and on FBS-coated hit 9 polymer for 7 days against Oct4 (green) and SSEA-4 (red). **c**, Karyotypic analysis of hES cells propagated on hit 9 polymer array for more than 2 months (>10 passages). A normal 46XY karyotype was maintained on the hit array. **d**, Gene-expression analyses through RT-PCR of various differentiation markers for the three germ layers generated through embryoid-body (EB) *in vitro* differentiation. Lane labels are as follows: 'M-EB' for EBs generated from hES cells cultured on mEFs, '9-EB' and '15-EB' for EBs generated from hES cells cultured on 9 and 15A-30% hit polymer arrays respectively and '9-ES' and '15-ES' for hES cells cultured on 9 and 15A-30% hit polymer arrays respectively. **e**, Teratoma formation in immunodeficient mice by cells cultured on 15A-30% hit arrays. H and E staining was carried out on the teratoma. The resulting teratoma contained tissues representing all three germ layers: ectoderm, epidermal and neural tissue (rosette); mesoderm, bone and cartilage; and endoderm, respiratory epithelium and intestinal-like epithelium. **f**, Fraction of adhered cells after 24 h of culture in mTeSR1 media on hit polymer arrays coated with either human serum (HuSerum) or human vitronectin (HuVitronectin) and with the specified integrin-blocking antibody. Cell numbers shown here are averages of 24 replicates of the following hit polymers: 15, 15B-10%, 15B-20%, 15B-25%, 15D-10 and 15D-20%. β_1 blocking had minimal effect either alone or in combination with $\alpha_v\beta_5$ and $\alpha_v\beta_3$ blocking, whereas both $\alpha_v\beta_5$ and $\alpha_v\beta_3$ blocking reduced adhesion.

from completely dissociated hES single cells (Fig. 6b), and similar behaviour was seen with other pluripotent cell lines: an hiPS cell line (Supplementary Fig. S15a,b) and a non-transgenic hES cell line (Supplementary Fig. S15c,d).

The hit arrays were further evaluated for their capacity to maintain pluripotency of hES cells after more than 2 months of culture (>10 passages). Cells after prolonged culture were found to maintain an undifferentiated state with robust expression of hES cell markers, Oct4, Nanog, Tra-1-60 and SSEA-4 (as in Fig. 6b). Clonal efficiency of cells after long-term culture remained ~20%. In addition, cells could be passaged immediately to mEFs (Supplementary Fig. 16a), and such immediate attachment and growth suggests that the hit polymers do not select for a rare subpopulation in parental culture. Additionally, a normal karyotype showed the capacity of both 'hit' polymers (9 and 15A-30%) to support genomic stability of hES cells after a long-term culture (Fig. 6c). Gene-expression results (Fig. 6d) confirmed robust differentiation of these hES cells into all three germ lineages after 13 days of EB cultivation, and derivatives of all three embryonic germ layers were seen in teratoma assays (Fig. 6e). These results demonstrate that hES cells cultured on the polymeric hits maintain their full pluripotent potential.

To develop a more clinically relevant, defined culture system for hES cells, long-term culture was conducted on human-serum-(HS-) coated hit polymer arrays in mTeSR1 medium, a completely chemically defined medium (Supplementary Fig. S17). HS-coated hit polymer arrays supported the expansion of dissociated hES cells in a similar manner to arrays coated with FBS (Fig. 6a). Further, the HS-coated hit arrays could support long-term culture for more than 1 month (>5 passages), with robust expression of hES cell markers (Supplementary Fig. S16b). Last, the HS-coated hit polymers could support the undifferentiated growth of hiPS and other hES cell lines (Supplementary Fig. S15).

To investigate the potential pathways through which human serum may be important for colony formation, we carried out blocking experiments of several highly expressed integrins on the hES/hiPS cell surface (Fig. 6f). Although there are multiple integrins on the hES cell surface that can interact in a complex manner to promote cell adhesion and self-renewal[40,41], blocking the vitronectin-binding integrins, $\alpha_v\beta_3$ and $\alpha_v\beta_5$, resulted in a significant decrease in day 1 cell adhesion, whereas blocking a matrigel-binding integrin[41], β_1, had no effect. Vitronectin is also abundantly present in serum[42], and we tested its capacity to support colony formation when coated on the hit polymers. Similar levels of hES cell adhesion at day 1 were observed on HS-coated and vitronectin-coated hit polymers (Fig. 6f). Colony-formation efficiency at day 7 on FBS/HS-coated hit polymer arrays was identical to the efficiency on vitronectin-coated hit polymers (Fig. 6a). The histogram of the cell number on the polymer spots at day 1 (Supplementary Fig. S5c) indicated that most colonies formed at day 7 are expanded from a single cell. Although vitronectin-coated TCPS was recently reported to support the culture of hES cells[9], these surfaces were not demonstrated to support hES cell clonal growth, and significant differentiation was observed during clonal growth (see, for example, Fig. 2d). Our results indicate that the surface chemistry of hit substrates interacts with vitronectin, which engages with proper hES cell surface receptors (integrin $\alpha_v\beta_3$ and $\alpha_v\beta_5$), to support the clonal, undifferentiated growth of human pluripotent cells.

The biological activities of polymeric substrates can be controlled by surface properties, which in turn can be determined by chemical moieties present on the polymer surface. However, it may be difficult to quantitatively predict the presence of certain chemical functional groups at the polymer surface from the monomer composition alone, as well as the effects of surface chemistry on biological performance[43]. Here we employed high-throughput material synthesis and analysis to rapidly establish quantitative models between polymer-surface chemical structures and hES cell clonal growth. Using surface chemical structure information of the adsorbed protein layer (for example, vitronectin) did not result in more predictive models of hES cell-colony formation (Supplementary Fig. S18), indicating that ToF-SIMS can resolve surface-chemistry differences on polymers leading to colony formation but is less likely to resolve protein structural information leading to colony formation. Our analysis suggests that adsorbed proteins from our initial protein coating (but potentially also from culture media or hES cell secretion) need to interact with the appropriate polymer surface chemistry to be in the proper conformation to optimally promote colony formation. In addition to surface chemistry, which was demonstrated to have a controlling effect on stem cell behaviour, the geometry of the spot may influence self-renewal of hES cells as well, potentially by enhancing autocrine or juxtacrine signalling[44,45]. Last, the combination of human-vitronectin-coated hit polymers and mTeSR1 medium provides an attractive platform to develop a fully chemically defined, xeno-free, feeder-free culture system, as the only animal component, bovine serum albumin from the mTeSR1 medium, can be replaced by human serum albumin. Together, these advances may permit the facile growth of clinically relevant hES/hiPS cells from fully dissociated single cells, thereby enabling more straightforward genetic manipulation.

Methods

Combinatorial array preparation. Polymers were printed in a humid Ar atmosphere on epoxy-monolayer-coated glass slides (Xenopore Xenoslide E, Hawthorne, New Jersey), which were first dip-coated in 4 w/v% poly(2-hydroxyethyl methacrylate), using modifications of robotic fluid-handling technology as described previously[23]. Spots were polymerized through 10 s exposure to long-wave ultraviolet radiation (365 nm), dried at <50 mtorr for at least 7 days and coated with 20% serum or other proteins. See also Supplementary Methods and Figs S1 and S11a for the composition of primary and secondary arrays.

Surface-roughness measurements. Using an atomic force microscope (Digital Instruments Dimensions 3000A) in tapping mode, 5-μm regions of each polymer were measured and the root mean squared roughness was calculated. See also Supplementary Methods.

Elastic-modulus measurements. Using a pendulum-based instrumented nanoindenter (NanoTest, Micro Materials), experiments were conducted in ambient air or using a modified platform for *in situ* liquid experiments[46]. Samples were indented with a spherical ruby indenter of radius $R = 500$ μm ($n = 3$ locations for each polymer spot) with loading and unloading rates of 0.5 mN s^{-1}, a dwell of 10 s and a maximum load of 3 mN or a maximum depth of 600 nm. Loading rates were chosen such that the reduced elastic modulus inferred from indentation could be calculated from the initial unloading slope[47,48]. The indentation elastic modulus presented in the manuscript was calculated from the measured reduced elastic modulus, assuming a Poisson's ratio of 0.49 for all polymers. See also Supplementary Methods.

Water-contact-angle measurements. Measurements were of the sessile-drop type and made using ultrapure water on a Kruss DSA 100 apparatus fitted with a piezo-doser head. See also Supplementary Methods.

Time-of-flight secondary-ion mass spectroscopy. A secondary-ion mass spectrometer (ION-TOF, IV, UK) was operated using a Bi_3^+ primary-ion source at 25 kV and in 'bunched mode'. A 1 pA primary-ion beam was rastered at an area of 100 × 100 μm. Secondary ions were collected from the same area of each polymer spot on the microarray over 10 s acquisition time. Ion masses were determined using a high-resolution TOF analyser, and the typical mass resolution (at m/z 41) was just over 6,000.

Cell culture. hES cell lines BG01 and WIBR3 were maintained on mitomycin C-inactivated mEF feeder layers in standard medium. hES BG01–Oct4–GFP cells were made by introducing a Oct4–GFP–puro construct into hES cells[25]. hiPS C1 cells were derived through lentiviral infection of Oct4, Sox2 and Klf4 and cultured on mEFs as described previously[49]. For EB-induced differentiation, hES cell colonies were cultured for 13 days in non-adherent suspension culture dishes (Corning) in DMEM supplemented with 15% FBS.

For fluorescence-activated cell sorting, hES or hiPS cell lines were cultured in 10 μM Rho-associated kinase inhibitor (Calbiochem; Y-27632) for 24 h in standard mEF conditions before sorting. Cells were harvested enzymatically with 1 mg ml^{-1} collagenase and then with 0.05% trypsin for 5 min at 37 °C. hiPS cells were immunostained using SSEA-4. Cells were collected in media with Rho-associated kinase inhibitor and sorted on a FACSAria Flow Cytometer (Becton Dickinson, San Jose, California). Cells were subsequently plated on various surfaces in medium supplemented with Rho-associated kinase inhibitor for the first 24 h of culture to reduce initial apoptosis of completely dissociated hES cells[50]. For HS-coated hit arrays, culturing occurred in mTeSR1 media (Stemcell Technologies). Long-term culture on hit arrays occurred in mTeSR1 media and by passaging 1:3 every 5–7 days using collagenase. See also Supplementary Methods.

NATURE MATERIALS DOI: 10.1038/NMAT2812 **ARTICLES**

Biological assays. For teratoma formation, hES cells were injected subcutaneously in the back of severe combined immunodeficient mice (Taconic) and tumours generally developed within 4–8 weeks. After sectioning, teratomas were diagnosed on the basis of haematoxylin and eosin staining. For karyotype analysis, chromosomal studies were carried out by Cell Line Genetics (Madison, Wisconsin) using standard protocols for high-resolution G-banding. For immunocytochemistry, cells were fixed in 4% paraformaldehyde in PBS and immunostained according to standard protocols. See also Supplementary Methods.

Numerical methods. Principal component analysis and PLS regression were carried out using the Eigenvector PLS Toolbox 3.5 using the SIMPLS algorithm. For linear and nonlinear least-squares regression, Excel (Microsoft) was used. Contour and three-dimensional plots were generated in MATLAB4 (Mathworks) using the v4 griddata method of data interpolation. See also Supplementary Methods.

Efficiency measurements. Using a low (1.6 cells mm^{-2}) seeding density, day 1 cell and day 7 colony numbers on mEFs and matrigel were normalized per scanned surface area and on 'hit' polymer spots as per 'hit' array (1,875 mm^2, 1,728 replicates). For experiments on TCPS, single cells were sorted individually directly into each well of a coated 96-well plate. Day 1 cell and day 7 colony numbers were measured by manually counting stained wells. For vitronectin-coated, matrigel-coated and HS-coated TCPS in mEF-conditioned media, no GFP^+ colonies were observed: in about 30% of cases we observed only differentiated cell growth. See also Supplementary Methods.

Received 21 September 2009; accepted 24 June 2010; published online 22 August 2010

References

1. Thomson, J. A. *et al.* Embryonic stem cell lines derived from human blastocysts. *Science* **282,** 1145–1147 (1998).
2. Takahashi, K. *et al.* Induction of pluripotent stem cells from adult human fibroblasts by defined factors. *Cell* **131,** 861–872 (2007).
3. Yu, J. *et al.* Induced pluripotent stem cell lines derived from human somatic cells. *Science* **318,** 1917–1920 (2007).
4. Daley, G. Q. & Scadden, D. T. Prospects for stem cell-based therapy. *Cell* **132,** 544–548 (2008).
5. Saha, K. & Jaenisch, R. Technical challenges in using human induced pluripotent stem cells to model disease. *Cell Stem Cell* **5,** 584–595 (2009).
6. Amit, M. *et al.* Clonally derived human embryonic stem cell lines maintain pluripotency and proliferative potential for prolonged periods of culture. *Dev. Biol.* **227,** 271–278 (2000).
7. Xu, C. *et al.* Feeder-free growth of undifferentiated human embryonic stem cells. *Nature Biotechnol.* **19,** 971–974 (2001).
8. Stojkovic, P. *et al.* Human-serum matrix supports undifferentiated growth of human embryonic stem cells. *Stem Cells* **23,** 895–902 (2005).
9. Braam, S. R. *et al.* Recombinant vitronectin is a functionally defined substrate that supports human embryonic stem cell self-renewal via alphavbeta5 integrin. *Stem Cells* **26,** 2257–2265 (2008).
10. Li, Y., Powell, S., Brunette, E., Lebkowski, J. & Mandalam, R. Expansion of human embryonic stem cells in defined serum-free medium devoid of animal-derived products. *Biotechnol. Bioeng.* **91,** 688–698 (2005).
11. Amit, M., Shariki, C., Margulets, V. & Itskovitz-Eldor, J. Feeder layer- and serum-free culture of human embryonic stem cells. *Biol. Reprod.* **70,** 837–845 (2004).
12. Yao, S. *et al.* Long-term self-renewal and directed differentiation of human embryonic stem cells in chemically defined conditions. *Proc. Natl Acad. Sci. USA* **103,** 6907–6912 (2006).
13. Ludwig, T. E. *et al.* Feeder-independent culture of human embryonic stem cells. *Nature Methods* **3,** 637–646 (2006).
14. Ludwig, T. E. *et al.* Derivation of human embryonic stem cells in defined conditions. *Nature Biotechnol.* **24,** 185–187 (2006).
15. Rodin, S. *et al.* Long-term self-renewal of human pluripotent stem cells on human recombinant laminin-511. *Nature Biotechnol.* **28,** 611–615 (2010).
16. Li, Y. J., Chung, E. H., Rodriguez, R. T., Firpo, M. T. & Healy, K. E. Hydrogels as artificial matrices for human embryonic stem cell self-renewal. *J. Biomedical Mater. Res. Part A* **79,** 1–5 (2006).
17. Gerecht, S. *et al.* Hyaluronic acid hydrogel for controlled self-renewal and differentiation of human embryonic stem cells. *Proc. Natl Acad. Sci. USA* **104,** 11298–11303 (2007).
18. Villa-Diaz, L. G. *et al.* Synthetic polymer coatings for long-term growth of human embryonic stem cells. *Nature Biotechnol.* **28,** 581–583 (2010).
19. Melkoumian, Z. *et al.* Synthetic peptide–acrylate surfaces for long-term self-renewal and cardiomyocyte differentiation of human embryonic stem cells. *Nature Biotechnol.* **28,** 606–610 (2010).
20. Zou, J. *et al.* Gene targeting of a disease-related gene in human induced pluripotent stem and embryonic stem cells. *Cell Stem Cell* **5,** 97–110 (2009).
21. Hockemeyer, D. *et al.* Efficient targeting of expressed and silent genes in human ESCs and iPSCs using zinc-finger nucleases. *Nature Biotechnol.* **27,** 851–857 (2009).
22. Zwaka, T. P. & Thomson, J. A. Homologous recombination in human embryonic stem cells. *Nature Biotechnol.* **21,** 319–321 (2003).
23. Anderson, D. G., Levenberg, S. & Langer, R. Nanoliter-scale synthesis of arrayed biomaterials and application to human embryonic stem cells. *Nature Biotechnol.* **22,** 863–866 (2004).
24. Mei, Y. *et al.* Mapping the interactions among biomaterials, adsorbed proteins, and human embryonic stem cells. *Adv. Mater.* **21,** 2781–2876 (2009).
25. Green, J. J. *et al.* Nanoparticles for gene transfer to human embryonic stem cell colonies. *Nano Lett.* **8,** 3126–3130 (2008).
26. Tamada, Y. & Ikada, Y. Effect of preadsorbed proteins on cell-adhesion to polymer surfaces. *J. Colloid Interf. Sci.* **155,** 334–339 (1993).
27. Underwood, P. A., Steele, J. G. & Dalton, B. A. Effects of polystyrene surface chemistry on the biological activity of solid phase fibronectin and vitronectin, analysed with monoclonal antibodies. *J. Cell Sci.* **104,** 793–803 (1993).
28. van Wachem, P. B. *et al.* The influence of protein adsorption on interactions of cultured human endothelial cells with polymers. *J. Biomed. Mater. Res.* **21,** 701–718 (1987).
29. Keselowsky, B. G., Collard, D. M. & Garcia, A. J. Integrin binding specificity regulates biomaterial surface chemistry effects on cell differentiation. *Proc. Natl Acad. Sci. USA* **102,** 5953–5957 (2005).
30. Luther, E., Kamentsky, L., Henriksen, M. & Holden, E. *Cytometry, New Developments,* Vol. 75, 4th edn, 185–218 (2004).
31. Urquhart, A. J. *et al.* High throughput surface characterisation of a combinatorial material library. *Adv. Mater.* **19,** 2486–2491 (2007).
32. Lipski, A. M. *et al.* Nanoscale engineering of biomaterial surfaces. *Adv. Mater.* **19,** 553–557 (2007).
33. Engler, A. J., Sen, S., Sweeney, H. L. & Discher, D. E. Matrix elasticity directs stem cell lineage specification. *Cell* **126,** 677–689 (2006).
34. Saha, K. *et al.* Substrate modulus directs neural stem cell behaviour. *Biophys. J.* **95,** 4426–4438 (2008).
35. Tamada, Y. & Ikada, Y. Effect of preadsorbed proteins on cell-adhesion to polymer surfaces. *J. Colloid Inter. Sci.* **155,** 334–339 (1993).
36. Thompson, M. T., Berg, M. C., Tobias, I. S., Rubner, M. F. & Van Vliet, K. J. Tuning compliance of nanoscale polyelectrolyte multilayers to modulate cell adhesion. *Biomaterials* **26,** 6836–6845 (2005).
37. Delcorte, A. *et al.* ToF-SIMS study of alternate polyelectrolyte thin films: Chemical surface characterization and molecular secondary ions sampling depth. *Surf. Sci.* **366,** 149–165 (1996).
38. Vickerman, J. C. *ToF-SIMS: Surface Analysis by Mass Spectrometry* (IM Publications, 2001).
39. Urquhart, A. J. *et al.* TOF-SIMS analysis of a 576 micropatterned copolymer array to reveal surface moieties that control wettability. *Anal. Chem.* **80,** 135–142 (2008).
40. Meng, Y. *et al.* Characterization of integrin engagement during defined human embryonic stem cell culture. *FASEB J.* **24,** 1056–1065 (2010).
41. Rowland, T. J. *et al.* Roles of integrins in human induced pluripotent stem cell growth on matrigel and vitronectin. *Stem Cells Dev.* doi:10.1089/scd.2009.0328 (2010).
42. Hayman, E. G., Pierschbacher, M. D., Ohgren, Y. & Ruoslahti, E. Serum spreading factor (vitronectin) is present at the cell surface and in tissues. *Proc. Natl Acad. Sci. USA* **80,** 4003–4007 (1983).
43. Neuss, S. *et al.* Assessment of stem cell/biomaterial combinations for stem cell-based tissue engineering. *Biomaterials* **29,** 302–313 (2008).
44. Guilak, F. *et al.* Control of stem cell fate by physical interactions with the extracellular matrix. *Cell Stem Cell* **5,** 17–26 (2009).
45. Peerani, R. *et al.* Niche-mediated control of human embryonic stem cell self-renewal and differentiation. *EMBO J.* **26,** 4744–4755 (2007).
46. Constantinides, G., Kalcioglu, Z. I., McFarland, M., Smith, J. F. & Van Vliet, K. J. Probing mechanical properties of fully hydrated gels and biological tissues. *J. Biomechanics* **41,** 3285–3289 (2008).
47. Oliver, W. C. & Pharr, G. M. An improved technique for determining hardness and elastic modulus using load and displacement sensing indentation experiments. *J. Mater. Res.* **7,** 1564–1583 (1992).
48. Cheng, Y-T. & Cheng, C-M. Scaling, dimensional analysis, and indentation measurements. *Mater. Sci. Eng.* **44,** 91–149 (2004).
49. Hockemeyer, D. *et al.* A drug-inducible system for direct reprogramming of human somatic cells to pluripotency. *Cell Stem Cell* **3,** 346–353 (2008).
50. Watanabe, K. *et al.* A ROCK inhibitor permits survival of dissociated human embryonic stem cells. *Nature Biotechnol.* **25,** 681–686 (2007).

Acknowledgements

We thank C. Beard, R. Alagappan, P. Xu, P. Wisniewski, C. Araneo, K. Wood, J. Daussman, R. Flannery, D. Fu, E. Luther and Compucyte for technical support.

ARTICLES NATURE MATERIALS DOI: 10.1038/NMAT2812

We thank all the members of the Langer laboratory and Jaenisch laboratory for helpful discussions and comments on the manuscript. R.J. was supported by NIH grants R37-CA084198, RO1-CA087869 and RO1-HD045022. R.L., R.J. and D.G.A. are advisors to Stemgent and R.L. and R.J. are cofounders of Fate Therapeutics. Financial support for J.Y. and A.H. is from the Wellcome Trust 085246. K.S. is supported by the Society in Science: the Branco Weiss Fellowship. D.G.A., R.L. and Y.M. are supported by NIH DE016516. Z.I.K. was supported by the US Army through the Institute for Soldier Nanotechnologies, under Contract W911NF-07-D-0004 with the US Army Research Office.

Author contributions

All of the authors developed experiments, participated in the generation and analysis of data and assisted in the writing of the manuscript.

Additional information

The authors declare no competing financial interests. Supplementary information accompanies this paper on www.nature.com/naturematerials. Reprints and permissions information is available online at http://npg.nature.com/reprintsandpermissions. Correspondence and requests for materials should be addressed to R.L., R.J. or D.G.A.

PERSPECTIVE

Three-dimensional biomaterials for the study of human pluripotent stem cells

Thomas P Kraehenbuehl[1,2], Robert Langer[1] & Lino S Ferreira[3,4]

The self-renewal and differentiation of human pluripotent stem cells (hPSCs) have typically been studied in flat, two-dimensional (2D) environments. In this Perspective, we argue that 3D model systems may be needed in addition, as they mimic the natural 3D tissue organization more closely. We survey methods that have used 3D biomaterials for expansion of undifferentiated hPSCs, directed differentiation of hPSCs and transplantation of differentiated hPSCs *in vivo*.

Human pluripotent stem cells (hPSCs), including human embryonic stem cells (hESCs)[1] and induced pluripotent stem cells (iPSCs)[2,3] are promising for research, drug discovery and regenerative medicine. The unique ability of hPSCs to self-renew in an undifferentiated state enables large-scale expansion *in vitro*, whereas their differentiation capacity allows the generation of all ~200 cell types of the human body.

During embryonic development, cells do not only interact with each other but also with the three-dimensional (3D) extracellular matrix (ECM), which comprises proteins such as collagen and laminin interlaced with proteoglycans[4]. For instance, in the mouse embryo, laminin and heparan sulfate proteoglycan are present on the cell surface at the two-cell stage, and collagen IV and fibronectin are first detected at the protein level in the blastocyst inner cell mass[5]. The ECM has multiple functions in cell growth, differentiation, cell maintenance and tissue morphogenesis[6]. Cells interact with the ECM via cell-surface receptors, including via integrins, which have mechanical roles (that is, anchoring) and also signal across the cell membrane.

There is an increasing interest in developing biomaterials that mimic the 3D architecture and biological role of the ECM to modulate the behavior of stem cells[7–9]. Recent reviews have discussed the design of artificial environments to study and instruct stem-cell fate generally[10,11]. In this Perspective, we consider the need for 3D scaffolds specifically for pluripotent stem cells (**Fig. 1**). Matrix requirements to control the self-renewal or differentiation of pluripotent stem cells are likely to differ from those required for adult stem cells. The 3D culture models could permit recapitulation of embryonic development *in vitro* to a degree of complexity not achievable in a 2D culture system. We summarize the achievements in this area thus far and discuss the limitations of existing 3D scaffolds for the culture, study and control of pluripotent stem cells.

Types of biomaterials

Natural materials (for example, collagen, hyaluronic acid, Matrigel (BD Biosciences), chitosan, dextran and alginate) can be obtained commercially and can be easily processed to prepare 3D scaffolds, but controlled incorporation of biomolecules requires specialized expertise. In most cases, cells are suspended in the polymeric solution, which is then cross-linked via a temperature transition (Matrigel and collagen)[12,13], pH transition (collagen, chitosan)[14], light exposure (hyaluronic acid or dextran-containing vinyl groups)[15,16] or addition of ions (alginate)[17]. Decellularized membranes such as amniotic membranes[18] also can be used to culture and transplant cells. Whereas natural materials have the advantage of providing biological signals to the cells cultured on them, they have the disadvantages of batch-to-batch variability and limited ability for controlled modification. In addition, clinical application may be limited by the risk of immunogenicity and disease transfer.

Synthetic materials allow easier modulation of the structural properties of the scaffold. Synthetic scaffolds can be prepared by microfabrication techniques[19] to display multiple gradients of immobilized chemical signals, which could in principle be useful to tailor the

[1]Department of Chemical Engineering, Massachusetts Institute of Technology, Cambridge, Massachusetts, USA. [2]Hoffmann-La Roche Inc., Translational Research Sciences, Nutley, New Jersey, USA. [3]Biocant–Center of Innovation in Biotechnology, Cantanhede, Portugal. [4]Center of Neurosciences and Cell Biology, University of Coimbra, Coimbra, Portugal. Correspondence should be addressed to L.S.F. (lino@biocant.pt) or R.L. (rlanger@mit.edu).

PUBLISHED ONLINE 30 AUGUST 2011; DOI:10.1038/NMETH.1671

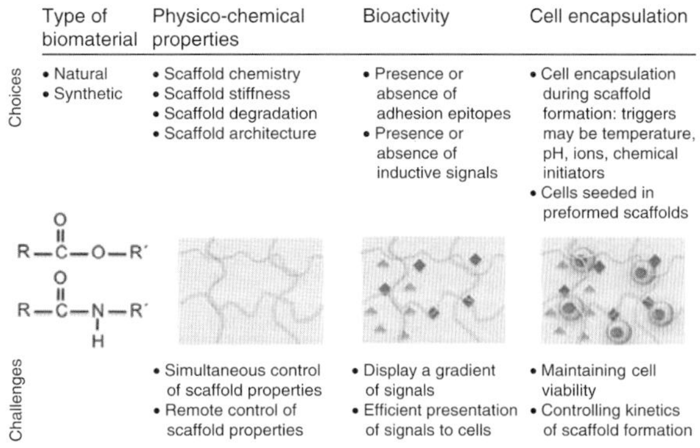

Figure 1 | Choices and challenges in the design of 3D environments for self-renewal, differentiation, *in vitro* culture or transplantation of hPSCs or hPSC-derived cells.

maturation process of successive progenitor cells. In addition, synthetic materials allow better control over the scaffold degradation profile and have a lower risk for immunogenic reactions. Synthetic materials used to build 3D scaffolds are typically derivatives of polyethylene glycol (PEG)[11,20–24], polyesters (poly(lactic acid) (PLA)[8,25], poly(lactic acid-*co*-glycolic acid) (PLGA)[8,25] and poly(glycerol sebacate) (PGS)[26]); these can be commercially obtained, but processing them into 3D scaffolds and modulating their physicochemical and biological properties requires specialized expertise. The inert PEG backbone can be tailored with tissue-specific bioactive elements (cell-adhesion ligands, growth factors and peptides that promote degradation) to render it cell-responsive[23,27,28]. Cells are immersed in the precursor gel solution containing a photoinitiator and the suspension is photo-cross-linked via UV-light exposure. It has been reported that hESCs[15] and their progeny cells[27] do not accumulate p53 upon short exposures to UV light (minutes range), suggesting that the exposure can be tolerated. Should UV-light toxicity be a concern, cells can be immersed in a precursor gel solution that can assemble and cross-link without the use of a photopolymerization reaction[24], although chemical groups in the cell membrane (for example, thiol groups[29]) may react with gel components (for example, vinyl-based macromonomers) and thus affect cell behavior.

The mechanical properties of ideal 3D scaffolds should be easily tailored. These properties can be controlled typically by the concentration and molecular weight of the polymer or the type or density of cross-linking between the polymer chains[21]. In addition, 3D scaffolds should be permissive to degradation by cells and migration of cells through the scaffold; the required degradation kinetics will depend on the particular study. Entirely synthetic polymers degrade at a constant rate: for instance, through hydrolysis of ester linkages in the case of polyesters[8,25]. Scaffolds of synthetic polymers modified with enzyme-sensitive peptides (collagenase-sensitive or metalloprotease (MMP)-sensitive peptides) degrade at multiple rates depending on the peptide density and the concentration of enzymes secreted by the cells[24,28]. Similarly, the degradation kinetics of natural hydrogels (collagen, hyaluronic acid and chitosan) depends on the amounts of secreted enzyme and on cross-linking density[15]. Ionically cross-linked scaffolds such as alginate normally undergo very slow dissolution[17,30], with the rate depending on the ionic environment. The type of scaffold should be selected based on its ultimate application (for instance, self-renewal, differentiation or transplantation), its capacity to replicate the chemical, mechanical, 3D architectural or degradative environment of a specific extracellular matrix, to the extent that this is known, and its availability.

3D biomaterials for hPSC self-renewal

hESCs had been first derived and cultured on mouse embryonic fibroblast feeder layers with medium containing animal components[1]. Since then, chemically defined hESC media[31] and 2D synthetic surfaces[32–35] have been developed that allow the expansion of several hPSC lines for more than 10 passages in an undifferentiated state. We propose that 3D systems may facilitate the long-term growth of hPSCs in the absence of feeder layers and without subculturing. At the blastocyst stage, the cells of the inner cell mass (the *in vivo* equivalent of hESCs) establish 3D cellular associations; although cells of the inner cell mass do not proliferate indefinitely *in vivo*, such associations may well be important to regulate the molecular circuitry of pluripotency. For instance, the signaling network formed by epithelial cadherin, nonmuscle myosin IIA and p120-catenin[36] and the interplay between epithelial cadherin and the small GTPase Rap1 (ref. 37) have been identified in 2D culture to be important in hESC self-renewal. But little is known about factors affecting proliferation of hESCs in 3D culture, and we believe that this aspect deserves additional study. So far, 3D scaffolds formed by alginate-chitosan complex, hyaluronic acid and PEG (**Table 1**) have been used to culture hPSCs for relatively short periods of time (typically < 1 month). It will be important to extend these studies for the long-term expansion of several hPSC lines. The development of artificial ECM preparations that allow long term *in vitro* culture of these cells could help to elucidate mechanisms of self-renewal. We anticipate that 3D scaffold degradation, chemistry and mechanical properties are important properties to modulate pluripotent stem cell self-renewal.

Scaffold mechanical properties. It is known that the mechanical properties of 2D substrates influence pluripotent stem cells, though the mechanisms are not as yet completely understood. In the mouse, for instance, soft 2D substrates (elastic Young's modulus of 0.6 kPa) that matches the intrinsic mouse ESC stiffness facilitate mESC self-renewal up to at least 15 passages in culture medium in the absence of leukemia inhibitory factor (LIF)[38]. It is thus tempting to speculate that engineering the 3D mechanical environment may enable modulation of stemness in hPSCs as well. So far, scaffolds tested for the encapsulation of hPSCs had an elastic modulus of around 10 kPa[15], which likely does not match the intrinsic stiffness of the cells.

Scaffold degradation. hESCs proliferate in 2D culture as compact colonies, and this architecture is linked to hESC self-renewal, pluripotency and survival[36,37,39]. Colony architecture that is most conducive to pluripotency in 3D culture is currently not known. It is likely that a successful 3D scaffold should allow cell expansion without perturbing colony integrity. This would require scaffold degradation in response to cellular enzymes[15] or other factors[14], or that the scaffold is sufficiently elastic to permit colony size increases. Scaffold degradation is also key for removal of cells from the matrix after expansion. Ideally, the 3D scaffold should

Table 1 | Biomaterials for hPSC self-renewal, differentiation and transplantation

Biomaterial	Application	Pore size or diameter (μm)	Young's modulus (kPa)	Reference
Alginate	Vascular tissue differentiation	50–100	500–1,136 (dry state)	17
	Transplantation into infarcted myocardium, rat	~100	NA	30
Alginate-chitosan complex	Self-renewal	~65	~800	14
Collagen	Liver tissue differentiation	NA	NA	13
Dextran	Vascular tissue differentiation	NA	1.5–6.0	16
Hyaluronic acid	Self-renewal	30–50	2–100	15
	Transplantation into cartilage, rat	NA	NA	55
Matrigel	Transplantation into infarcted myocardium, rat	NA	NA	12
	Transplantation subcutaneously, mouse	NA	NA	63
	Intestinal differentiation and organization	NA	NA	54
Poly(glycerol sebacate)	Vascular differentiation	20–200	40–60	26
PEG	Cartilage tissue differentiation	0.01–0 1	0.3–4	27
	Self-renewal, mouse	NA	0.02–1	23
	Transplantation into infarcted myocardium, rat	0.1	4	24
PLLA mixed with PLGA	Neuronal, cartilage, liver & skeletal muscle tissue differentiation	225–500	65–500	8,25
	Transplantation intramuscular, mouse and rat	225–500	65–500	25
	Cardiomyocyte differentiation and 3D organization	212–600	NA	58

NA, not available.

support the expansion of the cells and then degrade rapidly when an external trigger is added. Triggers that have been used for scaffold degradation include enzymes (for example, hyaluronidase enzyme in hyaluronic acid gels[15]) or small chemicals (for example, EDTA in alginate gels[14]).

Scaffold biochemical properties. The attachment of undifferentiated iPSCs or hESCs to 2D substrates requires functionalization to engage adhesion molecules expressed by these cells (for instance, the integrins $\alpha_V\beta_5$ (refs. 35,40), $\alpha_V\beta_3$ (refs. 35,41), $\alpha_2\beta_1$ (ref. 41), α_6 (refs. 41,42) and β_1 (refs. 41,42)). The involvement of integrins $\alpha_v\beta_5$ and $\alpha_v\beta_3$, which mediate cell adhesion to vitronectin, has yielded good results in terms of self-renewal and pluripotency in three independent hESC lines[40] and for hESC clonal growth[35]. However, the importance of engaging hPSC surface integrins in 3D scaffolds remains largely untested. Notably, hPSCs can self-renew in 3D biomaterials for which they have poor adhesion. For example, hESCs attached poorly to hyaluronic acid but maintained their pluripotency for at least 20 days[15].

3D biomaterials for hPSC differentiation

Many existing differentiation protocols for hPSCs use 2D culture systems[43–47]. Differentiation in two dimensions yields, in many cases, the desired cell lineages; it is easy to characterize the cells during the differentiation process as they are directly accessible; and it does not require nonbiological expertise. However, we believe that 2D approaches may have limitations. The 2D systems may promote unnatural interactions with external factors. As only the basal region of a cell in 2D culture interacts with the ECM and with neighboring cells, it is conceivable that the distribution of cell-surface integrins and the organization of intracellular signaling machinery is unnatural, thus changing the cellular response[44,45]. Low efficiencies of some directed differentiation protocols may be because 2D culture does not adequately reproduce important *in vivo* features. For example, the differentiation of hESCs into pancreatic endocrine cells requires multiple differentiation steps (~20 days) and involves several inductive agents to derive ~7% of differentiated cells[46]. It should be noted, however, that 2D culture allows rapid (~5 days) and efficient (>80%) differentiation of hPSCs into neurons in the presence of SMAD inhibitors[47]. Thus, 3D culture conditions are not an absolute requirement for the efficient differentiation of hPSCs. Although hPSC differentiation is often carried out in 3D aggregates termed embryoid bodies, these may not recapitulate all relevant aspects of embryonic development, in particular complex processes that involve substantial structural organization or maturation.

Most studies of hPSC differentiation in three dimensions use cells that have been predifferentiated via embryoid body formation. Either enzymatically dissociated single cells[8,25,48] or entire embryoid bodies are then encapsulated into the biomaterial[16] for 3D, tissue-specific differentiation (**Fig. 2**). Several 3D scaffolds have been used for hPSC differentiation, including scaffolds formed of alginate, collagen, dextran, hyaluronic acid, PLA mixed with PLGA, PEG and poly(glycerol sebacate) (**Table 1**). We anticipate that the mechanical, chemical and architectural properties of the scaffold are all likely to be important for modulating hPSC differentiation, although there is still much to be learned about the biochemical and biophysical factors that regulate hPSC differentiation and subsequent morphogenesis in three dimensions.

Scaffold mechanical properties. It has been shown that pluripotent stem cell differentiation can be influenced by the stiffness of 2D cell culture substrates[49]. Increasing substrate stiffness (polydimethylsiloxane) from 41 kPa to 2.7 MPa promoted mesendoderm differentiation of mouse embryonic stem cells. We anticipate that the effects of 3D scaffold mechanical properties on encapsulated hPSCs will be examined in the future. Recent advances in the development of techniques to measure the mechanical traction exerted by cells in 3D matrices[50] will contribute to this area.

Scaffold biochemical properties. It has been shown that cell differentiation can be modulated by growth factors covalently bound or physically entrapped in 3D scaffolds[16,51], or by ECM proteins or peptides covalently coupled to 3D scaffolds[16,24]. For instance, scaffolds functionalized to engage with cellular integrins β_1 (ref. 22) or α_6/β_4 (ref. 52) promote the differentiation

PERSPECTIVE

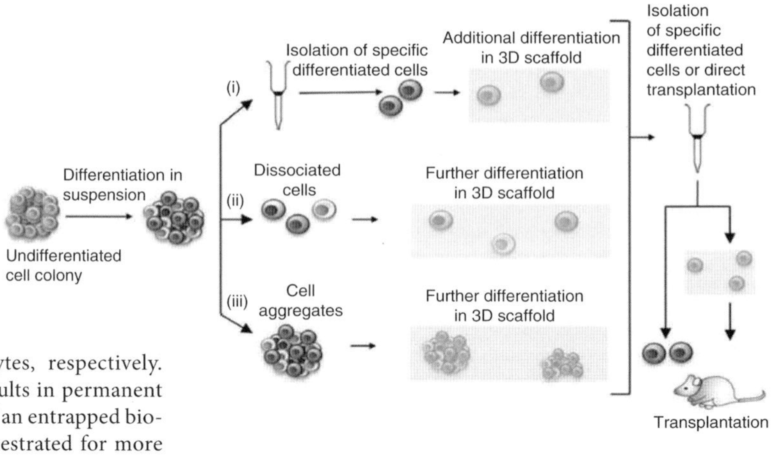

Figure 2 | Three-dimensional matrices for directed hPSC differentiation. Spontaneously differentiated hESC aggregates can be seeded into scaffolds in three different ways: isolation of specific target cells by cell sorting before seeding (i), enzymatic dissociation of aggregates followed by direct seeding without purification (ii) and direct seeding into the scaffold (iii). After cell cultivation in the biomaterials, the constructs can be either implanted directly into animal models or dissociated with subsequent selection of specific target cells for implant.

of hESCs into chondrocytes or keratinocytes, respectively. Covalent immobilization of biomolecules results in permanent stimulation of cells[16,22], whereas the release of an entrapped biomolecule can be spatially or temporally orchestrated for more controlled activation of a specific receptor[16,51].

Scaffold 3D architecture. 3D structures with characteristics of developing neuronal[48], cartilage[22,27], vascular[16,51] and liver[13] tissues have been reported for hESCs cultured in 3D scaffolds. Notably, the extent of differentiation[16], organization[27] and functionality[13,27] has been reported to be higher in cells cultured in 3D scaffolds than in embryoid bodies or cells cultured in 2D culture systems. However, efficient maturation of hPSC-derived tissues has proven difficult even in 3D culture. Typically, cells in the 3D constructs differentiate to variable degrees, which could be due to diffusion limitations in the construct, cell heterogeneity before seeding into the scaffold[53] or an insufficient understanding of the factors required for differentiation. In some cases, a rational combination of 2D and 3D systems may prove beneficial, as was recently demonstrated for the development of intestinal tissue from hPSCs[54]. The cells were first differentiated into hindgut cells in 2D culture and then placed in 3D Matrigel scaffolds for the induction of intestinal tissue. Remarkably, the cell aggregates developed into 3D intestinal tissue consisting of a polarized epithelium patterned into villus-like structures and crypt-like proliferative zones that expressed intestinal stem cell markers[54]. Depending on the tissue, a 3D scaffold may only be needed for the full maturation and 3D organization of the cells and not at early stages of hPSC differentiation.

3D biomaterials for *in vivo* transplantation of hPSC-derived cells

One of the goals of hPSC research is to develop new therapeutic approaches for regenerative medicine, for which 3D scaffolds could prove important. Scaffolds could be used to retain the cells at the desired location, while serving as a template for 3D cell assembly, survival and engraftment[12,24]. Several 3D scaffolds have been used for this purpose including PEG-based hydrogels, Matrigel, PLA mixed with PLGA and hyaluronic acid (**Table 1**). Scaffold-mediated improvement in cell retention and short-term survival does not necessarily indicate long-term survival of transplanted cells[55]. For instance, only 4% of the initial hESC-derived chondrocytes transplanted in a rat osteochondral defect model (rats received daily injections of cyclosporin A to prevent cell rejection) survived after 12 weeks. However, short-term surviving cells may be able to secrete pro-regenerative factors before they die[55]. 3D scaffolds may also contribute to the *in vivo* maturation of transplanted cells; this is likely more of a challenge for cells differentiated from hPSCs than from adult stem cells. A strategy that has been used is to seed immature hPSC-derived cells into 3D scaffolds and transplant them *in vivo*[12,27,55] (**Fig. 2**). For example, *in vitro*, hESC-derived cardiomyocytes express smooth muscle actin and MLC2A but not MLC2V, indicating their immaturity[56]; although *in vivo* a substantial number of grafted cells express MLC2V, a marker associated with a mature phenotype[12]. In our view, the properties of 3D biomaterials relevant for *in vivo* hPSC-derived cell transplant are scaffold degradation and the ability to present cell modulators (for example, growth factors, specific inhibitors and ECM peptides) in a spatiotemporally controlled manner.

Presentation of cell modulators. The *in vitro* differentiation state of hPSCs that leads to the best survival and functionality of hPSC-derived cells *in vivo* is poorly understood. Cells that are not terminally differentiated typically survive better, but their plasticity may result in undesired phenotypes. A strategy that is used to prevent undesired differentiation *in vivo* is to incorporate bioactive molecules in the scaffold for *in situ* control of the survival and maturation of encapsulated progenitor cells[16,21,24,27]. For example, inhibition of transforming growth factor β maintains the proliferation and identity of hESC-derived endothelial cells over 8 days[57]. When these cells are transplanted in a Matrigel plug into nonobese/severe combined immunodeficient mice they anastomose with recipient microvasculature. Combined culture in 3D scaffolds is another strategy that has been used to mature hPSC-derived cells. For example, *in vitro*, hESC-derived cardiomyocytes mature in 3D scaffolds when cultured together with endothelial cells and embryonic fibroblasts[58].

Scaffold degradation. Scaffolds for *in vivo* transplant should promote the structural organization of the attached cells and permit cell migration for their ultimate integration into the recipient tissue. Synthetic hydrogels that mimic key degradation characteristics of ECM may meet these requirements[11,59]. In our own work, MMP-sensitive hydrogels that contained pro-survival factors and that could release encapsulated hESC-derived vascular cells 'on demand' preserved the ejection fraction of the left ventricle in immunocompromised rats for 6 weeks after myocardial infarction[24].

Important challenges that remain to be addressed in the transplantation of hPSC-derived cells in 3D scaffolds are the vascularization of the tissue constructs (which is important for cell survival), the prevention of fibrous capsule formation and the prevention of adverse inflammatory responses.

Limitations of 3D scaffolds

Despite the possible advantages of 3D over 2D culture systems, some general limitations need to be considered. Because 3D *in vitro* models are not connected to the blood stream, cellular oxygenation, nutrition and waste removal occur by diffusion. To overcome potential mass diffusion limitations, the constructs should be millimeter-sized, unless they are maintained in bioreactors, which facilitate the exchange of oxygen and nutrients. Furthermore, it is often difficult to change one of the properties of the scaffold without affecting others. For example, the preparation of gel-based scaffolds with increased stiffness requires more cross-linking, which reduces scaffold pore size and thus could affect the diffusion of oxygen and nutrients in the scaffold. Effective control of hPSCs likely requires scaffolds with variable properties including stiffness, gradients of bioactive elements, responsiveness to enzymatic degradation and porosity. Although considerable progress has been made in the last years to design sophisticated scaffolds (for example, 3D scaffolds in which cell-material interactions are dynamically and remotely controlled)[11,20], most of these biomaterials are not yet commercially available. Systematic modulation of scaffold properties to identify factors important for a process of interest remains a challenge.

Future directions

The development of microscale technologies opens new opportunities for hPSC research[19]. Scaffolds can be fabricated at microscale and nanoscale by the use of 3D printing, microsyringe deposition and electrospinning of nanofibers. These methods allow control of the spatial location of biomolecules as well as scaffold topography at microscale and nanoscale, which may enable cellular control that is currently not possible. Some of these microtechnologies can be acquired commercially and are inexpensive, and we expect that they will be implemented in hPSC biology laboratories in the next years. Furthermore, the coupling of high-throughput screening methodologies with polymer libraries will continue to contribute to the design of new 3D scaffolds for the expansion and differentiation of hPSCs, the *in vivo* transplantation of their derivatives[35,60,61], and the tracking and manipulation of stem cells in real time[62].

We note that most of the 3D work done so far with hPSCs involves hESCs. It remains to be seen whether the challenges in biomaterial design for iPSCs are similar. Future studies will also be required to unravel the contribution of several variables—for instance, of hypoxia, mechanotransduction or cell-cell interactions—during differentiation and morphogenesis in 3D culture models. It is likely that numerical and computation techniques will be needed to interpret the data, prioritize factors for experimental testing and then to design scaffolds with improved structure and bioactivity. Finally, in-depth exchange between stem cell researchers and materials scientists will be key for development of hPSC-based clinical applications and for reliable and integrated use of these cells in pharmacological testing.

ACKNOWLEDGMENTS

This work was supported in part by US National Institutes of Health (grant HL060435). T.P.K. was supported by the Swiss National Science Foundation (grant PBELP3-127902), a Rotary Ambassadorial Scholarship and a Medicus Exchange Scholarship. L.S.F. was supported by a Marie Curie–Reintegration grant (FP7-People-2007-4-3-IRG; contract 230929), Massachusetts Institute of Technology–Portugal program and the Portuguese Foundation for Science and Technology (PTDC/SAU-BEB/098468/2008; PTDC/CTM/099659/2008).

COMPETING FINANCIAL INTERESTS

The authors declare no competing financial interests.

Published online at http://www.nature.com/naturemethods/.
Reprints and permissions information is available online at http://www.nature.com/reprints/index.html.

1. Thomson, J.A. *et al.* Embryonic stem cell lines derived from human blastocysts. *Science* **282**, 1145–1147 (1998).
2. Yu, J. *et al.* Induced pluripotent stem cell lines derived from human somatic cells. *Science* **318**, 1917–1920 (2007).
3. Takahashi, K. *et al.* Induction of pluripotent stem cells from adult human fibroblasts by defined factors. *Cell* **131**, 861–872 (2007).
4. Griffith, L.G. & Swartz, M.A. Capturing complex 3D tissue physiology *in vitro*. *Nat. Rev. Mol. Cell Biol.* **7**, 211–224 (2006).
5. Zagris, N. Extracellular matrix in development of the early embryo. *Micron* **32**, 427–438 (2001).
6. Hynes, R.O. The extracellular matrix: not just pretty fibrils. *Science* **326**, 1216–1219 (2009).
7. Adamo, L. *et al.* Biomechanical forces promote embryonic haematopoiesis. *Nature* **459**, 1131–1135 (2009).
8. Levenberg, S. *et al.* Differentiation of human embryonic stem cells on three-dimensional polymer scaffolds. *Proc. Natl. Acad. Sci. USA* **100**, 12741–12746 (2003).
9. Wozniak, M.A. & Chen, C.S. Mechanotransduction in development: a growing role for contractility. *Nat. Rev. Mol. Cell Biol.* **10**, 34–43 (2009).
10. Discher, D.E., Mooney, D.J. & Zandstra, P.W. Growth factors, matrices, and forces combine and control stem cells. *Science* **324**, 1673–1677 (2009).
11. Lutolf, M.P., Gilbert, P.M. & Blau, H.M. Designing materials to direct stem-cell fate. *Nature* **462**, 433–441 (2009).
12. Laflamme, M.A. *et al.* Cardiomyocytes derived from human embryonic stem cells in pro-survival factors enhance function of infarcted rat hearts. *Nat. Biotechnol.* **25**, 1015–1024 (2007).
13. Baharvand, H., Hashemi, S.M., Kazemi Ashtiani, S. & Farrokhi, A. Differentiation of human embryonic stem cells into hepatocytes in 2D and 3D culture systems *in vitro*. *Int. J. Dev. Biol.* **50**, 645–652 (2006).
14. Li, Z., Leung, M., Hopper, R., Ellenbogen, R. & Zhang, M. Feeder-free self-renewal of human embryonic stem cells in 3D porous natural polymer scaffolds. *Biomaterials* **31**, 404–412 (2010).
15. Gerecht, S. *et al.* Hyaluronic acid hydrogel for controlled self-renewal and differentiation of human embryonic stem cells. *Proc. Natl. Acad. Sci. USA* **104**, 11298–11303 (2007).
16. Ferreira, L.S. *et al.* Bioactive hydrogel scaffolds for controllable vascular differentiation of human embryonic stem cells. *Biomaterials* **28**, 2706–2717 (2007).
17. Gerecht-Nir, S., Cohen, S., Ziskind, A. & Itskovitz-Eldor, J. Three-dimensional porous alginate scaffolds provide a conducive environment for generation of well-vascularized embryoid bodies from human embryonic stem cells. *Biotechnol. Bioeng.* **88**, 313–320 (2004).
18. Ueno, M. *et al.* Neural conversion of ES cells by an inductive activity on human amniotic membrane matrix. *Proc. Natl. Acad. Sci. USA* **103**, 9554–9559 (2006).
19. Khademhosseini, A., Langer, R., Borenstein, J. & Vacanti, J.P. Microscale technologies for tissue engineering and biology. *Proc. Natl. Acad. Sci. USA* **103**, 2480–2487 (2006).
20. Kloxin, A.M., Kasko, A.M., Salinas, C.N. & Anseth, K.S. Photodegradable hydrogels for dynamic tuning of physical and chemical properties. *Science* **324**, 59–63 (2009).
21. Kraehenbuehl, T.P. *et al.* Three-dimensional extracellular matrix-directed cardioprogenitor differentiation: systematic modulation of a synthetic cell-responsive PEG-hydrogel. *Biomaterials* **29**, 2757–2766 (2008).
22. Hwang, N.S., Varghese, S. & Elisseeff, J. Derivation of chondrogenically-committed cells from human embryonic cells for cartilage tissue regeneration. *PLoS ONE* **3**, e2498 (2008).

PERSPECTIVE

23. Lee, S.T. *et al.* Engineering integrin signaling for promoting embryonic stem cell self-renewal in a precisely defined niche. *Biomaterials* **31**, 1219–1226 (2010).
24. Kraehenbuehl, T.P. *et al.* Human embryonic stem cell-derived microvascular grafts for cardiac tissue preservation after myocardial infarction. *Biomaterials* **32**, 1102–1109 (2011).
25. Levenberg, S. *et al.* Engineering vascularized skeletal muscle tissue. *Nat. Biotechnol.* **23**, 879–884 (2005).
26. Gerecht, S. *et al.* A porous photocurable elastomer for cell encapsulation and culture. *Biomaterials* **28**, 4826–4835 (2007).
27. Hwang, N.S. *et al.* In vivo commitment and functional tissue regeneration using human embryonic stem cell-derived mesenchymal cells. *Proc. Natl. Acad. Sci. USA* **105**, 20641–20646 (2008).
28. Kraehenbuehl, T.P., Ferreira, L.S., Zammaretti, P., Hubbell, J.A. & Langer, R. Cell-responsive hydrogel for encapsulation of vascular cells. *Bioma* **30**, 4318–4324 (2009).
29. Stephan, M.T., Moon, J.J., Um, S.H., Bershteyn, A. & Irvine, D.J. Therapeutic cell engineering with surface-conjugated synthetic nanoparticles. *Nat. Med.* **16**, 1035–1041 (2010).
30. Leor, J. *et al.* Human embryonic stem cell transplantation to repair the infarcted myocardium. *Heart* **93**, 1278–1284 (2007).
31. Ludwig, T.E. *et al.* Derivation of human embryonic stem cells in defined conditions. *Nat. Biotechnol.* **24**, 185–187 (2006).
32. Melkoumian, Z. *et al.* Synthetic peptide-acrylate surfaces for long-term self-renewal and cardiomyocyte differentiation of human embryonic stem cells. *Nat. Biotechnol.* **28**, 606–610 (2010).
33. Villa-Diaz, L.G. *et al.* Synthetic polymer coatings for long-term growth of human embryonic stem cells. *Nat. Biotechnol.* **28**, 581–583 (2010).
34. Klim, J.R., Li, L., Wrighton, P.J., Piekarczyk, M.S. & Kiessling, L.L. A defined glycosaminoglycan-binding substratum for human pluripotent stem cells. *Nat. Methods* **7**, 989–994 (2010).
35. Mei, Y. *et al.* Combinatorial development of biomaterials for clonal growth of human pluripotent stem cells. *Nat. Mater.* **9**, 768–778 (2010).
36. Li, D. *et al.* Integrated biochemical and mechanical signals regulate multifaceted human embryonic stem cell functions. *J. Cell Biol.* **191**, 631–644 (2010).
37. Li, L. *et al.* A unique interplay between Rap1 and E-cadherin in the endocytic pathway regulates self-renewal of human embryonic stem cells. *Stem Cells* **28**, 247–257 (2010).
38. Chowdhury, F. *et al.* Soft substrates promote homogeneous self-renewal of embryonic stem cells via downregulating cell-matrix tractions. *PLoS ONE* **5**, e15655 (2010).
39. Chen, G., Hou, Z., Gulbranson, D.R. & Thomson, J.A. Actin-myosin contractility is responsible for the reduced viability of dissociated human embryonic stem cells. *Cell Stem Cell* **7**, 240–248 (2010).
40. Braam, S.R. *et al.* Recombinant vitronectin is a functionally defined substrate that supports human embryonic stem cell self-renewal via alphavbeta5 integrin. *Stem Cells* **26**, 2257–2265 (2008).
41. Meng, Y. *et al.* Characterization of integrin engagement during defined human embryonic stem cell culture. *FASEB J.* **24**, 1056–1065 (2010).
42. Xu, C. *et al.* Feeder-free growth of undifferentiated human embryonic stem cells. *Nat. Biotechnol.* **19**, 971–974 (2001).
43. Hoffman, L.M. & Carpenter, M.K. Characterization and culture of human embryonic stem cells. *Nat. Biotechnol.* **23**, 699–708 (2005).
44. Tibbitt, M.W. & Anseth, K.S. Hydrogels as extracellular matrix mimics for 3D cell culture. *Biotechnol. Bioeng.* **103**, 655–663 (2009).
45. Gieni, R.S. & Hendzel, M.J. Mechanotransduction from the ECM to the genome: are the pieces now in place? *J. Cell Biochem.* **104**, 1964–1987 (2008).
46. D'Amour, K.A. *et al.* Production of pancreatic hormone-expressing endocrine cells from human embryonic stem cells. *Nat. Biotechnol.* **24**, 1392–1401 (2006).
47. Chambers, S.M. *et al.* Highly efficient neural conversion of human ES and iPS cells by dual inhibition of SMAD signaling. *Nat. Biotechnol.* **27**, 275–280 (2009).
48. Levenberg, S., Burdick, J.A., Kraehenbuehl, T. & Langer, R. Neurotrophin-induced differentiation of human embryonic stem cells on three-dimensional polymeric scaffolds. *Tissue Eng.* **11**, 506–512 (2005).
49. Evans, N.D. *et al.* Substrate stiffness affects early differentiation events in embryonic stem cells. *Eur. Cell Mater.* **18**, 1–14 (2009).
50. Legant, W.R. *et al.* Measurement of mechanical tractions exerted by cells in three-dimensional matrices. *Nat. Methods* **7**, 969–971 (2010).
51. Ferreira, L. *et al.* Human embryoid bodies containing nano- and microparticulate delivery vehicles. *Adv. Mater.* **20**, 2285–2291 (2008).
52. Guenou, H. *et al.* Human embryonic stem-cell derivatives for full reconstruction of the pluristratified epidermis: a preclinical study. *Lancet* **374**, 1745–1753 (2009).
53. Sachlos, E. & Auguste, D.T. Embryoid body morphology influences diffusive transport of inductive biochemicals: a strategy for stem cell differentiation. *Biomaterials* **29**, 4471–4480 (2008).
54. Spence, J.R. *et al.* Directed differentiation of human pluripotent stem cells into intestinal tissue *in vitro*. *Nature* **470**, 105–109 (2011).
55. Toh, W.S. *et al.* Cartilage repair using hyaluronan hydrogel-encapsulated human embryonic stem cell-derived chondrogenic cells. *Biomaterials* **31**, 6968–6980 (2010).
56. van Laake, L.W. *et al.* Human embryonic stem cell-derived cardiomyocytes survive and mature in the mouse heart and transiently improve function after myocardial infarction. *Stem Cell Res.* **1**, 9–24 (2007).
57. James, D. *et al.* Expansion and maintenance of human embryonic stem cell-derived endothelial cells by TGFβ inhibition is Id1 dependent. *Nat. Biotechnol.* **28**, 161–166 (2010).
58. Caspi, O. *et al.* Tissue engineering of vascularized cardiac muscle from human embryonic stem cells. *Circ. Res.* **100**, 263–272 (2007).
59. Lutolf, M.P. & Hubbell, J.A. Synthetic biomaterials as instructive extracellular microenvironments for morphogenesis in tissue engineering. *Nat. Biotechnol.* **23**, 47–55 (2005).
60. Anderson, D.G., Levenberg, S. & Langer, R. Nanoliter-scale synthesis of arrayed biomaterials and application to human embryonic stem cells. *Nat. Biotechnol.* **22**, 863–866 (2004).
61. Yang, F. *et al.* Combinatorial extracellular matrices for human embryonic stem cell differentiation in 3D. *Biomacromolecules* **11**, 1909–1914 (2010).
62. Ferreira, L., Karp, J.M., Nobre, L. & Langer, R. New opportunities: the use of nanotechnologies to manipulate and track stem cells. *Cell Stem Cell* **3**, 136–146 (2008).
63. Ferreira, L.S. *et al.* Vascular progenitor cells isolated from human embryonic stem cells give rise to endothelial and smooth muscle like cells and form vascular networks in vivo. *Circ. Res.* **101**, 286–294 (2007).

New Drug Development (e.g. Angiogenesis Inhibitors) and Enzyme Technology

Isolation of a Cartilage Factor That Inhibits Tumor Neovascularization

Abstract. *A cartilage fraction isolated by guanidine extraction and purified by affinity chromatography inhibits tumor-induced vascular proliferation and consequently restricts tumor growth. This fraction contains several different proteins; the major one has a molecular weight of about 16,000. The fraction strongly inhibits protease activity.*

Tumor-induced neovascularization is inhibited by neonatal cartilage and tumor growth is thereby restricted (*1*). We now report isolation and partial purification of a cartilage fraction that inhibits the vascularization of tumors.

Segments of cartilage 1.5 by 8 cm were excised from the distal tips of scapulae of calves less than 2 weeks old and extracted in lactated Ringer solution or in 1*M* guanidine hydrochloride (0.02*M* MES, *p*H 6) for 24 hours at room temperature. Sorgente *et al.* showed that a 1*M* guanidine extract has protease inhibitory activity, and they suggest that this activity may be responsible for cartilage avascularity (*2*).

Extracts were dialyzed exhaustively against water at 4°C and centrifuged. The supernatant was lyophilized and fractionated. Sixteen fractions were obtained. The first was an aqueous extract from lactated Ringer solution. The remaining 15 were obtained from guanidine extracts. Five of these came from Sephadex G-100, three from G-200, six from ammonium sulfate precipitation, and one from the trypsin affinity column. All fractions were dialyzed exhaustively against water at 4°C, passed through 0.45-μm Millipore filters, and lyophilized. Sterile technique was followed thereafter.

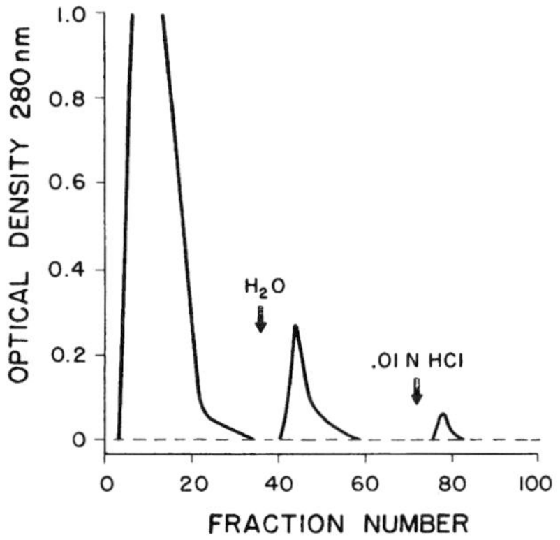

Fig. 1. Purification of protease inhibitor from cartilage by a trypsin-Sepharose affinity column. The column (1.5 by 15 cm) was equilibrated with buffer (0.05*M* tris, 0.5*M* NaCl, *p*H 8). The sample (1.45 g of guanidine extract of cartilage in 43 ml of buffer) was added at fraction 1, and buffer was added at fraction 7. The flow rate was 23 ml/hour at 4°C. Fractions of 7 ml were collected. High protease inhibitory activity was found in the HCl eluate.

Affinity chromatography with trypsin-Sepharose was used to purify the cartilage protease inhibitor. The affinity column was prepared as described by March *et al.* (*3*). We used 0.5 g of trypsin and 30 ml of packed Sepharose. Guanidine extract was passed over the affinity column, and the column was washed with buffer, distilled water, and 0.01*N* HCl (Fig. 1). The HCl eluate contained very high trypsin inhibitory activity. It migrated as one major and several minor bands on acrylamide (15 percent) slab gel electrophoresis. The major band fell between markers of myoglobin (molecular weight, 17,800) and egg white lysozyme (molecular weight, 14,400) on the gel. Approximately 500 g of cartilage produced 1 mg of this fraction.

Slow release polymers were made by mixing 5 mg of each dry fraction with 100 μl of 10 percent Elvax 40 (*4*) (an ethylene–vinyl acetate copolymer, 40 percent vinyl acetate by weight) in methylene chloride. The resulting dispersion was dried under vacuum, cut into 1-mm^3 pellets, and coated with a thin film of pure polymer. Polymer pellets containing lysozyme, alkaline phosphatase, and soybean trypsin inhibitor were prepared in identical fashion and tested in vitro to calibrate release rates and to ascertain that biological activity was not destroyed. Using agar gel diffusion assays (*5*), we found that polymers containing each of the three compounds released more than 100 ng/day of biochemically active material into aqueous media for periods longer than 100 days.

Implants of V2 carcinoma produce a neovascular response in the normally avascular rabbit cornea (*6*). Using this as a bioassay for tumor-induced vascularization, we assessed the inhibitory effect of 16 different fractions. Pellets of polymer and pieces of V2 carcinoma were placed into corneal pockets (Fig. 2A). The tumors grew as thin plaques, inducing vessels to sprout from the edge of the cornea 4 to 6 days after implantation. Vessel length and tumor diameter were measured four times each week by

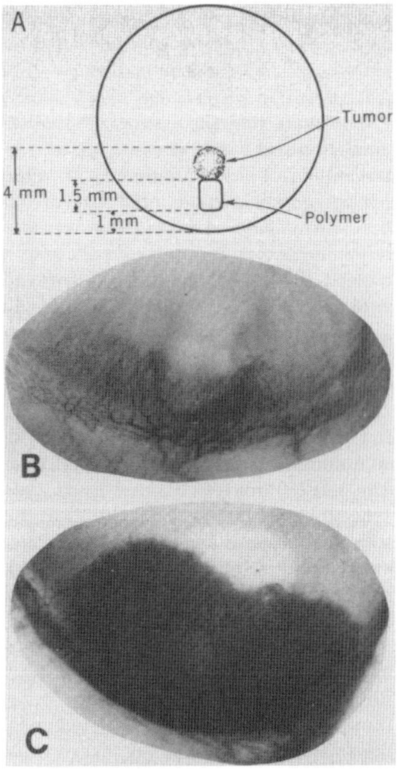

Fig. 2. (A) Schematic diagram of rabbit cornea with tumor and polymer. (B) Photograph of the lower third of a rabbit cornea containing tumor and polymer [12 days with inhibitor (the HCl eluate described in Fig. 1)]. Vessels are sparse and fail to grow in a zone surrounding the polymer. The tumor has not become vascularized. (C) Cornea containing tumor and polymer (12 days without inhibitor). Vessels appear as a dense carpet sweeping over the polymer. At this time, the tumor has already vascularized and is growing rapidly compared to that shown in (B) (~ × 7).

means of a calibrated slit-lamp stereomicroscope (accuracy, ± 0.1 mm). For each test fraction at least 12 corneas and an equal number of controls with identical tumor and empty polymer were used.

Tumor that was placed 2.5 mm from the edge of the cornea induced vessel growth rates nearly identical to those resulting from tumor and empty polymer. Empty polymer caused neither inflammation nor vessel growth. Of the 16 fractions tested, 8 caused severe inflammation, thereby preventing assessment of inhibitory activity. The remaining fractions were noninflammatory, and all except the HCl eluate from the affinity column had the same vessel growth rates as their controls.

Results obtained with the HCl eluate were highly significant. Vessel growth rates for 16 control and 21 experimental corneas with this fraction (Fig. 3) were nearly identical during the first week after the onset of neovascularization. During the second week, the rate of vessel growth in eyes with inhibitor was less than one-half that of the control group. By the third week of exposure to inhibitor, the tumor-induced vessels grew at one-third the rate of the controls and some vessels regressed (a *t*-test comparing controls to inhibitors in both the second and third weeks gave $P < .01$).

When polymers were empty, vessels appeared as a dense carpet sweeping over the polymer toward the tumor (Fig. 2C). When vessels penetrated the tumor, it grew rapidly into a large protruding mass occupying nearly the entire cornea, and necessitated killing the rabbits. In three rabbits, this stage was reached as early as the second week, and all rabbits with control polymers were dead by the third week. Very similar results were obtained when polymers containing fractions without inhibitory activity were tested.

By contrast, when inhibitor was present, vessels were sparse, grew slowly, and failed to grow in a zone surrounding the polymer (Fig. 2B). By the fourth week many vessels were regressing, and ten of the 21 tumors never became vascularized. Of those which were vascularized, vessels penetrated the tumor at a site remote from the inhibitor zone.

These studies show that a partially purified cartilage factor isolated by affinity chromatography will inhibit tumor-induced vascular proliferation and thereby restrict tumor growth. These data support the general concept (*7, 8*) that (i) solid tumors pass through an avascular and a vascular phase and (ii) preventing tumors from achieving the vascular phase may be an effective approach by which to bring about tumor dormancy. This approach to inhibition of tumor growth is unique because it operates at the level of tumor vessels rather than tumor parenchyma, and because the inhibitor is isolated from a normal tissue.

The use of a noninflammatory polymer capable of releasing macromolecules over a prolonged period was critical to these experiments. This was true because fractions diffused away from the implant site rapidly and because minute quantities of inhibitor were available for testing.

Some tumors eventually became vascularized even in the presence of inhibitor. This occurred at a site remote from the polymer. One might hypothesize that the inhibitor diffused into a small zone around the polymer. A vessel growing through an area remote from this zone could reach the tumor successfully, although more time would be required because of the greater distance. Furthermore, over the time course of the experiment, the polymer is releasing a diminishing, or at best constant, amount of inhibitor, and is confronted by a gradually spreading tumor with an increasing capacity to stimulate vessels.

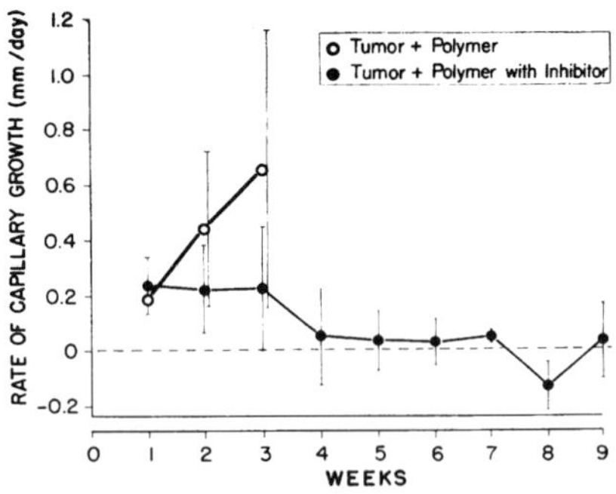

Fig. 3. Rate of capillary growth in 21 eyes with inhibitor (HCl eluate, Fig. 1) and 16 controls. Rates were determined by weekly differences in the length of the longest vessels. Calculation of rates began at the onset of neovascularization, which occurred 5 ± 1 days after implantation for each of the 37 corneas. ○, Vessel growth rates in the absence of inhibitor. All rabbits were dead by the end of the third week. ●, Vessel growth rates when inhibitor was present. By the fourth week some vessels were regressing.

Slit-lamp observations showed no adverse effect of the polymer-inhibitor combination on the avascular tumor itself. Histologic sections showed healthy tumor cells at the interface between tumor and polymer. Furthermore, the polymer-inhibitor did not retard growth of either V2 carcinoma or virus-transformed NIH 3T3 cells in culture.

The mechanism of inhibition of neovascularization by this cartilage material is unknown (*9*). These data do not provide any evidence that this inhibitor blocks tumor angiogenesis factor (*7*) directly. Furthermore, there is no evidence that this inhibitor is specific for tumor-induced vessels. Other forms of neovascularization (inflammation or wound healing) may also be inhibited. It is unclear whether antiprotease activity plays a role in this inhibition. It remains to be seen whether systemic administration of this inhibitor will suppress tumor neovascularization in a manner similar to that demonstrated here by local delivery.

ROBERT LANGER, HENRY BREM
KENNETH FALTERMAN
MICHAEL KLEIN, JUDAH FOLKMAN
Department of Surgery,
Children's Hospital Medical Center
and Harvard Medical School,
Boston, Massachusetts 02115

References and Notes

1. H. Brem and J. Folkman, *J. Exp. Med.* **141**, 427 (1975).
2. N. Sorgente, K. E. Keuttner, L. W. Soble, R. Eisenstein, *Lab. Invest.* **32**, 217 (1975); N. Sorgente *et al.* [in *Protides of the Biological Fluids*, H. Peeters, Ed. (Pergamon, New York, 1976), vol. 23, pp. 227–230)] report purification of a protease inhibitor from cartilage.
3. S. C. March, I. Parikh, P. Cuatrecasas, *Anal. Biochem.* **60**, 149 (1974).
4. The Elvax 40 was supplied through the courtesy of Alza Corporation, Palo Alto, California. Crude polymer was washed exhaustively in absolute ethanol to remove inflammatory impurities.
5. L. Carlsson and B. Karlsson, *Experientia* **28**, 990 (1972); W. B. Schill and G. F. B. Schumacher, *Anal. Biochem.* **46**, 502 (1972); G. F. B. Schumacher and W. B. Schill, *ibid.* **48**, 9 (1972).
6. M. A. Gimbrone, Jr., R. S. Cotran, S. B. Leapman, J. Folkman, *J. Natl. Cancer Inst.* **52**, 413 (1974).
7. J. Folkman, E. Merler, C. Abernathy, G. Williams, *J. Exp. Med.* **133**, 275 (1971).
8. J. Folkman, *Adv. Cancer Res.* **19**, 331 (1974); *Ann. Intern. Med.* **82**, 96 (1975).
9. Since this work was completed, Eisenstein *et al.* [*Am. J. Pathol.* **81**, 337 (1975)] have reported that 1*M* guanidine extracts of cartilage inhibit aortic endothelial cells (and to a lesser extent fibroblasts) in vitro.

10. We thank Ivan Preis, Germaine Grant, Susan Clark, Paul Wesley, Steven Fleit, and Dr. David Tapper for valuable technical assistance.

17 February 1976; revised 30 March 1976

Proc. Natl. Acad. Sci. USA
Vol. 77, No. 7, pp. 4331–4335, July 1980
Medical Sciences

Control of tumor growth in animals by infusion of an angiogenesis inhibitor

(cartilage/neovascularization/antiangiogenesis/B16 melanoma/V2 carcinoma)

ROBERT LANGER*†‡, HOWARD CONN*, JOSEPH VACANTI*, CHRISTIAN HAUDENSCHILD§, AND JUDAH FOLKMAN*¶

*Department of Surgery, Children's Hospital Medical Center and ¶Harvard Medical School, Boston, Massachusetts 02115; †Department of Nutrition and Food Science, Massachusetts Institute of Technology, Boston, Massachusetts 02139; and the §Mallory Institute of Pathology, Boston University School of Medicine, Boston, Massachusetts 02118

Communicated by Earl P. Benditt, April 21, 1980

ABSTRACT **Angiogenesis and tumor growth were inhibited in two different animal models by regional infusion of a partially purified cartilage extract. In rabbits bearing corneal implants of V2 carcinoma and receiving the inhibitor, vascular growth rates were <3% of those in control animals receiving either Ringer's solution or bovine trypsin inhibitor (Trasylol). Subconjunctival B16 melanoma implants in mice receiving the inhibitor weighed <2.5% of implants in mice receiving Ringer's solution, Trasylol, or albumin. Histologic study of major organs and standard blood tests revealed no toxic effects in any of the animals. The inhibitor did not retard the growth of either tumor cell type in tissue culture at concentrations as high as 1 mg/ml. These results suggest that the cartilage factor does not interfere with the growth of the tumor cell population directly but that it prevents tumor growth by inhibiting angiogenesis.**

The capacity to induce the growth of new capillaries in the host is a characteristic common to most solid malignant tumors. This phenomenon has been called "tumor angiogenesis." It has been hypothesized (1, 2) that until a tumor nodule is penetrated by new capillaries the limitations imposed by diffusion of oxygen and nutrients prevent it from growing beyond a few millimeters in diameter. This hypothesis implies that inhibition of angiogenesis, by whatever means, might control tumor growth. For this reason, "antiangiogenesis" was proposed as a potential therapeutic approach (3).

Indirect evidence for this hypothesis was drawn from experiments with tumors in isolated perfused organs (4), tumor spheroids grown in soft agar (2), and tumor implants floating in the anterior chamber of the rabbit eye (5).

Sorgente *et al.* (6) showed that when cartilage was extracted with 1 M guanidine it lost its resistance to vascular invasion. Eisenstein *et al.* (7) reported that a 1 M guanidine extract of cartilage inhibited the growth of aortic endothelial cells in culture. Brem and Folkman (8) showed that an implant of cartilage placed adjacent to a V2 carcinoma in the rabbit cornea inhibited growth of new blood vessels toward the tumor. Langer *et al.* (9) isolated a partially purified factor from cartilage that, when administered locally by a sustained-release polymer (10), inhibited angiogenesis induced by V2 carcinoma implanted in the rabbit cornea.

We now show that this angiogenesis inhibitor limits tumor growth by stopping neovascularization when the inhibitor is infused regionally in either mice or rabbits. Furthermore, the inhibitor appears to have no toxic effect on tumor cells and no short-term toxicity for the host.

In one set of experiments, the tumor (V2 carcinoma) was separated from the nearest vascular bed by a relatively large distance (approximately 1.0 mm) by implanting the tumor in the rabbit cornea. This system permitted daily measurements of vessel growth rate and demonstrated that vascular proliferation was inhibited and, subsequently, tumor growth was controlled.

In a second set of experiments, the tumor (B16 melanoma) was implanted directly onto a dense capillary bed in the mouse conjunctiva. This contiguous relationship between tumor and vascular bed more closely resembles most human and animal tumors. In this experimental design, tumor growth was measured directly and the vascular response was determined histologically. Regional infusion of the angiogenesis inhibitor again prevented neovascularization and markedly limited tumor growth.

MATERIALS AND METHODS

Rabbit V2 Carcinoma Tests. *Animals.* Male New Zealand white rabbits weighing 1.8–2.3 kg were used.

Tumor. V2 carcinoma (11) was maintained by serial intramuscular and subcutaneous injection (8). Tumors were harvested before they reached a diameter of 2.0 cm, and 1.5 × 1.5 mm pieces were implanted into the cornea of the rabbit 1.0 mm central to the limbus as described (8, 9).

Infusion system. Rabbits were infused through the right common carotid artery at a rate of 14.4 ml per day with the infusion system described by Conn and Langer (12). In this way, the inhibitor was delivered preferentially to the right eye; the left eye received inhibitor at a lower effective dose (Fig. 1).

Measurements. Rabbit corneas were examined with a Zeiss slit-lamp microscope at ×6 to ×10 magnification. The length of the longest vessel (accuracy, ±0.1 mm), measured from the limbus, was recorded.

Mouse Melanoma Experiments. *Animals.* Male C57BL/6J mice weighing 25–30 g were used.

Tumor. B16 melanoma was obtained from Arthur D. Little Company and serially passaged in mice as subcutaneous axillary implants. Fragments (1.0 × 1.0 mm) of tumor were excised 14 days after the subcutaneous implantation, kept in lactated Ringer's solution for <35 min, and implanted onto the conjunctiva as described below.

Mice were anesthetized by intraperitoneal injection of 3.6% chloral hydrate at 0.1 ml/10 g of body weight. Ether inhalation anesthesia was used as necessary for supplementation.

The eye was proptosed and held in position with a small hemostat at the lateral canthus. Under ×15 magnification, an incision was made into the bulbar conjunctiva at the limbus superiorly by using a glass micropipette (Pyrex capillary tubes,

The publication costs of this article were defrayed in part by page charge payment. This article must therefore be hereby marked "*advertisement*" in accordance with 18 U. S. C. §1734 solely to indicate this fact.

‡ To whom reprint requests should be addressed at: Department of Nutrition and Food Sciences, M.I.T., Cambridge, MA 02139.

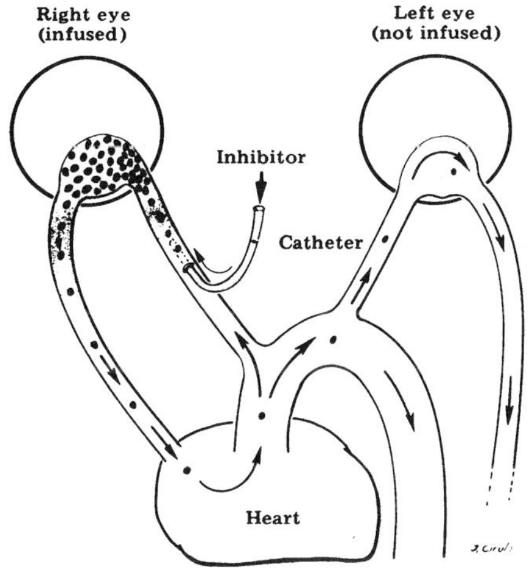

FIG. 1. Diagram of infusions through the right common carotid artery in both rabbits and mice. Technetium pertechnate scans demonstrated that the concentration reaching the right eye was 15- to 60-fold greater than that reaching the left eye.

1.0 mm outside diameter, 0.2 mm wall width; Fredrick Dimmock, Inc., Millville, NJ) with a sharp beveled edge approximately 70 μm in diameter. With microsurgical forceps, this incision was widened and a subconjunctival pocket was developed peripheral to the limbus, 2.0 mm in length and 1.0 mm in width. The tumor implant was then placed into the pocket. Tumor in a subconjunctival pocket rests in a vascular bed, in contrast to the rabbit corneal pocket where the tumor lies remote from the vascular bed.

Infusion system. Mice were infused at a rate of 0.96 ml/24 hr with a system modified from the technique of Brown and Goffinet (13). By positioning a silicone catheter in the right common carotid artery, the infusion system was thus arranged to deliver high concentrations of inhibitor to the right eye and orbital tissues and low concentrations to the left eye and orbital tissues (Fig. 1). The other end of the catheter was tunneled subcutaneously and brought out through the flank and taped to the tail and to a rubber band swivel apparatus. Animals were individually housed and had access to food and water ad lib. The fluid was delivered by a Harvard infusion pump (Harvard Apparatus Co., Millis, MA) that was modified to hold 1.0-ml syringes.

Measurements. Tumors were measured (animals under ether anesthesia) starting on day 5 after implantation. A Nikon stereomicroscope was used with an accuracy of ±0.1 mm. Mice were sacrificed on day 7 and the tumors were excised and weighed. Measurements were made without knowledge of how the animals had been treated.

Preparation of Angiogenesis Inhibitor. The angiogenesis inhibitor was prepared by a modification of our earlier method (9). Cartilage was excised from veal scapular bones within 48 hr after slaughter and all connective tissue was carefully removed. Cartilage slices (2 kg) were then extracted in 20 liters of 1.0 M guanidine·HCl/0.02 M sodium malate, pH 6, for 24 hr at 25°C. This solution was filtered and then concentrated to 500 ml in a Millipore ultrafiltration unit containing a PTGC cassette (cutoff, M_r 10,000). The retentate was dialyzed against water for 12 hr at 4°C and then desalted by using the same Millipore unit. The solution was judged to be effectively desalted when a drop of it added to a drop of 1% $AgNO_3$ gave no precipitate. It was then lyophilized, subjected to trypsin-Sepharose affinity chromatography, and sterilized as described (9).

Infusion Solutions. Angiogenesis inhibitor, Trasylol (a bovine trypsin inhibitor, Sigma), and bovine serum albumin (Sigma) were dissolved in lactated Ringer's solution (Abbott) and heparin was added to a concentration of 6 units/ml. Protein contents of all solutions as determined by the Lowry method (14) were identical.

In each set of tests, three experimental animals and up to 95 controls were used. The number of experimental animals was limited because of the scarcity of the inhibitor. To obtain 1 mg of the partially purified angiogenesis inhibitor, 500 g of cartilage was required.

The doses and patterns of administration used in both sets of experiments were determined by preliminary screening experiments. Because of the limited amount of inhibitor available, our experiments were designed to determine whether we could affect the vascular and tumor growth rates rather than to determine median lethal doses or to obtain a dose–response curve.

Mice received approximately 100 times the dose of inhibitor given to rabbits on a weight basis. This was because mouse infusions were started after the rabbit study was complete, and our first goal was to establish the toxic or lethal concentration for the inhibitor. Our original dose of 4 mg/day caused no host toxicity, yet angiogenesis inhibition was so striking (see *Results*) that we continued the series at this level.

Histology and Blood Tests. Rabbits and mice were sacrificed by using anesthesia at a lethal level, and the eyes including cornea and sclera were removed along with other major organs and fixed in 10% buffered formalin. Histologic sections were embedded in paraffin or Epon and were stained with hematoxylin/eosin or methylene blue. Blood was drawn from rabbits before sacrifice, and standard determinations of liver and renal functions, electrolyte balance, and hematologic functions were performed.

Cell Culture. Pieces of V2 carcinoma removed during tumor passage were minced and incubated in trypsin/EDTA solution (GIBCO, calcium- and magnesium-free) for 30 min at 37°C and washed with Eagle's medium (Flow Laboratories, Rockville, MD) supplemented with 10% calf serum (GIBCO), penicillin (50 units/ml), and streptomycin (50 μg/ml). The cells were plated in 60-mm culture dishes and allowed to grow to confluence.

Pieces of B16 melanoma removed during tumor passage were treated in the same way and also were grown to confluence.

B16 melanoma and V2 carcinoma cells were trypsinized from confluent plates and placed in the wells of a 24-well plate (Falcon, no. 3001) at a density of 5000 cells per well as determined by a Coulter particle counter (Coulter Electronics, model 2F, Hialeah, Florida). To ensure satisfactory plating, after 8 hr the cells were incubated in Dulbecco's medium with 10% calf serum containing Trasylol, bovine serum albumin, or angiogenesis inhibitor at protein concentrations from 1 μg/ml to 4 mg/ml. The controls consisted of cells incubated in medium with 10% serum alone. The cells were counted by observing five high-power fields across the center of each well by phase-contrast microscopy on days 1, 2, and 3 after plating. At the end of day 3, the cells were fixed in methanol and stained with 1% crystal violet. The cells in each well were counted again in five fields with the aid of a grid.

RESULTS

Rabbit Infusions. *Vascular Growth.* Three days after new blood vessels began growing toward the tumor, infusion of the cartilage-derived inhibitor was started in three rabbits. Controls consisted of 21 corneas infused with Ringer's solution, 4 infused with Trasylol, and the 3 untreated eyes of the experimental animals. In the control animals, the mean corneal vessel growth

Medical Sciences: Langer *et al.* *Proc. Natl. Acad. Sci. USA 77 (1980)* 4333

rates were 0.32 mm/day (Ringer's), 0.35 mm/day (untreated eyes in experimental group), and 0.46 mm/day (Trasylol) over the 6-day infusion period. In contrast, the average growth rate of vessels in the right corneas of the animals receiving the cartilage-derived inhibitor was 0.01 mm/day (Fig. 2).

The angiogenesis inhibitor affected the density of vessels as well as their rate of growth. Although all controls showed an increase in vessel number from approximately 4 to >100 vessels per cornea during the 6-day infusion period, 2 of the 3 treated corneas showed no increase in vessel density. The third cornea showed only a small increase in vessel density and a 0.1-mm increase in length. Yet, even this cornea was markedly less vascularized than the most mildly vascularized cornea of the controls (Fig. 3).

For 2 days, the rabbits were infused at a slightly higher level of inhibitor (3 mg/day) to determine if this increased dose would inhibit vascular growth in the contralateral eye (Fig. 2). We were not able to observe any quantitative differences compared to controls.

Tumor growth. All 28 controls developed large three-dimensional vascularized tumors. By contrast, the tumors in the treated eyes grew much more slowly and did not become vascularized during the infusion. New vessels resumed growth toward the tumor implant after the inhibitor infusion was discontinued (although in some cases a lag phase was observed) (Fig. 2). When the vessels reached the edge of the tumor, tumor growth also resumed. However, in two of the three treated animals, the tumors eventually regressed after vascularization.

Toxic effects. Gross and microscopic findings in all organs were within normal limits in the treated animals as well as in control animals with tumor and control animals without tumor. Also, leukocyte count, hematocrit value, and results of tests of liver and kidney function were within normal limits (15, 16). Alkaline phosphatase activity of the treated animals was below normal values (16) (and that of tumor-bearing and non-tumor-bearing controls) by a factor of 3.

Mouse Infusions. ***Tumor growth (B16 melanoma).*** Accurate measurements of tumor growth were attainable only after day 4 following implantation. Twenty-four hours after implantation, little evidence of either the implant or of the procedure itself was evident. Between days 3 and 4 a small flat black plaque appeared in the area of the subconjunctival pocket. The tumor cells rapidly grew into a three-dimensional mass. In control animals, focal hemorrhages occurred on the tumor surface, and large dilated vessels entered its base on day 5 to day 6. This was not observed in treated animals. Measurement of tumor length was the only serial determination possible. Tumor width and height could not be accurately quantitated until after removal because the tumor grew behind the orbital margin and the conjunctival reflection.

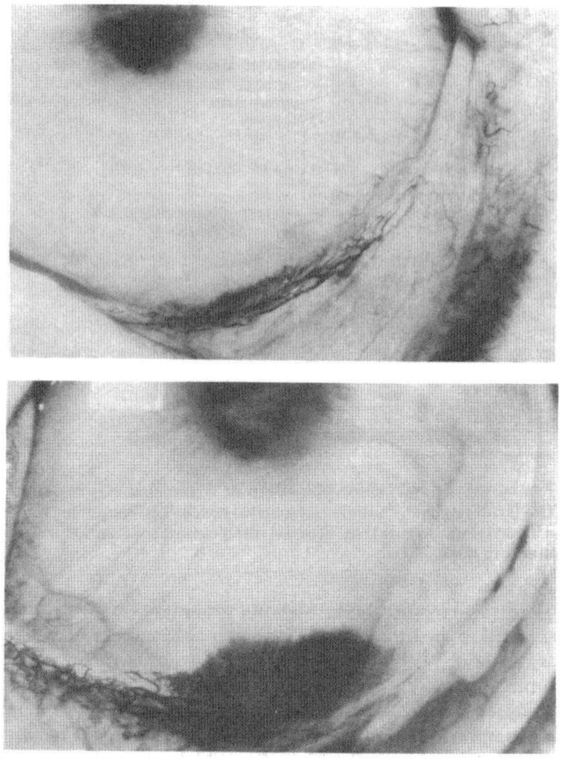

FIG. 3. Corneas in treated (*Upper*) and untreated (*Lower*) rabbits after 6-day infusion period (day 10). Note the much larger and broader vascular front in the untreated cornea. Before the infusion started (day 4), the vessels were shorter and less dense in the untreated cornea than in the treated cornea. (×10.)

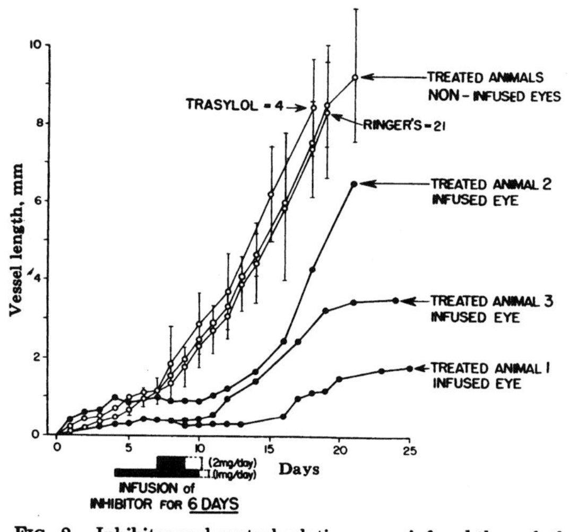

FIG. 2. Inhibitor and control solutions were infused through the right carotid artery in rabbits carrying V2 carcinoma in each cornea. Rabbits were sacrificed when tumors became large, protruding masses. The vessel length at this time averaged 9 mm. Results are shown as mean ± SD.

The rate of tumor growth was similar in all of the control animals. In non-infused mice, the tumor length increased at a rate of 0.69 mm/day (Fig. 4). In mice infused with lactated Ringer's, the rate was 0.64 mm/day. In mice infused with albumin or Trasylol, tumors grew at rates of 0.60 and 0.65 mm/day, respectively.

In contrast, the tumors in the right eyes of the animals infused with the cartilage-derived inhibitor showed growth rates of only 0.06 mm/day. The left eyes of these animals served as internal controls and had tumor growth rates similar to those of the other controls—i.e., 0.74 mm/day (Fig. 4).

Because the length of the tumor does not accurately reflect its three-dimensional growth, all animals were sacrificed on day 7 and the tumors were weighed. Control tumors weighed an average of 7.60 mg and no statistically significant difference in tumor weights was observed between any of the control groups. By contrast, the average tumor weight in the treated eyes was 0.18 mg, <2.5% that of controls (Fig. 5 *a* and *b*).

Histologic study of tumors in control animals showed densely

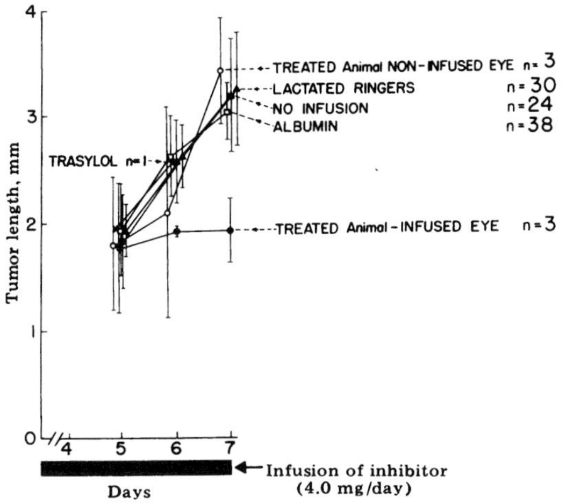

FIG. 4. Maximal tumor dimension of B16 melanoma implants in C57BL/6J mice in the subconjunctival space. The reason for only one Trasylol infusion is that Trasylol often caused coagulation in the catheter in mice.

packed pleomorphic cells. Melanin granules were found in about half of the tumor cells. The center of large tumor masses was necrotic. Within the densely packed viable tumor cells, prominent capillary blood vessels were present. They showed a collapsed lumen and were lined with endothelial cells similar in appearance to proliferating endothelium (Fig. 5c). The endothelial cell density in these vessels was higher than the density of the flat endothelium in adjacent tissue not exposed to tumor.

In contrast, animals treated with the angiogenesis inhibitor showed only a few melanin-producing tumor cells which did not form a solid tumor mass but were scattered as individual pleomorphic, often spindle-shaped, cells within a rather loose, edematous stroma. Spaces between the tumor cells were wide and contained extravasated erythrocytes in some areas and necrotic tumor cells in others. The newly branching tumor vessels and active, dense, endothelial cells observed in the controls were not noted in the treated animals. Instead, adjacent tissues showed dilated vessels with flat, sparse, quiescent endothelium. Occasionally, individual tumor cells were adjacent to these vessels but no directional growth of vascular tissue into the tumor implant was detectable (Fig. 5d).

Toxic effects. The infusion apparatus did not appear to affect tumor growth rates or the general well-being of the animal. Animals infused with the inhibitor did not demonstrate any signs of illness. They ate and drank normally throughout the infusion and were normally active. Histological features of the major organs were within normal limits. In two of the three treated animals, a widened femoral epiphyseal plate with increased numbers of chondrocytes was observed.

Cell Culture. B16 melanoma cells incubated with angiogenesis inhibitor, Trasylol, or bovine serum albumin at concentrations of 1–1000 μg/ml grew at approximately the same rate as controls (Fig. 6). At 4 mg/ml, all three solutions caused slight inhibition. The inhibitor caused no reduction in the growth rate of V2 carcinoma cells at concentrations as high as 1 mg/ml. At 4 mg/ml, the inhibitor caused a 10% reduction in growth rate of V2 carcinoma cells compared to controls. In all cases, including those in which cell growth rates were less than in controls, the tumor cells proliferated and appeared healthy.

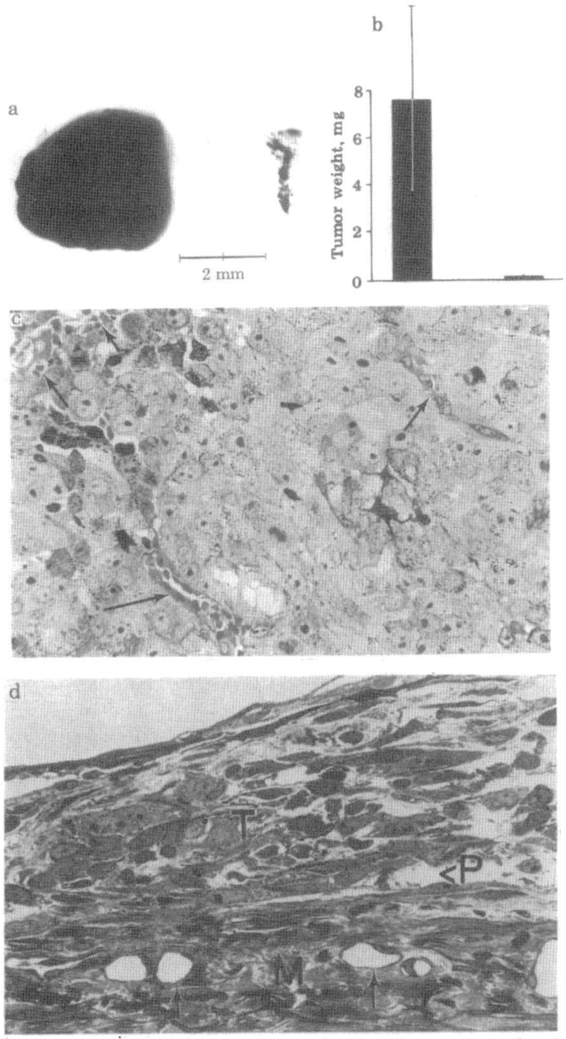

FIG. 5. (*a*) Excised B16 melanoma implants from untreated and treated mice at day 7. (*b*) Weights (mean ± SD) of excised B16 melanoma implants from untreated (*Left*) and treated (*Right*) mice at day 7. The untreated group includes all 15 controls—Ringer's solution, albumin, Trasylol, and the noninfused eyes of the treated animals—run in the same series as the treated tumors. (*c*) Section of B16 melanoma from untreated mouse, day 7. (×160.) The tumor contains densely packed melanoma cells interspersed with new capillary blood vessels (arrows) that penetrate throughout the tumor mass. Most tumor cells appear healthy. (*d*) Section through the whole tumor mass (T) of B16 melanoma from treated mouse, day 7. (×160.) The tumor implant has not been penetrated by new vessels and appears essentially avascular. There are only sparse melanoma cells. The vessels (arrows) in the contiguous vascular bed of muscle (M) are somewhat dilated and contain flat, nonactive, endothelial cells and there is no evidence of capillary proliferation or neovascularization. The edge of the avascular tumor is at plane P.

DISCUSSION

This study demonstrates that regional infusion of an angiogenesis inhibitor derived from cartilage will stop tumor vascularization and prevent subsequent tumor growth in two different species without toxicity to the animal host. This potential therapeutic approach appears to have no toxic effect on tumor cells *in vitro*. Furthermore, the inhibitor is a natural tissue product.

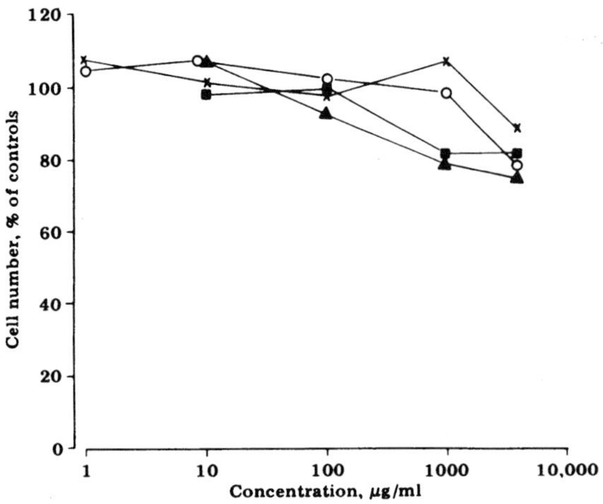

FIG. 6. Effect of the inhibitor, Trasylol, and albumin, on tumor cells in culture. Cells were fixed and counted 72 hr after exposure. X, Inhibitor and V2 carcinoma; ■, albumin and melanoma; O, inhibitor and melanoma; ▲, Trasylol and melanoma.

Implantation of V2 carcinoma in the rabbit cornea at a distance from the vascular bed permitted the quantitative measurement of inhibition of capillary proliferation and the subsequent limitation of tumor growth. The inhibition of the growth of B16 mouse melanoma further demonstrated that an aggressive solid tumor could be halted by an angiogenesis inhibitor, despite the fact that the tumor was implanted contiguous to a normal vascular bed.

The finding that inhibition of angiogenesis led to limitation of tumor growth was confirmed by including three different types of controls.

(*i*) To control for nonspecific inhibition due to infusion of protein in general or antitrypsin activity in particular, other protein solutions were infused including Trasylol, a bovine trypsin inhibitor. These infusions had no angiogenesis inhibitory effect.

(*ii*) To control for nonspecific inhibition of tumor growth due to "sickness" or debilitation of the animal, the left eyes of all treated animals received the inhibitor but at a substantially lower concentration than the right eye. If the animals were "sick," tumor suppression should have occurred in the left eye as well as the right eye. In fact, all tumors in the left eyes grew and became vascularized at the same rate as did tumors in untreated eyes.

(*iii*) To control for the possibility that the inhibited vessels might have stopped growing "spontaneously" or for some nonspecific reason, the behavior of the tumors in the treated eyes of the rabbits before and after infusion was studied carefully. Vessel growth essentially stopped in all eyes treated with the high dose of angiogenesis inhibitor; blood vessel growth always resumed when the angiogenesis inhibitor treatment was discontinued (Fig. 2).

These data suggest but do not prove that the inhibitor acts on growing new blood vessels rather than directly on the tumor itself. Before it becomes vascularized, the V2 carcinoma grows slowly in two dimensions within the collagen layers of the cornea. The treated tumor exhibited this same behavior under the stereomicroscope, suggesting that the V2 cells themselves were not affected by the inhibitor. In addition, histologic sections showed non-necrotic tumor cells present in the treated mice and rabbits. Furthermore, the angiogenesis inhibitor did not suppress the growth of B16 melanoma and V2 carcinoma cells *in vitro*.

One limitation of the tumor models used is that quantitative measurements of both tumor growth and angiogenesis cannot be made in the same animal system. In the mouse, the tumor is contiguous to the vascular bed and the new vessels are rapidly surrounded by growing tumor, so it is not possible to measure vessel length. However, tumor weight is an accurate measure of tumor growth and the melanoma is easily identified. By contrast, in the rabbit cornea, vessel length can be accurately measured because the tumor is separated from its vascular bed by a significant distance. However, measurement of tumor growth by weight is subject to large error, because the tumor infiltrates the cornea and the edge of the tumor is difficult to identify accurately.

Scarcity of the inhibitor precludes the possibility of testing it on tumors at more advanced stages of growth and vascularization. The point we wish to stress is that this angiogenesis inhibitor derived from cartilage may be considered as a prototype. It is possible that other angiogenesis inhibitors will be found or synthesized and that they may be even more potent (17). These results provide support for the concept of antiangiogenesis (3, 18, 19)—that control of tumor growth may be possible through control of neovascularization.

We thank Dr. James Schuck for his excellent work in developing the scale-up procedure for the first-stage extraction of bulk quantities of cartilage, Germaine Grant, Deborah Stark, and Ted Sisley for technical assistance, Carl Cobb for editorial assistance, and Mrs. Pauline Breen and Jo-Anne Hutchinson for typing. This work was supported by Grant CA 14019-06 from the National Cancer Institute and by a grant to Harvard University from Monsanto Company.

1. Folkman, J. (1974) in *Advances in Cancer Research*, eds. Klein, G. & Weinhouse, S. (Academic, New York), Vol. 19, pp. 331–358.
2. Folkman, J. & Hochberg, M. (1973) *J. Exp. Med.* **138**, 745–753.
3. Folkman, J. (1972) *Ann. Surg.* **175**, 409–416.
4. Folkman, J., Cole, P. & Zimmerman, S. (1966) *Ann. Surg.* **164**, 491–502.
5. Gimbrone, M. A., Jr., Leapman, S., Cotran, R. S. & Folkman, J. (1973) *J. Natl. Cancer Inst.* **50**, 219–228.
6. Sorgente, N., Kuettner, K. E., Soble, L. W. & Eisenstein, R. (1975) *Lab. Invest.* **32**, 217–222.
7. Eisenstein, R., Kuettner, K. E., Neopolitan, C., Soble, L. W. & Sorgente, N. (1975) *Am. J. Pathol.* **81**, 337–349.
8. Brem, H. & Folkman, J. (1975) *J. Exp. Med.* **141**, 427–439.
9. Langer, R., Brem, H., Falterman, K., Klein, M. & Folkman, J. (1976) *Science* **193**, 70–72.
10. Langer, R. & Folkman, J. (1976) *Nature (London)* **263**, 797–800.
11. Gimbrone, M. A., Jr., Leapman, S., Cotran, R. S. & Folkman, J. (1972) *J. Exp. Med.* **136**, 261–276.
12. Conn, H. & Langer, R. (1978) *Lab. Anim. Sci.* **28**, 598–602.
13. Brown, J. & Goffinet, D. (1970) *J. Lab. Clin. Med.* **76**, 175–186.
14. Lowry, O., Rosebrough, N. J., Farr, A. L. & Randall, R. J. (1951) *J. Biol. Chem.* **193**, 265–275.
15. Schalm, D. W. (1965) *Veterinary Hematology* (Lea & Febiger, Philadelphia), p. 30.
16. Melby, E. C. & Altman, N. M. (1974) in *Handbook of Laboratory Animal Sciences* (CRC, Cleveland, OH), p. 288.
17. Patz, A., Brem, S., Finkelstein, D., Chen, C. H., Lutty, G., Bennett, A., Coughlin, W. R. & Gardner, J. (1978) *Ophthalmology (Rochester)* **85**, 626–637.
18. Folkman, J. & Cotran, R. (1976) *Int. Rev. Exp. Pathol.* **16**, 207–248.
19. Folkman, J. (1976) *Sci. Am.* **234**, 58–73.

THE JOURNAL OF BIOLOGICAL CHEMISTRY

Vol. 260, No. 3, Issue of February 10, pp. 1849–1857, 1985
Printed in U.S.A.

Purification and Characterization of Heparinase from *Flavobacterium heparinum**

(Received for publication, October 9, 1984)

Victor C. Yang[a], Robert J. Linhardt[a,b], Howard Bernstein[c,d], Charles L. Cooney[a,c], and Robert Langer[a,d,e,f]

From the [a]Department of Nutrition and Food Science, Massachusetts Institute of Technology, Cambridge, Massachusetts 02139, the [b]College of Pharmacy, University of Iowa, Division of Medicinal Chemistry, Iowa City, Iowa 52242, the [c]Department of Chemical Engineering, Massachusetts Institute of Technology, Cambridge, Massachusetts 02139, the [d]Whitaker College of Health Sciences, Management and Technology, Massachusetts Institute of Technology, Cambridge, Massachusetts 02139, and the [e]Department of Surgery, Children's Hospital Medical Center, Boston, Massachusetts 02115

Heparinase (EC 4.2.2.7) isolated from *Flavobacterium heparinum* was purified to homogeneity by a combination of hydroxylapatite chromatography, repeated gel filtration chromatography, and chromatofocusing. Homogeneity was established by the presence of a single band on both sodium dodecyl sulfate and acid-urea gel electrophoretic systems. Amino acid analysis shows that the enzyme contains relatively high amounts of lysine residues (9%) consistent with its cationic nature (pI 8.5) but contains only 4 cysteine residues/polypeptide. The molecular weight of heparinase was estimated to be 42,900 ± 1,000 daltons by gel filtration and 42,700 ± 1,200 daltons by sodium dodecyl sulfate-polyacrylamide gel electrophoresis. The enzyme is very specific, acting only on heparin and heparan monosulfate out of 12 similar polysaccharide substrates tested. It has an activity maximum at pH 6.5 and 0.1 M NaCl and a stability maximum at pH 7.0 and 0.15 M NaCl. The Arrhenius activation energy was found to be 6.3 kcal/mol. However, the enzyme is very sensitive to thermal denaturation and loses activity very rapidly at temperatures over 40 °C. Kinetic studies of the heparinase reaction at 37 °C gave a K_m of 8.04×10^{-6} M and a V_m of 9.85×10^{-5} M/min at a protein concentration of 0.5 μg/ml. By adapting batch procedures of hydroxylapatite and QAE (quaternary aminoethyl)-Sephadex chromatography, gram quantities of heparinase that is nearly free of catalytic enzyme contaminants can be purified in 4–5 h.

Heparinase (EC 4.2.2.7) from *Flavobacterium heparinum* is an eliminase (1) which cleaves heparin at α-glycosidic linkages in heparin's major repeating unit: →4)-2-deoxy-2-sulfamino-α-D-glucopyranose 6-sulfate-(1→4)-α-L-idopyranosyluronic acid 2-sulfate-(1→ (2). Heparinase acts in a random endolytic fashion (3) to produce heparin fragments that are chain-shortened segments of heparin with minor-end group modification. The elimination reaction results in a Δ-4,5 site of unsaturation in the terminal uronic acid residue (4, 5), leaving the chemical structure of the glucosamine reducing end unaltered (6).

Heparinase has found use in numerous applications including structural determination of heparin (2, 6, 7), new bioassays for heparin (8, 9), investigation of the anticoagulant mechanism (10), preparation of low molecular weight heparin anticoagulants (11) and anti-tumor agents (12), and the development of immobilized enzyme filters for blood deheparinization (13). Despite its importance, heparinase has never been purified to homogeneity nor have its activity, stability, or kinetic characteristics been well characterized. In this report we present two methods for heparinase purification: (i) a method conducive to producing gram quantities of nearly catalytically pure heparinase in several hours using batch ion-exchange chromatography and (ii) a method for producing heparinase that is homogeneous as judged by single band purity on two different gel electrophoretic systems. We have also characterized the purified heparinase with respect to its reaction kinetics, stability, and activity.

* This work was supported by Grant GM 25810 from the National Institutes of Health. The costs of publication of this article were defrayed in part by the payment of page charges. This article must therefore be hereby marked "*advertisement*" in accordance with 18 U.S.C. Section 1734 solely to indicate this fact.

[f] To whom reprint requests should be addressed at MIT.

EXPERIMENTAL PROCEDURES

Materials

The sodium salt of heparin from porcine intestinal mucosa (157 USP units/mg) was obtained from Hepar Industries. Chondroitin 4- and 6-sulfates, dermatan sulfate, hyaluronic acid, pectin, chitin, dextran sulfate, alginic acid, polyvinyl sulfate, polygalacturonic acid, *p*-nitrocatechol sulfate, and protamine sulfate were purchased from Sigma. Heparan monosulfate was generously supplied by Dr. Cifonelli of Wyler Children's Hospital, Chicago. *N*-Acetyl heparin was prepared from *N*-desulfated heparin (14), according to the procedure of Roseman and Ludowieg (15). Azure A dye (A-970; total dye content, 70%) and Folin reagent were from Fisher. Sephadex G-100, QAE (quaternary aminoethyl)-Sephadex, Sephadex G-100 superfine, Polybuffer exchanger PBE 94, Polybuffer 96, and protein markers were obtained from Pharmacia Fine Chemicals. Hydroxylapatite (Bio-Gel HTP), Bio-Rad dye reagent, and chemicals for SDS[1]-polyacrylamide gel electrophoresis were purchased from Bio-Rad. Other chemicals were reagent grade, and water was twice distilled.

Methods

Protein Assays

Protein concentrations were measured by the Lowry method (16) or by the Bio-Rad protein assay (17).

Enzyme Assays

Heparinase activity was measured by the increase in ultraviolet absorption at 232 nm (5), according to the procedure of Linhardt *et al.* (3). However, this UV 232-nm assay was utilized only on relatively pure preparations of heparinase because contaminating enzymes such as glycuronidase (1, 4) can act on degradation products resulting in a

[1] The abbreviations used are: SDS, sodium dodecyl sulfate; PAGE, polyacrylamide gel electrophoresis.

loss of the chromophore being measured. In the less pure preparations, heparinase activity was assayed by the metachromatic shift of azure A from blue to red in the presence of heparin (18), according to the procedure of Galliher *et al.* (19). The change in absorbance was measured at 620 nm throughout the entire linear range of the assay and compared with a standard curve of 0 to 8.3 mg of heparin/ml in assay broth. One international unit of heparinase activity is defined as that amount which causes 1 μmol of double bonds formed/min, based on a molar extinction coefficient of 5.1×10^3 cm^{-1} M^{-1} at 232 nm for the degradation products, the α,β-unsaturated uronic acids (5). However, in the less pure preparations where the azure A assay is used, the international unit is no longer suitable. One unit of enzyme activity is, therefore, defined as the amount of enzyme which degrades 1 mg of heparin/h. Because of the cross-reaction of heparinase on heparan monosulfate, heparitinase activity was assayed at 43 °C, a temperature at which the heparitinase activity was maximized but the heparinase activity was minimized (4). Sulfatase activity was assayed by the procedure of Ototani *et al.* (20) using *p*-nitrocatechol sulfate as the substrate. Glycuronidase activity was measured by monitoring the rate of disappearance of absorption at 232 nm in the assay solution which had been previously used to measure haparinase activity, according to the method of Warnick and Linker (21).

Crude Enzyme Preparation

F. heparinum was grown in a complex protein digest medium as described previously (19), with heparin as the inducer. Approximately 100 g (dry weight) of cells were obtained in a 10-liter fermentation. The cells were harvested by centrifugation and resuspended in 3 volumes of cold 0.01 M sodium phosphate buffer, pH 6.8. The suspended microorganisms were homogenized at 5000 p.s.i. in 3–4 passages through a Gaulin model 15 M homogenizer thermostatically controlled at 4 °C by circulating ice water. After the addition of protamine sulfate to the homogenate (125 mg of protamine sulfate/g of protein) the cell debris and precipitated nucleic acids were removed by centrifugation at 10,000 × *g* for 15 min. The supernatant fluid thus obtained (the "crude extract") was subjected to further purification. Unless otherwise stated, all the enzyme preparations and purifications were carried out at 4 °C.

Large Scale Production of Heparinase

This includes batch procedures of hydroxylapatite and QAE-Sephadex chromatography.

Hydroxylapatite—The crude extract was batch loaded onto hydroxylapatite pre-equilibrated with 0.01 M phosphate buffer (pH 6.8), using 4 g of hydroxylapatite/g of protein. The hydroxylapatite-bound protein was then washed stepwise with buffers of increasing concentrations of sodium phosphate and NaCl. The solutions used for the sequential stepwise elutions were prepared by mixing 0.01 M sodium phosphate buffer, pH 6.8, with 0.25 M sodium phosphate, 0.5 M NaCl buffer, pH 6.8, according to a ratio (v/v) of 8/0, 7/1, 6.5/1.5, 6/2, 5.5/2.5, 5/3, 4/4, 3/5, 2/6, 1/7, and 0/8, respectively. The hydroxylapatite particles were removed by centrifugation at 3000 × *g* for 3 min, and the supernatant was assayed for protein concentration and heparinase activity.

QAE-Sephadex Adsorption—A negative adsorption on QAE-Sephadex was carried out. The resulting enzyme preparation (0.2–0.4 mg/ml), obtained from the hydroxylapatite chromatography in a 0.07 M sodium phosphate and 0.125 M NaCl wash (*i.e.* the 6/2 solution), was dialyzed for at least 4 h against 100 volumes of 0.01 M phosphate buffer at pH 6.8 and then added to QAE-Sephadex (5 mg of QAE-Sephadex/ml of protein solution). The mixture was kept at 4 °C for 10 min, with occasional and gentle vortexing, and the gel beads were removed by centrifugation at 3000 × *g* for 3 min. The supernatant was assayed for protein concentration and heparinase activity.

Polyacrylamide Gel Electrophoresis

The enzyme preparations obtained during the successive purification stages were subjected to SDS-PAGE according to a modified procedure of Laemmli (22). The separation gel contained 10% acrylamide and 0.1% SDS in 0.375 M Tris-HCl buffer (pH 8.8). The stacking gel contained 4% acrylamide and 0.1% SDS in 0.125 M Tris-HCl buffer (pH 6.8). Protein samples were dissolved in 0.0625 M Tris-HCl buffer (pH 6.8) containing 2% SDS, 10% 2-mercaptoethanol, and 12.5% glycerol, and then heated in a boiling water bath for 2 min prior to electrophoresis. Electrophoresis was carried out at room temperature at a constant voltage of 40 V until the bromphenol blue marker reached the bottom of the gel. The gels were fixed in 50% trichloroacetic acid overnight, stained for 3 h at room temperature with 0.25% Coomassie Blue R-250 in 45% methanol and 9% glacial acetic acid, and then diffusion destained by repeated washing with 15% glacial acetic acid in 10% methanol. To demonstrate the homogeneity of the purified enzyme, the urea-acetic acid system of Panyim and Chalkley (23) was also employed. The slab gel contained 10% polyacrylamide and 6.25 M urea in 0.9 M acetic acid (pH 3.2). The sample buffer contained 6.25 M urea and had a pH of 2.6. The electrode buffer was 0.9 M acetic acid (pH 2.6). The gel was stained and destained by the same procedures used in the SDS-PAGE system.

Elution of Heparinase from SDS-Polyacrylamide Gels

The concentrated QAE-purified heparinase was subjected to electrophoresis on a slab SDS-polyacrylamide gel as described above. After electrophoresis, a slot containing protein markers together with a slot containing the QAE-purified sample was cut out of the slab, fixed for 20 min, stained for 20 min, and destained for 15 min while the rest of the gel was stored at 4 °C in a refrigerator. The stained and unstained gels were lined up with each other, and the regions containing polypeptide bands were excised in slices. The gel slices were incubated overnight at 4 °C in separate tubes containing 3 ml of 1% (v/v) Triton X-100 in water. The eluate was dialyzed against 100 volumes of 0.01 M phosphate buffer (pH 7) for 10 h to remove the Triton since it produced a high background in the UV 232-nm assay. The preparations were assayed for heparinase activity.

Gel Filtration Chromatography

Sephadex G-100 gels (515 ml) and Sephadex G-100 superfine gels (190 ml) were packed in glass Pharmacia columns, equilibrated, and eluted with 0.01 M sodium phosphate buffer, pH 6.8. Samples (6 ml) were injected through a three-way valve and eluted at a flow rate of 48 ml/h for the Sephadex G-100 column and of 3.6 ml/h for the Sephadex G-100 superfine column. Protein markers used were albumin (M_r = 67,000), ovalbumin (M_r = 43,000), chymotrypsinogen A (M_r = 25,000), and ribonuclease A (M_r = 13,700).

Chromatofocusing

Chromatofocusing was performed with a 0.9-cm diameter column containing 15 ml of anion exchanger PBE 94. The anion exchanger was equilibrated with 0.25 M ethanolamine-CH_3COOH buffer at pH 9.4 and eluted with Polybuffer 96 diluted 1:10 with deionized water and adjusted to pH 6.0 with acetic acid. Proteins were monitored at 280 nm since Polybuffer interferes with the Lowry assay.

Amino Acid Analysis

The amino acid composition of purified heparinase was determined with a Beckman 121 *M* amino acid analyzer. About 20 μg of sample were hydrolyzed at 110 °C for 24, 48, and 72 h in 6 N HCl in an evacuated sealed tube. The sample was then dried, dissolved in 150 μl of 0.2 N sodium citrate buffer (pH 2.2) containing 0.5% thioglycol and 0.1% phenol, and loaded onto the analyzer. Tryptophan was destroyed under the conditions for hydrolysis. Asparagine and glutamine were converted to aspartic acid and glutamic acid, respectively. Cysteic acid was determined by hydrolysis of protein oxidized with performic acid (24).

Kinetics and Substrate Specificity

The kinetic constants of the heparinase reaction were determined at 37 °C by measuring the initial rates of heparin degradation as a function of heparin concentrations ranging from 9.1 μM to 2.27 mM. The substrate specificity of the purified heparinase on various polysaccharides including heparin, *N*-acetyl heparin, heparan monosulfate, chondroitin 4- and 6-sulfates, hyaluronic acid, dermatan sulfate, pectin, chitin, dextran sulfate, polygalacturonic acid, and alginic acid were measured at 30 °C, pH 7, by the UV 232-nm assay. The reaction mixture contained 10 mM phosphate, a protein concentration of 28 μg/ml, and a substrate concentration of 8.3 mg/ml. All substrates above were tested for their inhibitory effect on heparinase at a concentration of 8.3 mg/ml and in the presence of a heparin concentration of 8.3 mg/ml.

RESULTS

Large Scale Production of Heparinase—A high demand for heparinase in blood deheparinization studies (13) led us to seek methods suitable for production of large quantities of highly purified enzyme with a minimum of expense and inconvenience. This has been achieved by converting two successive ion-exchange chromatographic steps from a column technique to a batch process. The first step was a batch hydroxylapatite separation, the results of which are shown in Fig. 1. Heparinase activity was eluted over a broad range of ionic strength ($I = 0.22$–0.42); however, the maximum activity was obtained in a $I = 0.29$ wash. These results are consistent with those reported by other authors using hydroxylapatite column chromatography (20, 25). Although batch procedures are less efficient in resolution than column techniques, they allow enzyme production to be scaled up from the milligram quantities produced by a hydroxylapatite column to hundreds of grams. Polyacrylamide gel electrophoresis showed the presence of about 15 major bands in this preparation (Fig. 1). Enzymatic assays indicated that this preparation was not "catalytically" pure and contained low amounts of activities attributed to enzymes such as chondroitinases (26, 27), heparitinase (4, 20), and hyaluronidase (20, 27) which act on heparin-like mucopolysaccharides as well as other enzymes, sulfatases (4, 28) and glycuronidase (4, 20, 29), that act on heparin degradation products.

Further purification was achieved through anion-exchange chromatography on QAE-Sephadex. This is a convenient and efficient method since over 70% of the proteins in the hydroxylapatite preparation are acidic (as judged by isoelectric focusing) while heparinase is a cationic molecule. After QAE-Sephadex purification, 85% of the total activity was recovered, and the specific activity was enriched 3- to 4-fold over that obtained from the hydroxylapatite purification. In addition, the resulting enzyme preparation was nearly free of the catalytic enzyme contaminants. The relative activities of these contaminating enzymes, as compared to the heparinase activity (100%) at 30 °C on the basis of absorbance change at 232 nm/h were: chondroitinase B, 0%; heparitinase (measured at 43 °C as described under "Methods"), <4.5%; chondroitinase AC, <4.0%; hyaluronidase, <4.2%. Only minor sulfatase and glycuronidase activities were detected. Over a period of 1 h, less than 2.5% of the sulfate groups were desulfated and 1–2.8% of the α,β-unsaturated uronide chromophores were degraded.

Elution of Heparinase after SDS-PAGE—When QAE-purified heparinase was subjected to SDS-PAGE and eluted from the gel, no heparinase activity was found in the eluate as the gel slices were eluted with 1% SDS and dialyzed thereafter. However, about 20% of the enzyme activity was recovered when the SDS gel slices were eluted with 1% Triton X-100. It appears that heparinase is denatured faster in SDS than in Triton X-100. This may be accounted for by the nonionic nature of the Triton X-100, such that conformation change of the highly charged heparinase is much slower in Triton X-100 than in SDS. Fig. 2 shows that heparinase activity was only found in the region corresponding to a molecular weight between 38,000 to 50,100 daltons.

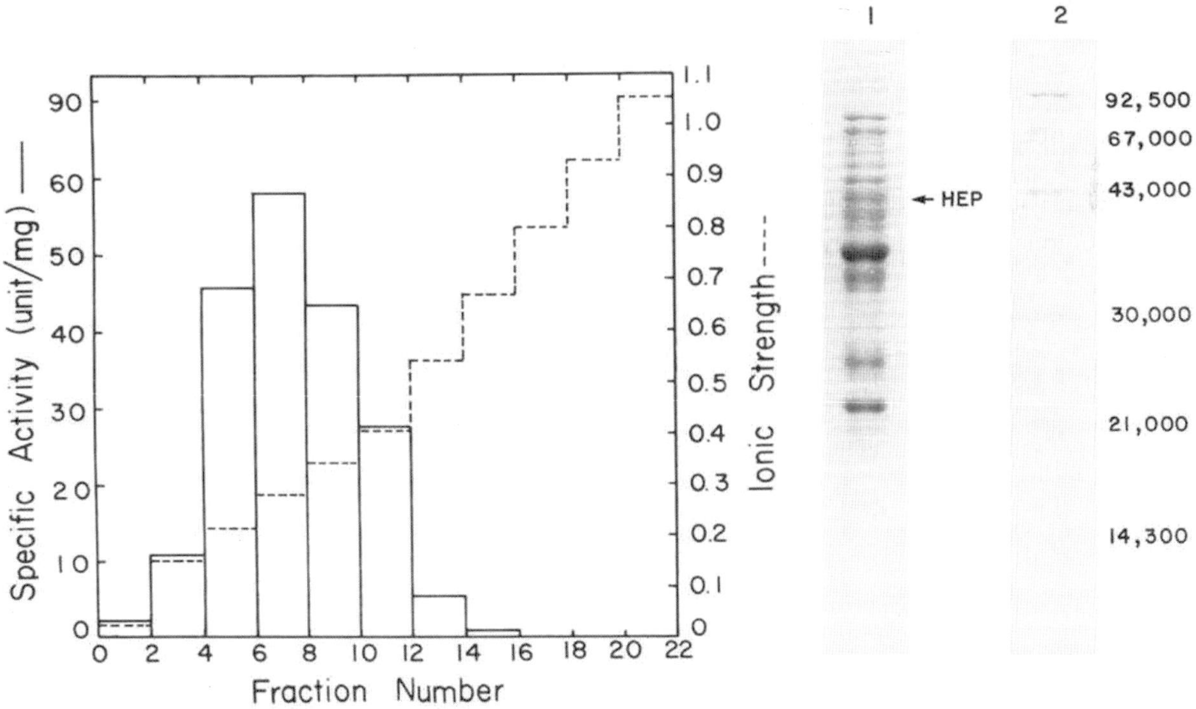

FIG. 1. **Elution of heparinase from batch hydroxylapatite chromatography.** One liter of the crude enzyme extract (30 mg/ml) was batch loaded onto 120 g of hydroxylapatite and then washed stepwise with buffers of increasing ionic strength as described under "Methods." Specific activity was measured by the azure A assay. One unit of heparinase activity is defined as the amount of enzyme which degrades 1 mg of heparin/h. Each fraction represents a volume of 1 liter of the eluate. *Right*, the active fractions resulting from an $I = 0.29$ wash were analyzed by SDS-polyacrylamide gel electrophoresis (*slot 1*). *HEP* designates heparinase. *Slot 2*, markers are phospholipase ($M_r = 94{,}000$), albumin ($M_r = 67{,}000$), ovalbumin ($M_r = 43{,}000$), carbonic anhydrase ($M_r = 30{,}000$), soybean trypsin inhibitor ($M_r = 21{,}000$), and lysozyme ($M_r = 14{,}300$).

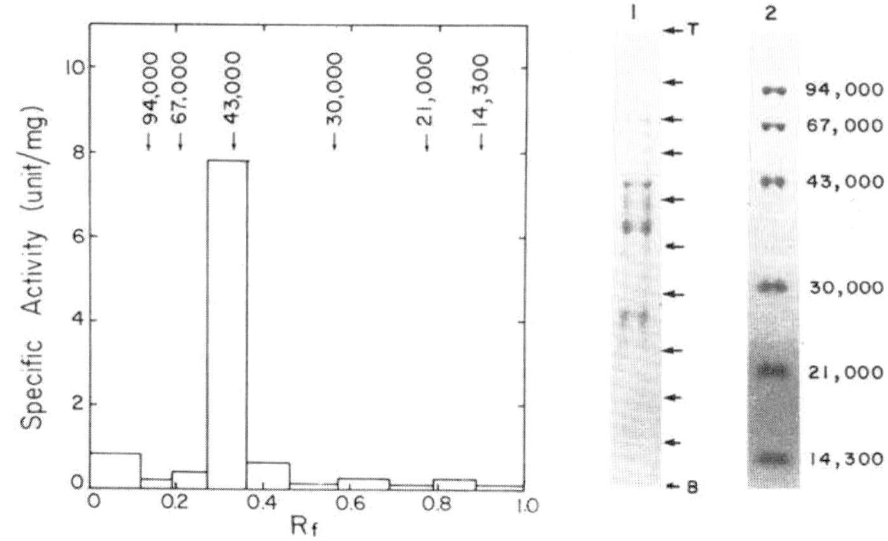

FIG. 2. **Elution of heparinase after SDS-polyacrylamide gel electrophoresis.** The experimental procedures are described under "Methods." R_f represents the distance from the top of the gel to the place where the gel is cut, as normalized with respect to the gel length. One unit of heparinase activity used here is defined as the amount of enzyme which can completely degrade 1 mg of heparin/h. *Right*, the QAE preparation was analyzed by SDS-polyacrylamide gel electrophoresis (*slot 1*). The *arrows* indicate the positions where the gel is cut. *Slot 2*, markers (see Fig. 1).

Purification of Heparinase to Homogeneity—Purification of heparinase to homogeneity was accomplished by additional column chromatography of the hydroxylapatite-purified heparinase through Sephadex G-100 gel filtration, chromatofocusing, and Sephadex G-100 superfine gel filtration.

Sephadex G-100 Gel Filtration Chromatography—Prior to gel filtration, the hydroxylapatite-purified heparinase was concentrated about 30-fold by ultrafiltration under a pressure of 40 p.s.i., using a Diaflo model 52 ultrafilter equipped with a hydrophilic and low nonspecific binding YM 30 membrane. More than 96% of activity was recovered in the retentate after the concentration. The concentrated enzyme solution (6 ml) was loaded onto a Sephadex G-100 column (2.6 × 97 cm) and eluted with 0.01 M sodium phosphate buffer (pH 6.8). Fig. 3 shows the elution pattern. The activity peak was found at the elution volume corresponding to a molecular weight of 42,900 ± 1,000 daltons. A significant fraction (about 20%) of the protein, possibly high molecular weight protein contaminants together with some protein aggregates formed from the effect of concentration polarization during the ultrafiltration process, was eluted at the void volume. However, no heparinase activity was found in this region. The resulting enzyme preparation showed no glycuronidase activity and had a specific activity of 1.4 IU/mg. SDS-PAGE showed 9 major bands in this preparation. As indicated in Fig. 2, the heparinase activity was eluted from a SDS-polyacrylamide gel in the region corresponding to a molecular weight between 38,000 to 51,000 daltons; the band with a molecular weight of 43,000 daltons, therefore, may be heparinase (Fig. 3).

Chromatofocusing—Active fractions indicated in Fig. 3 by cross-hatching were combined, concentrated to a volume of 6 ml by ultrafiltration, and then applied to a chromatofocusing column. As shown in Fig. 4, the heparinase activity was eluted as a broad symmetrical peak centered at a pI of 8.5. SDS-PAGE of the chromatofocused material showed the presence of two major bands with molecular weights of 42,700 and 25,000 daltons, respectively, as well as three very minor bands. Apparently the M_r 42,700 band is heparinase, because it is the only band lying in the region where heparinase was recovered after SDS-PAGE (Fig. 2).

Sephadex G-100 Superfine Gel Filtration Chromatography—Active fractions 25 to 40 from the chromatofocusing column (see Fig. 4) were combined, concentrated to a volume of 6 ml by ultrafiltration (as described earlier), and applied to a Sephadex G-100 superfine column (1.6 × 95 cm) pre-equilibrated with 0.01 M phosphate buffer at pH 6.8. The fractions eluting from the column were analyzed for both protein concentration and heparinase activity (Fig. 5). The activity peak coincided with the first protein peak in fraction 47. The second protein peak, which emerged at fraction 57, was presumably the M_r 25,000 contaminant seen in Fig. 4. The two shoulders appearing before the first and the second protein peak may be proteins attributed to the three minor bands seen in the gel of Fig. 4. The homogeneity of this preparation was demonstrated as a single band shown on both the SDS-PAGE system which separated polypeptides on the basis of molecular weight and the acid-urea gel system which separated polypeptides mainly on the basis of charge (Fig. 6). Densitometer traces of both gels also indicated the presence of a single polypeptide (>99% pure) in the preparation. The molecular weight of the purified heparinase was estimated to be 42,700 ± 1,200 daltons by the SDS-PAGE system.

Recovery and Yield—Table I summarizes the results of the heparinase purification. The enzyme was purified 280-fold over the cell homogenate. The specific activity of the purified enzyme was 26.7 IU/mg of protein. The overall mass yield of the purified protein was only 0.003% starting with 1 g of protein in the cell homogenate. The recovery of heparinase activity was about 1%.

Amino Acid Analysis—The amino acid composition of completely purified heparinase is shown in Table II. The enzyme is characterized by a low content of sulfur-containing amino acids (<2%) and a relatively high content of basic amino acids (14%) such as lysine, arginine, and histidine. The relatively high amount of basic amino acids is consistent with the cationic nature of the heparinase (pI 8.5).

Kinetics and Substrate Specificity—The kinetics of the heparinase reaction are shown in a Lineweaver-Burk plot (Fig. 7). The values of K_m and V_m, as obtained at 37 °C, pH 7, a protein concentration of 1 μg/ml, and from the average of three separate experiments, were 8.04×10^{-6} M and 9.85×10^{-5} M/min, respectively.

Substrate specificity studies of the purified heparinase show that the enzyme is very specific, acting only on heparin and heparan monosulfate out of 12 polysaccharides tested (see "Methods."). There was 72% less activity on heparan mono-

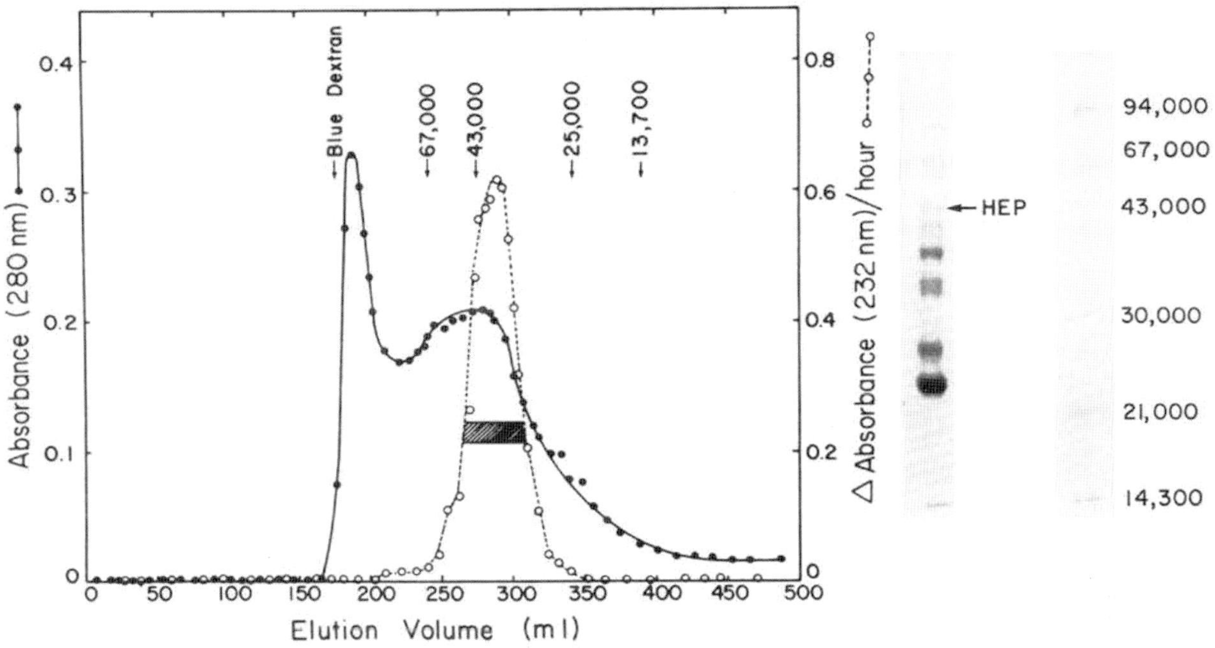

FIG. 3. **Gel filtration of heparinase on Sephadex G-100.** Two hundred milliliters of the hydroxylapatite-purified heparinase obtained in an $I = 0.29$ wash in Fig. 1 were concentrated to 6 ml by ultrafiltration and applied onto a Sephadex G-100 column (2.6 × 97 cm) equilibrated with 0.01 M sodium phosphate buffer, pH 6.8. The column was then eluted with the same buffer. The molecular weight markers are blue dextran ($M_r = 2 \times 10^6$), bovine serum albumin ($M_r = 67{,}000$), ovalbumin ($M_r = 43{,}000$), chymotrypsinogen A ($M_r = 25{,}000$), and ribonuclease A ($M_r = 13{,}700$). The flow rate was 48 ml/h, and the sample size was 6 ml of the protein solution with a concentration of 10 mg/ml. *Right*, the active fractions indicated by *cross-hatching* were pooled and analyzed by SDS-polyacrylamide gel electrophoresis (*slot 1*). *HEP* designates heparinase. *Slot 2*, markers (see Fig. 1).

sulfate compared with heparin. The lack of activity on *N*-acetyl heparin suggests that the purified heparinase contains no heparitinase activity. The activity on heparan monosulfate, therefore, is due to the cross-reaction of heparinase on heparan monosulfate rather than that there is heparitinase contaminant present. The other compounds were not active as substrates nor did they inhibit heparin degradation when present at concentrations as high as 8.3 mg/ml.

Effects of Salt, pH, and Temperature on Activity and Stability—The effects of salt concentration and pH on heparinase activity are shown in Fig. 8. The optimum salt concentration was around 0.1 M NaCl (Fig. 8*A*), and the optimum pH occurred at pH 6.5 (Fig. 8*B*). As shown in Fig. 9*A*, the stable salt concentration was 0.15 M NaCl where 93% of the enzyme activity remained after 96 h of incubation at 30 °C. The stable pH, after 72 h of incubation at 30 °C, was found to be at pH 7.0 (Fig. 9*B*).

The effect of temperature on enzyme activity was illustrated by an Arrhenius plot (Fig. 10). Initial rates were measured at pH 7 under conditions in which the enzyme was saturated with substrate. As is true for most enzymatic reactions, the reaction rate increases with temperature. The initial rates of degradation of heparin at 17, 31, and 41 °C were 0.17, 0.29, and 0.38 mM/min, respectively. The temperature coefficient (Q_{10}) was estimated to be 1.45, *i.e.* the reaction rate increased 1.45-fold for each 10 °C rise in temperature. The Arrhenius activation energy was calculated to be 6.3 kcal/mol. Although heparinase was very stable at 30 °C, it was sensitive to thermal denaturation above 38 °C. While over 90% of activity remained after 72 h of incubation at 30 °C, a loss of more than 25% of activity was observed after 20 min of incubation at 41 °C.

Storage Conditions—The hydroxylapatite-purified heparinase was stable when stored either frozen or lyophilized at −20 °C with a protein concentration of 0.2–0.4 mg/ml. More than 90% of activity was recovered after storage in the freezer for 6 months. Chromatofocused heparinase was also stable on storage and maintained 80% of the activity after 2 months of storage in the Polybuffer at −20 °C and a protein concentration of 50–100 μg/ml. However, in the absence of high salt concentration (0.1–0.2 M NaCl) or ampholytes, the enzyme is more stable when stored at −20 °C in the presence of 7.5% glycerol or 5% sucrose. An approximate 40% increase in stability was found for a preparation containing low ionic strength (10 mM sodium phosphate) and no ampholytes, when the sample was stored frozen for 2 months in the presence of 5% sucrose.

DISCUSSION

Heparin-induced *F. heparinum* contains a variety of enzymes acting on glycosaminoglycans by an eliminase mechanism (1, 28), as well as several enzymes acting on heparin degradation products by a hydrolysis mechanism (1, 29). The eliminases are heparinase (25, 29, 30), heparitinase (25), chondroitinases (26, 31), and hyaluronidase (26, 32). The hydrolases are sulfatases (sulfamidase and sulfoesterase) (30, 33, 34) and glycuronidase (1, 29, 35). Among these enzymes heparinase has gained increased attention with its potential therapeutic application for blood deheparinization in an immobilized enzyme reactor (13), as well as its importance in the elucidation of heparin structure (2, 6, 7). For the former studies it is essential to develop a method conducive to rapidly producing large quantities (grams) of catalytically pure hep-

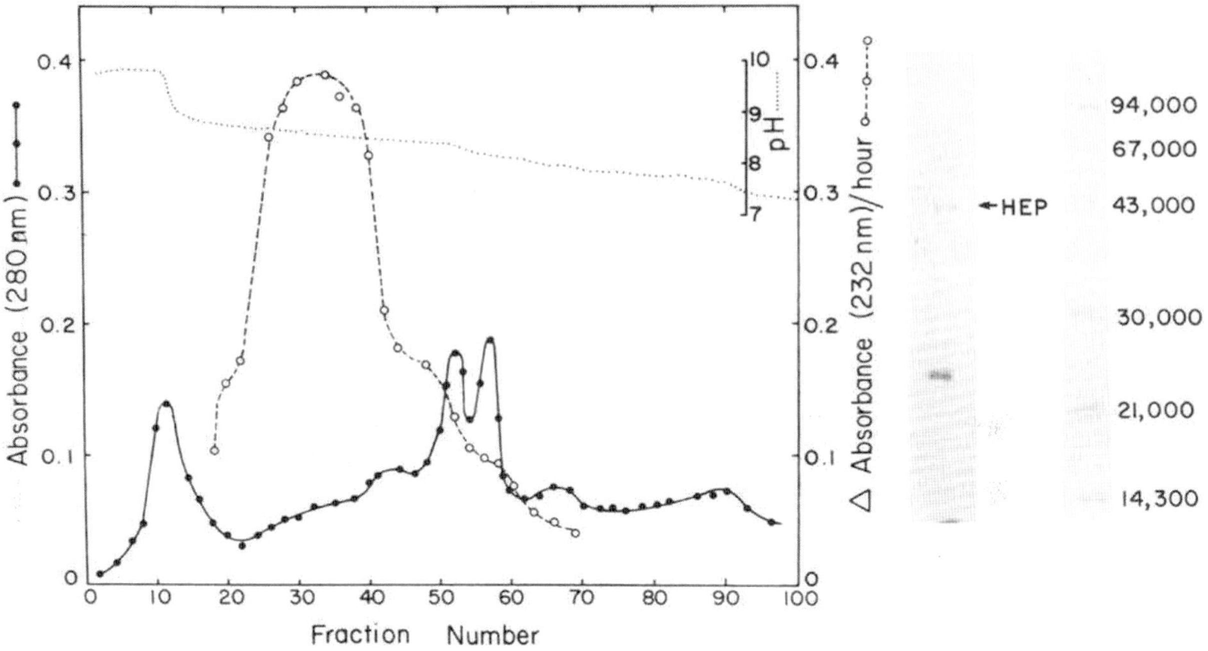

FIG. 4. **Chromatofocusing of heparinase on anion exchanger PBE 94.** Active fractions indicated in Fig. 3 by *cross-hatching* were concentrated to 6 ml by ultrafiltration, applied to the PBE 94 chromatofocusing column (0.9 × 24 cm), and eluted with Polybuffer 96-CH_3COOH (pH 6.0) as described under "Methods." Fractions of 1 ml are collected. The flow rate was 20 ml/h, and the sample size was 6 ml of the protein solution with a concentration of 1 mg/ml. *Right,* active fractions (25–40) were combined and analyzed by the SDS-polyacrylamide gel electrophoresis (*slot 1*). *HEP* designates heparinase. *Slot 2,* markers (see Fig. 1).

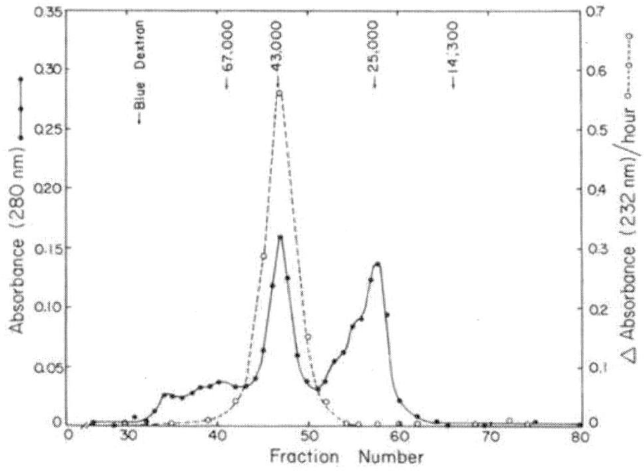

FIG. 5. **Gel filtration of heparinase on Sephadex G-100 superfine.** Active fractions (25–40) in Fig. 4 were pooled, concentrated to 6 ml by ultrafiltration, and applied to the Sephadex G-100 superfine column (1.6 × 95 cm) equilibrated with 0.01 M sodium phosphate buffer, pH 6.8. The column was then eluted with the same buffer. The flow rate was 3.6 ml/h, and the sample size was 6 ml of the protein solution with a concentration of 50 μg/ml.

a completely purified heparinase and to characterize its properties with regard to reaction kinetics, activity, and stability. Moreover, a homogeneous heparinase might eventually be used to produce antibodies which could subsequently be useful to facilitate heparinase purification through immunoaffinity chromatography.

Linker and Hovingh isolated heparinase from the other eliminases by a combination of hydroxylapatite and cellulose phosphate column chromatography (4). Dietrich and co-workers separated heparinase from heparitinase and sulfatases by preparative electrophoresis on agarose (7). Ototani *et al.* reported that by using repeated column chromatographic steps with hydroxylapatite, cellulose phosphate, and AH (aminohexyl)-Sepharose 4B coated with dermatan sulfate, heparinase can be purified to a point where all the contaminating activities are no longer present (20). However, these foregoing methods are time consuming and are not well suited for the preparation of large quantities of the enzyme. We have shown that by adapting a batch procedure of ion-exchange chromatography including hydroxylapatite and QAE-Sephadex, a facile large scale purification can be accomplished within a few hours. The preparation obtained after the QAE-purification is nearly free of sulfatases and glycuronidase and contains less than 5% of each of the other catalytic enzyme contaminants such as chondroitinase AC, hyaluronidase, and heparitinase. It is possible that the chondroitinase AC and hyaluronidase activities may be attributed to a single protein acting on both chondroitin sulfate and hyaluronic acid, since several authors (20, 26, 32) have reported the presence of co-purification of both activities throughout all their preparations of these enzymes from *F. heparinum.*

Heparinase has been purified to homogeneity by several successive chromatographic procedures including hydroxylapatite ion exchange, Sephadex G-100 gel filtration, chromatofocusing with PBE 94 and Polybuffer 96-CH_3COOH, and Sepahdex G-100 superfine gel filtration. The homogeneity of the purified heparinase is demonstrated by the presence of a single polypeptide band in both the SDS and the acid-urea polyacrylamide gel electrophoretic systems. The molecular weight was estimated to be 42,900 ± 1,000 daltons by gel

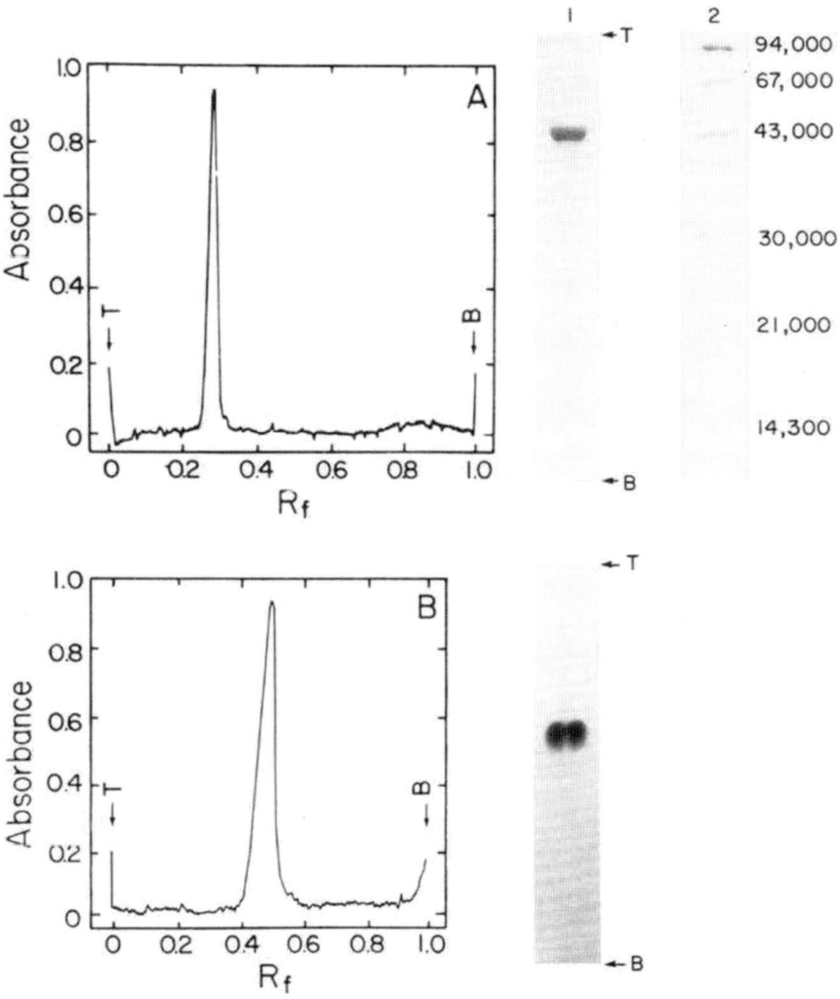

FIG. 6. **Homogeneity of the purified heparinase.** Active fractions (46–48) in Fig. 5 were combined and analyzed by the SDS-polyacrylamide gel electrophoresis and acid-urea polyacrylamide gel electrophoresis. *A*, densitometry trace of the SDS-polyacrylamide gel (*slot 1* on the *right*); *slot 2*, markers (see Fig. 1). *B*, densitometry trace of the acid-urea polyacrylamide gel (*right*). *T*, top of the gel; *B*, bottom of the gel. R_f represents the mobility of the polypeptide band.

TABLE I

Summary of purification of heparinase

Except in the less pure preparations where the glycuronidase is present (*i.e.* azure A assay is used), the international unit is used throughout the entire work.

Purification step	Total protein	Specific activity	Specific activity	Total activity	Recovery of activity
	mg	*units[a]/mg*	*IU[b]/mg*	*units[a]*	*%[a]*
Cell homogenate	1000	7.4		7400	
Protamine sulfate precipitation	670	10.0		6700	90
Hydroxylapatite	50	35.0		1750	23.8
Sephadex G-100	5.31	104.3	1.4	553.8	7.5
Chromatofocusing	0.23	619.3	8.1	142.4	1.9
Sephadex G-100 superfine	0.03	2045.8	26.6	61.4	0.8

[a] One unit of heparinase activity is defined as the amount of enzyme which degrades 1 mg of heparin/h.

[b] One international unit of activity is defined as the amount of enzyme which causes 1 μmol of double bonds formed/min.

filtration chromatography and 42,700 ± 1,200 daltons by SDS-PAGE. Furthermore, enzymatic assay of the polypeptides eluted from the polyacrylamide gel after SDS-PAGE showed that heparinase activity was present only in the region corresponding to a molecular weight between 38,000 to 50,100 daltons. These results suggest that heparinase is a single polypeptide chain with a molecular weight around 43,000 daltons.

Amino acid analysis of the purified heparinase indicates that the enzyme contains only 4 cysteine residues/polypeptide. In contrast, it contains relatively high amounts of lysine residues (9%). Preliminary results (data are not included) showed that heparinase was not inactivated by 2-mercaptoethanol. This resistance to the sulfhydryl-reducing reagent may be rationalized in terms of the low content of cysteine residues. The high level of lysine residues is consistent with the cationic nature of heparinase (pI 8.5). The lysine residues may be responsible for the coupling in heparinase immobili-

TABLE II
Amino acid composition of heparinase

Amino acid	Moles	Nearest integer[a]
	%	
Half-cystine[b]	0.89	4
Aspartic acid[c]	12.82	52
Threonine[d]	5.59	23
Serine[d]	6.17	25
Glutamic acid[c]	10.12	41
Proline	5.23	21
Glycine	8.62	35
Alanine	12.21	50
Valine	3.68	15
Methionine	0.74	3
Isoleucine	3.87	16
Leucine	6.93	28
Tyrosine	3.93	16
Phenylalanine	5.01	20
Lysine	9.04	37
Histidine	2.65	11
Arginine	2.50	10
Tryptophan	ND[e]	ND
Total		407

[a] The values are calculated assuming that the molecular weight of heparinase is 43,000 daltons.
[b] Represents moles (%) of cysteic acid generated by performic acid oxidation.
[c] Asparagine and glutamine were converted to aspartic acid and glutamic acid, respectively.
[d] Based on extrapolation of data obtained by hydrolysis of 24, 48, and 72 h.
[e] ND, not determined.

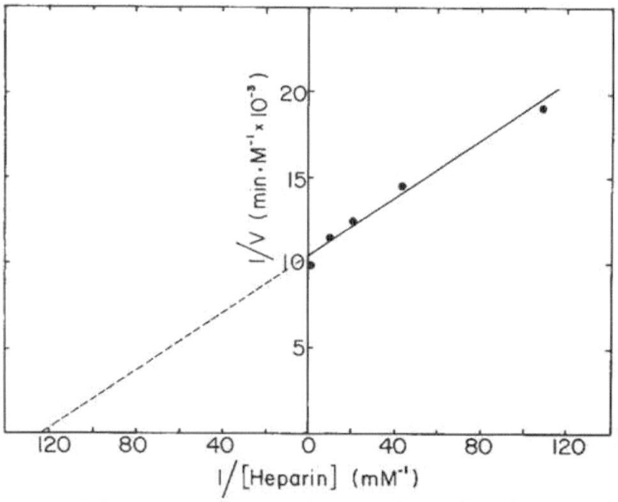

FIG. 7. **Lineweaver-Burk plot of initial rates of the heparinase reaction as a function of heparin concentration.** Initial rates (*V*) were measured by the UV 232-nm method at 37 °C, pH 7, and the indicated concentrations of heparin. Protein concentration in the reaction mixture was 0.5 μg/ml. An average molecular weight of 11,000 daltons was used for heparin to calculate its molar concentration.

zation to CNBr-activated Sepharose 4B (3) through its primary amino groups. However, if the active site is not protected with heparin during immobilization the enzyme loses activity (3). This suggests that certain lysine residues may be located either at or near the active site of heparinase.

The activity maximum of purified heparinase occurs at pH 6.5 and 0.1 M NaCl, while the stability maximum occurs at slightly higher pH (pH 7) and salt concentration (0.15 M NaCl). In order to maintain the stability of the heparinase, it

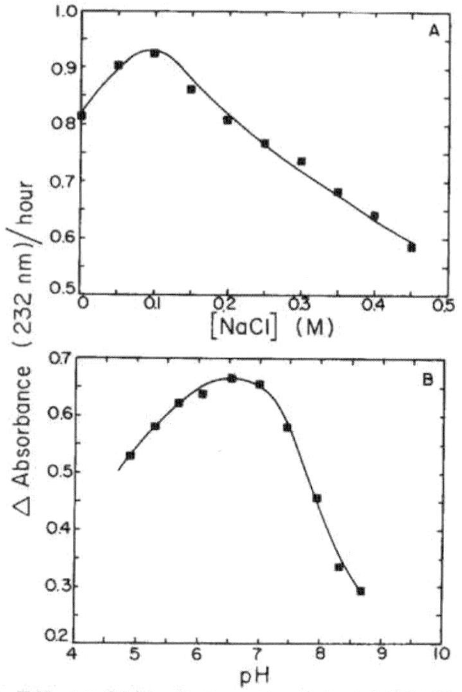

FIG. 8. **Effects of (*A*) salt concentration and (*B*) pH on heparinase activity.** Heparinase activity was assayed by the UV 232-nm method. In *A*, activity was determined at 30 °C, pH 7, a protein concentration of 12.8 μg/ml, and at the indicated salt concentrations. In *B*, activity was determined at 30 °C, a protein concentration of 11.2 μg/ml, and at the indicated pH. Reaction mixtures also contained 10 mM sodium phosphate.

is suggested to store the enzyme at −20 °C, pH 7, and in the presence of 0.15 M NaCl. The Arrhenius plot shows that the heparinase reaction has an activation energy of 6.3 kcal/mol, indicating that the degradation of heparin with heparinase proceeds faster at higher temperatures. However, the rate of thermal denaturation of heparinase also increased with temperature. The apparent temperature optimum is thus dependent upon the processing time of the activity assay. During an assay period of 40 min or less, the temperature optimum was found to be 37 °C. Prolonged incubation shifted the temperature optimum to a lower temperature. Dietrich (29) reported that the temperature optimum for heparinase was 30 °C. However, their assays included a 3-h incubation at each temperature.

Heparinase is very specific, acting only on heparin and heparan monosulfate. The activity toward heparan monosulfate may be rationalized in terms of the presence of heparin-like regions in the heparan monosulfate polymer (20, 25). Heparan monosulfate was a relatively poor substrate when compared to heparin, and the extent of reaction was only 28% that of heparin. This value is consistent with that reported by Hovingh and Linker on a relatively pure enzyme preparation (25) but is lower than that reported by Dietrich *et al.* (7). The difference may be attributed to the difference in sulfate content in the heparan monosulfate preparations (7, 25, 36). Sulfate groups appear to be an important structural feature required for recognition by the active site of the heparinase molecule. Preliminary results (data are not included) show that polyvinyl sulfate, a synthetic heparin analogue (37) which bears no obvious structural resemblance to heparin except the high sulfate content, is a competitive inhibitor of heparinase with a K_i of 3.0×10^{-8} M. The inhibitory effect declines as

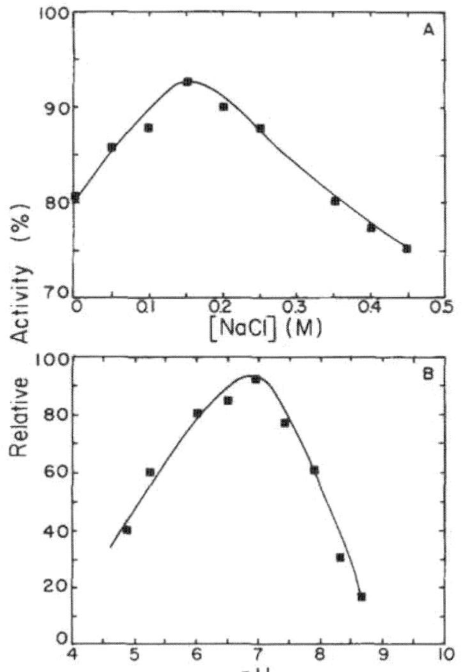

FIG. 9. **Effect of (*A*) salt concentration and (*B*) pH on heparinase stability.** In *A*, aliquots of the enzyme solution were incubated at 30 °C, pH 7, and a protein concentration of 12.8 μg/ml for 96 h. In *B*, aliquots were incubated at 30 °C and a protein concentration of 11.2 μg/ml for 72 h. The residual activity after incubation was assayed at 30 °C by the UV 232-nm method and was plotted as the relative activity. Reaction mixtures also contained 10 mM sodium phosphate.

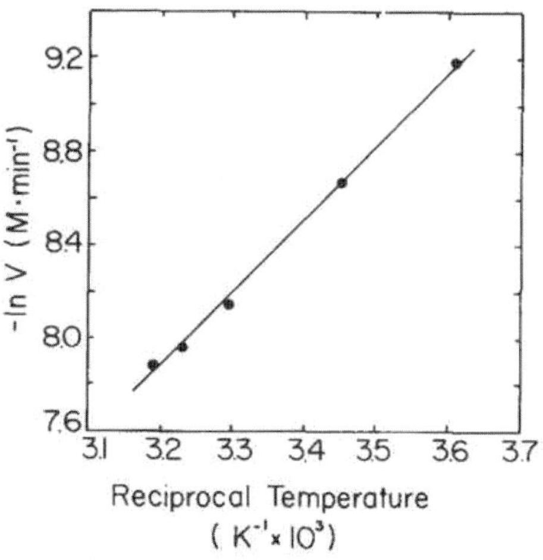

FIG. 10. **Arrhenius plot of the temperature effect on heparinase activity.** Initial rates (*V*) of the heparinase reaction were measured by the UV 232-nm method at temperatures of 4-41 °C. Reactions were conducted under conditions in which the enzyme (5.2 μg/ml) was saturated with substrate (heparin concentration, 24 mg/ml). An average molecular weight of 11,000 daltons was used for heparin to calculate its molar concentration. Reaction mixtures contained 10 mM sodium phosphate at pH 7.

the compound is desulfated by hydrolysis. The necessity of the sulfate group for either binding or reactivity becomes more evident as studies by several research groups show that heparinase cleaves the α-glycosidic linkages between a trisulfated disaccharide (2, 6). It is also interesting to note that heparinase possesses some properties resembling those discovered in a *Salmonella typhimurium* sulfate-binding protein (38). First, synthesis of both proteins is repressed by sulfate in the fermentation medium (38, 39). Second, the heparinase has a very low level of sulfur-containing amino acids (<2%), and the sulfate-binding protein has no such amino acids at all. It is not yet known whether other proteins acting upon sulfated substrates have characteristics and properties comparable to those of heparinase.

Acknowledgments—We thank Dr. Alfred Linker (Veterans Administration Hospital, Salt Lake City, UT) for generously providing the agar slants of *F. heparinum*, Dr. Ross Feldberg of Tufts University for reviewing the manuscript, Andre Dauphinais of Children's Hospital Medical Center in Boston for assistance in the amino acid analysis, and Cynthia Zannetos and Margaret Flanagan for their technical assistance.

REFERENCES

1. Linker, A., and Hovingh, P. (1965) *J. Biol. Chem.* **240**, 3724–3728
2. Silva, M. E., and Dietrich, C. P. (1975) *J. Biol. Chem.* **250**, 6841–6846
3. Linhardt, R. J., Fitzgerald, G. L., Cooney, C. L., and Langer, R. S. (1982) *Biochim. Biophys. Acta* **702**, 192–203
4. Linker, A., and Hovingh, P. (1972) *Methods Enzymol.* **28**, 902–911
5. Linker, A., and Hovingh, P. (1972) *Biochemistry* **11**, 563–568
6. Linker, A., and Hovingh, P. (1979) in *Heparin: Structure, Cellular Functions and Clinical Applications* (McDuffie, N. M., ed) pp. 3–24, Academic Press, New York
7. Dietrich, C. P., Silva, M. E., and Michelacci, Y. M. (1973) *J. Biol. Chem.* **248**, 6408–6415
8. Hutt, E. D., and Kingdom, H. S. (1972) *J. Lab. Clin. Med.* **79**, 1027–1034
9. Kanwar, Y. S., and Farquhar, M. G. (1979) *Proc. Natl. Acad. Sci. U. S. A.* **76**, 1303–1307
10. Lindahl, U., Backstrom, G., Hook, M., Thunberg, L., Fransson, L., and Linker, A. (1979) *Proc. Natl. Acad. Sci. U. S. A.* **76**, 3198–3202
11. Linhardt, R. J., Grant, A., Cooney, C. L., and Langer, R. (1982) *J. Biol. Chem.* **257**, 7310–7313
12. Folkman, J., Langer, R., Linhardt, R. J., Haudenschild, C., and Taylor, S. (1983) *Science (Wash. D. C.)* 221, 719–725
13. Langer, R., Linhardt, R., Hoffberg, S., Larsen, A. K., Cooney, C. L., Tapper, D., and Klein, M. (1982) *Science (Wash. D. C.)* 217, 261–263
14. Danishefsky, I., Eiber, H. B., and Carr, J. J. (1960) *Arch. Biochem. Biophys.* **90**, 114–121
15. Roseman, S., and Ludowieg, J. (1954) *J. Am. Chem. Soc.* **76**, 301–302
16. Lowry, O. H., Rosebrough, N. J., Farr, A. L., and Randall, R. J. (1951) *J. Biol. Chem.* **193**, 265–275
17. Bradford, M. M. (1976) *Anal. Biochem.* **72**, 248–254
18. Jaques, L. B. (1979) *Science (Wash. D. C.)* 206, 528–533
19. Galliher, P. M., Cooney, C. L., Langer, R. S., and Linhardt, R. J. (1981) *Appl. Environ. Microbiol.* **41**, 360–365
20. Ototani, N., Kikuchi, M., and Yosizawa, Z. (1981) *Carbohydr. Res.* **88**, 291–303
21. Warnick, C. T., and Linker, A. (1972) *Biochemistry* **11**, 568–572
22. Laemmli, U. K. (1970) *Nature (Lond.)* **227**, 680–685
23. Panyim, S., and Chalkley, R. (1969) *Arch. Biochem. Biophys.* **130**, 337–346
24. Hirs, C. H. W. (1967) *Methods Enzymol.* **11**, 197–199
25. Hovingh, P., and Linker, A. (1970) *J. Biol. Chem.* **245**, 6170–6175
26. Yamagata, T., Saito, H., Habuchi, O., and Suzuki, S. (1968) *J. Biol. Chem.* **243**, 1523–1535
27. Ototani, N., and Yosizawa, Z. (1979) *Carbohydr. Res.* **70**, 295–306
28. Linker, A., Hoffman, P., Meyer, K., Sampson, P., and Korn, E. D. (1960) *J. Biol. Chem.* **235**, 3061–3065
29. Dietrich, C. P. (1969) *Biochemistry* **8**, 2089–2094
30. Korn, E. D., and Payza, A. N. (1956) *J. Biol. Chem.* **223**, 859–864
31. Michelacci, Y. M., and Dietrich, C. P. (1974) *Biochem. Biophys. Res. Commun.* **56**, 973–980
32. Michelacci, Y. M., and Dietrich, C. P. (1976) *J. Biol. Chem.* **251**, 1154–1158
33. Dietrich, C. P. (1969) *Biochem. J.* **111**, 91–95
34. Lloyd, A. G., Law, B. A., Fowler, L. J., and Emberg, G. (1968) *Biochem. J.* **110**, 54P
35. Warnick, C. T., and Linker, A. (1970) *Fed. Proc.* **29**, 675
36. Dietrich, C. P., Nader, H. B., Britto, L. R. G., and Salva, M. E. (1971) *Biochim. Biophys. Acta* **237**, 430–441
37. Henry, D. (1974) *Clinical Chemistry: Principles and Techniques*, 2nd Ed., Harper and Row, Hagerstown, MD
38. Pardee, A. B. (1966) *J. Biol. Chem.* **241**, 5886–5892
39. Galliher, P. M., Linhardt, R. J., Conway, L. J., Langer, R., and Cooney, C. L. (1982) *Eur. J. Appl. Microbiol. Biotechnol.* **15**, 252–257

Adipose tissue mass can be regulated through the vasculature

Maria A. Rupnick*[†‡§], Dipak Panigrahy[†], Chen-Yu Zhang[¶], Susan M. Dallabrida[†], Bradford B. Lowell[¶], Robert Langer[‡], and M. Judah Folkman[†]

*Division of Cardiovascular Medicine, Brigham and Women's Hospital, [†]Division of Surgical Research, Children's Hospital, and [¶]Division of Endocrinology, Beth Israel Deaconess Medical Center, Boston, MA 02115; and [‡]Department of Chemical Engineering, Massachusetts Institute of Technology, Cambridge, MA 02139

Contributed by M. Judah Folkman, June 11, 2002

Tumor growth is angiogenesis dependent. We hypothesized that nonneoplastic tissue growth also depends on neovascularization. We chose adipose tissue as an experimental system because of its remodeling capacity. Mice from different obesity models received anti-angiogenic agents. Treatment resulted in dose-dependent, reversible weight reduction and adipose tissue loss. Marked vascular remodeling was evident in adipose tissue sections, which revealed decreased endothelial proliferation and increased apoptosis in treated mice compared with controls. Continuous treatment maintained mice near normal body weights for age without adverse effects. Metabolic adaptations in food intake, metabolic rate, and energy substrate utilization were associated with anti-angiogenic weight loss. We conclude that adipose tissue mass is sensitive to angiogenesis inhibitors and can be regulated by its vasculature.

Substantial evidence has established that tumor growth is angiogenesis-dependent (1). Neovascularization promotes tumor growth (2), whereas angiogenesis inhibition prevents it, and can regress the lesions (3). This raises the question of whether the endothelium functions similarly in nonneoplastic tissue growth (4).

The challenges in addressing this question are that most adult tissues normally do not grow, their mass is stable, and the supporting vasculature is quiescent (5). Exceptionally, adipose tissue can grow and regress throughout adulthood. It is highly vascularized and has angiogenic properties (6, 7). An extensive capillary network surrounds each adipocyte, and there are few other resident cell types (8). Therefore, adipose tissue is uniquely suited to study the role of angiogenesis in nonneoplastic adult tissue growth.

The potential for adipose tissue to grow and regress is substantial. One would expect the vasculature to have a commensurate capacity for remodeling. We hypothesized that adipose tissue growth is angiogenesis dependent and, therefore, may be inhibited by anti-angiogenesis agents. This would suggest that adipose tissue mass may be regulated via the vascular endothelium.

We primarily used *ob/ob* mice because they rapidly accumulate adipose tissue (9). This strain develops spontaneous obesity because of a lack of functional leptin, a protein secreted by adipocytes that acts on the hypothalamus to regulate appetite and metabolism (9–11). Without leptin, animals eat excessively, expend less energy, and become morbidly obese. Leptin replacement induces weight loss in *ob/ob* mice by specifically reducing adipose tissue (9), unlike diet restriction or appetite suppression (fenfluramine), which decreases both lean and fat mass (12, 13).

We treated *ob/ob* mice with various angiogenesis inhibitors. We primarily used TNP-470 (14) because of availability. TNP-470 inhibits endothelial cell proliferation *in vitro* (15) and angiogenesis *in vivo* (16). At significantly higher concentrations, TNP-470 can also suppress nonendothelial cell proliferation (14, 17). Angiostatin (kringle 1–4 domains of plasminogen) (18) and endostatin (a C-terminal fragment of collagen XVIII) (19) are angiogenesis inhibitors that act exclusively on endothelium. Bay-129566, a matrix metalloproteinase inhibitor (20), and thalidomide (21) were also studied.

We show that adipose tissue growth is angiogenesis dependent. Body and adipose tissue weights of treated mice were significantly less than controls. Loss of adipose tissue mass was similar to that resulting from leptin replacement. We observed decreased endothelial cell proliferation and increased apoptosis in the adipose tissue of treated animals compared with controls, suggesting that adipose endothelial cells were sensitive to angiogenesis inhibitors. We have also begun exploring the metabolic consequences of anti-angiogenic-induced weight reduction and have found adaptations in food intake, metabolic rate, and preferred energy substrate. These studies suggest that adipose tissue mass can be regulated through the vascular endothelium.

Methods

Animal Studies. Male C57BL/6J-Lep^{ob}, C57BL/6J, agouti (A^y), and fat (Cpe^{fat}) mice (The Jackson Laboratory) were used according to institutional guidelines. We housed mice individually with free access to water and standard chow unless noted. We measured body weights and remaining chow daily. We anesthetized animals with isoflurane (Baxter) and killed them by continuous inhalation of CO_2.

Solutions were prepared daily. TNP-470 (Takeda, Osaka; complexed with beta-cyclodextrin) was dissolved in saline and administered s.c. in rotating sites along the dorsum. Recombinant murine endostatin and angiostatin were produced in *Escherichia coli*, precipitated as nonrefolded proteins (18, 19), and administered s.c. in saline at 50 mg/kg per 12 h (endostatin and angiostatin) or 20 mg/kg per day (angiostatin). Bay 12-9566 (Bayer, Elkhart, IN) was suspended in 0.5% carboxymethylcellulose/0.2% Tween 80 and delivered via gastric lavage (100 mg/kg per day). We injected thalidomide (Entremed) in 0.5% carboxymethylcellulose i.p.(150 mg/kg per day). We administered leptin at 150 μg/kg per day (s.c. osmotic pump). Controls received the appropriate vehicle. We used the Kruskal–Wallis test to evaluate the overall treatment effect. Duncan's multiple range test on ranked data was used to compare results among treatment groups.

Cell Proliferation Assay. We washed confluent bovine capillary endothelial cell cultures with PBS and dispersed cells in 0.05% trypsin. Cells were resuspended in DMEM containing 10% bovine calf serum and 1% L-glutamine/penicillin/streptomycin and cultured on gelatinized plates (12,500 cells per well) (24 h). Media were replaced with assay media (DMEM/5% bovine calf serum/1% L-glutamine/penicillin/streptomycin) containing TNP-470 at 1 pg/ml to 100 μg/ml for 30 min, and basic fibroblast growth factor (1 ng/ml) was added. After 72 h, cells were dispersed in trypsin, resuspended in Hematall, and counted

Abbreviation: vWF, von Willebrand factor.

§To whom reprint requests should be addressed. E-mail: rupnick@hub.tch.harvard.edu.

 www.pnas.org/cgi/doi/10.1073/pnas.162349799

using a Coulter Counter. Following the above procedure, subconfluent 3T3-L1 cells were resuspended in DMEM containing 10% bovine calf serum, plated cells (8,000 cells per well), and were incubated in assay media without L-glutamine/penicillin/streptomycin.

Serum Glucose. Blood was collected (100 μl) from tail veins of anesthetized 9-month-old *ob/ob* mice following 115 d of treatment with TNP-470 (10 mg/kg per day) (n = 5) or saline (n = 13). Children's Hospital clinical chemistry laboratory (Boston) performed the measurements, which were compared using a one-tailed Student's t test.

Body Compositions. Bone-free lean body mass and fat mass were measured using a Lunar PIXImus J densitometer (Lunar, Madison, WI). Calibration was conducted using an aluminum/lucite phantom (TBMD, 0.0594 g/cm^2, percent fat 12.4%). Mice were fasted (3 h), anesthetized (80 mg/kg ketamine/10 mg/kg xylazine), and positioned with limbs extended. The imaging area was too small to contain the entire carcass, so the heads extended outside the field. We placed an exclusion region over any captured portion of the head to ensure that measurements represented only subcranial body composition. We scanned each mouse three times with intermittent repositioning.

Immunohistochemistry. Double-fluorescent labeling enabled simultaneous identification of proliferating cells (BrdUrd) and endothelial cells [von Willebrand factor (vWF)] in epididymal adipose tissue sections. Tissue was fixed in cold Carnoy's fixative (4 h), transferred to 100% ethanol, and paraffin-embedded. Sections (5 μm) were permeabilized with 10 μg/ml proteinase K in 0.1 M Tris-HCl buffer (pH 7.4) at 37°C (20 min) and washed with PBS. We labeled proliferating cells with anti-BrdUrd/nuclease. To label endothelium, sections were blocked with 5% horse serum (30 min) and incubated with rabbit anti-human vWF (Dako; 1:250) in 5% horse serum in PBS at 4°C overnight and washed. Sections were incubated with biotinylated anti-rabbit IgG (1:200) and Hoechst 33258 (1.0 μg/ml; 30 min) and washed. BrdUrd was detected with FITC-anti mouse IgG (1:200), and vWF was detected by Texas Red Avidin-conjugated goat anti-rabbit IgG (1:200; Vector Laboratories).

Similarly, double-fluorescent labeling enabled simultaneous detection of apoptotic cells (TUNEL assay) and endothelial cells. Apoptotic cells were identified using ApopTag plus *in situ* apoptosis detection kit (Integen) with a fluorescein tag. Sections were stained for endothelial cells (vWF) and counterstained (Hoechst 33258) as described.

Sections were preserved with Fluoromount G (Southern Biotechnology) and analyzed using a Zeiss Axiophot fluorescent microscope with a dual filter (FITC 485 nm, Texas Red, 510–560 nm). Sections were photographed at 40× magnification with Ektachrome P1600 film.

We did not quantify proliferating or apoptotic adipocytes because cell diameters can exceed 150 μm, and the frequency of visualizing adipocyte nuclei in thin sections was relatively low.

VO$_2$ and Respiratory Exchange Ratio. Serial measurements were made in 10-week-old *ob/ob* mice treated with TNP-470 (10 mg/kg per day) or vehicle (n = 3/group) using an OXYMAX System (Columbus Instruments, Columbus, OH). Conditions were settling time (100 s), measuring time (50 s), and room air (reference). Animals were placed individually in 0.3-liter chambers, and resting measurements were collected (4 h).

Results

Body Weight Responses of Obese Mice to Angiogenesis Inhibitors. TNP-470-treated *ob/ob* mice dose-dependently ($P < 0.05$) lost weight, whereas controls gained ($P \leq 0.003$) (Fig. 1*A*). To determine whether suppression of preadipocyte proliferation contributes to this effect, we treated 3T3-L1 preadipocytes (22) with TNP-470. A five logs greater TNP-470 concentration (10 μg/ml) was required for half-maximal cytostatic inhibition of 3T3-L1 preadipocytes compared with endothelial cells (100 pg/ml, $P < 0.05$; data not shown). Although this supports an endothelial-mediated mechanism, we cannot exclude preadipocyte suppression. Therefore, we tested specific angiogenesis inhibitors. Obese mice treated with angiostatin at 20 mg/kg per

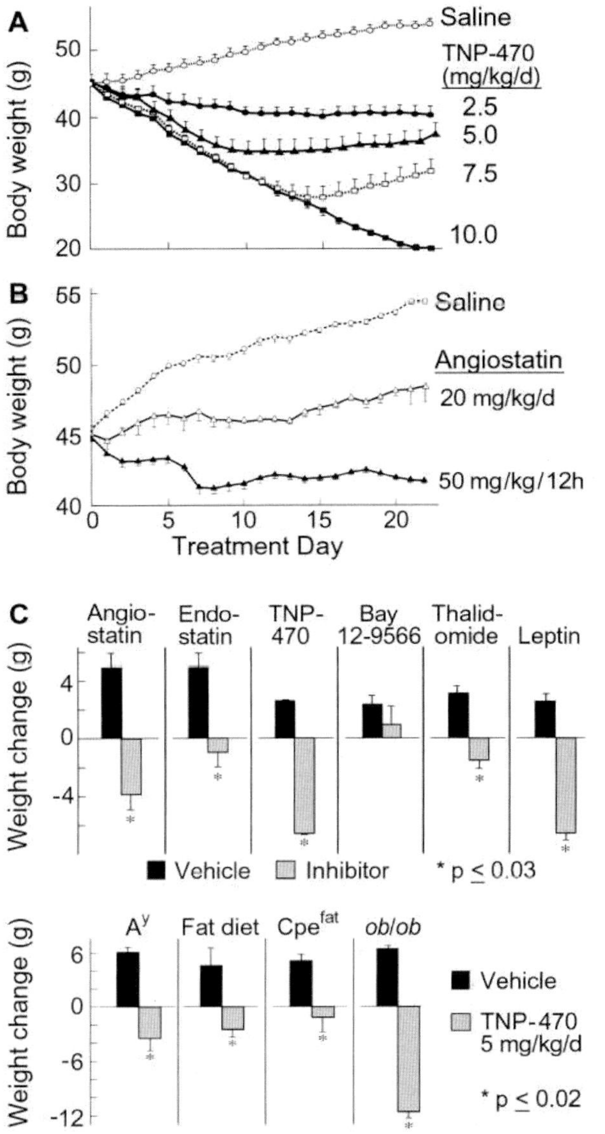

Fig. 1. Body weight responses to angiogenesis inhibitors. We treated 8-week-old *ob/ob* mice with various doses of TNP-470 (*A*) or angiostatin (*B*) for 21 days (n = 4 per group). (*C*) Responses of *ob/ob* mice treated with various angiogenesis inhibitors (hatched bars) for 7 d were compared with those of the corresponding vehicles (solid bars) and leptin (n = 3 per group). (*D*) Twelve-week-old mice from various obesity models were treated with vehicle (solid bars) or TNP-470 (hatched bars) for 7 days (n = 3 per group). These included agouti mice C57BL/6J-A^y (A^y), fat mice C57BLKS-Cpe^{fat}/J (Cpe^{fat}), *ob/ob* mice C57BL/6J-Lep^{ob}, and wild-type C57BL/6J mice fed a high-fat diet consisting of 40% fat by calorie (n = 3 per group).

day gained one-third that of controls ($P < 0.05$) (Fig. 1*B*), whereas those receiving 50 mg/kg per 12 h lost weight ($P < 0.05$). Angiostatin (up to 1 mg/ml) did not effect 3T3-L1 cell proliferation (data not shown).

Comparatively, endostatin ($P = 0.02$), Bay12–9566, thalidomide ($P = 0.01$), angiostatin ($P < 0.002$), and TNP-470 ($P = 0.01$) treated *ob/ob* mice gained less or lost weight relative to controls (Fig. 1*C*). TNP-470 (5 mg/kg per day) and leptin (150 μg/kg per day) yielded similar results. Mice from other obesity models (A^y, Cpe^{fat}, C57BL/6J mice on high-fat diet) treated with TNP-470 also weighed less than controls ($P \leq 0.02$) (Fig. 1*C*).

Effects of Long-Term or Cycled Anti-Angiogenic Treatment of Obese Mice. Daily TNP-470 treatment was well tolerated by *ob/ob* mice for over one-fourth of their life spans (Fig. 2*A*). Appetites decreased early but normalized despite continued therapy. Activity and grooming were maintained, and obesity-associated hyperglycemia was avoided (23). Controls had elevated serum glucose levels (224 ± 16 mg/dl), whereas levels from TNP-470-treated *ob/ob* mice (158 ± 12.6 mg/dl) were within normal range (62–175 mg/dl) (24).

Aged, relatively weight-stable *ob/ob* mice with negligible adipose endothelial cell proliferation (data not shown) lost weight when treated with TNP-470, whereas controls slightly gained ($P < 0.000002$) (Fig. 2*A*). This suggests adipose vasculature is susceptible to inhibitors, even when not proliferating.

To test reversibility, we intermittently treated *ob/ob* mice with TNP-470 until they reduced to the weight of age-matched C57BL/6 mice, and then suspended treatment until mice regained (Fig. 2*B*). Mice reduced weight while on TNP-470 and regained weight off treatment. At the bottom of the fourth cycle, control *ob/ob* mice were morbidly obese, whereas treated ones were similar in weight to wild-type mice (Fig. 2*C*).

Body Composition of Obese Mice Treated with Angiogenesis Inhibitors. We compared body compositions (fat and bone-free lean mass) by using dual-energy x-ray absorptiometry in TNP-470- and leptin-treated *ob/ob* mice (Fig. 3*A*). Dual-energy x-ray absorptiometry yields results similar (25) or well correlated (26) to chemical carcass analysis. Mice lost similar weight with TNP-470 and leptin, whereas controls gained ($P < 0.05$). Food intakes in treated groups were comparable and less than controls ($P < 0.0005$) (Fig. 3*B*). TNP-470 (30.4 ± 1.9 g) and leptin (29.0 ± 1.3 g) similarly reduced fat mass relative to controls (35.0 ± 0.9 g) ($P \leq 0.01$) (Fig. 3*C*), resulting in similar decreases in percent body fat (TNP-470 2.0 ± 1.1%; leptin 3.0 ± 0.6%). This was supported by significantly lower adipose tissue bed weights in both treated groups. For example, epididymal fat pad weights decreased from 1.73 ± 0.27 g in control mice to 1.35 ± 0.13 g with TNP-470 and to 1.33 ± 0.17 g with leptin ($P \leq 0.005$). Subcutaneous fat pad weights decreased from 1.42 ± 0.14 g in control mice to 1.25 ± 0.19 g with TNP-470 and to 1.27 ± 0.20 g with leptin ($P \leq 0.05$). Although insignificant, TNP-470-treated mice had the lowest average lean mass (control 18.5 ± 0.4 g; TNP-470 17.1 ± 0.8 g; leptin 18.15 ± 0.5), suggesting modest protein loss.

Vascular Remodeling in Adipose Tissue During Weight Shifts. We assessed cell proliferation (Fig. 4*A Left*) and apoptosis (Fig. 4*A Right*) in epididymal fat sections from treated and control *ob/ob* mice. TNP-470 and leptin significantly decreased median percents of endothelial cell proliferation (Fig. 4*B Left*) and increased apoptosis (Fig. 4*B Right*) compared with controls. Wide ranges reflect heterogeneity. Endothelial apoptosis also predominated in adipose tissue of diet-restricted *ob/ob* mice (data not shown).

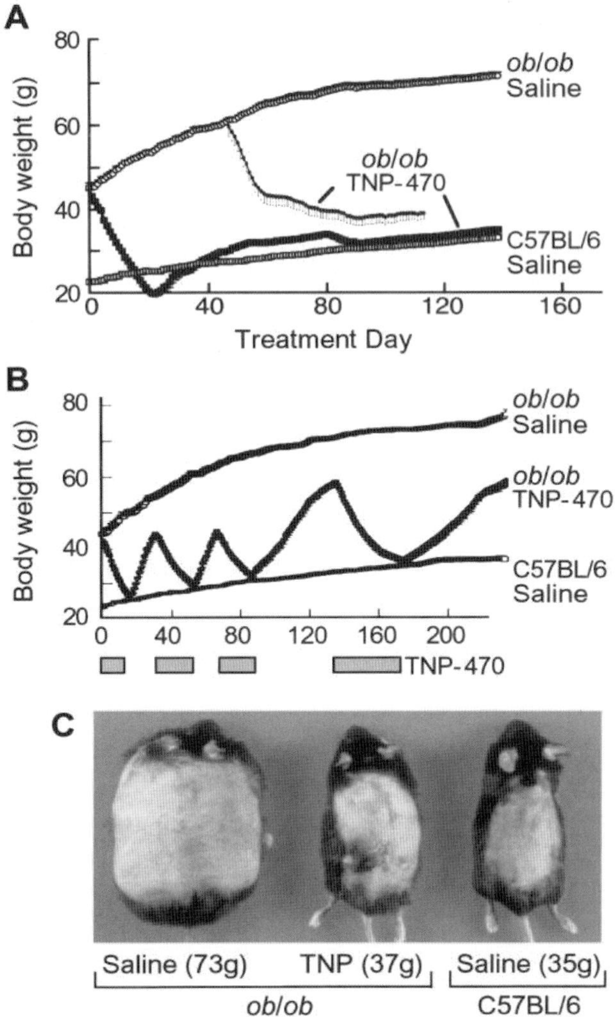

Fig. 2. Continuous and intermittent treatment of *ob/ob* mice with TNP-470. (*A*) *ob/ob* mice were treated with vehicle ($n = 4$) or TNP-470 (10 mg/kg/day; $n = 6$) starting at 8 weeks of age for 138 days, or at 6 months of age for 65 days. For comparison, the graph includes the growth curve of wild-type, age-matched C57BL/6 mice ($n = 5$). (*B*) Mice were treated with TNP-470 (10 mg/kg/day; $n = 15$) until they reduced to the weight of age-matched C57BL/6 mice. We then discontinued treatment, and the mice were permitted to regain before treatment was restarted. We cycled treatment four times. (*C*) Photographs show a representative mouse from each group on 173 days (bottom of cycle 4). Except for superficial scarring, daily TNP-470 treatment was well tolerated.

Metabolic Studies on Obese Mice Treated with Angiogenesis Inhibitors. Transiently decreased appetites in TNP-470-treated *ob/ob* mice were reflected by weight loss in paired-fed animals ($P < 0.05$) (Fig. 5*A*). Despite continued treatment, appetites normalized, as gauged by food intakes of C57BL/6 mice of similar weight. At lower doses, appetite was proportionately less affected (data not shown). Angiostatin-treated *ob/ob* mice also had decreased food intake ($P < 0.05$) (Fig. 5*B*). Reduced appetite in both groups coincided with about 7% loss of initial body weight. Endostatin-treated *ob/ob* mice had about 5% loss and no significant appetite effect (Fig. 5*C*).

Basal metabolic rate, as indicated by oxygen consumption (VO_2), increased in TNP-470-treated *ob/ob* mice by 4 days and

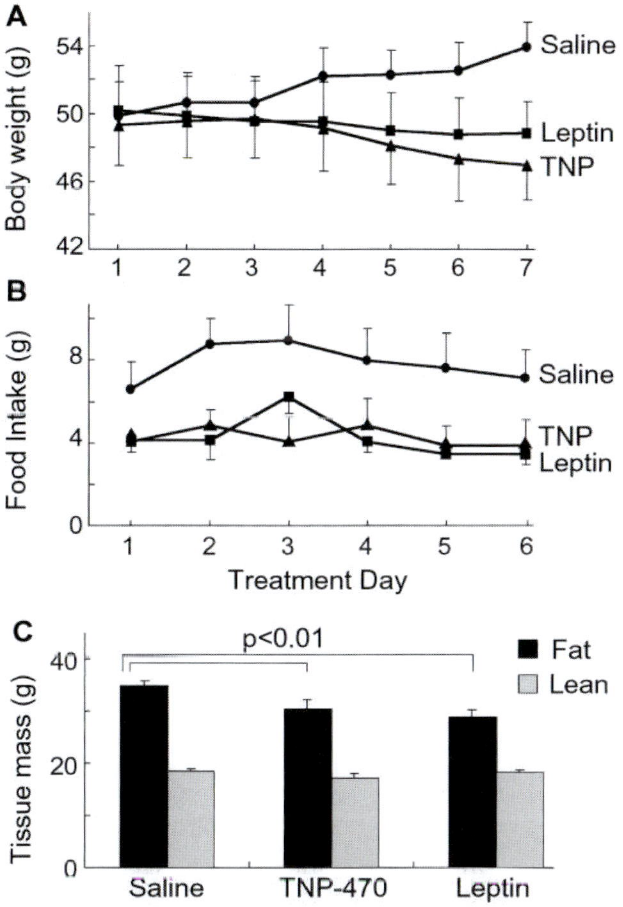

Fig. 3. Body compositions of *ob/ob* mice treated with TNP-470 or leptin. Nine-week-old female *ob/ob* mice were treated with TNP-470 (5 mg/kg/day) or leptin (150 μg/kg/day) for 7 days to achieve similar weight reduction (n = 4 per group). Control *ob/ob* mice received vehicle. (*A*) Body weight curves. (*B*) Daily food intakes. (*C*) Fat mass (black bars) and lean body mass (gray bars) components of body composition were measured using dual-energy x-ray absorptiometry.

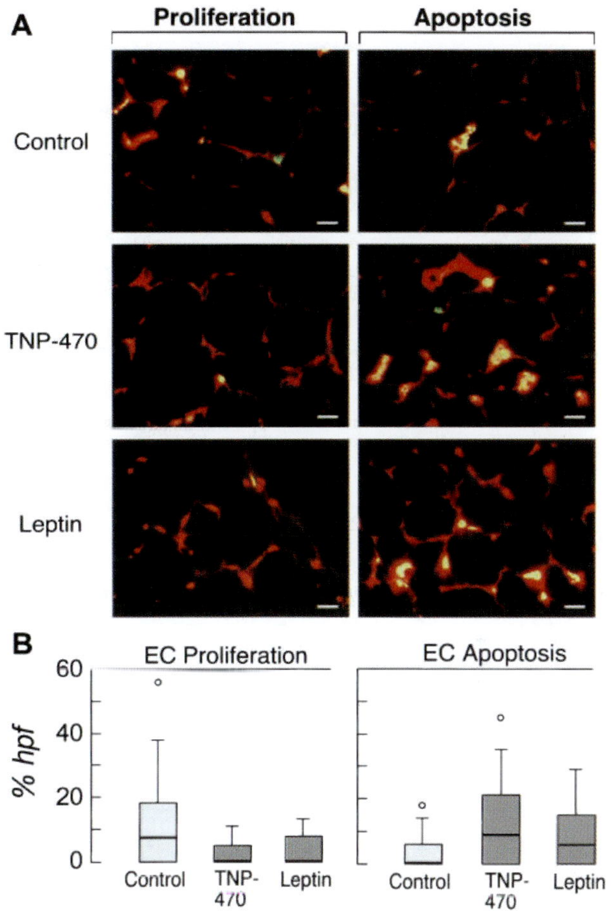

Fig. 4. Adipose tissue endothelial cell proliferation and apoptosis in *ob/ob* mice. (*A*) Endothelial cell proliferation and apoptosis were evaluated in epididymal adipose tissue sections from *ob/ob* mice following 7 days of treatment with vehicle, TNP-470, or leptin. We identified endothelial cells (red) by using vWF. We used fluorescein (green) with BrdUrd to identify proliferating cells (*Left*) or with the TUNEL assay to identify apoptotic cells (*Right*). Cells stained with both markers appear yellow. Adipose tissue from control *ob/ob* mice contained proliferating endothelial cells (yellow, top left), occasional proliferating nonendothelial cells (green), and few apoptotic endothelial cells (yellow, top right). In contrast, adipose tissue from TNP-470- and leptin-treated mice contained numerous apoptotic endothelial cells (yellow, middle and bottom right). (Bar = 25 μm.) (*B*) Median percents of endothelial cell proliferation and apoptosis were quantified for 30 high power fields per section (n = 5 per group). We used a box and whisker plot to illustrate the median and range of the measurements.

in C57BL/6 mice by 8 days ($P < 0.05$) (Table 1). VO_2 at 4, 12, and 24 h were unaltered (data not shown), although weight loss occurred by 24 h. Average body temperatures measured twice daily for 12 days via s.c. probes were similar in the control (95.3 ± 1.1°F) and TNP-470 (94.8 ± 0.8°F) treated mice (n = 4/group). Respiratory exchange ratio (respiratory exchange ratio = VCO_2/VO_2) decreased with TNP-470 by 8 days in both strains, suggesting the preferred energy substrate shifted from carbohydrates to fatty acids during weight loss ($P < 0.05$). Regressing fat is the likely fatty acid source and may contribute to transient appetite suppression (27).

Discussion

These studies demonstrate that adipose tissue mass can be regulated through the vasculature and that metabolic changes accompany anti-angiogenic-induced weight loss, which may contribute to weight reduction.

Different angiogenesis inhibitors significantly decreased body and adipose tissue weights (Figs. 1 and 3). We observed this effect in mice from distinct obesity models (Fig. 1). The magnitude of reduction varied by strain and angiogenesis inhibitor. This may reflect differences in bioavailability, tissue or cell specificity, or mechanisms of actions. Genetic differences in the animals' angiogenic potential may also contribute (28). Differences in percent body fat among the strains may also be a determining factor in their responsiveness. For example, *ob/ob* mice have greater fat content, which if proportionately reduced by angiogenesis inhibitors, would result in greater weight loss.

Adipose tissue's capacity to grow throughout life suggests it maintains the ability to recruit new capillaries. In the cornea assay, fat induced neovascularization, whereas liver and muscle did not (29). Angiogenesis and adipogenesis are spatially and temporally coupled during development (6), and endothelial cell proliferation occurs in expanding adult adipose tissue (6). This coordinated growth may be mediated by paracrine interactions between endothelial cells and adipocytes. Adipocytes produce

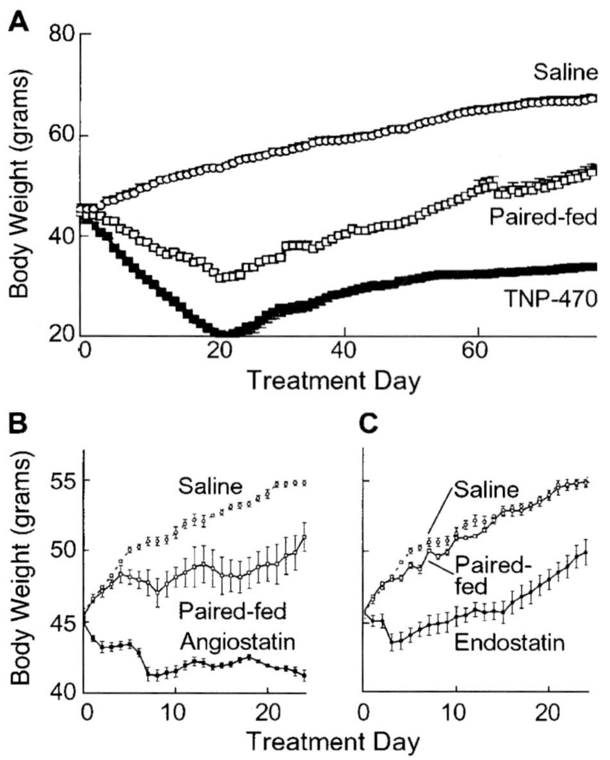

Fig. 5. Appetite of *ob/ob* mice treated with angiogenesis inhibitors. Eight-week-old *ob/ob* mice ($n = 4$ per group) with free access to chow were treated with (*A*) TNP-470 (10 mg/kg/day), (*B*) angiostatin (50 mg/kg/12 h), or (*C*) endostatin (50 mg/kg per 12 h). We fed the amount consumed per day by each treated mouse to a paired mouse in a second, untreated group ($n = 4$). We measured body weights of treated, paired-fed, and control mice daily.

endothelial cell-specific mitogens and angiogenic factors (6, 30), such as monobutyrin (31), vascular endothelial growth factor (32), and leptin (7). Conversely, endothelial cells produce soluble and matrix-bound preadipocyte mitogens (33, 34).

Weight reduction from angiogenesis inhibitors, leptin (Figs. 4 and 5), or diet restriction (data not shown) was associated with adipose endothelial cell apoptosis. Leptin-induced endothelial cell (35) and adipocyte (36) apoptosis during weight loss is recognized. Shifts from endothelial cell proliferation to apoptosis under various weight loss conditions suggest that vascular remodeling is a critical and common event during adipose tissue growth or regression. Whether weight loss leads to vascular regression or *vice versa* is likely a function of the initiating mechanism (37). Preliminary studies showing reductions in the endothelial cell to adipocyte ratios in TNP-470-treated mice (data not shown) are consistent with vascular regression as the initial event, whereas, with caloric restriction and leptin, vascular remodeling is probably a secondary response.

Cycling and aged mice studies (Fig. 3) illustrate that adipose vessels are capable of extensive remodeling and remain susceptible to anti-angiogenic agents, even when not growing. These findings may be explained by results from subsequent studies in which we determined that regulation of vessel maturation through shifts in angiopoietin-1 is coordinated with the extent of adipose tissue remodeling (S.M.D., D. Zurakowski, J.F., K. S. Moulton, and M.A.R., unpublished results). We propose that adipose tissue vessels may be maintained in a relatively immature state compared with that of other organs that are weight stable. Because anti-angiogenic agents target growing or newly formed vessels (whereas mature vessels are not responsive), adipose tissue vasculature may remain susceptible, even when not actively growing. If so, adipose tissue may be selectively reduced by angiogenesis inhibitors, thus, resulting in weight loss.

Mechanisms governing metabolic adaptations in food intake, VO_2, and respiratory exchange ratio (Table 1) during anti-angiogenic-induced weight loss are unknown. Food intake decreased with TNP-470 and angiostatin in leptin-deficient mice (Fig. 5). This may be the result of a direct central effect, which can be investigated through intrathecal administration of each drug. Alternatively, it may be related to the shared action of the inhibitors and perhaps an indirect response to adipose tissue loss. Factors released from regressing adipose tissue may inhibit feeding, analogous to appetite suppression seen with glucose (38) or fatty acid infusions (27). Treated animals preferentially used fatty acids as energy substrates during weight reduction. Despite continued TNP-470 treatment, body weights stabilize. Appetites increased, approximating that of wild-type mice of similar weight (data not shown). This suggests that the processes operating during weight reduction and weight maintenance may differ. We speculate that a leptin-independent feedback mechanism appropriately reduces appetite when anti-angiogenic agents prevent adipose tissue expansion.

Anti-angiogenic weight loss can occur without appetite changes as evidenced by endostatin (Fig. 5) and by low doses of TNP-470 or angiostatin (data not shown). Furthermore, treated mice weighed less than paired-fed mice having similar caloric intake. An increased metabolic rate in treated mice assists in accounting for these calories. A direct action on the adipocytes may also contribute.

Weight loss and appetite suppression are common nonspecific responses, thus necessitating consideration of drug toxicity. However, several observations favor a nontoxic mechanism. Weight reduction occurred with different angiogenesis inhibitors (Fig. 1). Numerous studies safely use TNP-470 to inhibit tumor growth (mice, refs. 16 and 39; clinical trials, ref. 40). Although toxicity can occur above the conventional anti-tumor dose (41), our maximum dose was two-thirds the anti-tumor dose. Mice treated daily for over 25% of their lifespan had no discernable ill effects, and food intake paralleled wild-type mice of compa-

Table 1. Body weight, VO_2, and respiratory exchange ratio in response to TNP-470

Mice	Time, day	Body weight, g		VO_2, ml/kg/min		Respiratory exchange ratio	
		Saline	TNP	Saline	TNP	Saline	TNP
C57BL6	0	23.0 ± 0.5	22.6 ± 0.2				
	4	22.8 ± 0.4	21.8 ± 0.4	44.4 ± 0.8	42.5 ± 3.2	0.70 ± 0.02	0.69 ± 0.01
	8	23.0 ± 0.4	21.1 ± 0.3	31.9 ± 0.8	38.3 ± 1.1*	0.71 ± 0.03	0.67 ± 0.03*
ob/ob	0	41.4 ± 0.7	41.7 ± 0.7				
	4	43.9 ± 0.6	41.1 ± 0.9	24.3 ± 1.1	30.0 ± 2.0*	0.74 ± 0.11	0.70 ± 0.01
	8	46.2 ± 0.5	36.7 ± 1.1	19.1 ± 0.5	24.0 ± 0.5*	0.76 ± 0.04	0.64 ± 0.05*

*$P < 0.05$.

rable weight. Long-term or repeated exposures did not impede mice from regaining weight off therapy. Mice regained at an accelerated rate, which resembled rebound weight gain after caloric restriction in healthy animals (42, 43). Body compositions of TNP-470- and leptin-treated mice were comparable following similar weight losses (Fig. 3). These observations argue against a toxic etiology. Instead, the data suggest TNP-470-induced weight reduction beginning with angiogenesis inhibition in adipose tissue, followed by subsequent effects on appetite, fuel utilization, and energy expenditure. The mechanism by which limiting vascular supply in fat affects these other parameters is presently unknown, but may be because of adipocyte-secreted factors.

Endothelium also mediates prostate growth (44) and supplies signals during organogenesis (45, 46). We propose angiogenesis mediators may be a means of regulating adipose tissue mass and that endothelium may have a role in regulating the mass of other adult tissues.

Generous gifts were provided by Takeda Chemical Company (TNP-470), Bayer Corporation (Bay 12-9566), and Entremed (thalidomide). Appreciation is extended to the veterinarian, Dr. Richard Hurley, for monitoring the health status of our animals during these studies. The valuable input of Dr. Rick Rodgers (Bio-Imaging Laboratory, Harvard School of Public Health) is also appreciated. Expertise in statistical analysis was provided by Elizabeth Allred (Children's Hospital Neuroepidemiology Unit, Boston). The technical assistances of Catherine Butterfield, Karen Keough, and Bin Shi are gratefully recognized. We also acknowledge photography services by Lori DeSantis and Kristin Gullage and graphic design by Advanced Medical Graphics (Boston). This study was supported in part by National Institutes of Health Grant K08 HL03698 (to M.A.R.).

1. Folkman, J. (1997) in *Cancer*, eds. DeVita, V., Jr., Hellman, S. & Rosenberg, S. (Lippincott, Philadelphia), pp. 3075–3085.
2. Rak, J., Filmus, J. & Kerbel, R. (1996) *Eur. J. Cancer* **32**, 2438–2450.
3. Boehm, T., Folkman, J., Browder, T. & O'Reilly, M. S. (1997) *Nature (London)* **390**, 404–407.
4. Folkman, J. (1998) *Endocrinology* **139**, 441–442.
5. Hobson, B. & Denekamp, J. (1984) *Br. J. Cancer* **49**, 405–413.
6. Crandall, D. L., Hausman, G. J. & Kral, J. G. (1997) *Microcirculation* **4**, 211–232.
7. Sierra-Honigmann, M. R., Nath, A. K., Murakami, C., Garcia-Cardena, G., Papapetropoulos, A., Sessa, W. C., Madge, L. A., Schechner, J. S., Schwabb, M. B., Polverini, P. J. & Flores-Riveros, J. R. (1998) *Science* **281**, 1683–1686.
8. Gersh, I. & Still, M. A. (1945) *J. Exp. Med.* **81**, 219–232.
9. Halaas, J. L., Boozer, C., Blair-West, J., Fidahusein, N., Denton, D. A. & Friedman, J. M. (1997) *Proc. Natl. Acad. Sci. USA* **94**, 8878–8883.
10. Zhang, Y., Proenca, R., Maffei, M., Barone, M., Leopold, L. & Friedman, J. M. (1994) *Nature (London)* **372**, 425–432.
11. Campfield, L. A., Smith, F. J. & Burn, P. (1996) *Horm. Metab. Res.* **28**, 619–632.
12. Carr, R. H., Ipaktchi, M. & Thenen, S. W. (1977) *Proc. Soc. Exp. Biol. Med.* **154**, 116–120.
13. Stein, L. J., Stellar, E., West, D. B., Greenwood, M. R., Foster, G. D., Feurer, I., Brown, J., Mullen, J. L. & Brownell, K. D. (1992) *Physiol. Behav.* **51**, 1–6.
14. Ingber, D., Fujita, T., Kishimoto, S., Sudo, K., Kanamaru, T., Brem, H. & Folkman, J. (1990) *Nature (London)* **348**, 555–557.
15. Wang, J., Lou, P. & Henkin, J. (2000) *J. Cell. Biochem.* **77**, 465–473.
16. Morita, T., Shinohara, N. & Tokue, A. (1994) *Br. J. Urol.* **74**, 416–421.
17. Kusaka, M., Sudo, K., Matsutani, E., Kozai, Y., Marui, S., Fujita, T., Ingber, D. & Folkman, J. (1994) *Br. J. Cancer* **69**, 212–216.
18. O'Reilly, M. S., Holmgren, L., Shing, Y., Chen, C., Rosenthal, R. A., Moses, M., Lane, W. S., Cao, Y., Sage, E. H. & Folkman, J. (1994) *Cell* **79**, 315–328.
19. O'Reilly, M. S., Boehm, T., Shing, Y., Fukai, N., Vasios, G., Lane, W. S., Flynn, E., Birkhead, J. R., Olsen, B. R. & Folkman, J. (1997) *Cell* **88**, 277–285.
20. Gatto, C., Rieppi, M., Borsotti, P., Innocenti, S., Ceruti, R., Drudis, T., Scanziani, E., Casazza, A. M., Taraboletti, G. & Giavazzi, R. (1999) *Clin. Cancer Res.* **5**, 3603–3607.
21. D'Amato, R. J., Loughnan, M. S., Flynn, E. & Folkman, J. (1994) *Proc. Natl. Acad. Sci. USA* **91**, 4082–4085.
22. Kuri-Harcuch, W. & Green, H. (1978) *Proc. Natl. Acad. Sci. USA* **75**, 6107–6109.
23. Coleman, D. L. & Hummel, K. P. (1973) *Diabetologia* **9**, 287–293.
24. Harkness, J. & Wagner, J. (1995) in *The Biology and Medicine of Rabbits and Rodents*, eds. Cahn, C. & Hunsberger, S. (Lea & Febiger, Philadelphia), p. 93.
25. Svendsen, O. L., Haarbo, J., Hassager, C. & Christiansen, C. (1993) *Am. J. Clin. Nutr.* **57**, 605–608.
26. Makan, S., Bayley, H. S. & Webber, C. E. (1997) *Can. J. Physiol. Pharmacol.* **75**, 1257–1261.
27. Scharrer, E. (1999) *Nutrition* **15**, 704–714.
28. Rohan, R. M., Fernandez, A., Udagawa, T., Yuan, J. & D'Amato, R. J. (2000) *FASEB J.* **14**, 871–876.
29. Silverman, K. J., Lund, D. P., Zetter, B. R., Lainey, L. L., Shahood, J. A., Freiman, D. G., Folkman, J. & Barger, A. C. (1988) *Biochem. Biophys. Res. Commun.* **153**, 347–352.
30. Castellot, J. J., Jr., Karnovsky, M. J. & Spiegelman, B. M. (1980) *Proc. Natl. Acad. Sci. USA* **77**, 6007–6011.
31. Dobson, D. E., Kambe, A., Block, E., Dion, T., Lu, H., Castellot, J. J., Jr., & Spiegelman, B. M. (1990) *Cell* **61**, 223–230.
32. Zhang, Q. X., Magovern, C. J., Mack, C. A., Budenbender, K. T., Ko, W. & Rosengart, T. K. (1997) *J. Surg. Res.* **67**, 147–154.
33. Lau, D. C., Schillabeer, G., Li, Z. H., Wong, K. L., Varzaneh, F. E. & Tough, S. C. (1996) *Int. J. Obes. Relat. Metab. Disord.* **20**, Suppl. 3, S16–S25.
34. Varzaneh, F. E., Shillabeer, G., Wong, K. L. & Lau, D. C. (1994) *Metabolism* **43**, 906–912.
35. Cohen, B., Barkan, D., Levy, Y., Goldberg, I., Fridman, E., Kopolovic, J. & Rubinstein, M. (2001) *J. Biol. Chem.* **276**, 7697–7700.
36. Qian, H., Azain, M. J., Compton, M. M., Hartzell, D. L., Hausman, G. J. & Baile, C. A. (1998) *Endocrinology* **139**, 791–794.
37. Hahnfeldt P., Panigrahy, D., Folkman, J. & Hlatky, L. (1999) *Cancer Res.* **59**, 4770–4775.
38. Grossman, B. M., Devore, M. L., Kelso, E. W. & Martin, R. J. (1997) *Physiol. Behav.* **61**, 169–173.
39. Yoshida, T., Kaneko, Y., Tsukamoto, A., Han, K., Ichinose, M. & Kimura, S. (1998) *Cancer Res.* **58**, 3751–3756.
40. Kruger, E. & Figg, W. (2000) *Expert Opin. Invest. Drugs* **9**, 1383–1396.
41. Kusaka, M., Sudo, K., Fujita, T., Marui, S., Itoh, F., Ingber, D. & Folkman, J. (1991) *Biochem. Biophys. Res. Commun.* **174**, 1070–1076.
42. Chlouverakis, C. (1970) *Experientia* **26**, 1262–1263.
43. Morimura, M., Ishiko, O., Sumi, T., Yoshida, H. & Ogita, S. (2001) *Int. J. Mol. Med.* **8**, 499–503.
44. Franck-Lissbrant, I., Haggstrom, S., Damber, J. E. & Bergh, A. (1998) *Endocrinology* **139**, 451–456.
45. Lammert, E., Cleaver, O. & Melton, D. (2001) *Science* **294**, 564–567.
46. Matsumoto, K., Yoshitomi, H., Rossant, J. & Zaret, K. S. (2001) *Science* **294**, 559–563.

nature biotechnology

ARTICLES

Oversulfated chondroitin sulfate is a contaminant in heparin associated with adverse clinical events

Marco Guerrini[1,7], Daniela Beccati[2,7], Zachary Shriver[2,3,7], Annamaria Naggi[1], Karthik Viswanathan[3], Antonella Bisio[1], Ishan Capila[2], Jonathan C Lansing[2], Sara Guglieri[1], Blair Fraser[4], Ali Al-Hakim[4], Nur Sibel Gunay[2], Zhenqing Zhang[5], Luke Robinson[3], Lucinda Buhse[4], Moheb Nasr[4], Janet Woodcock[4], Robert Langer[3,6], Ganesh Venkataraman[2,3], Robert J Linhardt[5], Benito Casu[1], Giangiacomo Torri[1] & Ram Sasisekharan[3]

Recently, certain lots of heparin have been associated with an acute, rapid onset of serious side effects indicative of an allergic-type reaction. To identify potential causes for this sudden rise in side effects, we examined lots of heparin that correlated with adverse events using orthogonal high-resolution analytical techniques. Through detailed structural analysis, the contaminant was found to contain a disaccharide repeat unit of glucuronic acid linked β1→3 to a β-*N*-acetylgalactosamine. The disaccharide unit has an unusual sulfation pattern and is sulfated at the 2-*O* and 3-*O* positions of the glucuronic acid as well as at the 4-*O* and 6-*O* positions of the galactosamine. Given the nature of this contaminant, traditional screening tests cannot differentiate between affected and unaffected lots. Our analysis suggests effective screening methods that can be used to determine whether or not heparin lots contain the contaminant reported here.

Heparin, a complex glycosaminoglycan polysaccharide, is widely used as an anticoagulant in a number of settings, including kidney dialysis and acute coronary syndromes[1–3]. The most serious adverse event associated with heparin, aside from a potential bleeding risk, is thrombocytopenia. Recently there has been a marked increase in serious adverse events associated with heparin therapy, with hundreds of individuals affected[4–7]. Although heparin therapy is generally well tolerated, recently patients presented—within several minutes after intravenous infusion of unfractionated heparin—with angioedema, hypotension, swelling of the larynx and related symptoms, which in some cases ended in death. Because heparin is a drug commonly used in the clinic, occurrence of these adverse events resulted in a crisis in the United States. Germany and other nations in the European Union have observed similar phenomena, turning this health problem into an international issue[4,5]. The rapid onset of these symptoms suggests an anaphylactic response, but the exact etiology is currently unknown. Given the clinical history of heparin, this spike in adverse events suggests the potential contamination of heparin. However, standard testing of heparin lots for typical biological contaminants, including protein, lipids and DNA (which, if present, may elicit such side effects), indicated that there is no difference in these regards between lots that elicit adverse events and those that do not[4]. Despite extensive analysis, no obvious differences were found with respect to other potential contaminants, including lead, dioxins and other molecular entities[8]. Definitive identification of how these heparin lots differ from the clinically approved heparin thus becomes imperative[9].

To understand the structure or structures of the contaminant(s) present within specific lots of heparin, we sought to identify these contaminants. This exercise required the use of multiple orthogonal techniques, including multidimensional NMR, to overcome the challenges inherent in the analysis of complex polysaccharides, including heparin, which in and of itself comprises a complex mixture of glycosaminoglycan chains. In doing so, we were able to determine definitively the structure of the contaminant, isolate it and confirm its structural identity by comparison to a chemically synthesized standard.

RESULTS

NMR shows unusual *N*-acetyl signals not seen in heparin

For this study, we examined lots that were associated with adverse clinical reactions (designated S1–S6) as well as four control lots of heparin not associated with adverse events (designated C1–C4). Initial analysis of S1–S6 by one-dimensional NMR indicated that all of these samples produced an unusual series of *N*-acetyl signals (**Fig. 1a**, **Supplementary Figs. 1** and **2** online). For example, particularly evident in the proton spectrum of S1 is the signal at 2.16 p.p.m. corresponding to an *N*-acetyl group different from that of heparin (2.04 p.p.m.). This *N*-acetyl signal is also distinct from that of

[1]Istituto di Ricerche Chimiche e Biochimiche "G. Ronzoni," Città Studi, via Giuseppe, Colombo 81, 20133 Milan, Italy. [2]Momenta Pharmaceuticals, Inc., 675 West Kendall Street, Cambridge, Massachusetts 02142, USA. [3]Department of Biological Engineering, Harvard-MIT Division of Health Sciences & Technology, Koch Institute for Integrative Cancer Research, Massachusetts Institute of Technology, 77 Massachusetts Ave., Cambridge, Massachusetts 02139, USA. [4]Center for Drug Evaluation and Research, US Food and Drug Administration, 10903 New Hampshire Ave., Silver Spring, Maryland 20993-0002, USA. [5]Center for Biotechnology and Interdisciplinary Studies, Rensselaer Polytechnic Institute, Troy, New York 12180, USA. [6]Department of Chemical Engineering, Massachusetts Institute of Technology, 77 Massachusetts Ave., Cambridge, Massachusetts 02139, USA. [7]These authors contributed equally to this work. Correspondence should be addressed to R.S. (rams@mit.edu).

Received 21 March; accepted 18 April; published online 23 April 2008; doi:10.1038/nbt1407

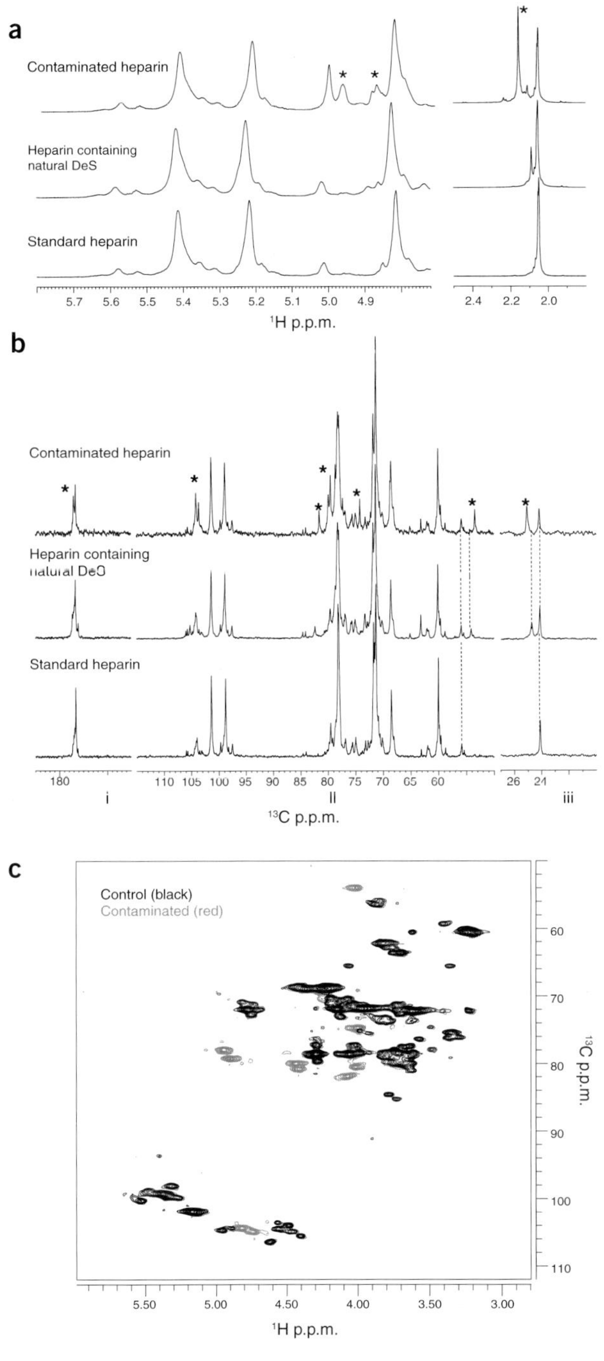

Figure 1 NMR analysis of contaminated heparin. (**a**) Comparison of anomeric and acetyl regions of the proton spectra of standard heparin, heparin containing natural dermatan sulfate (DeS) and contaminated heparin. (**b**) Comparison of carbonyl (i), sugar (ii) and *N*-acetyl regions (iii) of the carbon spectra of standard heparin, heparin containing natural dermatan sulfate, and contaminated heparin. Signals due to the contaminant are highlighted by asterisks. (**c**) HSQC spectrum of the contaminated sample S1 overlaid on control sample C1.

dermatan sulfate, a known impurity in heparin samples (2.08 p.p.m)[10]. To complement and extend the proton analysis, carbon (100 MHz) NMR spectroscopy was performed. Comparison of the carbon spectra of S1 and C1 indicates the presence of several additional signals not normally associated with heparin structural signatures (**Fig. 1b**). The acetyl signal at 25.6 p.p.m together with the signal at 53.5 p.p.m are indicative of the presence of an *O*-substituted *N*-acetylgalactosamine residue of unknown structure, but again distinct from the *N*-acetylgalactosamine contained within dermatan sulfate, with corresponding signals at 24.8 p.p.m and 54.1 p.p.m, respectively. Other signals are visible in the ring and anomeric regions of the carbohydrate moiety. The latter signals (103–105 p.p.m) agree with a beta configuration of glycosidic linkages for the contaminant.

To further identify the number and type of any major contaminants, we collected multidimensional heteronuclear single quantum coherence (HSQC) spectra on S1 and C1 to separate the observed signals in two dimensions—the carbon and proton signals. Ten major signals observed in sample S1 were not seen in C1. These same signals were observed in samples S2–S6 but not in samples C2–C4 (**Supplementary Figs. 3** and **4** online). In addition, these results and those from other two-dimensional experiments, including total correlation spectroscopy (TOCSY), correlation spectroscopy (COSY) and rotating frame nuclear Overhauser effect spectroscopy (ROSEY), were also consistent with the basic findings outlined above, namely that the principal contaminant consists of a polymeric repeat of *N*-acetylgalactosamine linked to glucuronic acid exclusively through beta linkages.

Identification of an impurity and a contaminant in heparin

Much of the analysis outlined above focused on sample S1, as it possessed signals in the *N*-acetyl region at 2.04 p.p.m (arising from heparin) and at 2.16 (arising from the unknown contaminant). In addition to this unusual signal at ~2.16 p.p.m, some samples (S2–S6 and C2–C4) possessed an additional *N*-acetyl peak (**Supplementary Fig. 5b** online), which had a chemical shift of 2.08 p.p.m, indicating the presence of another species that is distinct from the contaminant. Given the observed chemical shifts, this species is most likely dermatan sulfate, a known natural impurity of heparin[10,11]. To confirm the identity of this species, we performed a two-dimensional (2D) ^{1}H-^{13}C HSQC experiment on C2 (**Supplementary Fig. 5d**) and compared the results to those obtained on C1, a sample that did not show this additional signal. The chemical shifts observed in C2 but not in C1 are similar to those reported in the literature[10], and this observation suggests that the additional peaks in the HSQC of C2 can be assigned to dermatan sulfate. This assignment was confirmed by comparison to ^{1}H and HSQC data obtained on a standard of dermatan sulfate (**Supplementary Fig. 5a,c**). Through an analysis similar to that completed on sample C2, we confirmed the presence of dermatan sulfate in those samples presenting a proton NMR signal at 2.08 p.p.m. Therefore, samples S2–S6, but not sample S1, contained both dermatan sulfate (an impurity typically found in heparin) and the unusual contaminant.

Finally, to confirm the findings from the NMR analysis of the samples, we conducted enzymatic digestion of S1–S6 and C1–C4 with either a combination of heparinases or heparinases plus Δ4,5 glycuronidase and 2-*O* sulfatase followed by separation and analysis by high-performance liquid chromatography (HPLC). Digestion with the heparinases reduces heparin to its component di-, tri- and tetrasaccharides and imparts a Δ4,5 bond that can be monitored at 232 nm; completed in conjunction with treatment with glycuronidase and sulfatase, this digest permits the identification of minor heparin species, including those disaccharides containing a modified galacturonic acid[12,13]. Thus, concomitant use of a matrix of enzymes, especially in conjunction with liquid chromatography–tandem mass spectrometry analysis, allows for the complete separation, identification and quantification of heparin components in the mixture[14]. We find that the total areas under the curve, calculated by integration of all heparin di-, tri- and tetrasaccharides observed in the HPLC and summation of the resulting areas, of digested S1–S6 are substantially less than those of the control samples, indicating that the major contaminant is not substantially digested by the heparinases (**Table 1**). Furthermore, the difference in the area under the curve is correlated with the amount of contaminant present within the sample; samples with larger amounts of contaminant (as measured by proton NMR) had a lower area under the curve as measured by HPLC. Finally, the relative quantities of the individual heparin components are similar between the controls and suspect samples, with only minor differences.

To identify the unknown compound present in the contaminated heparin samples, we attempted to isolate the contaminant using a variety of methods. Given the overall properties of the contaminant, elucidated through NMR, capillary electrophoresis and HPLC analysis, we reasoned that this material could be differentially precipitated upon addition of an organic solvent. Partial purification of the contaminant was indeed achieved through the addition of increasing amounts of ethanol to an aqueous solution of S1. Similarly, because the contaminant contains *N*-acetylhexosamine (and not *N*-sulfohexosamine), it was also purified by degradation of heparin by nitrous acid and isolation of the remaining components.

Table 1 Total area under the curve for heparinase digests of S1–S6 and C1–C4

Sample	Total area
C1	2.52×10^7
C2	2.70×10^7
C3	2.33×10^7
C4	2.39×10^7
S1	1.74×10^7
S2	9.50×10^6
S3	1.94×10^7
S4	1.59×10^7
S5	1.91×10^7
S6	1.96×10^7

Isolation of the contaminant allowed definitive identification

After isolation, the proton spectrum of the isolated contaminant reveals a residual heparin content of ~10–30% (depending on the isolation method used) as determined by one-dimensional NMR (data not shown). We carried out additional, detailed NMR studies on this isolated sample. To facilitate an understanding of the experiments and their interpretation, the disaccharide repeat structures of heparin, chondroitin sulfate, dermatan sulfate and heparin, together with positional nomenclature, are presented in **Figure 2**. In the carbon spectrum, at neutral pH values, one apparent signal arises at 177.6 p.p.m, characteristic of carbonyl groups. Acidification of a solution of the product from pH 6.5 to pH 4 (**Fig. 3a**) reveals two distinct carbonyl signals consistent with the carbonyl group of an *N*-acetyl function and the protonated form of carboxylic acid, respectively. Similar shifts were not observed for any other signals except C-5 of uronic acid (U5), for which this carbon's chemical shift is sensitive to the ionization state of the carboxylic acid as well as the identity of the counterion present (**Fig. 3b**). To further characterize the isolated sample, we used homonuclear (COSY and TOCSY, data not shown) and heteronuclear (HSQC, HSQC-TOCSY and heteronuclear multiple bond correlation (HMBC)) 2D-NMR spectroscopy (**Fig. 3c,d**). These analyses indicated the presence of two types of residues. Chemical shift patterns were in agreement with one type of monosaccharide being a 4,6-*O*-sulfo-*N*-acetylgalactosamine and the other being a 2,3-*O*-sulfoglucuronic acid. In addition to confirming the assignments of the sugar moieties, the HMBC spectrum demonstrated the correlation across the glycosidic linkages, indicating the presence of a β-1,4 linkage between galactosamine and glucuronic acid and a β-1,3 linkage between the glucuronic acid and the galactosamine. Confirmatory evidence of this structure was provided by the long-range correlations of H1 of the galactosamine and H5 of glucuronic acid with two different carbonyl groups (177.5 and 177.8 p.p.m) owing to the *N*-acetyl of galactosamine and the carboxylic group of glucuronic acid.

Synthetic oversulfated chondroitin matches contaminant

The identification of the contaminant as an oversulfated chondroitin sulfate containing a tetrasulfated disaccharide unit consisting of glucuronic acid linked to *N*-acetylgalactosamine was surprising, as this is an unusual structure. To ensure the accuracy of this assignment, we prepared a standard by oversulfation of chondroitin sulfate using well-established chemistry. We analyzed this

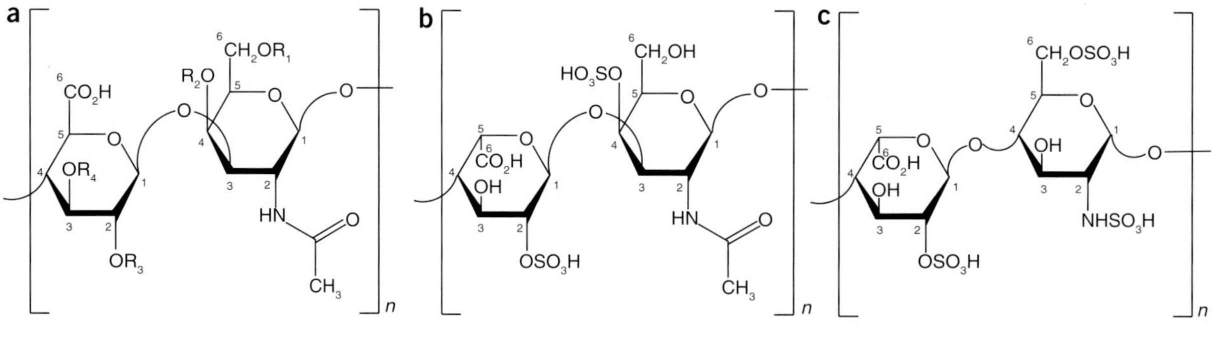

Figure 2 Chemical structures of major repeat units. (**a–c**) Schematic of major disaccharide repeat units of chondroitin sulfate (**a**), dermatan sulfate (**b**) and heparin (**c**). The carbon atoms are numbered for each monosaccharide of the disaccharide repeat. For chondroitin sulfate, R1-R4 can be either sulfated or unsubstituted.

standard by 2D NMR and carefully compared the HSQC spectra to both the literature[15] and the HSQC spectrum obtained for the isolated contaminant. Comparison of the HSQC spectrum of the synthesized standard with that of the isolated contaminant (**Fig. 4**) confirmed that the major contaminant consists of per-*O*-sulfated chondroitin sulfate, with all of the hydroxyl groups of both the uronic acid and galactosamine residues bearing sulfate substituents. Furthermore, the proton chemical shifts of the contaminant (**Table 2**) are in agreement with those assigned to fully sulfated chondroitin (degree of sulfation = 4)[15]. In addition, mass spectrometry completed after enzymatic digestion of a chemically desulfated[16] version of the contaminant was consistent with the proposed structure (**Fig. 5**). The final derived structure of the major contaminant present in heparin is shown in **Figure 2a**, with R_1, R_2, R_3 and R_4 all sulfated.

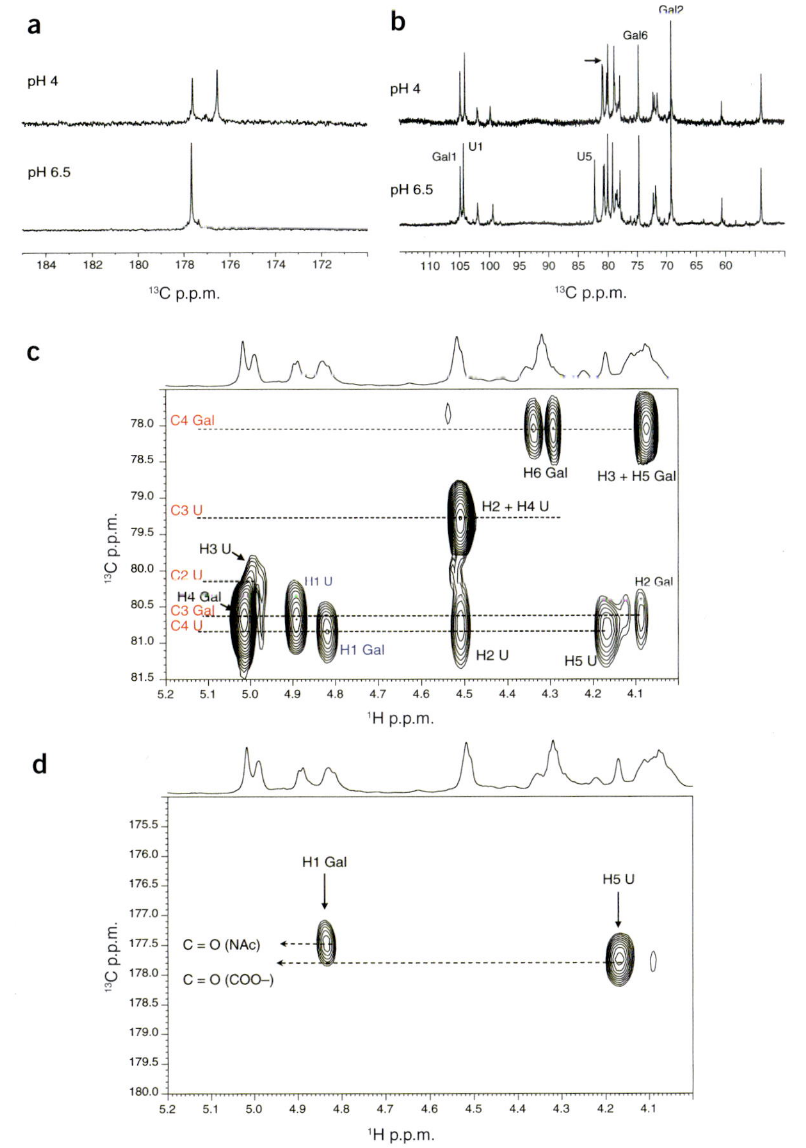

Figure 3 Detailed NMR structural analysis of the contaminant. (**a**) Carbonyl region of the carbon spectrum of the contaminant measured at pH 6.5 and 4. (**b**) Sugar region of the carbon spectrum of the contaminant measured at pH 6.5 and 4. The arrow points to the shift in the C5 of glucuronic acid (U5) upon pH adjustment. (**c**) A portion of the 600-MHz HMBC spectrum of the contaminant. Intramolecular two- and three-bond proton-carbon correlations are shown in black; interglycosidic proton-carbon correlations are indicated in blue; carbon connectivity shown in red. (**d**) Portion of the 600-MHz HMBC spectrum of the contaminant. Long-range correlation between the H1 of *N*-acetylgalactosamine (Gal) with the C=O of the acetyl group and H5 of glucuronic acid (U) with the carboxylic group are shown.

DISCUSSION

Several factors required the use of multiple approaches to ensure an accurate structural determination of the oversulfated chondroitin contaminant. First, heparin is a chemically complex polydisperse mixture of saccharides, making careful interpretation of results necessary to avoid misinterpretation. Detailed analysis is especially important when addressing the multiple isomeric possibilities within the chains of a complex mixture, and it necessitates the use of orthogonal techniques, including an enzyme matrix and multidimensional NMR. Second, oversulfated materials, such as the chondroitins, are resistant to enzymatic digestion techniques and copurify with heparin, rendering their isolation challenging[15,17].

These complexities manifest themselves in a number of ways: for example, structurally distinct species may have overlapping signals and properties, thereby 'masking' them within a single analysis. A nonintegrated approach can therefore potentially lead to false conclusions, especially when attempting to differentiate between heparin, dermatan sulfate and oversulfated chondroitin sulfate in any given sample. In this study, we present a set of experimental techniques that discriminate between dermatan sulfate (a known impurity of heparin) and oversulfated chondroitin sulfate.

The structure of the contaminant, which contains a tetrasulfated disaccharide repeat, is highly unusual. First, the presence of a 3-*O*-sulfated glucuronic acid is rare, only observed in specific contexts within certain organisms[18]. In addition, a tetrasulfated disaccharide repeat unit has not been isolated to date from animal tissues. Consequently, it is highly unlikely that the contaminant reported here is produced naturally. Finally, chemically synthesized tetrasulfated disaccharide repeat units of chondroitin sulfate are known to exhibit a high degree of anti–factor IIa activity[15], which could explain how contaminated heparin would pass an activity screen, such as a whole-blood coagulation test. Further investigation is warranted to understand how this contaminant was introduced.

With respect to the potential biological ramifications of this finding, it is possible that the presence of an oversulfated chondroitin sulfate within heparin preparations

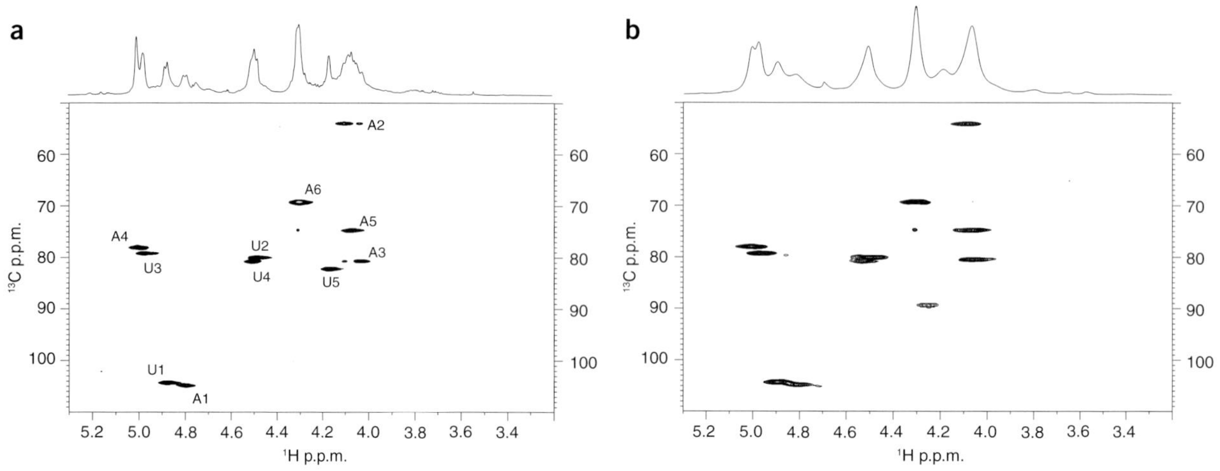

Figure 4 Structural assignment of the contaminant by NMR using isolated and synthetic materials. (**a**) ^{1}H-NMR and HSQC spectra of the isolated contaminant. Signals for 4,6-*O*-sulfo-*N*-acetylgalactosamine (A) and of 2,3-*O*-sulfo-glucuronic acid (U) are labeled. (**b**) ^{1}H-NMR and HSQC spectra of the chemically synthesized oversulfated chondroitin.

can provoke increased side effects. In the case of chondroitins, chondroitin sulfates A and C are expressed by human cells and are generally nonimmunogenic. Highly sulfated polysaccharides, however, such as oversulfated chondroitins, have been shown to be potent mediators of the immune response[19–22]. Indeed, complications associated with administration of highly sulfated chondroitins have been observed in humans. Arteparon, an oversulfated chondroitin that is structurally identical to the contaminant[23] (**Fig. 2**), injected intramuscularly in humans[24], was marketed for the treatment of degenerative joint disease in Europe. It was demonstrated that this product can induce an allergic-type response[25]. Because of patient deaths, most likely due to thromboembolic complications, the product was rapidly withdrawn from the European market[26]. The structural determination of the contaminant described here enables further investigation into the biological roles and potential pharmacological effects of oversulfated chondroitins, present within heparin preparations, in the recently reported adverse events.

Taken together, our orthogonal analytical experiments provide strong support for the structure we have assigned to the contaminant. This study also provides a set of screening methods that could be used to monitor the heparin supply and ensure the absence of oversulfated chondroitin sulfate contamination. For example, using the structural information presented here, it is now possible to (i) design reference standards that ensure accuracy, quantification and specificity of analysis for a given analytical method and (ii) devise an experimental protocol to clearly define the nature and extent, if any,

Table 2 Chemical shifts observed for contaminant by two independent laboratories and comparison to literature data on oversulfated chondroitin sulfate

Monosaccharide	Lab 1[a]		Lab 2[b]		^{1}H (literature)[c], fully *O*-sulfonated CS
	^{1}H (observed)	^{13}C (observed)	^{1}H (observed)	^{13}C (observed)	
4,6-*O*-Sulfo-*N*-acetyl-galactosamine					
H1/C1	4.77	105.0	4.79	104.8	4.86
H2/C2	4.06	54.0	4.10	54.0	4.10
H3/C3	4.05	80.7	4.03	80.7	4.10
H4/C4	4.98	78.0	5.00	78.0	5.02
H5/C5	4.05	74.8	4.07	74.7	4.06
H6,6'/C6	4.28	69.3	4.30	69.2	4.29
N-Acetyl	2.12[d]	25.6	2.16[e]	25.6	2.16
2,3-*O*-Sulfo-glucuronic acid					
H1/C1	4.87	104.5	4.87	104.3	4.97
H2/C2	4.47	80.0	4.49	80.0	4.53
H3/C3	4.95	79.3	4.98	79.2	4.94
H4/C4	4.46	80.9	4.51	80.8	4.55
H5/C5	4.12	82.0	4.17	82.2	4.20

[a]Chemical shifts are measured at 303 K and referenced to external 2,2-dimethyl-2-silapentane-5-sulfonate sodium salt (DSS). [b]Chemical shifts are measured at 308 K and referenced to external sodium trimethylsilylpropionate (TSP) (which resonates 0.12 p.p.m upfield of DSS[29]). [c]Measured at 303 K and referenced to TSP (ref. 15). CS, chondroitin sulfate. [d]Additional minor signals observed at 2.20 and 2.07 p.p.m. [e]Additional minor signals observed at 2.23 and 2.11 p.p.m.

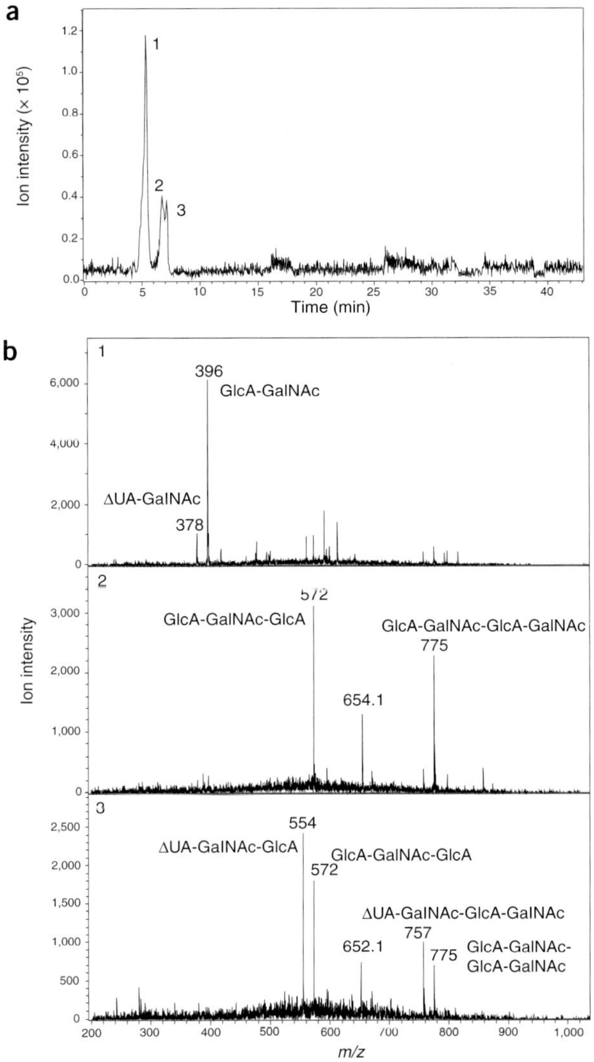

Figure 5 Liquid chromatography–mass spectrometry (LC-MS) analysis of the contaminant. (**a**) Total ion chromatogram with detection of three peaks (1–3). (**b**) Specific ions and associated structures observed in each peak. After desulfation, the contaminant was digested with chondroitinase ABC and analyzed by LC-MS. The presence of saturated and unsaturated oligosaccharides indicates that the backbone of the impurity could be digested with chondroitinase ABC and suggests that the backbone is a glucuronic acid–*N*-acetylgalactosamine repeat (UA-GalNAc).

of contamination in a given lot of heparin. Finally, the ramifications of these findings extend beyond scientific considerations and include clinical and health policy issues.

METHODS

Materials. Chondroitin sulfate type A from whale cartilage, pyridine–sulfur trioxide complex, tributylamine, dry *N,N*-dimethylformamide, pyridine, methanol, dimethylsulfoxide, $NaNO_2$, $NaBH_4$, 2,2-dimethyl-2-silapentane-5-sulfonate sodium salt and sodium trimethylsilylpropionate were purchased from Sigma-Aldrich. Tetrabutylammonium chloride was purchased from Fluka. 99.9% D_2O and 99.96% D_2O were obtained from Cambridge Isotope laboratories. The following ten lots of unfractionated heparin were tested: S1, S2, S3, S4, S5, S6, C1, C2, C3 and C4. All heparin samples were supplied by the FDA.

NMR analysis. Samples for ^{1}H-NMR or 2D NMR analysis were dissolved in 0.7 ml of D_2O (99.9%) and were either freeze-dried repeatedly to remove exchangeable protons or directly measured without any treatment to preserve potential volatile components. The thoroughly dried samples were dissolved in 0.7 ml of D_2O (99.96%). Before spectrum acquisition, samples were sonicated for 60 s to remove air bubbles. Spectra were obtained at 303 K or 308 K using a 600-MHz Varian VNMRS spectrometer or a 600-MHz Bruker Avance 600 spectrometer, both equipped with a 5-mm triple-resonance inverse cryoprobe. Monodimensional ^{1}H spectra were obtained with presaturation of residual HOD, for 32–128 scans. DQF-COSY and 2D-TOCSY spectra were acquired using 32 scans per series of 2,048 × 512 data points. HSQC and HSQC-TOCSY spectra were recorded with carbon decoupling during acquisition with 320 increments for 12–64 scans. The polarization transfer delay was set with a $^1J_{C\text{-}H}$ coupling value of 155 Hz. HMBC spectrum was recorded with 320 increments of 64 scans for each, without carbon decoupling and with twofold low-pass J-filter to suppress one-bond correlations[27]. The delay for evolution of long-range couplings was set with a J_{lr} of 8 Hz. Samples for ^{13}C-NMR analysis were dissolved in D_2O (99.9%) at 40 mg/ml and analysis was performed at 303 K with a 400-MHz Bruker spectrometer equipped with a multinuclear probe.

Composition analysis by ion-pair RPHPLC. Samples were constituted in capillary electrophoresis–grade water and digested under two different conditions. The first digest used an enzyme cocktail of heparinases I (500 mIU), II (400 mIU) and III (500 mIU) at 30 °C, for 16 h. A portion of this digest was further treated with 2-*O*-sulfatase (1,000 mIU) and Δ4,5-glycuronidase (2,000 mIU) for 6 h at 30 °C to obtain the second digest. Each digest was passed through a Ni^{2+} spin column (Qiagen) and analyzed by ion-pair RPHPLC similar to as previously described[14]. Elution was monitored by UV detection at 232 nm.

Isolation of major contaminant. Absolute ethanol was added to sample S1 (300 mg), dissolved in 1.5 ml of water until a white precipitate appeared (ethanol 23%, vol/vol). Precipitated material was separated by centrifugation for 5 min at 5,000 r.p.m. (~4,000g) on a Labofuge 200 (Heraeus). In addition to the above, sample S1 was treated with nitrous acid[28]. A solution of the sample (500 mg) was dissolved in 20 ml of H_2O and cooled at 4 °C. Also at 4 °C, 140 mg of $NaNO_2$ dissolved in 1 ml of water was added to the sample and the pH was adjusted to 1.7 with 0.1 M HCl. The solution was stirred at 4 °C for 20 min; an additional 100 mg $NaNO_2$ was added and the solution stirred for another 20 min. The solution pH was then adjusted to 7 with NaOH and the resulting reaction was brought to room temperature. Solid $NaBH_4$ (200 mg) was added in several portions with stirring. After 2 h, the pH was adjusted to 4 with 0.1 M HCl and the solution was neutralized with 0.1 M NaOH. The product, obtained by precipitation with four volumes of methanol, was recovered by centrifugation, dissolved in water and freeze-dried.

Chemical sulfonation of chondroitin sulfate. Fully sulfated chondroitin sulfate was prepared from chondroitin sulfate, as described[15]. Chondroitin sulfate (108 mg) was converted into its tributylamine salt and dissolved in dry *N,N*-dimethylformamide (1 ml). After addition of 159 mg of pyridine–sulfur trioxide complex, the solution was heated for 1 h at 40 °C . The reaction was interrupted by addition of 2 ml of water and the product was precipitated at 4 °C by addition of 35 ml of an ethanol solution saturated with sodium acetate. The product recovered by centrifugation was dissolved in water, dialyzed at room temperature and recovered by freeze-drying.

Note: Supplementary information is available on the Nature Biotechnology website.

ACKNOWLEDGMENTS

The authors would like to thank Sucharita Roy for work on the chemical synthesis of oversulfated chondroitin sulfate standards, Scott Bailey for assistance in the analysis of selected samples by one- and two-dimensional NMR and Laura Citterio for work in various separation and analytical steps. We also thank James Anderson, Josh Sorafine and Andre Jones for design of composition experiments and HPLC analysis of the heparinase-digested samples. Finally, we thank Ada Ziolkowski for help with manuscript preparation. This work was supported in part by US National Institute of General Medical Sciences grants GM57073 (R.S.) and GM38060 (R.J.L).

COMPETING INTERESTS STATEMENT
The authors declare competing financial interests: details accompany the full-text HTML version of the paper at http://www.nature.com/naturebiotechnology/.

Published online at http://www.nature.com/naturebiotechnology/
Reprints and permissions information is available online at http://npg.nature.com/reprintsandpermissions/

1. Capila, I. & Linhardt, R.J. Heparin-protein interactions. *Angew. Chem. Int. Edn. Engl.* **41**, 391–412 (2002).
2. Lepor, N.E. Anticoagulation for acute coronary syndromes: from heparin to direct thrombin inhibitors. *Rev. Cardiovasc. Med.* **8** (suppl. 3), S9–S17 (2007).
3. Fischer, K.G. Essentials of anticoagulation in hemodialysis. *Hemodial. Int.* **11**, 178–189 (2007).
4. Contaminant detected in heparin material of specified origin in the USA and in Germany; serious adverse events reported; recall measures initiated. World Health Organization Alert No. 118 (7 March 2008). <http://www.who.int/medicines/publications/drugalerts/Alert_118_Heparin.pdf>
5. Notice of Recall from Rotexmedica to Bfarm (German Regulatory Authorities). Rotexmedica/Bfarm Notice (7 March 2008). <http://www.akdae.de/20/40/Archiv/2008/20080310.pdf>
6. Acute allergic-type reactions among patients undergoing hemodialysis—multiple states, 2007–2008. *Morbid. Mortal. Wkly Rept.* **57** (1 February 2008).
7. 2008 Heparin Recall Information. Baxter Investigation Updates (5,14,19 March 2008). <http://www.baxter.com/products/biopharmaceuticals/heparin.html>
8. Baxter provides update on investigation. Baxter Investigation Update (14 March 2008). <http://www.baxter.com/products/biopharmaceuticals/downloads/heparin_03-14-08.pdf>
9. Communication. Information on heparin sodium injection. US Food and Drug Administration. <http://www.fda.gov/cder/drug/infopage/heparin/default.htm>
10. Holme, K.R. & Perlin, A.S. Nuclear magnetic resonance spectra of heparin in admixture with dermatan sulfate and other glycosaminoglycans. 2-D spectra of the chondroitin sulfates. *Carbohydr. Res.* **186**, 301–312 (1989).
11. Guerrini, M., Bisio, A. & Torri, G. Combined quantitative ^{1}H and ^{13}C nuclear magnetic resonance spectroscopy for characterization of heparin preparations. *Semin. Thromb. Hemost.* **27**, 473–482 (2001).
12. Myette, J.R. *et al.* Molecular cloning of the heparin/heparan sulfate delta 4,5 unsaturated glycuronidase from *Flavobacterium heparinum*, its recombinant expression in *Escherichia coli*, and biochemical determination of its unique substrate specificity. *Biochemistry* **41**, 7424–7434 (2002).
13. Myette, J.R. *et al.* The heparin/heparan sulfate 2-*O*-sulfatase from *Flavobacterium heparinum*. Molecular cloning, recombinant expression, and biochemical characterization. *J. Biol. Chem.* **278**, 12157–12166 (2003).
14. Kuberan, B. *et al.* Analysis of heparan sulfate oligosaccharides with ion pair-reverse phase capillary high performance liquid chromatography-microelectrospray ionization time-of-flight mass spectrometry. *J. Am. Chem. Soc.* **124**, 8707–8718 (2002).
15. Maruyama, T., Toida, T., Imanari, T., Yu, G. & Linhardt, R.J. Conformational changes and anticoagulant activity of chondroitin sulfate following its *O*-sulfonation. *Carbohydr. Res.* **306**, 35–43 (1998).
16. Solvolytic desulfation of glycosaminoglycuronan sulfates with dimethyl sulfoxide containing water or methanol. Nagasawa, K., Inoue, Y. & Kamata, T. *Carbohydr. Res.* **58**, 47–55 (1977).
17. Linhardt, R.J. Analysis of glycosaminoglycans with polysaccharide lyases. in *Current Protocols in Molecular Biology* (eds. Ausubel, F.A. *et al.*) 17.13B.1–17B.13.16 (Wiley, New York, 1994).
18. Kinoshita, A. *et al.* Novel tetrasaccharides isolated from squid cartilage chondroitin sulfate E contain unusual sulfated disaccharide units GlcA(3-*O*-sulfate)β1-3GalNAc(6-*O*-sulfate) or GlcA(3-O-sulfate)β1-3GalNAc. *J. Biol. Chem.* **272**, 19656–19665 (1997).
19. Hadding, U. *et al.* Ability of the T cell-replacing polyanion dextran sulfate to trigger the alternate pathway of complement activation. *Eur. J. Immunol.* **3**, 527–529 (1973).
20. Kang, O.H. *et al.* Suppressive effect of non-anaphylactogenic anti-IgE antibody on the development of dextran sulfate sodium-induced colitis. *Int. J. Mol. Med.* **18**, 893–899 (2006).
21. Siebeck, M. *et al.* Dextran sulfate activates contact system and mediates arterial hypotension via B2 kinin receptors. *J. Appl. Physiol.* **77**, 2675–2680 (1994).
22. Hojima, Y., Cochrane, C.G., Wiggins, H.C., Austen, K.F. & Steven, R.L. In vitro activation of the contact (Hageman factor) system of plasma by heparin and chondroitin sulfate E. *Blood* **63**, 1453–1459 (1984).
23. Naggi, A. *et al.* Sulfamino-galactosaminoglycans, a new class of semi-synthetic polysaccharides. in *Biomedical and Biotechnological Advances in Industrial Polysaccharides* (eds. Crescenzi, V., Dea, I.C.M. & Paoletti, S.) 101–108 (Gordon & Breach Science, New York, 1989).
24. Siegmenth, W. & Radi, I. Comparison of glycosaminoglycan polysulfate (Arteparon) and physiological saline solution in arthrosis of the large joints. Results of a multicenter double-blind study. *Z. Rheumatol.* **42**, 223–228 (1983).
25. Greinacher, A., Michels, I., Schäfer, M., Kiefel, V. & Mueller-Eckhardt, C. Heparin-associated thrombocytopenia in a patient treated with polysulphated chondroitin sulphate: evidence for immunological crossreactivity between heparin and polysulphated glycosaminoglycan. *Br. J. Haematol.* **81**, 252–254 (1992).
26. Weimann, G. *et al.* Glucosamine sulfate does not crossreact with the antibodies of patients with heparin-induced thrombocytopenia. *Eur. J. Haematol.* **66**, 195–199 (2001).
27. Cicero, O., Barbato, G. & Bazzo, R. Sensitivity enhancement of a two-dimensional experiment for the measurement of heteronuclear long-range coupling constants, by a new scheme of coherence selection by gradients. *J. Magn. Reson.* **148**, 209–213 (2001).
28. Cifonelli, J.C. Reaction of heparitin sulfate with nitrous acid. *Carbohydr. Res.* **8**, 233–242 (1968).
29. Wishart, D.S. *et al.* ^{1}H, ^{13}C and ^{15}N chemical shift referencing in biomolecular NMR. *J. Biomol. NMR* **6**, 135–140 (1995).

The NEW ENGLAND JOURNAL of MEDICINE

ORIGINAL ARTICLE

Contaminated Heparin Associated with Adverse Clinical Events and Activation of the Contact System

Takashi Kei Kishimoto, Ph.D., Karthik Viswanathan, Ph.D., Tanmoy Ganguly, Ph.D., Subbiah Elankumaran, Ph.D., Sean Smith, B.S., Kevin Pelzer, Ph.D., Jonathan C. Lansing, Ph.D., Nammalwar Sriranganathan, Ph.D., Ganlin Zhao, M.D., Zoya Galcheva-Gargova, Ph.D., Ali Al-Hakim, Ph.D., Gregory Scott Bailey, B.S., Blair Fraser, Ph.D., Sucharita Roy, Ph.D., Thomas Rogers-Cotrone, M.S., Lucinda Buhse, Ph.D., Mark Whary, Ph.D., James Fox, Ph.D., Moheb Nasr, Ph.D., Gerald J. Dal Pan, M.D., Zachary Shriver, Ph.D., Robert S. Langer, Sc.D., Ganesh Venkataraman, Ph.D., K. Frank Austen, M.D., Janet Woodcock, M.D., and Ram Sasisekharan, Ph.D.

ABSTRACT

BACKGROUND

There is an urgent need to determine whether oversulfated chondroitin sulfate (OSCS), a compound contaminating heparin supplies worldwide, is the cause of the severe anaphylactoid reactions that have occurred after intravenous heparin administration in the United States and Germany.

METHODS

Heparin procured from the Food and Drug Administration, consisting of suspect lots of heparin associated with the clinical events as well as control lots of heparin, were screened in a blinded fashion both for the presence of OSCS and for any biologic activity that could potentially link the contaminant to the observed clinical adverse events. In vitro assays for the activation of the contact system and the complement cascade were performed. In addition, the ability of OSCS to recapitulate key clinical manifestations in vivo was tested in swine.

RESULTS

The OSCS found in contaminated lots of unfractionated heparin, as well as a synthetically generated OSCS reference standard, directly activated the kinin–kallikrein pathway in human plasma, which can lead to the generation of bradykinin, a potent vasoactive mediator. In addition, OSCS induced generation of C3a and C5a, potent anaphylatoxins derived from complement proteins. Activation of these two pathways was unexpectedly linked and dependent on fluid-phase activation of factor XII. Screening of plasma samples from various species indicated that swine and humans are sensitive to the effects of OSCS in a similar manner. OSCS-containing heparin and synthetically derived OSCS induced hypotension associated with kallikrein activation when administered by intravenous infusion in swine.

CONCLUSIONS

Our results provide a scientific rationale for a potential biologic link between the presence of OSCS in suspect lots of heparin and the observed clinical adverse events. An assay to assess the amidolytic activity of kallikrein can supplement analytic tests to protect the heparin supply chain by screening for OSCS and other highly sulfated polysaccharide contaminants of heparin that can activate the contact system.

From Momenta Pharmaceuticals (T.K.K., T.G., S.S., J.C.L., G.Z., Z.G.-G., G.S.B., S.R., Z.S., G.V.), the Harvard–Massachusetts Institute of Technology Division of Health Sciences and Technology, Koch Institute for Integrative Cancer Research (K.V., Z.S., R.S.L., G.V., R.S.), and the Massachusetts Institute of Technology (M.W., J.F.) — all in Cambridge, MA; Virginia–Maryland Regional College of Veterinary Medicine, Virginia Polytechnic Institute and State University, Blacksburg (S.E., K.P., N.S., T.R.-C.); the Center for Drug Evaluation and Research, Food and Drug Administration, Silver Spring, MD (A.A.-H., B.F., L.B., M.N., G.J.D.P., J.W.); and Brigham and Women's Hospital and Harvard Medical School, Boston (K.F.A.). Address reprint requests to Dr. Sasisekharan at the Department of Biological Engineering, Harvard–MIT Division of Health Sciences and Technology, Koch Institute for Integrative Cancer Research, Massachusetts Institute of Technology, 77 Massachusetts Ave., 16-561, Cambridge, MA 02139, or at rams@mit.edu.

This article (10.1056/NEJMoa0803200) was published at www.nejm.org on April 23, 2008.

N Engl J Med 2008;358:2457-67.

In January 2008, health authorities in the United States began receiving reports of clusters of acute hypersensitivity reactions in patients undergoing dialysis that had been occurring since November 2007. Symptoms included hypotension, facial swelling, tachycardia, urticaria, and nausea. Although initial investigations focused on dialysis equipment, an investigation by the Centers for Disease Control and Prevention identified the receipt of heparin sodium for injection (1000 U per milliliter, in 10-ml and 30-ml multidose vials), manufactured by Baxter Healthcare, as a common feature of the cases.[1] This finding led Baxter Healthcare to recall, on January 17, 2008, nine lots of heparin sodium for injection. As of April 13, 2008, there were 81 reports of death that involved at least one sign or symptom of an allergic reaction or hypotension in patients receiving heparin since January 1, 2007. However, the fact that allergic symptoms or hypotension were reported does not mean that these were the cause of death in all cases.

After this initial recall, there were continuing reports of allergic-type reactions, including some deaths, after injection of bolus heparin not only in patients undergoing dialysis but also in patients in other clinical settings, such as those undergoing cardiac procedures. On February 28, 2008, Baxter Healthcare recalled all remaining lots and doses of its multidose and single-dose vials of heparin sodium for injection and HEP-LOCK heparin flush products. Since that recall, monitoring by the Food and Drug Administration (FDA) has indicated that, in March 2008, the number of deaths reported in association with heparin usage had returned to baseline levels.[2]

However, on March 6, a heparin recall was announced in Germany because of a cluster of reactions in patients undergoing dialysis that were linked to a different manufacturer's heparin. On the same day, the FDA posted descriptions of analytic methods on its Web site and recommended that all manufacturers and regulatory authorities screen for a contaminant in heparin.[3] This screening revealed widespread contamination of the heparin supply in at least 12 countries.

The contaminant was recently identified as an unusual oversulfated form of chondroitin sulfate (OSCS) representing up to approximately 30% wt/wt in suspect lots of heparin; no other contaminants were observed.[4] In addition, dermatan sulfate, a known impurity of heparin, was found in selected samples. Both heparin and chondroitin sulfate are members of the glycosaminoglycan family of complex polysaccharides; heparin contains a disaccharide repeat unit of glucuronic–iduronic acid linked to glucosamine, and chondroitin sulfate contains a disaccharide repeat unit of glucuronic acid linked to galactosamine. Analysis of the contaminant unexpectedly revealed an unusual type of sulfation not found in any natural sources of chondroitin sulfate and indicated that OSCS, containing four sulfates per disaccharide unit, is structurally similar to heparin (see the Supplementary Appendix, available with the full text of this article at www.nejm.org).

However, the biologic link between the presence of the OSCS in heparin and the adverse clinical events remained to be established. Highly charged polyanionic polymers are known to modulate various enzymatic cascades in plasma, affecting coagulation, fibrinolysis, inflammation, and vasculature function.[5,6] Bradykinin, a potent vasoactive peptide mediator, is generated through the activation of the contact system of coagulation, which is initiated upon contact of factor XII with a negatively charged surface in the presence of prekallikrein and high-molecular-weight kininogen. Highly sulfated polysaccharides have been shown to serve as a negatively charged surface that can initiate fluid-phase activation of the contact system.[5,7] However, initial attempts to recapitulate the adverse responses in experimental models were unsuccessful.[8] Without a definitive link between the contaminant and the clinical reactions, concerns remain that the screening tests currently in place may not be adequate to prevent further cases. We therefore set out to identify a biologic basis for a link between OSCS and allergic-type reactions.

CASE REPORT

A representative case involved a 63-year-old woman with a complex medical history, including end-stage renal disease treated with the use of hemodialysis for 7 years, who received heparin intravenously during hemodialysis (5000-U loading dose and 500 U per hour during the procedure) three times weekly. In mid-January 2008, the development of "low blood pressure" was reported, along with nausea and dyspnea, during dialysis. She was treated with normal saline and oxygen (2 liters per minute), and the rates of ultrafiltration and blood flow were slowed. She recovered after 30 minutes, and dialysis was contin-

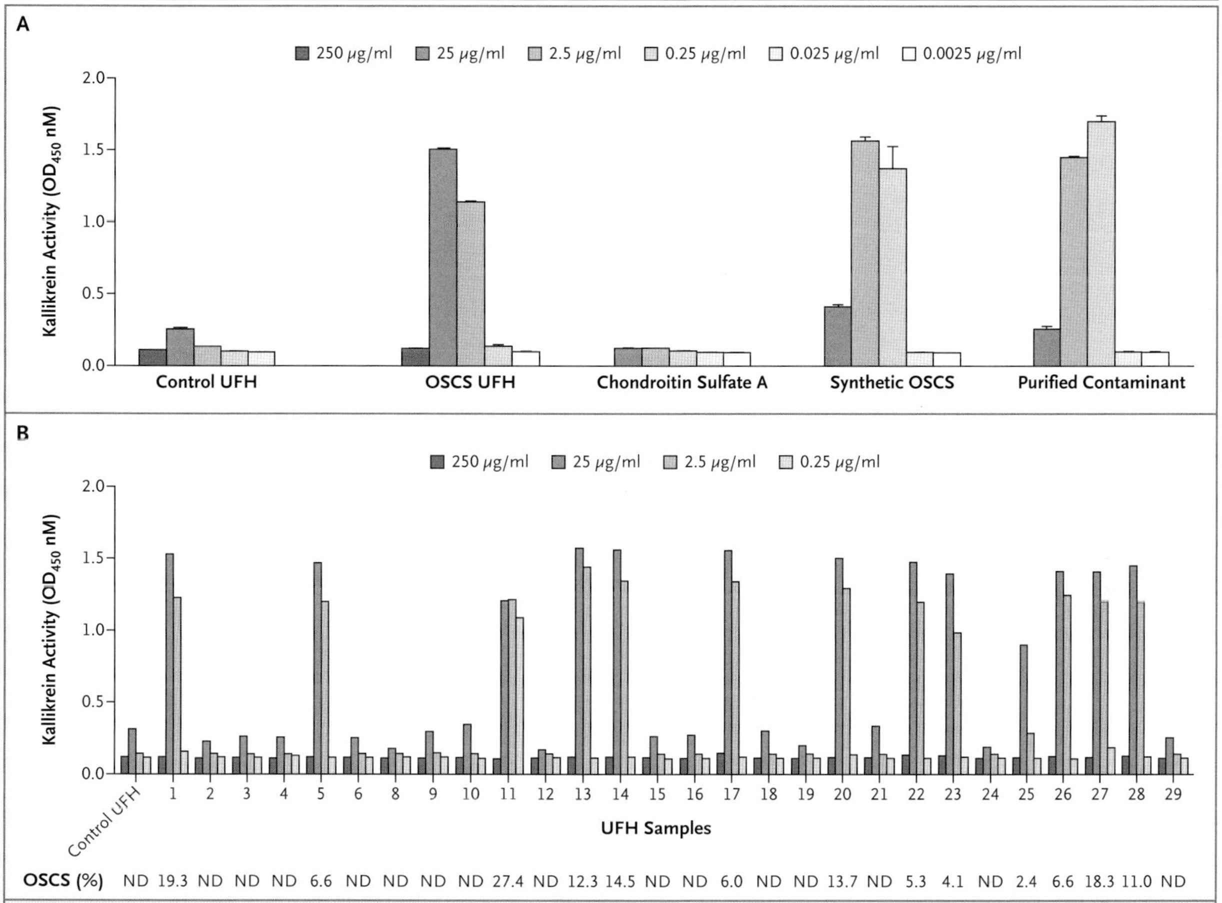

Figure 1. Effect of OSCS on Kallikrein Activity.

Pooled human plasma samples were treated with control unfractionated heparin (UFH) or OSCS-contaminated heparin (0.025 to 250 μg per milliliter) or with chondroitin sulfate A, synthetic OSCS, or purified OSCS contaminant (0.0025 to 25 μg per milliliter). Amidolytic activity was assessed by the addition of the S-2302 chromagenic substrate (D-Pro-Phe-Arg-p-nitroaniline); the effect on kallikrein amidolytic activity is shown (Panel A). The presence of OSCS in heparin was associated with the induction of kallikrein activity. Twenty-nine samples of heparin, representing both suspect heparin lots and control lots, were analyzed in a blinded fashion for both the presence of OSCS and the ability to activate kallikrein (Panel B). The presence of OSCS was detected and quantified by one-dimensional nuclear magnetic resonance spectroscopy (see Figure 2 in the Supplementary Appendix). The percentage of each sample that was OSCS is shown below the plot. Kallikrein amidolytic activity was assessed at various concentrations of heparin; Sample 7 was not analyzed for kallikrein activity owing to the limited quantity available. ND denotes not detectable, and OD optical density. In Panel A, T bars indicate standard deviations of replicate measurements.

ued. Two days later, she again received intravenous heparin (5000-U loading dose and 500 U per hour) from the same lots of heparin from the same manufacturer (Baxter Healthcare). Immediately after dialysis was initiated, the patient had an anaphylactoid reaction, with a sudden drop in blood pressure (to 65/34 mm Hg), dyspnea, nausea, vomiting, and constitutional symptoms. She was treated with a bolus of normal saline and oxygen (2 liters per minute). Hemodialysis was continued for another hour. The patient continued to feel ill, was admitted to the hospital, and was discharged 2 days later, after recovery. Further dialysis was performed with the use of heparin from another manufacturer.

METHODS

TEST SAMPLES

Twenty-nine clinical lots of heparin, including 13 associated with clinical adverse events, were procured from the FDA and coded as unknown samples 1 through 29. A laboratory lot of heparin was included as a control. For all analytic and biologic tests, samples were dosed on a weight basis; specific activity of heparin is typically approximately 180 U per milligram. OSCS was purified to homogeneity from a lot of heparin that was known to be contaminated, as previously described.[4] Briefly, OSCS-contaminated heparin was subjected to anion-exchange chromatography followed by alcohol precipitation to isolate the contaminant.[4] The identity of the contaminant was confirmed by means of multiple orthogonal techniques, including multidimensional nuclear magnetic resonance (NMR), enzymatic digestion followed by high-performance liquid chromatography, and liquid chromatography–mass spectrometry.[4] After identification of the contaminant as OSCS, a synthetic standard was generated through chemical sulfonation of chondroitin sulfate A and was exhaustively characterized to ensure authenticity, as previously described.[4] The synthetic OSCS was used in spiking experiments to qualify the analytic procedures (especially one-dimensional proton NMR, described below) to determine limits of detection and to establish accurate quantification.[4] The limit of detection for this assay was determined to be 0.3% on a weight basis for both dermatan sulfate and OSCS.

ANALYTIC METHODS

To ensure accurate identification and quantification of any contaminants and impurities, the 29 coded test samples were subjected to orthogonal analytic techniques. Proton NMR, anion-exchange chromatography, and capillary electrophoresis were used to screen the samples for the presence of OSCS, dermatan sulfate, and other nonheparin components. The levels of OSCS and dermatan sulfate were quantified with the use of a 600-MHz NMR instrument to ensure peak resolution. The details of quantification, as well as a representative spectrum, are given in Figure 1 and Table 1 in the Supplementary Appendix. For samples with unusual patterns, the identity of contaminants or impurities, including OSCS, was confirmed by means of detailed characterization, including the use of multidimensional NMR.[4]

AMIDOLYTIC ACTIVITY OF KALLIKREIN

Pooled human plasma or factor XII–depleted plasma (American Diagnostica) was treated with various concentrations of coded test samples of heparin, chondroitin sulfate A, or synthetic OSCS for 5 minutes at 37°C. The amidolytic activity of kallikrein (with a small contribution of factor XII)[9] was assessed by adding the S-2302 chromagenic substrate (D-Pro-Phe-Arg-p-nitroaniline [pNA]) for 30 minutes at 37°C, followed by spectrophotometric measurement of the absorbance at 450 nM.

GENERATION OF C3a AND C5a

Pooled human EDTA plasma or factor XII–depleted plasma (American Diagnostica) was treated with various concentrations of OSCS-contaminated heparin, control heparin, chondroitin sulfate A, or synthetic OSCS for 30 minutes at 37°C. C3a and C5a activation products of the complement cascade were assayed by means of a sandwich enzyme-linked immunosorbent assay (ELISA), as specified in the manufacturer's instructions (Becton Dickinson and Integrated Biotech Laboratories for C3a and C5a, respectively).

IN VIVO STUDIES

The swine were handled and treated in compliance with the Public Health Service Policy on Humane Care and Use of Laboratory Animals and the federal Animal Welfare Act. The experimental procedures were performed according to the Institutional Animal Care and Use Committee–approved protocol of the Virginia Polytechnic Institute and State University, Blacksburg. Domestic Yorkshire crossbred swine were of either sex (Virginia Polytechnic Institute and State University) and ranged in weight from 10 to 25 kg. They were initially anesthetized with an intravenous injection of 6 mg of tiletamine hydrochloride per kilogram of body weight and 2.2 mg of xylazine per kilogram, and then a single-lumen silicone catheter was implanted in the left jugular vein of each animal. Adequate anesthesia was maintained throughout the procedure with the administration of supplemental tiletamine. After a 5-minute stabilization period, each pig received an intravenous bolus infusion of 5 mg of the test substance per kilogram (three to six pigs per test substance). All the pigs were continuously monitored for vital signs with the use of an oscillometric blood pressure monitor (Cardell 9401/9403, CAS Medical Systems) for systolic, diastolic, and mean arterial blood pressures, pulse oximetry for pulse and respiratory rates, and a rectal probe for body temperature. At the end of the 60-minute observation period, the animals were euthanized with the use of an intravenous infusion of Fatal Plus

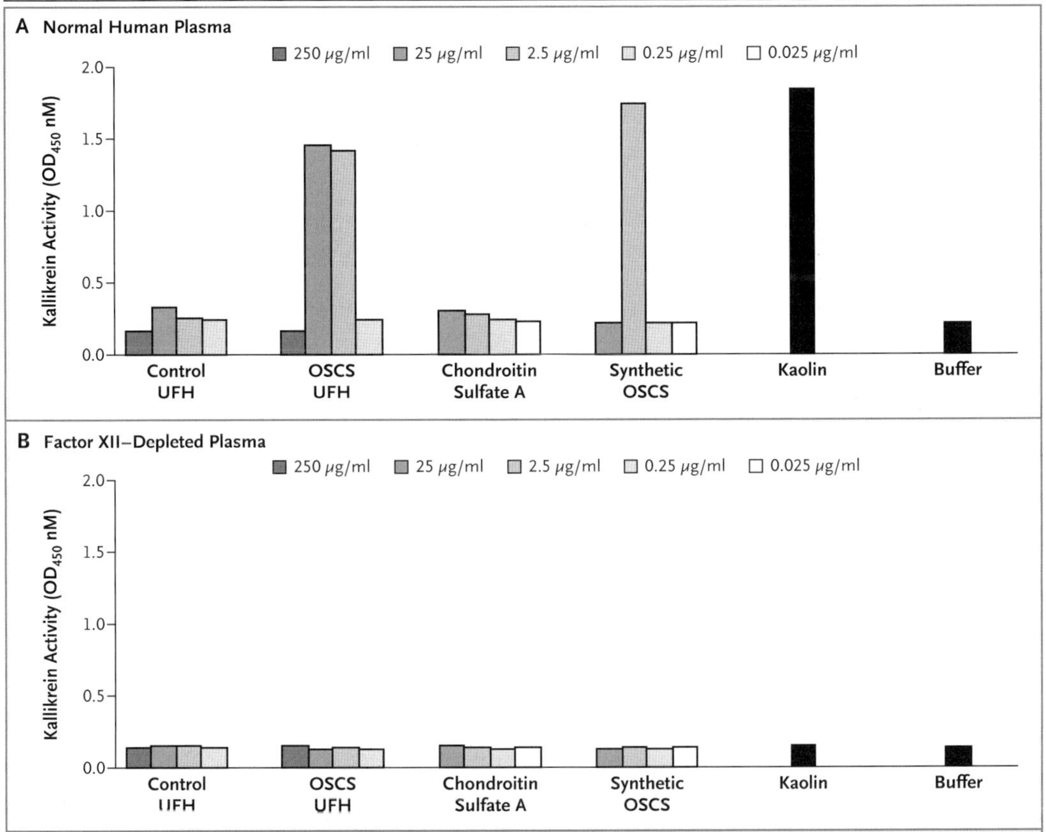

Figure 2. **OSCS-Induced Activation of Kallikrein in Normal or Factor XII–Depleted Plasma.**
Control unfractionated heparin (UFH), OSCS-contaminated UFH, chondroitin sulfate A, and synthetic OSCS were incubated with normal plasma (Panel A) or with factor XII–depleted plasma (Panel B); OSCS-induced activation of kallikrein is dependent on factor XII, as indicated. Kaolin-containing buffer was evaluated as a positive control activator of the contact system. Buffer alone was included as a negative control. Amidolytic activity was assessed by the addition of the S-2302 chromagenic substrate (D-Pro-Phe-Arg-p-nitroaniline). OD denotes optical density.

(Vortech Pharmaceuticals) at a dose of 0.22 ml per kilogram. Blood samples were collected at baseline and at 5, 10, 20, 40, and 60 minutes and were kept in 5 mM EDTA. Plasma was isolated after centrifugation at 4°C and flash-frozen on dry ice. Frozen samples were thawed at 4°C and assayed for amidolytic activity of kallikrein with the addition of the S-2302 chromagenic substrate (D-Pro-Phe-Arg-pNA), as described above.

RESULTS

Given the association of activation of the contact system with negatively charged polysaccharides, we sought to elucidate whether an in vitro biologic response could be correlated with the identity or levels of contaminant within heparin lots. To this end, we examined the ability of a sample of OSCS-contaminated heparin, containing 19.3% wt/wt OSCS (Table 1 in the Supplementary Appendix), to activate kallikrein amidolytic activity in human plasma (Fig. 1A). The contaminated heparin showed a bell-shaped dose response, which is typical of glycosaminoglycan-mediated responses.[5,10] At 2.5 and 25 μg per milliliter, robust activation of kallikrein was found with the contaminated heparin sample but not with a control sample of uncontaminated heparin. These concentrations are in the range of a clinically efficacious concentration of heparin of approximately 1 to 5 μg per milliliter, based on a specific activity of about 180 U per milligram. High concentrations of the OSCS-contaminated heparin (250 μg per milliliter) induced little or no amidolytic activity of kallikrein, suggesting that at this

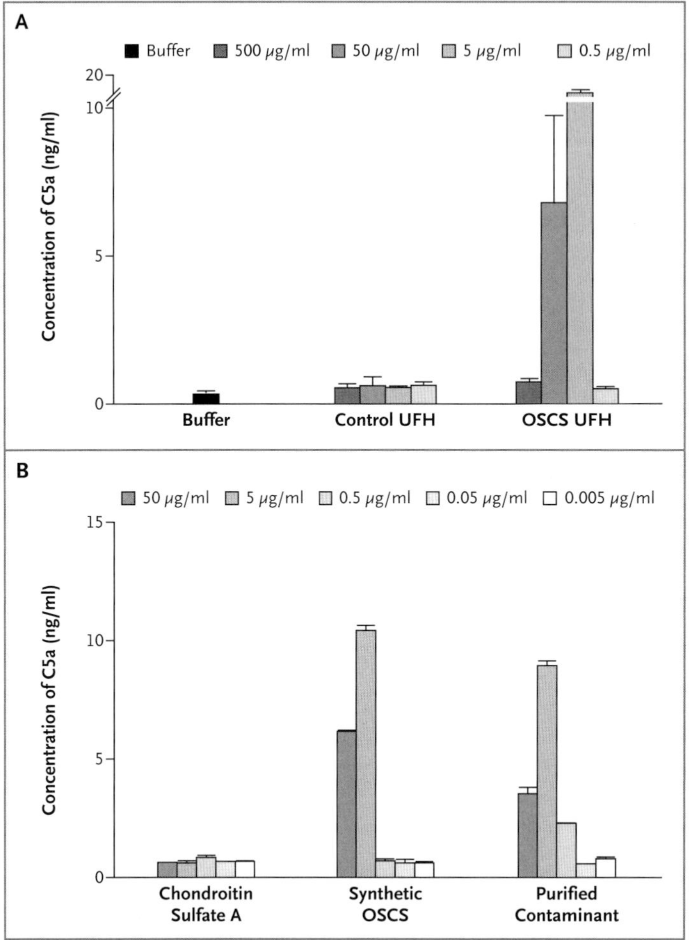

Figure 3. OSCS and the Generation of Complement-Derived C5a Anaphylatoxin in Human Plasma.

OSCS-contaminated unfractionated heparin (UFH) and control UFH (0.5 to 500 μg per milliliter) (Panel A) or chondroitin sulfate A, synthetic OSCS, and purified OSCS contaminant (0.005 to 50 μg per milliliter) (Panel B) were incubated with normal plasma anticoagulated with EDTA. The generation of C5a was assessed by means of ELISA. T bars indicate standard deviations of replicate measurements.

concentration, heparin may inhibit or cause depletion of factor XII, as previously described.[7,11,12] This high concentration of heparin also prevented activation of the contact system in response to kaolin, a potent activator (data not shown).

To further verify that the contaminant was responsible for the activation of the contact system, OSCS was purified to homogeneity by means of anion-exchange chromatography followed by alcohol precipitation. In addition, an OSCS standard was created through chemical sulfonation of chondroitin sulfate A, to form OSCS.[4] The purified contaminant and the OSCS standard were identical, as judged by several orthogonal analytic techniques, including two-dimensional NMR.[4] Both the purified contaminant and the synthetic OSCS showed robust activation of kallikrein activity at 0.25 μg and 2.5 μg per milliliter (Fig. 1A). The peak activity of the purified contaminant and the synthetic OSCS standard were observed at a level that was approximately an order of magnitude lower than that found for the contaminated heparin sample. This is consistent with the observation that the OSCS constituted approximately 20% of the contaminated sample. Chondroitin sulfate A showed no induction of amidolytic activity.

These results are in good agreement with the observations of Hojima et al.,[5] who demonstrated that oversulfated chondroitin, but not chondroitin A, B, or C, can activate the kinin pathway. Heparin also activated the contact system in an in vitro model system involving purified protein components[5,13] but did not in plasma,[13] suggesting that negative-regulatory factors present in plasma may prevent activation of the contact system by heparin. One such mechanism is the fact that heparin is known to enhance antithrombin III–mediated inhibition of factor XII. Our results indicate that OSCS, in contrast to heparin but similar to dextran sulfate,[13] can activate the contact system in plasma.

The 29 heparin samples procured from the FDA, consisting of both suspect heparin lots associated with clinical events as well as control heparin lots, were screened in a blinded fashion for both the presence of OSCS and the ability to activate the contact system (Fig. 1B). There was complete correspondence between the presence of detectable amounts of OSCS by one-dimensional proton NMR and the ability of a sample to induce robust amidolytic activity of kallikrein (Fig. 1B). The biologic activity was generally characterized as an all-or-none response, with all 13 samples containing detectable levels of OSCS having a positive response at 25 μg or 2.5 μg per milliliter. Sample 11, which contained the highest level of contaminant (27.4%), also showed activity at 0.25 μg per milliliter, whereas Sample 25, which contained the lowest level of contaminant (2.4%), showed only modest activity at 2.5 μg per milliliter. In contrast, there was no association between the level of inducible kallikrein activity and the level of dermatan sulfate (Fig. 2 in the Supplementary Appendix), an impurity found in many heparin preparations.

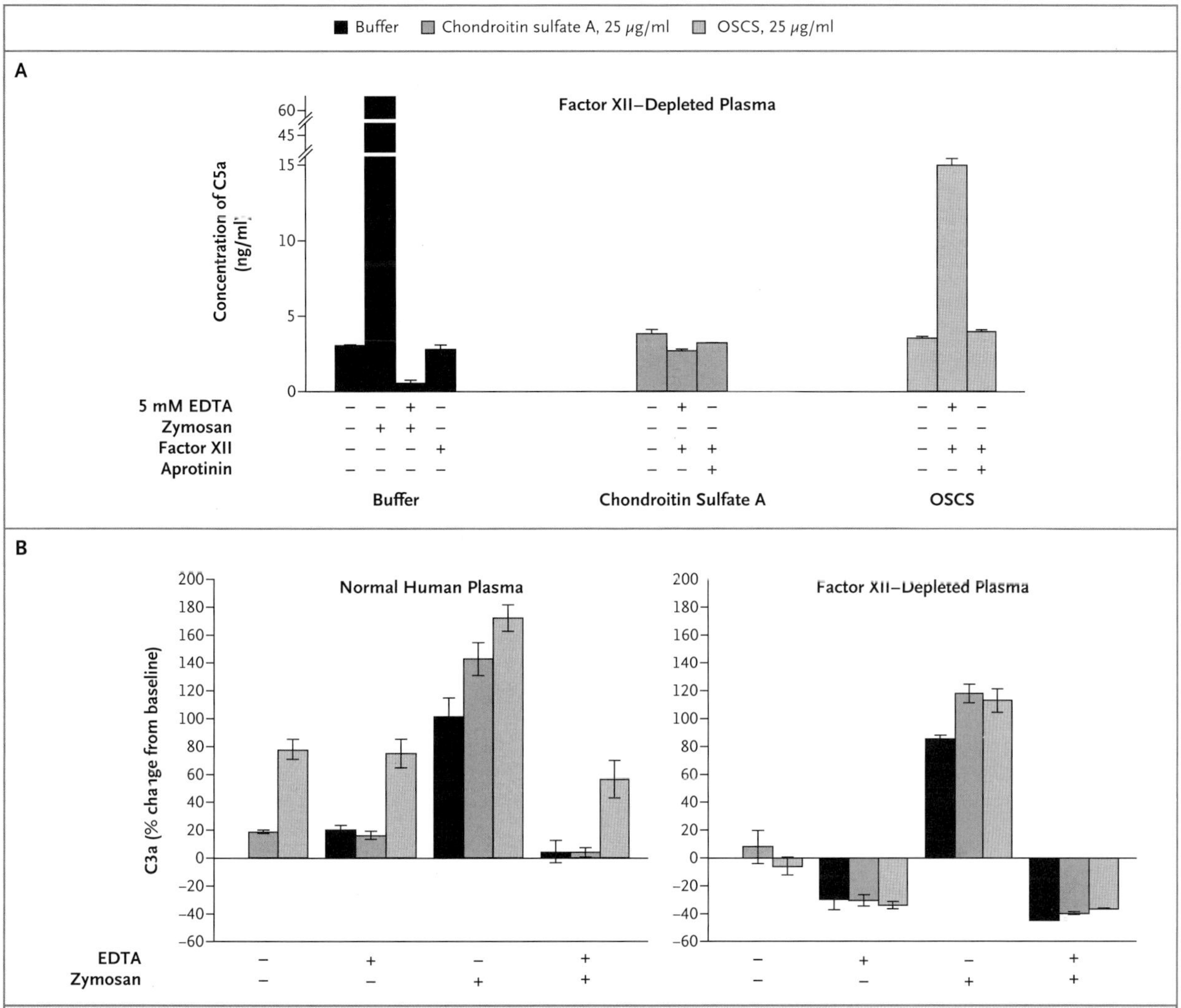

Figure 4. OSCS Induction of Complement-Derived C3a and C5a Anaphylatoxins and Its Relationship to the Contact System.

Factor XII–depleted plasma was incubated with chondroitin sulfate A, OSCS, or control buffer in the presence or absence of 5 mM EDTA, zymosan (1 mg per milliliter), or aprotinin (400 IU per milliliter) (Panel A). Specific samples were reconstituted with purified factor XII, as indicated. Normal human plasma or factor XII–depleted plasma was incubated with chondroitin sulfate A, OSCS, or control buffer in the presence or absence of 5 mM EDTA, zymosan (1 mg per milliliter), or both (Panel B). C3a and C5a generation was assessed by means of ELISA. T and I bars indicate standard deviations of replicate measurements.

Direct activation of the contact system by the contaminated heparin and the synthetic OSCS standard was confirmed through the use of human plasma depleted of factor XII, the upstream activator of prekallikrein[14] (Fig. 2). The contaminated heparin, the synthetically derived OSCS, and the positive control (the kaolin-containing reagent) all failed to induce the amidolytic activity of kallikrein in factor XII–deficient plasma.

We next examined the ability of contaminated heparin to generate C3a and C5a, potent anaphylatoxins derived from complement proteins. Exposure of human plasma to the contaminated heparin, but not to control heparin, induced the production of C5a (Fig. 3). OSCS-induced C5a generation showed a bell-shaped dose response similar to that found for kallikrein activation. Peak C5a activity was observed at 50 μg and 5 μg per milliliter of heparin containing 19.3% OSCS. At 500 μg per milliliter, significant generation of C5a was not observed. Similar results were obtained with the purified OSCS isolated from contaminated heparin and the synthetic OSCS standard, but not with chondroitin sulfate A.

Surprisingly, the generation of C5a by OSCS-contaminated heparin was more robust in the presence of EDTA, a Ca^{2+}- and Mg^{2+}-chelating agent, than in the absence of EDTA. The classic and alternative pathways of complement activation are known to be dependent upon Ca^{2+} and Mg^{2+}, respectively. As expected, EDTA blocked C3a and C5a generation in response to zymosan, a potent activator of the alternative pathway (Fig. 4). These results suggested the possibility that OSCS induces the generation of C3a and C5a in a manner that bypasses the C3 and C5 convertases. To determine whether the generation of C3a and C5a was linked to the activation of the contact system, we next examined C3a and C5a generation in factor XII–depleted plasma (Fig. 4). As expected, zymosan induced the generation of C3a and C5a in factor XII–depleted plasma, and this activity was inhibited by EDTA. In contrast, neither C3a nor C5a was generated in factor XII–depleted plasma activated with OSCS, suggesting that OSCS bypasses the normal pathways for complement activation in a manner that is dependent on contact activation through factor XII. The generation of C5a could be restored by reconstituting the factor XII–depleted plasma with purified factor XII (Fig. 4A). This finding is further supported by the observation that C5a generation induced by OSCS-contaminated heparin can be inhibited by aprotinin, a protease inhibitor of kallikrein but not of factor XIIa (Fig. 4A). Crosstalk between the contact system and the complement cascade has been suggested previously.[15-18] For example, factor XII has been shown to activate the classical pathway by activating C1.[15] It has also been proposed to substitute for factor D in activating the alternative pathway.[16] However, in these cases, activation of the complement cascade still occurs through divalent cation–dependent pathways. Kallikrein has been shown to act directly on C5 to generate C5a-like biologic activity.[17] Both kallikrein and factor XII can activate the plasminogen pathway leading to the activation of plasmin, which has also been implicated in complement activation.[18] Preliminary data suggest that OSCS is unable to induce C5a generation in plasminogen-depleted plasma (data not shown).

To identify an appropriate species for in vivo testing of OSCS, a panel of plasma samples were screened for amidolytic activity in response to OSCS-contaminated heparin (Fig. 5A). Only swine plasma supported robust amidolytic activity of kallikrein in response to kaolin and OSCS-contaminated heparin but not control heparin. In contrast, rabbit, horse, and rat plasma showed moderate-to-robust amidolytic activity in response to kaolin but not to OSCS-contaminated heparin. These findings are consistent with a report that initial attempts to provoke an allergic response with suspect lots of heparin were unsuccessful.[8] Similarly, we found that rabbits treated with 5 mg of intravenous OSCS-contaminated heparin per kilogram showed no change in temperature, blood pressure, or heart rate as compared with rabbits treated with control heparin (data not shown). Wiggins[19] demonstrated previously that dextran sulfate can induce hypotension in rabbits, but only at a high dose (20 mg per kilogram) and in a manner independent of complement or kinin activation. In contrast, moderate doses of dextran sulfate (5 mg per kilogram) induced a robust hypotensive response in pigs that was dependent on activation of the contact system.[20]

To test the in vivo activity of OSCS, pigs were treated with a single intravenous dose (5 mg per kilogram) of OSCS-contaminated heparin, control heparin, synthetic OSCS, or chondroitin sulfate A and were monitored for 60 minutes. Animals treated with control heparin and those treated with OSCS-contaminated heparin showed similar anti-Xa activity during the entire 60-minute

Figure 5 (facing page). **In Vitro and In Vivo Activity of OSCS.**
Human, rat, rabbit, pig, and horse plasma samples were incubated with various concentrations of OSCS-contaminated unfractionated heparin (UFH) or control UFH (Panel A). Kaolin-containing buffer was tested as a positive control. Buffer alone was included as a negative control. Kallikrein amidolytic activity was assessed by the addition of the S-2302 chromogenic substrate; OSCS induces hypotension and kallikrein activity in swine (Panels B and C). Anesthetized Yorkshire crossbred pigs (three to six pigs per group) were treated with a single intravenous bolus (5 mg per kilogram) of control UFH, OSCS-contaminated UFH, chondroitin sulfate A, or synthetic OSCS. Representative data for the heart rate, the mean arterial pressure, the systolic blood pressure, and the diastolic blood pressure are shown (Panel B). EDTA-anticoagulated plasma was collected at baseline and at 5, 10, 20, 40, and 60 minutes after infusion of test samples (Panel C). OD denotes optical density. In Panels A and C, T and I bars indicate standard deviations of replicate measurements.

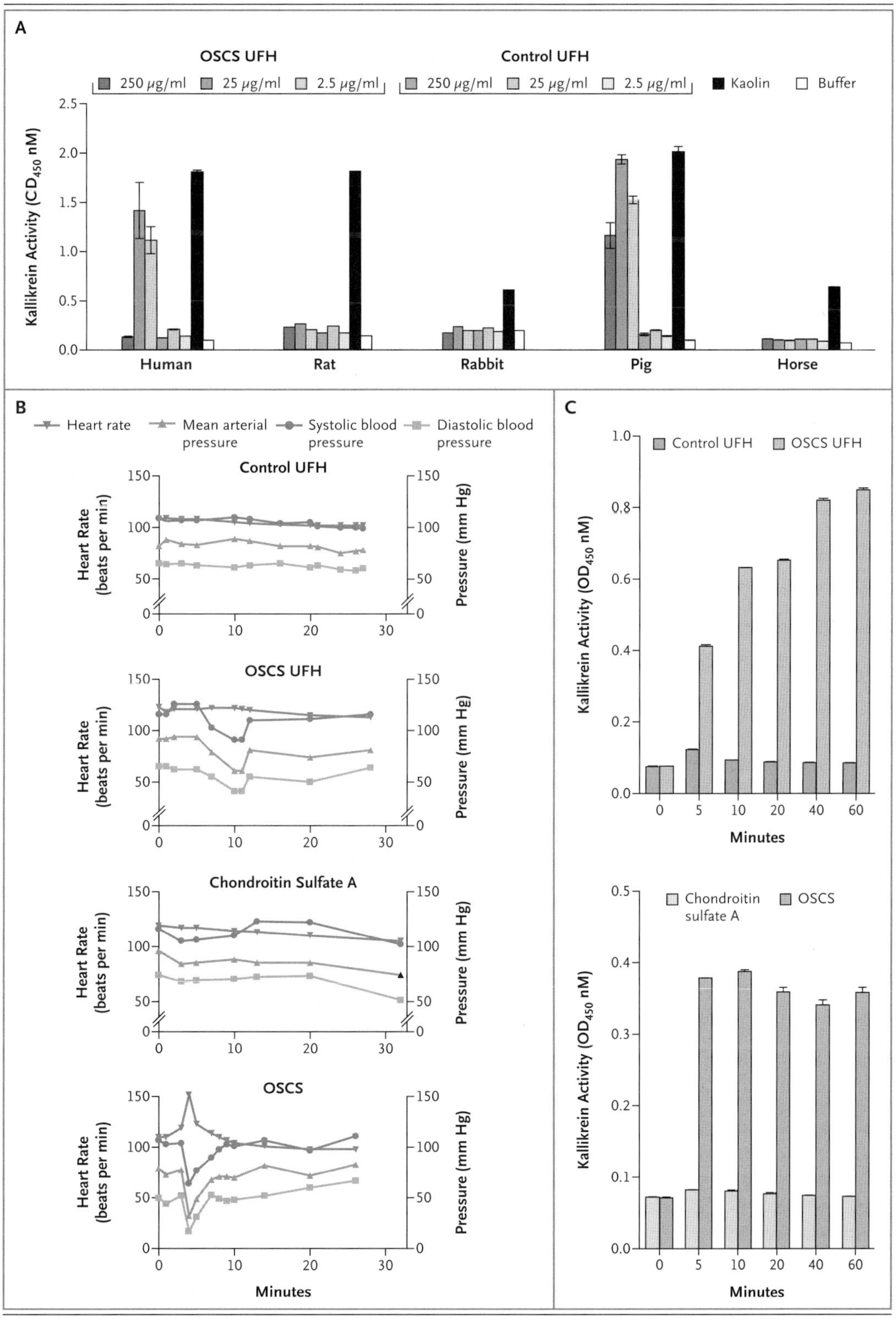
A
OSCS UFH
Control UFH
250 µg/ml
25 µg/ml
2.5 µg/ml
Kaolin
Buffer
Kallikrein Activity (CD450 nM)
Human
Rat
Rabbit
Pig
Horse
B
Heart rate
Mean arterial pressure
Systolic blood pressure
Diastolic blood pressure
Control UFH
OSCS UFH
Chondroitin Sulfate A
OSCS
Heart Rate (beats per min)
Pressure (mm Hg)
Minutes
C
Control UFH
OSCS UFH
Kallikrein Activity (OD450 nM)
Minutes
Chondroitin sulfate A
OSCS

observation period (activity at 5 minutes, approximately 3 to 4 IU per milliliter) (Fig. 4 in the Supplementary Appendix). Animals treated with chondroitin sulfate A or synthetic OSCS showed no anti-Xa activity. These results suggest that any anticoagulant activity of OSCS is mediated through a non–antithrombin III–dependent mechanism. Two of six animals treated with OSCS-contaminated heparin had at least a 30% drop in blood pressure over the first 30 minutes after infusion (Fig. 5B). One animal remained in a hypotensive state for more than 15 minutes. In contrast, none of the four animals treated with control heparin showed any substantive changes in blood pressure. The adverse events were more severe in pigs treated with the synthetic OSCS, a result consistent with the greater exposure to OSCS in animals treated with pure OSCS as compared with contaminated heparin containing approximately 20 to 30% OSCS. All three pigs treated with synthetic OSCS showed a profound drop in blood pressure (maximal decrease, 45 to 59%) and a concurrent increase in heart rate within minutes after infusion. One animal had difficulty breathing and became cyanotic after a precipitous drop in blood pressure. The heart rate of a second animal increased from 114 beats per minute to 196 beats per minute within 4 minutes after the infusion of OSCS. The third pig showed a transient but pronounced spike in heart rate with a corresponding drop in blood pressure (Fig. 5B). In contrast, none of the three pigs treated with chondroitin sulfate A showed any significant changes in blood pressure or heart rate within the first 30 minutes after drug infusion. Thus, intravenous infusion of OSCS is capable of recapitulating the hallmark cardiovascular features of the reaction in swine. The changes in physiological parameters were mirrored by rapid induction of the amidolytic activity of kallikrein (Fig. 5C). Kallikrein activity remained high throughout the observation period, even after the vital functions returned to normal, suggesting depletion of high-molecular-weight kininogen and inactivation of bradykinin by kininases in vivo, as previously shown with dextran sulfate.[20] Induction of kallikrein activity was evident in all animals that received OSCS-contaminated heparin, even when no substantive changes in blood pressure were observed. These findings suggest that activation of kallikrein does not always manifest as clinical symptoms, perhaps because of individual variation in control mechanisms that regulate bradykinin activity. Nonetheless, these results also suggest that swine may be an appropriate species in which to assess the potential consequences of OSCS contaminant in cardiovascular and dialysis models as well as in heparin-coated devices.

DISCUSSION

The recent reports of allergic-type serious adverse events in patients receiving heparin and the subsequent detection of widespread contamination have caused intense international concern about the safety of this essential drug. Urgent problems included an immediate and unknown risk to patients' lives, a threat to the supply of a widely used, essential drug, and the need for international cooperation in managing the integrity of a global supply chain. This crisis necessitated an urgent need to both understand the basis for these clinical events and to prevent future occurrences. The development of an analytic assay for OSCS, coupled with the rapid response of manufacturers and regulatory authorities around the world, has undoubtedly limited the harm. However, in the absence of a biologic link between the OSCS contaminant and the adverse events, the adequacy of screening heparin lots to prevent a recurrence is a concern.

Determining whether a link exists between the presence of OSCS and a biologic response required the convergence of two distinct analyses. First, there was a requirement to develop analytic techniques of sufficient sensitivity and specificity to ensure accurate identification and quantification of contaminants or impurities that are present within heparin. Second, there was a requirement to develop a sensitive, clinically appropriate biologic test to determine at what levels, if any, the OSCS would have relevant biologic activity.

With regard to the analytic techniques, a tiered approach was required to ensure effective translation to biologic characteristics. Screening methods were developed to rapidly identify whether heparin lots were contaminated or impure. Then, methods were further developed to enable quantification of the contamination levels. Finally, more sophisticated techniques, such as multidimensional NMR, enabled complete characterization of the contaminant or impurity. This tiered approach was necessitated by the fact that heparin is a polydisperse mixture of glycosaminoglycan chains; orthogonal techniques were therefore required to ensure capture of the other nonheparin components.

Here, we demonstrate that the OSCS present in suspect heparin lots, as well as a synthetic OSCS standard, can directly activate the contact system and induce the generation of C3a and C5a anaphylatoxins in vitro. Moreover, OSCS activates kallikrein in vivo and can induce a profound hypotensive response in pigs, thus providing a potential biologic link between the contaminant and the anaphylactoid reactions seen in affected patients. The finding that hypotension did not develop in all animals treated with OSCS-contaminated heparin, even at a relatively high dose, is consistent with the observation that the majority of patients who received contaminated heparin did not experience an adverse event. However, it is important to note that all animals treated with OSCS-contaminated heparin showed evidence of kallikrein activation in vivo, even in the absence of clinical signs. Patients undergoing dialysis who are also receiving heparin therapy are already at high risk for hypotension because of their exposure to the dialysis membrane, which can also activate the contact system, and their treatment with angiotensin-converting–enzyme inhibitors, which inhibit bradykinin degradation. Exposure to OSCS-contaminated heparin may further increase the risk and could potentially trigger an adverse event. Finally, these findings also suggest that a simple in vitro bioassay could complement the previously described analytic tests[4] to help protect the global supply chain of heparin, by allowing the screening of heparin lots for the presence not only of OSCS but also of other polysulfated contaminants that may have unintended pharmacologic consequences.

Supported by grants from the National Institute of General Medical Sciences and the National Heart and Lung Institute (GM57073 and HL080278-01, respectively, to Dr. Sasisekharan).

Drs. Kishimoto, Ganguly, Lansing, Zhao, Galcheva-Gargova, Roy, Shriver, and Venkataraman, Mr. Smith, and Mr. Baily report being employees of Momenta Pharmaceuticals and holding equity in the company, which has technology on the analysis and characterization of complex mixtures, including heparin. Drs. Sasisekharan and Langer report receiving consulting fees from Scientific Protein Labs and Momenta Pharmaceuticals and holding equity in Momenta Pharmaceuticals. Dr. Austen reports receiving consulting fees from Momenta Pharmaceuticals. No other potential conflict of interest relevant to this article was reported.

We thank Dr. Allen P. Kaplan for helpful discussions; Pete Jobst, Animal Resource Manager, and Andrea Aman, Animal Care Technician, at the Virginia–Maryland Regional College of Veterinary Medicine for their excellent technical help during the in vivo experiments; and Ms. Alison Long, Mr. Chris Honan, Dr. Cedric Hubeau, Mr. Erick Moy, and the staff of the Massachusetts Institute of Technology Division of Comparative Medicine for help with the animal studies.

REFERENCES

1. Acute allergic-type reactions among patients undergoing hemodialysis — multiple states, 2007–2008. MMWR Morb Mortal Wkly Rep 2008;57:124-5.
2. Information on adverse event reports and heparin. Rockville, MD: Food and Drug Administration, 2008. (Accessed May 12, 2008, at http://www.fda.gov/cder/drug/infopage/heparin/adverse_events.htm.)
3. Information on heparin sodium injection. Rockville, MD: Food and Drug Administration, 2008. (Accessed May 12, 2008, at http://www.fda.gov/cder/drug/infopage/heparin/default.htm#screening.)
4. Guerrini M, Beccati D, Shriver Z, et al. Oversulfated chondroitin sulfate is a major contaminant in heparin associated with adverse clinical events. Nat Biotechnol (in press).
5. Hojima Y, Cochrane CG, Wiggins RC, Austen KF, Stevens RL. In vitro activation of the contact (Hageman factor) system of plasma by heparin and chondroitin sulfate E. Blood 1984;63:1453-9.
6. Henry SP, Giclas PC, Leeds J, et al. Activation of the alternative pathway of complement by a phosphorothioate oligonucleotide: potential mechanism of action. J Pharmacol Exp Ther 1997;281:810-6.
7. Silverberg M, Diehl SV. The autoactivation of factor XII (Hageman factor) induced by low-Mr heparin and dextran sulphate: the effect of the Mr of the activating polyanion. Biochem J 1987;248:715-20.
8. Baxter provides update on the heparin investigation. Deerfield, IL: Baxter, March 19, 2008. (Accessed May 12, 2008, at http://www.baxter.com/products/biopharmaceuticals/downloads/heparin_03-19-08.pdf.)
9. Silverberg M, Dunn JT, Garen L, Kaplan AP. Autoactivation of human Hageman factor: demonstration utilizing a synthetic substrate. J Biol Chem 1980;255:7281-6.
10. Verhamme IM, Bock PE, Jackson CM. The preferred pathway of glycosaminoglycan-accelerated inactivation of thrombin by heparin cofactor II. J Biol Chem 2004;279:9785-95.
11. Stead N, Kaplan AP, Rosenberg RD. Inhibition of activated factor XII by antithrombin-heparin cofactor. J Biol Chem 1976;251:6481-8.
12. Olson ST, Sheffer R, Francis AM. High molecular weight kininogen potentiates the heparin-accelerated inhibition of plasma kallikrein by antithrombin: role for antithrombin in the regulation of kallikrein. Biochemistry 1993;32:12136-47.
13. Pixley RA, Cassello A, De La Cadena RA, Kaufman N, Colman RW. Effect of heparin on the activation of factor XII and the contact system in plasma. Thromb Haemost 1991;66:540-7.
14. Kaplan AP, Austen KF. A prealbumin activator of prekallikrein. II. Derivation of activators of prekallikrein from active Hageman factor by digestion with plasmin. J Exp Med 1971;133:696-712.
15. Ghebrehiwet B, Silverberg M, Kaplan AP. Activation of the classical pathway of complement by Hageman factor fragment. J Exp Med 1981;153:665-76.
16. DiScipio RG. The activation of the alternative pathway C3 convertase by human plasma kallikrein. Immunology 1982;45:587-95.
17. Wiggins RC, Giclas PC, Henson PM. Chemotactic activity generated from the fifth component of complement by plasma kallikrein of the rabbit. J Exp Med 1981;153:1391-404.
18. Schaiff WT, Eisenberg PR. Direct induction of complement activation by pharmacologic activation of plasminogen. Coron Artery Dis 1997;8:9-18.
19. Wiggins RC. A different cleavage site for high molecular weight kininogen in vivo following intravenous injection of dextran sulfate in the rabbit. Circ Res 1986;58:595-604.
20. Siebeck M, Cheronis JC, Fink E, et al. Dextran sulfate activates contact system and mediates arterial hypotension via B2 kinin receptors. J Appl Physiol 1994;77:2675-80.

The NEW ENGLAND JOURNAL of MEDICINE

ORIGINAL ARTICLE

Outbreak of Adverse Reactions Associated with Contaminated Heparin

David B. Blossom, M.D., Alexander J. Kallen, M.D., M.P.H., Priti R. Patel, M.D., M.P.H., Alexis Elward, M.D., M.P.H., Luke Robinson, B.S., Ganpan Gao, Ph.D., Robert Langer, Sc.D., Kiran M. Perkins, M.D., Jennifer L. Jaeger, M.D., Katie M. Kurkjian, D.V.M., M.P.H., Marilyn Jones, R.N., M.P.H., Sarah F. Schillie, M.D., M.P.H., Nadine Shehab, Pharm.D., Daniel Ketterer, M.D., Ganesh Venkataraman, Ph.D., Takashi Kei Kishimoto, Ph.D., Zachary Shriver, Ph.D., Ann W. McMahon, M.D., K. Frank Austen, M.D., Steven Kozlowski, M.D., Arjun Srinivasan, M.D., George Turabelidze, M.D., Ph.D., Carolyn V. Gould, M.D., Matthew J. Arduino, Dr.P.H., and Ram Sasisekharan, Ph.D.

ABSTRACT

BACKGROUND

In January 2008, the Centers for Disease Control and Prevention began a nationwide investigation of severe adverse reactions that were first detected in a single hemodialysis facility. Preliminary findings suggested that heparin was a possible cause of the reactions.

METHODS

Information on clinical manifestations and on exposure was collected for patients who had signs and symptoms that were consistent with an allergic-type reaction after November 1, 2007. Twenty-one dialysis facilities that reported reactions and 23 facilities that reported no reactions were included in a case–control study to identify facility-level risk factors. Unopened heparin vials from facilities that reported reactions were tested for contaminants.

RESULTS

A total of 152 adverse reactions associated with heparin were identified in 113 patients from 13 states from November 19, 2007, through January 31, 2008. The use of heparin manufactured by Baxter Healthcare was the factor most strongly associated with reactions (present in 100.0% of case facilities vs. 4.3% of control facilities, $P<0.001$). Vials of heparin manufactured by Baxter from facilities that reported reactions contained a contaminant identified as oversulfated chondroitin sulfate (OSCS). Adverse reactions to the OSCS-contaminated heparin were often characterized by hypotension, nausea, and shortness of breath occurring within 30 minutes after administration. Of 130 reactions for which information on the heparin lot was available, 128 (98.5%) occurred in a facility that had OSCS-contaminated heparin on the premises. Of 54 reactions for which the lot number of administered heparin was known, 52 (96.3%) occurred after the administration of OSCS-contaminated heparin.

CONCLUSIONS

Heparin contaminated with OSCS was epidemiologically linked to adverse reactions in this nationwide outbreak. The reported clinical features of many of the cases further support the conclusion that contamination of heparin with OSCS was the cause of the outbreak.

From the Epidemic Intelligence Service Program (D.B.B., A.J.K., J.L.J., K.M.K., S.F.S.), the Office of Workforce and Career Development (D.B.B., A.J.K., K.M.P., J.L.J., K.M.K., S.F.S., D.K.), and the Division of Healthcare Quality Promotion (D.B.B., A.J.K., P.R.P., S.F.S., N.S., A.S., C.V.G., M.J.A.), Centers for Disease Control and Prevention, Atlanta; St. Louis Children's Hospital (A.E.), BJC Healthcare (M.J.), and the Missouri Department of Health and Senior Services (G.T.) — all in St. Louis; the Departments of Biological Engineering (L.R., G.G., R.L., R.S.) and Chemical Engineering (R.L.), the Harvard–MIT Division of Health Sciences and Technology (R.L., R.S.), and the Koch Institute for Integrative Cancer Research (L.R., G.G., R.L., R.S.), Massachusetts Institute of Technology; and Momenta Pharmaceuticals (G.V., T.K.K., Z.S.) — both in Cambridge, MA; the Center for Drug Evaluation and Research, Food and Drug Administration, Silver Spring, MD (A.W.M., S.K.); and Brigham and Women's Hospital and Harvard Medical School, Boston (K.F.A.). Address reprint requests to Dr. Patel at the Centers for Disease Control and Prevention, 1600 Clifton Rd., MS-A31, Atlanta, GA 30333, or at ppatel@cdc.gov.

This article (10.1056/NEJMoa0806450) was published at www.nejm.org on December 3, 2008.

N Engl J Med 2008;359:2674-84.

UNFRACTIONATED HEPARIN IS AN ANTIcoagulant medication that is used to prevent or treat thromboembolic disorders. Heparin is also commonly used to prevent clotting of extracorporeal blood during hemodialysis and cardiac surgery, as well as to maintain the patency of intravenous devices. Chemically, heparin is a heterogeneous mixture of sulfated polysaccharides; its main anticoagulant activity is mediated through activation of antithrombin. Commercially available heparin is derived from animal tissues; only porcine-derived heparin is approved for the U.S. market. Although heparin-induced thrombocytopenia is a well-described immune-mediated phenomenon among patients receiving heparin, immediate hypersensitivity reactions (e.g., exanthemas, bronchospasm, angioedema, and anaphylaxis) that are directly attributable to heparin have rarely been reported.[1,2]

On January 7, 2008, the Missouri Department of Health and Senior Services notified the Centers for Disease Control and Prevention (CDC) about a cluster of allergic-type reactions among patients undergoing hemodialysis at a pediatric hospital.[3] Symptoms occurred within minutes after the initiation of a dialysis session, and manifestations included facial edema, tachycardia, hypotension, urticaria, and nausea. The CDC's national case-finding effort identified, in multiple states, additional clusters of similar reactions among patients undergoing hemodialysis, and subsequently among patients undergoing photopheresis or treatment for cardiac conditions. A common feature that preceded many of the reactions was the receipt of heparin produced by Baxter Healthcare. On January 17, 2008, nine lots of vials of heparin manufactured by Baxter were voluntarily recalled.[4] A more extensive recall of heparin products manufactured by Baxter occurred on February 28, 2008.[5]

In March 2008, the Food and Drug Administration (FDA) announced that a "heparin-like" compound had been identified as a contaminant in the active pharmaceutical ingredient used in heparin manufactured by Baxter. This contaminant was identified as oversulfated chondroitin sulfate (OSCS),[6] and the ability of OSCS in the active pharmaceutical ingredient to activate the contact and complement systems was shown.[7] In this report, we describe the CDC epidemiologic investigation that was undertaken to establish the cause of the allergic-type reactions among patients undergoing dialysis, provide a clinical description of the reactions that occurred after the administration of heparin, and report on laboratory tests for the presence of OSCS in finished-product heparin vials that were related to these reactions.

METHODS

EPIDEMIOLOGIC INVESTIGATION OF REACTIONS AMONG PATIENTS UNDERGOING DIALYSIS

Case Finding

After being notified about the Missouri cluster, the CDC began active case finding. Inquiries about allergic-type reactions were circulated by various methods, including the use of e-mail distribution lists targeting providers in the field of nephrology and the CDC's Epidemic Information Exchange. We used a standard form to collect information about the adverse events, as well as information about the demographic and clinical characteristics of the patients who had reactions and exposures of those patients to medication and medical devices. Since the original cluster and the majority of reactions that were reported subsequently occurred among patients undergoing dialysis, our initial case definition was restricted to reactions associated with dialysis. For this aspect of the investigation, we characterized a definite case as the sudden onset of angioedema (i.e., facial edema) or urticaria in a patient within 1 hour after the initiation of a hemodialysis session that occurred after November 1, 2007. A probable case was characterized by the development, within the same period after initiation of hemodialysis, of hypotension, loss of consciousness, or signs and symptoms from at least two of the following categories: sensation of burning, warmth, or flushing; numbness or tingling; difficulty swallowing; shortness of breath, audible wheezing, or chest tightness; tachycardia; and nausea, vomiting, or diarrhea.

Facility-Based Case–Control Study

Because a clustering of cases was noted in specific dialysis facilities, and patients within a facility had relatively uniform exposures to medical products, we conducted a case–control study to identify risk factors, using the facility as the unit of analysis. Dialysis facilities that completed a case-report form for at least one definite or probable case (21 facilities in 11 states) were considered to be case facilities and were compared with control facilities that reported no such reactions (23 facilities in 9 states). Control facilities were identified with the use of the Centers for Medicare and Medicaid Services Dialysis Facility Compare Web site.[8] These

Table 1. Characteristics of Facilities Evaluated in a Facility-Based Case–Control Study.

Characteristic	Facilities with at Least One Case of Adverse Reaction (N=21)	Facilities with No Case of Adverse Reaction (N=23)	P Value
	no. (%)		
Manufacturer of heparin used			
Baxter Healthcare	21 (100.0)	1 (4.3)	<0.001
APP Pharmaceuticals	2 (9.5)	20 (87.0)	<0.001
Other	0	2 (8.7)	0.49
Manufacturer of dialysis machines used			
Gambro	16 (76.2)	7 (30.4)	0.03
Fresenius Medical Care	9 (42.9)	13 (56.5)	0.55
B. Braun	1 (4.8)	3 (13.0)	0.61
Other	1 (4.8)	0	0.48
Manufacturer of dialyzers used			
Gambro	10 (47.6)	8 (34.8)	0.54
Fresenius Medical Care	9 (42.9)	14 (60.9)	0.37
Other	7 (33.3)	6 (26.1)	0.75
Reused dialyzers	15 (71.4)	9 (39.1)	0.04
Saline priming solution delivered to patient*	11 (52.4)	13 (56.5)	0.76
Patient census >70 patients	10 (47.6)	12 (52.2)	1.00

* Saline solution is used to clear the tubing and dialyzer of air and residual disinfectant. In some instances, some of the priming solution that has passed through the circuit is delivered to the patient.

facilities were selected randomly among the pooled dialysis facilities in the 11 states in which case facilities were located. Owing to the pooling of potential controls, two states with case facilities were not represented among the control facilities. A representative of each case and control facility was contacted by telephone and surveyed between January 28, 2008, and February 8, 2008. Oral consent was obtained, and a clinical manager was asked to identify the medical products and supplies that had been used at the facility in the period after December 15, 2007, including heparin products and dialysis equipment (e.g., machines, tubing, and dialyzers), and to describe the facility's practices with respect to dialyzer reuse and reprocessing. Proportions were compared with the use of Fisher's exact test and Stata software, version 9.0 (Stata). All reported P values are two-sided.

ASSESSMENT OF REACTIONS, HEPARIN PRODUCTS, AND EXPOSURE TO CONTAMINATED LOTS

Clinical Description of Heparin Reactions

After preliminary findings suggested that heparin was strongly associated with cases and similar reactions were reported among patients who were not undergoing hemodialysis, we expanded our case definition to include reactions in a broader population of patients. For the purposes of describing heparin reactions in this broader population, a probable or definite adverse reaction associated with heparin met the same clinical and temporal criteria as those in the case definition for patients undergoing dialysis but was defined as occurring within 1 hour after the administration of heparin (rather than within 1 hour after the initiation of a hemodialysis session).

Analytic and In Vitro Evaluation of Heparin Products

Unopened finished-product vials of heparin were solicited from health care facilities that reported cases. Heparin vials received by the CDC were categorized by lot number; 10 lots of heparin manufactured by Baxter Healthcare were tested. Samples from each unique lot number and three controls were tested for the presence of OSCS and were assessed for their effect on the amidolytic activity of kallikrein in human plasma as a measure of the

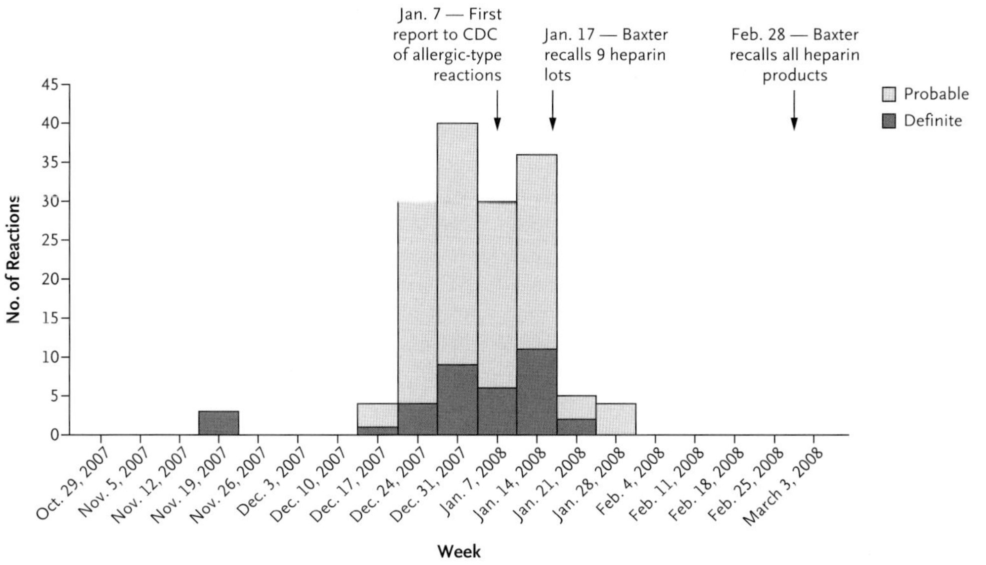

Figure 1. **Cases of Adverse Reactions Associated with Heparin, According to Week of Onset.**
The figure shows a total of 152 cases of adverse reactions associated with heparin that were reported to the Centers for Disease Control and Prevention from November 19, 2007, through January 31, 2008.

activation of the kinin–kallikrein pathway. The quantification of OSCS levels and the measurement of amidolytic activity were performed with the use of previously described methods.[6,7]

Exposure to Contaminated Heparin Lots

Although the specific lot number of heparin that had been administered in each case patient was requested, most facilities did not routinely record this information. Instead, many facilities reported likely exposures on the basis of the heparin lots that were present in the facility at the time of the reaction. Exposure to heparin lots contaminated with OSCS was described in the following manner. A known exposure to an OSCS-contaminated lot was defined as either documented receipt by a patient of heparin from a lot that tested positive for OSCS or exposure of a patient to one of several heparin lots in the facility, all of which tested positive for OSCS.

RESULTS

DIALYSIS CASES

A total of 131 adverse events met our initial case definition. Of these cases, 128 (97.7%) occurred after the administration of heparin; in 122 of those cases (95.3%), the heparin that was used was manufactured by Baxter Healthcare. The only other factors identified in more than 50% of the cases were the use of acid concentrate manufactured by Minntech (59.5%), the use of dialysis machines (53.8%) and dialyzers (53.5%) manufactured by Gambro, and the practice of reusing hemodialyzers (52.0%).

FACILITY-BASED CASE–CONTROL STUDY

Twenty-one case facilities in 11 states had been identified by late January 2008 and were included in the study. The mean number of cases at case facilities was 4.2 (range, 1 to 11). Fifty-two facilities were contacted to obtain 23 control facilities in nine states. Of the 29 facilities that were contacted but not included as controls, 16 refused to participate, 7 did not respond, 5 reported a possible allergic-type reaction, and 1 was not a dialysis facility.

In univariate analysis, the use of Gambro dialysis machines and the administration of Baxter heparin were significantly associated with the presence of adverse reactions at the facility (Table 1). The factor with the strongest association was the use of Baxter heparin, which was reported by all case facilities and only one control facility (100.0% vs. 4.3%, P<0.001).

ADVERSE REACTIONS ASSOCIATED WITH HEPARIN

A total of 152 heparin reactions that met our case definition were identified from November 19, 2007, through January 31, 2008 (Fig. 1), among the 194 events that were reported to the CDC. The 152 cases occurred in 113 patients from 13 states and included 130 reactions in 100 patients undergoing hemodialysis, 8 reactions in 6 patients undergoing treatment for cardiac conditions, and 14 reactions in 7 patients undergoing photopheresis. The average age of the 113 case patients was 53 years; 57 (50.4%) of the case patients were women.

Table 2 shows the clinical characteristics of the 152 heparin reactions in 113 case patients. The mean time to a reaction after exposure to heparin was 5.1 minutes among patients undergoing hemodialysis. The most common manifestations were hypotension (50.0%), nausea (48.7%), and shortness of breath (37.5%). Thirty-six reactions (23.7%) involved facial swelling. Although urticaria was reported in several of the cases in the initial cluster, this feature was infrequent (3.3%) among all cases. Fever (0.7%), chills (1.3%), wheezing (0%), and difficulty swallowing (0%) were also rare

Table 2. Clinical Characteristics of the 152 Adverse Reactions after Administration of Heparin.*

Characteristic	All Reactions (N=152)	Probable Reactions (N=116)	Definite Reactions (N=36)
Time from administration of heparin to reaction — min			
During hemodialysis	5.1±9.1	5.3±10.0	4.5±6.2
During treatment for cardiac conditions	14±17.2	15.9±17.7	1±0
During photopheresis†	30±13.1	30±13.1	—
Manifestation — no. (%)			
Facial edema			
Any	36 (23.7)	0	36 (100)
Lips	23 (15.1)	0	23 (63.9)
Eyelids	17 (11.2)	0	17 (47.2)
Throat	12 (7.9)	0	12 (33.3)
Tongue	11 (7.2)	0	11 (30.6)
Mouth	10 (6.6)	0	10 (27.8)
Urticaria	5 (3.3)	0	5 (13.9)
Low blood pressure	76 (50.0)	69 (59.5)	7 (19.4)
Systolic pressure <80 mm Hg	17 (11.2)	14 (12.1)	3 (8.3)
Nausea	74 (48.7)	56 (48.3)	18 (50.0)
Shortness of breath	57 (37.5)	38 (32.8)	19 (52.8)
Vomiting	37 (24.3)	27 (23.3)	10 (27.8)
Tingling	36 (23.7)	28 (24.1)	8 (22.2)
Flushing	35 (23.0)	31 (26.7)	4 (11.1)
Tachycardia	33 (21.7)	29 (25.0)	4 (11.1)
Diaphoresis	23 (15.1)	23 (19.8)	0
Abdominal pain	17 (11.2)	12 (10.3)	5 (13.9)
Diarrhea	8 (5.3)	7 (6.0)	1 (2.8)
Loss of consciousness	6 (3.9)	6 (5.2)	0
Chills	2 (1.3)	2 (1.7)	0
Fever	1 (0.7)	1 (0.9)	0
Difficulty swallowing	0	0	0
Wheezing	0	0	0

Table 2. **(Continued.)**

Characteristic	All Reactions (N=152)	Probable Reactions (N=116)	Definite Reactions (N=36)
Follow-up care — no./total no. (%)			
Blood cultures obtained‡	32/144 (22.2)	26/109 (23.9)	6/35 (17.1)
Evaluation in an emergency department§	22/144 (15.3)	9/109 (8.3)	13/35 (37.1)
Hospitalization§	13/144 (9.0)	5/109 (4.6)	8/35 (22.9)
Use of intravenous heparin — no. (%)	149 (98.0)	115 (99.1)	34 (94.4)
Brand of heparin — no. (%)			
Baxter Healthcare	141 (92.8)	110 (94.8)	31 (86.1)
APP Pharmaceuticals	6 (3.9)	4 (3.4)	2 (5.6)
Baxter or APP	2 (1.3)	1 (0.9)	1 (2.8)
Not reported	3 (2.0)	1 (0.9)	2 (5.6)

* Plus–minus values are means ±SD.
† Although heparin is administered at the beginning of photophoresis, the patient is not exposed to it until later in the process.
‡ The total numbers exclude eight reactions (one definite and seven probable) owing to missing data.
§ The total numbers exclude the eight reactions (one definite and seven probable) that occurred in patients undergoing treatment for cardiac conditions because these patients were hospitalized at the time of the adverse reactions.

or absent. In 15.3% of the cases, the reaction required further evaluation in the emergency department, and in 9.0% of the cases, required hospitalization. A total of 149 reactions (98.0%) occurred after intravenous administration of heparin. The other three involved exposure to heparin through other means (e.g., a dialysis circuit primed with heparin). The brand of heparin most commonly used (in 92.8% of the cases) was manufactured by Baxter Healthcare.

None of the 113 patients with adverse reactions that met our case definition died immediately after the reaction. Three deaths among patients undergoing treatment for cardiac conditions were reported to the CDC, but the reactions in these patients did not meet our case definition because they occurred between 8 and 11 hours after administration of the heparin.

ANALYTIC AND IN VITRO FINDINGS

The samples tested included 10 different lot numbers of heparin vials collected from facilities and 3 control samples (Fig. 2). Of the lots collected from facilities, Lots A through G and Lot J represent eight of the nine lots of Baxter heparin that were recalled on January 17, 2008. A high level of activation of kallikrein was observed with samples from Lots A through D and F through J at concentrations of 2.5 and 25 μg per milliliter. These concentrations are in the range of a clinically efficacious concentration of heparin of approximately 1 to 5 μg per milliliter, based on a specific activity of approximately 180 U per milligram. There was little activation of kallikrein from heparin samples that did not contain OSCS, including the control samples and Lot E.

EXPOSURE TO CONTAMINATED HEPARIN LOTS

Information on the lot of Baxter heparin was reported for 130 of the 152 heparin reactions. In 128 of the 130 (98.5%), OSCS-contaminated heparin was present in the facility; in 106 of the 130 (81.5%), facility records indicated that only OSCS-contaminated heparin could have been received by the patient (either because all heparin lots present in the facility tested positive for OSCS or because the single lot received by the patient was known and tested positive). Furthermore, of 54 heparin reactions for which the specific lot number of administered heparin was known, 52 (96.3%) resulted from an OSCS-contaminated lot.

The 106 reactions for which there was known exposure to OSCS-contaminated heparin occurred in 77 case patients. Of these case patients, 40 (51.9%) had a reported allergy to medication, most commonly to antibiotics. Six patients (7.8%) had a reported food allergy. An angiotensin-converting–enzyme (ACE) inhibitor was prescribed for 20 of these case patients (26.0%) at the time of the reaction.

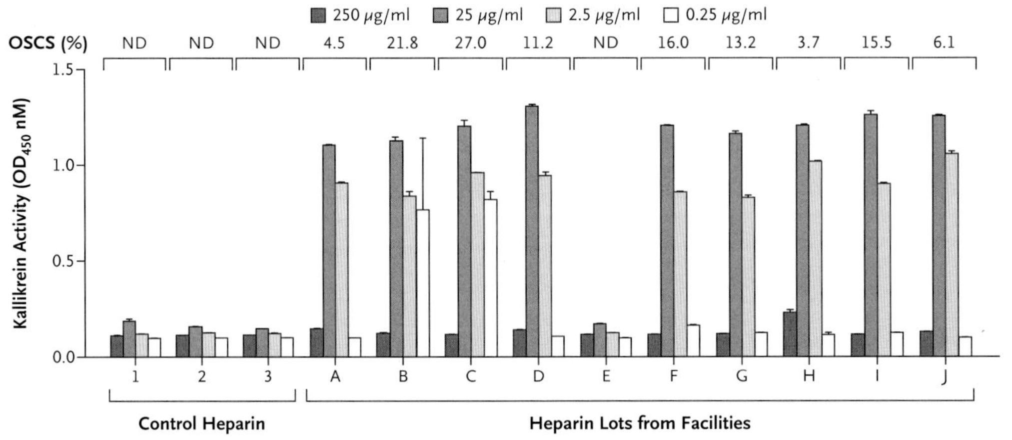

Figure 2. Association of Oversulfated Chondroitin Sulfate (OSCS) in Unfractionated Heparin with Induction of Kallikrein Activity.

Thirteen samples of heparin, including one sample of non–clinical-grade heparin (control heparin 1), representing both suspect heparin lots and control lots, were analyzed in a blinded fashion for both the presence of OSCS and the ability to activate kallikrein. The presence of OSCS was detected and quantified as described elsewhere.[6] The amidolytic activity of kallikrein was assessed at various concentrations of heparin, as indicated. Lots A through D and F through J contained OSCS. Samples from Lot E, as well as controls 1 through 3, contained no detectable OSCS. Lots A through G and Lot J were recalled on January 17, 2008. T bars indicate standard deviations of replicate measurements. ND denotes not detected.

Table 3 shows the clinical characteristics of the heparin reactions in case patients who were known to have received OSCS-contaminated heparin (106 reactions) and heparin that contained more than 20% OSCS (18 reactions). Low blood pressure was documented in 58.5% of the reactions; a systolic pressure lower than 80 mm Hg, however, was reported for only 9.4% of the reactions. Other common symptoms in case patients included nausea (46.2% of reactions), vomiting (28.3%), shortness of breath (25.5%), flushing (25.5%), tingling (24.5%), and tachycardia (24.5%). Facial swelling was associated with 17 reactions (16.0%). The 18 reactions involving heparin contaminated with more than 20% OSCS showed similar manifestations, although nausea was more frequent in this subgroup of cases (occurring in 72.2% of reactions), as was shortness of breath (38.9%).

DISCUSSION

We describe findings from a national investigation of adverse reactions among patients who received heparin and characterize 152 cases that occurred between November 19, 2007, and January 31, 2008. Our initial evaluation of reactions among patients undergoing hemodialysis suggested that heparin was a potential cause, and a facility-based case–control study confirmed a strong epidemiologic association between the receipt of heparin produced by Baxter Healthcare and reactions. Reports to the CDC of heparin reactions declined after the initial Baxter recall of nine lots of heparin vials on January 17, 2008 (Fig. 1). By February 28, 2008, when all Baxter heparin products had been recalled, the CDC was no longer receiving reports of cases. Our findings confirm the presence of OSCS in finished-product vials of heparin manufactured by Baxter Healthcare that were used by facilities that reported cases, as well as the ability of these vials to induce kallikrein activation, and provide evidence that the vast majority of the patients with reactions had been exposed to heparin vials contaminated with OSCS.

Kishimoto et al.[7] previously demonstrated kallikrein activation from OSCS in the active pharmaceutical ingredient of contaminated heparin. Since finished-product vials may contain heparin from more than one lot of the active pharmaceutical ingredient, the vials tested in this investigation best represent the heparin product that case patients received. We found similar biologic activity among multiple vials of heparin that were known to result in adverse reactions, and the clinical picture described among the outbreak cases nation-

Table 3. **Clinical Characteristics of the 106 Adverse Reactions in Patients Confirmed to Have Received Heparin Contaminated with OSCS and with >20% OSCS.***

Characteristic	Reaction with OSCS-Contaminated Heparin (N=106)	Reaction with Heparin Contaminated with >20% OSCS (N=18)
Case status — no. (%)		
Probable	89 (84.0)	16 (88.9)
Definite	17 (16.0)	2 (11.1)
Patient population — no. (%)		
Dialysis	85 (80.2)	18 (100)
Cardiac	7 (6.6)	0
Photopheresis†	14 (13.2)	0
Time from administration of heparin to reaction — min		
During dialysis	4.8±9.0	3.7±1.7
During treatment for cardiac conditions	15.9±17.7	—
During photopheresis	30±13.2	—
Manifestations — no. (%)		
Facial edema		
Any	17 (16.0)	2 (11.1)
Eyelids	11 (10.4)	0
Lips	11 (10.4)	1 (5.6)
Tongue	5 (4.7)	1 (5.6)
Mouth	5 (4.7)	2 (11.1)
Throat	3 (2.8)	1 (5.6)
Urticaria	2 (1.9)	0
Low blood pressure	62 (58.5)	10 (55.6)
Systolic pressure <80 mm Hg	10 (9.4)	0
Nausea	49 (46.2)	12 (66.7)
Vomiting	30 (28.3)	4 (22.2)
Shortness of breath	27 (25.5)	7 (38.9)
Flushing	27 (25.5)	4 (22.2)
Tachycardia	26 (24.5)	0
Tingling	26 (24.5)	7 (38.9)
Diaphoresis	17 (16.0)	5 (27.8)
Loss of consciousness	4 (3.8)	0
Difficulty swallowing	0	0

* Forty-six reactions are not included: 2 were in patients who did not receive OSCS-contaminated heparin, and 22 were in patients who may have received OSCS-contaminated heparin but for whom receipt could not be confirmed; for the remaining 22 reactions, we did not receive lot information and were thus unable to make a determination. Plus–minus values are means ±SD. OSCS denotes oversulfated chondroitin sulfate.

† Although heparin is administered at the beginning of photopheresis, the patient is not exposed to it until later in the process.

ally is consistent with the biologic mediators previously identified in response to OSCS.

The adverse reactions that were reported to the CDC encompassed a constellation of signs and symptoms, initially among patients undergoing hemodialysis. Some manifestations were overt (e.g., facial edema), others were more subtle (e.g., flushing), and many were not uncommon for patients undergoing hemodialysis or treatment for cardiac conditions (e.g., hypotension and dyspnea). Our initial case definition was intentionally broad to accommodate the many presentations and potential sources of allergic-type reactions in the absence of a clear cause or mechanism.

Similar adverse reactions have been documented among patients undergoing dialysis and have, in the past, been attributed to many causes, including dialyzer membranes, water impurities, residual disinfectants, and medications such as ACE inhibitors.[9-11] A systemic inflammatory response has also been described in the setting of cardiopulmonary bypass and has been attributed to activation of the contact system resulting from interaction of blood with the artificial surfaces of the bypass circuit and other mechanisms that activate the kinin–kallikrein pathway, complement system, and other systems.[12]

Heparin alone rarely causes the symptoms, such as angioedema, that were observed in this investigation.[13] Heparin has antiinflammatory properties, and the use of heparin-coated devices is thought to decrease the risk of an inflammatory response.[14-17] Although there have been occasional adverse reactions associated with heparin that have been attributed to animal proteins or allergens,[2] most adverse events reported in association with heparin products have been the result of either intrinsic or extrinsic microbial contamination.[18-20]

Various clinical manifestations were observed in patients with adverse reactions who received OSCS-contaminated heparin. Although hypotension was the most common, a large proportion of case patients had nausea, shortness of breath, vomiting, tingling, flushing, and diaphoresis. Urticaria was not a prominent feature among the case patients. This finding is consistent with reactions that are not mediated by mast cells and supports the role of bradykinin and other mediators instead.[21] In addition, there was no substantial difference between reactions that occurred as a result of heparin contaminated with OSCS and those that occurred as a result of heparin contaminated with a high concentration of OSCS (>20%). This finding may be consistent with data showing that at clinical concentrations of heparin (2.5 and 25 μg per milliliter), the level of kallikrein activation is similar, regardless of the concentration of OSCS. Furthermore, the relatively infrequent need for case patients to be evaluated in the emergency department or hospitalized supports the notion that the clinical manifestations of these adverse reactions were mostly transient.

Twenty-six percent of case patients were taking an ACE inhibitor when they received OSCS-contaminated heparin and had the reaction. ACE inhibitors cause an accumulation of bradykinin and, thus, might be expected to predispose patients to a reaction or to worsen the reaction. The prevalence of ACE-inhibitor use among patients who received OSCS-contaminated heparin at these facilities but did not have reactions is unknown. In two studies of patients undergoing hemodialysis, the prevalence of ACE-inhibitor use was 24% and 51%,[22,23] suggesting that the prevalence among case patients in this investigation was not greater than expected.

On the basis of available evidence, an allergic mechanism seems to be an unlikely cause of these reactions. Thus, it is interesting that a preexisting allergy to medication was identified among 51.9% of case patients, although it is possible that these reports of a history of intolerance to medication did not truly indicate allergic phenomena, and the prevalence of preexisting allergies among patients who did not have a reaction is unknown.

It should be recognized that the cases described in this report do not represent the entire outbreak. As of May 31, 2008, the FDA Adverse Event Reporting System had received reports of more deaths after heparin administration (238 deaths) than we had reports of cases.[24] No determination of a causal association between these deaths and heparin administration has been reported, and many deaths occurred among patients with severe underlying or life-threatening conditions.

We could not calculate attack rates because we did not know the total number of patients in the United States who received heparin during this period. We suspect that OSCS-contaminated heparin was used more broadly than in only those facilities that reported cases. According to national sales distribution data from the IMS National Sales Perspectives (IMS Health), approximately 7,163,700 single-dose and 3,339,400 multidose vials of Baxter heparin were sold by the manufacturer to distributors in the United States during the period from November 2007 through January 2008 (data abstracted by the FDA from IMS Health: IMS National Sales Perspectives: Retail and Non-Retail. Nov '07–Jan. '08. Extracted September 4, 2008). These data do not provide a direct estimate of use, but they do provide a national estimate of the number of vials sold by the manufacturer into retail and nonretail channels of distribution. The lack of reports from other facilities may represent underreporting or underrecognition of reactions, evidence of intermittent contamination, patterns of distribution and use, or other factors.

There were several challenges to this investigation. First, health care facilities rarely recorded lot numbers of the heparin that was administered to patients. In many cases, we could determine only the heparin lots that a patient might have received. Second, our case descriptions are limited by the accuracy of information provided by health care personnel at reporting facilities. However, the fact that health care personnel reported reactions with the use of standard case-report forms probably increased the overall quality of the clinical information we obtained, as compared with information obtained by other methods. Finally, it was difficult to build precise case definitions owing to inherent uncertainties in this investigation. We attempted to reduce misclassification by establishing limits with respect to the time of onset and the symptoms that were required to be classified as a case, but some true cases may have been misclassified as noncases.

This report of a nationwide outbreak attributed to a newly discovered contaminant in heparin products contributes to our understanding of the epidemiology and biology of the adverse reactions that occurred. It also underscores the importance of a public health mechanism to address serious noninfectious adverse events in health care settings, the pivotal role of clinicians who recognize and report clusters of unusual events to public health authorities, and the need for ongoing collaboration among public health agencies, clinicians, basic-science researchers, and industry to prevent future safety threats associated with medications.

The CDC investigation had no external funding. The laboratory analysis was supported in part by grants from the National Institutes of Health (GM57073 and HL080279, to Dr. Sasisekharan).

Dr. Venkataraman reports being an employee of Momenta Pharmaceuticals and holding equity in the company, which performs the analysis and characterization of complex mixtures, including heparin, and being an inventor on patent applications for Momenta Pharmaceuticals; Dr. Kishimoto, being an employee of, holding equity in, and being an inventor on patent applications for Momenta Pharmaceuticals; Dr. Langer, receiving consulting fees from Momenta Pharmaceuticals and Scientific Protein, having equity in Momenta Pharmaceuticals, and serving on a paid board of Momenta Pharmaceuticals; Dr. Shriver, being an employee of Momenta Pharmaceuticals and holding equity in the company; Dr. Sasisekharan, holding equity in Momenta Pharmaceuticals and receiving consulting fees or serving on a paid board of Momenta Pharmaceuticals and Scientific Protein Labs; Dr. Austen, receiving consulting fees from Momenta Pharmaceuticals; Dr. Elward, receiving consulting fees or serving on a paid advisory board for Trinity and receiving grant support from SAGE Products; and Dr. Schillie, holding equity in Pfizer, Lilly, Monsanto, Walgreens, General Electric, and Abbott. No other potential conflict of interest relevant to this article was reported.

The findings and conclusions in this report are those of the authors and do not necessarily represent the views of the Centers for Disease Control and Prevention. Use of trade names and commercial sources is for identification only and does not imply endorsement by the U.S. Department of Health and Human Services.

We thank the many health care professionals at facilities at which adverse events occurred for reporting to public health authorities and for sending heparin vials for laboratory testing; the state epidemiologists and representatives of the End Stage Renal Disease Network for their assistance in facilitating contact between facilities with adverse events and the CDC; Laura Governale (IMS Health), Gary Warns (Gambro), and Bill Singley (Minntech) for the timely information they provided; and representatives of the FDA for their close collaboration, especially in the early stages of this outbreak investigation, in particular, Karen Deasy and Tarun Mallick, who served as important links between the FDA and the CDC and furthered our collective efforts.

REFERENCES

1. Bircher AJ, Harr T, Hohenstein L, Tsakiris DA. Hypersensitivity reactions to anticoagulant drugs: diagnosis and management options. Allergy 2006;61:1432-40.

2. Bottio T, Pittarello G, Bonato R, Fagiolo U, Gerosa G. Life-threatening anaphylactic shock caused by porcine heparin intravenous infusion during mitral valve repair. J Thorac Cardiovasc Surg 2003;126:1194-5.

3. Acute allergic-type reactions among patients undergoing hemodialysis — multiple states, 2007–2008. MMWR Morb Mortal Wkly Rep 2008;57:124-5.

4. Recall — firm press release: Baxter issues urgent nationwide voluntary recall of heparin 1,000 units/ml 10 and 30ml multidose vials. Rockville, MD: Food and Drug Administration, 2008. (Accessed November 24, 2008, at http://www.fda.gov/oc/po/firmrecalls/baxter01_08.html.)

5. Recall — firm press release: Baxter to proceed with recall of remaining heparin sodium vial products. Rockville, MD: Food and Drug Administration, 2008. (Accessed November 24, 2008, at http://www.fda.gov/oc/po/firmrecalls/baxter02_08.html.)

6. Guerrini M, Beccati D, Shriver Z, et al. Oversulfated chondroitin sulfate is a contaminant in heparin associated with adverse clinical events. Nat Biotechnol 2008;26:669-75.

7. Kishimoto TK, Viswanathan K, Ganguly T, et al. Contaminated heparin associated with adverse clinical events and activation of the contact system. N Engl J Med 2008;358:2457-67.

8. Information on Dialysis Facility Compare. Baltimore: Centers for Medicare & Medicaid Services, 2008. (Accessed November 24, 2008, at http://www.medicare.gov/Dialysis/Include/DataSection/Questions/SearchCriteria.asp?version=default&browser=IE%7C6%7CWinXP&language=English&defaultstatus=0&pagelist=Home.)

9. Arduino MJ. CDC investigations of noninfectious outbreaks of adverse events in hemodialysis facilities, 1979-1999. Semin Dial 2000;13:86-91.

10. Ebo DG, Bosmans JL, Couttenye MM, Stevens WJ. Haemodialysis-associated anaphylactic and anaphylactoid reactions. Allergy 2006;61:211-20.
11. Pegues DA, Beck-Sague CM, Woollen SW, et al. Anaphylactoid reactions associated with reuse of hollow-fiber hemodialyzers and ACE inhibitors. Kidney Int 1992; 42:1232-7.
12. Zakkar M, Taylor K, Hornick PI. Immune system and inflammatory responses to cardiopulmonary bypass. In: Gravlee GP, Davis RF, Stammers AH, Ungerleider RM, eds. Cardiopulmonary bypass: principles and practice. 3rd ed. Philadelphia: Lippincott Williams & Wilkins, 2007:321-7.
13. Berkun Y, Haviv YS, Schwartz LB, Shalit M. Heparin-induced recurrent anaphylaxis. Clin Exp Allergy 2004;34: 1916-8.
14. Walenga JM. Non-anticoagulant effects of unfractionated and low-molecular weight heparins. Clin Adv Hematol Oncol 2007;5:759-60.
15. Ludwig RJ, Alban S, Boehncke WH. Structural requirements of heparin and related molecules to exert a multitude of anti-inflammatory activities. Mini Rev Med Chem 2006;6:1009-23.
16. Maharaj C, Laffey JG. New strategies to control the inflammatory response in cardiac surgery. Curr Opin Anaesthesiol 2004;17:35-48.
17. Wan S, LeClerc JL, Antoine M, DeSmet JM, Yim AP, Vincent JL. Heparin-coated circuits reduce myocardial injury in heart or heart-lung transplantation: a prospective, randomized study. Ann Thorac Surg 1999;68:1230-5.
18. Pseudomonas bloodstream infections associated with a heparin/saline flush — Missouri, New York, Texas, and Michigan, 2004–2005. MMWR Morb Mortal Wkly Rep 2005;54:269-72.
19. Souza Dias MB, Habert AB, Borrasca V, et al. Salvage of long-term central venous catheters during an outbreak of Pseudomonas putida and Stenotrophomonas maltophilia infections associated with contaminated heparin catheter-lock solution. Infect Control Hosp Epidemiol 2008;29:125-30.
20. Vonberg RP, Gastmeier P. Hospital-acquired infections related to contaminated substances. J Hosp Infect 2007;65: 15-23.
21. Schwartz LB. Heparin comes clean. N Engl J Med 2008;358:2505-9.
22. Bishu K, Gricz KM, Chewaka S, Agarwal R. Appropriateness of antihypertensive drug therapy in hemodialysis patients. Clin J Am Soc Nephrol 2006;1:820-4.
23. Leypoldt JK, Cheung AK, Delmez JA, et al. Relationship between volume status and blood pressure during chronic hemodialysis. Kidney Int 2002;61:266-75.
24. Information on adverse event reports and heparin. Rockville, MD: Food and Drug Administration, 2008. (Accessed November 24, 2008, at http://www.fda.gov/cder/drug/infopage/heparin/adverse_events.htm.)

List of Patents

Drug Delivery (Including Nanotechnology)

1. 4,391,797
   Systems for the Controlled Release of Macromolecules
   Folkman et al.

2. 4,591,496
   Process for Making Systems for the Controlled Release of Macromolecules
   Cohen et al.

3. 4,757,128
   High Molecular Weight Polyanhydride and Preparation Thereof
   Domb et al.

4. 4,767,402
   Ultrasound Enhancement of Transdermal Drug Delivery
   Kost et al.

5. 4,863,735
   Biodegradable Polymeric Drug Delivery System with Adjuvant Activity
   Kohn et al.

6. 4,891,225
   Bioerodible Polyanhydrides for Controlled Drug Delivery
   Langer et al.

7. 4,952,406
   Feedback Controlled Release
   Brown et al.

8. 5,019,034
   Control of Transport of Molecules Across Tissue Using Electroporation
   Weaver et al.

9. 5,330,768
   Controlled Drug Delivery Using Polymer/Pluronic Blends
   Park et al.

10. 5,543,158
    Biodegradable Injectable Nanoparticles
    Gref et al.

11. 5,565,215
    Biodegradable Injectable Particles for Imaging
    Gref et al.

12. 5,578,325
    Nanoparticles and Microparticles of Non-linear Hydrophilic-hydrophobic Multiblock Copolymers
    Domb et al.

13. 5,618,563
    Biodegradable Polymer Matrices for Sustained Delivery of Local Anesthetic Agents
    Berde et al.

14. 5,626,862
    Controlled Local Delivery of Chemotherapeutic Agents for Treating Solid Tumors
    Brem et al.

15. 5,783,567
    Microparticles for Delivery of Nucleic Acid
    Hedley et al.

16. 5,797,898
    Microchip Drug Delivery Devices
    Santini, Jr. et al.

17. 5,874,064
    Aerodynamically Light Particles for Pulmonary Drug Delivery
    Edwards et al.

18. 6,254,854 B1
    Porous Particles for Deep Lung Delivery
    Edwards et al.

19. 7,550,441 B2
    Controlled Release Polymer Nanoparticle Containing Bound Nucleic Acid Ligand for Targeting
    Farokhzad et al.

20. 7,829,112 B2
    Methods and Devices for the Sustained Release of Multiple Drugs
    Ron et al.

21. 8,193,334 B2
    Polymer-encapsulated Reverse Micelles
    Radovic-Moreno et al.

22. 8,343,497 B2
    Targeting of Antigen Presenting Cells with Immunonanotherapeutics
    Shi et al.

23. 8,442,611 B2
    Medical Device with Reservoir-based Sensors
    Santini, Jr. et al.

24. 8,450,298 B2
    Aminoalcohol Lipidoids and Uses Thereof
    Mahon et al.

25. 8,591,905 B2
    Nicotine Immunonanotherapeutics
    von Andrian et al.

26. RE41,996 E
    Composition and Methods for Enhancing Receptor-mediated Cellular Internalization
    Deaver et al.

Novel Biomaterials

27. 4,753,652
    Biomaterial Implants which Resist Calcification
    Langer et al.

28. 4,806,621
    Biocompatible, Bioerodible, Hydrophobic, Implantable Polyimino Carbonate Article
    Kohn et al.

29. 5,128,420
    Method of Making Hydroxamic Acid Polymers from Primary Amide Polymers
    Domb et al.

30. 6,149,864
    Supercritical Fluid Sterilization Method
    Dillow et al.

31. 6,174,952 B1
    Monolithic Polymer Composition having a Water Absorption Material
    Hekal et al.

32. 6,388,043 B1
    Shape Memory Polymers
    Langer et al.

33. 7,427,394 B2
    Biodegradable Poly (β-Amino Esters) and Uses Thereof
    Anderson et al.

34. 7,722,894 B2
    Biodegradable Polymer
    Wang et al.

35. 8,287,849 B2
    Biodegradable Poly(Beta-amino Esters) and Uses Thereof
    Langer et al.

36. 8,303,625 B2
    Biodegradable Shape Memory Polymeric Sutures
    Lendlein et al.

Tissue Engineering and Stem Cells

37. 5,041,138
    Neomorphogenesis of Cartilage in vivo from Cell Culture
    Vacanti et al.

38. 5,512,600
    Preparation of Bonded Fiber Structures for Cell Implantation
    Mikos et al.

39. 5,759,830
    Three-dimensional Fibrous Scaffold Containing Attached Cells for Producing Vascularized Tissue in vivo
    Vacanti et al.

40. 5,770,193
    Preparation of Three-dimensional Fibrous Scaffold for Attaching Cells to Produce Vascularized Tissue in vivo
    Vacanti et al.

41. 5,985,320
    Materials and Methods for Enhancing Cellular Internalization
    Edwards et al.

42. 6,306,819 B1
    Method for Regulating Size of Vascularized Normal Tissue
    Rupnick et al.

43. 6,537,567 B1
Tissue-engineered Tubular Construct having Circumferentially Oriented Smooth Muscle Cells
Niklason et al.

New Drug Development (e.g. Angiogenesis Inhibitors) and Enzyme Technology

44. 4,373,023
Process for Neutralizing Heparin
Langer et al.

45. 4,443,545
Process for Producing Heparinase
Langer, Jr. et al.

46. 4,994,443
Inhibition of Angiogenesis
Folkman et al.

47. 5,389,539
Purification of Heparinase I, II, and III from *Flavobacterium Heparinum*
Sasisekharan et al.

48. 5,569,600
Purification, Composition and Specificity of Heparinase I, II, and III from *Flavobacterium Heparinum*
Sasisekharan et al.

49. 5,714,376
Heparinase Gene from Flavobacterium Heparinum
Sasisekharan et al.

50. 8,580,253 B2
Immunomodulating Compositions and Methods of Use
Rubin-Bejerano et al.

The Lemelson-MIT Awards Program
Robert S. Langer
THE NATIONAL MEDAL OF TECHNOLOGY AND INNOVATION
Commissioner's Special Citation
IEEE
Drexel University
Wolf Prize

Appendix

Awards

2016 Raymond and Beverly Sackler Award for Sustained National Leadership (Research! America)

2016 Benjamin Franklin Medal in Life Science (Franklin Institute)

2015 Queen Elizabeth Prize for Engineering

2015 Hoover Medal (American Society of Mechanical Engineers)

2015 Scheele Award (Swedish Pharmaceutical Society)

2015 Kazemi Award for Research Excellence in Biomedicine

2015 Cornell Entrepreneur of the Year

2014 Kyoto Prize in Advanced Technology

2014 Breakthrough Prize in Life Sciences

2014 Biotechnology Heritage Award (Chemical Heritage Foundation)

2014 Chemical Pioneer Award (American Institute of Chemists)

2014 ETH Zurich Chemical Engineering Medal

2014 Elected a Foreign Corresponding Member of the Austrian Academy of Sciences

2014 Mack Memorial Award (Ohio State University)

2013 Wolf Prize in Chemistry

2013 United States National Medal of Technology and Innovation (2011)

2013 RUSNANOPRIZE International Prize in Nanotechnology

2013 Fellow, American Association for the Advancement of Science

2013 Distinguished Investigator Award (American College of Clinical Pharmacology)

2013 Honorary Fellow Award (American College of Clinical Pharmacology)

2013 Julio Palmaz Award for Innovation in Healthcare and the Biosciences (BioMed SA)

2013 Industrial Research Institute Medal

2013 IEEE Medal for Innovations in Healthcare Technology

2013 Founders Award (Society of Biomaterials)

2013 MDEA Lifetime Achievement Award

2012 Scientist of the Year (R&D Magazine)

2012 Elected to National Academy of Inventors (Charter Fellow)
2012 Priestley Medal (American Chemical Society)
2012 Perkin Medal (Chemical Heritage Foundation)
2012 Terumo International Prize (Japan)
2012 Wilhelm Exner Medal (Austria)
2012 Fellow, American Institute of Chemical Engineers
2012 Feodor Lynen Award (Nature Biotechnology)
2011 Economist Innovation Award for Bioscience
2011 Warren Alpert Foundation Prize (Warren Alpert Foundation/Harvard Medical School)
2011 Walker Prize (Boston Museum of Science)
2011 Apple Award (American Spinal Injury Association/Shepherd Center/Thomas Land Publishers Inc.)
2011 Frontier of Science Award (Society of Cosmetic Chemists)
2011 Fellow (American Chemical Society)
2010 Elected International Fellow (Royal Academy of Engineering)
2010 Founders Award (National Academy of Engineering)
2010 Robert Flectcher Award (Thayer School of Engineering, Dartmouth College)
2010 EMBS Academic Career Achievement Award (IEEE Engineering in Medicine and Biology Society)
2010 Elected to the Controlled Release Society College of Fellows
2010 Founding POLY Fellow (Division of Polymer Chemistry, American Chemical Society)
2009 Distinguished Chemist Award (New England Institute of Chemists)
2009 Biomedical Research Leaders Award (Massachusetts Society for Medical Research)
2008 Millennium Technology Prize
2008 Max Planck Research Award
2008 Prince of Asturias Award for Technical and Scientific Research
2008 Elected Foreign Academic Member (Royal Academy of Pharmacy of Spain)
2008 Founders Award (American Institute for Chemical Engineers)
2008 Innovation in Health and Technology Award (Boston History and Innovation Collaborative)
2008 Acta Biomaterialia Gold Medal (Acta Materialia)
2008 Named one of the 100 Top Chemical Engineers of the 20th Century (AIChE)
2007 United States National Medal of Science (2006)
2007 Chemistry of Materials Award (American Chemical Society)
2007 Herman F. Mark Award (American Chemical Society, Polymer Chemistry Division)

2007 Wenig Memorial Award for Outstanding Achievement in Drug Delivery
2007 First Prize of Scientific Award (BMW Group)
2007 Elected to the Biotechnology Hall of Fame
2006 Elected to the National Inventors Hall of Fame
2006 Bailey Award (American Institute of Chemical Engineers)
2005 Von Hippel Award (Materials Research Society)
2005 Albany Medical Center Prize in Medicine and Biomedical Research
2005 Dan David Prize (Materials Science)
2005 Lifetime Achievement Award (Society for In Vitro Biology)
2005 Rainer Hoffmann Product through Science Award (Controlled Release Society)
2005 Technology Innovation and Development Award (Society of Biomaterials)
2005 Washington Award (Western Society of Engineers)
2004 Charles F. Kettering Prize (General Motors Cancer Research Foundation)
2004 Nelson Taylor Award (Pennsylvania State University)
2003 Heinz Award for Technology, Economy and Employment
2003 Harvey Prize in Science and Technology and Human Health
2003 John Fritz Medal (American Association of Engineering Societies)
2003 Elected to the Academy of Achievement (Golden Plate Award)
2002 Dickson Prize for Science (Carnegie Mellon University)
2002 Charles Stark Draper Award (National Academy of Engineering)
2002 Othmer Gold Medal (Chemical Heritage Foundation)
2002 Nagai Innovation Award (Controlled Release Society)
2002 Herman Schwan Award (University of Pennsylvania)
2001 Harrison Howe Award (American Chemical Society)
2000 Glaxo Wellcome International Achievement Award (Royal Pharmaceutical Society of Great Britain)
2000 Millennial Pharmaceutical Scientist Award (Millennial World Congress of Pharmaceutical Sciences)
2000 First Pierre Galletti Award (American Institute of Medicine & Biological Engineering)
2000 Wallace Carothers Award (American Chemical Society, Delaware Section)
1999 American Chemical Society Award in Polymer Chemistry
1999 Esselen Award (American Chemical Society, Northeast Section)
1999 Ebert Prize (American Pharmaceutical Association)
1999 G.N. Lewis Medal (University of California at Berkeley)
1998 Outstanding Pharmaceutical Paper Award (Controlled Release Society)
1998 Lemelson–MIT Prize for Invention and Innovation
1998 The Nagai Foundation Tokyo International Prize

1997 Killian Faculty Achievement Award (MIT)
1997 Wiley Medal (U.S. Food and Drug Administration)
1996 Gairdner Foundation International Award
1996 William Walker Award (American Institute of Chemical Engineers)
1996 Society of Plastics Engineers International Award
1996 Ebert Prize (American Pharmaceutical Association)
1996 Elected a Fellow of Biomaterials Science and Engineering
1996 Avis Distinguished Visiting Professor (University of Tennessee)
1995 International John W. Hyatt Service to Mankind Award (Society of Plastics Engineers)
1995 Ebert Prize (American Pharmaceutical Association)
1995 Elected a Fellow (American Association of Pharmaceutical Scientists)
1995 PEL Associates Award (PEL Associates, Groton, Connecticut)
1994 Elected to the American Academy of Arts and Sciences
1994 Elected a Fellow, Society of Biomaterials
1993 Distinguished Pharmac. Scientist Award (Highest Honor of the Amer. Assoc. of Pharm. Scient.)
1992 Elected to the National Academy of Sciences
1992 Elected to the National Academy of Engineering
1992 American Chemical Society Award for Applied Polymer Science (Phillips Award)
1992 Elected a Founding Fellow, American Institute of Medical and Biological Engineering
1992 Outstanding Pharmaceutical Paper Award (Controlled Release Society)
1992 Perlman Memorial Award (American Chemical Society, Biochemical Technology Division)
1991 Organon Teknika Award (European Society for Artificial Organs)
1991 Charles M.A. Stine Award in Materials Science and Eng. (American Institute of Chemical Engineers)
1990 Professional Progress Award (American Institute of Chemical Engineers)
1990 Clemson Award for Basic Research (Society for Biomaterials)
1990 Outstanding Pharmaceutical Paper Award (Controlled Release Society)
1989 Elected to the National Academy of Medicine
1989 Creative Polymer Chemistry Award (American Chemical Society, Polymer Division)
1989 Outstanding patent in Massachusetts and one of the twenty outstanding patents in the U.S. (Intellectual Property Owners, Inc.)
1989 Founders Award for Outstanding Research (Controlled Release Society)
1988 Elected to the Gordon Conference Research Council

2005 John and Donna Hall Lectureship (Vanderbilt University)

2005 Keewaunee Lecturer (Duke University)

2005 Gavel Clinical Research Lecturer (The Forsyth Institute)

2005 Talamo Lecturer (Harvard Medical School)

2004 Presidential Lecturer (University of Texas Health Science Center, San Antonio)

2004 Pirkey Lectureship in Chemical Engineering (University of Texas, Austin)

2004 Evans Memorial Award Lectureship (Ohio State University)

2004 Benjamin Zweifach Distinguished Lecturer (City College of New York)

2004 Nelson Leonard Lecturer (University of Illinois)

2004 Donald Katz Lecturer (University of Michigan)

2003 Founders Lecturer (University of Wisconsin)

2003 Rohm and Hass Lecturer (Stanford University)

2003 Tripathy Endowed Memorial Lecture (University of Massachusetts, Lowell)

2003 Skinner Memorial Lecture (Northwestern University)

2003 Maurice and Yetta Glicksman Lecturer (Brown University)

2003 FMC Lecturer (Princeton University)

2003 Seymour J. Kreshover Lecturer (National Institutes of Health)

2003 Whitaker Lecturer (American Society of Artificial Organs)

2002 Feigenbaum–Levine Lecturer (Beth Israel Hospital at Harvard Medical School)

2002 Distinguished Lecturer (University of Louisville)

2002 Institute Lecturer (American Institute of Chemical Engineers)

2001 Ullyot Lecturer (Chemical Heritage Foundation)

2001 Clapp Lecturer (Brown University)

2001 Julian Smith Lecturer (Cornell University)

2001 Mason Lecturer (Stanford University)

2001 Distinguished Lecturer (Carnegie Mellon)

2000 Herman Beerman Lecturer (Society for Investigative Dermatology)

2000 Millennial Lecturer (University of Liverpool)

2000 Bayer Lecturer (University of Pittsburgh)

2000 Bayer Stein Honorary Lecture (University of Massachusetts at Amherst)

2000 William G. Lowrie Lectureship (The Ohio State University)

2000 Frank T. Gucker Lecturer (Indiana University)

2000 First Patten Distinguished Lectureship (University of Colorado at Boulder)

1999 Beckman Lecturer (University of Illinois at Urbana)

1999 Reilly Lecturership (Notre Dame University)

1998 Wagner Lectureship (University of Michigan)

1998 Ewing Halsell Foundation Lectureship (University of Texas Health Center, San Antonio)

1998 Robert R. Linton Distinguished Lecture; New England Society for Vascular Surgery

1998 Marcus Memorial Lecturer (Washington University, St. Louis)

1998 Joseph Stokes, Jr. Lectureship (University of Pennsylvania)

1997 William J. Rashkind Memorial Lecture (American Heart Association)

1997 Rohm and Haas Lecturer in Materials Chemistry (University of North Carolina)

1996 The Berkeley Lecturer (University of California, Berkeley)

1995 Distinguished Medical Scientist Lecturer (Ohio State University)

1995 Lacy Lecturer (California Institute of Technology

1995 Ralph Peck Memorial Lecturer (Illinois Institute of Technology)

1994 Whitaker Distinguished Lecturer (Biomedical Engineering Society)

1994 Miles Lecturer (Cornell University)

1994 Feigenbaum Memorial Lecturer (Beth Israel Hospital, Harvard Medical School)

1993 Kurt Wohl Memorial Lecturer (University of Delaware)

1993 Priestley Lecturer (Penn State University)

1992 Kelly Distinguished Lecturer (Purdue University)

1992 Miles Distinguished Lecturer (University of Pittsburgh)

1991 Louis W. Busse Lecturer (University of Wisconsin)

1991 Sidney Riegelman Lecturer (University of California, San Francisco)

1991 Ashton-Cary Lecturer (Georgia Institute of Technology)

1991 Sandoz-Dorsey Lecturer (Ohio State University)

1989 Walter F. Enz Lecturer (University of Kansas)

1988 Robert Rushmer Lecturer (University of Washington, Seattle)

1988 1st Presidential Lecturer, Controlled Release Society (Basel, Switzerland)

1987 Biomedical Research Council Lecturer (University of Michigan)

1986 Elmer L. Linseth Lecturer (Case Western Reserve University)

1983 Merck, Sharpe and Dohme Lecturer (University of Puerto Rico)

Honorary Degrees

2016 Karolinska Institutet, Sweden

2016 Hong Kong University of Science and Technology, Hong Kong

2016 Carnegie Mellon University

2015 University of Maryland

2015 Hanyang University, South Korea

2015 University of New South Wales, Australia

2014 Drexel University
2014 University of Western Ontario, Canada
2013 Boston University
2013 Tel Aviv University, Israel
2013 Ben Gurion University, Israel
2011 Bates College
2010 Rensselaer Polytechnic Institute
2010 Willamette University
2009 Harvard University
2009 Mount Sinai School of Medicine
2009 University of California at San Francisco Medal
2007 Yale University
2006 Northwestern University
2006 Albany Medical College
2005 Pennsylvania State University
2005 Uppsala University, Sweden
2005 University of Nottingham, England
2003 University of Liverpool, England
2002 Hebrew University of Jerusalem, Israel
2000 The Catholic University of Louvain, Belgium
1997 The Technion, Israel
1996 Eidgenossische Technische Hochschule (ETH), Switzerland

Founder, Start-up Companies

Enzytech (merged to become Alkermes, spun out Opta foods which went public and was acquired by Sun)

Echo Therapeutics (public)

Advanced Inhalation Research (merged with Alkermes, spun out as Civitas, which was acquired by Accorda)

MicroCHIPS Biotech Transform Pharmaceuticals (acquired by Johnson & Johnson)

Momenta (public)

Pervasis (acquired by Shire)

Arsenal Medical (spun out 480 biomedical)

BIND Therapeutics (public)

T2 Biosystems (public)

Semprus (acquired by Teleflex)

Focal (acquired by Genzyme)

Living Proof

Combinent BioMedical Systems (acquired by Juniper)

EnzyMed (acquired by Albany Medical)

Neomorphics (acquired by Advanced Tissue Sciences which was acquired by Smith and Nephew, which spun out Advanced Biohealing and was acquired by Shire)

Pulmatrix Inc. (public)

PureTech Ventures (public)

Reprogenesis Inc. (merged with Creative Biomolecules and Ontogeny to form Curis) (public)

Taris

Selecta Biosciences

Seventh Sense

Gecko Biomedical

Blend Therapeutics

SQZ

Lyndra

In Vivo Therapeutics (public)

Xtuit

Arsia

Moderna Therapeutics

Siglion

Board Member, Non-profit

U.S. Food and Drug Administration

McGovern Institute

Whitehead Institute

Ragon Institute

HST Governing Council

Boston Childrens Hospital

Cornell University

LAM Foundation